Reliability

Reliability

Modeling, Prediction, and Optimization

WALLACE R. BLISCHKE

D. N. PRABHAKAR MURTHY

A Wiley-Interscience Publication

JOHN WILEY & SONS, INC.

New York • Chichester • Weinheim • Brisbane • Singapore • Toronto

This text is printed on acid-free paper. ⊗

Copyright © 2000 by John Wiley & Sons, Inc.

All rights reserved. Published simultaneously in Canada.

For ordering and customer service, call 1-800-CALL-WILEY.

Library of Congress Cataloging in Publication Data is available.

ISBN 0-471-18450-0

Printed in the United States of America

10 9 8 7 6 5 4 3 2 1

Dedicated to the memory of our fathers
Walter Henry Blischke
and
Dodderi Narsihma Murthy

CONTENTS

PART B: BASIC RELIABILITY METHODOLOGY

PART C: RELIABILITY MODELING, ESTIMATION, AND PREDICTION

PREFACE

In the real world, all products and systems are unreliable in the sense that they degrade with age and/or usage and ultimately fail. This has important consequences for both the producer and user of such products. For both, the end result of unreliability is increased cost—increased support cost and potential loss of sales for the producer; increased cost of ownership for the buyer.

The reliability of a product depends on a complex interaction of the laws of physics, engineering design, manufacturing processes, management decisions, random events, and usage. Increasing the reliability of a product is often a complex process, involving one or more of a number of activities, including redesign, upgrading of materials and process improvements, as well as additional elements such as handling, storage, and shipping.

Reliability theory is concerned with the study of the various aspects of product reliability. These range from improving product reliability at the design stage, to effective control (through testing) during the manufacturing stage, to proper usage and effective maintenance at the operational stage. Reliability studies involve concepts and techniques from many disciplines, including engineering, mathematics, material science, statistics, and operations research. The knowledge in each discipline is ever increasing. In addition, many new products are being developed and sold and many existing products are becoming more complex, involving new technologies and materials. Limited experience and lack of information about new materials, products, and so forth, are a source of considerable uncertainty. As a result, the study of product reliability has become both more important and more complex.

In applications, interaction between the disciplines involved is necessary for successful and effective implementation of a reliability program. Results of reliability studies are important inputs to decisions regarding design, manufacturing, marketing, strategy, and postsale support requirements, all of which impact costs. Significant involvement of management in this activity is therefore also essential. This is a convergent process oriented to solve specific problems or achieve specified goals.

One of the primary objectives of this book is to integrate the theoretical concepts and methodologies from diverse disciplines in an effective manner for dealing with the many diverse issues involved with product reliability in the real world. The emphasis is on providing concepts for understanding and tools needed for reliability study.

To this end, we look at the different stages in a product life cycle and the reliability issues—assessment, prediction, improvement, and so forth—at each stage. The systems approach provides framework for analysis of the various technological, scientific and commercial issues in product reliability in a unified and integrated manner. This is achieved by providing bridges between concept, theory and application, as well as an integration of the various disciplines needed for the effective study and management of product reliability. A key element of the systems approach is the use of mathematical models. The mathematical results needed for modeling and analysis are discussed in detail. Mathematical derivations and proofs, on the other hand, are generally omitted, with references given for the interested reader.

Applications are emphasized throughout. Cases based on real applications and using real data are introduced early on (in Chapter 2) and used frequently in subsequent chapters to illustrate the methodology. The relevance of the study in the context of reliability management is often discussed in some detail. To illustrate the integrated approach that is required in the real world, the capstone chapter includes a detailed analysis and discussion of two of the cases.

Aspects of reliability discussed in the book cover a broad range of topics, including product design, marketing, maintenance, warranty costing, design of complex systems, life-cycle costing, service contracts, engineering changes, bidding strategies, and others. Cases have been selected to cover a broad range of applications, including basic materials (various types of fibers), simple hardware items (ball bearings, electric leads), relatively simple consumer and commercial products (light bulbs, audio speakers), complex consumer products (automobiles), commercial items (jet engines, aircraft windshields), medical devices (heart pacemakers), military systems (helicopters, submarine diesel engines), and others.

The book is organized into the following five parts:

- Part A. Context of Reliability Analysis
- Part B. Basic Reliability Methodology
- Part C. Reliability Modeling, Estimation, and Prediction
- Part D. Reliability Management, Improvement, and Optimization
- Part E. Epilogue

Part A is an introduction to reliability and its applications. In Part B, we look at some of the basic mathematical and statistical tools and concepts needed for reliability analysis. Part C deals with more advanced topics in modeling and data analysis and the use of these in modeling and analyses of reliability from the part to the system level. Although much of the emphasis is on hardware, a chapter on software reliability is included as well. Additional management issues, including effective management of reliability, product support, optimal selection of testing procedures, price and warranty terms, and many related topics are discussed in Part D. Part E includes the detailed case studies and an extensive list of sources of technical information, data, and computer programs for reliability analysis.

An outline of the chapters in each of the five Parts is given in Chapter 1. Each chapter includes a moderate to extensive set of exercises, ranging from relatively straightforward to challenging. These relate to most or all of the topics covered in the chapter and are intended to both provide practice in application of the methodology as well as to supplement the material covered. (A word of advice to the student regarding the exercises: many exercises in later chapters refer to results of earlier ones, so keep a copy of your homework.)

The literature on reliability—books, journal articles, conference proceedings, and so forth—is huge. No attempt has been made to survey this literature. We have, however, included numerous references in each chapter, both as sources of the results given and to indicate sources for further reading and additional results. Many of the latter are discussed briefly in the notes at the end of each chapter. The collected references appear at the back of the book.

Since there are a large number of computer programs available for data analysis and reliability analysis (see Chapter 20) and these are frequently modified, we have made no attempt to illustrate their use. Most of the statistical analyses in the book were done on Minitab, and Minitab output is frequently given. Other computations were done using standard mathematical programs or programs written by us (or our students!).

Finally, this book is intended for use by:

- Practicing engineers
- Managers involved with the design, development, operation, and maintenance of products and systems
- Production engineers and managers
- Quality engineers and managers
- Managers involved with postsale product support
- Cost analysts and managers
- Applied statisticians involved with reliability
- Graduate and advanced undergraduate students in industrial engineering, operations research, and statistics.

The book is intended both as a reference book and as a text in graduate and advanced undergraduate level courses. Because of its integrative nature and applications orientation, we hope that it will appeal to a broad audience in reliability and related fields.

We express our appreciation to our wives, Carol Blischke and Jayashree Murthy, and to our families, for their patience and understanding during the preparation of this book. The many constructive comments of our colleagues and students and of the reviewers of the text are very much appreciated. Finally, we gratefully acknowledge the help and encouragement of the editors, Steve Quigley and Andrew Prince, in the production of this book.

W. R. Blischke
Sherman Oaks, California, USA

D. N. P. Murthy
Brisbane, Queensland, Australia

PART A

Context of Reliability Analysis

CHAPTER 1

An Overview

1.1 INTRODUCTION

The twentieth century was characterized by rapid changes in technology, with many changes occurring at an exponential rate, and this will certainly continue into the twenty-first century as well. Technology is, according to a standard dictionary definition, "the totality of the means employed to provide objects necessary for human sustenance and comfort." A more appropriate definition for our purposes is "the totality of goods, tools, processes, methods, techniques, procedures, and services that are invented and put into some practical use." (Bayraktar, 1990). Our interest in this book is on "objects" in the first definition and "goods" in the second. They can vary from relatively simple products, such as light bulbs, tires, toasters, or articles of clothing, to complex systems such as bridges, factories, power generating systems, or communications networks. Items such as these are engineered and manufactured to perform in some specified manner when operated under normal operating conditions.

By and large, engineered objects such as these perform satisfactorily, but occasionally they fail. A dictionary definition of failure is "falling short in something expected, attempted, or desired, or in some way deficient or lacking." From an engineering point of view, it is useful to define failure in a broader sense. Witherell (1994) elaborates as follows: "It [failure] can be any incident or condition that causes an industrial plant, manufactured product, process, material, or service to degrade or become unsuitable or unable to perform its intended function or purpose safely, reliably, and cost-effectively." Accordingly, "the definition of failure should include operations, behavior, or product applications that lead to dissatisfaction, or undesirable, unexpected side effects."

When a failure occurs, no matter how benign, its impact is felt. For example, failure of an air conditioner or a flat tire in a car causes a certain degree of inconvenience. In contrast, failure of the brakes on a car can result in personal injury, damage to property, and a significant economic loss. When the failure is catastrophic (as for example, in the crash of an aircraft, collapse of a bridge, or failure of a nuclear reactor), the total economic damage and loss of life can be very dramatic, affecting society as a whole.

Failures occur in an uncertain manner and are influenced by factors such as design, manufacture or construction, maintenance, and operation. In addition, the human factor is important. Improper operation of a machine, unsafe driving habits, and unpredictable human behavior generally lead to accidents and other types of failures in many situations.

There is no way that failures can be totally eliminated. Every engineered object is unreliable in the sense that it will fail sooner or later, even with the best design, construction, maintenance,

3

and operation. The reason for this is that there are limits to everything and as a result all objects, whether engineered and manufactured or natural (living organisms) must fail eventually. What can be done is reduce the chance of occurrence of failures within a limited time frame. This requires effective integration of good engineering with good management so that the failures and their consequences are minimized and the object can fulfill its intended purpose.

As we move into the twenty-first century, engineered objects are becoming more and more complex. This, combined with the use of new materials and new construction methods, often increases the risk of failure and the possible damage that may result. Civilized society has always taken a dim view of the damage suffered by its members that is caused by someone or some activity and has demanded a remedy or retribution for offenses against it. Consequently, manufacturers are required to provide compensation for any damages resulting from failures of an object. This has serious implications for manufacturers of engineered objects. Product-liability laws and warranty legislation are signs of society's desire to ensure fitness of products for their intended use and compensation for failures.

Similarly, the actions of user–owners (e.g., operations and maintenance) of engineered objects may have an impact on failure, and individuals and businesses need to understand the implications of this. For example, operating an engine at a higher load than that for which it is rated might lead to increased output but hasten its failure and hence lead to loss rather than gain. A bridge that is not properly maintained or is allowed to be overloaded may collapse even though it was properly engineered.

This book deals with the study of various aspects related to failures of engineered objects, the consequence of such failures, and techniques for their avoidance. The study of these topics requires that we begin with a good and clear conceptual understanding and have a framework that allows us to integrate the various issues involved in an effective manner. The systems approach provides the framework needed. An important feature of this approach is the use of mathematical models to obtain solutions to a variety of problems of interest to manufacturers and user–owners.

This chapter discusses the basic concepts needed and defines the scope and outline of the book. The structure of the chapter is as follows. We commence, in Section 1.2, with a description of a few engineered objects ranging from simple products to complex systems. We discuss characterization of a product or system in terms of its various parts. This is essential to the analysis, since the failure of a product or system is related to failure of one or more parts. Following this, we clarify the basic concepts of failure, failure causes, and failure modes in Section 1.3. We then introduce the concept of reliability and discuss very briefly the evolution of reliability theory as a discipline in Section 1.4. In Section 1.5, we discuss the reliability of a product over the life cycle of the product. This sets the scene for discussion of some of the many different problems of interest to manufacturers and buyer–users. These are discussed in Sections 1.6 and 1.7, respectively. A framework and the use of the systems approach to solve these problems are discussed in Section 1.8. This provides the background for discussion of the aims of the book, which is done in Section 1.9. The book takes an approach that links theory with practice, as discussed in Section 1.10. Finally, we conclude the chapter with an outline of the book in Section 1.11.

1.2 ILLUSTRATIVE EXAMPLES OF PRODUCTS AND SYSTEMS

Broadly speaking, products can be categorized into the following three groups:

1. Consumer durables
2. Industrial and commercial products
3. Specialized defense-related products

Consumer durables (e.g., televisions sets, appliances, automobiles, PCs) are consumed by society at large as well as by commercial users and government agencies. They are characterized by a large number of consumers for the product. The complexity of the product can vary considerably and the typical small consumer is often not sufficiently well informed to evaluate product performance, especially in the case of complex products.

Industrial and commercial products (e.g., large-scale computers, cutting tools, pumps, X-ray machines, commercial aircraft, hydraulic presses) are characterized by a relatively small number of consumers and manufacturers. The technical complexity of such products and the mode of usage can vary considerably. The products can be either complete units such as cars, trucks, pumps, etc., or product components needed by a manufacturer, such as batteries, drill bits, electronic modules, turbines, etc.

Specialized defense products (for example, military aircraft, ships, rockets) are characterized by a single consumer and a relatively small number of manufacturers. The products are usually complex and expensive and involve "state-of-art" technology with considerable research and development effort required of the manufacturers. These products are usually designed and built to consumer specifications. In contrast, systems (for example, power stations, computer networks, communication networks, chemical plants) are collections of several interlinked products. As such, they are inherently complex and almost always custom designed.

Both products and systems can be decomposed into a hierarchy of levels, with the product or system at the top level and parts at the lowest level.[1] A decomposition of this type is necessary in a detailed and thorough analysis of reliability. There are many ways of describing this hierarchy. We shall use the following eight-level description:

Level	Characterization
0	System
1	Subsystem
2	Major Assembly
3	Assembly
4	Subassembly
5	Component
6	Part
7	Material

As mentioned in the Introduction, the complexity of products has been increasing with technological advances. The following example of a farm tractor is from Kececioglu (1991). The numbers of critical components are given as:

Model year	1935	1960	1970	1980	1990
Number of components	1200	1250	2400	2600	2900

For more complex products, the number of parts may be orders of magnitude larger. The success of the Mariner/Mars spacecraft required the satisfactory performance of some 138,000 components over the nine months of its mission in space [*Engineering Opportunities,* October 1964, pp. 14–15]. The Boeing 747 has 4.5 million parts (Appel, 1970). Other very large systems in terms of parts counts are the space shuttle and its launch system, large naval ships such as carriers, telecommunications systems, and so forth. Parts counts, in fact, can provide a crude notion of relative reliability; the more parts, the lower the reliability, all other things being equal, simply because there are more things that can go wrong.

Systems of the type indicated above are very complex, indeed, requiring huge charts and schematics for design and analysis. We illustrate in detail the decomposition of products and systems into parts with one or more intermediate levels through a few examples, beginning with much simpler products and extending to complex systems. These examples will be discussed several times later in the chapter. We shall see that even for quite simple products there are many possible causes of failure. Theoretically, any part, even some not explicitly shown, such as adhesives, could fail and lead to item failure.

Example 1.1   Incandescent Electric Bulb

The components of a typical incandescent light bulb are shown in Figure 1.1. The light-emitting part is the filament. When heated to 2000 to 3000°C, it emits light due to incandescence. The source of heating is the resistance of the filament to the electrical current flowing through it. The filament is made by first pressing tungsten ingots and sintering them. The ingot is shaped into round rods and drawn through a die to produce thin wire. The lead-in wires are usually made of nickel, copper, or molybdenum and the support wires are made of molybdenum. The base is made of aluminum. The bulb is filled with an inert gas, usually a mixture of nitrogen and argon.

Example 1.2   Hydraulically Operated Fail-Safe Gate Valve

The components of a typical hydraulically operated fail-safe gate valve are shown in Figure 1.2. The valve is held open by hydraulic pressure and is spring compressed. When the pres-

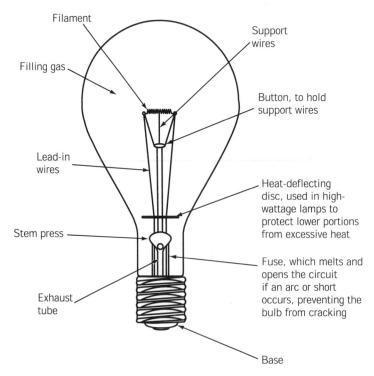

Figure 1.1. Components of a typical light bulb.

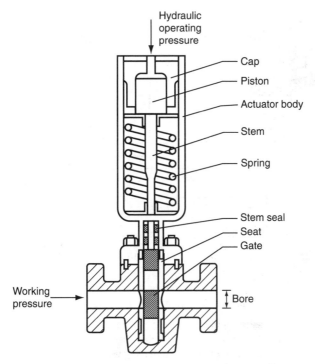

Figure 1.2. Hydraulically operated fail-safe gate valve. Reprinted from *Reliability Engineering and System Safety,* Vol. 53, p. 75, Copyright 1996, with permission from Elsevier Science.

sure is released (or bled off), the spring reverts to its normal (uncompressed) state and the valve closes.

These valves are used extensively in many industries to control fluid flow. The essential function of this type of valve is to stop the flow of the fluid at the appropriate time. The valve can be viewed as being an assembly of a complex processing system, for example an oil refinery.

Example 1.3   Pneumatic Pump

In an aquarium, it is essential that there be a continuous supply of dissolved oxygen to ensure the health of the aquatic animals and plants inside the tank. This is achieved by pumping air using a pneumatic pump.

The components of such a pneumatic pump are shown in Figure 1.3. The pump operating mechanism is fairly simple. When an alternating voltage (from the main supply) is applied across the coil, an alternating magnetic field is set up. This field interacts with the permanent magnet to produce an alternating motion (in a vertical plane) of the magnet. This motion is transmitted through the lever to produce a vertical motion of the bellows. When the bellows move up, air (at atmospheric pressure) is drawn through the inlet valve (with the outlet valve closed), and when it moves down, air is pumped (at a pressure higher than the atmospheric pressure) through the outlet valve (with the inlet valve closed).

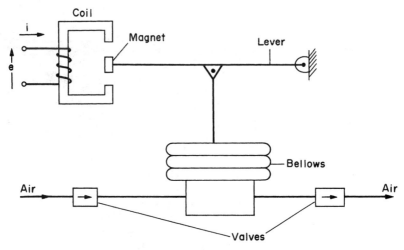

Figure 1.3. Pneumatic pump.

Example 1.4   Brake Mechanism for Train Wagons

In its simplest characterization, a train wagon (e.g., a boxcar used for transporting goods) can be viewed as consisting of three assemblies—body, bogey, and axle–wheel combinations. We focus on the brake mechanism used to bring a moving wagon to rest. The details of the brake rigging are shown in Figure 1.4(a). The brakes can be operated either by the engine operator (air brake) or manually (hand brake) as shown in Figure 1.4(a). The air brakes use pressurized air from a reservoir for operation. Charging of the reservoir, application, and release of the air brake are done using a triple valve, shown schematically in Figure 1.4(b).

The basic principle of operation of the air brake is as follows. The brake cylinder provides the force for the air brake. The force is transmitted through the levers and rods of the brake rigging to each brake block in order to obtain the pressure desired between it and the wheel. In contrast, the hand brake is applied by rotating the hand brake wheel. A chain is wound around a spindle, thus pulling the cylinder lever. The force is then transmitted through the rigging in a manner similar to that for the air brake. Braking is achieved by the pads of the brake block pressing against the wheels. This generates a retarding frictional force that decelerates the train and finally brings it to a stop.

Example 1.5   Liquid Rocket Engine

Rocket propulsion systems consist of liquid rocket engines, which use liquid fuel and oxidizers, solid rocket motors (SRM's), or a combination of both. For example, the booster propulsion system of the space shuttle consists of three liquid rocket engines (the space shuttle main engines, or SSMEs) built by the Rocketdyne Division of Rockwell International (now North American Boeing), and two strap-on SRM's built by Thiokol. The entire system consists of the shuttle itself, the propulsion system, including a large fuel tank containing liquid oxygen and liquid hydrogen, and the guidance systems, comprising, in all, several hundred thousand parts.

Predicting the reliability of a system such as this is a monumental undertaking. This is done by analyzing major subsystems, components, etc., on down to the part level when necessary. The SSME is one of the major subsystems that has been analyzed in great detail by

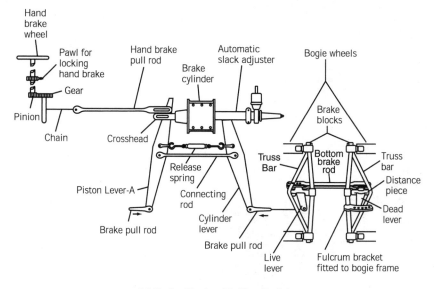

(*a*) Brake Rigging (Trailing Bogie)

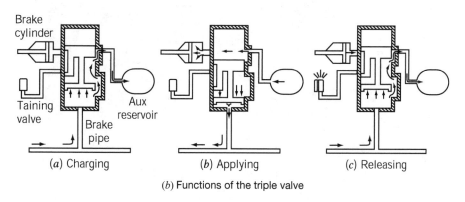

(*b*) Functions of the triple valve

Figure 1.4. Components of trian wagon brake system.

the builder, the purchaser (NASA), and a number of independent contractors and analysts. It is a remarkable machine, consisting of some thousands of parts, and generating 471,500 lb of thrust at sea level.

A schematic of the propellant flow of an SSME is given in Figure 1.5(a). The engine weighs about 7000 pounds, is 14 feet long and 8 feet wide. Its major components include low- and high-pressure hydrogen and oxygen turbopumps, the main fuel injector, main combustion chamber, and nozzle. All of these include numerous parts. As an example, a schematic of the high pressure fuel turbopump is given in Figure 1.5(b). Other components include a cooling system, preburners, and many valves and other essential parts. Many of these undergo extreme stress in operation. Thermal stresses include temperatures ranging from –423 (the temperature of liquid hydrogen) to over 6000°F, the temperature reached in the combustion

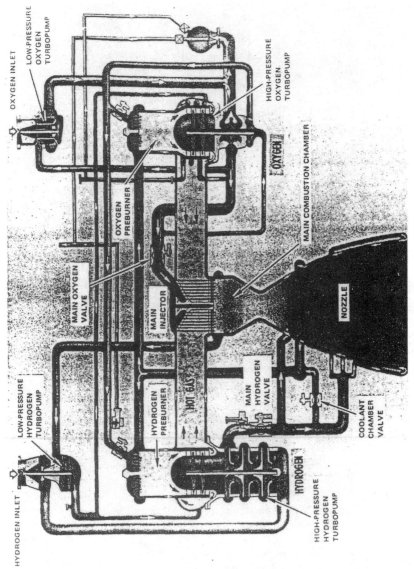

(a) SSME propellant flow schematic

10

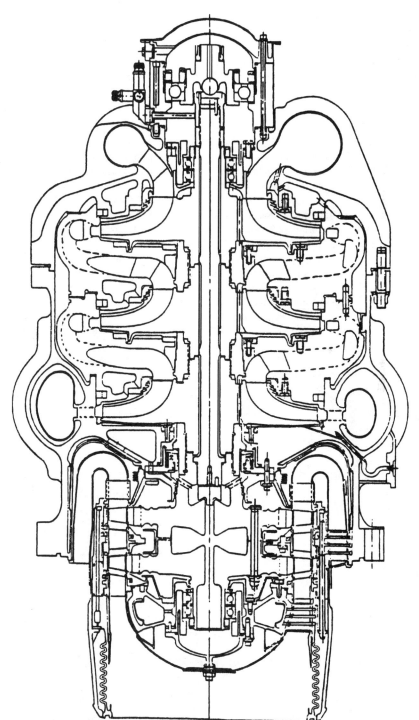

(b) High Pressure Fuel Turbopump

Figure 1.5. Space shuttle main engine.

chamber. Other stresses include extreme levels of vibration and very high performance requirements. The high pressure fuel turbopump spins at 30,000 rpm, and has a discharge pressure of nearly 6900 psi at maximum operating condition. The oxidizer turbopump has a discharge pressure of nearly 8000 psi.

All in all, this engine is a significant engineering achievement. No SSME has ever failed catastrophically in flight.

Example 1.6   Electric Power System

Industrial nations require electrical energy for use in homes as well as in commerce and industry. This electricity is generated by power plants and transmitted to demand centers (which include domestic and/or commercial and industrial consumers) using a network of high- and low-voltage transmission lines. In schematic form, an electric power system can be represented as a network, as shown in Figure 1.6(a). The network consists of two types of nodes—square nodes representing power plants and round nodes representing demand centers—and connecting arcs representing transmission lines that transfer the power from power plants to demand centers.

Each power plant is a complex subsystem consisting of several elements. The main elements of a thermal power plant are shown schematically in Figure 1.6(b). The basic process in a thermal power station is as follows. The chemical energy contained in the fuel is converted into heat energy through combustion in the boiler. This energy is used to generate steam from water. The steam is used to drive a turbine that converts the thermal energy into mechanical energy. Finally, the generator transforms the mechanical energy into electrical energy for transmission over high-voltage lines.

Each element of a thermal power plant consists of several components and these in turn can be decomposed into various parts. A partial list of the components for the different elements are as follows:

- Boiler
  Water tubes
  Drum
  Headers
  Superheater/reheater tubes
- Turbine
  High-, intermediate-, and low-pressure units
    Rotor disk
    Bladings
    Inner casings, Steam chests
- Generator
  Stator
  Rotor
    Retaining rings
    Coils

1.3 SYSTEM AND PRODUCT DETERIORATION AND FAILURE

In the Introduction, we defined failure in an intuitive manner. In this section we refine this concept and discuss some related notions in order to define clearly what is meant by deterioration and failure in a system.[2]

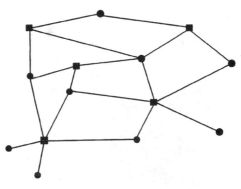

■   **Generating Station**

●   **Demand Center**

—   **High-Voltage Transmission Line**

(a) Power Network

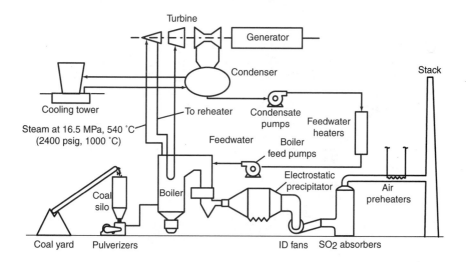

(b) Schematic of a Coal -Fired Steam Power Plant

Figure 1.6. Schematic representation of an electric power system.

1.3.1 Failures and Faults

We begin with some definitions of *failure*.

- "Failure is the termination of the ability of an item to perform a required function." [International Electronic Commission, IEC 50(191)]
- "Equipment fails, if it is no longer able to carry out its intended function under the specified operational conditions for which it was designed." (Nieuwhof, 1984)

- "Failure is an event when machinery/equipment is not available to produce parts at specified conditions when scheduled or is not capable of producing parts or perform scheduled operations to specification. For every failure, an action is required." (Society of Automotive Engineers, "Reliability and Maintainability Guideline for Manufacturing Machinery and Equipment")
- "Recent developments in products-liability law has given special emphasis to expectations of those who will ultimately come in direct contact with what we will do, make or say or be indirectly affected by it. Failure, then, is any missing of the mark or falling short of achieving these goals, meeting standards, satisfying specifications, fulfilling expectations, and hitting the target." (Witherell, 1994)

As can be seen, the key term in the above definitions is the inability of the system or product to function as required. Rausand and Oien (1996) suggest a classification of functions for items of a complex system. The various functions in their classification are as follows:

1. *Essential functions:* This defines the intended or primary function. In Example 1.1, the primary function is to provide light. In Example 1.3, it is to provide air into the tank. In Example 1.5, the primary function is to provide the required amount of thrust as and when needed. In Example 1.6, it is to provide electric power on demand to the consumers who are part of the network.

2. *Auxiliary functions:* These are required to support the primary function. In Example 1.6, transport of power from areas of low demand to areas of high demand, storage of excess capacity, and sale to other networks are examples of auxiliary functions.

3. *Protective functions:* The goal here is to protect people and the environment from damage and injury. In Example 1.4, the brakes provide the protective function. Failure of the brakes to function when needed can lead to a major accident. Similarly, in Example 1.6, relays in the network serve the primary role of offering protection against current surges, and scrubbers on smokestacks remove particulate matter to protect the environment.

4. *Information functions:* These comprise condition monitoring, gauges, alarms, etc. In the liquid rocket engine of Example 1.5, an electronic controller evaluates performance 50 times per second and adjusts valves for peak performance during a shuttle launch. In Example 1.6, the main control panel displays various bits of information about the different subsystems, e.g., voltage and current output of generators, pressure and temperature of steam in various parts of a power generating plant, and so on.

5. *Interface functions:* This deals with the interface between the item under consideration and other items. In Example 1.5, the connecting wires and cables provide this function. If the connection is broken, then the performance of the system is affected. Many interfaces between the components of the propulsion system and other major system of the vehicle are in operation prior to and during a launch and for the entire mission as well.

6. *Superfluous functions:* These are superfluous to the system. They occur due to modifications to a system that make an item no longer necessary.

A *fault* is the state of the system characterized by its inability to perform its required function. (Note, this excludes situations arising from preventive maintenance or any other intentional shutdown period during which the system is unable to perform its required function.) A fault is hence a state resulting from a failure.

It is important to differentiate between failure (fault) and error. According to the International Electrotechnical Commission [IEC 50(191)], an error is a "discrepancy between a computed, observed or measured value or condition and the true, specified or theoretically correct value or condition." As a result, an error is not a failure because it is within the acceptable limits of deviation from the desired performance (target value). An error is sometimes referred to as an incipient failure. [See, Rausand and Oien (1996) for further discussion.]

1.3.2 Failure Modes

A *failure mode* is a description of a fault. It is sometimes referred to as fault mode [for example, IEC 50(191)]. Failure modes are identified by studying the (performance) function of the item. Blache and Shrivastava (1994) suggest a classification scheme for failure modes; this is shown in Figure 1.7.

A brief description of the different failure modes is as follows:

1. *Intermittent failures:* Failures that last only for a short time. A good example of this is software faults that occur only under certain conditions that occur intermittently.
2. *Extended failures:* Failures that continue until some corrective action rectifies the failure. They can be divided into the following two categories:

   a) *Complete failures,* which result in total loss of function
   b) *Partial failures,* which result in partial loss of function

   Each of these can be further subdivided into the following:

   a) *Sudden failures:* Failures that occur without any warning
   b) *Gradual failures:* Failures that occur with signals to warn of the occurrence of a failure

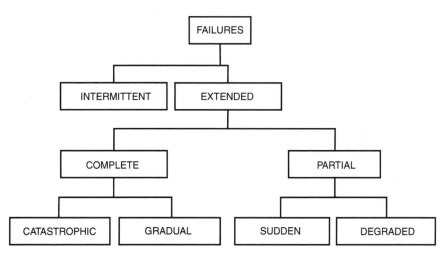

Figure 1.7. Failure classification (from Blache and Shrivastava, 1996).

A complete and sudden failure is called a *catastrophic failure* and a gradual and partial failure is designated a *degraded failure.*

In Example 1.1, failure of the item is sudden and catastrophic in the sense that failure is essentially instantaneous and after failure the bulb no longer emits light. In Example 1.3, if the valve does not shut properly, the flow is not reduced to zero and this can be viewed as a partial failure. However, if the valve fails to operate when the pressure is released (due, for example, to the spring not functioning properly), then the failure is a complete failure. The valve usually wears out with usage and this corresponds to a gradual failure. In a power station (Example 1.6), the bearings (for turbine and generator) often fail due to wear, resulting in gradual deterioration. In contrast, failure due to a lightning strike is sudden and can lead to either partial or complete failure of the network.

1.3.3 Failure Causes and Severity

According to IEC 50(191), *failure cause* is "the circumstances during design, manufacture or use which have led to a failure." Failure cause is useful information in the prevention of failures or their reoccurrence. Failure causes may be classified (in relation to the life cycle of the system) as shown in Figure 1.8. We briefly describe each of these failure causes.

1. *Design failure:* Due to inadequate design
2. *Weakness failure:* Due to weakness (inherent or induced) in the system so that the system cannot stand the stress it encounters in its normal environment
3. *Manufacturing failure:* Due to nonconformity during manufacturing
4. *Aging failure:* Due to the effects of age and/or usage
5. *Misuse failure:* Due to misuse of the system (operating in environments for which it was not designed)
6. *Mishandling failure:* Due to incorrect handling and/or lack of care and maintenance

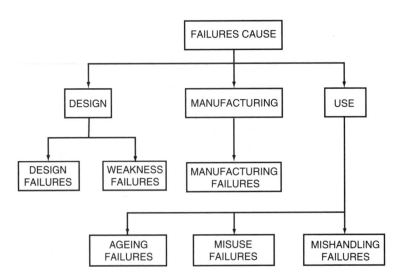

Figure 1.8. Failure cause classification [from IEC 50(191)].

Note that the various failure causes shown in Figure 1.8 are not necessarily disjoint. Also, one can differentiate between primary (or root) cause and secondary and other levels of failures that result from a primary failure.

In Example 1.6, blades in the steam turbine can fail due to excessive thermal stress resulting from poor design. In Example 1.3, aging is often the cause of leaks developing in the bellows. A good example of misuse failure is cooking in a microwave oven using metallic cookware. A mishandling failure in the case of Example 1.1 occurs when the filament breaks loose under a mechanical impact. Damage in shipping or in installation of the SSME would also constitute a mishandling failure. In Example 1.6, bearings of the turbine may fail due to improper lubrication (mishandling), aging, or, in fact, for any of the other reasons.

Finally, the severity of a failure mode signifies the impact of the failure mode on the system as a whole and on the outside environment. A severity ranking classification scheme (MIL-STD 882) is as follows:

1. *Catastrophic:* Failures that result in death or total system loss
2. *Critical:* Failures that result in severe injury or major system damage
3. *Marginal:* Failures that result in minor injury or minor system damage
4. *Negligible:* Failures that result in less than minor injury or system damage

Another classification is given in the reliability centered maintenance (RCM) approach (see Moubray, 1991), where the following severity classes (in descending order of importance) are used:

1. Failures with safety consequences
2. Failures with environmental consequences
3. Failures with operational consequences
4. Failures with nonoperational consequences.

1.3.4 Deterioration

Failure is often a result of the effect of deterioration. The deterioration process leading to a failure is a complicated process, and this varies with the type of product and the material used. Failure mechanisms may be divided into two broad categories (Dasgupta and Pecht, 1991): (i) overstress failures, and (ii) wear-out failures.

Overstress failures are those due to brittle fracture, ductile fracture, yield, buckling, large elastic deformation, and interfacial deadhesion. Wear-out failures are those due to wear, corrosion, dendritic growth, interdiffusion, fatigue crack propagation, diffusion, radiation, fatigue crack initiation, and creep.

As an illustrative example, consider fatigue failure. When cyclic stress is applied to a mechanical component (e.g., the lever rod in Example 1.3), failure of the material occurs at stresses much below the ultimate tensile strength of the material because of the accumulation of damage. Fatigue failure begins with the initiation of a small, microscopic crack. The crack typically develops at a point of discontinuity or at a defect in the material that can cause local stress and plastic strain concentration. This is termed fatigue crack initiation. Once a fatigue crack has been initiated, the crack can propagate in a stable fashion under cyclic stress, until it becomes unstable under applied stress amplitude. The crack propagation rate is a material property. We will discuss the other failure mechanisms in Chapter 6.

The rate at which the deterioration occurs is a function of time and/or usage intensity. The following example from Kordonsky and Gertsbakh (1995) will illustrate this.

Example 1.7 Deterioration in an Aircraft

An aircraft is a complex system consisting of many subsystems. We confine our attention to (i) body, and (ii) engine. The different deterioration processes that lead to failures and the time scales and usage factors that affect the rate of deterioration are as follows:

Subsystem	Time scale/usage intensity	Deterioration
Body	Calendar year	Corrosion
	Time in air	Wear
		Accumulation of fatigue
	Number of landings	Accumulation of fatigue
	(take offs) or flights	High amplitude loading
Jet engine	Calendar year	Corrosion
	Operation time	Wear
		Accumulation of fatigue
	Number of operation cycles	Temperature cycling
	Time in take-off regime	High temperature

The final failure is often due to a cumulative effect of these different types of deterioration.

1.4   CONCEPTS AND HISTORY OF RELIABILITY

1.4.1   Basic Concepts

Reliability of a product (system) conveys the concept of dependability, successful operation or performance, and the absence of failures. Unreliability (or lack of reliability) conveys the opposite. Since the process of deterioration leading to failure occurs in an uncertain manner, the concept of reliability requires a dynamic and probabilistic framework. We use the following definition:

The *reliability* of a product (system) is the probability that the product (system) will perform its intended function for a specified time period when operating under normal (or stated) environmental conditions.

Reliability theory deals with the interdisciplinary use of probability, statistics, and stochastic modeling, combined with engineering insights into the design and the scientific understanding of the failure mechanisms, to study the various aspects of reliability. As such, it encompasses issues such as (i) reliability modeling, (ii) reliability analysis and optimization, (iii) reliability engineering, (iv) reliability science, (v) reliability technology, and (vi) reliability management.

Reliability modeling deals with model building to obtain solutions to problems in predicting, estimating, and optimizing the survival or performance of an unreliable system, the impact of the unreliability, and actions to mitigate this impact.

Reliability analysis can be divided into two broad categories: (i) qualitative and (ii) quantitative. The former is intended to verify the various failure modes and causes that contribute to the unreliability of a product or system. The latter uses real failure data in conjunction with suitable mathematical models to produce quantitative estimates of product or system reliability.

Reliability engineering deals with the design and construction of systems and products, taking into account the unreliability of its parts and components. It also includes testing and programs to improve reliability. Good engineering results in a more reliable end product.

Reliability science is concerned with the properties of materials and the causes for deterioration leading to part and component failures. It also deals with the effect of manufacturing processes (e.g., casting, annealing) on the reliability of the part or component produced.

Reliability management deals with the various management issues in the context of managing the design, manufacture, and/or operation of reliable products and systems. Here the emphasis is on the business viewpoint, as unreliability has consequences in cost, time wasted, and, in certain cases, the welfare of an individual or even the security of a nation.

"The soundness of management is reflected in the quality of products produced and in customer satisfaction. Reliability is merely one quality of the product; others might be performance, style, convenience, economy and so on." (Lloyd and Lipow, 1962, p. 13).

1.4.2   A Brief History of Reliability

Prior to World War II, the notion of reliability was largely intuitive, subjective, and qualitative. The use of actuarial methods (involving statistical techniques) to estimate survivorship of railroad equipment began in the early part of the twentieth century (Nelson, 1982, p. 2). In the late 1930s, extreme value theory was used to model fatigue life of materials and was the forerunner of later probabilistic developments.

A more quantitative (or mathematical) and formal approach to reliability grew out of the demands of modern technology and particularly out of the experiences in World War II with complex military systems (Barlow and Proschan, 1965, p. 1). Since the appearance of this classic book, the theory of reliability has grown at a very rapid rate, as can be seen by the large number of books and journal articles that have appeared on the subject. Many of these will be referred to later; still others are listed in the bibliography given in the final chapter of the book.

Barlow (1984) deals with a historical perspective of mathematical reliability theory up to that time. Similar perspectives on reliability engineering in electronic equipment can be found in Coppola (1984); on space reliability technology in Cohen (1984); on nuclear power system reliability in Fussel (1984) and on software reliability in Shooman (1984).[3]

1.5   PRODUCT LIFE CYCLE AND RELIABILITY

A product life cycle (for a consumer durable or an industrial product), from the point of view of the manufacturer, is the time from initial concept of the product to withdrawal of the product from the marketplace. It involves several stages, as indicated in Figure 1.9.[4]

The process begins with an idea to build a product to meet some customer requirements, such as performance (including reliability) targets. This is usually based on a study of the market and potential demand for the product being planned. The next step is to carry out a feasibility study. This involves evaluating whether it is possible to achieve the targets within specified cost limits. If this analysis indicates that the project is feasible, an initial product design is undertaken. A prototype is then developed and tested. It is not unusual at this stage to find that achieved performance levels of the prototype product are below the target values. In this case, further product development is undertaken to overcome the problem. Once this is

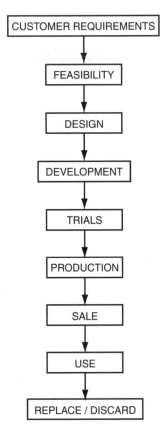

Figure 1.9. Product life cycle.

achieved, the next step is to carry out trials to determine performance of the product in the field and to start a preproduction run. This is required because the manufacturing process must be fine tuned and quality control procedures established to ensure that the items produced have the same performance characteristics as those of the final prototype. After this, the production and marketing efforts begin. The items are produced and sold. Production continues until the product is removed from the market because of obsolescence and/or the launch of a new product.

The life cycle for other products (defense or specialized industrial products) is similar. Here, the product requirements are supplied by the customer and the manufacturer builds the product to these specifications.

We focus our attention on the reliability of the product over its life cycle. Although this may vary considerably, a typical scenario is as shown in Figure 1.10 (based on Figure 9 of Court, 1981). A feasibility study is carried out using the specified target value for product reliability. During the design stage, product reliability is assessed in terms of part and component reliabilities. Product reliability increases as the design is improved. However, this improvement has an upper limit. If the target value is below this limit, then the design using available parts and components achieves the desired target value. If not, then a devel-

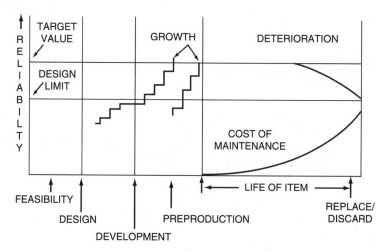

Figure 1.10. Reliability over the product life cycle. Reprinted from *Reliability Engineering,* Vol. 2, p. 256, Copyright 1981, with permission from Elsevier Science.

opment program to improve the reliability through test–fix–test cycles is necessary. Here the prototype is tested until a failure occurs and the causes of the failure are analyzed. Based on this, design and/or manufacturing changes are introduced to overcome the identified failure causes. This process is continued until the reliability target is achieved.

The reliability of the items produced during the preproduction run is usually below that for the final prototype. This is caused by variations resulting from the manufacturing process. Through proper process and quality control, these variations are identified and reduced or eliminated and the reliability of items produced is increased until it reaches the target value. Once this is achieved, full-scale production commences and the items are released for sale.

The reliability of an item in use deteriorates with age. This deterioration is affected by several factors, including environment, operating conditions, and maintenance. The rate of deterioration can be controlled through preventive maintenance, as shown in Figure 1.10.

It is worth noting that if the reliability target values are too high, they might not be achievable with development. In this case, the manufacturer must revise the target value and start with a new feasibility study before proceeding further.

The changing nature of reliability of a product over its life cycle has implications for both the manufacturer and the buyer (owner–user) as will be indicated in the next two sections.

1.6   BUYER'S PERSPECTIVE

As mentioned earlier, buyers can be divided into three categories—individuals (buyers of consumer durables), businesses (buyers of commercial and industrial products/systems as well as consumer durables), and government agencies (buyers of the above plus specialized products/systems). In this section, we discuss reliability related issues from the buyer's perspective for each of the three types of buyers, in each case listing a sample of the types of problems that may be encountered and decisions that must be made.

1.6.1 Individuals

Individuals buy products either for obtaining certain benefits (a refrigerator for extending the life of perishable products, a washing machine to reduce the effort needed to wash clothes, tools for various purposes, etc.), for pleasure (television, stereo, recreational vehicle), or both (cars, personal computers, sports equipment). The performance of the product has a major impact on consumer satisfaction. The decision to buy a product is influenced by this factor. Product performance is in turn affected by the reliability of the product, usage pattern, and operational environment.

One way for manufacturers to inform buyers of high product reliability is through attractive warranty terms. A warranty is a contract that requires the manufacturer to fix problems that occur within the warranty period subsequent to the sale. Over the last few years, manufacturers (and other businesses such as credit card agencies and retailers) have been offering extended warranties. These cover failures beyond the normal warranty period but at an extra cost to the buyer. In this context, two considerations of great interest to a buyer are:

1. Is the extended warranty worth the price?
2. If warranty options are offered, which extended warranty policy best suits his/her needs?

1.6.2 Businesses

In order to function, businesses require equipment of many types—computers and related items; photocopiers, lathes and power tools in a factory, extractors and pumps in a processing plant, tractors and other machines on a farm, engines and vehicles for transport. The performance of such equipment depend on its reliability as well as on other factors, such as usage intensity and maintenance.

In this context, the reliability of equipment and machines becomes critical. When a failure occurs, the impact can be significant—not only economic loss but also damage to property and persons. The deterioration in reliability and its impact on business performance can be reduced by increased maintenance efforts. This improvement is achieved at the expense of increasing costs. This raises a variety of problems of interest to a company, involving the trade-off of the cost of maintenance against the benefits derived. We list a few of these problems.

3. What maintenance policies are feasible and which policy is optimal?
4. How many spares are needed for different parts over the life of an item and what is the optimal purchase strategies for these spares?

In many cases, it is uneconomical for maintenance to be carried out by in-house staff for a variety of reasons (e.g., the need for specialized maintenance equipment and trained staff). In this case, outsourcing is an option. The consideration here is

5. How does one evaluate different service contracts and select the most economical alternative?

1.6.3 Government

Government agencies, especially the military, regularly buy specialized systems—ships, planes, radar equipment, armaments, and so forth. These often involve new technologies and must meet very demanding performance criteria. Such systems are not only very expensive to purchase, they are also expensive to operate and to maintain. (Government agencies, of course, purchase a great deal of more mundane items as well—tires, uniforms, paper goods, copy machines, and other typical consumer and commercial goods—which do not require special design, development, maintenance and operation.)

In the case of specialized systems, bids to develop and build the system are requested from (usually) a small group of manufacturers. A contract is awarded to the successful bidder. This raises several problems related to system reliability and the award of the contract, including

6. Economic decisions must be based on life cycle cost (LCC), which includes not only the acquisition cost, but also the cost of maintaining and operating the system over its life cycle and the cost of disposal at the end of its life cycle. How does one determine the LCC for different options and system reliability requirements?

7. How is the actual reliability of the system delivered evaluated and compared with the reliability stated in the contract? What type of testing should be used?

8. How are the reliability aspects of such programs effectively managed and administered?

1.7 MANUFACTURER'S PERSPECTIVE

From a manufacturer's point of view, the reliability of a product is influenced by several technical factors—design, materials, manufacturing, distribution, and so forth. Product reliability, in turn, affects the commercial side of the business—sales, warranty costs, profits—as shown in Figure 1.11. Poor reliability implies low buyer satisfaction and this in turn affects sales and results in higher warranty costs. This implies that the manufacturer must find solutions to a range of reliability related problems in order to manage reliability issues from the overall business point of view. We indicate a few of these problems.

9. What is the expected cost of servicing warranty for different types of warranties as a function of the reliability of the product?

10. What are the optimal warranty terms and pricing strategy, given the reliability of the product?

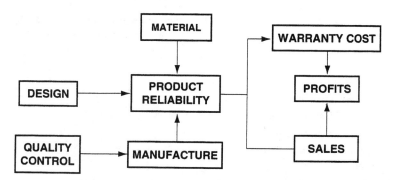

Figure 1.11. Effect of product reliability for a manufacturer.

11. How is the optimal reliability determined, taking into account product development, if the marketing strategy requires offering some specific warranty terms (e.g., a longer warranty period)?

12. What is the number of spares needed to service a specific warranty? How does this change with the warranty duration?

13. What is the impact of process variations on the reliability of the product? How can these be controlled?

The solutions to these problems require the evaluation of product and system reliability. This involves testing items to failure, which often must be done in an accelerated mode (i.e., under higher stress than will be encountered in normal operating conditions) in order to reduce the total time required for testing. Accelerated testing raise several additional problems:

14. How is the reliability of parts or components estimated based on different types of data (failure or nonfailure of items under test) obtained from testing?

15. What is the appropriate relationship between reliability at normal and accelerated testing conditions? How should results be adjusted so that correct inferences under normal conditions can be drawn based on the results of the accelerated tests?

In some cases, a recall of the product to replace a part might be the most economical strategy when there are excessive failures or potential safety issues. The concern here is:

16. When should a manufacturer decide to recall a product with a reliability related defect?

The above list is a small illustrative sample of problems that a manufacturer may need to address. Other problems of relevance relate to issues such as marketing implications, data collection and analysis, product liability and other legal issues, and so on.

1.8   FRAMEWORK FOR SOLVING RELIABILITY RELATED PROBLEMS

1.8.1   Reliability Issues

As can be seen from the list of problems of interest to buyers and to manufacturers, the study of product reliability requires a framework that incorporates many interrelated technical, operational, commercial and management issues. We list some important issues in each of these areas.

Technical issues:

- Understanding of deterioration and failure (material science)
- Effect of design on product reliability (reliability engineering)
- Effect of manufacturing on product reliability (quality variations and control)
- Testing to obtain data for estimating part and component reliability (design of experiments)
- Estimation and prediction of reliability (statistical data analysis)

Operational issues:

- Operational strategies for unreliable systems (operations research)
- Effective maintenance (maintenance management)

Commercial issues:

- Cost and pricing issues (reliability economics)
- Marketing Implications (warranties, service contracts)

Management issues:

- Administration of reliability programs (engineering management)
- Impact of reliability decisions on business (business management)
- Risk to individuals and society resulting from product unreliability (risk theory)
- Effective management of risks from a business point of view (risk management)

The uncertain nature of deterioration and failure implies the need for a suitable framework in which to formulate and solve these problems. The systems approach provides an integrated framework for effectively addressing the issues raised. Figure 1.12 shows many of the important issues and the disciplines involved in their analysis in a diagrammatic manner.

1.8.2 The Systems Approach

The systems approach to problem solving in the real world involves several stages. These are shown in Figure 1.13. The execution of each stage requires a good understanding of concepts and techniques from many disciplines.[5]

The key step is characterization of the system in such a way that the details of the system that are relevant to the problem being addressed are made apparent and appropriately modeled. The variables used in the system characterization and the relationships between them depend on the problem. If the problem is to understand system failures, then the variables of

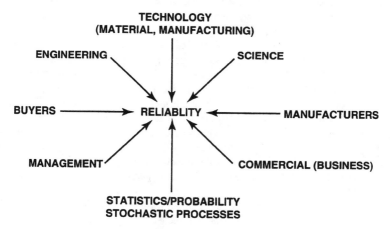

Figure 1.12. Framework for the study of reliability.

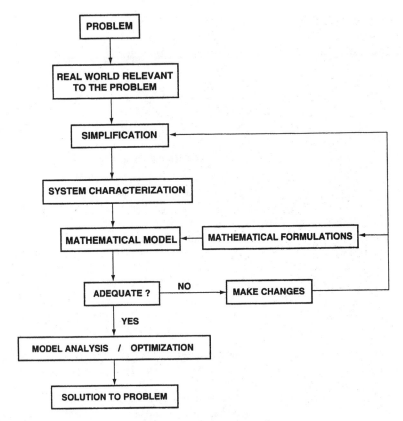

Figure 1.13. Systems approach.

the system characterization are from the relevant engineering sciences; if the problem is to study the impact of reliability on sales, then one would use variables from the theory of marketing and economics in the system characterization; and so forth.

For reliability related problems, most of the variables used in the system characterization are dynamic (changing with time) and stochastic (changing in an uncertain manner). The mathematical formulations needed for modeling reliability are obtained from statistics, probability theory, and stochastic processes. It is important to ensure that the model used is adequate for solving the real problem and that adequate and relevant data can be obtained. If not, then the analysis will yield results that are of limited use for solving the problem. In general, obtaining an adequate model requires an iterative approach, wherein changes are made to the simplification and/or the mathematical formulation during each iteration. Adequate and relevant data are obtained by proper testing as well as from other sources. Statistical methods are used in test design and for both parameter estimation and model validation.

Once an adequate model is developed, techniques from statistics, probability theory and stochastic processes and optimization theory are needed for analysis and optimization. Reliability theory provides the concepts and tools for this purpose.

1.9   OBJECTIVES OF THIS BOOK

The aims of this book are as follows: (i) to discuss a variety of concepts and techniques; (ii) to develop a framework that integrates the technological, operational, commercial, and management issues; and (iii) to use the systems approach to find effective solutions to reliability related problems.

As noted, an important feature of the systems approach is the use of mathematical models in solving problems. In this book, we focus on the building of mathematical models for solving reliability problems. We commence with the modeling of reliability at the part and system levels. Following this, a variety of models for solving many different reliability related problems are developed. Both theoretical issues and real applications are discussed throughout. Our goal is to bridge the gap between reliability theory and its application in real-life problems.

1.10   APPROACH

The approach used in this book is to begin with a look at the real world of reliability by introducing cases and data sets at the outset. In each case, the real-world reliability problem is discussed and the relevant data are presented and described. Concepts and tools for modeling and analysis of systems as well as for analysis of data are then discussed. The applications of these are illustrated by use of appropriate case data.

The cases cover a broad range of topics, including basic materials properties, product design, marketing, maintenance, warranty costing, design of complex systems, life-cycle costing, service contracts, engineering changes, and bidding strategies. Applications cover a broad range of items, including simple hardware items (e.g., composite fibers), relatively simple consumer products (light bulbs), complex consumer products (automobiles), highly complex systems (rocket propulsion systems), commercial items (locomotives, jet engines, aircraft), medical devices (heart pacemakers), highly reliable parts (turbine blades), and others.

Two additional major cases are presented at the end of the book. They are more complex than those used in the earlier chapters. These cases deal with multiple aspects of reliability—design, production, costing, reliability assessment and growth, maintenance, life-cycle costing, and so forth, as appropriate. Each case is discussed in detail, including description, analysis, and interpretation of the data; approaches to reliability modeling and analysis; managerial aspects; and alternative business decisions and their possible consequences.

1.11   OUTLINE OF THIS BOOK

This book is organized into five parts, with a total of 20 chapters. The outlines of the five parts are as follows:

Part A: Context of Reliability Analysis

Part A consists of two chapters and serves as an introduction to reliability and its application in practical situations. In Chapter 1 we have dealt with the issues involved in, and the overall framework needed for effective analysis of reliability, and discussed the scope and outline of the book. Chapter 2 provides a set of illustrative real-world cases covering a range of areas of application. These are used throughout the text to link various reliability issues in an integrative manner.

Part B: Basic Reliability Methodology
Part B consists of three chapters, which deal with the simplest reliability scenario—simple random samples, part-level data, and a single-failure mode. Chapters 3 through 5 deal with, respectively, the collection and preliminary analysis of part-level (or, more generally, single-level) data, basic techniques for modeling reliability in this context, and statistical methods for further analysis of the data and estimation of part reliability.

Part C: Reliability Modeling, Estimation, and Prediction
The six chapters in Part C deal with analysis of more complex systems. We first deal with data and reliability modeling at the component level in Chapter 6, and then extend the results to the level of multicomponent systems in Chapter 7. Various approaches to analysis at these levels are considered, including methods for dealing with highly complex systems. In Chapter 8, additional statistical methods for analysis of reliability data are presented. Chapter 9 provides an introduction to the modeling of software reliability. Chapter 10 deals with data collection at the component and system levels, including some of the trade-offs that might be made in formulating test plans and emphasizing estimation and prediction of system reliability. Finally, Chapter 11 is concerned with model selection and validation.

Part D: Reliability Management, Improvement, and Optimization
The seven chapters in this section cover management decisions that must be made by both consumer and manufacturer in dealing with unreliable systems. Chapter 12 develops the framework for managing unreliable systems in the context of the product life cycle and integrates the technological and commercial issues involved. Chapter 13 deals with reliability engineering. This is of particular significance during the design and development stages of the life cycle. In Chapter 14, we look at reliability assessment and prediction over the life cycle, emphasizing techniques for aggregating data from various sources and updating reliability predictions as additional data are obtained. Chapter 15 deals with alternate approaches to reliability improvement and optimal development strategies. Chapter 16 deals with the operation and maintenance of unreliable systems and examines different maintenance strategies. In Chapter 17, we look at various postsale service issues, including warranties and service contracts. These issues are of importance in the marketing of unreliable products in situations where postsale services are bundled with the product and sold as a package. Finally, Chapter 18 deals with various aspects of optimization of reliability, including optimal design, development, manufacturing, and maintenance strategies.

Part E: Epilogue
The final two chapters (Chapters 19 and 20) of the book provide two case studies and an annotated bibliography. The cases illustrate the integration of the results of the earlier chapters and emphasize interpretation of the results and management issues. The References section includes a large number of sources, covering both theory and applications, where additional resource material on the topics covered in the book may be found.

NOTES

1. The number of levels needed in the decomposition of a system is problem-specific. It depends on the complexity of the system and the goal (or focus) of the study.

2. For an interesting collection of case studies dealing with failures and their consequence, see Schlagar (1994).

3. For more on the history of reliability, see Knight (1991) and Villemeur (1992). The 50th Anniversary Special Publication of the *IEEE Transactions on Reliability* (Volume 47, Number 3-SP, September 1998) contains several historical papers dealing with various aspects of reliability (e.g., reliability prediction, component qualification) and applications, including space systems reliability, reliability physics in electronics, and communications reliability.

4. Product life cycle is discussed in more detail in Chapters 12 and 13.

5. There are several books on the systems approach and mathematical modelling. Murthy, Page, and Rodin (1990) deals with various issues involved in mathematical modeling at an introductory level. It also contains an annotated literature review of books on mathematical modeling. Mathematical modeling for study of reliability requires concepts and tools from probability theory and stochastic processes. A brief summary of the relevant material needed is given in the appendices at the end of the book.

EXERCISES

1.1. Describe the functioning of the following systems in terms of their components:

  (a) An electric toaster

  (b) A bicycle

  (c) An automobile

  (d) A personal computer

  (e) A large communications network

1.2. Describe the different notions of failure for each of the systems listed in Exercise 1.

1.3. For each of the failures identified in Exercise 2, identify the failure causes and the severity of each failure.

1.4. The tire is a major component of a bicycle. Describe the different physical processes that may lead to its degradation and failure.

1.5. Discuss the implications of failure of traffic signals at a busy city intersection from the following three perspectives:

  (a) Users—pedestrians and cars

  (b) The Highway Department

  (c) Law enforcement

1.6. From a user's point of view, the life cycle of an item (e.g., a car) is the time from purchase until disposal at the end of its useful life. Discuss the impact of equipment reliability on the total life-cycle cost to the user.

1.7. List some of the elements of concern in the management of reliability in a manned space system.

1.8. The concept of reliability is applicable to all biological systems as well. Discuss this in the context of humans, listing the various modes of failure and their causes.

1.9. Discuss the impact of unreliability to buyer and seller in the context of an electric power network.

CHAPTER 2

Illustrative Cases and Data Sets

2.1 INTRODUCTION

The purpose of this chapter is to present a sample of real-life cases, each of which deals with problems involving one or more aspects of reliability and which collectively span a number of different types of applications. This is done by describing, where possible, the management and engineering objectives of the study and presenting a set of data on which the relevant decisions will be based, at least in part. In some instances, these objectives may not be known in detail or at all; here the interest is only in the reliability aspects of the problem.

These cases and the accompanying data will be used to illustrate many of the concepts and analyses to be presented in subsequent chapters. The goals are therefore both to motivate the reader and to provide a data-based foundation for the reliability analyses to follow. In addition, some of cases will be revisited a number of times as we progress through increasingly complex models and analytical techniques, and will thereby provide a unifying theme.

Most readers should find some data sets of interest in the collection selected for this volume. The breadth of applications of reliability should also become apparent: cases range from consumer goods to heavy industry, medical appliances, aerospace, and electronics. Later analyses will be undertaken at the part level, the level of simple components, subsystems, and so forth, and range to the level of complex systems. In the final analysis of each set of data (and along the way, as appropriate), the managerial implications of the results will be briefly discussed.

The cases given in this chapter require reliability analysis from many points of view. Managerial and engineering decisions cover many aspects—design decisions, simple part reliability assessment, reliability prediction for a complex system, life-cycle costing, maintenance, warranty cost analysis, and so forth.

Many additional applications and data sets will be presented as illustrative examples and in the exercises that appear at the end of each chapter. Still others will be given in Chapter 19, where two quite complex cases will be discussed in detail. This is intended to be a capstone summarization of the methods of reliability analysis, in which the key data analytical and mathematical tools dealt with in the book are illustrated, complete with interpretation and caveats.

Having previously stated our intent with regard to breadth of coverage, we hasten to note that we have, in fact, provided only a small sample of the nearly endless list of products and decision situations in which reliability analysis is an integral and important aspect of decision making.

As will quickly become apparent as the cases are described, many types of data are encountered in reliability applications. A brief description of the basic types is given in the next section. Data structures will be discussed in detail in Chapter 3.

2.2 RELIABILITY DATA

From a scientific point of view, the best data are those from a carefully designed experiment in which all of the major factors that could influence the response are controlled, proper experimental techniques (e.g., randomization) are used, and the response is precisely measured. In reliability experiments, response is typically time to failure or some related quantity, and this may be affected by many factors, including type of material used, manufacturing process, design factors (such as geometry of a part, design of components, subsystems, and so forth), usage rate, stresses, environmental factors, and many others.

In some instances, failure data are obtained from carefully designed experiments. This is particularly true of data at the part level. On the other hand, data are often available at many levels, and as the level increases up to the system itself, the items being tested become increasingly complex and so do the structure of the experiment, reliability modeling, and the data analysis. Furthermore, there are often many other data sources available to the engineer and reliability analyst. These include basic information on materials properties, data on similar parts or components used in other applications, data from suppliers, operational, field, and warranty data, and so forth. Caution must be exercised in using many such data sets. Most nonexperimental data sets are not well structured, cannot be considered a proper (random) sample, and cannot be verified as to validity and reliability (in the data reliability sense).

There are many other difficulties in collecting and analyzing reliability and related data. In many cases, it would be too costly or infeasible to continue an experiment until all items have failed. In such cases, experimentation is often curtailed prior to the failure of all items on test. The resulting data are "incomplete" in the sense that the experimenter has actual failure times for only some of the items, and times of operation without failure ("service times") for the rest. Correspondingly, field data will may include failure times for failed items and service times for those that had not failed. In both cases, the service times cannot be ignored in assessing item reliability.

Many data sets, most commonly warranty data, will include only failure times (with varying degrees of accuracy) for failed items. It may not even be known how many other such items are in service, and service times will almost never be known. In fact, many failure times will also not be known, namely, those that fail out of warranty and those that fail during the warranty period for which no warranty claim is made.

All in all, there are many types of incomplete data, both experimental and nonexperimental. These will be discussed in more detail in the next chapter.

Another difficulty in obtaining failure data is encountered in dealing with very highly reliable items, i.e., items that rarely fail in certain applications (e.g., turbine blades, bolts, many electronic components). In these instances, it is never possible to obtain complete data, and in fact, it is often the case that no failures at all will be observed, even in extensive testing. There are situations, however, for example, in aerospace and nuclear reactors, where the possibility of failure of such items must be considered and the likelihood of failure assessed. One approach to obtaining data on highly reliable items is accelerated testing, that is, testing under conditions of stress much higher than those that will be encountered in actual operation and then using appropriate models to relate the results to normal operating conditions. Experiments of this type will be discussed in Chapter 14.

The remainder of this chapter is devoted to brief descriptions of many reliability related data sets. Included are true experimental data (from planned, structured experiments) as well as nonexperimental data of a number of types. Complete data as well as several types of incomplete data are represented in the selected sets.

2.3 CASES

The cases that follow will serve to illustrate the types of problems that require reliability analysis. Each is a real situation and the data given in each case are real data, though sometimes disguised at the request of the provider in the interest of confidentiality. Many of the data sets are new to this volume, while some have been published previously. Because of confidentiality in a number of instances, references to the sources of the data are not provided, except for those taken from the literature.

In discussing the cases, it is inevitable that a number of technical terms not yet precisely defined will be used. Most will be familiar to the reader. For those that are not, an intuitive understanding is adequate at this point. More rigorous definitions will be given as needed.

Case 2.1   Reliability of a Simple Part

The company is a large manufacturer of components used widely in audio systems of all types and is well known for the high quality sound and high reliability of its audio equipment. A key component of the system in question consists of three parts. Two of these are metallic parts used in creating sound. The third "part" is an adhesive binding these two parts. The adhesive is the part of interest in the study.

Special adhesives for binding metals are available from a number of sources. In selecting a supplier of the adhesive, the company conducts some fairly extensive tests of each adhesive product under consideration. Here we are concerned with two such tests of a specific adhesive. The first involves components tested after storage under "normal" conditions of usage of the product, basically conditions typical of a normal office environment, say 23°C and 50% relative humidity (RH). The second study is a test of components stored under stress conditions, that is, much higher temperatures and relative humidities, up to 40°C and 90% RH or more, such as might be encountered in a warehouse during hot, humid summer months. The response variable is bond strength in pounds per square inch.

If the part in question—the bond—is not done properly, it can cause sound distortion. Audio engineers designing these sound systems are of the opinion that this will eventually occur only at extremely low bond strengths, less than 0.5 pounds. Because of the stringent quality control (QC) procedures used (all components are tested to 25 pounds bond strength), this virtually never happens when the components are shipped without undue stress, stored in a benign environment, and used under normal operating conditions. In fact, in one application, many millions of these components have been sold without a single failure of this part under any conditions.

Note, incidentally, that here we are looking at the reliability of the bond strength, which is not the same as operational reliability of the component. The latter would involve time to failure of the component as an acceptable producer of audio signals. The two—bond strength and component performance—are related, but in a complex and poorly understood fashion. All of the components tested in this study had bond strengths in excess of that required for high audio performance and nearly all had strengths very substantially in excess of that amount.

Testing of this part is typically done primarily when it is to be used in a new application, which was the case here. Tests may also be done for monitoring purposes, if quality control detects an increase in failure rates, or for special purposes, such as tests under stress conditions, e.g., storage for significant periods of time in areas of high heat and humidity. In the new application, the item may be used in conditions of higher stress, so tests of this type were undertaken as well.

The data are basically field data in that nominal conditions were, in fact, not maintained in either case. Rather, the parts were warehoused under "normal" environments, for varying amounts of time. The objectives of this study were simply to characterize bond strength under the random set of conditions that might be encountered prior to delivery of an audio system to a customer in various locations and at various times of the year and to investigate the effect of additional stress after such storage.

Table 2.1(a) contains the set of data on bond strength obtained for 85 specimens stored for

TABLE 2.1(a). Length of storage (days) and bond strength (pounds) for audio components stored under normal conditions.

Days	Strength	Days	Strength	Days	Strength
3	296	11	278	52	264
3	337	11	271	52	306
3	266	11	316	52	290
3	299	11	248	52	210
3	310	11	273	52	196
4	197	12	312	68	212
4	312	12	308	68	268
4	317	12	309	68	290
4	332	12	327	68	132
4	341	12	256	68	230
5	309	13	289	75	318
5	297	13	212	75	260
5	296	13	249	75	246
5	242	13	273	75	252
5	253	13	309	75	302
6	307	14	298	80	290
6	298	14	252	80	271
6	320	14	326	80	360
6	277	14	235	80	224
6	244	14	297	80	228
7	323	17	332	82	294
7	287	17	247	82	356
7	344	17	311	82	284
7	306	17	259	82	268
7	299	17	352	82	204
10	252	49	237		
10	235	49	278		
10	290	49	217		
10	303	49	268		
10	290	49	235		

varying lengths of time, ranging from three days to 82 days after manufacture. Data for parts stored under various conditions ranging from mild to moderate stress are given in Table 2.1(b). The test conditions for the four data sets were:

Test 1—27°C, 50% RH (a warm room)

Test 2—27°C, 70% RH (warm with somewhat elevated humidity)

Test 3—32°C, 70% RH (hot, somewhat humid)

Test 4—27°C, 100% RH (hot, extremely humid)

The sample sizes were 32, 32, 29, and 10, respectively.

TABLE 2.1(b).   Bond strength for audio components under test conditions.

Test 1	Test 2	Test 3	Test 4
345.0	210.0	378.0	44.5
230.0	272.0	254.0	3.5
26.0	210.0	277.5	11.5
91.0	247.0	352.5	4.0
222.0	223.0	359.0	2.5
325.0	263.0	275.5	132.0
251.0	9.0	245.0	7.5
131.0	265.0	281.5	1.0
322.0	214.0	307.5	48.0
237.0	32.0	125.5	3.5
8.0	282.0	265.0	
306.0	1.0	244.5	
45.0	276.0	266.0	
272.0	202.0	289.0	
264.0	231.0	175.5	
277.0	251.0	272.5	
332.0	5.0	352.5	
100.0	75.0	320.0	
7.0	325.0	139.0	
254.0	50.0	266.0	
157.0	240.0		
260.0	68.0		
204.0	233.0		

Case 2.2   Reliability of a Major Component, Incomplete Data

In this case we are concerned with a commercial product, specifically an aircraft windshield. The company that produces the product is a leader in advanced composite technology for production of windshields and canopies for high-performance commercial, military, and general aviation aircraft. The company has been producing this product line for over fifty years. Its products are used in over 100 different models of aircraft.

The windshield on a large aircraft is a complex piece of equipment, comprised basically of several layers of material, including a very strong outer skin with a heated layer just beneath it, all laminated under high temperature and pressure. Failures of these items are not structural failures. Instead, they typically involve damage or delamination of the nonstructural outer ply, or failure of the heating system. These failures do not result in damage to the aircraft, but do result in replacement of the windshield.

Data on all windshields are routinely collected and analyzed. At any specific point in time, these data will include failures to date of a particular model as well as service times of all items that have not failed. Data of this type are incomplete in that not all failure times have as yet been observed.

The data in the study with which we are concerned were originally used in analysis and prediction of warranty costs. Several alternative warranty policies were being considered. Management issues involved the relative potential future cost of different warranties and warranty periods. These were to be predicted based on the existing data set. Details of the warranty analysis and a more complete description of the data are given in Blischke and Murthy (1994).

Data on failure and service times for a particular model windshield are given in Table 2.2. The data consist of 153 observations, of which 88 are classified as failed windshields and the remaining 65 are service times of windshields that had not failed at the time of observation.

TABLE 2.2. Aircraft windshield failures, failure times, and service times of unfailed items (thousands of hours).

Failure times				Service times		
0.040	1.866	2.385	3.443	0.046	1.436	2.592
0.301	1.876	2.481	3.467	0.140	1.492	2.600
0.309	1.899	2.610	3.478	0.150	1.580	2.670
0.557	1.911	2.625	3.578	0.248	1.719	2.717
0.943	1.912	2.632	3.595	0.280	1.794	2.819
1.070	1.914	2.646	3.699	0.313	1.915	2.820
1.124	1.981	2.661	3.779	0.389	1.920	2.878
1.248	2.010	2.688	3.924	0.487	1.963	2.950
1.281	2.038	2.823	4.035	0.622	1.978	3.003
1.281	2.085	2.890	4.121	0.900	2.053	3.102
1.303	2.089	2.902	4.167	0.952	2.065	3.304
1.432	2.097	2.934	4.240	0.996	2.117	3.483
1.480	2.135	2.962	4.255	1.003	2.137	3.500
1.505	2.154	2.964	4.278	1.010	2.141	3.622
1.506	2.190	3.000	4.305	1.085	2.163	3.665
1.568	2.194	3.103	4.376	1.092	2.183	3.695
1.615	2.223	3.114	4.449	1.152	2.240	4.015
1.619	2.224	3.117	4.485	1.183	2.341	4.628
1.652	2.229	3.166	4.570	1.244	2.435	4.806
1.652	2.300	3.344	4.602	1.249	2.464	4.881
1.757	2.324	3.376	4.663	1.262	2.543	5.140
1.795	2.349	3.385	4.694	1.360	2.560	

Failure times in the data correspond to several categories of failure, including delamination, upper-sill delamination, coating burn-out, outer-ply breakage, and accidental damage (which includes damage due to human causes as well as that due to natural causes). Failure causes that are not associated with normal use (e.g., shattered during maintenance) were treated as service times in the data. Failures due to faulty installation, and similar nonusage-related failures were excluded from the data entirely.

Case 2.3 Heart Pacemaker Data

The U.S. Food and Drug Administration (FDA) requires that failures of implantable medical devices be reported to the Center for Devices and Radiological Health (CDRH) on a regular basis. Reports are submitted by manufacturers, importers, distributors, medical practitioners, and user facilities. In this case, failures are defined to be malfunctions that could cause serious injury or death. The data given in Table 2.3 were obtained from CDRH Device Experience Network, which contains over 600,000 reports submitted through August 1, 1996. The data selected for this case are data submitted on faults of heart pacemakers. The data are available on the web at http://www.fda.gov/cdrh/mdrfile.html.

Table 2.3 contains failure data for two manufacturers of heart pacemakers (Companies "A" and "B"). The data are on similar products for the two companies and were reported in

TABLE 2.3. Heart pacemaker faults/failures.

Fault number	Fault	Frequency Company A	Company B
1	Sensing anomaly	28	29
2	Malfunction observed in testing	40	0
3	Battery problem	28	15
4	Elective replacement	0	30
5	Loss/anomaly of capture and sensing	33	39
6	Battery voltage	7	0
7	Improper pacing/no output	26	15
8	Other medical problems	10	24
9	Telemetry problem	12	5
12	Programming problem	6	2
13	Other pacer problem	12	21
15	Increasing lead impedance	2	2
19	High lead impedance	14	9
30	Low lead impedance	26	5
31	Faulty lead	2	9
32	Broken lead	8	29
33	Insulation	2	7
34	Mechanical problem	1	0
35	Lead(s) dislodged	7	1
36	Twiddler's syndrome	1	1
37	Rising thresholds	7	10
38	Leads incorrectly connected	0	1

January, 1996. During the period covered, Company A reported 272 faults and Company B, 255 faults. Neither dates of production nor dates of failure of specific items are known, though most failures are likely to have taken place within a month or so prior to the report date. Production, however, could have been as long as seven years earlier. Production rate at Company A is 150 to 200 units per day; at Company B it is about 300 to 400 per day.

Many types of failures are indicated. Some of these suggest a possible design problem, some suggest production problems, some are medical problems during implant, and some are medical problems having nothing to do with implantation of the device. In addition, some are indeterminate. The category "Other pacer problem" includes a number of faults not clearly described and/or reported only once.

The important considerations with products of this type are determining the most frequently occurring failure modes, and determining how to correct them through improved design and/or production. It is apparent from Table 2.3 that the products produced by the two companies can fail in many ways. A difficulty in comparing the two companies is that they may define failure types differently.

Case 2.4   Automobile Warranty Data

Automobiles are probably the most complicated pieces of equipment most people buy. The combination of number of parts and number of ways they can go wrong number into the tens of thousands. Auto manufacturers have many sources of data appropriate to reliability analysis—test data, data on identical or very similar parts on previous year's models, data from suppliers, warranty data, and so forth. Here we look at data resulting from warranty claims involving the engine of a particular model. This is a very small subset of the vast amounts of warranty data routinely collected by an automobile manufacturer.

Warranty data are used by the manufacturer for many purposes, the most important of which is usually prediction of future warranty costs (or, more precisely, updating such predictions). There are many difficulties in using warranty data, and caution must be exercised. One of the problems is that there is very often a delay in reporting a failure; unless it is disabling, a motorist will wait until it is convenient to bring a car in to the dealer for warranty service. Thus the reported mileage at failure overstates the lifetime of the failed item. In addition, some claims are not legitimate and some failures go unreported.

The data in this example were collected by a distributor of 4-wheel drive vehicles imported into Australia and are for a single model year. A total of about 5000 units were sold. These data are incomplete because failures that occur after the end of the warranty period, in this case 40,000 km, are not included in the data base. Another difficulty is that no indication of the environment in which the vehicles were driven is given. Driving conditions in cities are quite different from those in the outback.

The data are given in Table 2.4. Included are usage (thousands of kilometers) at the time of a warranty claim because of an engine problem and cost ($) of the resulting repair. Note that the engine is a very reliable component and most of the failures are relatively minor problems, as is evident from the cost data. The data were collected on a total of 329 automobiles of this make and model year for which a warranty claim of some type had been made. Of these, 297 experienced no engine problems up to 40,000 km and hence had no warranty claims relating to the engine. Given in the table are data on time to first failure for the remaining 32 automobiles, each of which experienced at least one engine failure within the warranty period.

TABLE 2.4. Warranty claims data for automobile engines
(*n* = 329). Warranty period 40,000 km.

Auto number	km at failure (000)	Cost of repair ($)
1	13.1	24.60
2	29.2	5037.28
3	13.2	253.50
4	10.0	26.35
5	21.4	1712.47
6	14.5	127.20
7	12.6	918.53
8	27.4	34.68
9	35.5	1007.27
10	15.1	658.36
11	17.0	42.96
12	27.8	77.22
13	2.4	77.57
14	38.6	831.61
15	17.5	432.89
16	14.0	60.35
17	15.3	48.05
18	19.2	388.30
19	4.4	149.36
20	19.0	7.75
21	32.4	29.91
22	23.7	27.58
23	16.8	1101.90
24	2.3	27.78
25	26.7	1638.73
26	5.3	11.70
27	29.0	98.90
28	10.1	77.24
29	18.0	42.71
30	4.5	1546.75
31	18.7	556.93
32	31.6	78.42

Case 2.5 Industrial Product Sold in Batches

A producer of industrial equipment sells the product to other commercial as well as government organizations. Depending on the customer's needs at the time of purchase, the product is sold in batches ranging from 1 to 100 items. The item is essential for operation of the facility. As a result, buyers will ordinarily purchase a large enough batch so that an adequate number of spares are available for replacements when items in service fail. Customers often store failed items for a period of time or until the number of reserves reaches too low a point, at which time the reserved failed items are returned to the manufacturer for repair or replacement.

Failure data for this product, reported by Lyons and Murthy (1996), are given in Table 2.5.

TABLE 2.5. Failure data, industrial product sold in batches.

Batch number	Batch size	Age of batch (days)	Number of claims	Claim times	Number of failures at claim time
1	2	679	1	642	1
2	2	636	0		0
3	2	623	0		0
4	2	588	0		0
5	2	588	2	255	1
				350	1
6	2	588	3	255	1
				354	1
				355	1
7	2	588	1	322	1
8	10	574	0		0
9	7	556	1	556	2
10	4	505	3	131	1
				172	1
				228	1
11	4	494	0		0
12	2	487	0		0
13	8	477	1	209	1
14	100	469	7	104	12
				154	4
				189	6
				272	23
				346	3
				350	12
				368	5
15	1	463	0		0
16	2	463	0		0
17	2	463	0		0
18	2	456	0		0
19	2	444	0		0
20	2	441	1	67	1
21	2	441	0		0
22	2	441	0		0
23	2	435	0		0
24	1	382	0		0
25	2	380	0		0
26	4	379	0		0
27	2	378	0		0
28	2	354	0		0
29	24	351	2	216	4
			218	2	
30	1	345	0		0
31	20	344	2	221	1
				243	1
32	2	275	0		0

TABLE 2.5. *Continued*

Batch number	Batch size	Age of batch (days)	Number of claims	Claim times	Number of failures at claim time
33	1	245	1	45	1
34	16	234	4	217	1
				95	2
				193	1
				219	3
35	2	224	0		0
36	1	224	0		0
37	4	221	0		0
38	2	220	0		0
39	38	205	0		0
40	4	182	0		0
41	2	182	1	102	1
42	4	155	0		0
43	12	142	0		0
44	20	126	0		0
45	10	101	0		0
46	4	63	0		0
47	2	10	0		0
48	2	9	0		

The data were collected over a period of 22 months. Shown are batch size, age of items at the time of observation, total number of claims for that batch, claim times, and number of failed items included in each claim. For example, three claims were filed for Batch 10, which consisted of 4 items. The age of the batch (time since sale) was 505 days, with one failure time having been reported at 131 days, one at 172 days, and one at 228 days. Batch sizes vary from 1 to 100.

The data are incomplete in several ways. First, the times given include storage times for those items not put into immediate use and no indication of the amount of storage time is given. Second, batches having no failures are included; these results are thus storage time, if any, plus service time. Finally, some of the claims are for multiple items at the same age, implying that these failures occurred at or before the stated age of the batch.

These data are to be used for evaluating product reliability and predicting future warranty costs. For the latter purpose, additional data not included in this data set are needed. This would include types of failure, repair cost for each, warranty terms, and so forth.

Case 2.6 Tensile Strength of Fibers

Composite materials are comprised of a matrix material and reinforcing elements, with the latter ordinarily being dominant in determining material strength. Zok et al. (1995) report the results of an experiment in which strengths of silicon carbide fibers were measured after extraction from a ceramic matrix. This experiment is an attempt to estimate fiber strength after incorporation into the composite, rather than modeling composite strength and estimating each component separately, which is the usual approach in estimating composite strength. The difficulty with the latter is that it is necessary to account for the interaction between the two materials in determining strength (see Fox and Walls, 1997).

Fiber strength is measured as stress applied until fracture failure of the fiber. One of the main objectives of the experiment was to determine the distribution of failures as a function of gauge length of the fiber. Another is to compare strengths after incorporation into the composite with that of a standard pristine fiber.

Gauge lengths used in the study and sample sizes (n) at each were 265 mm ($n = 50$), 25.4 mm ($n = 64$), 12.7 mm ($n = 50$), and 5.0 mm ($n = 50$). The results of the tensile tests are given in Table 2.6

TABLE 2.6. Tensile strengths of fibers (SiC).

Fiber length (mm)							
265		25.4		12.7		5.0	
0.36	1.93	1.25	2.81	1.96	3.29	2.36	3.81
0.50	1.96	1.50	2.82	1.98	3.30	2.40	3.88
0.57	1.97	1.57	2.90	2.06	3.36	2.54	3.93
0.95	1.99	1.85	2.92	2.07	3.39	2.67	3.94
0.99	2.04	1.92	2.93	2.07	3.39	2.68	3.94
1.09	2.06	1.94	3.02	2.11	3.41	2.69	3.94
1.09	2.06	2.00	3.11	2.22	3.41	2.70	3.698
1.33	2.08	2.02	3.11	2.25	3.43	2.77	4.04
1.33	2.11	2.13	3.14	2.39	3.52	2.77	4.07
1.37	2.26	2.17	3.20	2.42	3.72	2.79	4.08
1.38	2.27	2.17	3.20	2.63	3.96	2.83	4.08
1.38	2.27	2.20	3.22	2.67	4.07	2.91	4.16
1.39	2.38	2.23	3.26	2.75	4.09	3.04	4.18
1.41	2.39	2.24	3.29	2.75	4.13	3.05	4.22
1.42	2.47	2.30	3.30	2.75	4.13	3.06	4.24
1.42	2.48	2.33	3.34	2.89	4.14	3.24	4.35
1.45	2.73	2.42	3.35	2.93	4.15	3.27	4.37
1.49	2.74	2.43	3.37	2.95	4.29	3.28	4.50
1.50		2.45	3.43	2.96		3.34	
1.56		2.49	3.43	2.97		3.36	
1.57		2.51	3.47	3.00		3.39	
1.57		2.54	3.61	3.03		3.51	
1.75		2.57	3.61	3.04		3.53	
1.78		2.62	3.62	3.05		3.59	
1.79		2.66	3.64	3.07		3.63	
1.79		2.68	3.72	3.08		3.64	
1.82		2.71	3.79	3.13		3.64	
1.83		2.72	3.84	3.20		3.66	
1.86		2.76	3.93	3.22		3.71	
1.89		2.79	4.03	3.23		3.73	
1.90		2.79	4.07	3.26		3.75	
1.92		2.80	4.13	3.27		3.78	

Case 2.7   Reliability of Hydraulic Systems

The hydraulic components of large machines used in moving ore and rock (called load–haul–dump, or "LHD" machines) comprise a critical subsystem with regard to the reliability of this piece of equipment. Kumar and Klefsjö (1992) discuss hydraulic subsystems used in LHD's in underground mines in Sweden. Data were collected on time between consecutive failures (TBF) for such a system during the development phase, over a period of two years.

Of concern is the formulation of maintenance policies for machines of this type. There are other critical components that will affect these decisions, but this study focused only on the hydraulic subsystem. Of concern are maintenance intervals, system deterioration, and related issues, including engineering modifications and reliability growth. In addition to the managerial and engineering aspects, there were a number of statistical questions that were addressed in the study. These include trend analysis, correlation, and modeling of the failure data. These issues are discussed in detail in the reference cited.

The data on which the various analyses were based are given in Table 2.7. Data are suc-

TABLE 2.7.   Time (operating hours) between successive failures of hydraulic systems of selected LHD machines

LHD1	LHD3	LHD9	LHD11	LHD17	LHD20
327	637	278	353	401	231
125	40	261	96	36	20
7	397	990	49	18	361
6	36	191	211	159	260
107	54	107	82	341	176
277	53	32	175	171	16
54	97	51	79	24	101
332	63	10	117	350	293
510	216	132	26	72	5
110	118	176	4	303	119
10	125	247	5	34	9
9	25	165	60	45	80
85	4	454	39	324	112
27	101	142	35	2	10
59	184	38	258	70	162
16	167	249	97	57	90
8	81	212	59	103	176
34	46	204	3	11	370
21	18	182	37	5	90
152	32	116	8	3	15
158	219	30	245	144	315
44	405	24	79	80	32
18	20	32	49	53	266
	248	38	31	84	
	140	10	259	218	
		311	283	122	
		61	150		
			24		

cessive times to failure (operating hours) for six hydraulic subsystems. The first two machines (LHD1 and LHD3) are the oldest of the machines observed, the second two (LHD9 and LHD11) are of medium age, and the last two (LHD 17 and LHD 20) are relatively new machines. The number of failures of the hydraulic subsystems over the period of observation ranged from 23 to 28.

Case 2.8   Bus Motor Failure Data

A well-known and often-analyzed data set in reliability applications is the record of times to initial and subsequent failures of bus motors in a large city fleet. The data were first reported and analyzed by Davis (1952). Subsequent analyses are reported by Mudholkar et al. (1995) and in the references cited therein. Failures considered are major breakdowns consisting of those which either caused the motor to be completely inoperative or to have insufficient power for operation of the bus. Minor failures were not included.

The data recorded are time to first through fifth failures for 191 buses, as measured by distance traveled (thousands of miles) between failures. The data, grouped into 20,000 mile intervals, are given in Table 2.8.

Note from the totals in the table that not all buses experienced more than one motor failure; only 104 of the 191 experienced a second failure, and so forth. The data also do not enable us to determine the individual failure sequences (that is, time to first failure for an individual motor, and then time interval between first and second failures for that motor, etc.). In addition, the fleet size is not given, so it is not known if there were buses that experienced no engine failures. Modes of failure are also not known (except that they were disabling), nor are repair actions, which could have an effect on types and times of succeeding failures.

Here the interest is in characterizing time between successive failures. Because of the limitations imposed by the structure of the data (individual units not having been followed through time), this cannot be done in detail, but at least average times to failure can be studied as a function of number of prior failures.

TABLE 2.8.   Bus motor failure data (miles between failures, in thousands).

| Miles (000) | Failure | | | | |
	First	Second	Third	Fourth	Fifth
< 20	6	19	27	34	29
20–40	11	13	16	20	27
40–60	16	13	18	15	14
60–80	25	15	13	15	8
80–100	34	15	11	12*	7*
100–120	46	18	16*		
120–140	33	7			
140–160	16	4*			
160–180	2				
> 180	2				
Total	191	104	101	96	85

*Indicates greater than lower bound of category.

Case 2.9 Lifetimes of Light Bulbs

Another of the data sets given by Davis (1952) consists of lifetimes of 40 Watt incandescent light bulbs obtained in monitoring product quality. Although these data are over fifty years old, they represent well the kinds of data obtained in this context and are illustrative of the product, a schematic of which is given in Figure 1.1.

The data, given in Table 2.9, are in groups of ten per week over a 42-week period in 1947 (with three observations missing, for reasons not explained). The main objective of the experiment was to characterize the production process through time, monitoring for quality changes. This is useful both in determining conformance to specifications and in predicting the consumer demand for replacements.

Case 2.10 Air Conditioning Unit Failures

Data on times between failures of air conditioning units on aircraft are reported by Proschan (1963). The data on thirteen aircraft are given in Table 2.10. The purpose of the study, as noted by Proschan, was to characterize the distribution of failure times. Information of this type is useful in analysis of many aspects of reliability, maintenance, and so forth. Since the aircraft is out of service if the air conditioning unit is inoperable, an effective maintenance policy is essential. Proschan provides additional discussion of some of these issues and a detailed analysis of the data.

Case 2.11 Electronic Connectors

In a medical application, it is necessary to be able to insert and withdraw leads to a connector with a minimum amount of force. In this application, 2.5 pounds is considered too high. An experiment was conducted using a sample of 30 leads. For each lead, force for insertion and withdrawal was recorded on each of two trials. The data are given in Table 2.11.

Here an observed value of 2.5 pounds pull strength or greater would be considered a failure. Note that the data set includes no failures. A key objective of the study is to determine the proportion of such leads that would be classified as failed in accordance with this standard. This is an illustration of a situation that is not uncommon in dealing with highly reliable parts. The analyst must assess the chances of failure based on data in which no failures have occurred.

A question that must be addressed in this regard is the appropriateness of various probability models on which this assessment is to be based. A secondary issue in the analysis is whether or not the data can be pooled in some meaningful way to provide an overall estimate of the chances of failure. If it cannot be determined with an acceptable level of confidence that the proportion of items with pull strengths greater than 2.5 pounds does not exceed a specified value, it may be necessary to redesign the lead.

Case 2.12 Laser Weld Strength

Tests of the strengths of laser welds bonding titanium parts were run at three settings of the welding equipment (schedules). Strengths are measured by means of pull tests, with strengths of less than 3 pounds considered unacceptable for the application in which this part is to be used. The data, consisting of results of pull tests from samples of size 30 from production runs at each of the three schedules, are given in Table 2.12.

TABLE 2.9. Time to failure for 417 40W light bulbs

Week	Item 1	Item 2	Item 3	Item 4	Item 5	Item 6	Item 7	Item 8	Item 9	Item 10
1	1067	919	1196	785	1126	936	918	1156	920	948
2	855	1092	1162	1170	929	950	905	972	1035	1045
3	1157	1195	1195	1340	1122	938	970	1237	956	1102
4	1022	978	832	1009	1157	1151	1009	765	958	902
5	923	1333	811	1217	1085	896	958	1311	1037	702
6	521	933	928	1153	946	858	1071	1069	830	1063
7	930	807	954	1063	1002	909	1077	1021	1062	1157
8	999	932	1035	944	1049	940	1122	1115	833	1320
9	901	1324	818	1250	1203	1078	890	1303	1011	1102
10	996	780	900	1106	704	621	854	1178	1138	951
11	1187	1067	1118	1037	958	760	1101	949	992	966
12	824	653	980	935	878	934	910	1058	730	980
13	844	814	1103	1000	788	1143	935	1069	1170	1067
14	1037	1151	863	990	1035	1112	931	970	932	904
15	1026	1147	883	867	990	1258	1192	922	1150	1091
16	1039	1083	1040	1289	699	1083	880	1029	658	912
17	1023	984	856	924	801	1122	1292	1116	880	1173
18	1134	932	938	1078	1180	1106	1184	954	824	529
19	998	996	1133	765	775	1105	1081	1171	705	1425
20	610	916	1101	895	709	860	1110	1149	972	1102
21	990	1141	1127	1181	856	716	1308	943	1272	917
22	1069	976	1187	1107	1230	836	1034	1248	1061	1550
23	1240	932	1165	1303	1085	813	1340	1137	773	787
24	1438	1009	1002	1061	1277	892	900	1384	1148	
25	1117	1225	1176	709	1485	1225	1011	1028	1227	1277
26	1222	912	885	1562	1118	1197	976	1080	924	1233
27	1135	623	983	883	1088	1029	1201	898	970	1058
28	1160	831	1023	1354	1218	1121	1172	1169	1113	1308
29	1166	1470	1635	1141	1555	1054	1461	1057	1228	1187
30	1016	744	1197	1122	666	1022	964	1085	612	1003
31	1235	942	1055	893	1235	1056	968	1056	1014	1096
32	1013	889	1430	926	1297	1033	1024	1103	1385	
33	1077	813	1121	960	1156	1033	1255	225	525	675
34	1211	995	924	732	935	1173	1024	1254	1014	
35	798	1080	862	1220	1024	1170	1120	898	918	1086
36	1028	1122	872	826	1337	965	1297	1096	1068	943
37	1490	918	609	985	1233	985	985	1075	1240	985
38	1105	1243	1204	1203	1310	1262	1234	1104	1303	1185
39	759	1404	944	1343	932	1055	1381	816	1067	1252
40	1248	1324	1000	984	1220	972	1022	956	1093	1358
41	1024	1240	1157	1415	1385	824	1690	1302	1233	1331
42	1109	827	1209	1202	1229	1079	1176	1173	769	905

TABLE 2.10. Time between failures of aircraft air conditioning systems.

						Plane						
7907	7908	7909	7910	7911	7912	7913	7914	7915	7916	7917	8044	8045
194	413	90	74	55	23	97	50	359	50	130	487	102
15	14	10	57	320	261	51	44	9	254	493	18	209
41	58	60	48	56	87	11	102	12	5		100	14
29	37	186	29	104	7	4	72	270	283		7	57
33	100	61	502	220	120	141	22	603	35		98	54
181	65	49	12	239	14	18	39	3	12		5	32
	9	14	70	47	62	142	3	104			85	67
	169	24	21	246	47	68	15	2			91	59
	447	56	29	176	225	77	197	438			43	134
	184	20	386	182	71	80	188				230	152
	36	79	59	33	246	1	79				3	27
	201	84	27	15	21	16	88				130	14
	118	44	153	104	42	106	46					230
	34	59	26	35	20	206	5					66
	31	29	326		5	82	5					61
	18	118			12	54	36					34
	18	25			120	31	22					
	67	156			11	216	139					
	57	310			3	46	210					
	62	76			14	111	97					
	7	26			71	39	30					
	22	44			11	63	23					
	34	23			14	18	13					
		62			11	191	14					
		130			16	18						
		208			90	163						
		70			1	24						
		101			16							
		208			52							

The objectives of the study were to compare schedules and characterize the results, including analysis of the proportion of welds that, on average, will be found to have strengths of less than three pounds.

Case 2.13 · Clerical Errors

Many applications involve the tracking of failures sequentially through time, recording the time between successive failures. One version of this occurs in situations wherein a task, performed either correctly or incorrectly, is repeated through a (fixed or random) number of trials. Here "time" is measured as the number of trials rather than as clock time, and "failure" is a task incorrectly performed.

A set of data of this type is given by Davis (1952). In this application, data are entered into

TABLE 2.11.   Pull strength (lb) for insertion and withdrawal of leads.

	Insertion		Withdrawal	
Lead	Trial 1	Trial 2	Trial 1	Trial 2
1	1.450	1.350	1.225	1.130
2	1.115	1.020	1.000	1.030
3	1.450	1.510	1.035	0.930
4	0.995	0.935	0.800	0.760
5	1.315	1.395	0.985	0.990
6	1.515	1.455	0.950	0.835
7	1.415	1.455	0.995	0.995
8	1.420	1.505	1.115	1.410
9	0.985	1.025	1.005	0.920
10	1.335	1.375	1.115	1.100
11	0.960	0.950	0.875	0.875
12	1.225	1.380	0.865	0.835
13	1.355	1.440	0.995	0.940
14	1.530	1.545	1.265	1.365
15	1.425	1.390	1.035	1.075
16	1.175	1.160	1.110	1.085
17	1.240	1.220	1.140	1.005
18	1.595	1.405	0.995	1.025
19	1.445	1.350	1.225	1.100
20	1.075	0.980	1.025	0.985
21	1.310	1.225	0.985	1.210
22	1.160	1.135	1.055	1.095
23	1.340	1.330	1.070	1.070
24	1.280	1.200	0.940	1.005
25	1.110	1.280	1.000	1.120
26	1.145	1.155	0.985	1.175
27	1.280	1.045	1.250	1.220
28	1.050	1.155	1.110	0.890
29	0.990	0.780	1.005	0.850
30	1.790	1.660	1.305	1.300

a file, entries in the data file are verified, and the number of correct entries until the next incorrect entry in the file is observed for five clerks over a ten-day period. (In counting trials, the trial in which the error occurs is included in the count.) Other types of possible errors, e.g., machine or verification errors, are not identified in the data set. The data are given in Table 2.13.

This data set is representative of many applications of this type, e.g., manual entry of data into computer files. Reliability of the data entry process has important implications with regard to the results of any analyses done on the data and their use in managerial decisions. Conclusions based on unreliable data can be misleading or even disastrous.

A question of interest in analysis of these data is whether or not errors occur randomly. This would be the case if the error rate for each clerk were constant. The primary interest here

TABLE 2.12.   Bond strengths of laser welds for three schedules.

	Schedule	
1	2	3
7.48	5.70	8.92
7.22	5.14	17.74
7.22	6.40	16.56
9.00	5.28	14.70
8.96	3.76	16.04
7.96	4.06	13.96
7.54	7.10	18.22
5.80	8.24	13.34
6.78	5.64	11.46
6.36	6.30	17.74
7.74	4.62	12.80
6.68	7.68	11.10
10.16	4.40	14.60
7.86	5.50	12.88
5.46	7.28	11.92
7.26	5.30	16.04
6.80	5.58	11.58
8.54	5.90	14.42
6.72	5.00	18.84
6.74	5.12	16.62
8.20	5.80	18.64
6.72	5.76	18.84
6.92	5.98	13.50
10.06	5.86	10.64
6.98	4.22	16.70
7.46	7.28	16.20
6.38	5.98	18.24
7.56	6.20	12.84
7.48	5.76	15.30
7.28	5.86	13.06

is in modeling the random error process, though the results could also be used to compare clerks and to estimate the number of errors in other data files.

Case 2.14   Ball Bearing Data

Another classic data set in reliability analysis is a set of measurements on fatigue failure of deep-groove ball bearings. The data are taken from Lieblein and Zelen (1956), who provide a detailed analysis on just under 5,000 bearings tested by four companies over the period 1936 to 1951. Results from hundreds of tests under various loads are reported. Good engineering requires data of this type.

TABLE 2.13.   Number of entries until an error occurs.

Clerk 1	Clerk 2	Clerk 3	Clerk 4			Clerk 5	
734	451	726	234	50	149	26	61
121	3	883	133	3	74	12	305
404	1116	142	14	115	170	190	267
646	1143	196	233	65	2	170	573
1072	447	14	46	295	129	201	282
148	630	1905	1327	235	3	481	414
312	37	456	259	781	65	157	48
773	2031	2565	115	343	44	165	150
43	1786	610	369	49	204	75	212
1102	659	1263	563	551	333	334	189
111	151	847	286	105	60	671	47
641	210	881	145	19	11	10	95
754	1426	1214	69	127	60	290	57
598	72	248	1822	77	20	321	160
86	699	195	686	82	608	6	99
2138	426	548	183	24	19	231	139
150	1040	234	146	64	64	204	18
1047	277	1096	268	941	113	45	609
907	72	530	115	1153	413	94	64
165	1286	338	494	147	75	176	211
166	235	356	215	67	22	488	35
6	625	217	539	878	403	76	249
94	493	195	467	339	299	233	81
1023	2	77	4	151	396	350	294
903	756	392	37	545	6	33	98
355	1460	3114	453	565	156	333	418
303			629	651	330	66	233
1378							
202							
343							
1266							

The specific data set that has been used in many studies is given in Table 2.14. The data consist of 23 observations on bearings containing nine balls each, taken from a worksheet used in the study and reproduced by Leiblein and Zelen (with two other points omitted for reasons not specified). The data are in millions of revolutions until failure. Failure modes are also indicated. These are data on a single product from a single manufacturer. The data are complete failure data and the objective of the experiment is to evaluate performance of the product. This is an example of a well-designed experiment, with controlled conditions.

Additional analyses of these data are discussed by Lawless (1982), Richards and McDonald (1987), and elsewhere. The data have been used to illustrate and fit various probabilistic models.

TABLE 2.14. Endurance life (millions of revolutions) of ball bearings.

Bearing number	Endurance	Type of failure
16	17.88	Ball
10	28.92	Ball
5	33.00	Ball
19	41.52	Inner Ring
9	42.12	Ball
11	45.60	Ball
15	48.48	Ball
12	51.84	Ball
20	51.96	Ball
18	54.12	Inner Ring
13	55.56	Inner Ring
1	67.80	Ball
4	68.64	Ball
6	68.64	Left Bore
25	68.88	Disc
22	84.12	Ball
17	93.12	Ball
7	98.64	Inner Ring
23	105.12	Inner Ring
24	105.84	Disc.
21	127.92	Ball
8	128.04	Outer Ring
14	173.40	Disc

Case 2.15. Helicopter Failure Data

Luxhoj and Shyur (1995) report data on time to failure of helicopter components and provide a reliability analysis of the data. Data on three parts for a particular model are given in Table 2.15. The data include aircraft (AC) flight hours as well as operational time to failure of each part.

These data, which account for 37% of system (i.e., helicopter) failures, consist of failure data on three parts. The objective of the study was to determine an appropriate reliability model. This involves a probabilistic analysis of the failure data.

Case 2.16. Compressor Failure Data

Large air compressors require "bleeding," that is, clearing of lines and the tank of any liquid build-up due to condensation, prior to operation. Data on units of this type at military bases are reported by Abernethy et al. (1983) for several environments. Table 2.16 contains data on failures at a base located near a seacoast, where salt air was thought to be a major contributor to the problem. The failures considered are those due to binding in the bleed system

Failure data are reported on 202 units, 192 of which had not failed prior to the time of ob-

TABLE 2.15. Aircraft flight hours and time to failure of three helicopter parts.

Part	AC hours	Part hours
1	3566.3	406.30
1	8266.2	644.60
1	4495.8	744.80
1	4479.3	213.40
1	4479.3	213.30
1	833.7	156.50
1	13825.3	265.70
1	13825.3	265.70
1	11699.1	1023.60
1	9139.1	337.70
1	9139.1	337.70
1	3523.3	774.80
1	2476.5	573.50
1	2476.5	573.50
2	4477.5	495.90
2	290.3	117.30
2	3652.9	16.90
2	506.5	207.53
2	506.5	207.53
2	906.6	410.10
2	906.6	410.10
2	3006.7	573.60
2	3006.7	573.60
2	1162.5	209.48
2	270.2	270.20
2	920.6	920.60
2	750.1	750.10
2	750.1	750.10
2	354.5	354.50
2	564.5	564.50
2	392.1	392.10
3	4057.0	4057.00
3	1500.9	1163.00
3	4096.6	607.40
3	1161.2	158.70
3	14241.8	1088.40
3	8402.8	1199.80
3	3271.3	420.00
3	11279.7	751.10
3	5387.0	838.00

Reprinted from *Reliability Engineering and System Safety,* Vol. 38, p. 230, Copyright 1995, with permission from Elsevier Science.

TABLE 2.16.

(a) Operating time until compressor failure	
Operating hours	Observation number
708	7
808	23
884	37
884	38
1013	72
1082	87
1105	92
1198	115
1249	127
1251	128

(b) operating time for 202 compressors (failed and unfailed units)	
Operating hours	Frequency
201–300	2
301–400	0
401–500	0
501–600	2
601–700	2
701–800	10
801–900	26
901–1000	27
1001–1100	22
1101–1200	24
1201–1300	24
1301–1400	11
1401–1500	11
1501–1600	20
1601–1700	8
1701–1800	4
1801–1900	2
1901–2000	3
2001–2100	3
2101–2200	1

servation. The original data included service times (times of operation without failure) for these units. All of the data, service and failure times, were ordered from smallest to largest and numbered in order of magnitude. Table 2.16(a) includes the failure times and the observation number for each failed unit. (Gaps in the numbers recorded indicate that unfailed units had service times between this and the prior entry in the table.) All of the remaining units had operating times in excess of 1275 hours without having experienced a failure. A frequency distribution (See Chapter 3) of the entire data set is given in Table 2.16(b).

The objective of the study is to characterize the probabilistic failure structure of this unit based on the incomplete information provided. Nothing is known about other variables that may provide information of relevance to failure, e.g., age of the equipment, frequency of operation, other environmental conditions, and so forth.

Case 2.17   Jet Engines

Many additional data sets are reported by Abernethy et al. (1983). Nearly all involve aircraft engines or components or other military equipment. The data in this case are another such example.

A jet aircraft engine consists of many parts and, though generally very reliable, can fail in many ways. One approach to analyzing data on multiple failure modes is to analyze each mode separately. Table 2.17 includes data on 31 military aircraft at a particular base. Of these, six had been removed and repaired prior to a scheduled maintenance action because of failure of a turbine airfoil. Data are flight hours until failure for the failed items and approximate flight hours for nonfailed items. (Data reported for nonfailed items were given in interval groupings; the data shown in Table 2.17 are the midpoints of the reported intervals.)

TABLE 2.17.   Failure data (flight hours) for jet engines

Failure times	Service times (nonfailures)	
684	350	1350
701	650	1450
770	750	1550
812	850	1550
821	850	1650
845	950	1750
	950	1850
	1050	1850
	1050	1950
	1150	2050
	1150	2050
	1250	2050
	1250	

Case 2.18   Offshore Oil Exploration Equipment

Assuring high reliability of equipment used in offshore oil drilling operations is a high priority in the design, operation, and maintenance of the equipment. The equipment is operated in extreme environments and is subject to high levels of stress. Unreliability has serious potential consequences, leading to hazards to crews, equipment, and the environment, and to significant loss of production.

A data collection project for reliability assessment and evaluation of various types of offshore equipment is described by Sandtorv et al. (1996). Data were collected in several phases, with increasingly broad objectives, ranging from risk and availability assessment to maintenance optimization. A summarization of the data, based on Sandtorv et al.(1996) is given in Table 2.18. Included are data on 14 categories of equipment. The data were collected from 1981 through 1992. The data are summary data only, giving number of units and number of

TABLE 2.18. Offshore oil equipment failure data summary.

Equipment class	Number of units	Number of failures
Gas turbines	154	5245
Compressors	112	4762
Electric generators	125	2111
Pumps	852	5159
Vessels	742	1180
Heat Exchangers	764	505
Valves	2202	1918
Fire and gas detectors	9511	5363
Process sensors/controls	4227	2078
Drilling equipment	880	1106
Electric power systems	1321	455
Subsea systems	77	85
Miscellaneous safety systems	1703	737
Miscellaneous utility systems	1035	2147

Reprinted from *Reliability Engineering and System Safety,* Vol. 51, p. 161, Copyright 1996, with permission from Elsevier Science.

failures observed in each category, with no indication of individual times to failure, number of failures per unit or other details that may be important in reliability analysis.

Case 2.19 Throttle Failures

Carter (1986) and Jiang and Murthy (1995a and 1995b) report data on throttle failures in prototype models of general purpose load carrying vehicles. The data consist of failure times for 25 units and service times for 25 units that had not failed at the time of observation. "Time" is measured in kilometers driven prior to either failure or cessation of observation. The data are given in Table 2.19. The objective of this study was to determine an appropriate model for the data that might give some insight into the nature of the failure mechanism.

TABLE 2.19. Failure and service times of throttles.

Failure times		Service times	
484	3791	478	2981
626	4443	583	3392
850	5900	753	3392
1071	6226	753	3904
1318	6711	801	4829
1472	6835	834	5328
1579	6947	944	5562
1610	7878	959	6122
1729	7884	1377	6331
1792	10263	1534	6531
1847	13103	2400	11019
2550	23245	2639	12986
2568	.	2944	.

Case 2.20   Breaking Strengths of Single and Bundled Fibers

Breaking strengths of materials in general are related to many factors, including the nature of the material, the geometry of the structure in which it is used, the stress to which it is subjected, and so forth. Individual fibers are one of the simplest geometrical structures and are used as the elementary unit in many primary analyses of materials' properties. A second-level structure is a bundle of such fibers, and one of the goals of testing is to determine the relationship of between the strengths of individual fibers and of fiber bundles of various sizes.

Many explanations have been offered by materials scientists for the relationship between size and strength of the fiber and size and strength of the fiber bundle, and much experimentation in this area has been done. Information of this type is very valuable in the engineering design of products that employ these items.

Experiments typically involve stressing fibers or bundles until failure occurs and recording the maximum stress prior to failure. Data of this type for carbon fibers are given by Watson and Smith (1985). The data consist of failure stresses (giga Pascals) for single carbon fibers of length 1, 10, 20, and 50mm, given in Table 2.20(a), and failures stresses of bundles of 1000 such fibers of lengths 20, 50, 150, and 300mm, given in Table 2.20(b).

These data and some similar data sets have been further analyzed by Smith (1991), and Durham and Padgett (1997).

Case 2.21   Breaking Strength of Carbon Fibers in Resin

Experiments involving single carbon fibers similar to those discussed in Case 2.20 are described by Wolstenholme (1995), except that in this case the fibers were embedded in resin. The experiments described were part of a study of the "weakest link principle," and are important in obtaining an understanding of failure mechanisms of fibers and of composites containing them.

In the experiments, segments of various lengths are cut from individual fibers and tested to determine breaking strength. Since the segments are from individual fibers rather than simply randomly selected from the material, the effect of variability of fiber diameter is eliminated. In one such experiment, stress at failure was observed for segments of lengths 5, 12, 30, and 75mm from each of 26 fibers. The data are given in Table 2.21 (Table 4 in Westenholme, 1995). Note that one observation is missing for Fiber 18 and two are missing for Fiber 21 due to breakage prior to testing.

Case 2.22   Stress Fatigue Failures, Tungsten Alloy

As a final set of data on fiber strength tests, we look at a second set of data analyzed by Smith (1991). The data are experimental results on bend strength of specimens of a brittle material, a tungsten carbide alloy containing 6% cobalt with ground surface finish. Specimens were subjected to five different stress rates in a logarithmic progression from 0.1 to 1000 MN per square meter per second, with 12 specimens tested at each stress rate. The data were also analyzed by Green (1984) using an alternative statistical approach. The observed fracture stresses are given in Table 2.22.

Case 2.23   Software Reliability

The production of software that is free of any significant errors is of obvious importance to software vendors, particularly in highly competitive markets. It can be of critical importance

TABLE 2.20.   Failure stresses (Gpa) of single carbon fibers and bundles (tows) of 1000 fibers.

(a) Failure stresses of single carbon fibers							
1 mm Fiber		10 mm Fiber		20 mm Fiber		50 mm Fiber	
2.247	4.519	1.901	3.139	1.312	2.478	1.339	2.308
2.640	4.542	2.132	3.145	1.314	2.490	1.434	2.335
2.842	4.555	2.203	3.220	1.479	2.511	1.549	2.349
2.908	4.614	2.228	3.223	1.552	2.514	1.574	2.356
3.099	4.632	2.257	3.235	1.700	2.535	1.589	2.386
3.126	4.634	2.350	3.243	1.803	2.554	1.613	2.390
3.245	4.636	2.361	3.264	1.861	2.566	1.746	2.410
3.328	4.678	2.396	3.272	1.865	2.570	1.753	2.430
3.355	4.698	2.397	3.294	1.944	2.586	1.764	2.431
3.383	4.738	2.445	3.332	1.958	2.629	1.807	2.458
3.572	4.832	2.454	3.346	1.966	2.633	1.812	2.471
3.581	4.924	2.454	3.377	1.997	2.642	1.840	2.497
3.681	5.043	2.474	3.408	2.006	2.648	1.852	2.514
3.726	5.099	2.518	3.435	2.021	2.684	1.852	2.558
3.727	5.134	2.522	3.493	2.027	2.697	1.862	2.577
3.728	5.359	2.525	3.501	2.055	2.726	1.864	2.593
3.783	5.473	2.532	3.537	2.063	2.770	1.931	2.601
3.785	5.571	2.575	3.554	2.098	2.773	1.952	2.604
3.786	5.684	2.614	3.562	2.140	2.800	1.974	2.620
3.896	5.721	2.616	3.628	2.179	2.809	2.019	2.633
3.912	5.998	2.618	3.852	2.224	2.818	2.051	2.670
3.964	6.060	2.624	3.871	2.240	2.821	2.055	2.682
4.050		2.659	3.886	2.253	2.848	2.058	2.699
4.063		2.675	3.971	2.270	2.880	2.088	2.705
4.082		2.738	4.024	2.272	2.954	2.125	2.735
4.111		2.740	4.027	2.274	3.012	2.162	2.785
4.118		2.856	4.225	2.301	3.067	2.171	2.785
4.141		2.917	4.395	2.301	3.084	2.172	3.020
4.216		2.928	5.020	2.339	3.090	2.180	3.042
4.251		2.937		2.359	3.096	2.194	3.116
4.262		2.937		2.382	3.128	2.211	3.174
4.326		2.977		2.382	3.233	2.270	
4.402		2.996		2.426	3.433	2.272	
4.457		3.030		2.434	3.585	2.280	
4.466		3.125		2.435	3.585	2.299	

(b) Failure stresses of tows of 1000 carbon fibers			
20 mm tow	50 mm tow	150 mm tow	300 mm tow
2.526	2.485	2.11	1.889
2.546	2.526	2.26	2.115
2.628	2.546	2.34	2.177
2.628	2.546	2.44	2.259

continued

TABLE 2.20.   *Continued*

(b) Failure stresses of tows of 1000 carbon fibers			
20 mm tow	50 mm tow	150 mm tow	300 mm tow
2.669	2.567	2.51	2.279
2.669	2.628	2.51	2.320
2.710	2.649	2.57	2.341
2.731	2.669	2.57	2.341
2.731	2.710	2.61	2.382
2.731	2.731	2.61	2.382
2.752	2.752	2.61	2.402
2.752	2.772	2.65	2.443
2.793	2.793	2.67	2.464
2.834	2.793	2.71	2.485
2.834	2.813	2.71	2.505
2.854	2.813	2.71	2.505
2.875	2.854	2.75	2.526
2.875	2.854	2.75	2.587
2.895	2.854	2.75	2.608
2.916	2.895	2.75	2.649
2.916	2.916	2.77	2.669
2.957	2.936	2.77	2.690
2.977	2.936	2.79	2.690
2.998	2.957	2.83	2.710
3.060	2.957	2.83	2.751
3.060	3.018	2.83	2.751
3.060	3.039	2.87	2.854
3.080	3.039	2.87	2.854
	3.039	2.90	2.875
	3.080	2.90	
		2.92	
		2.94	

in commercial and industrial operations, where computerized systems are used in a vast and growing number of applications, and in military and space applications, where mission success and safety of personnel and equipment are vitally dependent on the correct operation of both hardware and software.

In assessing software performance, the interpretations of "failure" and "reliability" are somewhat different from those of hardware reliability. For example, failure may be defined as the existence of a bug in a line of code (and note that there are many types of bugs) or as failure of the program to perform one of its intended functions correctly, or as production of improper output, and so forth. Similarly, time to failure may be variously defined as actual time to detection of a bug in the debugging process, number of trials until a bug is found, etc. In addition, the testing of software proceeds along quite different lines from the testing of hardware. For obvious reasons, rather different models and data analytical techniques are used, and the results are sometimes used for quite different purposes (e.g., to estimate the number of remaining bugs, rather than to redesign a component).

TABLE 2.21. Stress at failure for carbon fibers in resin.

	Fiber length		
5 mm	12 mm	30 mm	75 mm
3.0720	3.2551	3.2272	2.9928
3.7711	3.7819	3.3454	3.1283
4.1194	3.4787	3.3540	3.4787
3.6335	3.4465	3.6206	3.5260
3.2164	3.1906	2.9369	2.8187
3.3884	3.2895	3.0982	2.7219
3.1476	3.0853	3.1433	2.6015
3.1476	2.9305	2.8015	2.8703
3.5260	3.3884	3.0229	2.9133
3.4572	3.5131	3.2422	3.1949
3.5991	3.6701	3.5260	3.0874
3.8098	3.4852	3.6701	3.1605
3.6701	3.8012	3.3218	3.0358
3.7303	3.7711	3.2788	3.1154
3.5884	3.7088	3.4099	3.0530
3.7819	3.6336	3.4787	2.9606
3.8829	3.5626	3.5991	3.2788
*	3.6228	3.1906	3.2895
4.0399	3.3390	3.4185	3.3347
3.5067	3.5841	3.4465	3.2143
*	3.8636	*	3.2229
3.8206	3.9367	3.4615	3.3218
3.6959	3.6572	3.5561	3.2465
3.9603	4.0012	3.7002	3.4465
3.9066	3.7453	3.7819	3.3949
3.9775	3.8980	3.5884	3.3583

*Missing value.

TABLE 2.22. Fracture stress of tungsten carbide alloy at five stress rates.

		Stress rate		
0.1	1	10	100	1000
1676	1895	2271	1997	2540
2213	1908	2357	2068	2544
2283	2178	2458	2076	2606
2297	2299	2536	2325	2690
2320	2381	2705	2384	2863
2412	2422	2783	2752	3007
2491	2441	2790	2799	3024
2527	2458	2827	2845	3068
2599	2476	2837	2899	3126
2693	2528	2875	2922	3156
2804	2560	2887	3098	3176
2861	2970	2899	3162	3685

An analysis of software failure data from the Naval Tactical Data System (NTDS) is given by Hossain and Dahiya (1993). The data, summarized in Table 2.23, are time in days between detection of errors in four phases of preparation of a software module—production phase, test phase, user phase, and a later test phase. The data have been used by various authors in analyses of software failure models.

Case 2.24   Aircraft Radar Component Failures

Since the early 1970's, federal law has required that warranties be offered on nearly all items purchased by the military, including weapons systems. The cost of warranty to both the government and the defense industry is strongly dependent on the reliability of the equipment, and many other factors, such as maintainability, availability, repairability, and so forth, also play a role. The warranties in many cases are quite complex, often involving reliability improvement efforts even after delivery of production items has begun, and are negotiated as a part of the procurement activity. Determining compliance with the warranty may also be a difficult problem.

As a result, a large amount of data is collected by both the suppliers of defense equipment and the customer. An analysis of field data for military aircraft data is discussed by Lakey (1991). The stated objective is to provide a basis for making important warranty decisions. The statistical techniques are illustrated by application to a number of electronic and mechanical items used in F-15 and F/A-18 aircraft. A part of the data (reconstructed from the charts in Lakey, 1991) is given in Table 2.24. The data are hours flying time until first failure of a component called a radar receiver exciter used in the F/A-18 aircraft. There were 241 aircraft in all, the time frame was 1982–1985, and each exciter experienced at least one failure in that period. The data are grouped into forty-hour intervals.

Case 2.25   Ship Engine Maintenance

Equipment reliability has a significant impact on maintenance cost, in an inverse relationship. As a result, initial expenditures on design and manufacturing to increase reliability can be expected to be reflected in reduced future maintenance costs. Another approach to reducing the total cost of maintenance for certain types of equipment is to schedule regular preventive maintenance procedures.

Lee (1980) provides an analysis of a set of data on time between "significant maintenance events" on the Number 4 main propulsion diesel engine on the submarine USS Grampus. The data (cumulative operating hours), given in Table 2.25, include unscheduled

TABLE 2.23.   Summary of software failure data.

Phase	Number of errors found	Average time to detection
1	26	9.615
2	5	58.000
3	1	258.000
4	2	25.500
Overall	34	24.971

TABLE 2.24. Time (flying hours) to first failure of radar receiver exciter WUC 742G2 F/A-19A, 1982–1985.

Hours	Frequency	Hours	Frequency
0-40	39	520-560	4
40-80	35	560-600	4
80-120	25	600-640	5
120-160	15	640-680	2
160-200	14	680-720	3
200-240	16	720-760	3
240-280	13	760-800	2
280-320	15	800-840	2
320-360	10	840-880	3
360-400	5	880-920	2
400-440	7	920-960	1
440-480	7	960-1000	1
480-520	8		

corrective maintenance actions as well as regularly scheduled engine overhauls. The data were used by the author and the references cited to study the effect of regular overhaul on engine reliability.

Case 2.26 Failures of Electronic Modules

Many large organizations that utilize large numbers of certain items in a variety of applications maintain computerized data bases of reliability-related information. This is particularly true of the military, NASA, and some other government agencies. In some instances, the data

TABLE 2.25. Cumulative operating hours until the occurrence of a significance maintenance event.

Failure times						Overhaul
860	3902	5755	9042	12368	14173	1203
1258	3910	6137	9330	12681	14173	3197
1317	4000	6221	9394	12795	14449	5414
1442	4247	6311	9426	13399	14587	7723
1897	4411	6613	9872	13668	14610	10594
2011	4456	6975	10191	13780	15070	14357
2122	4517	7335	11511	13877	22000	15574
2439	4899	8158	11575	14007	22575	
3203	4910	8498	12100	14028		
3298	5676	8690	12126	14035		

are quite complete, giving the usage conditions, failure history, manufacturer, application, and many other pertinent facts. More usually, the data are sparse, incomplete, and perhaps inaccurate. This is particularly true of field (operational) data. Nonetheless, data of this type can be useful in providing some reliability information in early design stages.

The data in Table 2.26 are records of failure data of amplifier and logic circuits used in military equipment. This is an illustration of a typical situation in which the data are provided with limited additional information. They are taken from a data bank as it existed in the mid-1970's and consist of only item and manufacturer identification, numbers of parts and numbers of failures, and total hours of operation (in millions). The data bank also provides partial information on some elements not given in Table 2.26, for example, application (aircraft,

TABLE 2.26.   Failure data for microelectronic circuits.

(a) Amplifier circuit			
Manufacturer	Total number of parts	Million part hours	Number of failures
1	624	0.097968	0
1	480	0.075360	0
1	240	0.037680	0
2	3708	0.582156	0
1	5	0.071885	0
3	9	0.129393	1
4	60	0.861780	3
8	1480	0.270840	5
5	10	0.027820	0
5	2	0.008556	1
1	12	0.030432	0
6	35	0.017500	1
6	210	0.105000	4
6	210	.0105000	6

(b) Logic circuit			
Manufacturer	Total number of parts	Million part hours	Number of failures
6	7872	1.235904	1
6	7968	1.250976	1
6	5520	0.866640	2
6	4320	0.678240	0
6	3216	0.504912	1
6	1488	0.233616	0
6	432	0.067824	0
6	384	0.060288	0
6	1632	0.256224	0
1	3	0.043131	0
1	8	0.115016	0
1	21	0.301476	2
1	23	0.330671	0
9	318	0.841428	1
7	48	0.127008	1

shipboard, etc.) for some items and electrical, thermal, and mechanical stress (low, medium, high) for some items.

A great deal of information is not given; for example, failure times for individual items, service times of unfailed items, types of failures, details of usage and environment, and many other relevant items. Furthermore, submission of information to this data bank was voluntary. Thus the data are nowhere nearly complete, nor can they be considered to be a scientific (random) sample of the manufactured items. This is a good illustration of the difference between the types of data that can be obtained from carefully designed scientific experiments (as in many of the previous cases) and what might be described as haphazard data collection. As will be seen in later chapters, the implications with regard to statistical analysis and interpretation and use of the results in engineering design and analysis are substantial.

Case 2.27   Nuclear Power Plant Cooling System

An important safety system in certain types of nuclear reactors is the low-pressure coolant injection system (LPCI), which provides coolant to the reactor in case of an accident that causes the pressure in the reactor vessel to be low. Martz and Waller (1990) describe a system of this type used in 1150 megawatt U.S. nuclear-powered, boiling-water electric power plants. The LPCI system normally operates in a stand-by mode and must be available on demand and on starting must operate for at least a specified length of time.

The system consists of pumps, check valves, and motor-operated valves. In order for the system to function, a minimal number of these components must be operable; the motors operating the pumps must start, the check valves must open, and the motor-operated valves must function properly. A schematic of such a system is given by Martz and Waller. They provide data on a system consisting of 4 pumps (A, . . . , D), two of which (C and D) are back-up units, a check valve (CV-48) for each pump, motor-operated valve (MOV-25) for each pair of pumps, and a check valve (CV-48) connected to each MOV. The data, given in Table 2.27, consist of the results of 240 success–fail type tests for each component.

These data are used by Martz and Waller along with data from other sources to assess reliability of the LPCI system.

NOTES

1. There are several databases that contain failure data information. References to some of these can be found in Chapter 20.

TABLE 2.27.   LPCI system component test data.

Component	Failure mode	Number of tests	Number of successes
Pump A	Failure to start	240	236
Pump B	Failure to start	240	240
Pumps C and D	Failure to start	240	238
CV-48 A,B,C,D	Failure to open	240	240
CV-46 A,B	Failure to open	240	240
MOV-25 A,B	Failure to operate	240	240

2. We have not given any data sets associated with biological systems. Appendix 1 of Kalbfleisch and Prentice (1980) contains many such data sets. See also Elandt-Johnson and Johnson (1980) for some additional data sets.

3. Books that deal with specific topics in reliability also contain many interesting real data sets. For example, Nelson (1990) contains several data sets relating to accelerated testing.

EXERCISES

2.1. Prepare a table that summarizes the data given in the cases discussed in this chapter. Include the following items in the table:

   (a) Case title
   (b) Variables for which data are given, including units of measurement
   (c) Identify which data sets are quantitative and which are qualitative
   (d) For the quantitative data sets, identify which are complete data and which are incomplete

2.2. True experimental data are data collected under controlled conditions, with proper randomization employed in the experimental procedure (or at least with data collected in such a way that the result can reasonably be assumed to constitute a random sample). Nonexperimental data are data that arise under conditions over which the experimenter or analyst has little or no control and the sample cannot be considered random. Add a column (e) to your table of case data in which you indicate whether you think the data are experimental or nonexperimental data. (Note that the distinction is not always obvious—some sets may be considered either, depending on the assumptions made regarding the data source. Furthermore, in some cases, not enough information may be given.)

2.3. Add a column (f) giving a brief description of the objectives of the study. In several of the cases, we have indicated that certain essential information is not given in the source from which the data were obtained. Summarize these in Column (g). Think about any other information that might be useful in addressing the objectives given in Column (f) and indicate this in an additional column (h).

2.4. Further background information may be obtained from the sources cited in many of the cases. Look up these sources where possible, and prepare brief summaries of the broader context and objectives of the study. Include any other information you think may be pertinent to a reliability analysis.

2.5. Many of the references cited were not the original sources of the data. In these cases, determine the original source, obtain a copy of the article, and write a brief summary of any additional relevant information.

2.6. A number of the sources mentioned in the previous exercises contain additional reliability data. Prepare a list of these additional data sets, describing each, and, if possible, including tables of the data. Indicate briefly the conclusions of the study with regard to data and reliability issues.

2.7. Collect one or more data sets that involve reliability applications from your own experience as an engineer, manager, analyst, or student. (Don't hesitate to ask colleagues, professors, etc.) Describe each in detail, including all the elements listed in the previous problems.

PART B

Basic Reliability
Methodology

CHAPTER 3

Collection and Preliminary Analysis of Failure Data

3.1 INTRODUCTION

In Chapter 1, the two concepts most important to this book, failure and reliability, were introduced. There are many aspects to failure—time to failure, failure mode, causes of failure, consequences and their severity, and so forth—and many dimensions of reliability—theory, modeling, test design, assessment, optimization, and so forth. In this context, the acquisition, analysis, and interpretation of data play a key role in wedding theory and practice, mathematics and the real world.

It is apparent from the examples in Chapter 1 and the cases in Chapter 2 that many types of data are encountered in reliability applications and that the range of applications is quite extensive, including electronics, medicine, aerospace, automotive, heavy industry, simple consumer products, power tools, appliances, and so forth, i.e., nearly every industrial and commercial enterprise. One of the objectives of this chapter is to categorize these data types and indicate other possible sources of information that may be useful to engineers, analysts, and managers in the analysis of reliability. These range from test data to operational data to subjective data ("engineering judgment").

In most cases, we will be dealing with data on time to failure. Such data are variously called "failure data," "lifetime data," "life data," and so forth. In more general applications (and sometimes in reliability), they may simply be time until occurrence of some event (e.g., a warranty claim, completion of a repair, and so forth). Furthermore, "lifetime" data may not, in fact, be lifetimes for all items observed. For example, if data collection is stopped at a point in time prior to the failure of all items being observed, observations will be actual lifetimes of failed items and service times of those that did not fail. Data of this type are called "incomplete." There are many types of incomplete data, as we shall see. This and many other factors must be taken into consideration in a proper analysis of the data. In other cases, we will be dealing with count data—number of failures per unit of time, number of defectives in a batch of items, etc.

The point of the above discussion is that the key to application of reliability techniques to real-world problems is acquisition, analysis, and interpretation of data. As a result, the field of statistics, which deals with effective and efficient collection of data (design of experiments) and description and analysis of data (descriptive statistics and statistical inference), plays a major role in applications of reliability.

The second key objective of this chapter is to give an introduction to some basic techniques that are essential in the proper summarization and presentation of data. This provides a means of extracting and expressing the information contained in the data, and forms the basis of later modeling and data analyses.

Most of the descriptive techniques described are appropriate in the context of simple data sets—data at a single level, which we will call "part," and relatively simple assumptions—samples of independent observations from a single population, with actual lifetimes of all elements in the sample recorded. Data of a more complex structure will be mentioned briefly, with more detailed discussions deferred to later chapters.

This chapter is organized as follows: Section 3.2 deals with types of data, data sources and levels of data. We also comment briefly on the use of the many types of data that may be available in the various applications and levels of reliability analysis. These concepts will be developed more fully in later chapters. In Sections 3.3 and 3.4, we look at the various types of test data that may be obtained in reliability analysis. Section 3.3 deals with "pass/fail" data, that is, data in which counts of items are recorded. Section 3.4 is concerned with failure or lifetime data. Both complete and incomplete data are discussed.

Section 3.5 discusses data structure and the notion of statistical inference. Basic concepts of statistical description, including both graphical and numerical techniques, are discussed in Section 3.6. Finally, we summarize some key statistical methods for preliminary analysis of part data, discussing both pass/fail data and time-to-failure data in Section 3.7. Further use of these techniques and introduction of basic methods of statistical inference in the reliability context will be presented in Chapter 5 and later chapters.

3.2   DATA-BASED RELIABILITY ANALYSIS

3.2.1   Levels of Data

Reliability assessments may be necessary at many levels. A hierarchy of levels is given in Chapter 1. We call the lowest level "part." This may be, for example, a capacitor, a lead to a capacitor, a formed piece of sheet metal, a glue joint, and so forth; in short, the lowest-level element that could fail. Collections or assemblies of parts are variously called components, subassemblies, assemblies, subsystems, systems, items or products, and so forth. All of these consist of parts connected in various ways to form an operational system. This is discussed in detail in Chapter 1.

In nearly all cases, it is the reliability of the system that is of primary concern. As noted in Chapter 1, by reliability we mean the probability that the system satisfactorily performs its intended function for at least a specified period of time. For simple systems, consisting of relatively few parts or components, it may be possible and cost-effective to assess reliability at the system level. For complex systems, system reliability is usually assessed by building up from the part level to the component level, and so forth, up to the system level. This requires an understanding of how the system functions, i.e., the way in which the parts and components are connected and how they operate together (e.g., independently or interacting). It also requires realistic models for expressing these relationships.

In this chapter, we are concerned mainly with data at a single level, whether it be part, component, system, etc. The term "item" will usually be used to identify the element under discussion, rather than specifying a particular level. Thus, by "part" we here mean the level at which reliability is being assessed and we will not be concerned with how this might be related to reliability at a different level.

The focus of the book as a whole with regard to test and other data is aggregation from parts to components, from components to subsystems, etc., in the context of reliability assessment, prediction, and optimization. In this sense, the data focus parallels the theoretical focus. The intent is to integrate the two in order to provide an understanding of the concepts and a basis for realistic application.

We turn next to a discussion of the types of relevant data that may be available to the reliability engineer or manager or may be obtained by appropriate experimentation or by acquisition from other sources.

3.2.2   Types of Data and Data Sources

Although most of what we will do will emphasize experimental data, by which is meant data from planned tests or random samples from well-defined populations, many other types of data and relevant information may also be available to the reliability analyst. In this section, these various types of information will be described. The focus will be on the structure of such data. The uses of this information and, to some extent, how and where it may be obtained will be discussed in Section 3.4 and in later chapters, particularly under Bayesian statistical analysis. Types of data include:

Historical data. In reliability estimation and prediction, particularly in the early stages of design and analysis, many types of historical data may be relevant. Test and field data on failure times of existing products can provide a useful starting point for reliability prediction for similar new or redesigned products. Even products that have undergone major design changes or are based on completely new designs may have some parts or even major components in common with existing products. We shall see that this data can be combined with new data to obtain estimates of reliability for the new product or system in a number of ways.

Vendor data. Few companies produce all of the parts and components that make up even simple products. Outside sources are nearly always involved and test data on the purchased items are often available from the vendor. These data, along with some internal test data for verification, if necessary, can be used in basically the same way as historical data.

Data obtained by simple random sampling. In a sense, all data are obtained as a result of sampling—rarely does one have access to an entire population or the resources to measure every individual. A simple random sample (SRS) is a sample selected in such a way that every sample of size n drawn from a population of size N is equally likely to be selected. This is done by use of a computer program or a random number table. When sampling from an infinite or conceptually infinite population, such as a production run, items are chosen at random by this process from the items that are available. Ideally, data are obtained by means of an SRS, but this is not always possible, and one often has to assume that something essentially equivalent to simple random sampling has produced the data at hand.

Test data. To assess product reliability, testing of parts and components should begin at an early stage of product development. This is particularly true of new items for which no historical or vendor data are available, but it is also true of existing items that are to be used in new applications. Ideally, test data are obtained from carefully designed experiments, conducted under carefully controlled conditions, with simple random samples of items. For relatively "simple" products such as light bulbs, computer diskettes, hand tools, and so forth, tests of the final assembled product may be adequate. For complex products such as television sets, automobiles, computers, radars, and so forth, tests are required at several levels—parts, components, systems. For exceedingly complex systems such as jet aircraft, space systems, and advanced electronics, very extensive testing is required. Accordingly, increasingly extensive and complex data sets are obtained and increasingly sophisticated statistical techniques are required for aggregation and analysis.

Handbook data. In theoretical assessments of reliability, for example at the part level, physical failure models are sometimes used. The models typically require inputs such as geometrical configurations, materials and their properties, environments in which the item will be operated, and measures of variability in all of these variables. Much of this information can be obtained from standard engineering, physics, and chemistry handbooks.

Operational data. Operational or usage data ("field data") are data collected under actual operating conditions after the product (or part or component from a supplier) has been acquired by the customer. Experimental data of this type (i.e., data collected under controlled conditions or by means of a truly random sample) are seldom available, for obvious reasons. Exceptions occur in special applications, for example, aircraft components such as windshields or engines, and space systems, where each item in the entire production may be tracked throughout its useful life. In these cases, the data may be used in updating reliability estimates as failures occur. More often, however, operational data are obtained haphazardly, for example, on items used under extreme conditions (and failing frequently) or as a part of data collected for other purposes, e.g., warranty claims data. In these cases, the items on which data are collected cannot be considered to be a random sample, which is an assumption of most statistical analyses.

Operational data are useful for reliability assessment in design and development efforts of an existing item only if they realistically represent the whole of operational conditions and then only in the case of long-term projects involving design updates. On the other hand, they may be very useful in formulating initial estimates of reliability for similar, later-generation, products. Operational data are also useful for nonreliability purposes, such as prediction of future warranty costs.

Judgmental data. When hard data are sparse or lacking altogether, it is not unusual to use "engineering judgment" in attempting to predict item characteristics, including reliability. Estimates of this type, based on quantified subjective or partially subjective information, enable the analyst to introduce engineering knowledge and experience into the reliability assessment process. This type of information may also form the basis of an important input element (the "prior distribution") to Bayesian analysis, which provides a method of incorporating this information with test data as it is obtained and using all of this information in refining and updating reliability predictions. Bayesian analysis will be discussed in Chapter 5.

Environmental data. The environment in which an item is expected to perform can have a profound effect on its reliability. Materials' properties can change, usually for the worse, under extreme conditions of temperature and humidity and other types of stress, such as vibration, or after prolonged exposure to mildly extreme stresses. Information on environmental effects can often be found in handbooks and is of importance in designing for reliability.

Cost data. Many types of cost data, both direct and indirect, are relevant to management decisions with regard to reliability. Rather than affecting reliability, these are factors that are affected by it and hence are of importance in reliability management. Relevant data include data on sales, service costs, and warranty claims, all of which are related to product performance, reliability, and other aspects of quality. Accounting and financial data are important in assessing the direct and indirect effects of product unreliability. If the data indicate a sufficiently adverse impact (for example, on sales or profitability), the ultimate result may be a management decision to redesign the product to improve reliability, or, if this does not appear to be cost effective, to discontinue the product.

3.2.3   Use of Data in Reliability Analysis

The usefulness of data from the various sources listed in the previous section depends on the relevance of the data to the problem at hand and on the reliability and validity of the data. Data relevance will depend on the stage of product design, development, production, and so forth, and on the types of reliability models being employed. A number of reliability models are developed in the next chapter and in later chapters on component and system reliability. The use of data for assessing reliability in the context of these models is discussed in Chapter 5 for single elements and in later chapters for more complex items.

In early stages of design, little or no test data may be available and considerable use may be made of data from other sources. Historical data, from previous testing of parts or components, either in-house or by vendors, can be highly relevant if the same parts are being used in a new product; less so if it is a "similar" part. Handbook data are most often information of a more basic type, e.g., materials properties, such as strength as a function of temperature for different alloys, conductivity and other electrical properties, and so forth. Information of this type is generally accepted as valid. Models that employ such data are usually quite complex, involving modeling of stresses at the part level along with environmental and other factors, and requiring computer simulation for evaluation of reliability. An approach based on models of this type used in simulation experiments in rocket engines is given by Moore et al. (1990) and applied to individual parts (turbine disk and heat exchanger coil) used in the space shuttle main engine (Newlin et al., 1990 and Sutharshana et al., 1990).

Extra caution must be exercised in using some types of data, particularly those having a subjective component and those over which the analyst has little or no control. This is true of vendor data, where complete information concerning the test procedure should be requested along with the data and analysis. It is certainly true of engineering judgment, which is always at least partially subjective. In order for information of this type to be useful in modeling, it is necessary to quantify it. An approach to accomplishing this is provided by Bayesian statistical analysis.

The Bayesian approach enables the analyst to combine information that may be available from various sources prior to experimentation with test data to provide a basis for reliability estimation. This is done by setting up an appropriate probabilistic model for reliability (the topic of the next chapter) and then using the available information to formulate a prior distribution. The prior distribution is a probability function that expresses uncertainty regarding the parameters of the model or regarding reliability per se.

Ideally, the prior distribution should embody all of the information one has at the time. When information is available from a number of sources, for example historical information on reliability of some parts, vendor information on others, engineering judgment on the item as a whole, and so forth, combining this into a single prior distribution becomes a challenging task. It is further complicated by the fact that some elements may be subjective. Opinions may vary and may change in time. This will be discussed in more detail in Chapter 9.

The essence of the Bayesian approach is to update previously acquired information as new data, usually as the result of testing, become available. In fact, this is done periodically as testing proceeds. At each stage, new reliability predictions are produced. Furthermore, information on part reliability may be used in formulating prior distributions for component reliability, aggregating with prior information at the component level, and so forth. These would be updated as test data at the component and higher levels were obtained.

In analysis of a complex system, the process will typically proceed through several stages: design, prototype, production, and operation, with many types of data obtained at each stage. A typical scenario for data generation and reliability prediction is as follows:

- *Design stage.* Available information may be limited, with much of the most relevant information at the part level. Prediction at higher levels is difficult. At the system level, it may be possible to predict only within orders of magnitude.

- *Prototype stage.* As test data on parts and components become available and prototypes are tested, more refined estimates of reliability at each level are obtained.

- *Production stage.* Significant testing will have been done, and much directly relevant data obtained. The role of preliminary information is much diminished as this information is entered into the model estimation.

- *Operational stage.* After a system becomes operational and field data, including data on warranty claims and perhaps data resulting from later tests, become available, the models are used to provide still more refined and precise estimates of reliability. As the amount of data collected increases, these later information sets will tend to overwhelm the prior data.

We shall elaborate on these ideas in Part C. It is important to note that to achieve realistic estimates and predictions of reliability, realistic models and valid data are important at every stage.

3.3 ATTRIBUTE DATA

3.3.1 Data Structure

In many tests, the result is simply whether or not an item is acceptable, i.e., whether or not it passes the test. This is particularly true when testing at the part level when parts are relatively simple. Depending on whether items pass or fail the test, they are classified as pass/fail, operative/inoperative, nondefective/defective, go/no go, and so forth. The generic term for such data is attribute data. The generic dichotomy is "yes/no," in accordance with whether or not the item possesses the attribute. Data structures involving attribute data are discrete in nature. This type of response is often obtained in acceptance testing of parts received from suppliers.

Attribute data may arise in two ways. The first is in cases where the dichotomous response is inherent in the measurement process. Examples are light bulbs that either burn or fail instantaneously, an ignitor that does or does not produce a spark, a switch or sensor that is or is not operable. The second source of data of this type is obtained when only pass/fail outcomes are recorded for responses that are inherently quantitative. This occurs in situations where it may be difficult, expensive, or even impossible to record the actual value of the quantitative response, but easy to record whether or not it exceeds a certain threshold. This occurs frequently in dealing with electrical parts or components, where the item is determined to be "in spec" or "out of spec." There are also many biological applications in which this type of data is encountered. For example, batches of n insects may be subjected to various dosages of an insecticide, with the number of survivors recorded in each case. The underlying unobservable continuous variable is the exact amount of insecticide corresponding to a lethal dose. An analogous situation occurs when batches of parts are subjected to increasing temperature levels and/or time periods of increasing length and then categorized as operative or inoperative. The exact time–temperature combination that will cause failure cannot be determined directly.

Data of this type are called *quantal* data. The occurrence or nonoccurrence of the phenomenon being studied (e.g., failure) is a function of an underlying related unobservable (or simply unobserved) continuous variable.

In dealing with quantal data, the individual pass/fail responses are seldom of interest (although the items themselves may be of considerable interest). Instead, the information is usually aggregated in some way, resulting in data in the form of a discrete variable, namely, a count of items possessing a certain characteristic. We look next at the various forms of discrete data that may result.

3.3.2 Count Data

Count data are aggregated attribute data, that is, counts of the number of individuals in a data set possessing the attribute. Counts may be obtained in various ways, depending on whether or not the number of items is fixed, whether or not the observation time is fixed, and how the sampling is done. We discuss several cases, which are classified based on the discrete distributions (to be discussed in Chapter 4) appropriate for modeling the data sets. The resulting data types are classified as binomial, hypergeometric, multinomial, Poisson, and negative binomial, based on the probability distributions.

Case (1): The presence or absence of a particular characteristic is observed for each of a fixed number n of items in a random sample. The count is the number of items out of n that possess the characteristic (failure, survivor, etc.). It is assumed that the sample is selected from an infinite population (or is selected with replacement from a finite population, which is seldom the case in reliability applications). Examples are number of defectives (e.g., dry-soldered joints) in a sample of n items from a production line; number of underweight bags of potato chips out of 50 chosen at random from the output of a packaging machine; number of cancer patients who survive for at least five years out of 138 given a particular treatment regime. In each of these examples, the population from which the sample is selected is conceptually infinite. The term "binomial" indicates "two names"—pass or fail, and so forth. Data of this type are modeled by the binomial distribution.

Case (2): The context here is basically the same as for Case (1), except that sampling is from a finite population (of size N, say) and sampling is done without replacement. The count is again the number of items possessing a specified characteristic in a sample of size n. This is common in acceptance testing, where a batch of items is received from a supplier and it is desired to test a sample of n distinct items. Note that if N is large relative to n, there will not be much difference between this and binomial sampling. That is, if we sample 200 items out of a batch of 1000, quite different samples can result depending on whether sampling is done with or without replacement. If we sample 200 out of a population of 2,500,000, replacement or not matters little; we will almost always get very nearly identical samples. The hypergeometric distribution is used to model this type of data.

Case (3): Suppose that each item in a random sample of size n is classified into one of k classes. Counts are numbers of items in each class. A common application involves classification of failures according to several failure modes. For example, glass or ceramic items tested under stress conditions may fail due to shattering, cracking, discoloring, or deterioration of finish. Manufactured items may be returned to the manufacturer for many reasons. As an example, items returned to a clothing factory may be classified by type of defective, with categories 1) buttons missing, 2) improper stitching, 3) material flaws, 4) improperly dyed, 5) sized incorrectly, 6) other. "Multinomial" indicates "many names." The appropriate model for data of this type is the multinomial distribution.

Case (4): Here a process is observed for a fixed period of time. The count is the number of occurrences of a particular phenomenon during the observation period. It is assumed that occurrences are independent and occur at a constant rate. (The notion of a constant rate of occurrence will be made more precise in Chapter 4.) This is illustrated by a situation in which an item is put into service, replaced instantly on failure, and replacement is continued through a fixed period of time T. The count is the number of failures in the period of length T. Examples would be number of failed light bulbs in a lamp over a five-year period and number of replacement batteries in a watch, medical device, single-cell flashlight, and so forth. The model used for data of this type is the Poisson probability function, first discovered by the French mathematician Simeon D. Poisson, who published his research on this topic in 1837.

Case (5): The situation is the same as for Case (1), except that here we fix the number of occurrences r of the phenomenon and sample until that number is reached. The count is the number of items required until r occurrences are observed. Suppose, for example, that we wish to select r defective items for analysis in a design improvement study. Sampling from production is done until the desired number of defective items is found. The count in question is the total number of items that must be tested. The appropriate model for data of this type is the negative binomial distribution.

3.4   TEST DATA ON PRODUCT LIFETIMES (FAILURE DATA)

The previous section dealt with discrete data. Many of the characteristics measured in reliability studies are continuous. Data of this type are often called *variables data*[1] (as opposed to *attribute data*). In general, variables data may assume values in the interval $(-\infty, \infty)$ or on any subinterval or set of subintervals of this. In reliability, we are concerned primarily with test data in which lifetimes of items that fail during the course of the test are recorded or with variables related in some way to item lifetimes. Lifetimes, of course, are necessarily in the interval $[0, \infty)$. Other variables may have different domains.

3.4.1   Complete and Incomplete Data

It is assumed that the data are obtained as the result of a designed experiment under more or less controlled conditions. One such situation is that in which a random sample of items is selected from those available and put on test. Data of this type could also arise in a number of other ways, for example, as operational lifetimes of a random sample of items put into service.

If the actual lifetime of every item in the sample is recorded, the data are called *complete* data. To obtain complete data, it is necessary to continue the experiment until the last item on test or in service has failed. In cases where even a few items in the sample may have very long lifetimes, experiments of this type can go on for very long periods of time, in fact, well beyond the point at which the results are no longer of any use. Under these circumstances, it is desirable to discontinue the study prior to failure of all items in the sample. (This is also the case for relatively short-lived items if answers to reliability questions are urgently needed.)

When observation is discontinued prior to all items having failed, the resulting data are called *incomplete* data. The recorded observations are failure times for those items that failed prior to cessation of testing and service or operating time for those that did not. The data are incomplete in the sense that we do not know the exact time to failure of the unfailed items, only that their failure times are greater than the recorded service time. The end result, then, is that we have two types of observations in the sample: failure times and bounds on failure times. Many of the data sets given in Chapter 2 are incomplete in this sense, as was pointed out in the case discussions.

Data of this type are also called *censored* data. There are a number of ways in which censored data may arise. The discussion to this point has focused on the testing of manufactured items. Here the type of censoring depends on the criterion used to determine when to conclude experimentation. In some cases, most commonly in medical studies, this decision is not under the control of the experimenter. For example, lifetimes of certain types of cancer patients given a particular drug regimen may be of interest. If a patient dies due to a cause other than the cancer being treated, the test time is censored. An analogous situation occurs in the case of product lifetimes, when products may fail by more than one failure mode. For example, an electronic module may fail because of weakening of one of its parts, failure of a connector, or burn-out of a part. Failure by any one mode acts as a censoring device for all others.

Operational data are often censored. For relatively expensive items, particularly those not produced in large quantities, e.g., radar units, windshields on large commercial aircraft, jet engines, diesel engines in locomotives, and so forth, manufacturers may track each unit in service. At any particular observation time, there may be units with no failures or no failures by a particular failure mode, resulting in censored data of one or more of the types mentioned.

Warranty claims data are another type of field data that are inherently censored. Legitimate claims are made only prior to the end of the warranty period. Lifetimes that exceed the length of the warranty period are very rarely recorded.

We see, then, that there are many underlying causes of censoring and a number of types of censored data. It is useful to categorize these.

3.4.2   Types of Censoring

The major reason for intentional censoring is to limit the time and/or cost of experimentation. This may be done in several ways, as will be seen. In other cases, censoring is unintentional or, at least, unscheduled. This is true in the medical illustration given previously. It is also true in some cases in product testing. As a result, there are several modes of censoring. These are:

Type I censoring. Censoring that occurs as a function of time is called Type I censoring. This is true in the illustrations given in the previous paragraphs, where the observation period stops at a particular point in time and the data recorded are lifetimes of items (products, patients, etc.) that failed prior to that time and service times of all other items. More specifically, data of this type are called *right censored.*

If all items are put on test or put in service at the same time and the observation period is the same for all items (that is, observation stops at a fixed time T), the data are *singly censored* (more precisely, *singly right censored*). In this case, all unfailed items have been in service for the same amount of time, namely, the observation period T.

Data for which unfailed items have been operating for variable amounts of time are called *multiply censored* (or *multiply right censored*). This often occurs when dealing with operational data, for example, when units are put into service at different times but all are observed at time T from the time of first installation. In this case, in fact, nearly all the data may be censored. This would occur, for example, if failed items are replaced or repaired and put back into service, so that nearly all item are in operation at the time of observation.

Data may also be censored on the left. *Left censored* data result when it is only known that failures have occurred prior to some time, say t. This may come about, for example, in multiple installations of items (fluorescent tubes in classrooms; heat sensors in warehouses, ships, etc.) which are checked periodically for failures. In this case, exact failure times cannot be determined; for a failed item, it is known only that the failure occurred since the previous inspection. This type of censoring also occurs when items are used intermittently and a failure occurs during a period when the item is not in use. The end result of left censoring is that the observation provides an upper bound on the lifetime of the item. As with right censoring, items may be singly or multiply left censored.

Both right and left censoring can occur. This would be the case, for example, in the fluorescent tube and heat sensor illustrations if observation were stopped prior to the failure of all items. At any observation time prior to complete failure, lower bounds on lifetimes would be observed for unfailed items and upper bounds for failed items.

Another distinction that is important is that between censoring at predetermined times and random censoring, that is, censoring at times that occur randomly rather than under the control of the experimenter or observer. This is the case is many of the illustrations given. Random censoring is typical of multifailure-mode data. It occurs frequently in biological studies and when operational data are used. It occurs less frequently in laboratory studies, where test conditions and observation times are more easily controlled.

Type II censoring. In Type I censoring, the censoring factor is time dependent. This may provide too limited an amount of information if the time of observation is (usually inadvertently) too short. An alternative is to continue observation until a predetermined number r of failures have occurred (with $r < n$). This also results in incomplete data, since the lifetimes of the remaining $n - r$ items are not observed. This type of censoring is called *Type II censoring*.

Type II censoring is used primarily in laboratory experimentation. It is rarely, if ever, used or encountered in biological data or field data on products. Under Type II censoring, multiply censored data and random censoring are unusual but occasionally occur, for example when multiple failure modes are involved.

An extension of this type of censoring is *progressive Type II censoring*. Under this censoring scheme, batches of items are removed from test as in Type II censoring in two or more stages, as follows: n items are initially put on test. After r_1 failures have occurred, an additional $n_1 - r_1$ items are removed from test. Testing is continued on the remaining $n - n_1$ items until an additional r_2 failures are observed, at which time an additional $n_2 - r_2$ items are removed, with testing continuing on the remaining $n - n_1 - n_2$ items, and so on, through k stages ($k > 1$).

Other types of censoring. Various other combinations of censoring mechanisms are sometimes used. For example, it may be desirable to observe a reasonable number of failures and yet limit test time. (Time is not limited under Type II censoring.) In this case, a combination of Type I and II censoring may be used, whereby testing continues until either r failures occur or test duration T is reached, whichever comes first.[2]

3.5 EFFECT OF DATA STRUCTURE ON STATISTICAL ANALYSIS

Proper analysis of a set of data requires an accurate and thorough understanding of the nature and structure of the data. This includes the type of data to be analyzed and how and under what conditions they were collected.

There are two basic objectives of data analysis. The first is descriptive, that is, description and summarization of the information contained in the sample data. This is accomplished by selection of appropriate descriptive measures and will be discussed in Section 3.6. The second is statistical inference, that is, the use of sample information to make inferences about the entire population from which the sample was drawn. Selection of appropriate methods for inference is basically a mathematical problem. This will be dealt with to some extent in

Chapter 5 and subsequent chapters and is treated in considerable depth in the references cited.

In the next section, we look more carefully at the types of data that one might encounter and the basic computations that may meaningfully be done with each type.

3.5.1 Scales of Measurement

In the preceding sections, data were categorized as discrete or continuous and as complete or incomplete. It is useful to refine these distinctions by types or levels of scales on which measurements may be recorded. There are four such scales—nominal, ordinal, interval, and ratio. These increase in order of complexity (meaning, basically, information content) and, correspondingly, in the number and types of arithmetic operations that can meaningfully be applied to the data. Nominal and ordinal data are inherently qualitative in nature, representing item attributes; interval and ratio data are inherently quantitative, representing amounts rather than types. In practice, numerical values are used for data on all four scales (especially in computer files). The important differences are in how the numerical values are interpreted and what can legitimately be done with the data. The scales are defined as follows:

Nominal data. Nominal data ("names"), or data on a nominal scale are purely categorical, with values for the variable recorded only as the category into which the item falls. Binomial and multinomial data are of this type. Examples are characteristics such as supplier of an outsourced component, whether or not an item is of acceptable quality, failure mode, plant at which an item was manufactured, and variables in a marketing study such as gender, brand preference, ethnicity, and so forth. It is usual to assign numbers to the categories and to enter nominal data into computer files in this form.

Since the assignment of numerical values to the categories is arbitrary, no arithmetic operations (sum, average, and so forth) give inherently meaningful results. (There is no meaning, for example, to the statement that the average failure mode is 2.7, although a computer can easily calculate many such results!) The only meaningful statistical calculations are counts, proportions or percentages in each category.

Ordinal data. Values on an ordinal scale indicate categories that have a rank order relationship, with increasing values on the scale indicating the presence of an increasingly larger amount of the characteristic being measured. Numerical values have an inherent meaning, but only insofar as their order is concerned. Data of this type are often subjective (or at least partially subjective). This is true, for example, in the case of rating scales, which are the most commonly used ordinal scales in most applications. For example, "Rate the quality of this product on the scale of 1 (High) to 7 (Low)." Quality of sound of a speaker system, taste tests of food products, subjective (or partially subjective) ratings of the severity of a disease, the seriousness of the consequences of a failure, and so forth, are other examples of data of this type.

Since the measurements on an ordinal scale indicate only order and not absolute amounts, the intervals between successive numbers on the scale cannot be assumed to have the same meaning. For example, the difference between categories 5 and 6 of severity may not be the same as that between categories 1 and 2. As a result, numerical operations are also not meaningful on an ordinal scale. It follows, incidentally, that it is not meaningful to compare scores given by two groups (e.g., males and females, different age groups, etc.) by comparing averages. Again, only counts and proportions or percentages are legitimate calculations.

Interval data. If the intervals between values on the scale are equal, the measurements are said to be on an interval scale. The classical example of an interval scale is temperature. Rating scales (e.g., sound quality of speakers on a scale of 1 to 5, subjective reliability evaluation of an item on a scale of 1 to 9) are interval scales only if it is possible to assume that the intervals between successive values are equal. This is a frequently made (though often dubious) assumption.

A feature of interval data is that there is not a well-defined zero on the scale. The zero value on the temperature scale, for example, is arbitrary (compare Fahrenheit and Centigrade). As a result, operations such as sums, difference, averages, and related quantities are meaningful, but multiplication and division are not. Thus it makes sense to state that the average noon temperature in Chicago in August is 82, but not that 82 is twice as hot as 41. A sound quality of 4 is not twice as good as 2 (and is probably not 2 units better either).

Ratio data. Data on a ratio scale have all of the characteristics of those on an interval scale with the additional feature of a well-defined zero. Height, weight, profit, number of failures, time to failure, breaking strength, and sound intensity in decibels are all examples of ratio data. For ratio data, all of the usual numerical operations, including multiplication and division, are meaningful. For example, it is meaningful to say that 8 cm is twice as long as 4 cm.

Note, incidentally, that the level of scale is separate from the distinction between discrete and continuous scales. Except for the nominal scale, which is always discrete, data on the remaining scales may be either discrete or continuous. For example, time to failure is a continuous ratio variable, whereas number of failures in a sample of size n is a discrete ratio variable.

3.5.2 Statistical Inference

In data analysis, the most important objective is to be able to draw inferences from data in a meaningful way concerning population characteristics. Traditionally, this is done through estimation of population parameters or other characteristics, and/or by testing hypotheses about parameter values. In reliability analysis, other characteristics of interest include population fractiles (directly related to product reliability), failure rates, mean time to failure, reliability functions, hazard functions, and related quantities. All of these characteristics will be defined precisely and methodologies for estimating them will be discussed in later chapters.

Selection of an appropriate method for analysis is based on three data-related criteria: (1) the scale of measurement of the data; (2) the structure of the data, including how it was collected and whether it is complete or incomplete; and (3) the assumptions made about the underlying random phenomenon. As will be seen, many important decisions in reliability analysis are data based. It is essential that the data be analyzed correctly.

In the previous section, the relationship of analysis to levels of scales of measurement was discussed. Most of the standard statistical techniques in common use assume interval or ratio data. The reason is that many inferences involve means or related quantities and these are only meaningful on those scales. Many procedures also assume continuous data, though this is not a problem in practice except for small samples. Other methods are appropriate for nominal and ordinal data.

How the data are collected is of critical importance in data analysis. In sampling, this includes whether the sample was drawn with or without replacement. We will ordinarily assume that samples are from very large or infinite populations, in which case this doesn't matter. If that is not the case, and sampling is done without replacement, the procedures given require modification. See sampling texts such as Thompson (1992) and Schaeffer et al. (1996) for appropriate methods of analysis.

In designed experiments, the structure of the analysis is directly related to the design of the experiment.[3] Thus the design drives the analysis, and it is essential to know exactly how the experimental procedure was carried out. If this is not known precisely, it is unlikely that the data will be analyzed properly. Worse, if the data are simply collected haphazardly, there may be no method of analysis that will provide information about the questions and populations of interest. This is particularly true of complex experiments involving multiple, possibly interacting, factors and special structures with regard to randomization. We will deal with experimental design and its relationship to data analysis in Chapter 10.

Because of the importance of data collection methodology in analyzing data, particular caution should be taken in using results from sources about which the analyst has no direct knowledge. Thus historical data, data from outside sources, and some of the other types of data mentioned previously should be carefully evaluated.

Probabilistic assumptions regarding the data also play an important role in analysis. Traditionally, it is assumed in data analysis that data follow the well-known normal distribution or at least that sample sizes are large enough so that the sample means are approximately normally distributed (which follows, for sufficiently large samples, from the Central Limit Theorem; see Appendix A.3.5). This tradition is not followed to such an extent in reliability analysis, where lifetime data are seldom normal, and many alternative assumptions are more tenable, though large-sample theory still plays a very important role. Alternatives to the normal probability models will be discussed in detail in Chapter 4. More on statistical methods for analysis of data under these assumptions will be given in Chapters 5, 9, and .11.

Finally, the effect of censoring on inference must also be taken into consideration, since this significantly changes the probabilistic representation of the data. For complete data (with a sample of size n), the probability structure is based on the n individual failure times, these being the random elements in the data. For Type I censoring, the number of failed units is random; for Type II censoring, the test duration is random. Including these random elements as a part of the probability structure is a significant complication.

3.6 BASIC DESCRIPTIVE STATISTICS

In statistics, a *parameter* is defined to be a population characteristic, for example the *mean time to failure* (MTTF), μ. The corresponding quantity in a sample, e.g., the *sample mean*, $\bar{x}$, is called a *statistic*. More generally, a statistic is any quantity calculated from sample data. In this context, the objective of descriptive statistics is to calculate appropriate statistics for purposes of description and summarization of the information in a set of data. The objective of inferential statistics is to determine appropriate statistics for effectively and efficiently making inferences concerning parameters.

In this section, we discuss descriptive statistics generally as well as specifically in the context of the types of data described previously. The emphasis will be on part-level data (or, more generally, single-level data at any level). Both graphical and computational measures will be presented. Modifications of these techniques for grouped data will be given in Chapter 9.

3.6.1 Frequency Distributions

A *frequency distribution* is a graphical or numerical description of an entire set of data. The objective is to present the data information in a concise form and in such a way that, if possible, the general shape of the distribution is displayed. This not only makes the data more comprehensible, it can sometimes give some insights into structural issues. For example, it may become apparent in looking at failure data in this way that data are concentrated in two or more areas, suggesting multiple failure modes.

Most standard statistical program packages include algorithms for numerical and graphical frequency distributions of various types, with many options. The examples in this chapter were analyzed using Minitab.

Numerical frequency distributions. A *numerical frequency distribution* is a grouping of data into distinct categories or classes. This is done by determining an appropriate number k of classes, determining appropriate class boundaries, and simply counting to determine the frequency in each class. Usually, k is chosen to be between 5 and 20. A rule of thumb is *Sturge's Rule,* which gives k as the nearest integer to $1 + 3.322 \log(n)$, where n is the sample size. Usually a value of k close to this choice will lead to a good representation of the data set. A rule of thumb for class boundaries is to use classes of equal width unless the data appear to be highly skewed, tailing off substantially on one or both ends, in which case open-ended end intervals may be preferable. In any case, the objective is to give an honest and informative picture of the entire data set.

Cumulative frequency distributions. *Cumulative frequency distributions* are obtained by forming partial sums, that is, by summing frequencies over all classes up to and including a given class. Thus, cumulative distributions give the number of values in the data set that are at or below a given value.

Relative frequency distributions. Division of the previous counts by the sample size n produces *relative frequency* and *cumulative relative frequency distributions*. These give, respectively, the proportion of values in a given class and the proportion in that class or below. Relative frequencies are often expressed in percents.

Example 3.1 (Case 2.1)
Data on bond strength for 85 assemblies are given in Table 2.1(a). For $n = 85$, Sturge's Rule gives the number of classes to be $1 + 3.322 \log(85) = 7.41$, suggesting that $k = 7$ or 8 classes are reasonable choices. Frequency, cumulative frequency, and relative frequency distributions for these data are given in Table 3.1 for $k = 7$.

Thus 25 (29.4%) of the bond strengths are between 240 and 279 pounds, while 47% are less than 280. If a minimum of 280 were required in a certain application, this would be a highly unreliable part (47% failures). If only 100 pounds were required, the part is highly reliable (no failures). ■

Graphical frequency distributions. There are many ways of presenting the numerical information in a frequency distribution in graphical form. The most common is a *histogram,* which is simply a bar chart with classes indicated on the horizontal axis and heights of the

TABLE 3.1. Frequency distributions for bond strength data of Table 2.1(a).

Class	Frequency	Cumulative frequency	Relative frequency	Cumulative relative frequency
120–159	1	1	0.0118	0.0118
150–199	2	3	0.0235	0.0353
200–239	12	15	0.1412	0.1765
240–279	25	40	0.2941	0.4706
280–319	33	73	0.3882	0.8588
320–359	11	84	0.1294	0.9882
360–399	1	85	0.0118	1.0000

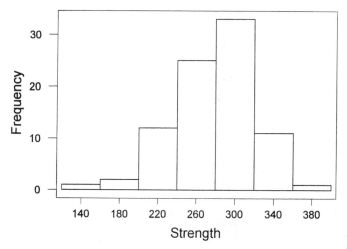

Figure 3.1. Histogram of bond strength data (Table 2.1a).

bars representing frequencies. A histogram of the bond strength data of Example 3.1 is given in Figure 3.1. It is apparent from the graph that the distribution of bond strengths is somewhat skewed left.

Cumulative frequencies can also be presented in this way. A cumulative histogram for the bond strength data is given in Figure 3.2. An alternative graphical representation is a *frequency polygon,* which is formed by connecting the midpoints of a histogram or cumulative histogram.

A plot similar to a histogram, but giving much more detail, is a *stem-and-leaf plot.* This is formed by splitting each numerical value in the data set into two parts, the stems, consisting of all digits but the last, with the leaves being the final digit. Data may be rounded, if neces-

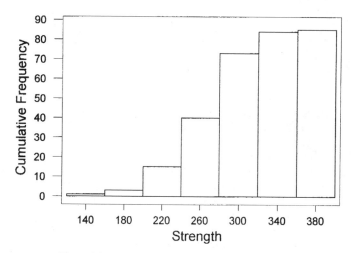

Figure 3.2. Cumulative histogram for bond strength data.

```
1 | 3
1 |
1 |
1 | 99
1 | 01111
2 | 2233333
2 | 444444555555
2 | 6666667777777
2 | 88899999999999999
3 | 0000000001111111
3 | 2222333
3 | 4455
3 | 6
```

Figure 3.3. Stem-and-leaf plot of bond strength data (Table 2.1(a)) (leaf unit = 10).

sary, to produce fewer stems. To produce more stems, the leaves may be divided into two groups (0–4 and 5–9) or 5 groups (0–1, etc.). The stems are then arranged in increasing order, as are the leaves on each stem.

Figure 3.3 is a stem-and-leaf plot of the bond strength data, rounded to two significant digits and using five stems for each leading digit. From the plot, the data can be reproduced completely to two digits. Note that in this plot the left skewness appears to be only slight, except for the smallest observation (the *minimum,* or *min*), read as 130 from the chart, which now becomes apparent as a possible *outlier* or unusual observation. Note also that the largest observation (the *max*) is read as 360.

3.6.2   Other Graphical Methods

Many additional graphical data displays are available. Some of these are used to show different features of a data set; some are for specialized types of data, different from the complete data on a single continuous variable used as an example in the previous section. We look at a few of these.[4]

Boxplots. A *boxplot* summarizes a set of data graphically on the basis of the min, the three quartiles, so called because they divide the data into quarters, and the max. The quartiles, Q_1, Q_2, and Q_3, are determined as follows: The data are first ordered from smallest to largest (with ties, if any, included as repetitions), resulting in the set $x_{(1)}, x_{(2)}, \ldots, x_{(n)}$, say. Q_2, also called the *median,* is taken to be $x_{((n+1)/2)}$ if n is odd, and is calculated as $[x_{(n/2)} + x_{(n/2)+1}]/2$ if n is even. Q_1 and Q_3 are calculated as the medians of all points lying below and above the median, respectively. (The set [min, Q_1, Q_2, Q_3, Max] is called the five-number summary of a set of data.)

To prepare a boxplot,[5] a rectangle of length $Q_3 - Q_1$ and any convenient width is drawn. $Q_3 - Q_1$ is called the *interquartile range* (IQR). The median is indicated by a line drawn across the box. Lines are drawn from the middle of the lower end of the box either to the min or of length 1.5IQR, whichever is shorter, and from the upper end to the max or of length 1.5IQR. These are the *whiskers.* Points beyond the whiskers, if any, are indicated by a "*" and are considered to be possible outliers. (Often, points more than 3IQR from the box are indicated by another symbol and are considered to be definite outliers.)

Another feature that may become apparent in a boxplot is skewness. If the distribution of

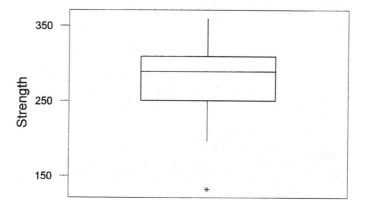

Figure 3.4. Boxplot of bond strength data (Table 2.1a).

the data is symmetrical, or nearly so, the median will fall in the center of the box and the whiskers will be of the same length.

A boxplot of the bond strength data is given in Figure 3.4. Note that the possible outlier mentioned earlier is flagged in the plot, and that left skewness is indicated by the location of the median.

Pareto charts. There are a number of ways of representing qualitative or categorical data graphically. One of the most useful in the context of failure data is the *Pareto chart*. The most common use is in displaying frequencies when several failure modes are involved. Counts of the numbers of failures of each type are made, the modes are ordered in accordance with their frequency of occurrence, and a bar chart of the frequencies, beginning with the largest, is prepared. Pareto charts often also include cumulative frequencies or proportions.

The importance of Pareto charts is in identifying the most frequently occurring, and hence usually most important, failure modes, separating out the "vital few from the trivial many" (the so called Pareto Principle). As noted by Kane (1989), 80% of defectives from a production process are often due to 20% of the errors.

Example 3.2 (Kane, 1989)
Quality checks on a day's production in an auto plant uncovered 70 defects that were classified as body anomalies (paint, dents, chrome, etc.), 51 accessory defects (air conditioning, sound system, power accessories, etc.), and 11 electrical, 8 transmission, and 5 engine defects. Figure 3.5 is a Pareto chart of these data. Note that the chart, prepared by use of Minitab, also includes cumulative percentages. This example is discussed in more detail in McClave and Benson (1994). ■

Example 3.3 (Case 2.3)
A set of data involving many faults reported on removed pacemaker implants is discussed in Case 2.3. Data from two manufacturers are included. A Pareto chart of the data for Company A is given in Figure 3.6. (Note: The category "Other" in the chart indicates a failure not in the list; "Others" identifies a set of categories with very small frequencies.) It is apparent from the figure that there are no faults that are the "vital few." The proportions tail off quite slowly. This is indicative of a difficult problem with regard to reliability improvement—the product is already highly reliable, and correction of no single fault will improve it significantly. ■

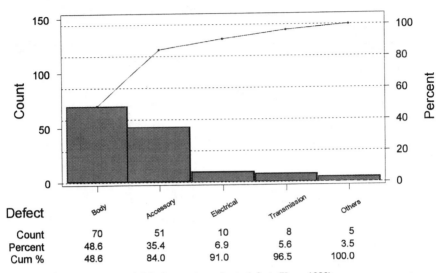

Figure 3.5. Pareto chart of auto defects (Kane, 1989).

Pie charts and other pictorial representations. There are many other ways of presenting qualitative data. One of the most common is a pie chart. This is constructed by dividing the 360° of a circle into segments proportional to the frequencies in each category of the data, and drawing wedges in the circle accordingly. A pie chart of the automobile defect data is given in Figure 3.7.

A more fanciful presentation is a pictograph, which is essentially a set of drawings of

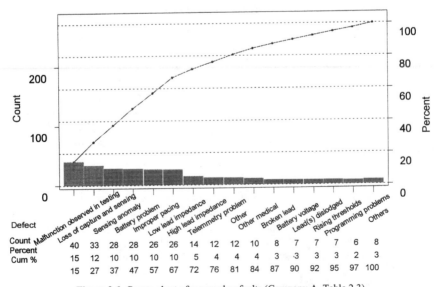

Figure 3.6. Pareto chart of pacemaker faults (Company A, Table 2.3).

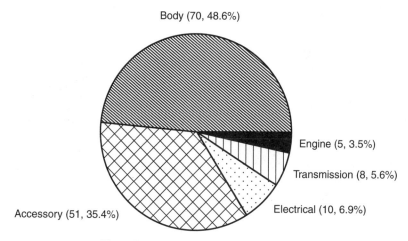

Figure 3.7. Pie chart of auto defects (Kane, 1989).

items, with size representing frequency (e.g., autos of various sizes representing sales of different brands or sales in different years, dollar signs of different sizes representing warranty costs or other costs of unreliability, and so forth).

In all of these graphical approaches, it is easy to give a misleading picture and care must be taken to avoid misrepresentation.

Time series charts. It often occurs that data are obtained sequentially in time. For example, in experimentation, appropriate test plans often require that tests be run in a random sequence, and these are run one at a time. In sampling to determine acceptability of a lot, individual items are selected at random and tested. In analyzing data such as these, it is useful to plot the resulting data against time (usually by means of an index indicating the temporal order rather than the actual time of measurement). Plots against time uncover certain anomalies in the experiment; for example equipment going out of calibration, technicians systematically varying the procedure (e.g., becoming either more skilled or more tired or bored), and so forth. They can also indicate failure of statistical assumptions, e.g., results becoming more variable or less variable through time; nonindependence of successive measurements.

Another use of such plots is in identifying possible outliers. A plot of bond strength versus time on test (days) for the data of Table 2.1(a) is given in Figure 3.8. No particular pattern is apparent through time, but note that the outlier previously identified on Day 68 is easily seen in the plot.

Higher dimensional plots. When two or more variables are recorded in a data set, data relationships are of interest. Plots of these variables against one another in two or three dimensions can be very useful in exploring and modeling these relationships, checking assumptions, identifying outliers, and suggesting further statistical analysis.

Care must be taken in interpreting two- and three-dimensional plots in situations involving many variables. Plots of raw data with variables taken two at a time can be very misleading. Apparent patterns can easily be artifacts due to relationships with other variables not appearing in the plot. Statistical adjustments to the data may be necessary to investigate the nature of the relationship free of such confounding variables.

A number of additional plotting techniques will be introduced as needed in data analysis

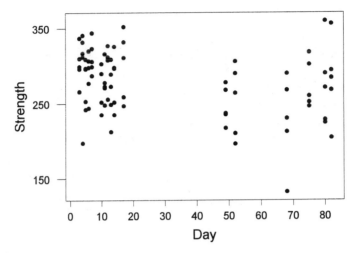

Figure 3.8. Bond strength versus days on test (Table 2.1a).

in the ensuing chapters. Most of the charts discussed here and elsewhere can be produced fairly easily by use of standard statistical software packages.

3.6.3   Measures of Center

Graphical techniques are useful for displaying data and, to some extent, for comparing different data sets. For many purposes (including statistical comparisons), summary statistics that convey essential characteristics of the data are required. Here we look at a few of the most common of such statistics, called variously measures of *center*, measures of *central tendency*, measures of *location*, or simply *means*.

The most commonly used measure of center is the sample mean, $\bar{x}$, given by

$$\bar{x} = \frac{1}{n} \sum_{i=1}^{n} x_i \tag{3.1}$$

Another common measure is the sample median, Q_2, defined in Section 3.6.2.

Although the mean and median are both measures of center, a comparison of the two can give some information on another feature of the distribution, namely skewness. If $\bar{x} > Q_2$, the distribution is skewed right; if $\bar{x} < Q_2$, the distribution is skewed left. For symmetric distributions, the two measures are equal.

In using these two statistics, it should be noted that the mean is quite sensitive to outliers, whereas the median is not. In fact, outliers can be a cause of apparent skewness. Failure data are often skewed, a feature that will be important later in modeling. It is important in this context to determine (insofar as possible) that the skewness is real and not the result of faulty data.

An approach to avoiding or mitigating the influence of outliers is to use a trimmed mean, obtained by omitting a fixed proportion of observations at each extreme (2.5% or 5% at each end are common choices), and averaging the remaining data.

Variable	N	Mean	Median	Trimmed Mean	S D	SE Mean
Strength	85	278.92	289.00	280.09	41.96	4.55

Variable	Min	Max	Q_1	Q_3
Strength	132.00	360.00	250.50	309.00

Figure 3.9. Descriptive statistics for bond strength data (Table 2.1a).

Summary statistics given by computer software packages generally include most if not all of the measures of this and the next section and may include many others as well. The standard output of this type from Minitab for the bond strength data of Example 3.1 is given in Figure 3.9.

From Figure 3.9, we find $\bar{x}$ to be 278.92 pounds. The median is 289 pounds, indicating some left skewness. As we saw from the boxplot and stem-and-leaf displays of the data, this is primarily due to an apparent outlier. The trimmed mean (trimming the lowest and highest 5% of the values) is found to be 280.09 pounds, very close to the median, confirming this conclusion. The output also includes the remaining elements of the five-number summary, as well as SD and SE Mean, which will be defined later.

For most purposes of statistical inference, the preferred measure of the center of a set of data is the sample mean.

3.6.4 Measures of Variability

The second commonly used characteristic of a set of data is a measure of *variability, spread, or dispersion.* Measures of this type provide additional information concerning the distribution and further facilitate comparisons of different data sets.

The most common measures of dispersion are the variance and standard deviation. The *sample variance,* s^2, is given by

$$s^2 = \frac{\sum_{i=1}^{n}(x_i - \bar{x})^2}{n-1} = \frac{\sum_{i=1}^{n}x_i^2 - \frac{1}{n}\left(\sum_{i=1}^{n}x_i\right)^2}{n-1} \tag{3.2}$$

The square root of this quantity is s, the *sample standard deviation.*

Another common measure of spread, used especially in quality control applications, is the sample range, calculated as max − min. The range, s^2 and s, are sensitive to outliers. A measure that is not is the *interquartile range* (IQR), defined previously as IQR = $Q_3 - Q_1$. The IQR is a measure of the spread of the middle 50% of the data.

Finally, the *coefficient of variation* (CV), calculated as $s/\bar{x}$ (often expressed as a percent), is a unit-free measure of spread that is useful in comparing variables that differ in order of magnitude, e.g., fuel consumption for large and small machines.

For purposes of statistical inference, the preferred measure of spread is usually the standard deviation. Some comments on interpretation of this measure will be given in the next section.

For the data of Example 3.1, $s = 41.96$ pounds is obtained from Figure 3.9. From this, s^2 is found to be 1760.64 pounds. The range is calculated as $360 - 132 = 228$ pounds. The IQR is $309 - 250.5 = 58.5$. Finally, the CV is $100(41.96)/278.92 = 15.04\%$.

3.6.5 Interpretation of the Mean and Standard Deviation

The mean and standard deviation can be used jointly to express the location and spread of a distribution. In particular, it is useful to describe spread by measuring distance from the mean in standard deviation units. Two results along these lines are:

The 68-95-99.7 rule. This rule is applicable to distributions that are "bell-shaped" or "mound-shaped" and are symmetric or nearly so. For such distributions,

- Approximately 68% of the observations lie in the interval $\bar{x} \pm s$
- Approximately 95% of the observations lie in the interval $\bar{x} \pm 2s$
- Approximately 99.7% of the observations lie in the interval $\bar{x} \pm 3s$

The Chebyshev inequality. The Chebyshev Inequality is applicable to any distribution having finite variance. It states that at least $100(1 - 1/k)\%$ of the distribution lies within k standard deviations of the mean, regardless of the shape of the distribution. Thus, at least 75% lies within 2 standard deviations from the mean, at least 89% lies within 3 standard deviations of the mean, and so forth Alternatively, not more than 25% is beyond 2 standard deviations, 11% beyond 3 standard deviations, and so forth. Note that this is a much cruder result than the previous one. On the other hand, it is much more widely applicable.

3.6.6 Other Data Characteristics

The mean and variance do not uniquely determine a distribution (nor, in general, do any of the other characteristics discussed in the previous sections). As a result, it is often useful to look at other characteristics as well. Here we list a few additional such measures.

Fractiles (percentiles). The *p-fractile* of a distribution is a value x_p having the property that at least a proportion p of the data lies at or below x_p. x_p is also called a *percentile* of the distribution. Examples are the three quartiles, Q_1, Q_2, and Q_3, which are, respectively, the .25-, .50-, and .75-fractiles of the distribution. Other fractiles are calculated in a similar manner to that of the quartiles.

Fractiles are particularly important in certain reliability applications. In some situations, for example, it is of interest to determine the proportion of a distribution of lifetimes that lies beyond a certain value. Conversely, we may wish to know the value beyond which a proportion p of the population lies. This is the $(1-p)$-fractile. For example, 99% of item lifetimes will be longer than $x_{.01}$.

Higher moments. The mean is also called the *first moment* of a distribution. The variance is the *second moment about the mean*. This notion can be extended to higher moments as well, and these provide additional descriptive measures. The rth moment about the mean, K_r, is given by

$$K_r = \frac{\sum_{i=1}^{n}(x_i - \bar{x})^r}{n} \tag{3.3}$$

The third and fourth moments measure, respectively, skewness and kurtosis (heaviness of the tails).

3.7 PRELIMINARY ANALYSIS OF RELIABILITY DATA

The descriptive measures discussed in the previous sections are not all appropriate for all types of data. In this section, a few guidelines for use of the various graphical and numerical methods for the preliminary analysis of pass/fail and complete and incomplete time-to-failure data are given. The basic objective here is summarization and presentation of the data. Additional techniques will be given as the various types of data are discussed in succeeding chapters. It should be noted that although this chapter has nominally dealt with part-level data, the techniques of statistical description and summarization given are applicable for preliminary analysis of reliability data at any level.

3.7.1 Description and Summarization of Pass/Fail Data

Case (1) and Case (2) data may be represented as a sequence $x_1, \ldots, x_n$, with each $x_i = 1$ or 0 (corresponding to "success" and "failure," respectively). The principal summary statistics are x = number of successes = Σx_i and the sample proportion $p = x/n$. The sample variance of x is given by $np(1-p)$ for Case (1) data and by $[n(N-n)/N]p(1-p)$ for Case (2), where N is the size of the population (or lot, etc.) from which the sample was drawn. If multiple samples are drawn under identical conditions, it is meaningful, since these are ratio data, to average the results, calculate any other useful summary statistics, and apply graphical techniques as appropriate.

Case (3) data, being an extension of the binomial to more than two classes, are handled similarly. The main statistics are counts and proportions in each class. For multiple samples, averages, etc., for each class are meaningful, but it is not meaningful to average over classes.

Case (4) data are generally a sample of observations, each of which is a count of the number of successes (or failures) in a fixed time period. These are ratio data and all of the descriptive methods discussed previously are applicable.

For Case (5) data, the basic statistic is the variable itself, namely x = number of trials until the rth success. For multiple observations, all of the descriptive statistics are meaningful.

For categorical data, averages, etc., are meaningless. These are usually best described by simple counts and Pareto charts.

3.7.2 Description and Summarization of Complete Failure Data

In this section, we restrict attention to complete data. It is assumed that we have available a random sample of n independent observations. For each individual in the sample, a measurement on a single variable X is recorded. In most reliability applications, X will be an item lifetime or some characteristic that is related to or is a proxy for time to failure (e.g., bond strength). As before, the sample values are denoted $x_1, x_2, \ldots, x_n$.

It is apparent that nearly all of the variables dealt with in this context are continuous ratio variables. It follows that all of the techniques of the previous section are applicable. Analysis should begin with exploratory techniques—boxplots, stem-and-leaf plots, and descriptive measures such as those given in Figure 3.9. The idea is to provide an overview of the data, identify possible outliers, summarize the results in a concise and comprehensible fashion, and lay the groundwork for modeling and further analyses.

3.7.3 Description and Summarization of Incomplete Failure Data

Special techniques are required for analysis of censored failure data. Clearly, it is not meaningful simply to average the observations, which consist of failure times for failed items and service times for censored items. It is only slightly more meaningful to average each group separately. The average time to failure of failed items may be interpretable as a conditional average, given that failure occurs before censoring, but this is not very meaningful except for Type I censoring with a common censoring time. Average censoring time in and of itself does not provide much useful information concerning reliability.

The important point to be made is that averaging and the other techniques of this chapter do not provide direct information concerning the MTTF, which is one of the characteristics of most interest in preliminary data analysis in the context of reliability. More complex methods are required. This is also true of graphical methods. A proper analysis of incomplete data requires probability modeling and special graphing techniques. These will be presented in the next three chapters.

NOTES

1. Both the terms *attribute data* and *variables data* have their origins in quality control applications.

2. A number of other censoring regimes have been employed. General censoring procedures and a thorough discussion of the theoretical problems associated with censoring are discussed by Efron (1977), Kalbfleisch and Prentice (1980), Lawless (1982), and in the references cited therein.

3. There are many texts on design of experiments. A few examples in which the topic is considered from somewhat different perspectives are Lorenzen and Anderson (1993), Montgomery (1997), and Winer (1971).

4. Many other graphical techniques for presenting information are discussed in great detail in the three excellent treatises by Tufte (1983, 1989, 1997). The many potential pitfalls in graphical presentations are also discussed in detail.

5. There are several approaches to the preparation of boxplots. The method given here is a fairly standard one and is the one used in Minitab.

EXERCISES

3.1. Determine frequency distributions and plot histograms for the breaking strength data of Table 2.1(a) with $k = 6, 8, 9,$ and 10, and compare the results with those of Table 3.1 and Figure 3.1.

3.2. Determine the scales of measurement for each variable of each of the data sets of Chapter 2. (Include the information in the chart prepared as a part of the Chapter 2 Exercises.)

3.3. For each data set in the cases of Chapter 2 that is incomplete, determine the type of censoring involved and add this information to the chart.

3.4. A Pareto chart for the pacemaker failures recorded by Manufacturer A is given in Figure 3.5. Prepare a Pareto chart for the data of Manufacturer B, given in Table 2.3. Compare the results of the two manufacturers. (Put both Manufacturers' data on the same chart, if possible.)

3.5. Prepare a time series chart of the data of Table 2.1(a). Does there appear to be a time trend? If so, what is the apparent shape of the relationship?

3.6. Prepare boxplots of the three sets of bond-strength test data given in Table 2.1(b). Plot these on the same graph and compare the results. Also prepare histograms and stem-and-leaf plots of the three data sets.

3.7. Prepare Pareto charts for the OREDA data on equipment failures on offshore oil rigs given in Table 2.18. Include charts for item types, numbers of failures, and number of failures per item (rounded). Discuss the results.

3.8. Calculate means, medians, coefficients of variation, and the five-number summary for each set of test data in Table 2.1(b). Use the results to compare the data sets.

3.9. Perform a preliminary analysis of the warranty claims data of Case 2.4. Include summary statistics for each variable, appropriate plots for each variable and a two-dimensional plot of cost versus km at failure. Does there appear to be a relationship between time of failure and cost?

3.10. Perform a preliminary analysis of the data on fiber strengths in Table 2.6. Do separate analyses for each fiber length. Include graphical frequency distributions and summary statistics and use the results to compare lengths.

3.11. Perform a preliminary analysis of the data on bus motor failures given in Table 2.8.

3.12. Perform a preliminary analysis of the data on lifetimes of light bulbs given in Table 2.9. Provide results for each of the 42 weeks as well as overall results for the entire data set. Plot each summary statistic as a function of time. Are there any apparent patterns?

3.13. Perform a preliminary analysis of the data of Table 2.10.

3.14. Perform a preliminary analysis of the data of Table 2.11. Compare trials and insertion/withdrawal.

3.15. Perform a preliminary analysis of the data of Table 2.12. Compare schedules.

3.16. Perform a preliminary analysis of the data of Table 2.13. Compare clerks.

3.17. Perform a preliminary analysis of the data of Table 2.14, including any appropriate summaries and comparisons.

3.18. Repeat Exercise 3.17 on the data of Table 2.15. Include histograms of the failure data ("Part Hours") for each part and for the aggregated failure data.

3.19. Repeat Exercise 3.17 on the data of Table 2.16.

3.20. Repeat Exercise 3.17 on the data of Table 2.18.

3.21. Repeat Exercise 3.17 on the data of Table 2.20(a).

3.22. Repeat Exercise 3.17 on the data of Table 2.20(b).

3.23. Repeat Exercise 3.17 on the data of Table 2.21.

3.24. Repeat Exercise 3.17 on the data of Table 2.22.

3.25. Analyze the interfailure times of Table 2.23.

3.26. Calculate and analyze the times between maintenance events given in Table 2.25.

3.27. For the binomial data of Table 2.27, calculate the mean p and the standard deviation for each component/failure mode.

3.28. Calculate the third and fourth moments for the data of Tables 2.6, 2.7, and 2.13. In each case, calculate these separately for each group.

3.29. Prepare a table of summary statistics for all of the data sets analyzed in the above exercises and append it to the summary table prepared in the Chapter 2 exercises. (Include at least the following for each set: $\bar{x}$, s, CV, and five-number summary. Prepare a file of all data plots.

3.30. Each of the following comparisons of data characteristics provides information concerning symmetry or lack thereof:

    (i) the mean versus the median
    (ii) max $- Q_3$ versus $Q_1 -$ min
    (iii) max $- Q_2$ versus $Q_2 -$ min
    (iv) $Q_3 - Q_2$ versus $Q_2 - Q_1$

(a) For each of these four measures, determine which condition would indicate left skewness, right skewness, and symmetry.

(b) Can you think of any other measures of skewness based on these statistics? Indicate how these would be interpreted.

3.31. For each of the data sets analyzed in the above exercises, calculate the measures of skewness defined in the previous exercise. Interpret the results.

3.32. The third moment K_3 is also a measure of skewness. The distribution is skewed left if $K_3 < 0$, symmetric if $K_3 = 0$, and skewed right if $K_3 > 0$. Calculate K_3 for each of the data sets analyzed above and interpret the results.

3.33. Add the results of the previous two exercises to your table of summary statistics for the data sets. Summarize your conclusions regarding skewness and add a comment on your findings to the summary table.

CHAPTER 4

Probability Distributions for Modeling Time to Failure

4.1 INTRODUCTION

In Chapter 1 we discussed a variety of reliability-related problems and mentioned the use of the systems approach to finding solutions to these problems. The systems approach involves the use of mathematical models. An important and critical factor in the building of these models is modeling the system (product) failures.

A system (product) is, in general, comprised of several parts and system failure is related to part failures. As such, the starting point is the modeling of part failures. For a nonrepairable part, we need to consider only the first failure. For repairable parts, it is necessary to differentiate first failure from subsequent failures, since the latter depend on the type of repair action taken. The failure of a part can be characterized in many different ways. These lead to the use of different mathematical formulations in modeling failures.

In this chapter, we focus our attention on the modeling of first failures based on the "black-box" characterization. Here, a part is characterized as being in one of two states—working or failed. We consider two cases—static and dynamic. In the static case, because of manufacturing defects a part produced can initially be in a failed state. When such a part is put into operation, its (failed) state is detected immediately. In the dynamic case, the part is in its working state to start with and fails after a certain length of time (called time to first failure). Note that the static case can be viewed as a special case of the dynamic case with the time to failure being zero. As seen from the cases of Chapter 2, the times to failure vary significantly from item to item. In the "black-box" approach, one models this uncertainty in the time to failure without directly considering the mechanisms responsible for failure. In contrast, in the "white-box" approach, the failure is characterized in terms of the underlying failure mechanism. As a result, failure models based on this approach are more complex, since the failure mechanisms are modeled explicitly. We shall deal with models based on this approach in Chapter 6, where we also discuss the modeling of subsequent failures in the case of repairable parts. In Chapter 7 we relate part failures to system failure. The models of these three chapters will be used extensively throughout the remainder the book.

In general, modeling involves linking the descriptive system characterization to a suitable mathematical formulation. Since the time to failure is uncertain, the mathematical formulation appropriate for modeling first failure is a distribution function, a concept developed in the theory of probability. In this chapter, we discuss a variety of distributions useful in modeling the time to first failure at the part level for a new item, based on the black-box characterization.

We note that whereas we generally discuss the notions in this chapter in the context of part reliability, in appropriate applications they are useful for modeling reliability at higher levels as well, and are widely used for that purpose. Thus "part" reliability may be interpreted as reliability of a part, component, etc., up to the product or system level. In fact, this nonspecificity is the nature of the black-box characterization.

The outline of this chapter is as follows. Our starting point is the concept of a random variable and distribution and density functions. This is done in Section 4.2. We focus primarily on the class of distributions useful in modeling failure times and we call them failure distributions. Some basic concepts from the theory of probability useful for a better understanding are given in Appendix A.1. We consider both discrete and continuous distribution functions. Also, we introduce the concept of a failure rate function associated with a continuous time failure distribution. This concept plays an important role in the study of unreliable systems. In Section 4.3, we discuss several discrete distribution functions useful for modeling the static case and for other purposes as well. The next three sections (Sections 4.4–4.6) deal with a variety of continuous distribution functions useful for modeling the dynamic case. Section 4.7 deals with the classification of continuous failure distributions based on various notions of aging and the relationships between these concepts. We also present some lower and upper bounds for a failure distribution function. The shapes of the density and failure rate functions depend on the parameters of the failure distribution function. In Sections 4.8 and 4.9 we discuss the parametric study of these shapes for a few failure distribution functions. Failures are often influenced by environmental factors. The mathematical modeling of this phenomenon is discussed in Section 4.10.

4.2 RANDOM VARIABLES AND PROBABILITY DISTRIBUTIONS

In this section, we introduce the notion of a random variable and its characterization through a distribution function. A more basic approach involves sets of outcomes and probabilities of these outcomes. This is discussed briefly (and in a nonrigorous manner) in Appendix A.1.

4.2.1 Random Variables

A random variable X can be viewed as a function which assumes values in the interval $(-\infty, \infty)$. The value assumed is unpredictable and is dependent on some chance mechanism.

Example 4.1
Due to uncertainty in manufacturing, items produced are unreliable in the static sense of reliability. As a result, the state X of an item produced is uncertain before the event of testing. It can be either in working state ($X = 1$) or in failed state ($X = 0$). These two states are also known as nondefective (or conforming) and defective (or nonconforming) states, respectively, and these terms are used interchangably throughout the book. Note that the value of X is unknown before inspection and belongs to the set $\{0, 1\}$. After testing, the true value of X is known. ∎

Example 4.2
An item is in working state when put in use. The time to failure T is uncertain when the item is put in use (which may occur for a variety of reasons, such as variability in raw material,

manufacturing, operating environment, and so forth). In this case, T can assume any value in the interval $[0, \infty)$. After failure, the value of T is known. ■

If the set of values that X can assume is a countable set $\{x_1, x_2, \ldots, x_n\}$, where n can be infinite, then the random variable is said to be a *discrete random variable*. On the other hand, if X can assume all real values in the interval (a, b), where a can be $-\infty$ and/or b can be ∞, then the random variable is said to be a *continuous random variable*.

4.2.2   Distribution and Density Functions

The uncertainty in the values that X can assume can be described through a distribution function $F(x; \theta)$ which characterizes the probability $P\{X \le x\}$ and is defined as

$$F(x; \theta) = P\{X \le x\} \tag{4.1}$$

for $-\infty < x < \infty$. $F(x; \theta)$ is a nondecreasing function with $F(-\infty; \theta) = 0$ and $F(\infty; \theta) = 1$. θ denotes the parameters (or parameter set) of the distribution function. For any $x_1 < x_2$, we have

$$P\{x_1 < X \le x_2\} = P\{X \le x_2\} - P\{X \le x_1\} = F(x_2; \theta) - F(x_1; \theta) \tag{4.2}$$

which gives the probability that the random variable assumes values in the interval $(x_1, x_2]$.

In the case of discrete random variable,

$$F(x; \theta) = \sum_{x_i \le x} p_i \tag{4.3}$$

where

$$p_i = P\{X = x_i\} \tag{4.4}$$

The distribution is called a *discrete distribution*. Discrete distributions have been used in modeling static and dynamic failures. An example of the latter is:

Example 4.3
Every landing causes a stress on the landing gear of an aircraft and a failure occurs when the stress exceeds the strength of the material. In this case, the number of landings before a failure is a discrete random variable assuming a value from the set of positive integers. ■

In the case of (absolutely) continuous random variables, the distribution is called *continuous distribution* and there exists a function $f(x; \theta)$ such that

$$F(x; \theta) = \int_{-\infty}^{x} f(x; \theta)dx \tag{4.5}$$

or equivalently,

$$f(x; \theta) = \frac{dF(x; \theta)}{dx} \tag{4.6}$$

over the interval $-\infty < x < \infty$. $f(x; \theta)$ is called the *density function* associated with the distribution function $F(x; \theta)$. From (4.2) it follows that

$$P\{x < X \le x + \delta x\} = f(x; \theta)\delta x + 0(\delta x^2) \tag{4.7}$$

where $0(\delta x^2)/\delta x \to 0$ as $\delta x \to 0$. This implies that $f(x; \theta)\,\delta x$ is approximately the probability that x will assume values in the interval $(x, x + \delta x]$ for very small δx. Note that in the continuous case $P(X = x)$ is zero.

Sometimes a continuous-valued random variable can assume a set of discrete values with nonzero probability. In this case, $F(x; \theta)$ has jumps at these discrete values where the derivative (in the usual sense) of $F(x; \theta)$ does not exist. Elsewhere, $F(x; \theta)$ is differentiable. Distributions of this type are called *mixed continuous and discrete* distributions. An example in the context of an unreliable item is the following:

Example 4.4

Because of manufacturing variability, an item produced is in failed state with probability p or in working state with probability $(1 - p)$. The time to failure for a working item is given by a continuous distribution function $H(x)$ with $H(x) = 0$ for $x \le 0$. In this case, we have

$$F(x; \theta) = p + (1 - p)H(x)$$

and the parameter set consists of the parameters of the distribution function $H(x)$ and p. ∎

4.2.3 Failure Distributions

We use the term *failure distribution* for a continuous distribution with $F(x; \theta) = 0$ for $x < 0$. Distributions of this type are used extensively for modeling failure times. In this case X corresponds to the age of an item at failure. Failure time is also called the lifetime of the item in the case of a nonrepairable item.

The *survivor function, $S(x; \theta)$,* characterizes the probability that $X > x$. In other words, it is the probability that the item will not fail before it reaches an age x. $S(x; \theta)$ is also often called the *reliability* of the item. It is related to $F(x; \theta)$ by

$$S(x; \theta) = P\{X > x\} = 1 - P\{X \le x\} = 1 - F(x; \theta) \tag{4.8}$$

$\overline{F}(x; \theta)$ is used to denote $[1 - F(x; \theta)]$. We shall use this and $S(x; \theta)$ interchangeably.

The conditional probability that the item will fail in the interval $(x, x + t]$ given that it has not failed at or prior to x is given by

$$F(t\,|\,x,\,\theta) = \frac{F(t + x;\,\theta) - F(x;\,\theta)}{1 - F(x;\,\theta)} \tag{4.9}$$

The *hazard function* (or *failure rate function*) $r(x; \theta)$ associated with $F(x; \theta)$ is defined as

$$r(x;\,\theta) = \lim_{t \to 0} \frac{F(t|x;\,\theta)}{t} = \frac{f(x;\,\theta)}{1 - F(x;\,\theta)} \tag{4.10}$$

The hazard function $r(x; \theta)$ can be interpreted as the probability that the item will fail in $(x, x + \delta x]$ given that it has not failed at or prior to x. In other words, it characterizes the effect of age on item failure more explicitly than $F(x; \theta)$ or $f(x; \theta)$.

The *cumulative failure rate function, $R(x; \theta)$,* is defined as:

$$R(x;\,\theta) = \int_0^x r(y;\,\theta)dy \tag{4.11}$$

and is also called the *cumulative hazard function*. These functions are related as follows:

$$F(x; \theta) = 1 - e^{-R(x;\theta)} \qquad (4.12)$$

$$f(x; \theta) = r(x; \theta)e^{-R(x;\theta)} \qquad (4.13)$$

and

$$R(x; \theta) = -\log(1 - F(x; \theta)) \qquad (4.14)$$

4.2.4   Moments

$M_j(\theta)$, the jth moment of a random variable (also called the jth moment of the probability distribution), $j \geq 1$, is given by

$$M_j(\theta) = E[X^j] \qquad (4.15)$$

where $E[\ ]$ is the *expectation* or *expected value* of the random variable. In the case of a discrete random variable, this is given by

$$M_j(\theta) = \sum_{i=1}^{n} (x_i)^j p_i \qquad (4.16)$$

where n may be ∞. In the case of a continuous random variable, it is given by

$$M_j(\theta) = \int_{-\infty}^{\infty} x^j f(x; \theta)dx \qquad (4.17)$$

The first moment $M_1(\theta)$ is also denoted μ and $M_j(\theta)$ is also denoted μ_j' for $j > 1$.
    $\mu_j(\theta)$, the jth central moment of a random variable X, $j \geq 1$, is given by

$$\mu_j(\theta) = E[(X - \mu)^j] \qquad (4.18)$$

This is calculated as

$$\mu_j(\theta) = \sum_{i=1}^{n} (x_i - \mu)^j p_i \qquad (4.19)$$

in the case of a discrete random variable, and as

$$\mu_j(\theta) = \int_{-\infty}^{\infty} (x - \mu)^j f(x; \theta)dx \qquad (4.20)$$

in the case of a continuous random variable.
    Note that $M_j(\theta)$ and $\mu_j(\theta)$ ($j \geq 1$) are functions of the parameters of the distribution function $F(x; \theta)$. This will be used in the next chapter in the estimation of the parameters based on failure data and an assumed failure distribution.

Some notions defined in terms of the first few moments are:

1. The first moment μ is called the *mean* (of the random variable or of the probability distribution). It is a measure of central tendency.

2. The second central moment μ_2 (θ) is called the *variance* and is also denoted σ^2. The square root of the variance (σ) is called the *standard deviation*. Both are measures of dispersion or spread of a probability distribution.

3. $\gamma_1 = \mu_3(\theta)/(\mu_2(\theta))^{1/2}$ is called the *coefficient of skewness*. It is measure of departure from symmetry of the density function $f(x; \theta)$. If $\gamma_1 = 0$, then $f(x; \theta)$ is symmetric, if $\gamma_1 > 0$, it is positively skewed (or skewed right), and if $\gamma_1 < 0$, it is negatively skewed (or skewed left).

4. $\gamma_2 = [\mu_4(\theta)/(\mu_2(\theta))^2 - 3]$ is called the *coefficient of excess* or *kurtosis*. It is a measure of the degree of the flattening of the density function. If $\mu_2 = 0$, the density function is called mesokurtic. If $\gamma_1 > [<] 0$, it is called leptokurtic [platykurtic].

5. σ/μ is called the *coefficient of variation*. It is a standardized (unit free) measure of dispersion.

The moments of a distribution are easily derived from the *moment generating function*. Another important function (closely related to the moment generating function and useful in analysis later on) is the *characteristic function*. These two functions are briefly discussed in Appendix A.2.

4.2.5 Fractiles and Median

For a continuous distribution, the α-fractile, x_α, for a given α, $0 < \alpha < 1$, is a number such that

$$P\{X \le x_\alpha\} = F(x_\alpha) = \alpha \tag{4.21}$$

The fractiles for $\alpha = 0.25$ and 0.75 are called first and third *quartiles* and the 0.50 fractile is called the *median*.

4.3 BASIC DISCRETE DISTRIBUTIONS

In this section we discuss some basic discrete distribution functions. We shall confine our attention to the case where X assumes integer values from the set $\{0, 1, 2, \ldots, n\}$ with n being either finite or infinite. Let p_i, $0 \le i \le n$ denote the probability that X assumes the value i, i.e., $p_i = P\{X = i\}$.

Some of the following distributions are useful for modeling static reliability at the part level. Others will be used in later chapters (especially Chapter 6 and 7). For each distribution, we give the mathematical expressions for p_i, indicate the parameter set of the distribution, and give expressions for the mean and variance of the distribution. In some cases, other important characteristics (coefficients of skewness and kurtosis) of the distribution are also given.[1]

4.3.1 Bernoulli Distribution

Here X takes on two values, 0 or 1. As a result $n = 1$ and p_i, $i = 0$ and 1, is given by

$$p_0 = p \quad \text{and} \quad p_1 = (1 - p) \tag{4.22}$$

The parameter set is $\theta = \{p\}$ with $0 \leq p \leq 1$. The mean and variance are

$$\mu = p \qquad \text{and} \qquad \sigma^2 = p(1-p) \tag{4.23}$$

This distribution can be used to model static reliability at the part level with $X = 0$ corresponding to a failed part and $X = 1$ corresponding to a working part. The uncertainty is due to manufacturing variations. p is the probability that an item is in the failed state. Smaller values of p correspond to more reliable items.

4.3.2 Binomial Distribution

Here X can assume values from 0 to n and p_i, $0 \leq i \leq n$, is given by

$$p_i = \frac{n!}{i!(n-i)!} p^i (1-p)^{(n-i)} \tag{4.24}$$

The parameter set is $\theta = \{n, p\}$ with $0 \leq p \leq 1$ and $0 < n < \infty$. The mean and variance are

$$\mu = np \qquad \text{and} \qquad \sigma^2 = np(1-p) \tag{4.25}$$

and the coefficients of skewness and kurtosis are

$$\gamma_1 = \frac{1 - 2p}{\sqrt{[np(1-p)]}} \qquad \text{and} \qquad \gamma_2 = \frac{1}{np(1-p)} - \frac{1}{n} \tag{4.26}$$

The distribution can be used to model the number of nonworking items in a lot of size n with X representing this number. p is the probability that an item is nonworking and $(1-p)$ is the probability that it is working.

Finally, it is worth noting that X can be viewed as the sum of n independent Bernoulli random variables, so that its distribution can be obtained as the n-fold convolution of the Bernoulli distribution. (See Appendix A.)

Example 4.5

Consider the unreliable component of Example 4.1. Let p denote the probability that an item is unreliable (in the sense of static unreliability). A sample of size n is selected from the output of a production run. X is the number of defective items in a sample.

Suppose that $p = 0.03$ and $n = 40$. From (4.25), we have the mean number of defective items in a sample of size 40 given by $\mu = 40(0.03) = 1.20$ and the standard deviation by $\sigma = \sqrt{40(0.03)(0.97)} = 1.0789$. This implies that on the average, one can expect to find 1.2 defectives in a sample of 40. From (4.24), the probability of finding 3 defective items in a sample is given by

$$P(X = 3) = \binom{40}{3}(0.03)^3(0.97)^{37} = 0.0864$$

Finally, the probability of finding 3 or more nonworking items in a sample is given by

$$P(X \geq 3) = 1 - P(X < 3) = 1 - P(X = 0) - P(X = 1) - P(X = 2)$$
$$= 1 - 0.2957 - 0.3658 - 0.2206 = 0.1179$$

This implies that 8.6% of the time one will find 3 defectives; about 12% of the time, one will find 3 or more defectives in a lot.

Note: When both $np > 5$ and $n(1 - p) > 5$, the binomial distribution can be approximated by a normal distribution (See Section 4.4.) with mean and variance given by (4.25). ■

4.3.3 Geometric Distribution

Here X can assume values from 0 to ∞ and p_i, $0 \leq i \leq \infty$, is given by

$$p_i = (1 - p)^i p \tag{4.27}$$

The parameter set is $\theta = \{p\}$ with $0 \leq p \leq 1$. The mean and variance are

$$\mu = \frac{(1 - p)}{p} \quad \text{and} \quad \sigma^2 = \frac{(1 - p)}{p^2} \tag{4.28}$$

This distribution can be used to model the number of nondefective parts produced during manufacturing before a defective part is produced with X representing this number. p is the probability that an item is defective and $(1 - p)$ is the probability that it is not.

Example 4.6

Consider the unreliable component of Example 4.1. As before, p is the probability that an item is defective. The manufacturer produces the items in continuous production. The items produced are inspected to detect defectives. Let X denote the number of items inspected before a defective item is detected.

Suppose that $p = 0.03$. From (4.28), the mean and standard deviation of X are $\mu = (0.97)/(0.03) = 32.33$ and $\sigma = \sqrt{(0.97)/(0.03)^2} = 32.83$. This implies that on the average 32.33 nondefective items are inspected before a defective item is detected. When p is 0.01, the average number inspected changes to 99. This is to be expected, since the items produced are more reliable and it should take longer to find a defective. The reverse occurs when p is 0.05; in this case the average number inspected is 19.

From (4.27), with $p = 0.03$, the probability that $X = 3$ is given by

$$P(X = 3) = (0.97)^3(0.03) = 0.0274.$$

The probability that $X \leq 3$ is given by

$$P(X \leq 3) = P(x = 0) + P(X = 1) + P(X = 2) + P(X = 3) = 0.1147. ■$$

4.3.4 Negative Binomial Distribution

For the negative binomial distribution, X assumes values from 0 to ∞. The distribution is given by

$$p_i = P(X = r + i) = \binom{r + i - 1}{r - 1} p^r (1 - p)^i \quad i = 0, 1, 2, \ldots \tag{4.29}$$

The parameter set is $\theta = \{r, p\}$ with $r \geq 1$ and $0 \leq p \leq 1$. The mean and variance are given by

$$\mu = \frac{r(1 - p)}{p} \quad \text{and} \quad \sigma^2 = \frac{r(1 - p)}{p^2} \tag{4.30}$$

This distribution is useful in modeling the number of items inspected until r defective items are found, with p being the probability that an item is defective.

Example 4.7
Suppose $p = 0.03$ and let $r = 10$. From (4.30) the mean and variance of X are $\mu = 10(0.97)/(0.03) = 323.33$ and $\sigma^2 = 103.82$. This implies that on the average it is necessary to test 323.33 items in order to obtain 10 defective items. ∎

4.3.5   Hypergeometric Distribution

Here X assumes integer values in the interval $\max(0, n - n + D) \leq i \leq \min(n, D)$ where N, D, and n are the three parameters of the distribution, with n, D, and N positive integers satisfying $n \leq N$ and $D \leq N$. The p_i are given by

$$p_i = P(X = i) = \frac{\dbinom{D}{i}\dbinom{N-D}{n-i}}{\dbinom{N}{n}} \tag{4.31}$$

The mean and variance are given by

$$\mu = \frac{nD}{N} \quad \text{and} \quad V(X) = \frac{(N-n)n}{N-1}\left(\frac{D}{n}\right)\left(1 - \frac{D}{N}\right) \tag{4.32}$$

This distribution is useful in modeling the number of defective items in a sample when sampling is done without replacement, as illustrated in the following example:

Example 4.8
The items produced by a manufacturer are unreliable and the (static) reliability is as indicated in Example 4.5. These are sold in batches of 200 units per batch to retailers who in turn sell them in smaller batches to customers. Suppose that there are 6 defective items in a batch bought by the retailer. He sells 40 items to a customer. Let X denote the number of defective items in the batch of 40 items sold to the customer. This can be modeled by a hypergeometric distribution with $N = 200$, $D = 6$, and $n = 40$. Note that X can take on the values $0, 1, \ldots, 6$. The probability that there are three defective items in the lot sold can be obtained from (4.31) and is given by

$$P(X = 3) = \frac{\dbinom{197}{37}\dbinom{6}{3}}{\dbinom{200}{40}} = \frac{194!6!40!160!}{37!157!3!3!200!} = \frac{5\cdot4\cdot40\cdot39\cdot38\cdot160\cdot159\cdot158}{200\cdot199\cdot198\cdot197\cdot196\cdot195} = 0.0803$$

On comparing this result with that of Example 4.4, we see that the probability in this case is slightly smaller. The reason for this is that in Example 4.4 the lot of 40 is drawn from an infinite population so that the probability of an item being defective is always 0.03. In contrast, here this changes as we draw items from the batch of 200 to fill the order for 40 items. The difference in the two types of sampling is also reflected in the mean and variance, which in this case, using (4.32), are $\mu = 40(6/200) = 1.2$ and $\sigma^2 = [40(200-40)/(199)][0.03(0.97)] = 0.9359$. For the binomial, the mean is the same, but the variance is larger (1.0789). There is less variability when sampling from a finite population is done without replacement. ∎

4.3.6 Poisson Distribution

Here X can assume values from 0 to ∞ and p_i, $0 \le i < \infty$, is given by

$$p_i = \frac{e^{-\lambda}\lambda^i}{i!} \tag{4.33}$$

The parameter set is $\theta = \{\lambda\}$ with $\lambda > 0$. The mean and variance are given by

$$\mu = \lambda \quad \text{and} \quad \sigma^2 = \lambda \tag{4.34}$$

and the coefficients of skewness and kurtosis are

$$\gamma_1 = \frac{1}{\sqrt{\lambda}} \quad \text{and} \quad \gamma_2 = \frac{1}{\lambda} \tag{4.35}$$

This distribution will be used later in the book in many different contexts.

Example 4.9
Let X denote the number of flaws (e.g., microcracks) in a component. We assume that the phenomenon can be modeled by a Poisson process with $\lambda = 0.20$. From (4.34), we have that the average number of flaws in a component is 0.2. From (4.33), the probability of finding no flaws and one flaw in a component are given by

$$P(X=0) = \frac{e^{-0.2}(0.2)^0}{0!} = 0.8187 \quad \text{and} \quad P(X=1) = \frac{e^{-0.2}(0.2)^1}{1!} = 0.1638$$

and the probability of finding more than 1 flaw in a component is given by

$$P(X>1) = 1 - P(X=0) - P(X=1) = 0.0175$$

The probability of finding more than 2 flaws is still smaller, namely $P(X \ge 2) = 0.0011$. Thus roughly 11 in 10,000 items will have 2 or more flaws. ∎

Example 4.10
Let X denote the number of failures of a component over a six month period and assume that the failure process can be modeled by a Poisson distribution with $\lambda = 2.92$. From (4.34) the mean and variance of X are given by $\mu = \sigma^2 = 2.92$. From (4.33), the probabilities for $X = 0$, 1, 2, 3, 4 and 5 are found to be:

i:	0	1	2	3	4	5
p_i:	0.0539	0.1575	0.2299	0.2238	0.1634	0.0954

From this one can see that the probability of no failure in 6 months is 0.0539 and the probability that there will be more than two failures in six months is $1 - 0.0539 - 0.1575 - 0.2299 = 0.5587.$ ∎

4.4 CONTINUOUS DISTRIBUTIONS—I (BASIC DISTRIBUTIONS)

In this section we first give a few basic failure distributions (with $F(x; \theta) = 0$ for $x < 0$) that are useful for modeling time to first failure at the part level. Later we give a few more general basic distributions. These cannot be used to model failure times unless $P(X < 0) \approx 0$. In the sections that follow, we derive other distributions that are based on these and which are useful for modeling failure times.[2]

4.4.1 Exponential Distribution

The distribution function for the exponential distribution is given by

$$F(x; \theta) = 1 - e^{-\lambda x} \tag{4.36}$$

for $x \geq 0$. The parameter set is $\theta = \{\lambda\}$, with $\lambda > 0$. The density function and the failure rate functions are given by

$$f(x; \theta) = \lambda e^{-\lambda x} \tag{4.37}$$

and

$$r(x; \theta) = \lambda \tag{4.38}$$

respectively. Note that the failure rate is constant and does not change with x (the age of the item).

The first two moments are given by

$$\mu = \frac{1}{\lambda} \quad \text{and} \quad \sigma^2 = \frac{1}{\lambda^2} \tag{4.39}$$

and the coefficients of skewness and kurtosis are

$$\gamma_1 = 2 \quad \text{and} \quad \gamma_2 = 6 \tag{4.40}$$

The median of the distribution is $0.6931/\lambda$.

Figure 4.1 shows the shapes of $f(x; \theta)$ for $\mu = 1/\lambda = 500, 1000, 1500, 2000,$ and 5000. $r(x)$ is not plotted; it is constant with value λ.

The exponential distribution has been used to model failures of electronic and electrical parts and in many other applications. In fact, it is one of the most widely used failure distributions. The distribution is appropriate whenever failures occur randomly and are not age dependent.

Comments
1. Define $\beta = 1/\lambda$. Then $F(x; \theta) = 1 - e^{-x/\beta}$. β is called the scale parameter.
2. There is a close link between the exponential and Poisson distributions. This will be discussed in Chapter 6.

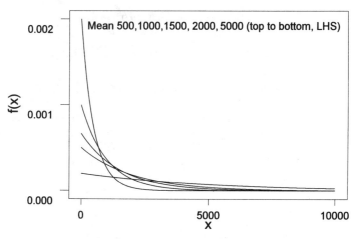

Figure 4.1. Exponential distribution.

3. A second reason for the widespread use of the exponential distribution is that it is mathematically tractable in most applications (or at least more so than nearly any other distribution). This is not a good reason for assuming exponentiality.

Example 4.11
The mean time to failure (MTTF) for a certain brand of light bulbs is $\mu = 1500$ hours. From (4.39) we have $\lambda = 1/1500 = 0.0006667$ per hour. Note that the median lifetime is $0.6931/\lambda = 1039.6$. Although these bulbs have an average lifetime of 1500 hours, half will burn out in 1040 hours.

The probability $S(x)$ that a light bulb will survive beyond x is given below for a range of values of x (in hours):

x	500	1000	1500	2000	3000	5000
$S(x)$	0.7164	0.5132	0.3679	0.2633	0.1356	0.0356

Results such as those of this example have implications in terms of maintenance, as will be discussed in Chapter 16. ∎

4.4.2 Gamma Distribution

The gamma density function is given by

$$f(x;\ \theta) = \frac{x^{\alpha-1}e^{-x/\beta}}{\beta^{\alpha}\Gamma(\alpha)} \tag{4.41}$$

for $x \geq 0$. The parameter set is $\theta = \{\alpha, \beta\}$ with $\alpha > 0$ and $\beta > 0$. Here $\Gamma(.)$ is the gamma function and is tabulated extensively by Abramowitz and Stegun (1964).

The failure distribution and failure rate functions are complicated functions involving confluent hypergeometric functions (Abramowitz and Stegun, 1964). The mean and variance are

$$\mu = \alpha\beta \qquad \text{and} \qquad \sigma^2 = \alpha\beta^2 \tag{4.42}$$

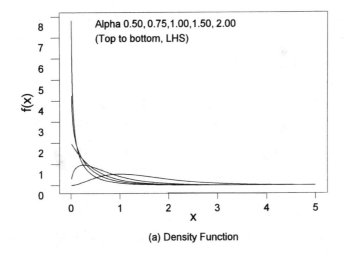

(a) Density Function

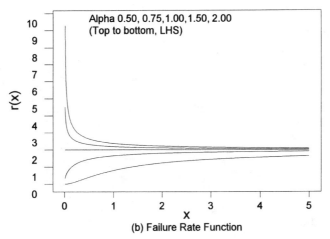

(b) Failure Rate Function

Figure 4.2. Gamma distribution $\beta = 0.50$.

and the coefficients of skewness and kurtosis are

$$\gamma_1 = \frac{1}{\sqrt{\alpha}} \qquad \text{and} \qquad \gamma_2 = \frac{6}{\alpha} \tag{4.43}$$

Figure 4.2 shows the shapes of $f(x; \theta)$, and $r(x; \theta)$ for $\alpha = 0.50, 0.75, 1.00, 1.50$, and 3.00 and $\beta = 0.50$. The gamma distribution has been used to model failure times for many different objects.

Comments
   1. β is the scale parameter of the distribution.
   2. When α takes on only integer values, the distribution is also known as the Erlangian distribution. In this case the distribution can be interpreted as the distribution of a sum of exponential random variables.

3. When $\alpha = 1$, the gamma distribution reduces to the exponential distribution with $\lambda = 1/\beta$.
4. When $\beta = 2$ and $\alpha = \nu/2$, where ν is a positive integer, the distribution is called the chi-square distribution with ν "degrees of freedom."

Example 4.12
A small hydraulic pump has a failure distribution that can be modeled by a gamma distribution with $\alpha = 2$ and $\beta = 2.5$ years. Then from (4.42) the mean life of the pump is 5 years and the variance is 12.5.

The probability that the pump will survive beyond x is given below for a range of values of x (in years):

x	1	2	3	4	5	6	7	8	9	10
$S(x)$	0.938	0.808	0.663	0.525	0.406	0.308	0.231	0.171	0.126	0.092

This implies that that the probability that the pump will fail before 5 years (the mean life) is $1 - 0.406 = 0.594$.

4.4.3   Weibull Distribution

The two-parameter Weibull distribution function is given by

$$F(x; \theta) = 1 - e^{-(x/\beta)^{\alpha}} \tag{4.44}$$

for $x \geq 0$. The parameter set is $\theta = \{\alpha, \beta\}$ with $\alpha > 0$ and $\beta > 0$. The failure density and failure rate functions are given by

$$f(x; \theta) = \frac{\alpha x^{(\alpha-1)} e^{-(x/\beta)^{\alpha}}}{\beta^{\alpha}} \tag{4.45}$$

and

$$r(x; \theta) = \frac{\alpha x^{(\alpha-1)}}{\beta^{\alpha}} \tag{4.46}$$

respectively. The mean and variance are

$$\mu = \Gamma\left(1 + \frac{1}{\alpha}\right)\beta \quad \text{and} \quad \sigma^2 = \left[\Gamma\left(1 + \frac{2}{\alpha}\right) - \left\{\Gamma\left(1 + \frac{1}{\alpha}\right)\right\}^2\right]\beta^2 \tag{4.47}$$

and the coefficients of skewness and kurtosis are given by

$$\gamma_1 = \frac{\mu_3}{\sigma} \quad \text{and} \quad \gamma_2 = \frac{\mu_4}{\sigma^4} - 3 \tag{4.48}$$

where μ_3 and μ_4 are given by

$$\mu_m = \beta^m \sum_{j=0}^{m} (-1)^j \binom{m}{j} \Gamma\left(\frac{m-j+\alpha}{\alpha}\right) \Gamma^j\left(\frac{1+\alpha}{\alpha}\right) \tag{4.49}$$

for $m = 3$ and 4 respectively. The median is $(0.6931^{1/\alpha})\beta$.

Figure 4.3 shows the shapes of $f(x; \theta)$ and $r(x; \theta)$ for a range of parameter values chosen so that the means and variances of the distributions are the same as those of the gamma distributions of Figure 4.2. The corresponding parameter sets for the Weibull are $(\alpha, \beta) = (0.721, 0.203), (0.869, 0.350), (1, 0.50), (1.23, 0.804),$ and $(1.80, 1.69)$. (The plots are in this order from top to bottom in both Figure 4.3(a) and 4.3(b).)

The Weibull distribution was first proposed by Weibull (1951). Harter and Moore (1976) give a comprehensive list of references dealing with the applicability of the Weibull distribution to model failure times for many different technical objects. (See Table IV of their paper.) Kececioglu (1991, p. 313) discusses the use of this distribution in modeling failure times for electron tubes, capacitors, ball bearings, leakage from batteries and many other applications.

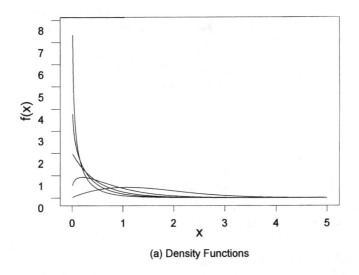

(a) Density Functions

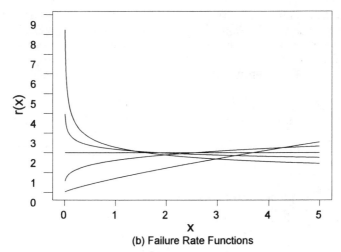

(b) Failure Rate Functions

Figure 4.3. Weibull distribution.

Comments

1. β is called the scale parameter and α is called the shape parameter. As α varies, the shape of the distribution varies distinctively, as shown in Figure 4.3. The values of α giving rise to the various shapes are $0 < \alpha < 1$, $\alpha = 1$, $1 < \alpha < 2$, and $\alpha \geq 2$.

2. When $\alpha = 2$, the distribution is called a Rayleigh distribution.

3. When $\alpha = 1$, the distribution reduces to an exponential distribution (with $\beta = 1/\lambda$).

4. If α is greater than 3, the distribution is approximately normal.

5. Because of its substantial flexibility (including having the exponential as a special case), the Weibull is considered to be the most widely used failure distribution in reliability applications.

Example 4.13

The lifetime of a medical implant can be modeled by a Weibull distribution with shape parameter $\alpha = 3$ and scale parameter $\beta = 22$ (in years). Since the shape parameter is greater than 1, it follows from Equation (4.46) that the hazard function is increasing with time. From (4.47) the mean time to failure is $\mu = 22\Gamma(1.33) = 19.64$ years and the variance is $\sigma^2 = 22^2(.9028) - 19.64^2 = 51.22$. [The values of the gamma function required for these calculations are found from the tables of Abramowitz and Stegun to be $\Gamma(1.333) = 0.8929$ and $\Gamma(1.667) = 0.9028$.]

The reliability of the implant for different time intervals (x, in years) is as follows:

x	5	10	15	20	25	30
$S(x)$	0.9884	0.9104	0.7284	0.4717	0.2305	0.0792

The median of the distribution is $(0.6931)^{1/3}(22) = 19.47$, i.e., half of the implants will fail prior to 19.47 years of service. Note that this value is nearly the same as the mean time to failure, indicating that the distribution is nearly symmetrical. ∎

Example 4.14

Suppose that the lifetime of an electronic component can be adequately modeled by a Weibull distribution with shape parameter $\alpha = 0.8$ and scale parameter $\beta = 17.335$ Since the scale parameter is less than 1, the hazard function is decreasing. From (4.42), the mean time to failure is given by $\mu = (17.335)\Gamma(2.25) = 19.64$ years [since $\Gamma(2.25) = 1.133$]. Note that these values were chosen so that the mean time for this component is the same as that for the implant in Example 4.13.

The reliability of the component for different time intervals (x, in years) is as follows:

x	5	10	15	20	25	30
$S(x)$	0.6908	0.5252	0.4104	0.3259	0.2617	0.2121

Here the median is 10.964 years, which is much smaller than the mean, indicating strong right skewness. ∎

4.4.4 Normal (Gaussian) Distribution

The density function for the normal distribution is given by

$$f(x;\ \theta) = \frac{e^{-(x-\mu)^2/2\sigma^2}}{\sigma\sqrt{2\pi}} \tag{4.50}$$

$-\infty < x < \infty$. The parameter set is $\theta = \{\mu, \sigma^2\}$ with $\sigma > 0$ and $-\infty < \mu < \infty$. It is not possible to give analytical expressions for the distribution function. In general, this distribution cannot be used to model failure times, since $f(x; \theta) > 0$ for $x < 0$. However, if $\mu \gg \sigma$, then $P\{X < 0\} \approx 0$ and this distribution can be used to model failure times. In addition, the distribution is used to model many variables that are related to reliability. The lognormal distribution, for which $f(x; \theta) = 0$ for $x < 0$, is derived from this and used in extensively in modeling failure times. This will be discussed in the next section. The mean and variance, μ and σ^2, are also the parameters of the distribution. Also, the coefficients of skewness and kurtosis are both 0.

Normal probabilities are calculated using the table for the standard normal distribution given in Appendix C, Table C1. The standard normal distribution has mean $\mu = 0$ and variance $\sigma^2 = 1$. A standard normal random variable is usually denoted Z. Table C1 gives $F_Z(z) = P(Z \le z)$. If X is an arbitrary normal random variable, $P(X \le x)$ is calculated as $P(X \le x) = F_Z[(x - \mu)/\sigma]$.

Example 4.15

Suppose that the breaking strength of leads used in heart pacemakers is normally distributed with mean $\mu = 20$ years and variance $\sigma^2 = 4$. The design requirement is that the strength be greater than L. This is calculated as $P(X > L) = 1 - P(X \le L)$. The probabilities that a lead will meet the design requirements for various values of L are as follows:

L	14	16	18	20	22	24	26
$P(X > L)$	0.9987	0.9772	0.8413	0.5000	0.1587	0.0228	0.0013	■

4.4.5 Inverse Gaussian Distribution

In this case $f(x; \theta)$ is given by

$$h(x; \theta) = \sqrt{\{\lambda/(2\pi x^3)\}}e^{-\lambda(x-\mu)^2/2\mu^2 x} \tag{4.51}$$

for $x \ge 0$. The parameter set is $\theta = (\lambda, \mu)$ with $\lambda > 0$ and $\mu > 0$. Inverse Gaussian distributions are also called Wald distributions, and were first used in analysis of Brownian motion. The distribution was named "inverse Gaussian" because certain characteristics of the distribution are inversely related to those of the normal (Gaussian) distribution. Inverse Gaussian distributions have been used as alternatives to the Weibull to model breaking strengths of fibers and in other applications. (See Durham and Padget, 1997).

It is not possible to give analytical expressions for the failure distribution and failure rate functions. The mean and variance are given by

$$M_1(\theta) = \mu \quad \text{and} \quad \sigma^2 = \frac{\mu^3}{\lambda} \tag{4.52}$$

Figure 4.4 indicates the shapes of $f(x; \theta)$ and $r(x; \theta)$ for a range of parameter values corresponding to those used for the gamma and Weibull distributions with the same means and variances. The parameter sets satisfying these conditions are $(\mu, \lambda) = (0.25, 0.125)$, $(0.375, 0.28125)$, $(0.50, 0.50)$, $(0.75, 1.125)$, and $(1.50, 4.50)$. The top to bottom curves in Figure 4.4 are in this order.

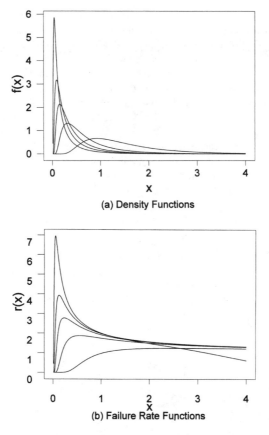

(a) Density Functions

(b) Failure Rate Functions

Figure 4.4. Inverse Gaussian distribution.

Folks and Chhikara (1978) review applications of the inverse Gaussian distribution. For more on this distribution, see Chhikara and Folks (1989).

4.4.6 Gumbel Distributions

The Gumbel (type I) distribution of the smallest extreme is given by

$$F(x; \theta) = 1 - e^{-e^{(x-\alpha)/\beta}} \tag{4.53}$$

for $-\infty < x < \infty$. The parameter set is $\theta = \{\alpha, \beta\}$ with $-\infty < \alpha < \infty$ and $\beta > 0$.

The Gumbel (type I) distribution of the largest extreme is given by

$$F(x; \theta) = e^{-e^{-(x-\alpha)/\beta}} \tag{4.54}$$

for $-\infty < x < \infty$. The parameter set is $\theta = \{\alpha, \beta\}$ with $-\infty < \alpha < \infty$ and $\beta > 0$. Again, since $f(x; \theta) > 0$ for $x < 0$, this distribution cannot be used to model failure times. However, two distributions derived from this (based on transformed and truncated variables) are used in modeling failure times. These will be discussed in the next section.

Comment

The Gumbel distribution is related to the extreme value distribution. We will encounter the extreme value distribution in Chapter 7.

4.5 CONTINUOUS DISTRIBUTIONS—II (DERIVED DISTRIBUTIONS)

In this section, we give several failure distributions derived from basic distributions. Let $F(x; \theta)$ $[f(x; \theta)]$ denote the basic distribution (density) function and X the random variable having the basic distribution. Note that X need not be constrained to be nonnegative. Let Y denote the random variable from the derived failure distribution (density) function, which will be denoted $H(y; \theta)$ $[h(y; \theta)]$. We can divide the derived distributions into the following two broad categories:

Category A: Y is related to X by some transformation.

Category B: $H(\cdot; \theta)$ is related to $F(\cdot; \theta)$ by some transformation.

4.5.1 Lognormal Distribution (Category A)

In this case $f(x; \theta)$ is given by (4.50) and Y is related to X by the transformation $X = \log(Y)$, with X normally distributed. The density function $h(y; \theta)$ is given by

$$h(y; \theta) = \frac{e^{-\{(\log(y)-\mu)^2/2\sigma^2\}}}{\sigma y \sqrt{2\pi}} \tag{4.55}$$

for $y \geq 0$. The parameter set is $\theta = \{\mu, \sigma\}$ with $\sigma > 0$ and $-\infty < \mu < \infty$. It is not possible to give analytical expressions for the failure distribution and failure rate functions. Probabilities are easily calculated, however, by use of the relationship with the normal distribution.

The mean and variance are

$$M_1 = e^{(\mu+\sigma^2/2)} \quad \text{and} \quad V(Y) = \omega(\omega-1)e^{2\mu} \tag{4.56}$$

where $\omega = e^{\sigma^2}$. The coefficients of skewness and kurtosis are

$$\gamma_1 = \frac{\omega^3 - 3\omega + 2}{(\omega - 1)^{3/2}} \quad \text{and} \quad \gamma_2 = \frac{\omega^6 - 4\omega^3 + 6\omega - 3}{(\omega - 1)^2} - 3 \tag{4.57}$$

Figure 4.5 shows the shapes of $f(x; \theta)$ and $r(x; \theta)$ for a range of parameter values that give the same means and variances as in the previous figures. For the lognormal, these are (μ, σ^2) = $(-1.9356, 1.0986)$, $(-1.4045, 0.8473)$, $(-1.0397, 0.6932)$, $(-0.5431, 0.5108)$, and $(0.2616, 0.2877)$ (top to bottom curves in Figure 4.5).

The lognormal distribution has been used to model the failure times in many semiconductor devices, such as LEDs, IMPATTT diodes, lasers, etc. (see Cheng, 1977). Crow and Shimizu (1988) discuss the application of this distribution in many other contexts. See also Aitchison and Brown (1957) for more on the lognormal distribution.

Example 4.16

Suppose that the hydraulic pump in Example 4.12 can be modeled by a lognormal distribution with $\mu = 0.62$ and $\sigma^2 = 4$. From (4.56) we have the mean time to failure = 5.0 years, the

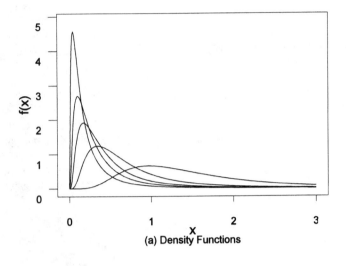

(a) Density Functions

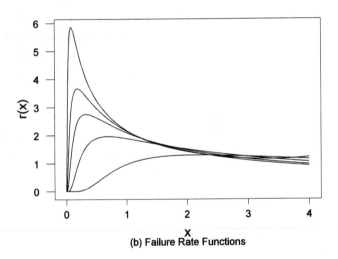

(b) Failure Rate Functions

Figure 4.5. Lognormal distribution.

same as that of the gamma distribution of Example 4.12. The reliability of the pump for the first ten years is given by

x	1	2	3	4	5	6	7	8	9	10
$S(x)$	0.6217	0.4854	0.4054	0.3508	0.3103	0.2790	0.2537	0.2328	0.2152	0.2001

Note that although this distribution and that of Example 4.12 have the same means, the reliabilities are quite different. The survivor function is much flatter and here 69% of the items fail prior to the mean of 5 years.

4.5.2 Three-Parameter Weibull Distribution (Category A)

In this case $F(x; \theta)$ is given by (4.41) and Y is related to X by a simple linear relationship $Y = X + \gamma$. As a result, $H(x; \theta)$ is given by

$$H(y; \theta) = 1 - e^{-\{(y-\gamma)/\beta\}^\alpha} \qquad (4.58)$$

for $y \geq \gamma$ (and $H(y; \theta) = 0$ for $y < \gamma$). In this case the parameter set is $\theta = \{\alpha, \beta, \gamma\}$ with $\alpha > 0$, $\beta > 0$ and $\gamma > 0$. This distribution is identical to the two-parameter Weibull distribution function except that it is shifted by an amount γ. Hence γ is called a shift parameter. As in the two-parameter case, α is the shape parameter and β is the scale parameter.

Comments:
1. $\gamma > 0$ implies that the item will have a guaranteed minimum life of γ. Although this would not be true for most products, the distribution is useful, in general, in modeling lifetimes for items for which there is essentially no chance of failure during an initial period of operation. An example would be the shaft of a large turbine, which if properly installed, will not fail for an extended period of time.
2. When $\alpha = 1$, the distribution reduces to the two-parameter exponential distribution.

4.5.3 Truncated Gumbel Distribution of the Smallest Extreme (Category B)

In this case $F(x; \theta)$ is given by (4.53). Hoyland and Rausand (1994) suggest $H(y; \theta)$ as a left-truncated version of $F(y; \theta)$, i.e.,

$$H(y; \theta) = \frac{F(y; \theta) - F(0; \theta)}{1 - F(0; \theta)} \qquad (4.59)$$

for $y \geq 0$ and is zero for $y < 0$. As a result, $H(y; \theta)$ is given by

$$H(y; \theta) = 1 - e^{-\psi\{e^{(y/\beta)}-1\}} \qquad (4.60)$$

where $\psi = e^{-(\alpha/\beta)}$. The parameter set is $\theta = \{\alpha, \beta\}$, with $-\infty < \alpha < \infty$ and $\beta > 0$. The failure rate is given by

$$r(y; \theta) = (\psi/\beta)e^{(y/\beta)} \qquad (4.61)$$

4.5.4 Distribution Derived from the Gumbel Distribution of the Largest Extreme (Category A)

In this case $r(x; \theta)$ is given (4.54). Lloyd and Lipow (1962, p. 139) suggest the nonlinear relationship between Y and X of the form

$$e^{\gamma X} - 1 = e^{-(Y/\beta)} \qquad (4.62)$$

with $\gamma > 0$. They show that in this case $H(y; \theta)$ is given by

$$H(y; \theta) = 1 - e^{-\psi\{e^{\gamma y}-1\}} \qquad (4.63)$$

for $y \geq 0$ with $\psi = e^{\alpha/\beta}$. The parameter set is $\theta = \{\alpha, \beta, \gamma\}$ with $-\infty < \alpha < \infty$, $\beta > 0$ and $\gamma > 1$. The failure rate is given by

$$r(y; \theta) = \psi \gamma e^{\gamma y} \tag{4.64}$$

Note that this is similar to (4.61), but the expressions for ψ are different.

4.5.5   Exponentiated Weibull Distribution (Category B)

In this case $H(y; \theta)$ is related to $F(x; \theta)$, given by (4.44), as

$$H(y; \theta) = [F(y; \theta)]^\eta \tag{4.65}$$

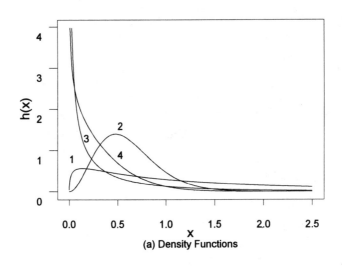

(a) Density Functions

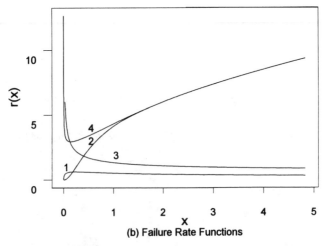

(b) Failure Rate Functions

Figure 4.6. Exponentiated Weibull distribution.

for $y \geq 0$ and $\eta > 0$. The parameter set is $\theta = \{\alpha, \beta, \eta\}$ with $\alpha > 0$, $\beta > 0$ and $\eta > 0$. The density function is given by

$$h(y; \theta) = \eta[F(y; \theta)]^{(\eta-1)}f(y; \theta) \tag{4.66}$$

For the Weibull distribution, as can be seen from (4.46), the failure rate function is continuously increasing for $\alpha > 1$, continuously decreasing for $\alpha < 1$. For $\alpha = 1$, it reduces to the exponential distribution and the failure rate is constant. In contrast, the exponentiated Weibull distribution allows for a more diverse range of shapes including the bathtub shape (discussed further in Section 4.9.2). Figure 4.6 shows plots of $h(y; \theta)$ and the failure rate $r(y; \theta)$ for a range of parameter values that demonstrate the various shapes possible. The parameter values used in the plots are $\beta = 0.50$ and $(\alpha, \eta) = (0.50, 3.00)$, $(1.50, 2.00)$, $(0.75, 0.50)$, and $(1.50, 0.50)$, with the curves numbered $1, \ldots, 4$, in this order. Note that density 1 gives rise to a failure rate function that is first increasing and then decreasing, density 2 has a strictly increasing failure rate, that of density 3 is strictly decreasing, and that of density 4 is a form of the bathtub curve. Mudholkar and Srivastava (1993) and Mudholkar et al. (1995) deal with this distribution and illustrate its use in modeling a diverse range of complex failure data.

4.6 CONTINUOUS DISTRIBUTIONS—III (INVOLVING SEVERAL BASIC DISTRIBUTIONS)

Often, the failure of a component is due to one of many different failure modes. As a result, the failure data of such components are complex and the empirical plots of the density and failure rates exhibit shapes that cannot be adequately modeled by the distributions discussed in Sections 4.5 and 4.6.

Distributions involving two or more basic distributions allow for a more diverse range of shapes for both density and failure-rate functions (for example, bimodal shapes for the density function and roller coaster shapes for the failure rate, as will be indicated later in the chapter). As a result, they are extremely useful for modeling complex data that cannot be adequately modeled by any of the standard distributions.

In this section, we discuss several additional classes of failure distributions derived from basic failure distributions. For notational ease, we omit the parameter θ so that $F(x; \theta)$ will be represented simply as $F(x)$, and so on. The new distributions, denoted $H(x)$, are obtained in terms of K (≥ 2) basic distributions $F_i(x)$, $1 \leq i \leq K$.

4.6.1 Mixtures of Distributions

A *finite mixture* of distributions is a weighted average of distribution functions. Specifically, $H(x)$ is a mixture of distribution functions $F_i(x)$ (called the *components* of the mixture), $1 \leq i \leq K$, if it is of the form

$$H(x) = \sum_{i=1}^{K} p_i F_i(x) \tag{4.67}$$

with $p_i > 0$, $i = 1, \ldots, K$, and $\Sigma_{i=1}^{K} p_i = 1$. Assuming differentiability of all of the components, the corresponding density function $h(x)$ is given by

$$h(x) = \sum_{i=1}^{K} p_i f_i(x) \tag{4.68}$$

where $f_i(x)$, $1 \leq i \leq K$, is the density function associated with $F_i(x)$.

An important application of mixture models is in situations involving multiple failure modes. In these cases, K is the number of modes, $F_i(x)$ is the distribution of time to failure for the ith mode, and p_i is the proportion of items that fail according to the ith mode. Table 2.1.3 of Titterington et al. (1985) provides references to various applications of mixtures of distributions. In the context of reliability modeling, these include failure times for valves, bus engine failure times, laser life times, and a number of others. This distribution is also useful in modeling component failure times with quality uncertainty. Because of this, an item produced can conform to design specification with probability p and not conform with probability $(1 - p)$. A conforming item has a failure distribution $F_1(x)$ and nonconforming one has a failure distribution $F_2(x)$. This issue is discussed further in Chapter 13.

It is easily seen that the the jth moment, $M_j(\theta)$, is the weighted sum of the jth moments for the K component distributions.

4.6.2 Competing Risk Models

A competing risk model is one in which $H(x)$ is of the form

$$H(x) = 1 - \prod_{i=1}^{K}(1 - F_i(x)) \tag{4.69}$$

The corresponding density function is given by

$$h(x) = \sum_{i=1}^{K}\left[\prod_{\substack{k=1 \\ k \neq i}}^{K}\{1 - F_k(x)\}\right]f_i(x) \tag{4.70}$$

and the failure rate is given by

$$r(x) = \sum_{i=1}^{K} r_i(x) \tag{4.71}$$

where $r_i(x)$, $1 \leq i \leq K$, is the failure rate associated with $F_i(x)$.

The reason that this is called a "competing risk" model is as follows. The system can fail due to one of K causes. If the failure is due to cause i, it occurs at time X_i, after being put into operation. X_i is a random variable from a distribution function $F_i(x)$. Since the failure causes are competing, the system failure occurs at time X given by

$$X = \min_{i}\{X_1, X_2, \ldots, X_K\} \tag{4.72}$$

From this, it follows that the distribution of X is given by (4.69). For further details on this distribution and its applications, see David and Moeschberger (1978).

4.6.3 Multiplicative Models

In contrast to the competing risk model, here we have

$$X = \max_{i}\{X_1, X_2, \ldots, X_K\} \tag{4.73}$$

As a result, $H(x)$ is of the form

$$H(x) = \prod_{i=1}^{K} F_i(x) \tag{4.74}$$

The density function is given by

$$h(x) = \sum_{i=1}^{K} \left\{ \prod_{\substack{k=1 \\ k \neq i}}^{K} F_k(x) \right\} f_i(x) \tag{4.75}$$

An application of this model is as follows. One way of improving the component reliability is to replace it with a module that consists of K ($K \geq 2$) identical replicas which are connected in parallel. As a result, the module fails only when the last working replica fails, so that the module failure time is given by (4.73). Note that in this case $F_i(x) = F(x)$ for all i. Redundancy and the use of this model are discussed further in Chapter 15.

4.6.4   Sectional Models

Let x_i, $1 \leq i \leq K$, be an increasing sequence with $x_0 = 0$ and $x_K < \infty$. A sectional model is one in which the distribution changes in each "section," so that $H(x)$ is given by

$$H(x) = \gamma_i F_i(x) \tag{4.76}$$

for $x_{i-1} \leq x < x_i$, $1 \leq i \leq K$. It is assumed that γ_i is positive and is selected so as to ensure that $H(x)$ and its derivatives (to some specified order) are continuous at x_k, $1 \leq i \leq K$. The density function is given by

$$h(x) = \gamma_i f_i(x) \tag{4.77}$$

for $x_{i-1} \leq x < x_i$, $1 \leq i \leq K$.

Comment

In contrast to the earlier models of the section, there is no meaningful physical justification for this model. The main reason for its use is that it offers greater flexibility in modeling complex data sets and as such has been used extensively in practice.

4.6.5   Special Case: $K = 2$ and Weibull Distributions

In this section we briefly discuss distributions derived from two Weibull distributions with

$$F_1(x) = 1 - e^{-(x/\beta_1)^{\alpha_1}} \quad \text{and} \quad F_2(x) = 1 - e^{-(x/\beta_2)^{\alpha_2}} \tag{4.78}$$

Mixture of Distributions

$$H(x) = p[1 - e^{-(x/\beta_1)^{\alpha_1}}] + (1 - p)[1 - e^{-(x/\beta_2)^{\alpha_2}}] \tag{4.79}$$

The distribution is characterized by a five-parameter set, $\theta = \{\alpha_1, \beta_1, \alpha_2, \beta_2, p\}$. If $\alpha_1 = \alpha_2$, then $\beta_1 \neq \beta_2$ must hold or the model reduces to a single Weibull distribution. The density function is given by

$$h(x) = p\frac{\alpha_1 x^{(\alpha_1-1)} e^{-(x/\beta_1)}}{(\beta_1)^{\alpha_1}} + (1-p)\frac{\alpha_2 x^{(\alpha_2-1)} e^{-(x/\beta_2)}}{(\beta_2)^{\alpha_2}} \tag{4.80}$$

The first two moments are given by

$$M_1 = p\beta_1 \Gamma\left(1 + \frac{1}{\alpha_1}\right) + (1-p)\beta_2 \Gamma\left(1 + \frac{1}{\alpha_2}\right) \tag{4.81}$$

and

$$M_2 = p\beta_1^2 \Gamma\left(1 + \frac{2}{\alpha_1}\right) + (1-p)\beta_2^2 \Gamma\left(1 + \frac{2}{\alpha_2}\right) \tag{4.82}$$

Example 4.17
Due to quality variations, the failure distribution of an item is given by $F_1(x)$, with probability p and $F_2(x)$, with probability $(1 - p)$. The parameters of $F_1(x)$ are $\alpha_1 = 1.25$ and $\beta_1 = 5.3659$, implying MTTF = 5 years. This corresponds to the item produced conforming to the design. The parameters of $F_2(x)$ are $\alpha_2 = 1.25$ (same as that for $F_1(x)$) and $\beta_2 = 1.0732$ implying MTTF = 1 year and corresponding to the item being nonconforming.

From (4.81), the MTTF for the item is $[5p + (1 - p)]$ years and is a function of p. As the manufacturing process deteriorates with age and usage, p decreases, implying that the quality (in terms of conformance) is going down. The effect of this on the component reliability is shown in Figure 4.7 for p varying from 0 to 1 in steps of 0.2. Not surprisingly, the effect is substantial; the reliability for a period of 3 years changes from 0.617 for $p = 1$ to 0.027 for $p = 0$.

Competing Risk Model:

$$H(x) = 1 - e^{-\{(x/\beta_1)^{\alpha_1} + (x/\beta_2)^{\alpha_2}\}} \tag{4.83}$$

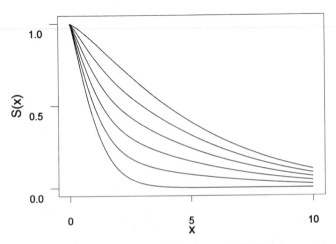

Figure 4.7. Survival function, mixed Weibull distribution. $p = 0.0(0.2)1.0$, bottom to top curves.

The distribution is characterized by four parameters. If $\alpha_1 = \alpha_2$, and $\beta_1 = \beta_2$, then the model reduces to a single Weibull distribution with $\beta = \beta_1 + \beta_2$. The density and failure rate functions are given by

$$h(x) = \left[\frac{\alpha_1 x^{(\alpha_1-1)}}{\beta_1^{\alpha_1}} + \frac{\alpha_2 x^{(\alpha_2-1)}}{\beta_2^{\alpha_2}} \right] e^{-\{(x/\beta_1)^{\alpha_1}+(x/\beta_2)^{\alpha_2}\}} \tag{4.84}$$

and

$$r(x) = \left[\frac{\alpha_1 x^{(\alpha_1-1)}}{\beta_1^{\alpha_1}} + \frac{\alpha_2 x^{(\alpha_2-1)}}{\beta_2^{\alpha_2}} \right] \tag{4.85}$$

Example 4.18
Brake failures in rolling stock can be due to normal wear, which can be modeled by $F_1(x)$. The second mode of failure is due to driver negligence, which can be modeled by $F_2(x)$. Let the parameters of $F_1(x)$ be $\alpha_1 = 1.25$ and $\beta_1 = 5.3659$, so that the mean time to failure for this component due to normal wear is 5 years. Let $\beta_2 = 1.0732$. The plots of $h(x)$ and $r(x)$ for $\alpha_2 = 0.50, 0.75, 1.0, 1.25,$ and 1.50 are given in Figure 4.8(a) and (b), respectively. As can be seen from Figure 4.8(a), when $\alpha_2 < 1$, the failure rate has a bathtub shape (i.e., initially decreasing and then increasing) and when $\alpha_2 > 1$, the failure rate is continuously increasing.

Multiplicative Model

$$H(x) = [1 - e^{-(x/\beta_1)^{\alpha_1}}] [1 - e^{-(x/\beta_2)^{\alpha_2}}] \tag{4.86}$$

and the distribution is characterized by four parameters. The density function is given by

$$h(x) = \frac{\alpha_1 x^{(\alpha_1-1)}}{\beta_1^{\alpha_1}} e^{-(x/\beta_1)^{\alpha_1}} [1 - e^{-(x/\beta_2)^{\alpha_2}}] + \frac{\alpha_2 x^{(\alpha_2-1)}}{\beta_2^{\alpha_2}} e^{-(x/\beta_2)^{\alpha_2}} [1 - e^{-(x/\beta_1)^{\alpha_1}}] \tag{4.87}$$

Sectional Model

$$H(x) = \begin{cases} [1 - e^{-(x/\beta_1)^{\alpha_1}}] & \text{for } 0 \le x < x_1 \\ \gamma[1 - e^{-(x/\beta_2)^{\alpha_2}}] & \text{for } x_1 \le x < \infty \end{cases} \tag{4.88}$$

The distribution has six parameters. The number of parameters reduces to five if $H(x)$ is required to be continuous at x_1 and to four if both $H(x)$ and $h(x)$ are required to be continuous at x_1.

   Jiang and Murthy (1995a,b, 1997a) and Murthy and Jiang (1997) illustrate the use of these models in modeling various failure data sets.[3]

4.7   CLASSIFICATION OF FAILURE DISTRIBUTIONS

Failure distributions can be classified into different categories based on different notions related to the aging properties of the distributions. In this section we discuss some of these no-

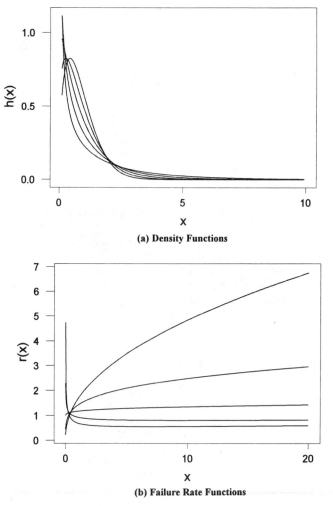

Figure 4.8. Density and failure-rate functions, competing risk model. $\alpha_1 = 1.25$, $\beta_1 = 5.3659$, $\beta_2 = 1.0732$, $\alpha_2 = 0.50$, 0.75, 1.00, 1.25, and 1.50 (bottom to top curves).

tions. A very useful result of these categorizations is that they enable one to derive bounds (upper and lower) for the reliability (or survivor function) $S(x)$. In addition, these results are useful in other aspects of reliability analysis, as will be seen in later chapters.

4.7.1   Definitions and Classification

We need the following concepts. Define $\gamma(x)$ and $L(x)$ as:

$$\gamma(x) = \frac{R(x)}{x} \tag{4.89}$$

and

$$L(x) = \frac{\int_x^\infty S(y)dy}{S(x)} \tag{4.90}$$

where $R(x)$ and $S(x)$ are given by (4.13) and (4.8) respectively (with the parameter θ omitted for notational ease). $\gamma(x)$ can be viewed as the average failure rate over $[0, x)$. $L(x)$ has an intuitive interpretation as the mean remaining lifetime given that the item has survived to age x. In other words,

$$L(x) = E[X - x | X > x] \tag{4.91}$$

It is easily seen (using integration by parts) that

$$M_1 = \mu = \int_0^\infty xf(x)dx = \int_0^\infty S(x)dx = L(0) \tag{4.92}$$

The classification is as follows. A failure distribution $F(x)$:

1. has a *decreasing failure rate* (DFR) if $r(x)$ is decreasing in x
2. has a *constant failure rate* (CFR) if $r(x)$ is constant in x
3. has an *increasing failure rate* (IFR) if $r(x)$ is increasing in x
4. has an *increasing failure rate on the average* (IFRA) if $\gamma(x)$ is increasing in x for $x \geq 0$
5. has a *decreasing failure rate on the average* (DFRA) if $\gamma(x)$ is decreasing in x
6. is *new better than used* (NBU) if

$$[1 - F(x + y)] \leq [1 - F(x)][1 - F(y)]$$

   for $x \geq 0$ and $y \geq 0$
7. is *new worse than used* (NWU) if

$$[1 - F(x + y)] \geq [1 - F(x)][1 - F(y)]$$

   for $x \geq 0$ and $y \geq 0$
8. is *new better than used in expectation* (NBUE) if

$$\int_0^\infty [1 - F(x + y)]dy \leq [1 - F(x)]\int_0^\infty [1 - F(y)]dy$$

   for $x \geq 0$ and $F(x) < 1$
9. is *new worse than used in expectation* (NWUE) if

$$\int_0^\infty [1 - F(x + y)]dy \geq [1 - F(x)]\int_0^\infty [1 - F(y)]dy$$

   for $x \geq 0$ and $F(x) < 1$
10. has an *increasing mean residual life* (IMRL) if $L(x)$ is increasing in x for $x \geq 0$
11. has a *decreasing mean residual* life (DMRL) if $L(x)$ is decreasing in x for $x \geq 0$

The chain of implications between these notions is shown in Figure 4.9. We omit the proofs, which can be found in the references cited in Shaked and Shantikumar (1990). As can be seen, a distribution function that is IFR is also IFRA, NBU, DMRL, and NBUE. On the other hand, a distribution that is NBU is also NBUE but need not be IFR or IFRA.

A characterization of some of the distributions discussed in Section 4.6 is as follows:

Weibull: IFR for $\alpha > 1$ and DFR for $\alpha < 1$

Gamma: IFR for $\alpha > 1$ and DFR for $\alpha < 1$

Lognormal: Neither IFR nor DFR

Inverse Gaussian: Neither IFR nor DFR.

4.7.2 Bounds on Reliability

The reliability function is given by the survivor function $S(x)$. Bounds (upper and lower) on reliability are often useful in reliability analysis, particularly when the actual failure distribution is unknown or leads to analytical difficulties. In this context, the sharpest possible upper and lower bounds are of great interest. In this section, we present some results for different classes of failure distributions in terms of the moments of the distribution. The results are given without proof; proofs may be found in Sengupta (1994).

Upper Bounds

The jth moment of the failure distribution, denoted M_j, is given by (4.17). The bounds given below hold for any $j \geq 1$. Upper bounds based on moments are:

(i) For IFR, IFRA, and NBU failure distributions

$$S(x) \leq \begin{cases} 1, & \text{if } x < (M_j)^{1/j} \\ \delta_x, & \text{if } x \geq (M_j)^{1/j} \end{cases}$$

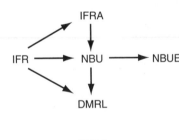

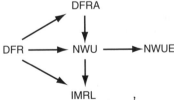

Figure 4.9. Relationship between different notions of aging.

where δ_x is given by

$$\int_0^1 jy^{(j-1)}\delta_x^y\,dy = \frac{M_j}{x^j}$$

(ii) For DFR distributions

$$S(x) \leq \begin{cases} e^{-jx/x_0}, & \text{if } x < x_0 \\ (x_0/x)^j e^{-j} & \text{if } x \geq x_0 \end{cases}$$

where x_0 is given by

$$x_0 = j\left[\frac{\mu_j}{\Gamma(j+1)}\right]^{(1/j)}$$

(iii) For DFRA distributions

$$S(x) \leq \delta_x$$

where δ_x is given by

$$\delta_x + \int_1^\infty jy^{(j-1)}\delta_x^y\,dy = \frac{M_j}{x^j}$$

(iv) For NWU distributions

$$S(x) \leq \delta_x$$

where δ_x is given by

$$\sum_{i=1}^\infty \delta_x^j[i^j - (i-1)^j] = \frac{M_j}{x^j}$$

Lower Bounds
(i) For IFR failure distributions

$$S(x) \geq \begin{cases} \inf_{0 \leq \beta \leq x} e^{-\alpha}, & \text{if } x < (M_j)^{1/j} \\ 0, & \text{if } x \geq (M_j)^{1/j} \end{cases}$$

where

$$\int_0^\infty \left\{\beta + \left[\frac{z(x-\beta)}{\alpha}\right]\right\}^j e^{-z}\,dz = M_j \quad \text{and} \quad \alpha = [\Gamma(j+1)/M_j]^{1/j}$$

(ii) For IFRA distributions

$$S(x) \geq \begin{cases} \delta_x, & \text{if } x < (M_j)^{1/j} \\ 0, & \text{if } x \geq (M_j)^{1/j} \end{cases}$$

where δ_x is given by

$$1 + \int_1^\infty jy^{(j-1)}\delta_x^y\,dy = \frac{M_j}{x^j}$$

(iii) For NBU distributions

$$S(x) \geq \begin{cases} \delta_x, & \text{if } x < (M_j)^{1/j} \\ 0, & \text{if } x \geq (M_j)^{1/j} \end{cases}$$

where δ_x is given by

$$\sum_{i=0}^\infty \delta_x^i[(i+1)^j - i^j] = \frac{M_j}{x^j}$$

(iv) For DFR, DFRA, and NWU distributions

$$S(x) \geq 0$$

For other inequalities, see Shaked and Shantikumar (1990) and the references cited therein.[4]

4.8   SHAPES OF DENSITY FUNCTIONS

We confine our attention to failure distributions discussed in Sections 4.4–4.6. The four different typical shapes of the density functions for these distributions are shown in Figure 4.10 and can be classified as follows:

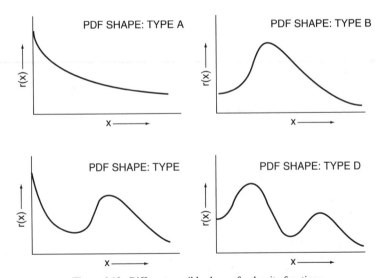

Figure 4.10. Different possible shapes for density functions.

Type A: Monotonically decreasing

Type B: Unimodal

Type C: Initially decreasing and then unimodal

Type D: Bimodal

Table 4.1 is a summary of the possible different shapes for some of the distributions discussed in the earlier sections of this chapter. This information is useful in deciding whether a particular failure distribution is appropriate for modeling a given failure data set based on a histogram or other plot of the data.

For a given distribution, the shape of the density function depends on its parameter values. A parametric characterization (identification of the different regions in the parameter space corresponding to the different shapes possible) is not too difficult for some of the basic distributions but, in general, is a difficult task for other distributions.

Weibull-Based Models

For the two- (or three-) parameter Weibull, the scale parameter β has no influence and the shape is type A for $\alpha \leq 1$ and type B for $\alpha > 1$. The parametric characterization is complex, as it involves identifying the different possible shapes and then finding the boundaries in the parameter spaces that separate one shape from another. The different possible shapes are listed in Table 4.1. The parametric study for the mixture of two Weibull distributions can be found in Murthy and Jiang (1997) and Jiang and Murthy (1997b); for the competing risk Weibull ($K = 2$) in Jiang and Murthy (1997c); for the multiplicative Weibull ($K = 2$) in Jiang and Murthy (1996d); for the sectional Weibull ($K = 2$) in Murthy and Jiang (1997); and for the exponentiated Weibull in Jiang and Murthy (1997e).

4.9 SHAPES OF FAILURE-RATE FUNCTIONS

4.9.1 Failure-Rate Classification

We again confine our attention to failure distributions discussed in Sections 4.4–4.6. There are ten different typical shapes of the failure-rate function for these distributions. These are:

Type A: Nonincreasing (includes montonically decreasing)

TABLE 4.1. Possible shapes of density functions

Distribution	Type			
	A	B	C	D
Exponential	Y	N	N	N
Weibull	Y	Y	N	N
Lognormal	Y	Y	N	N
Mixture (Weibull $K = 2$)	Y	Y	Y	Y
Competing risk (Weibull $K = 2$)	Y	Y	Y	Y
Multiplicative (Weibull $K = 2$)	Y	N	Y	Y
Sectional (Weibull $K = 2$)	Y	Y	Y	Y
Exponentiated Weibull	Y	Y	N	N

Type B: Nondecreasing (includes monotonically increasing)

Type C: Constant

Type D: Bathtub

Type E: Reverse bathtub (unimodal)

Type F: Decreasing followed by unimodal

Type G: Unimodal followed by increasing

Type H: Unimodal with $r(x) > 0$ as $x \to \infty$

Type J: Bimodal with $r(x) > 0$ as $x \to \infty$

Type K: Bimodal with $r(x) \to 0$ as $x \to \infty$

Type L: Bimodal with $r(x) \to \infty$ as $x \to \infty$

These forms for $r(x)$ are illustrated in Figure 4.11.

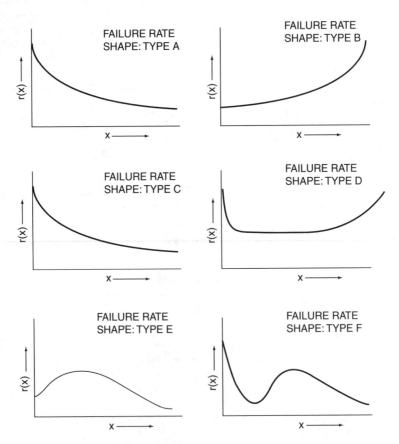

Figure 4.11. Possible shapes for failure-rate functions.

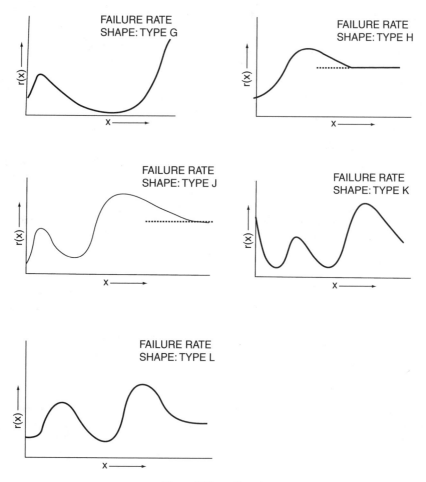

Figure 4.11. *continued.*

The results of Glaser (1980) allow one to determine if a failure distribution is type A, type B, type D, or type E. The analysis involves $\eta(x)$ and $\eta'(x)$ $(= d\eta(x)/dx)$ where

$$\eta(x) = -\frac{f'(x)}{f(x)}$$

The result is given in the following theorem:

Theorem (Glaser, 1980)

1. If $\eta'(x) > 0$ for all $x > 0$, then the distribution has a failure rate of type B.
2. If $\eta'(x) < 0$ for all $x > 0$, then the distribution has a failure rate of type A.
3. Suppose there exists an x_0 such that $\eta'(x) < 0$ for all $x \in (0, x_0)$, $\eta'(x_0) = 0$, and $\eta'(x) > 0$ for all $x > x_0$. Then

   a) If $\lim_{x \to 0^-} f(x) = 0$, then the distribution has a failure rate of type B.

   b) If $\lim_{x \to 0^-} f(x) \to \infty$, then the distribution has a failure rate of type D.

4. Suppose there exists an $x_0 > 0$ such that $\eta'(x) > 0$ for all $x \in (0, x_0)$, $\eta'(x_0) = 0$, and $\eta'(x) < 0$ for all $x > x_0$. Then

   a) If $\lim_{x \to 0^-} f(x) = 0$, then the distribution has a failure rate of type E.

   b) If $\lim_{x \to 0^-} f(x) \to \infty$, then the distribution has a failure rate of type D.

Table 4.2 is a summary of the possible different shapes for the failure rate functions for the distributions discussed in Sections 4.4–4.6. As will be seen in later chapters, this information is useful in deciding whether a particular failure distribution is appropriate for modeling a given failure data set based on the empirical failure rate plot of the data set.

For a given distribution, the shape of the failure-rate function often depends on the parameter values. The identification of the different possible shapes and their characterization in the parametric space (defining the boundaries that separate regions with different shapes) is, in general, a difficult task for all but few simple distributions.

Weibull-Based Models

For the two (or three) parameter Weibull, the scale parameter β has no influence and the shape is type A for $\alpha < 1$, type B for $\alpha > 1$, and type C for $\alpha = 1$. For models involving two Weibull distributions, the different possible shapes are as indicated in Table 4.2. Parametric characterizations for these models are fairly involved and can be found in Jiang and Murthy (1997b) for the mixture model; in Jiang and Murthy (1997c) for the competing risk model; in Jiang and Murthy (1997d) for the multiplicative model; in Jiang and Murthy (1997a) and Murthy and Jiang (1997) for the sectional model, and in Mudholkar et al. (1995) for the exponentiated Weibull model.

Finally, the type F shape for the failure rate is also called "roller coaster" failure rate by Wong (1989), who deals with data that fit a distribution having this shape for the failure-rate function.

4.9.2 Bathtub Failure Rate

The bathtub failure rate is of special significance in modeling item failure. The shape of the failure rate is characterized by three regions. In the first region ($0 < x < x_1$) the failure rate is

TABLE 4.2. Possible shapes of failure-rate functions

Distribution	Type										
	A	B	C	D	E	F	G	H	J	K	L
Exponential	N	N	Y	N	N	N	N	N	N	N	N
Weibull	Y	Y	Y	N	N	N	N	N	N	N	N
Truncated Gumbel	N	Y	N	N	N	N	N	N	N	N	N
Mixture (Weibull $K = 2$)	Y	Y	Y	Y	N	Y	Y	Y	Y	Y	Y
Competing risk (Weibull $K = 2$)	Y	Y	N	N	Y	N	N	N	N	N	N
Multiplicative (Weibull $K = 2$)	Y	Y	N	N	Y	N	Y	N	N	N	N
Sectional (Weibull $K = 2$)	Y	Y	N	Y	Y	N	N	N	N	N	N
Exponentiated Weibull	Y	Y	Y	Y	Y	N	N	N	N	N	N

decreasing, in the second region ($x_1 \leq x \leq x_2$) it is roughly constant, and in the third region ($x > x_2$) it is increasing. The first region corresponds to infant mortality, where failures occur due to poor manufacturing. The failures in the second region are purely chance and age has no effect (as the failure rate is essentially constant). Finally, failures in the third region are due to the effect of aging.

A variety of failure distribution models with a bathtub failure rate have been studied. The starting point for constructing these failure distributions is the mathematical form of the failure rate. Several forms have been proposed and we indicate three of them.

Quadratic:	$r(x) = \alpha + \beta x + \gamma x^2$
Exponential:	$r(x) = e^{(\alpha + \beta x + \gamma x^2)}$
Exponential power:	$r(x) = \beta \alpha x^{(\alpha - 1)} e^{x^\alpha}$

For a more comprehensive list, see Rajarshi and Rajarshi (1988). Xie and Lai (1996) propose an additive Weibull distribution model that has a bathtub-shaped failure rate. There is very little theoretical justification for the forms that have been studied. This is basically an empirical approach in which the most appropriate form would be determined by plotting the failure rate.

4.10   EFFECT OF ENVIRONMENTAL FACTORS

In practice, the time to failure (X) depends on environmental factors. The effect of the environment can be captured through explanatory variables (given by a vector $\mathbf{z}$) which affect the failure distribution. We discuss two such formulations.

(i) Accelerated Testing. In this type of testing, the item is subjected to a harsher environment than normal, which hastens the time to failure. Usually, several elevated stress levels are used in the experiment. It is assumed that the changing failure rates can be captured through a time scaling factor θ_i corresponding to stress level i. θ_i increases with i. If the time to failure under the normal (or base) condition is X, then under stress level i it is assumed to be given by X/θ_i, with $\theta_i > 1$, implying that the time to failure decreases as θ_i increases.

As a result, the survivor function under stress level i, $S_i(x)$, is related to that under normal conditions, $S_0(x)$, by the relationship

$$S_i(x) = S_0(\theta_i x) \tag{4.93}$$

This implies that

$$r_i(x) = \theta_i r_0(\theta_i x) \tag{4.94}$$

One form that links θ_i to the explanatory variables $Z_j, j = 1, 2, \ldots, N$, is

$$\theta_i = e^{\phi_i} \tag{4.95}$$

where ϕ_i is a linear function of the explanatory variables given by

$$\phi_i = a_{1i} Z_1 + a_{2i} Z_2 + \cdots + a_{Ni} Z_N \tag{4.96}$$

where $a_{ji}, j = 1, 2, \ldots, N$ are constants.

(ii) Multiplicative Failure Rate. Here, the effect of the explanatory variables on the failure rate is through a multiplicative factor affecting the base failure rate, so that

$$r_i(x) = g(\phi_i)r_0(x) \tag{4.97}$$

with ϕ_i given by (4.96). Note that the Z_j can be random variables, in which case, conditional on the Z_j, the failure rate is given by (4.97). As a result, the conditional survivor function is given by

$$S_i(x) = \{S_0(x)\}^{g(\phi_i)} \tag{4.98}$$

We shall discuss this further in Chapter 6. Here our emphasis is on the form of distributions used to model failure times, taking into account the effect of environmental factors. For further details, see Oakes (1983).

4.11   CONCLUSIONS

In this chapter, we have presented several distribution functions useful for modeling first failure at the part level. We again note that the results may be applicable at a higher level as well and are often used in this context. Models of this form are simplifications in that they do not take into account the relationship between reliabilities at the different levels (part, component, subsystem, etc.) in calculating reliability at a given (higher) level. In this approach, the unit being analyzed is considered as an entity in itself. Again, this is the black-box approach, and it has been found very useful in modeling, analysis, and prediction in a very large number of applications.

In applying this approach, the selection of an appropriate distribution and values for its parameters is an important initial step. This should be based on a careful preliminary analysis of the failure data (and any other data available), as discussed in Chapter 3, and the more advanced data analysis techniques given in Chapters 5 and 8. When no such analysis is possible, the choice must be based on past experience in relation to the item under consideration— e.g., the exponential distribution has been used extensively in modeling the failures of electrical components and parts and the Weibull distribution in modeling failures of mechanical components.

One can use either graphical or statistical methods to determine if a given failure data set can be adequately modeled by one of the failure distributions discussed in this chapter. This topic is the focus of the next chapter and of Chapter 11. The "white-box" approach to modeling failures at part level is done in Chapter 6.

NOTES

1. There are many other basic discrete distributions. For a comprehensive list, see Johnson and Kotz (1969). The authors also provide expressions for higher moments, moment generating functions, and other characteristics for the distributions discussed in this section as well as for many other distributions.

2. Many other continuous distributions (for example, Pareto, Fisk, Lomax, generalized gamma, Burr, to name a few) can be used for modeling failure times. Details of these can be found in Johnson and Kotz (1970a,b) and Patel et al. (1976). McDonald and Richards

(1987a,b) and Richards and McDonald (1987) propose a family tree wherein several of these distributions (along with the ones discussed earlier) are special cases of the generalized beta type I and type II distributions. They give expressions for the density, failure rate, and distribution functions and the moments.

3. For additional references on models involving two Weibull distributions along with the application of these distributions in modeling failure times, see Jiang (1996).

4. Another property for continuous distributions that has not been discussed is the *ordering* property between two distributions. One such property is stochastic ordering. Let X and Y be two random variables from distribution functions $F(\ )$ and $G(\ )$, respectively. X is said to be stochastically larger than Y, written $x \geq_{st} Y$, if

$$P\{X > a\} \geq P\{Y > a\} \qquad \text{for all } a.$$

This is equivalent to $\overline{F}(a) \geq \overline{G}(a)$ for all a. For more on this and other results with regard to ordering (e.g., hazard rate ordering, likelihood ratio ordering) and their applications in reliability analysis, see Ross (1983).

An inequality relationship for probability, involving the first two moments, is the well-known Chebyshev inequality, given by

$$P\{|X - \mu| \geq k\sigma\} \leq \frac{1}{k^2}$$

For other inequalities, see Chapter 2 of Patel, Kapadia, and Owen (1976).

EXERCISES

1. Let Z denote the number of defective items in a lot of size $n = 20$. Let p denote the probability that an item is defective. Suppose that $p = 0.05$.

   (i) Calculate the probability of finding 0, 1, and 2 defective items in a lot.
   (ii) What is the probability that there are more than 2 defective items in a lot?
   (iii) Calculate these probabilities for $p = 0.02$.

2. A computer manufacturer intends to purchase 100 chips from a supplier. He expects that some of the chips can be defective but is willing to tolerate only 4 defectives in the batch. He checks the quality by drawing a random sample of 3 chips without replacement from the lot. If one or more of the chips sampled is defective, the lot is rejected. If a lot contains 4 defectives, what is the probability that the lot is rejected? What is this probability if the lot contains 5 defectives? 8 defectives?

3. The number of incidents requiring an action by the Highway Patrol for a section of a highway over a 24 hour period can be modeled by a Poisson process with $\lambda = 10$.

   (i) Find the probability that there are 0, 1, 2, and 3 incidents over the 24 hour period. Find the probability that there are more than 15.
   (ii) Each incident costs the Highway Patrol $100. One way of recovering this cost is through a toll per car. Suppose that 1000 cars use the highway each day. What should the toll be so that, on the average, the costs are recovered?

4. The failure distribution of bearings for a small electric motor is given by a Weibull distribution with shape parameter $\alpha = 1.25$ and scale parameter $\beta = 8.0$ years. What is the probability that a bearing will fail within 5 years? Suppose that the bearing has not failed at the end of its first year. What is the probability that it will not fail in the next 4 years?

5. Repeat Problem 4 with $\alpha = 0.90$ and β chosen so that the mean time to failure is the same as in Problem 4. Compare the results with those of Problem 4.

6. A component has two failure modes. One failure mode, due to external conditions, has a constant failure rate of 0.08 failures per year. The second, due to wearout, can be modeled by a Weibull distribution with shape parameter 1.25 and a mean life of 10 years. As a result, the time to failure can be modeled by a competing risk model. Calculate the reliability of the system for a operational life of 3 years. How does this change if the operational life changes to 5 years?

7. A cutting tool wears out with time and is deemed to have failed when the wear exceeds some specified value. In the normal mode of operation, the time to failure can be modeled by a normal distribution with a mean of 20 days and a standard deviation of 3 days.

   (i) Find the reliability of the tool life if the tool is replaced every three days.
   (ii) Determine the probability that the tool will last for 3 days, given that it has not failed after being in use for one day.
   (iii) Determine the time for replacement that ensures that the tool will not fail with probability 0.98.

8. The failure distribution for a VCR can be modeled by a lognormal distribution with $\mu = 1.5$ years and $\sigma = 1$ year.

   (i) Find the mean time to failure.
   (ii) If the VCR is sold with a warranty for 2 years, what is the probability that there is no claim within the warranty period?
   (iii) How does the probability in (ii) change if the warranty is increased to 3 years?

9. A bus travels between two cities, A and B, which are 100 miles apart. If the bus has a breakdown, then the distance X of the point of breakdown from city A has a uniform distribution over the interval $[0, 100]$.

   (i) A tow truck is sent from the city nearest to the point of breakdown. What is the probability that the tow truck has to travel more than 20 miles?
   (ii) Suppose that a tow truck is positioned at the midpoint in addition to one each at the two cities. What is the probability that the tow truck has to travel more than 20 miles? How would you decide if this is a better strategy than that of (i)?

10. Due to quality variations in manufacturing, an item produced can have a failure characteristic that is different from the design specification. The failure distribution of such an item is given by an exponential distribution with mean $\mu = 3$ years. When the item meets the design specification, its failure time is also given by an exponential distribution, but in this case with $\mu = 10$ years. The probability that an item fails to meet the design specification is p.

   (i) Calculate the mean and variance of the time to failure for $p = 0.05$ and 0.10
   (ii) Suppose that the items are tested (in an accelerated test mode) for a period of 0.5 year. What is the probability that a item that does not conform to the design specification will fail during the testing period? What is this probability for an item that meets the design specification?

11. The mode of a probability density function is a point where the density function has a local maximum. Show that for a Weibull density function with scale parameter $\alpha > 1$, there is a unique mode. Characterize the mode as a function of the scale and shape parameters.

12. If $f(x)$ is a failure density function, show that $g_j(x)$, $j = 1, 2, 3, \ldots$, given by

$$g_j(x) = \frac{x^j f(x)}{M_j}$$

is also a failure distribution function, where M_j is the jth moment of X with density function $f(x)$.

13. Suppose that $f(x)$ in Exercise 12 is an exponential distribution with parameter λ. Plot $g_j(x)$ for $j = 1$ and 2. Obtain the mean, variance, and mode for the density functions.

14. Prove that the failure rate is constant if and only if the failure distribution is exponential.

15. Let $F(x)$ be a probability distribution function. Show that $G(x)$ given by

$$G(x) = \begin{cases} 0 & \text{for } x < 0 \\ e^{-\alpha\{1-F(x)\}} & \text{for } x \geq 0 \end{cases}$$

is also a probability distribution function if $\alpha > 0$.

16. Suppose that X is distributed according to a Weibull distribution with shape parameter α. Show that the random variable

$$Y = X^\alpha$$

is distributed according to an exponential distribution.

17. An illuminated advertising board has 500 identical bulbs. The failure times for the bulbs is given by an exponential distribution with mean time to failure = 5 years.
   (i) What is the probability that 3 bulbs will fail within the first year?
   (ii) Suppose that no bulb has failed in the first 6 months. What is the probability that no failure will occur in the next 6 months?

18. The mean time to failure (MTTF) is a function of the parameters of the failure distribution and can be displayed as a curve in the parameter plane if the number of parameters is 2. Plot contours of constant MTTF (1, 2, 3, 4, and 5 years) in the parameter plane for the Weibull distribution.

19. Repeat Problem 18 for the lognormal distribution.

20. In Section 4.7.2, expressions for various upper and lower bounds are given. Select the bound given under (i) for both cases and compute the resulting bounds for the Weibull distribution with $\beta = 5$ and $\alpha = 1.25$ involving (a) the first and (b) the second moments. Compare the error between the actual reliability and the bounds.

CHAPTER 5

Basic Statistical Methods for Data Analysis

5.1 INTRODUCTION

Chapter 2 dealt with applications of reliability and included many examples of data sets of various types that might be obtained in reliability studies. In Chapter 3, we looked more closely at the structure of these many types of data sets and discussed methods of description and summarization of data. Chapter 4 dealt with probability models for representing failure data as well as other types of data that may be relevant in reliability studies.

The topics covered in this chapter are all in the general category of *statistical inference,* which in a very real sense relates all of the concepts of the previous three chapters. In probability, we model uncertainty (randomness), for example, through the distribution function, and can use this model to make statements about the nature of the data that may result if the model is correct. The principal objective of statistical inference is to use data to make statements about the probability model, either in terms of the probability distribution itself or in terms of its parameters or some other characteristics. Thus probability and statistical inference may be thought of as inverses of one another:

$$\text{Probability:} \quad \text{Model} \rightarrow \text{Data}$$

$$\text{Statistics:} \quad \text{Data} \rightarrow \text{Model}$$

The statistical inference procedures we will look at here will be appropriate for part or item data, i.e., data at a single level. Multi-level data will be considered in later chapters. We will assume that the data have been obtained as a result of random sampling and that the observations in the sample are independent. We also assume complete data, i.e., no censoring.

Procedures of statistical inference are the basic tools of data analysis. Most are based on quite specific assumptions regarding the nature of the probabilistic mechanism that gave rise to the data. Thus the distributions of Chapter 4 play a key role in determining an appropriate method of analysis. Furthermore, most employ one or more of the descriptive statistics given in Chapter 3. In this chapter, we will look at the use of the descriptive measures of Chapter 3 in the context of the probability models of Chapter 4 and apply the methods to some of the data sets of Chapter 2.

The basic questions to be addressed are the following:

1. Suppose that the form of the distribution function $F(x)$ (= $F(x; \Theta)$, where Θ is the set of parameters) is specified (e.g., exponential, Weibull, Erlangian) based on an understanding of the failure mechanism. What are the values of the parameter set Θ? More specifically, how do we use the data to "determine" them?

2. Suppose that both F and Θ are specified (hypothesized), based, for example, on some assumption concerning the failure mechanism. How do we use data to determine if the hypothesized value can reasonably be assumed to be the true value?

Question 1 deals with problems in *estimation;* Question 2 with *hypothesis testing.* Entire books have been written on each of these topics. We give highlights of some basic procedures in each area. In most cases, mathematical derivations and details will be omitted, with references given to sources for more in-depth coverage.

A third question, which actually would have to be dealt with prior to either of the above is:

3. Suppose that neither F nor Θ is known. How do we determine these, i.e., how do we check to determine if a specified F can reasonably be assumed to be the true F?

Question 3 deals with *goodness-of-fit tests.* This topic will be covered in Chapter 11.

The outline of this chapter is as follows: In Section 5.2 we cover basic procedures of statistical estimation, including moment and maximum likelihood estimators, and provide a list of parameter estimation procedures for most of the important life distributions given in Chapter 4. Confidence intervals for these parameters will be given in Section 5.3. In Section 5.4, the results are extended to estimation of functions of parameters. Section 5.5 deals with tolerance intervals, that is, confidence interval statements for regions that cover a specified proportion of a population. In Section 5.6, we give a brief introduction to hypothesis testing.

5.2   ESTIMATION

5.2.1   Overview

We assume complete data resulting from a random sample of size n. As before, the sample is denoted $X_1, X_2, \ldots, X_n$ when considered as a set of random variables and $x_1, x_2, \ldots, x_n$ when considered as a set of data (i.e., numerical values of the X_i's). The sample is assumed to have been drawn from a population represented by distribution function $F(x) = F(x; \Theta)$, where $\Theta = (\theta_1, \ldots, \theta_k)$ is a k-dimensional parameter. Observations having this structure are said to be independent and identically distributed (IID).[1]

The form of $F(.)$ is assumed to be known; the objective of statistical estimation is to assign numerical values to the parameter based on the sample data. In classical estimation, there are two approaches to this problem: *point estimation* and *interval* (or *confidence interval*) *estimation.* In point estimation, a numerical value for each θ_i is calculated. In interval estimation, a k-dimensional region is determined in such a way that the probability that the region contains the true parameter Θ is a specified, predetermined value. (If $k = 1$, this region is almost always an interval, hence the name for this type of inference.) This section of the chapter will be devoted to point estimation. Confidence intervals are the topic of Section 5.3.

The symbol "^" placed over the symbol for a parameter is used to denote an estimate of that parameter, i.e., $\hat{\Theta}$ estimates Θ. $\hat{\Theta}$ is a function of the observations and may also be a function of other quantities, for example, parameters whose values are known. A distinction is made between $\hat{\Theta}$ as a function of $X_1, X_2, \ldots, X_n$ and $\hat{\Theta}$ as a function of $x_1, x_2, \ldots, x_n$. $\hat{\Theta} = \hat{\Theta}(X_1, X_2, \ldots, X_n)$ is called an *estimator* and is a random variable; $\hat{\Theta} = \hat{\Theta}(x_1, x_2, \ldots, x_n)$ is

called an *estimate* and is the numerical value calculated from the data. Another way of looking at this is that an estimator is a procedure or formula for calculating an estimate, which is
a numerical value resulting from application of the formula to a set of data. When it is clear
from the context what is meant, the dependency of $\hat{\Theta}$ on the X's or x's will be suppressed. $\hat{\Theta}$
may also be written as $\hat{\Theta}_n$ to indicate its dependence on the sample size n.

We also use $\hat{\theta}_i$ to denote an estimator or estimate of θ_i (omitting the subscript if $k = 1$), and
symbols, $\hat{\mu}$, $\hat{\lambda}$, etc., in estimation of the corresponding parameters.

In many cases estimators of parameters or other populations characteristics are based on
logically connected descriptive statistics. For example, to estimate a mean μ, it would appear
to be reasonable to use the sample mean $\overline{X}$, that is, the estimator

$$\hat{\mu} = \overline{X} = \frac{1}{n} \sum_{i=1}^{n} X_i \qquad (5.1)$$

On the other hand, if the distribution is symmetrical, then the mean and median are equal and
the sample median is an equally logical choice as an estimator of μ . Other possibilities
might be a weighted average of the mean and the median, the average of the max and the
min, the average of the first 20% of the (unordered) data, and so forth. In short, there are an
infinite number of possible estimators. To choose an estimator, a number of properties that
assess the performance of the procedure in the context of the assumed distribution function
are considered. We look at a few of these in the next section.

5.2.2 Properties of Estimators

Many methods of evaluating estimators have been developed. Here we look at a few of the
more commonly used approaches. These will include small sample properties as well as asymptotic properties (as the sample size $n \rightarrow \infty$). To evaluate estimators, we look at their
properties, i.e., at how well they perform in estimating parameters in the long run. The objective is to find a "best" estimator, that is, one that outperforms all others with regard to all of
the properties of estimators that are considered important.

This leads to a number of difficulties, among which are: first, the notion of what is "best"
is a subjective one, although many analysts would agree that some basic properties should be
included in the set. Second, for some distributions, there may exist no estimators having a
given set of properties, or there may exist an infinite number of such estimators. A third important problem is that of finding *any* estimator having a given desired property. The approach to be taken here is as follows: We define several properties of estimators, then look at
two estimation procedures: the method of moments and the method of maximum likelihood.
Both of these are constructive in nature, that is, they lead to specific formulas or equations
for solution that lead to numerical estimates of the parameters. We then list the general properties of these estimators and the assumptions required for these properties to hold. Finally,
moment and maximum likelihood (ML) estimators will be given for the parameters of many
of the distributions discussed in Chapter 4, with the results applied to some of the data sets of
Chapter 2.

The following are some of the properties that are considered in evaluating and comparing
estimators:[2]

Unbiasedness. Definition: An estimator $\hat{\theta}_i$ of θ_i is said to be *unbiased* if $E(\hat{\theta}_i) = \theta_i$ for all
possible values of θ_i. $\hat{\Theta}$ is an unbiased estimator of Θ if $\hat{\theta}_i$ is unbiased for $i = 1, \ldots, k$.

The interpretation of this is that the average value of the estimates obtained by use of the
estimator $\hat{\theta}_i$, averaging over all possible samples, is the true value, θ_i.

It is easily shown that the sample mean $\overline{X}$ is an unbiased estimator of the population mean μ for any distribution for which μ is finite. The sample variance s^2, given by Equation (3.5), as an estimator of the population variance σ^2 is also an unbiased estimator (for any distribution for which $E(X^2)$ is finite), which is one of the reasons for dividing by $(n-1)$ rather than by n. Finally, the sample proportion in binomial sampling is an unbiased estimator of the population proportion. Most other parameters or other population characteristics must be dealt with on a case by case basis.

An estimator for which $E(\hat{\theta}_i) \neq \theta_i$ is said to be *biased*. The *bias* $b(\hat{\theta}_i)$ of an estimator $\hat{\theta}_i$ is given by

$$b(\hat{\theta}_i) = E(\hat{\theta}_i) - \theta_i \tag{5.2}$$

Asymptotic Unbiasedness. Definition: An estimator $\hat{\theta}_i$ of θ_i is said to be *asymptotically unbiased* if $E(\hat{\theta}_i) \rightarrow \theta_i$ as $n \rightarrow \infty$ for all possible values of θ_i. $\hat{\Theta}$ is an asymptotically unbiased estimator of Θ if $\hat{\theta}_i$ is asymptotically unbiased for $i = 1, \ldots, k$.

Replacing $(n-1)$ by n in the formula for the sample variance yields an estimator that is biased but is asymptotically unbiased for any distribution for which $E(X^2) < \infty$. The sample standard deviation s is a biased estimator of σ for any finite n, but is asymptotically unbiased.

Consistency. Definition: An estimator $\hat{\theta}_i$ of θ_i is said to be consistent if for any $\varepsilon > 0$ and all possible values of θ_i, $P(|\hat{\theta}_i - \theta_i| > \varepsilon) \rightarrow 0$ as $n \rightarrow \infty$. $\hat{\Theta}$ is a consistent estimator of Θ if $\hat{\theta}_i$ is consistent for $i = 1, \ldots, k$.

The interpretation is that for a consistent estimator the probability that the estimate will deviate from the true value by any amount, no matter how small, approaches zero as the sample size becomes increasingly large; i.e., for large samples the estimate will be very close to the true parameter value with high probability. Consistency in this sense is also called weak consistency. Other forms of consistency will be discussed later.

Sufficiency. Definition: An estimator $\hat{\Theta}$ of Θ is *sufficient* if the conditional distribution of $X_1, X_2, \ldots, X_n$ given $\hat{\Theta}$ does not depend on Θ. This implies that the estimator contains all of the sample information about the parameter. It is sufficient to know its value; given that, no additional information about Θ is contained in the data. All of the estimators presented in the following sections are sufficient.

Efficiency. The efficiency of an estimator is measured in terms of its variability, which in turn may be assessed in a number of ways. The rationale is that use of an inefficient estimator requires more data to do as well and hence it costs less to use an efficient estimator. Efficiency of an estimator may be assessed relative to another estimator or estimators (*relative efficiency*) or relative to an absolute standard. We look at two methods of assessing efficiency, restricting attention to $k = 1$. The concept is more complex in the multidimensional parameter case; see the references on estimation cited previously. The results given below involve the variance of the estimator, denoted $V(.)$.

Minimum Variance Unbiased Estimators. Definition: An unbiased estimator $\hat{\theta}$ of a parameter θ is minimum variance unbiased if $V(\hat{\theta}) \leq V(\theta^*)$ for any other unbiased estimator θ^* and for all possible values of θ.

Estimators that are efficient in this sense are also called UMVUE (for Uniformly Minimum Variance Unbiased Estimator), and, if $\hat{\theta}$ is a linear function of the X_i's, BLUE (for Best Linear Unbiased Estimator.) All of the above definitions extend in an obvious way to estimating a function of θ, say $\tau(\theta)$. A result that is useful in determining whether or not an estimator is UMVUE is the following:

Cramér-Rao Inequality. Suppose that $T = T(X_1, X_2, \ldots, X_n)$ is an unbiased estimator of $\tau(\theta)$. Then

$$V(T) \geq \frac{[\tau'(\theta)]^2}{E\left(\dfrac{\partial \log[L(X_1, \ldots, X_n; \theta)]}{\partial \theta}\right)^2},$$

(5.3)

where

$$L(X_1, \ldots, X_n; \theta) = \prod_{i=1}^{n} f(X_i; \theta)$$

(5.4)

with $f(.; \theta)$ the density function of the X's.

Notes: (1) This result requires that certain regularity conditions be met. These involve the existence of second derivatives and interchangeability of integration and differentiation. See Stuart and Ord (1991). These should be checked in any application, especially for distributions involving a shift parameter, as the conditions are often not met in such cases. Examples are the three-parameter Weibull distribution of Equation (4.48) and the two-parameter exponential distribution. These are problems in nonregular estimation and require special treatment.

(2) If $\tau(\theta) = \theta$, the inequality provides a lower bound on the variance of unbiased estimators. An estimator θ^* is efficient if its variance achieves this bound.

(3) Other variance bounds, applicable in nonregular cases and extensions to multiparameter distributions, have been developed.

(4) In many estimation problems, the bounds may not be achievable for small samples but are achievable asymptotically. In this case, the estimators are called *asymptotically efficient.*

Minimum MSE Estimation. Definition: The *mean square error* (MSE) of an estimator $t(X_1, X_2, \ldots, X_n)$ of $\tau(\theta)$ is given by

$$\text{MSE}(t) = V(t) + [b(t)]^2$$

(5.5)

where $b(t)$ is the bias of the estimator, given by Equation (5.2) An estimator is *MSE efficient* if no other estimator has smaller MSE. Note that if an estimator is unbiased, $\text{MSE}(t) = V(t)$.

MSE Consistency. Definition: An estimator $\hat{\theta}_i$ of θ_i is said to be *mean square error consistent* if $\text{MSE}(\hat{\theta}_i) \to 0$ as $n \to \infty$ for all possible values of θ_i. $\hat{\Theta}$ is an MSE consistent estimator of Θ if $\hat{\theta}_i$ is MSE consistent for $i = 1, \ldots, k$.

MSE consistency is also called *strong* consistency. Strong consistency implies weak consistency but not conversely. MSE consistent estimators are asymptotically unbiased.

5.2.3 Moment Estimators

Here we assume that the values of all components of Θ are unknown. (If not, the dimensionality of the problem is reduced in an obvious way.) The *method of moments* is based on expressing population moments in terms of the parameters of the assumed distribution, solving the resulting equations to express the parameters in terms of the moments, and then replacing these by the corresponding sample moments. For this purpose, either the moments $\mu'_r = E(X^r)$, the moments about the mean, $\mu_r = E(X - \mu)^r$, or a combination of the two are used. In some cases, notably for some discrete distributions, it is convenient to use factorial moments, $E[X(X-1) \ldots (X-r+1)]$, instead. We use the symbol $\hat{\theta}$ to denote the moment estimator.

The sample moments M'_r and moments about the mean M_r are defined by

$$M'_r = \frac{1}{n} \sum_{i=1}^{n} x_i^r$$

(5.6)

and

$$M_r = \frac{1}{n} \sum_{i=1}^{n} (x_i - \bar{x})^r \qquad (5.7)$$

If $k = 2$, it is common to use the mean $\mu = \mu_1'$ and the variance σ^2 and to estimate these by the sample mean $M_1' = \bar{X}$ and the sample variance s^2 given by Equation (3.2).

In most applications, moment estimators are consistent and asymptotically unbiased. Since the moments M_r' are unbiased estimators of the μ_r', moment estimators that are linear functions of the moments will also be unbiased. Under quite general conditions, moment estimators are asymptotically (as $n \to \infty$) normally distributed, and standard errors of the estimators can often be obtained without difficulty. We will return to this briefly in the next section.

Moment estimators are usually not efficient, and ordinarily should not be used if efficient estimators are available. They are of used for several reasons: (1) as preliminary estimates; (2) for simplicity in cases where efficient estimators are intractable; (3) as initial guesses in numerical solution of efficient estimation equations; (4) for very large data sets where efficiency is not an important issue.

To illustrate the technique, we consider the gamma distribution, with density given by Equation (4.34). The moments are given in Equation (4.36) as $\mu = \mu_1' = \alpha\beta$ and $\sigma^2 = \alpha\beta^2$, so that $\alpha = \mu^2/\sigma^2$ and $\beta = \sigma^2/\mu$. The moment estimates are $\hat{\alpha} = \bar{x}^2/s^2$ and $\hat{\beta} = s^2/\bar{x}$.

Example 5.1 (Case 2.1, Gamma Distribution)

Plots of the data on bond strength given in Table 2.1b show evidence of skewness, so the gamma distribution may be a reasonable assumption. We estimate the parameters of the gamma distribution for the data of Test 1. For these data, $n = 32$, $\bar{x} = 204.9$, and $s = 104.0$. Thus $\hat{\alpha} = (204.9/104.0)^2 = 3.882$, and $\hat{\beta} = 104.0^2/204.9 = 52.787$. The estimated value of the shape parameter α indicates an increasing failure rate for this part, i.e., bond strength appears to decrease with age. ■

Moment estimators for selected distributions. Derivation of moment estimators requires solution of a system of equations. In some cases, straightforward algebraic solutions yield analytical expressions for the estimators. In other cases, the solutions may not be expressible in closed form and numerical techniques may be necessary to obtain the estimates. The following list gives formulas for moment estimators for many of the distributions discussed in Chapter 4. In each case, reference is given to the distribution and moment estimates are given explicitly where possible and as solutions to equations where necessary.

Binomial distribution (Equation (4.24)):

$$\hat{p} = x/n \qquad (5.8)$$

the sample proportion.

Geometric distribution (Equation (4.27)):

$$\hat{p} = \frac{1}{1 + \bar{x}} \qquad (5.9)$$

Poisson distribution (Equation (4.33)):

$$\hat{\lambda} = \bar{x} \qquad (5.10)$$

Negative binomial distribution (Equation (4.29)): In some applications, r is unknown and may take on any positive value. In reliability applications, however, r is a known integer (the number of items having a certain characteristic) and we need only estimate p. The moment estimate is

$$\hat{p} = \frac{r}{r + \bar{x}} \tag{5.11}$$

For the case of r also unknown, see Johnson and Kotz (1969), Chapter 5.

Hypergeometric distribution (Equation (4.31)): In reliability, we wish to estimate D, the number of defectives in the lot. (From this, we can also estimate the proportion of defectives, $p = D/N$, if desired.) The moment estimate is

$$\hat{D} = \frac{Nx}{n} \tag{5.12}$$

Exponential distribution (Equation (4.37)):

$$\hat{\lambda} = 1/\bar{x}. \tag{5.13}$$

Gamma distribution (Equation 4.41):

$$\hat{\alpha} = \bar{x}^2/s^2 \tag{5.14}$$

and

$$\hat{\beta} = s^2/\bar{x}. \tag{5.15}$$

Weibull distribution (Equation (4.45)): $\hat{\alpha}$ is obtained as the solution of

$$\frac{s^2}{\bar{x}^2} = \frac{\Gamma(1 + 2/\hat{\alpha})}{\Gamma^2(1 + 1/\hat{\alpha})} - 1 \tag{5.16}$$

(See Blischke and Scheuer (1986).) The estimate of β is then given by

$$\hat{\beta} = \frac{\bar{x}}{\Gamma(1 + 1/\hat{\alpha}\}} \tag{5.17}$$

Normal distribution (Equation (4.50)):

$$\hat{\mu} = \bar{x}, \hat{\sigma}^2 = s^2 \tag{5.18}$$

Lognormal distribution (Equation (4.55)): Transform the data to $z_i = \log(x_i)$. The corresponding random variable Z_i is normally distributed. Use (5.18) with x replaced by z.

Inverse Gaussian distribution (Equation (4.51)):

$$\hat{\mu} = \bar{x}, \hat{\lambda} = \frac{\bar{x}^3}{s^2} \tag{5.19}$$

Two-parameter exponential distribution (Equation (4.58) with $\alpha = 1$ and $\lambda = 1/\beta$):

$$\hat{\gamma} = \bar{x} - s, \hat{\lambda} = \frac{1}{s} \qquad (5.20)$$

Three-parameter Weibull distribution (Equation (4.58)): The moment estimation equations may be set up in a number of ways, none of which admits of an easy solution. One possibility is:

$$\bar{x} = \hat{\gamma} + \hat{\beta}\Gamma(1 + 1/\hat{\alpha}) \qquad (5.21)$$

$$s^2 = \hat{\beta}^2[\Gamma(1 + 2/\hat{\alpha}) - \Gamma^2(1 + 1/\hat{\alpha})] \qquad (5.22)$$

$$M_3' - 3\bar{x}M_2' + 2\bar{x}^3 = \hat{\beta}^3[\Gamma(1 + 3/\hat{\alpha}) - 3\Gamma(1 + 1/\hat{\alpha})\Gamma(1 + 2/\hat{\alpha}) + 2\Gamma^3(1 + 1/\hat{\alpha})] \qquad (5.23)$$

The moment estimates are obtained as the solution to these three equations. Another possibility is to estimate γ by $\hat{\gamma} = \min(x_i)$ or $\hat{\gamma} = \min(x_i) - 1/n$, translate to $y_i = x_i - \hat{\gamma}$, and apply Equations (5.16) and (5.17) to the y_i.

Mixture of two exponential distributions (Equation (4.55) with $K = 2$ and $F_k(x)$ given by (4.29)): The parameters are p, λ_1 and λ_2. We solve, instead for p and $\mu_i = 1/\lambda_i$. This is most easily done in terms of the moment equations

$$z_1 = \bar{x} = \hat{p}\hat{\mu}_1 + (1 - \hat{p})\hat{\mu}_2 \qquad (5.24)$$

$$z_2 = \frac{1}{2}M_2' = \hat{p}\hat{\mu}_1^2 + (1 - \hat{p})\hat{\mu}_2^2 \qquad (5.25)$$

$$z_3 = \frac{1}{6}M_3' = \hat{p}\hat{\mu}_1^3 + (1 - \hat{p})\hat{\mu}_2^3 \qquad (5.26)$$

Let $a = (z_3 - z_1z_2)/(z_2 - z_1^2)$. The solution of Equations (5.24)–(5.26) for $\hat{\mu}_1$, $\hat{\mu}_2$ is

$$\hat{\mu}_1, \hat{\mu}_2 = \frac{a \pm \sqrt{a^2 - 4(az_1 - z_2)}}{2} \qquad (5.27)$$

with $\hat{\mu}_2$ taken to be the larger value; The moment estimate of the mixing parameter p is given by

$$\hat{p} = \frac{\hat{\mu}_2 - \bar{x}}{\hat{\mu}_2 - \hat{\mu}_1} \qquad (5.28)$$

Example 5.2 (Case 2.1, Normal Distribution)
The bond strength data given in Table 2.1a were collected under basically random conditions. From the various plots of these data given in Figures 3.1–3.3, it is apparent that the data are nearly normal, though slightly skewed left. (In a later chapter, we will see that a statistical test

does not reject normality.) In addition to skewness, there appears to be an outlier at the low end. Under the assumption of normality, moment estimates of the parameters are given, from (5.18), by $\hat{\mu} = \bar{x} = 278.92$ and $\hat{\sigma}^2 = s^2 = 1760.64$. If the outlier is omitted, these values become $\hat{\mu}^2 = 280.67$ and $\hat{\sigma}^2 = 1518.66$. Note that the outlier has a substantial effect on the variance, but not on the mean. If we accept normality and eliminate the outlier, we estimate to average bond strength to be 281 pounds and the standard deviation to be $\sqrt{1518.66} = 38.97$. The 68-95-99.7 Rule (See Section 3.6.5.) tells us that under the conditions under which the data were collected, approximately 95% of the bond strengths will lie between 203 and 359 pounds. ∎

Example 5.3 (Case 2.20, Weibull and Inverse Gaussian Distributions)
Data on breaking strengths of individual fibers and of tows of 1000 fibers are given in Table 2.20. In analyses of these data, Smith (1991) and Durham and Padget (1997) use both the Weibull and inverse Gaussian distributions. We calculate moment estimates for the parameters of both distributions for individual fibers of length 20 mm. Moment estimates for the inverse Gaussian are given in (5.19). For the selected data set the mean and standard deviation are $\bar{x} = 2.4497$ and $s = 0.4917$. The estimates are $\hat{\mu} = 2.4497$, and $\hat{\lambda} = 2.4497^3/0.4917^2 = 60.80$.

For the Weibull distribution, moment estimates require solution of (5.16) for $\hat{\alpha}$. Here $s^2/x^2 = 0.0403$. The solution of (5.16), which may be obtained by use of the tables in Blischke and Scheuer (1986) or by trial and error using tables of $\Gamma(1 + x)$, is $\hat{\alpha} = 5.77$. From (5.17), the moment estimate of the scale parameter is found to be $\hat{\beta} = 2.4497/\Gamma(1.173) = 2.65$. The large value of $\hat{\alpha}$ indicates an increasing failure rate. ∎

Example 5.4 (Case 2.13, Exponential and Mixed Exponential Distributions)
Table 2.13 includes data on number of correct entries until an error is made for five clerical workers. If entries are made at essentially a constant rate, the tabulated values are a good proxy for time to failure. If *errors* occur at a constant rate, the data should follow an exponential distribution. Finally, if the error rate is the same for all clerks, the composite data should be exponentially distributed. In the composite, there are 218 observations, with a mean of $\bar{x} = 406$, which is the moment estimate of μ, the mean number of correct entries between errors. The moment estimate of λ is $\hat{\lambda} = 1/\bar{x} = 0.00246$.

Suppose that errors can occur in two ways (e.g., skilled and less-skilled clerks or error-prone and less-error-prone entries), and that each type is exponentially distributed. The resulting distribution of errors is then a mixture of two exponential distributions, and we estimate the proportion of each type and the parameters of each component by use of (5.27) and (5.28). The required sample moments are $\bar{x} = 406.0$, $M_2' = 391,702$, and $M_3' = 596,350,032$, from which we obtain $z_1 = 406$, $z_2 = 196,851$, $z_3 = 99,391,672$, and $a = 640.850$. The roots of the quadratic give the estimates $\hat{\mu}_1 = 124.60$ and $\hat{\mu}_2 = 516.23$. The individual rates are estimated as the reciprocals of these. The estimated proportion of component 1 (having the smaller mean) is $\hat{p} = (516.23 - 406.00)/(516.23 - 124.60) = 0.2815$. We estimate that 28% of the errors occur with a MTTF of 124.60 and the remaining 72% with a MTTF of 516.23. ∎

5.2.4 Maximum Likelihood Estimators

The *likelihood function L* is given by (5.4) with θ replaced by Θ. The *maximum likelihood estimator* (MLE) of Θ is that value Θ^* that maximizes L. In most cases, the MLE is obtained by differentiation, equating to zero and solving the resulting equations, with the process applied to log L rather than L for ease of solution.[3]

Under regularity conditions similar to those mentioned previously, maximum likelihood estimators are consistent, asymptotically unbiased, asymptotically efficient and asymptotically normally distributed. Asymptotic efficiency here means that as $n \to \infty$, the covariance matrix of the estimators achieves the Cramér-Rao bound. This is an important result in that it leads to methods for obtaining confidence intervals (to be discussed in Section 5.3) that are valid, at least for large samples, in situations where the exact theory/derivation is cumbersome, if not impossible to implement. For a rigorous treatment of the theory of maximum likelihood, asymptotic results and related topics, see Stuart and Ord (1991).

To illustrate the procedure, we derive the MLE for the exponential distribution. We assume a sample of size n from the distribution given in (4.37). The likelihood function is

$$L(x_1, \ldots, x_n) = \prod_{i=1}^{n} (\lambda e^{-\lambda x_i}) = \lambda^n \exp\left\{-\lambda \sum_{i=1}^{n} x_i\right\} \tag{5.29}$$

so that $\log L$ becomes

$$\log L = n \log(\lambda) - \lambda \sum_{i=1}^{n} x_i \tag{5.30}$$

Differentiation with respect to λ leads to the likelihood equation

$$\frac{n}{\lambda^*} - \sum_{i=1}^{n} x_i = 0 \tag{5.31}$$

with solution

$$\lambda^* = \frac{n}{\sum_{i=1}^{n} x_i} = \frac{1}{\bar{x}} \tag{5.32}$$

Note that the MLE in this case is the same as the moment estimator. (This is not true in general.) The asymptotic variance of λ^* is

$$\left\{-E\left(\frac{\partial^2 \log L}{\partial \lambda^2}\right)\right\}^{-1} = \frac{\lambda^2}{n} \tag{5.33}$$

Maximum likelihood estimators for selected distributions. The following list gives formulas for MLEs for many of the distributions discussed in Chapter 4. In each case, reference is given to the distribution and MLEs are given explicitly where possible and as solutions to equations where necessary. Variances of the estimators are also given in many cases. The notation $V(.)$ is used as before for exact variances and we use $V_a(.)$ for asymptotic variances. (Note that these are, in fact, large sample approximations based on the asymptotic variances of the normalized estimators.) This is useful in that the variance can be estimated as well (e.g., by substituting MLEs for the parameters in the variance formulas). The square root of the result is an estimate of the standard error of the MLE and is a measure of the uncertainty in the estimate.

Binomial Distribution (Equation (4.24)): The MLE is

$$p^* = x/n \tag{5.34}$$

with variance (exact, not asymptotic)

$$V(p^*) = \frac{p(1-p)}{n} \tag{5.35}$$

Geometric Distribution: (Equation (4.27)): The MLE is

$$p^* = 1/(1 + x) \tag{5.36}$$

with asymptotic variance

$$V_a(p^*) = \frac{p^2(1-p)}{n} \tag{5.37}$$

Poisson Distribution (Equation (4.33)): The MLE is

$$\lambda^* = \bar{x} \tag{5.38}$$

with variance

$$V(\lambda^*) = \frac{\lambda}{n} \tag{5.39}$$

Negative Binomial Distribution (Equation (4.29)): The MLE is

$$p^* = \frac{r}{r + \bar{x}} \tag{5.40}$$

with asymptotic variance

$$V_a(p^*) = \frac{p^2(1-p)}{nr} \tag{5.41}$$

Hypergeometric Distribution (Equation (4.31)): The MLE of D is

$$D^* = \left[\frac{x(N+1)}{n} \right] \tag{5.42}$$

where $[c]$ denotes the greatest integer less than or equal to c. The variance is

$$V(D^*) = \frac{(N+1)^2(N-n)}{n(N-1)} \left(\frac{D}{N} \right) \left(1 - \frac{D}{N} \right) \tag{5.43}$$

Exponential Distribution (Equation (4.37)): See Equations (5.32) and (5.33) for estimation of λ. In reliability studies, one often wishes to estimate μ = MTTF as well. The MLE is $\mu^* = \bar{x}$, with variance $V(\mu^*) = 1/(n\lambda^2)$.

Two-Parameter Exponential Distribution (Equation (4.36) with x replaced by $(x - a)$): The MLEs are

$$a^* = \min(x_1, \ldots, x_n) \tag{5.44}$$

$$\lambda^* = \frac{1}{\bar{x} - a^*} \tag{5.45}$$

with asymptotic variances

$$V_a(a^*) = \frac{1}{\lambda^2 n^2} \tag{5.46}$$

$$V_a(\lambda^*) = \frac{1}{\lambda^2}\left(\frac{1}{n} + \frac{1}{n^2} - \frac{2}{n^3}\right) \tag{5.47}$$

Gamma Distribution (Equation (4.41)): For the two-parameter case, the MLE of the shape parameter α is obtained as the solution of the equation

$$\frac{1}{n}\sum_{i=1}^{n} \log(x_i) - \log(\bar{x}) = \psi(\alpha^*) - \log(\alpha^*) \tag{5.48}$$

where ψ is the digamma function defined as

$$\psi(x) = \frac{\partial}{\partial x}\log[\Gamma(x)] \tag{5.49}$$

with $\Gamma(.)$ the gamma function. The MLE of the scale parameter β is then obtained as

$$\beta^* = \frac{\bar{x}}{\alpha^*} \tag{5.50}$$

Tables and methods for evaluation of the gamma and digamma function can be found in Abramowitz and Stegun (1964). Computer algorithms are also available. The asymptotic variances of the estimators are

$$V_a(\alpha^*) = \frac{\alpha}{n[\alpha\psi'(\alpha) - 1]} \tag{5.51}$$

$$V_a(\beta^*) = \frac{\beta^2 \psi'(\alpha)}{n[\alpha\psi'(\alpha) - 1]} \tag{5.52}$$

Here $\psi'(x)$ is the trigamma function, also evaluated as indicated in the previous references. See Johnson and Kotz (1970a) for additional details, including various methods of solution and evaluation of the large sample variances, as well as the correlation between the estimators. The three-parameter gamma distribution is also considered in detail.

Weibull Distribution (Equation (4.45)): The MLE of the shape parameter α is obtained as the solution of

$$\frac{\sum_{i=1}^{n}(x_i^{\alpha^*}\log x_i)}{\sum_{i=1}^{n}x_i^{\alpha^*}} - \frac{1}{\alpha^*} - \frac{1}{n}\sum_{i=1}^{n}\log x_i = 0 \tag{5.53}$$

(In many standard statistical program packages, it is not difficult to write a Macro for solution of this equation.) The scale parameter is then estimated as

$$\beta^* = \left(\frac{1}{n}\sum_{i=1}^{n}x_i^{\alpha^*}\right)^{1/\alpha^*}$$ (5.54)

The large sample variances are

$$V_a(\alpha^*) = \frac{1.1087\beta^2}{n\alpha^2}$$ (5.55)

$$V_a(\beta^*) = \frac{0.6079\alpha^2}{n}$$ (5.56)

See Johnson and Kotz (1970a) for additional discussion and approaches to the three-parameter case.

Normal Distribution (Equation (4.50)): The MLEs are

$$\mu^* = \bar{x}$$ (5.57)

$$\sigma^* = \left(\frac{1}{n}\sum_{i=1}^{n}(x_i - \bar{x})^2\right)^{1/2}$$ (5.58)

The variances of the estimators are

$$V(\mu^*) = \frac{\sigma^2}{n}$$ (5.59)

$$V(\sigma^*) = \sigma^2\left[1 - \frac{1}{n} - \frac{2\Gamma^2\left(\frac{n}{2}\right)}{n\Gamma^2\left(\frac{n-1}{2}\right)}\right]$$ (5.60)

The actual distribution of σ^* is related to the chi-square distribution and this fact is usually used in statistical inference problems regarding σ and σ^2.

Lognormal Distribution (Equation (4.55)): Again, transform to $y_i = \log(x_i)$ and the results for the normal distribution are directly applicable. The three-parameter version of the lognormal distribution (including a shift parameter) is much more difficult; see Johnson and Kotz (1970a).

Inverse Gaussian Distribution (Equation (4.51)): The MLEs are

$$\mu^* = \bar{x}$$ (5.61)

$$\lambda^* = n\left[\sum_{i=1}^{n}\left(\frac{1}{x_i} - \frac{1}{\bar{x}}\right)\right]^{-1}$$ (5.62)

The variance of μ^* is

$$V(\mu^*) = \frac{\mu^3}{n\lambda}$$ (5.63)

The distribution of λ^* is proportional to a chi-square distribution, which is thus used for inference problems regarding λ.

Example 5.5 (Case 2.1, Gamma Distribution)
We again consider Test 1, with data given in Table 2.1b. The moment estimates were calculated in Example 5.1. The maximum likelihood estimates are obtained from (5.48) and (5.49). For the Test 1 data, we have $n = 32$, $\bar{x} = 204.9$, and $[\Sigma \log(x_i)]/32 = 5.0071$, so the LHS of (5.48) is -0.315 and α^* is obtained as the solution of

$$\psi(\alpha^*) - \log(\alpha^*) = -0.315.$$

The solution, found by trial and error using the tables of Abramowitz and Stegun (1964), is $\alpha^* = 1.735$. From (5.50), $\beta^* = 204.9/1.735 = 118.1$. Note that these values differ substantially from the moment estimates, which is not unusual for small or even modest sample sizes. Approximate variances may be estimated by substituting these estimates into (5.51) and (5.52). Using the tables of Abramowitz and Stegun, we find $\psi'(1.735) = 0.77262$, so that $\hat{V}(\alpha^*) = 0.1592$ and $\hat{V}(\beta^*) = 989.0$, giving estimated standard deviations of 0.3990 and 31.45, respectively. (These may not be accurate because of the small n and caution should be used in interpretation of such results.) ■

Example 5.6 (Case 2.13, Exponential Distribution)
If we assume that the composite data (with $n = 218$) are exponentially distributed, the MLE of λ is the reciprocal of the overall mean. We obtain $\lambda^* = 1/406.0 = 0.00246$. From (5.33), the estimated variance of the MLE is $\hat{V}(\lambda^*) = (\lambda^*)^2/n = 2.776 \times 10^{-8}$. The estimated standard deviation of λ^* is 0.0001666.

If clerks are thought to have different error rates, they are analyzed separately. For Clerk 1, the results are $n = 31$, $\bar{x} = 580.4$, $\lambda^* = 0.001732$, and $\hat{V}(\lambda^*) = 9.677 \times 10^{-8}$. ■

Example 5.7 (Case 2.20, Weibull and Inverse Gaussian Distributions)
We continue the analysis of the data of Case 2.20 for 20 mm fibers. Moment estimates of the parameters under Weibull and inverse Gaussian assumptions were calculated in Example 5.3. Here we calculate MLEs for both distributions. For the Weibull, we begin with the solution of (5.53). For trial and error or more sophisticated search routines, a good strategy is to begin with the moment estimate of α. Alternatively, a number of computer programs (e.g., Minitab) will calculate the MLEs. The solution is found to be $\alpha^* = 5.524$. From (5.54) we obtain $\beta^* = (15122/70)^{1/5.52} = 2.648$. Here the MLEs are nearly in agreement with the moment estimates. From (5.55) and (5.56), the estimated variances of the MLEs are $\hat{V}(\alpha^*) = 0.003640$ and $\hat{V}(\beta^*) = 0.2650$.

For the inverse Gaussian, we again have $\mu^* = \bar{x} = 2.45$. From (5.62) we obtain $\lambda^* = 91.54$, somewhat different from the moment estimate. The estimated variance of μ^* is obtained from (5.63) as $\hat{V}(\mu^*) = (\mu^*)^3/n\lambda^* = 0.002295$. The estimated standard deviation of μ^* is 0.0479. ■

5.3 CONFIDENCE INTERVAL ESTIMATION

5.3.1 Basic Concepts

In point estimation of a parameter or other population characteristic, we use a single number to estimate a parameter or a set of k numbers to estimate a k-dimensional parameter. For

MLEs, we also gave variances (exact or asymptotic) of the estimators. The square root of this variance, called the *standard error,* is a measure of uncertainty in the estimate. A *confidence interval* (or *confidence interval estimator*) takes this uncertainty into account by providing an estimate in the form an interval of numbers along with a measure of the "confidence" one has that the interval will, in fact, contain the true value of the parameter or characteristic being estimated. For k parameters, a separate confidence interval may be calculated for each, or a k-dimensional *confidence region* may be defined.

In the case of a single parameter, say θ, a confidence interval based on a sample of size n, $X_1, X_2, \ldots, X_n$, is formally defined as follows:

Definition: A *confidence interval* for a parameter θ is an interval defined by two limits, the lower limit $L_1(X_1, X_2, \ldots, X_n)$ and the upper limit $L_2(X_1, X_2, \ldots, X_n)$, having the property that

$$P(L_1(X_1, \ldots, X_n) < \theta < (L_2(X_1, \ldots, X_n)) = \gamma \qquad (5.64)$$

where γ (the *confidence coefficient*) is a constant with $0 < \gamma < 1$.

Confidence is usually expressed in percent; e.g., if $\gamma = 0.95$, the result is a "95% confidence interval" for θ. Note that the random variables in expression (5.64) are L_1 and L_2, not θ, i.e., this is not a probability statement about θ, but about L_1 and L_2. Hence we use the term "confidence" rather than "probability" when discussing this as a statement about θ. The proper interpretation is that the procedure gives a correct result $100\gamma\%$ of the time.

It is desirable in practice that the width of a confidence interval be small, i.e., that the result be precise in the sense that we can have high confidence that the true value of the parameter lies within a relatively narrow interval (or small region, in the multiparameter case). In general, the width of the interval depends on the data and on the desired confidence. Specifically, the width of the confidence interval (1) decreases as n increases; (2) increases as the confidence γ increases; and (3) decreases as the variability in the data decreases. Thus in theory, the width of the confidence interval can be controlled, but in practice this is not always easy. In particular, it usually means incurring the expense of obtaining large samples.

Confidence intervals for functions $\tau(\theta)$ are defined by replacing θ by $\tau(\theta)$ in (5.64). If $\tau(\theta)$ is a monotonic function, the interval is simply $(\tau(L_1), \tau(L_2))$. For example, a confidence interval for a standard deviation may be obtained by taking square roots of the confidence limits for the variance. One of the most important functions for which we may wish to calculate a confidence interval is item reliability. This will be discussed in detail in Chapter 11.

The confidence interval in Equation (5.64) is a *two-sided interval;* part of the remaining probability, $(1 - \gamma)$, is that below L_1 and part is that above L_2 (usually $(1 - \gamma)/2$ on each side). In some applications, a *one-sided confidence interval* is desired. A *lower* (*upper*) one-sided confidence interval is obtained by omitting L_2 in (5.64) and modifying $L_1(L_2)$ accordingly; the interpretation is that we are $100\gamma\%$ confident that the true value is at least L_1 (at most L_2). These are also called lower and upper *confidence limits.* Lower one-sided confidence intervals are particularly important in estimating reliability and in estimating the MTTF in reliability applications.

Construction of confidence intervals, i.e., derivation of the limits L_1 and L_2, may be done in several ways (See Mood et al. (1974)). Usually the interval is based on the distribution of the best point estimator, if such exists. If not, it is based on the distribution of a "good" point estimator. If the distribution is unknown or is mathematically intractable, approximations may be used. Another approach is to use asymptotic results, most commonly those discussed in the previous section in connection with the MLE. We turn next to some specific results.

5.3.2 Confidence Intervals for the Parameters of Distributions Used in Reliability

In this section, we list confidence intervals (CI) for the parameters of some important failure distributions as well as other distributions of importance in reliability and related applications. We begin with confidence intervals for the mean of the normal distribution. Although this distribution is not often used as a life distribution, the result is important in this context because it is frequently used in obtaining asymptotic confidence intervals, as will be discussed below.

In estimating the mean of a normal distribution, we distinguish two cases: (1) the standard deviation σ known, and (2) σ unknown. The results are:

CI for μ, normal distribution, σ known. The (two-sided) interval is

$$\bar{x} \pm z_{1-(\alpha/2)} \frac{\sigma}{\sqrt{n}} \tag{5.65}$$

where $\alpha = 1 - \gamma$ and z_p is the p-fractile of the standard normal distribution, given in Table C1 in Appendix C. Situations in which the standard deviation is known or may assumed to be known usually involve either large samples or large amounts of historical data, or theory that provides a specific value of σ. For small- or modest-size data sets, we require the following:

CI for μ, normal distribution, σ unknown. In this case, we substitute s for σ and fractiles of the Student-t distribution for z_p in Equation (5.65). The result is

$$\bar{x} \pm t_{1-(\alpha/2)} \frac{s}{\sqrt{n}} \tag{5.66}$$

The parameter of the Student-t distribution is a quantity called *degrees of freedom* (df). In this application, df $= (n - 1)$. Fractiles of the Student-t distribution for selected values of df are given in Table C2. For df not in the table, one may (1) interpolate; (2) use the tabulated value corresponding to the largest df less than the desired df (a slightly conservative approach); or (3) use a computer program (e.g., Minitab) to calculate the desired value.

One-sided limits are obtained by replacing $1 - \alpha/2$ by $1 - \alpha$ in (5.65) and (5.66) and using only the positive sign for an upper limit and the negative sign for a lower limit.

Example 5.8 (Case 2.1, Confidence Interval for μ)
We again consider the data of Table 2.1a. Here the standard deviation is not known and the sample standard deviation s is used in (5.66). We had $\bar{x} = 278.92$, $s = 41.96$, and $n = 85$. Suppose we wish to calculate a 95% CI for the mean bond strength μ. We need $t_{.975}$ with 84 df. This value is not found in the table so we use 80 df; the corresponding tabulated value is 1.990. (The exact value for 84 df, calculated by Minitab, is 1.989.) The 95% CI for μ is $278.92 \pm 1.99(41.96)/\sqrt{85}$, giving (269.86,287.98). We are 95% confident that the true mean breaking strength is between 270 and 288 psi. ∎

Example 5.9 (Case 2.1, One-Sided Confidence Interval)
Suppose we wish to calculate a lower 95% confidence interval for μ in the previous example. In this case, we replace $t_{1-(\alpha/2)}$ in equation (5.66) by $t_{1-\alpha}$. For $\alpha = 0.05$ ($\gamma = 0.95$) with 84 df, the tabulated value is 1.663. The lower one-sided confidence bound is $\bar{x} - t_{.95}s/\sqrt{n} = 278.92 - 1.663(41.963)/\sqrt{85} = 271.35$, so the confidence interval is (271.35,∞). We are 95% confident that the mean bond strength is *at least* 271.35 pounds. ∎

We now list exact and/or approximate confidence intervals for the parameters of the distributions considered in the previous sections. The approximations suggested are based on the fact that the MLEs are (under the required regularity conditions) asymptotically normally distributed with variances as given in Section 5.2.4. The procedure is to calculate an asymptotic confidence interval using the results for the mean of a normal distribution but with $\bar{x}$ replaced by θ^* and the standard deviation replaced by an estimate of the standard error of θ^*. Thus to estimate a parameter θ, we use either

$$\theta^* \pm z_{1-(\alpha/2)} \hat{V}(\theta^*) \tag{5.67}$$

or

$$\theta^* \pm t_{1-(\alpha/2)} \hat{V}(\theta^*) \tag{5.68}$$

It is suggested that Equation (5.68) be used for small to moderate n (say $n \leq 100$). In any case, neither approximation should be used for small n. Sample sizes of at least 20 to 30 are almost always required for the approximation to be adequate, and n's of 50 to 100 are often required. The number required depends on the skewness of the actual distribution of the estimator. If this distribution is symmetrical or nearly so, the normal approximation will be adequate for relatively modest sample sizes. If the distribution is highly skewed, hundreds or thousands of observations may be required.

The following are suggested confidence interval procedures for parameters of selected distributions:

Binomial Distribution: If n is large enough so that $n\hat{p} \geq 5$ and $n(1-\hat{p}) \geq 5$, then a normal approximation to the binomial distribution may be used to obtain a confidence interval for p. The result is

$$\hat{p} \pm z_{1-(\alpha/2)} \sqrt{\frac{\hat{p}(1-\hat{p})}{n}} \tag{5.69}$$

If the conditions on n are not satisfied, procedures based on the exact binomial distribution should be used. Tables and charts for this purpose are available in some elementary statistics texts, for example, Dixon and Massey (1965).

Geometric Distribution: Use (5.67) or (5.68) with MLE given in (5.36) and variance in (5.37).

Poisson Distribution: The MLE is $\lambda^* = \bar{x}$. Let $k = n\bar{x}$. The exact CI is

$$\left(\frac{1}{2n} \chi^2_{2k,(\alpha/2)}, \frac{1}{2n} \chi^2_{2k+2,1-(\alpha/2)} \right) \tag{5.70}$$

where $\chi^2_{a,p}$ is the p-fractile of the chi-square distribution with a df, which may be gotten from Table C3 or calculated by use of a statistical program package. If λ is expected to be at least 5 or if n is sufficiently large, the following normal approximation may be used:

$$\bar{x} + \frac{z^2_{1-(\alpha/2)}}{2n} \pm z_{1-(\alpha/2)} \sqrt{\frac{\bar{x}}{n} + \frac{z^2_{1-(\alpha/2)}}{n^2}} \tag{5.71}$$

If n is large enough so that terms of order $1/n$ can be ignored, this reduces to

$$\bar{x} \pm z^2_{1-(\alpha/2)} \sqrt{\frac{\bar{x}}{n}} \tag{5.72}$$

which agrees with the asymptotic normal approximation based on the MLE.

Negative Binomial Distribution: Use (5.67) or (5.68) with MLE given in (5.40) and variance in (5.41).

Hypergeometric Distribution: Use (5.67) or (5.68) with MLE given in (5.42) and variance in (5.43).

Exponential Distribution: To estimate the MTTF, we use $\mu^* = \bar{x}$. The CI is

$$\left(\frac{2n\bar{x}}{\chi^2_{2n,1-(\alpha/2)}}, \frac{2n\bar{x}}{\chi^2_{2n,(\alpha/2)}} \right) \tag{5.73}$$

A confidence interval for λ may be obtained by calculating reciprocals of the limits on μ. Alternatively, for large n, the normal approximation based on Equations (5.32), (5.33), (5.67), and (5.68) may be used.

Two-Parameter Exponential Distribution: Use (5.67) or (5.68) with MLEs given in (5.44) and (5.45) and variances in (5.46) and (5.47)

Gamma Distribution: Use (5.67) or (5.68) with MLEs obtained from (5.48) and (5.50) and variances given in (5.51) and (5.52).

Weibull Distribution: Use (5.67) or (5.68) with MLEs obtained from (5.53) and (5.54) and variances given in (5.55) and (5.56).

Normal Distribution: Confidence intervals for μ are given in Equations (5.58) and (5.59). A confidence interval for the variance σ^2 is given by

$$\left(\frac{(n-1)s^2}{\chi^2_{n-1,1-\alpha/2}}, \frac{(n-1)s^2}{\chi^2_{n-1,\alpha/2}} \right) \tag{5.74}$$

To obtain a confidence interval for σ, take square roots of the limits on σ^2.

Example 5.10 (Case 2.1, Confidence Interval for σ^2)

For the data of Table 2.1a, $n = 85$ and we had $s = 41.96$, so $s^2 = 1760.64$. We calculate 99% confidence intervals. The required tabulated fractiles of the of the χ^2-distribution, obtained by computer, are $\chi^2_{84,.995} = 121.12$ and $\chi^2_{84,.005} = 54.368$. The 99% confidence interval for σ^2 is obtained from Equation (5.85) as $[84(1760.64)/121.12,\ 84(1760.64)/54.368] = (1221.05, 2720.24)$. The 99% confidence interval for σ is $(34.94, 52.15)$. ∎

Lognormal Distribution. Again, analyze log x, using the results for the normal distribution.

Inverse Gaussian Distribution. Use (5.67) or (5.68) for a confidence interval for μ with MLE given by (5.63). The confidence interval for λ is

$$\left(\frac{\lambda^*}{n} \chi^2_{n-1,\alpha/2},\ \frac{\lambda^*}{n} \chi^2_{n-1,1-\alpha/2} \right) \tag{5.75}$$

Example 5.11 (Case 2.27, Binomial Distribution, Confidence Interval for p)

For Pump A, 236 successes were observed in 240 tests. The estimate of p, the reliability of the pump under the test conditions, is $\hat{p} = 236/240 = 0.99167$. Since $n(1 - \hat{p}) < 5$, the normal approximation is not appropriate. To illustrate the procedure, suppose that 6 failures and 234 successes had been observed. then $\hat{p} = 234/240 = 0.9750$, the conditions for use of the approximation are satisfied, and we obtain a 95% confidence interval for p as $0.975 \pm 1.96[0.975(0.025)/240]^{1/2} = 0.9750 \pm 0.020$, or $(0.955, 0.995)$. ∎

Example 5.12 (Case 2.13, Exponential Distribution, Confidence Intervals for λ and μ)
We consider Clerk 1 and calculate 90% confidence intervals. From Example 5.6, we have n = 31, $\mu^* = 580.4$, and $\lambda^* = 0.001723$. For calculation of confidence intervals we need $\chi^2_{62,0.05}$ = 44.8890 and $\chi^2_{62,0.95} = 81.3907$. The 90% confidence interval for λ is

$$\frac{44.8890}{62(580.4)}, \frac{81.3907}{62(580.4)} \text{ or } (0.001248, 0.002262)$$

We are 90% confident that the true error rate for Clerk 1 lies between 0.001248 and 0.002262. A 90% confidence for the mean "time" between errors is calculated as the reciprocals of these values. We are 90% confident that the true mean μ lies between 442.1 and 801.6.

Note that the normal approximation gives an estimated standard deviation of λ^* as $\lambda^*/\sqrt{n} = 0.0003095$, which results in an approximate 90% confidence interval of $0.001723 \pm 1.645(0.0003095) = (0.001214, 0.002232)$, in fairly close agreement with the exact result. ■

Example 5.13 (Case 2.20, Confidence Intervals for the Parameters of the Weibull and Inverse Gaussian Distributions)
For the 20 mm fibers analyzed previously, $n = 70$, which should be adequate for use of the asymptotic results and the normal approximation. For the Weibull distribution, the MLEs of the parameters were found in Example 5.7 to be $\alpha^* = 5.52$ and $\beta^* = 2.65$, with estimated standard deviations of 0.008519 and 0.5144, respectively. We calculate 95% confidence intervals using (5.68). $t_{.975}$ with 69 df is found from Minitab to be 1.995. The 95% confidence interval for α is $5.52 \pm (1.995)(.008519)$, giving (5.52,5.56). Similarly, the 95% confidence interval for β is (1.62,3.68). Note from (5.55) and (5.56) that when α is large and β is not small, there is much uncertainty in the estimate of β, and this is reflected in the width of the confidence intervals.

Finally, we calculate confidence intervals for μ and λ in the Inverse Gaussian distribution using the same data. From Example 5.7, we have $\mu^* = 2.45$, with an estimated standard deviation of 0.0479. The 95% confidence interval for μ is $2.45 \pm 1.995(0.0479)$, or (2.35,2.55). The 95% confidence interval for λ is obtained from (5.75). The required values of χ^2 with 69 df are 47.92 and 93.86. The interval is (62.7,122.7). ■

5.4 ESTIMATION OF FUNCTIONS OF PARAMETERS

5.4.1 Basic Approach

In reliability analysis, it is often of interest to estimate functions of parameters in addition to the parameters themselves. In later chapters, we will encounter a number of such functions; examples are item reliability as a function of time, MTTF as a function of estimated parameters, system reliability as a function of component reliabilities, and so forth. A common practice is to simply substitute parameter estimates in the functional form to estimate the function. In fact, if the MLE is used to estimate a function $\tau(\theta)$, i.e., we calculate $\tau(\theta^*)$, the result is the MLE of τ. If the moment estimator is used instead, the result is the moment estimator of τ. Under fairly general conditions, estimators of this type are asmptotically normally distributed. This fact will be used in the next section to construct asymptotic confidence intervals for $\tau(\theta)$. What is needed for this purpose is a method for determining the variance in this asymptotic distribution.

Results that are useful in this regard are given in Appendix A.5. Equations (A5.1) and (A5.2) of the appendix give the approximate mean and variance of a function of a single random variable (e.g., a sample mean). Extensions to more than one variable (e.g., two or more sample moments) are given in (A5.4) and (A5.5).

We illustrate the use of these results by application to the *coefficient of variation, C* = σ/μ. This quantity is a standardized measure of variability often used in the physical sciences because it is nondimensional and tends to remain stable over measures of different orders of magnitude. Suppose that we have a sample from a normal distribution. The MLE of C is C^* = $\tau(\mu^*, \sigma^*) = \sigma^*/\mu^*$, where σ^* and μ^* are given by Equations (5.57) and (5.58), with variances given by (5.59) and (5.60). In the case of the normal distribution, the estimators are independent, so the covariance is zero. The required derivatives are

$$\frac{\partial \tau}{\partial \mu} = \frac{-\sigma}{\mu^2} \qquad \frac{\partial \tau}{\partial \sigma} = \frac{1}{\mu} \qquad \frac{\partial^2 \tau}{\partial \mu^2} = \frac{2\sigma}{\mu^3} \qquad \frac{\partial^2 \tau}{\partial \sigma^2} = 0 \qquad (5.76)$$

Example 5.14 (Case 2.1, Coefficient of Variation)

Suppose that it is reasonable to assume that the bond strength data given in Table 2.1a comprise a sample from a normal distribution. Summary statistics for these data are given in Table 3.9, from which we find that $\bar{x}$ = 278.92 and s = 41.96, with n = 85. The MLE of σ is $\sigma^* = s\sqrt{(n-1)/n}$ = 41.71. It is common to express the coefficient of variation in percent; the result is c^* = 100(41.71)/278.92 = 14.95%, indicating a modest amount of variability. The derivatives of (5.76), evaluated at the MLEs, are, respectively, −0.0005361, 0.003585, 0.000003844, and 0. The estimated variances are $\hat{V}(\bar{x})$ = 41.71²/85 = 20.47 and

$$\hat{V}(\sigma^*) = 41.71^2\left[1 - \frac{1}{85} - \frac{2\Gamma^2(42.5)}{85\Gamma^2(42)}\right]$$

The gamma functions in this expressions may be evaluated by table look-up (Abramowitz and Stegun, 1964, p. 272). This gives $\Gamma(42.5)$ = 2.16153E50, $\Gamma(42)$ = 3.34525E49, and $\hat{V}(\sigma^*)$ = 10.20. The estimator C^* is approximately unbiased; the final terms in (5.79) reduce to $2\sigma^{*3}/n\bar{x}^3$ = 0.00008. The approximate variance of C^* is

$$V(C^*) \approx \frac{\sigma^2}{\mu^4} V(\bar{x}) + \frac{1}{\mu^2} V(\sigma^*)$$

which is estimated by substitution of sample values as

$$V(C^*) \approx \frac{41.71^2}{278.92^4}(20.47) + \frac{1}{278.92^2}(10.20) = 0.0001370$$

The standard error is $\sqrt{.0001370}$ = 0.01170, indicating a reasonably precise estimate of the true coefficient of variation. ■

5.4.2 Functions of Random Variables

The most commonly occurring, and hence most important, functions of random variables encountered in analyzing data are sums, differences, products and quotients. Examples are the sample mean, the difference between two sample means, the product of the estimated reliabilities of two components, and the coefficient of variation. Results concerning exact and asymptotic distributions, means, and variances for these are given in Appendix A5.

In cases where it is possible to determine the distribution of a function, moment or maximum likelihood methods may be applied as in Section 5.2. Alternatively, approximations such as that of the previous section may be derived.

5.5 TOLERANCE INTERVALS

5.5.1 Basic Concept

A *tolerance interval* is essentially a confidence interval for a population fractile. It is a statement of the form "We are $100\gamma\%$ confident that at least $100P\%$ of the population lies within the interval (a,b)." As before, γ is called the confidence; P is called the coverage. Again, the interval may be two-sided or one-sided, either lower or upper bounds (in which case either a is $-\infty$ or b is ∞). Factors for calculating tolerance intervals have been tabulated for the normal distribution and for nonparametric tolerance intervals (not assuming a particular distributional form). These will be discussed briefly in Sections 5.5.2 and 5.5.4, respectively. They can also be calculated for the exponential distribution. This is discussed in Section 5.5.3.

Parametric tolerance intervals for distributions other than the normal can be obtained as well. The analysis is straightforward for the exponential distribution. Other distributions can, in principle, be analyzed, but the mathematics is difficult and little else has been done in this area. One approach, particularly in cases where closed form expressions for the CDF are available, would be to substitute MLEs for the parameters and use the methods for estimating functions of parameters and asymptotic normality to derive the tolerance limits. In most cases, this will also lead to analytical difficulties and the nonparametric approach is suggested as a preferable alternative.

This concept is an important one in reliability applications. In analysis of manufactured parts, tolerance intervals yield intervals in which the measured characteristic of a specified large proportion of parts will lie with high probability (assuming that P and γ are close to one). If this interval is sufficiently small, the part is acceptable. Secondly, tolerance intervals on failure times provide bounds on item lifetimes. In this case, we are usually interested is lower one-sided tolerance intervals, since these provide a number beyond which a specified proportion of lifetimes will lie with stated confidence.

Note, incidentally, that tolerance intervals will always be much wider than confidence intervals for single parameters such as the mean. Here we are attempting to estimate an entire segment of a population rather than simply a single parameter value.

5.5.2 Normal Tolerance Limits

Tolerance limits for the normal distribution are based on the sample mean $\bar{x}$ and sample standard deviation s. Two-sided limits are of the form $\bar{x} \pm Ks$. The factor K depends on n, P, and γ. The width of the tolerance interval increases as either P or γ is increased and decreases as n increases. K-factors for two-sided tolerance intervals with coverage $P = 0.75, 0.90, 0.95$ and 0.99 and confidence $\gamma = 0.90, 0.95,$ and 0.99 are given in Table C4 for selected values of n.

Lower and upper tolerance limits for the normal distribution are of the form $\bar{x} - K's$ and $\bar{x} + K's$, respectively. The factors K' are given in Table C5 for the same P, γ combinations. More extensive tables for both one- and two-sided tolerance intervals may be found in Owen (1962).

Example 5.15 (Case 2.1, Normal Tolerance Intervals)
We calculate both two- and one-sided tolerance intervals using the data of Table 2.1a. Suppose we wish to determine a 95% two-sided tolerance interval with coverage $P = .90$. With $n = 85$, the factor is found from Table C4 to be $K = 1.897$. The tolerance interval is $\bar{x} \pm Ks = 278.92 \pm 1.897(41.96)$ or $(199.32, 358.52)$. We are 95% confident that 95% of the bond strengths will be between 199.3 and 358.5 pounds. Suppose we instead want a 95% lower tolerance limit with 99% coverage. From Table C5, we find (by interpolation) $K' = 2.720$, so that the tolerance limit is $278.92 - 2.720(41.96) = 164.79$. We are 95% confident that 99% of the bond strengths are at least 164.79 pounds. ∎

5.5.3   Tolerance Limits for the Exponential Distribution

In many reliability applications, the exponential distribution is used to model item lifetimes. In this case, what is usually of interest is a lower tolerance limit L. This may be calculated as

$$L = \frac{2n\bar{x} \log(1/P)}{\chi^2_{2n,\gamma}} \tag{5.77}$$

where $\chi^2_{2n,\gamma}$ is the γ-fractile of the χ^2-distribution with $2n$ df. We have confidence γ that a proportion P of items will have lifetimes of at least L.

Similarly, a two-sided tolerance interval is calculated as

$$\left(\frac{2n\bar{x} \log(2/(1+P))}{\chi^2_{2n,1-(\alpha/2)}}, \ \frac{2n\bar{x} \log(2/(1-P))}{\chi^2_{2n,(\alpha/2)}} \right) \tag{5.78}$$

Example 5.16 (Case 2.13, Exponential Tolerance Interval)
We calculate a 95% lower tolerance interval with confidence 90% for Clerk 1. We had $\bar{x} = 580.4$, with $n = 31$. From Example 5.12, $\chi^2_{62,0.095} = 81.3807$. The tolerance limit is

$$\frac{62(580.4) \log(1/0.9)}{81.3807} = 46.58. \qquad ∎$$

5.5.4   Nonparametric Tolerance Intervals

Nonparametric or *distribution-free* tolerance intervals are based on the ordered observations, which we denote $X_{(1)}, X_{(2)}, \ldots, X_{(n)}$, with $X_{(i)} \leq X_{(i+1)}$, $i = 1, \ldots, (n-1)$. The $X_{(i)}$ are called *order statistics*. It can be shown that the coverage provided by an interval of the form (X_i, X_{i+j}), with $1 < i < i + j < n$, is at least an amount P with probability a calculated value γ, the result being true for any continuous distribution. The problem is to determine the values of i and j so as to obtain the desired coverage and confidence. The theory on which the choice is based is discussed by Stuart and Ord (1991).

Aids for determining distribution-free tolerance intervals have been tabulated in a number of ways—coverage given n, i, and j; n required for a given coverage and confidence; and so forth. Owen (1962) provides a number of tables of various types. Here we give tables of i or i,j values that provide one- or two-sided intervals for specified coverage P and confidence γ. These are given in Tables C6 and C7 for samples of size 50 or greater. Note that, as one might expect, it is not possible to have high confidence and high coverage unless the sample size is substantial, and not much can be done if the sample size is small.

Example 5.17 (2.1, Nonparametric Tolerance Interval)
We calculate a two-sided 95% distribution-free tolerance interval with coverage $P = 0.90$ analogous to that of Example 5.15. From Table C6, the values satisfying this are $(i,j) = (2,2)$, i.e., we require the second smallest and second largest values. The data are given in Table 2.1a, from which we find $x_{(2)} = 196$ and $x_{(84)} = 356$. We are 95% confident that at least 90% of the bond strengths lie in the interval (196,356). Note that the result is not much different from that obtained under the normality assumption. This is not always the case; the distribution-free intervals are often found to be substantially wider. ■

5.6 HYPOTHESIS TESTING

In the previous sections, we looked at estimation, which involves making statements about the numerical values of population parameters and other characteristics based on sample information. We now give a brief introduction to and discussion of another important aspect of data analysis, called *hypothesis testing,* whereby we formulate various hypotheses about the population parameters or other characteristics and then use the data to select one of these as the "correct" hypothesis. The concepts will be illustrated in the case of the normal distribution and extended to other distributions by use of asymptotic results and the relation between hypothesis testing and confidence interval estimation.

5.6.1 Basic Concepts

The Null and Alternate Hypotheses
In the classical approach to hypothesis testing, two hypotheses are formulated, the *null hypothesis,* denoted H_0, and the *alternate hypothesis,* H_a. By definition, H_0 is the hypothesis that is tested; H_a is the hypothesis that is accepted if H_0 is rejected. The basic idea in testing H_0 is to look at the data obtained and evaluate the likelihood of occurrence of this sample *given that* H_0 is true. If the conclusion is that this is a highly unlikely sample under H_0, H_0 is rejected and H_a is accepted. If not, we say that we "fail to reject H_0" (not that we accept it). The philosophy here is that we can "disprove" H_0 if the evidence against it is strong enough, but if the evidence against it is not strong, this does not "prove" that it is true. For example, a larger sample may lead to rejection of H_0. (In practice, however, we often proceed as though H_0 is true when it is not rejected, or at least that it is a reasonable approximation of the true situation.)

In setting up a hypothesis testing problem, the first step is determining the appropriate null and alternate hypotheses. This must be done in the context of the experimental objectives. In fact, different objectives often lead to different formulations of the hypotheses and seemingly different conclusions based on the same set of data. The process begins with a determination of what is to be demonstrated or "proven," or, in a managerial context, what conclusion will lead to some action being taken. This is made the alternate hypothesis. A rule of thumb is that the burden of proof is put on H_a.

We illustrate these concepts by means of an example. The product to be considered is a lead used in an electronic device, for example, a heart pacemaker. The manufacturer's specification is that the lead has a breaking strength of 20 pounds, and has designed the production process accordingly. In monitoring the quality of the product, samples of leads are randomly selected on a regular basis and tested for breaking strength. The production manager wishes to know if the average strength is different from nominal. The parameter in question is the population mean, "different from nominal" becomes the alternate hypothesis, and we have

$$H_0: \mu = 20 \qquad H_a: \mu \neq 20 \qquad (5.79)$$

Here H_a is called a *two-sided alternative.*

Now consider the point of view of the purchaser of this item, the manufacturer of the pacemaker. The lead is acceptable as long as the breaking strength is *at least* 20 pounds, which is what is to be demonstrated. In setting up hypotheses, "=" is always made a part of H_0. (The reason is a mathematical one, involving the selection of a "best" test .) Thus we actually formulate H_a as *more than 20,* and have

$$H_0: \mu \leq 20 \qquad H_a: \mu > 20 \qquad (5.80)$$

Finally, suppose that it is suspected that the strength of the lead is actually less than 20 pounds. (This may, for example, be the subject of a lawsuit or an investigation by a regulatory agency.) The burden of proof now is on this hypothesis and we have

$$H_0: \mu \geq 20 \qquad H_a: \mu < 20 \qquad (5.81)$$

The alternate hypotheses in (5.80) and (5.81) are called *one-sided alternatives.*

Note: The generic symbol for the hypothesized mean is μ_0. In (5.79)–(5.81), $\mu_0 = 20$. In general, we use the subscript "0" to indicate the null hypothesized value of a parameter (e.g., $\lambda_0, \theta_0, \Theta_0$, etc.).

Test Statistics
In performing tests, we rarely look at the entire sample to determine its likelihood of occurrence. Instead, we calculate *test statistics,* which summarize the sample information about the characteristic in question. These are often based on the "best" point estimator of the characteristic. The requirement is that we can calculate probabilities that assess "likelihood of occurrence." This means that (1) the test statistic must have a known (or determinable) distribution; and (2) neither the statistic nor its distribution depend on any unknown parameters. It is sometimes difficult to find appropriate statistics having these (and other desirable) properties, and, as in estimation, we often resort to asymptotic results based on the normal distribution to obtain approximate tests. Results for the important distributions previously considered will be given below.

Type I and Type II Errors and Error Rates
In hypothesis testing as formulated here, there are two possible decisions: (1) fail to reject H_0, and (2) reject H_0 and accept H_a . It follows that there are two types of errors that can be made: rejecting H_0 when it is true, and failing to reject it when it is not true. These are called *Type I* and *Type II Errors.* We have the following situation:

	State of Nature	
Decision	H_0 True	H_a True
Do Not Reject H_0	OK	Type II Error
Reject H_0 and Accept H_a	Type I Error	OK

"OK" in the two cells means that we have not made an error. Type I and II errors are as indicated. Type I and Type II *error rates* are the probabilities of making the two types of errors. Error rates depend on the sample size n, the true parameter value Θ, and the procedure used. Specifically, the Type I and II error rates, denoted $\alpha(n, \Theta)$ and $\beta(n, \Theta)$, respectively, are

$$\alpha(n, \Theta) = P\{\text{Reject } H_0 | H_0 \text{ true}\} \tag{5.82}$$

and

$$\beta(n, \Theta) = P\{H_0 \text{ not rejected} | H_a \text{ true}\} \tag{5.83}$$

In order to explicitly specify a test procedure, it is necessary to state what is meant by an "unlikely" sample. This is done in terms of the *level of significance* of the test, denoted α and given by

$$\alpha = \max_{\Theta} P\{\text{Reject } H_0 | H_0 \text{ true}\} \tag{5.84}$$

Usually α is stated as a percent, e.g., if $\alpha = 0.05$, we say that we are testing at the 5% level of significance, or simple "testing at the 5% level."

The quantity $1 - \beta(n, \Theta)$ is called the *power* of the test. Note that this is the probability of not making a Type II error. Tests are usually considered optimal if they are *uniformly most powerful*, that is, if, for all possible values of Θ, no other test has higher power. Note, incidentally, that Type I and Type II error rates are inversely related; as one increases, the other decreases. Selection of an appropriate level of significance is based, at least in part, on this relationship. A standard choice of level of significance is $\alpha = 0.05$. If it is more important to protect against Type I errors, the level of significance should be chosen to be a smaller value, e.g., 0.01 or less. In this case, the power of the test against any specific alternative will be relatively small, i.e., the Type II error rate will be relatively large. If a small Type II error rate is desired, testing should be done at a higher level of significance, say at the 10% or 20% level.

The Use of p-*Values in Hypothesis Testing*
Most statistical program packages do not require the specification of a level of significance when setting up a hypothesis testing problem. Rather, they require specification of the alternate hypothesis (upper, lower, or two-sided alternative) and then calculate a "*p*-value" and provide this as a part of the output. This rule for using this *p*-value for testing H_0 at the $100\alpha\%$ level is as follows: *Reject H_0 if $p \le \alpha$. If $p < \alpha$, do not reject H_0.*

5.6.2   Testing Hypotheses About the Mean of a Normal Distribution

Recall that the best estimator of the mean μ of a normal distribution is the sample mean $\overline{X}$. This forms the basis of the test statistic. As in the case of confidence interval estimation, the procedure is based on the standard normal (Z) distribution if σ is known, and on the Student-t distribution if it is not. The hypotheses may be one- or two-sided, analogous to (5.79)–(5.81). The test statistic is

$$z = \frac{\overline{x} - \mu_0}{\sigma / \sqrt{n}} \tag{5.86}$$

if σ is known. If σ is not known, it is replaced by s in the formula and we have

$$t = \frac{\bar{x} - \mu_0}{s/\sqrt{n}} \tag{5.86}$$

The distribution of the z-statistic is the standard normal under H_0. The t-statistic has a Student-t distribution with $(n-1)$ df under H_0. These results are used to test H_0 in the three situations (one- and two-sided alternatives). The test procedures (which use the fractiles of the z and t distributions as previously defined) are as follows:

z-tests: (1) H_0: $\mu = \mu_0$ vs. H_a; $\mu \neq \mu_0$—Reject H_0 if $|z| > z_{1-\alpha/2}$.

(2) H_0: $\mu \leq \mu_0$ vs. H_a; $\mu > \mu_0$—Reject H_0 if $z > z_{1-\alpha}$.

(3) H_0: $\mu \geq \mu_0$ vs. H_a; $\mu < \mu_0$—Reject H_0 if $z < z_\alpha$.

t-tests: Use calculated t in place of calculated z in tests (1) – (3), and use fractiles of the Student-t distribution with $(n-1)$ df in place of the normal fractiles.

Terminology: The test of the two-sided alternative in (1) is called a *two-tailed test*. When the alternate is ">" as is (2), the test is an *upper-tail test*; if the alternate is "<" as in (3), the test is a *lower-tail test*. The latter two are also called *one-tailed tests*, for obvious reasons. This terminology is used as well for parameters other than μ.

Example 5.18 (Case 2.1, t-test)
We continue analysis of the data of Table 2.1a. Suppose that a particular application requires an average bond strength of more than 275 pounds (along with additional requirements, e.g., on σ). In testing for the requirement on the mean, the burden of proof is on showing that the requirement on μ is met, so this is the alternate hypothesis and we test H_0: $\mu \leq 275$ versus H_a: $\mu > 275$. Suppose that we test at the 5% level. Based on our previous information, the appropriate test is a t-test and we have

$$t = \frac{\bar{x} - \mu_0}{s/\sqrt{n}} = \frac{278.92 - 275}{41.96/\sqrt{85}} = 0.86$$

The appropriate tabulated value with 84 df is $t_{.95} = 1.663$. Since the calculated value is less than the tabulated value, we fail to reject H_0 and conclude that there is no evidence, testing at the 5% level, that the product meets the required specification. Note that here we do not conclude that μ exceeds 275, even though the sample mean is nearly four pounds higher. The reason is that a value of $\bar{X} > 279$ is not that unlikely when $\mu = 275$.

The following is a listing of typical computer output for this analysis (in this case from Minitab):

t-Test of the Mean

```
            Test of mu = 275.00 vs mu > 275.00

Variable      N      Mean     StDev    SE Mean      T        P
Strength      85    278.92    41.96      4.55      0.86     0.20
```

Note that the hypotheses are specified and the calculated p-value is given as 0.20. Since 0.20 > 0.05, we do not reject H_0 at the 5% level. The p-value tells us that we can expect to obtain a sample mean of 278.92 or higher 20% of the time when the true mean is $\mu = 275$. ■

5.6.3   Relationship between Hypothesis Testing and Confidence Interval Estimation

As is apparent in a comparison of hypothesis testing and confidence interval estimation, there is a relationship between these two techniques of data analysis. In particular, a confidence interval with confidence coefficient $\gamma = 1 - \alpha$ is the set of null hypothesized values that would not be rejected when testing at level of significance α, with two-sided confidence intervals corresponding to two-tail tests and one-sided confidence intervals corresponding to one-tail tests. This relationship may be used to form tests of hypotheses about a parameter θ based on calculated confidence intervals for θ as follows:

1. To test H_0: $\theta = \theta_0$ *versus* H_a: $\theta \neq \theta_0$ at level of significance α, calculate a two-sided confidence interval for θ with confidence coefficient $\gamma = 1 - \alpha$. If θ_0 is contained in the interval, *do not reject H_0*; if θ_0 is *not* in the interval, *reject H_0*.
2. To test H_0: $\theta \leq \theta_0$ *versus* H_a: $\theta > \theta_0$ at level α, calculate a $(1 - \alpha)$ *lower* confidence limit. If θ_0 exceeds the lower confidence limit (i.e., is contained in the resulting confidence interval), *do not reject H_0*; if not, *reject H_0*.
3. To test H_0: $\theta \geq \theta_0$ *versus* H_a: $\theta < \theta_0$ at level α, calculate a $(1 - \alpha)$ *upper* confidence interval. If θ_0 is contained in the interval, *do not reject H_0*; if not, *reject H_0*.

Note: This provides a method of testing hypotheses about the parameters of all of the distributions for which confidence intervals were given in Section 5.3.2. In many cases, these will be asymptotic tests, in the sense that they are based on the asymptotic normality of the point estimator. Asymptotic approximations can also be used to test hypotheses about functions of parameters (e.g., other population characteristics such as reliability), using the results of Section 5.2.5.

Example 5.19 (Case 2.1, One-Tailed Test)

To test the hypothesis of Example 5.18, we use a lower 95% confidence interval. This was calculated in Example 5.9. The result is $(271.35, \infty)$. Since the null-hypothesized value of 275 is contained in this interval, we do not reject H_0. This conclusion is in agreement with that of the formal hypothesis testing approach, as will always be the case. ∎

5.6.4   Suggested Approach to Testing Hypotheses

The following steps are necessary in setting up and solving hypothesis testing problems with regard to parameters or other characteristics of specified distributions:

1. Determine the context and the engineering/managerial objectives.
2. Determine the appropriate variable to be measured and an appropriate probability distribution. (Some aids for selecting a distribution, if it is not known, are given later in the text.)
3. Express the objectives of the study in terms of the parameter(s) of the specified probability distribution and set up the appropriate null and alternate hypotheses. (Remember that the burden of proof is on H_a and equality is what is actually tested, i.e., must be included in H_0.)
4. Select a level of significance α based on the relative importance of Type I and Type II errors.
5. Determine the appropriate test statistic or corresponding confidence interval procedure (as indicated in the previous sections or as specified in subsequent chapters).

6. Obtain an appropriate sample (either an existing data set or by conducting an experiment as indicated in Chapter 8 and later chapters).

7. Determine the statistical conclusion by one of the following methods:

   (i) Compare the calculated value of the test statistic with the tabulated value as indicated in Section 5.6.2.

   (ii) Calculate a p-value and reject H_0 if $p < \alpha$.

   (iii) Determine whether or not θ_0 lies in the calculated confidence interval and arrive at the conclusion as indicated in Section 5.6.3.

8. Determine the engineering/managerial conclusion corresponding to the statistical conclusion and initiate actions as appropriate.

NOTES

1. Strictly speaking, this is true when sampling from an infinite population or sampling *with replacement* from a finite population. The assumption is valid in the cases considered here except for the hypergeometric distribution.

2. For a more thorough discussion of the methods of estimation discussed here and many other methods, see Mood et al. (1974), Stuart and Ord (1991), or any other standard text on theoretical statistics. Alternative variance bounds can also be found in Stuart and Ord (1991). Additional details on estimation of the parameters of the distributions discussed in this section and many others can be found in Johnson and Kotz (1969, 1970a,b, 1972).

3. Since $\log(f(x))$ is a monotonic transformation, the maximum of $\log(f(x))$ and $f(x)$ occur at the same value of x. $\log L$ is easier to deal with because we are differentiating a sum rather than a product.

EXERCISES

Note: Most of the following exercises involve analysis of data. One of the many computer programs available for this purpose should be used in these analyses wherever possible. See Chapter 20 for a list of some of the available programs.

1. Calculate moment estimates $\hat{\alpha}$ and $\hat{\beta}$ for the parameters of the gamma distribution for the data of Tables 2.1a and 2.1b (each data set separately).

2. Calculate maximum likelihood estimates of the parameters of the gamma distribution for each data set in Tables 2.1a and 2.1b. Estimate the standard deviation of α^* and β^* in each case.

3. Use the results of Exercise 2 to calculate approximate 95% confidence intervals for α and β for each data set.

4. Use the results of Exercise 3 to test the null hypothesis H_0: $\alpha = 1$ versus the alternative H_a: $\alpha \neq 1$ at the 5% level for each data set. State your conclusion. Do the data appear to follow an exponential distribution? Does the scale parameter appear to be the same for each test?

5. Assuming that the data for Tests 1 through 4 given in Table 2.1b are samples from normal distributions, (a) calculate point estimates and 99% confidence intervals for μ for each test, (b) calculate point estimates and 95% confidence intervals for σ^2 for each test, and (c) calculate point estimates and 95% confidence intervals for σ.

6. Calculate 95% lower normal tolerance intervals with coverage $P = 0.90$ for each data set of Case 2.1.

7. Repeat Exercise 6 using nonparametric tolerance intervals and compare the results.

8. Suppose that the results for Test 4 (Table 2.1b) are thought to be lognormally distributed. Calculate point and 95% confidence intervals for the parameters of this distribution.

9. Costs of warranty claims in Case 2.4 are found to have a distribution that is highly skewed. For these data, calculate moment and maximum likelihood estimates assuming (a) a lognormal distribution, and (b) an inverse Gaussian distribution.

10. Repeat Exercise 9 omitting observations that appear to be extreme outliers. Compare the results.

11. For Case 2.6, assume that tensile strength is normally distributed and calculate point and 95% confidence intervals for μ and σ for each of the four fiber lengths. Does there appear to be a relationship between μ and length? Between σ and length?

12. Repeat Exercise 11 assuming a Weibull distribution. Estimate α and β for each length and look at how these may be related to length.

13. For Case 2.7, assume that time between failures is lognormally distributed. Give point and 90% confidence interval estimates of the parameters separately for old, medium-aged, and new trucks. Do the parameters appear to change with age?

14. Each item in a batch of 400 circuit boards is tested. Eighteen of these are found to be defective. (a) Estimate the proportion p of defectives for the production process for these circuit boards. (b) Verify that the normal approximation to the binomial is appropriate. (c) Calculate a 95% confidence interval for p.

15. (Continuation) Suppose that the production process is considered to be in control only if there is evidence that the defective rate is less than 5%. Formulate appropriate null and alternate hypotheses for this problem and test at the 5% level. Is the process in control?

16. For Case 2.13, assume that time between errors is exponentially distributed, with possibly different error rates for each clerk. Estimate λ for each clerk and calculate 95% confidence intervals for each λ. Estimate the MTTF for each clerk and calculate 95% confidence intervals for each μ.

17. (Continuation) If time between errors is exponentially distributed, then number of errors per unit of time follows a Poisson distribution. Count the number of errors per 2,000 entries for each clerk. (For Clerk 1 the results are 4, 3, 4, 4, 2, 3, 4, 3, 4.) Estimate λ for each clerk and calculate 95% confidence intervals. Compare the results with those of the previous exercise.

18. Items are sampled from a production process until four defectives are found. The number samples is 535. Estimate the proportion p of defectives and the asymptotic variance of the estimate.

19. For the bus motor failure data of Case 2.8, for time to first failure, estimate parameters and calculate 90% confidence intervals assuming (a) exponential time to failure, and (b) a Weibull distribution. Does the exponential assumption appear to be justified? Give a reason for your answer.

20. For Case 2.9, combine data over all periods and calculate MLEs of parameters assuming (a) a normal distribution, (b) an exponential distribution, and (c) a Weibull distribution.

21. (Continuation) Prepare plots of the normal, exponential, and Weibull distributions hav-

ing as parameters the values calculated in the previous exercise. Compare the results with a histogram of the data. Do the data appear to be adequately represented by one or more of these distributions? If not, why not?

22. For the data of Case 2.10, estimate λ for each of the 13 aircraft, assuming time between air conditioning failures to be exponentially distributed. Do the λ's for different aircraft appear to differ?

23. (Continuation) Assuming that aircraft do not differ and that time between failures is exponentially distributed, combine data over aircraft and (a) calculate an estimate of the common λ. (b) calculate a 99% confidence interval for the true air conditioner MTBF, (c) calculate a lower tolerance interval with coverage 0.95 and confidence 90%. Interpret the result.

24. Calculate normal tolerance intervals for each of the data sets in Case 2.11. Use $P = 0.90$ and $\gamma = 0.99$.

25. Calculate two-sided nonparametric tolerance intervals with $P = \gamma = 0.95$ for each of the three samples in Case 2.12.

26. For Case 2.14, calculate moment and maximum likelihood estimates of the parameters, assuming Weibull, gamma, and lognormal distributions. Compare the moment and MLE results in each case.

27. For Case 2.15, assume exponential time to failure for each part, estimate each λ, and calculate values beyond which 90% of the part lifetimes will lie with confidence 95%.

28. The data for 20 mm single fibers in Case 2.20 were analyzed in Examples 5.3, 5.7, and 5.13. Repeat these analyses the remaining three fiber lengths, using both the Weibull and inverse Gaussian distributions. Do the parameters of each of these distributions appear to be related to fiber length? If so, how are they related.

29. Repeat Exercise 28 for bundles (tows) of fibers. Is there any apparent relationship between results for single fibers and those for tows?

30. Estimate parameters and give 95% confidence intervals for the mean strength of fibers in resin (Case 2.21) assuming Weibull distributions for each length. Do the shape parameters of the Weibull distributions appear to be the same for each length?

31. For Case 2.22, logarithm of fracture stress is assumed to follow a Weibull distribution for each stress rate. Analyze the data under this assumption.

32. Analyze the data of Case 2.23, assuming that time to detection of a software defect is lognormally distributed.

33. Analyze the data of Case 2.25, assuming that time to failure follows a Weibull distribution and that time to failure is measured as time from last overhaul.

34. Calculate the sample coefficient of variation for each of the data sets in Case 2.6 and calculate estimated standard deviations for each length. Do the coefficients of variation appear to be the same?

35. Repeat Exercise 34 for each of the groups of trucks (i.e., the three ages) in Case 2.7.

36. Repeat Exercise 34 for each part in Case 2.15.

37. Repeat Exercise 34 for each of the 8 samples (4 lengths for individual fibers and 4 for tows) in Case 2.20. Make all comparisons of interest.

38. Repeat Exercise 34 for the four data sets of Case 2.21.

39. Repeat Exercise 34 for the five data sets of Case 2.22.

40. Combine the data sets in Exercise 2.20. Assume that the composite data comprise a sam-

ple from a mixture of two exponential distributions. Estimate the parameters of the mixed distribution and the MTBF for each component.

41. For the GIDEP data of Case 2.26, the only information available is total number of parts, total operational time, number of failures, and manufacturer. Analyze amplifier circuits, assuming an exponential distribution. Do a separate analysis for each line in the table and include in your analysis: (a) a point estimate of the failure rate λ, and (b) a 95% upper confidence interval for λ.

42. Repeat Exercise 41, combining data for each manufacturer and combining all data. Do the failure rates of the different manufacturers appear to be the same?

43. Repeat Exercises 41 and 42 for logic circuits.

44. Could the data of Case 2.26 be analyzed under assumptions other than the exponential (e.g., the Weibull, gamma, or lognormal distributions)? Explain.

PART C

Reliability Modeling and Estimation

CHAPTER 6

Modeling Failures at the Component Level

6.1 INTRODUCTION

As discussed in Chapter 1, a system is a collection of parts and failure of the system is related to failure of its parts. In Chapter 4, we discussed the modeling of part failure based on the black-box characterization, under which a part is characterized in terms of two states—working or failed. Typically, a part starts in its working state and changes to a failed state after a certain time. The time to failure is a random variable and in Chapter 4 we looked at modeling of the time to failure by a failure distribution function.

The failure of a part occurs due to a complex set of interactions between the material properties and other physical properties of the part and the stresses that act on the part. The process through which these interact and lead to a part failure is complex and is different for different types of parts (for example, failure mechanisms that lead to failure of mechanical parts are different from those that lead to failure of electrical parts). In this chapter, we briefly discuss the different mechanisms that lead to failure at the part level. We then develop models for first failure based on the physics of failure.

Subsequent failures of a part depend on the type of rectification actions taken. These in turn depend on whether the part is repairable or not. In the case of a nonrepairable part, a failed part needs to be replaced by a new one. In the case of a repairable part, subsequent failures depend on the type of repair made, since one can often subject a failed part to different types of action to make it operational.

Modeling of the first failure, and of subsequent failures as well, based on the mechanisms of failure, requires the use of many different types of stochastic process formulations. A brief introduction to stochastic formulations that will be used in modeling in this chapter is given in Appendices B.1–B.4.

The outline of the chapter is as follows. We commence in Section 6.2 with a discussion of the various physical mechanisms that can lead to part failure. The emphasis here is on a qualitative explanation of the different failure mechanisms. In Section 6.3, we deal with the case of static reliability and discuss the stress–strength model. Section 6.4 deals with models for dynamic reliability. We look at two classes of models—(1) overstress models and (2) wearout models. Modeling based on the mechanisms of failures leads to models which, in general, are too complex to handle. In Section 6.5, we look at multistate models that are a compromise between the black-box models of Chapter 4 and the failure mechanism models

169

of this chapter. In Section 6.6, we discuss failure modeling of repaired items. Section 6.7 deals with modeling of failures over time. In this context, the failures depend on whether the part is repairable or not and we consider the modeling of both of these cases. Lastly, in Section 6.8 we look at modeling of the environmental effects on item failures.

6.2   MODELING FAILURE MECHANISMS

According to Dasgupta and Pecht (1991)[1], one can divide the mechanisms of failure into two broad categories—(1) overstress mechanisms and (2) wearout mechanisms. In the former case, an item fails only if the stress to which the item is subjected exceeds the strength of the item. If the stress is below the strength, the stress has no permanent effect on the item. In the latter case, however, the stress causes damage that usually accumulates irreversibly. The accumulated damage does not disappear when the stress is removed, although sometimes annealing is possible. The cumulative damage does not cause any performance degradation as long as is it below the endurance limit. Once this limit is reached, the item fails. The effects of stresses are influenced by several factors—geometry of the part, constitutive and damage properties of the materials, manufacturing and operational environment.

Following Dasgupta and Pecht, we list the many different failure mechanisms under each of these two groups.

1. Overstress Failures
   - Brittle fracture
   - Ductile fracture
   - Yield
   - Buckling
   - Large elastic deformation
   - Interfacial deadhesion
2. Wearout Failures
   - Wear
   - Corrosion
   - Dendritic growth
   - Interdiffusion
   - Fatigue crack propagation
   - Diffusion
   - Radiation
   - Fatigue crack initiation
   - Creep

We will briefly discuss each of these later in the chapter.

The stresses that trigger the failure mechanism can be mechanical, electrical, thermal, radiation and/or chemical. This forms the basis for the following alternate categorization:

- Mechanical failures as a result of
   elastic and plastic deformation
   buckling

      brittle and ductile fracture
      fatigue crack initiation and crack growth
      creep and creep rupture
- Electrical failures as a result of
      electrostatic discharge
      dielectric breakdown
      junction breakdown in semiconductor devices
      hot electron injection
      surface and bulk trapping
      surface breakdown
- Thermal failures as a result of
      heating beyond critical temperatures (e.g., melting point)
      thermal expansions and contractions
- Radiation failures as a result of
      radioactive containment
      secondary cosmic rays
- Chemical failures as a result of
      corrosion
      oxidation
      surface dendritic growth

Often an item failure can be the result of interactions among these various types of stresses. For example, a mechanical failure due to a thermal expansion mismatch or a stress-assisted corrosion.

Martin et al. (1983) suggest a different classification for failure mechanisms based on the concept of physical degradation. They suggest the following coarse classification.

Cracking	Mechanical degradation
Deformation	↑
Wear	↕
Corrosion	↓
Fouling	Chemical degradation

There are several subclasses for each of these. For example, the cause–consequence-style diagram shown in Figure 6.1 summarizes the failure mechanisms in the case of corrosion.

A proper understanding of the response of materials to the stresses that may be encountered, as well as the effect of other factors (geometry of the part, raw materials, manufacturing, etc.) is essential for the modeling of part reliability based on the mechanism of failure. In the next two sections, we give brief descriptions of the various failure mechanisms leading to overstress and wearout failures.

6.3   OVERSTRESS FAILURE MECHANISMS

In this section, we give a qualitative characterization of the various mechanisms that may lead to overstress failures.

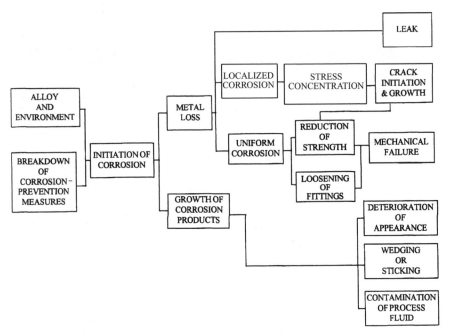

Figure 6.1. Corrosion failure modes. Reprinted from *Reliability Engineering,* Vol. 3, p. 32, Copyright 1983, with permission from Elsevier Science.

6.3.1 Large Elastic Deformation

Elastic deformation typically occurs in slender items. Failure is due to excessive deformation under overstress. Being elastic, the deformations are reversible and therefore do not cause any permanent change in the material. These types of failures occur in structures such as long antennas and solar panels, where large deformations can trigger unstable vibration modes and thereby affect the performance. The overstress is due to external factors and failure occurs when the stress exceeds some safe limit. For further details, see Dasgupta and Hu (1992a) and the references cited therein.

6.3.2 Yield

A component that is stressed past its yield strength results in an irreversible plastic strain and leads to a permanent deformation due to a permanent change in the material. This phenomenon most often occurs in components made of metals that are ductile and are crystalline materials (as opposed to brittle materials, which fracture with no plastic deformation). Plastic deformation is an instantaneous deformation occurring due to a slip motion (called dislocation) of planar crystal defects in response to an applied stress. In contrast, creep (to be discussed later) is a time-dependent deformation. The two main types of dislocations are edge and screw dislocations. For further discussion, see Dasgupta and Hu (1992b).

Permanent deformation may or may not constitute a failure, depending on the context. In the case of a spring, a yield failure occurs when the item fails to operate as a spring after being subjected to an overstress. Another example is a connector losing its clamp pressure.

6.3.3   Buckling

Buckling is a phenomenon that occurs in slender structures under compressive overstress. Deformation in the direction of the compressive load can suddenly change at a critical point, resulting in an instantaneous and catastrophic deformation in a direction perpendicular to the loading direction. The load at which buckling occurs is called the critical load and the new deformation mode is termed postbuckling. Buckling is a structural rather than a material failure mechanism. Postbuckling deformation often involves large deformation and/or finite rotation of the structure. For further details, see Dasgupta and Haslach (1993).

6.3.4   Brittle Fracture

In brittle materials (such as glass and ceramics), high stress concentration can occur at local microscopic flaws under overstress. This excessive stress can cause a failure by sudden catastrophic propagation of the dominant microflaw. The failure is not only related to the applied stress on the component but also depends on the flaw size. Failure resulting from brittle fracture is also referred to as cracking.

Brittle fracture typically occurs at preexisting microscopic flaws as a result of nucleation and sudden propagation of cracks. Cleavage fracture is the most common brittle type of fracture and it occurs by direct separation along crystallographic planes and is due to tensile breaking of molecular bonds. For further details, see Dasgupta and Hu (1992c).

6.3.5   Ductile Fracture

In ductile fracture, the failure is due to sudden propagation of a preexisting crack in the material under external stress. It differs from brittle fracture in the sense that there is large-scale yielding at the tip of the crack, which preceded the crack propagation.

Ductile fracture is dominated by shear deformation, and occurs by nucleation and coalescence of micro voids due to pileups of dislocation at defects such as impurities and grain boundaries. For further details, see Dasgupta and Hu (1992d) and the references cited therein.

6.3.6   Interfacial Deadhesion

Interfacial de-adhesion occurs at the interface between two adhering materials. One of the factors enhancing interfacial deadhesion between two dissimilar materials is interdiffusion. The interfacial strength depends on the chemical and mechanical properties of the interface. Typical examples of such interfacial adhesion are laminated composite materials and bonded joints. Failure due to interfacial deadhesion is delamination in laminated materials or failure of a bonded joint in others.

6.4   WEAR-OUT FAILURE MECHANISMS

In this section we give a qualitative characterization of the different failure mechanisms leading to wear-out failures.

6.4.1   Fatigue Crack Initiation and Growth

When a component is subjected to a cyclic stress, failure usually occurs at stresses much below the ultimate tensile strength and is due to accumulation of damage. Such failures are a result of incremental damage that occurs during each load cycle and accumulates with the number of cycles. Failures of this type are termed fatigue failures.

Fatigue failure is comprised of two stages—crack initiation and crack propagation. A crack typically develops at a point of discontinuity (such as a bolt hole or a defect in the material grain structure) because of local stress concentration. Once initiated, a crack can propagate stably under cyclic stress until it becomes unstable and leads to overstress failure. Fatigue is the leading cause of wear-out failures in engineering hardware. Typical examples are fatigue cracking in airframes (the cyclic loading being the result of landings and take-offs), rotating shafts, reciprocating components, and large structures such as buildings and bridges. For further details, see Dasgupta (1993).

6.4.2 Diffusion and Interdiffusion

Diffusion is a mass transport process at the atomic level and interdiffusion is a mutual mass transfer process that occurs when two species intermingle at an interface. Interdiffusion has been exploited in the fabrication of interfaces in heterogeneous structures (e.g., silicon oxide layers and other thin films in electronic devices).

Failure mechanisms such as corrosion, creep, and electromigration are all driven by diffusion phenomena. In mechanical components, diffusion is not the primary mechanism of failure, rather it plays a role in other types of failures, as mentioned earlier. In electronic devices, loss of control over the interdiffusion process can cause a failure of the device by degrading either electrical, chemical, mechanical or thermomechanical properties of one or more components of the device.

Diffusion occurs due to atomic or molecular motion. In the case of solids, it is the migration of atoms or molecules from one lattice site to another site. Diffusion mechanisms fall into several categories—lattice diffusion, dislocation diffusion, grain boundary diffusion, and surface diffusion. For further details of these, see Li and Dasgupta (1994).

6.4.3 Creep

Creep is a time-dependent deformation. It differs from plastic and elastic deformation in terms of the time scales for the deformations to occur. In creep, the time scales are orders of magnitude larger. The deformation mechanism in polymers is different from those in metals and ceramics because of their special microstructure. In polymers, the deformation is due to a polymer chain reorientation. In metals it is due to grain boundary sliding and intergranular or transgranular void migration. The mechanism is strongly influenced by temperature. At high temperatures, the mobility of atoms increases rapidly so that they can diffuse through the lattice of the material. For further details, see Li and Dasgupta (1993).

6.4.4 Corrosion and Stress Corrosion Cracking

Corrosion is the process of chemical or electrochemical degradation of materials. The three common forms of corrosion are uniform, galvanic, and pitting corrosion.

As the name suggests, in uniform corrosion, the reactions occurring at the metal–electrolyte interface are uniform over the surface of the item. Continuation of the process depends on the nature of the corrosion product and the environment. If the corrosion product is washed off or otherwise removed, fresh metal is exposed for further corrosion.

Galvanic corrosion occurs when two different metals are in contact. In this case one acts as a cathode (where a reduction reaction occurs) and the other as an anode (where corrosion occurs as a result of oxidation).

Pitting corrosion occurs at localized areas and results in the formation of pits. The corrosive conditions inside the pit accelerate the corrosion process.

Stress corrosion cracking is an interaction between the mechanisms of fracture (e.g., resulting from fatigue) and corrosion that occurs because of the simultaneous action of mechanical stress and corrosion phenomena. The corrosion reduces the fracture strength of the material. The process is synergistic—one process assists the other in leading to item failure.

Corrosion is a very pervasive problem in some engineering environments, e.g., offshore rigs operating in a salty environment and chemical plants.

6.4.5   Wear

Wear is the erosion of material resulting from the sliding motion of two surfaces under the action of a contact force. Erosion can be due to physical and chemical interactions between the two surfaces. The various microscopic physical processes, by which the particles are removed as wear debris, are called wear mechanisms and the engineering science dealing with the study of such contacts is called tribology.

Wear mechanisms can be broadly classified into five categories—adhesive, abrasive (when a hard material is sliding against a soft material), surface fatigue, corrosive (chemical), and thermal (mainly in polymers). Wear erosion can be uniform (e.g., wear-away of piston rings in an internal combustion engine) or nonuniform (e.g., pitting in gear teeth and cam surfaces). For further details, see Engel (1993) and the references cited therein.

6.4.6   Other Mechanisms

There are many other failure mechanisms that we have not discussed.[2] We list some these along with references where further details can be found:

Failure due to electromigration [Young and Christou (1994)]
Failure due to radiation [Al-Sheikhly and Christou (1994)]

Temperature has a strong effect on the failure of electronic components. Lall (1996) discusses the effect of temperature on the reliability of microelectronics.

6.5   STATIC RELIABILITY: STRESS–STRENGTH MODELS

Because of manufacturing variability, the strength of a part, X, may vary significantly and unpredictably, i.e., X must be modeled as a random variable. When the part is put into use, it is subjected to a stress, Y. If X is smaller than Y, then the part fails immediately (time to failure is negligible) because its strength is not sufficient to withstand the stress to which it is subjected. If Y is smaller than X, then the strength of the part is sufficient to withstand the stress, and the part is functional. In this section, we focus our attention on models for evaluating the (static) reliability of a part in the context of the stress–strength framework. We first consider the case where Y is a deterministic variable and later the case where it is random.[3]

In real life, both X and Y (whether deterministic or random) are variables assuming values in $[0, \infty)$. However, we will treat them as variables assuming values in the interval $(-\infty, \infty)$ as this is needed for certain cases—for example, when both X and Y are modeled as being normally distributed. When the strengths are modeled as being nonnegative, then the density function is zero for negative values. The models that do not assume this (such as the normal), are used in situations where the parameters are such that the probability assigned to $(-\infty, 0)$ is essentially zero (true for the normal if $\mu \gg \sigma$).

6.5.1   Deterministic Stress and Random Strength

In this case Y is deterministic and X is a random variable with distribution function $F_X(x)$. The reliability R is given by

$$R = P\{X > Y\} \tag{6.1}$$

and it is easily seen that

$$R = 1 - F_X(Y) \tag{6.2}$$

Note that as Y increases, the reliability of the part decreases, as expected.

6.5.2   Random Stress and Strength

In this case Y is a random variable with distribution [density] function $F_Y(y)$ $[f_Y(y)]$ and X is a random variable with distribution [density] function $F_X(x)$ $[f_X(x)]$. We assume that X and Y are independent. The reliability R is given by (6.1). There are two approaches to obtaining R.

  Approach 1: Define $Z = X - Y$. Let $F_Z(z)$ $[f_Z(z)]$ denote the distribution [density] function for Z. Then, since $Z = X + (-Y)$, from (A3.14) of Appendix A.3, we have

$$f_Z(z) = \int_{-\infty}^{\infty} f_X(y) f_Y(z + y) dy \tag{6.3}$$

and R is given by

$$R = P\{Z > 0\} = 1 - F_Z(0) \tag{6.4}$$

  Approach 2: The result may also be obtained by use of a conditional approach. First, conditional on $Y = y$, we have

$$P\{X > Y | Y = y\} = \int_{y}^{\infty} f_X(x) dx \tag{6.5}$$

and, on removing the conditioning,

$$P\{X > Y\} = \int_{-\infty}^{\infty} f_Y(y) \left\{ \int_{y}^{\infty} f_X(x) dx \right\} dy \tag{6.6}$$

which can also be written as

$$P\{Y < X\} = \int_{-\infty}^{\infty} f_X(x) \left\{ \int_{-\infty}^{x} f_Y(y) dy \right\} dx \tag{6.7}$$

Model 6.1 (Stress and Strength Normally Distributed)

Let X be normally distributed with mean μ_X and variance σ_X^2. Similarly, let Y be normally distributed with mean μ_Y and variance σ_Y^2. Then it is well known (and easily proved by use of Approach 1) that Z is also normally distributed with mean $\mu_Z = \mu_X - \mu_Y$ and $\sigma_Z^2 = \sigma_X^2 + \sigma_Y^2$. As a result, from (6.4) we have

$$R = 1 - \Phi_Z(0) = \Phi(\mu_Z/\sigma_Z) \qquad (6.8)$$

where $\Phi_Z(\ )$ is the distribution function for Z and $\Phi(\ )$ is the distribution function for the standard normal variable with mean 0 and variance 1.

Example 6.1
Due to uncertainty in the load, the stress to which a mechanical component is subjected is normally distributed with a mean of 10 and a standard deviation of 3. Due to uncertainty in the manufacturing process, the strength of the component is also normally distributed with mean 25 and standard deviation σ_X. From (6.8), the reliability of the component, for $\sigma_X = 5$, is given by $R = \Phi(15/\sqrt{34}) = 0.9950$.
  The following are values of R for a range of σ_X:

σ_X	3	4	5	6	7	8
R	0.9998	0.9987	0.9950	0.9873	0.9756	0.9604

As can be seen, the reliability decreases as σ_X increases, as is to be expected. ∎

Model 6.2 (Stress and Strength Exponentially Distributed)

Let $f_X(x)$ and $f_Y(x)$ be exponential distributions with respective means $1/\lambda_X$ and $1/\lambda_Y$. Then based on Approach 2, it can be shown that

$$R = \frac{\lambda_Y}{\lambda_X + \lambda_Y} \qquad (6.9)$$

Note that in this case, we have an analytical expression for R.

Example 6.2
The electrical stress that an electronic component is subjected to is exponentially distributed with parameter λ_Y and the component strength is exponentially distributed with parameter λ_X. As can be seen from (6.9), the reliability R is a function of the ratio of λ_X/λ_Y. Reliability as a function of the ratio of the two parameters is as follows:

λ_X/λ_Y	1.0	0.9	0.8	0.7	0.6	0.5	0.4	0.3	0.2	0.1
R	0.5	0.53	0.56	0.59	0.63	0.67	0.71	0.77	0.83	0.91

∎

6.6 MODELING FAILURE BASED ON THE FAILURE MECHANISM

As mentioned in Section 6.2, although the failure mechanisms vary, they essentially fall into two broad categories—failures due to (1) overstress and (2) wear-out. In this section we consider some generic models for first failure modeling appropriate for failures in these two categories.

A rigorous approach to modeling requires a deep understanding of stochastic processes and is beyond the scope of this book. Instead, we use an intuitive approach to highlight the use of different stochastic process formulations in the modeling of failure mechanisms leading to part failures. We omit most of the derivations so as not to digress from the main aim of the chapter, which is to present a few simple models based on failure mechanisms. We give relevant references where interested readers can find the mathematical details of the derivation of the results given as well as related results.

6.6.1   Overstress Failure Modeling

We assume that when an item is put into use, its strength exceeds the stress it encounters, so that it is in its operational mode. At some subsequent point in time, the stress exceeds the strength and this results in item failure. We look at two different model formulations. In the first case (Case A), the stress does not change with time but the strength decreases with time as a result of weakening of the item. This would be appropriate for fatigue type failure. In the second (Case B), the strength does not change with time and the stresses occur as random points along the time axis. This is appropriate where the item experiences shocks (for example, power surge in a network, impact load in the case of a mechanical component) that result in stresses on the item. The severity of the shock determines the stress level. If the stress is less than the strength, there is no damage to the item. If the stress exceeds the strength, then the outcome is an instantaneous failure.

Case A
As before, let Y denote the stress on the item. This may be either deterministic or random. It does not change with time. Let $X(t)$ denote the strength of the item at time t. Note that the assumption that we do not have an instantaneous failure implies that $X(0) > Y$. $X(t)$ is a decreasing function and failure occurs at the first instant in time when $X(t)$ becomes less than Y. This is illustrated in Figure 6.2.

$X(t)$ decreases as a result of material degradation. In general, this occurs in an uncertain manner, i.e., $X(t)$ is a stochastic process. One can model $X(t)$ in many different ways and we discuss two subcases.

Sub-Case (1): $X(t)$ is modeled by a function $h(t; \theta)$, where θ is a random parameter. For a given θ, $h(t; \theta)$ is a deterministic function with $dh(t; \theta)/dt < 0$. One can select many different forms for $h(t; \theta)$.

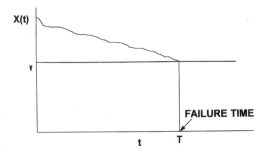

Figure 6.2. Failure due to strength reduction with time.

Model 6.3

Let $h(x; \theta)$ be an exponential function of the form $h(t; \theta) = Ae^{-\theta t}$, with θ a random variable distributed uniformly over the interval $[a, b]$ with $b > a > 0$. Note that we must have $A > Y$. For a given θ, the time to failure, T, is given by

$$T = \frac{\log(A/Y)}{\theta} \tag{6.10}$$

and this is a random variable since θ is random. The distribution of T can be obtained (in principle) from the distribution of θ. When Y is a random variable (assuming values in the interval $[c, d]$ with $0 < c < d < A$), then the distribution of T depends on the distributions of both θ and Y.

Example 6.3

Let the density function for θ be exponential with mean β and Y a deterministic stress. Then the density function of T is given by

$$f(t) = \left(\frac{K}{\beta t^2} \right) e^{-(K/\beta t)}$$

where $K = \log(A/Y)$. Note that in this case $1/T$ is exponential with parameter $\lambda = K/\beta$. ∎

Sub-Case (2): In this case $X(t)$ is modeled by a stochastic process with $X(0) > Y$ and $X(t)$ decreasing with time. The time to failure can be viewed as a level crossing problem for the stochastic process. The modeling of this is very similar to modeling the wear process, which is discussed in the next section. Hence we omit any further discussion except to note that, in general, these types of problems are extremely difficult to solve analytically and hence are of limited use in modeling for our purposes.

Case B

In this case, X does not change with time and it may be either deterministic or a random variable. $Y(t)$ is modeled as a marked point process (see Section 2.4.1 of Appendix B.2). The points correspond to the time instants at which the item is subjected to shocks and the mark associated with each point denotes the stress on the item resulting from that shock. Let t_i denote the time instant for the ith shock and let Y_i denote the mark indicating the corresponding stress on the item. In between shocks, the stress is zero, so that the failure can only occur at the instance of a shock occurrence.

Model 6.4

The point process, $N(t)$, is given by a stationary Poisson process (see Section 2.1 of Appendix B.3) with intensity λ. The resulting shocks are random variables with distribution $G(y)$ and X is deterministic. Let T denote the time to failure and $H(t)$ denote its distribution function. Then, conditional on $N(t) = n$, we have no failure only when each of n shocks cause a stress that is below the item strength X. As a result

$$P\{T > t | N(t) = n\} = \{G(X)\}^n \tag{6.11}$$

On removing the conditioning, we have

$$P\{T > t\} = \sum_{n=0}^{\infty} \{G(X)\}^n P\{N(t) = n\} = \sum_{n=0}^{\infty} [\{G(X)\}^n \frac{e^{-\lambda t}(\lambda t)^n}{n!} = e^{-\lambda t[1 - G(X)]} \tag{6.12}$$

Define $\theta = G(X)$. Note that $0 < \theta < 1$ and since $H(t) = 1 - P\{T > t\}$, we have

$$H(t) = 1 - e^{-\lambda(1-\theta)t} \tag{6.13}$$

In other words, the time to failure is given by an exponential distribution. The mean time to failure is

$$E[T] = \frac{1}{\lambda(1-\theta)}$$

Note that as X increases (i.e., the component becomes stronger), θ increases and as a result, the mean time to failure also increases, as would be expected. Similarly, as λ decreases (shocks occur less frequently), the mean time to failure decreases, as expected.

Example 6.4
A tall TV tower is designed to withstand cyclonic winds of speeds up to 130 mph. The hurricane winds can be modeled as being normally distributed with a mean of 100 mph and a standard deviation of 10 mph. Cyclones occur according to a Poisson process with an intensity function of five per year.

This implies that $G(y)$ is a normal distribution with mean $\mu = 100$ and standard deviation $\sigma = 10$ and $X = 130$. As a result, $\theta = \Phi((130 - 100)/10) = \Phi(3.0) = 0.99865$. The reliability (or survivor function) of the tower is given by

$$H(t) = 1 - e^{-0.00675t}$$

This implies that the mean life of the tower is $(1/0.00675) = 148$ years. The probability that it will not fail in 50 years is given by $(1 - e^{-0.3375}) = 0.2864$. When X is a random variable assuming values in $[0, \infty)$ with a distribution [density function] $F(x)$ $[f(x)]$, then conditional on $X = x$, we have

$$H(t|X = x) = e^{-\lambda(1-G(x))t}$$

and on removing the conditioning, we have

$$H(t) = \int_0^\infty e^{-\lambda(1-G(x))t} f(x)dx$$

Depending on the form of $G(x)$ and $f(x)$, one might or might not be able to derive an analytical expression for $H(t)$. ■

A slightly more general model is one where $\overline{P}_n$ is the probability that the item will survive the first n shocks with $1 = \overline{P}_0 \geq \overline{P}_1 \geq \dots$. [Model 6.3B is a special case, with $\overline{P}_n = \{G(X)\}^n$.] In this case, we have

$$H(t) = \sum_{n=1}^\infty \overline{P}_n \frac{e^{-\lambda t}(\lambda t)^n}{n!} \tag{6.14}$$

and the following proposition:

Proposition 6.1: $H(t)$ given by (6.14) is

(i) IFR [DFR] if $(\overline{P}_n/\overline{P}_{(n-1)})$ is decreasing [increasing] in $n = 1, 2, 3, \ldots$

(ii) IFRA [DFRA] if $(\overline{P}_n)^{1/n}$ is decreasing [increasing] in $n = 1, 2, 3, \ldots$

(iii) NBU [NWU] if $\overline{P}_n \overline{P}_m \geq [\leq] \overline{P}_{n+m}$, $n, m = 0, 1, 2, \ldots$

(iv) NBUE [NWUE] if $\overline{P}_n \sum_{j=0}^{\infty} \overline{P}_j \geq [\leq] \sum_{j=n}^{\infty} \overline{P}_j$, $n = 0, 1, 2, \ldots$

For a proof, see Esary et al. (1973).

6.6.2 Wear-out Failure Modeling

Wear is a phenomenon whereby the effect of damage accumulates with time, ultimately leading to item failure. Typical examples are crack growth in a mechanical part, a tear in a conveyor belt, or bearings wearing out. These can be modeled by a variable $X(t)$, which increases with time in an uncertain manner, and failure occurs when the value of $X(t)$ reaches some threshold level x^*. One can model the changes in $X(t)$ in two ways—(1) occurring at discrete points in time as a result of some external shocks and (2) occurring continuously over time (appropriate for corrosion or fatigue-type failures). In the former case, $X(t)$ is a jump process where the jumps coincide with the occurrence of shocks. In this section, we discuss the modeling of both of these cases.

Case A [X(t) changes at discrete time points]

Here changes in $X(t)$ are modeled by a marked-point process formulation (see Section 2.4.1 of Appendix B.3). The point process models the occurrence of shocks and the attached mark characterizes the resulting damage.

Model 6.5

Let T_1 denote the time to first shock and T_i $(i = 2, 3, \ldots)$ denote the time between shocks. We assume that these are independent and identically distributed random variables with distribution $F(t)$. Let X_i $(i = 1, 2, \ldots)$ denote the damage caused by shock i. These are assumed to be independent and identically distributed random variables with distribution $G(x)$. Let $N(t)$ denote the number of shocks received in $[0, t)$. Then the total damage at time t (assuming that the item has not failed) is given by $X(t) = 0$ if $N(t) = 0$ and by

$$X(t) = \sum_{i=1}^{N(t)} X_i \tag{6.15}$$

for $N(t) = 1, 2, \ldots$. Note that $X(t)$ is a cumulative process (see Section 2.4.2 of Appendix B.3).

Let T denote the time to failure. T is given by

$$T = \min\{t: X(t) > x^*\} \tag{6.16}$$

Let $H(t)$ denote the distribution function of T.

The item will not have failed up to time t if the total damage $X(t) \leq x^*$. In other words,

$$P\{T > t\} = 1 - H(t) = P\{X(t) \leq x^*\} \tag{6.17}$$

We obtain $P\{X(t) \leq x^*\}$ by conditioning on $N(t)$. The result is

$$P\{X(t) \le x^*\} = \sum_{n=0}^{\infty} P\{X(t) \le x^*|N(t) = n\}\, P\{N(t) = n\} \tag{6.18}$$

Since $X(t)$, conditional on $N(t) = n$, is a sum on n independent variables, we have (from Appendix A.3)

$$P\{X(t) \le x^*|N(t) = n\} = G^{(n)}(x^*) \tag{6.19}$$

where $G^{(n)}(x)$ is the n-fold convolution of $G(x)$ with itself. From (B3.12) of Appendix B.3, we have

$$P\{N(t) = n\} = F^{(n)}(t) - F^{(n+1)}(t) \tag{6.20}$$

where $F^{(n)}(t)$ is the n-fold convolution of $F(t)$ with itself. Using (6.18)–(6.20) in (6.17) we have

$$H(t) = 1 - \sum_{n=0}^{\infty} G^{(n)}(x^*)[F^{(n)}(t) - F^{(n+1)}(t)]$$

(with $G^{(0)}(x) = F^{(0)}(x) = 1$ for all x). This can be rewritten as

$$H(t) = \sum_{n=0}^{\infty} F^{(n+1)}(t)[G^{(n)}(x^*) - G^{(n+1)}(x^*)] \tag{6.21}$$

The mean time to failure is given by

$$E[T] = \int_0^{\infty} t\, h(t)dt$$

From (6.21), after some simplification, we have

$$E[T] = E[T_i] \sum_{n=0}^{\infty} G^{(n)}(x^*) \tag{6.22}$$

For the special case where both $F(t)$ and $G(x)$ are exponential, analytical expressions exist for both $H(t)$ and $E[T]$. Let

$$F(t) = 1 - e^{-\lambda t} \qquad \text{and} \qquad G(x) = 1 - e^{-\upsilon x}$$

Then from Cox (1962), we have

$$H(t) = 1 - e^{-\lambda t}[1 + \sqrt{\lambda \upsilon t} \int_0^{x^*} e^{-\upsilon x} x^{-1/2} I_1(2\sqrt{\lambda \upsilon t x})dx] \tag{6.23}$$

where $I_1(W)$ is the Bessel function of order 1 (see Abromowitz and Stegun, 1964) and

$$E[T] = \frac{\lambda x^*}{\upsilon} \tag{6.24}$$

The failure rate associated with $H(t)$ is given by

$$r(t) = \frac{\lambda e^{-\lambda t - vx^*} I_0(2\sqrt{\lambda vt x^*})}{1 + \sqrt{\lambda vt x^*} \int_0^{x^*} e^{-vx} x^{-1/2} I_1(2\sqrt{\lambda vt x^*}) dy]} \tag{6.25}$$

For most other forms of $F(t)$ and $G(x)$, it is not possible to obtain $H(t)$ and $E[T]$ analytically.

This model has been extended in several ways. Nakagawa and Osaki (1974) deal with two other models. In the first, each shock does damage with probability p and no damage with probability $(1 - p)$. The second looks at x^* as a random variable. Hameed and Proschan (1973) consider the case where shocks occur according to a nonstationary Poisson process. (See Appendix B3, Section 2.2.) Shaked (1984) deals with various issues relating to the aging properties of $H(t)$.

Case B [X(t) continuously changing]

In this case, the changes in $X(t)$ are modeled by an appropriate stochastic differential equation of the form

$$dX(t) = \psi_1(X(t))dW(t) + \psi_2(X(t))dt \tag{6.26}$$

where $\psi_1(\ )$ and $\psi_2(\ )$ are functions of $X(t)$ and $W(t)$ is a stochastic process. $X(0) = 0$ and failure occurs when $X(t)$ reaches a level x^*. Two different forms of $W(t)$ have been used in modeling. These are

1. The Weiner process with positive drift (See Appendix B.4)
2. The Gamma process

Model 6.6

Here $W(t)$ in (6.22) is a Weiner process (see Section 1 of Appendix B.4) with positive drift. In this case $W(0) = 0$, $W(t)$ has stationary independent increments and, for $t_2 > t_1$, $(W(t_2) - W(t_1))$ is normally distributed with mean $(t_2 - t_1)$ and variance $(t_2 - t_1)$. Note that in this case, $X(t)$ is no longer monotonically increasing.

The time to failure, T, is the first passage time for the process $X(t)$ (i.e., the first time $X(t)$ reaches the level x^*). Let $H(t)$ denote the distribution function for T. For the special case where $\psi_1(x) = \sigma^2 x$ and $\psi_2(x) = \mu x$ (with $\mu > 0$), $H(t)$ is given by the inverse Gaussian distribution (Equation (4.44)). We omit the proof; it can be found in, for example, Cox and Miller (1965) and Barndorff-Nielsen et al. (1978).

Birnbaum and Saunders (1969) modeled the growth in a crack with each cycle by a normal distribution and assumed that the item fails (as a result of fatigue) after a random number of cycles. They view the number of cycles (a discrete variable) as a continuous variable to obtain the resulting failure distribution, which is similar to the inverse Gaussian. For a comparison of the two distributions, see Bhattacharyya and Fries (1982), where it is shown that the distribution of Birnbaum and Saunders can be viewed as an approximation to the inverse Gaussian distribution.

Model 6.7

Here $W(t)$ is Gamma process. It has the following properties:

1. $W(0) = 0$
2. For $t_2 > t_1$, $(W(t_2) - W(t_1))$, has stationary independent increments and is distributed according to a Gamma distribution (Equation (4.33)) with scale parameter 1 and shape parameter $(t_2 - t_1)$.

As in the previous model, the time to failure, T, is the first passage time for the process $X(t)$. In this case, it is difficult to obtain an analytical expression for the first passage time distribution. For further discussion, see Sobczyk (1987) and Singpurwalla (1995a).[4]

6.6.3   Wear-out Modeling with Sudden Failure

In this section we consider two models where the failure can either occur as a result of wear-out ($X(t)$ reaching x^*) or earlier due to the occurrence of some hazardous event (e.g., external shock). As such, this approach combines the features of the two models discussed in the previous sections.

Model 6.8

The item can fail due to one of two causes. The first cause is a purely random mechanism with the time to failure (T_1) given by an exponential distribution with mean ($1/\lambda_0$). The second cause is wearout with time to failure (T_2) having distribution function given by $P\{T_2 < t\}$. The item failure time is given by $T = \min\{T_1, T_2\}$.

The time to failure due to wear-out is given by the first passage time for a stochastic process $X(t)$, which starts with $X(0) = 0$ and occurs on reaching a value x^*. At time t ($< T$) the amount of wear is given by $X(t)$ ($< x^*$). Conditional on $X(t) = x$, the item can fail in a small interval $[t, t + \delta t)$ with probability $k(x)\delta t$. Lemoine and Wencour (1985) call $k(x)$ the *killing rate* associated with $X(t) = x$. Let $H(t)$ denote the distribution for the time to failure. Then, we have

$$\overline{H}(t) = P\{T > t\} = e^{-\lambda_0 t}E[e^{-\int_0^t k(X(s))ds}] \tag{6.27}$$

If the killing rate $k(x)$ does not depend on x, i.e., is a constant given by λ_1, say, then (6.27) reduces to

$$\overline{H}(t) = P\{T > t\} = e^{-(\lambda_0 + \lambda_1)t} \tag{6.28}$$

Model 6.9

In this model the changes to $X(t)$ occur in a continuous manner as a result of wear and in a discontinuous manner as a result of external shocks. As a consequence, $X(t)$ has jumps at the time instants at which the item experiences shocks. A typical form of $X(t)$ is shown in Figure 6.3. The continuous wear component is modeled by a Weiner process and the jumps by a marked point process. As a result, if $X(t) = x$, then the change in $(X(t+\delta t) - X(t))$ in a small time increment δt is given by (1) a normal distribution with mean $b(t, x)\delta t$ and variance $2a(t, x)\delta t$ and (2) by Y with probability $p(t, x)\delta t$ with y assuming values in $[y, y + \delta y)$ with probability $q(y)\delta t$. Finally, the probability that the item will fail in the interval $[t, t + \delta t)$, given that $X(t) = x$, is given by $c(t, x)\delta x$.

It follows that this model is a generalization of some of the earlier models. The model is

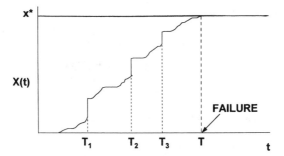

Figure 6.3. Failure due to shock damage.

fairly complex and it is not possible to give an analytical expression for the distribution function for time to failure. For further discussion, see Giglmayr (1987).

6.6.4   Some Other Models

Taylor (1987) deals with the failure of polymer material under static load. In this case, the item is a bundle of several elements and as each element fails, it increases the load on the remaining nonfailed elements of the bundle.

Lloyd and Lipow (1962) develop an interesting model for failure due to corrosion and we discuss this model next.

Model 6.10 [Lloyd and Lipow (1962)]
This model deals with the depth of microscopic pits on the surface of a tube. When the tube is manufactured, there are N pits with the depth of pit i being D_i ($i = 1, 2, \ldots, N$). The thickness of the tube is D. As time progresses, the depth of pits increase due to corrosion and a failure occurs when one of the pit depths equals D. The initial depths are modeled by a truncated exponential distribution so that

$$P(D_i \geq d\} = \frac{e^{-\lambda d} - e^{-\lambda D}}{1 - e^{-\lambda D}} \tag{6.29}$$

for $0 \leq d \leq D$. Let $T_i = k(D - d_i)$ be the time to failure for the ith pit. Then the distribution function for T_i is given by

$$G(t) = P\{T_i \leq t\} = P\left\{d_i \geq D - \frac{t}{k}\right\} = \frac{e^{-\lambda t/k} - 1}{e^{\lambda D} - 1} \tag{6.30}$$

for $0 \leq t \leq kD$.

The time to failure, T, is given by

$$T = \min_i \{T_i\} \tag{6.31}$$

Let $H(t)$ denote the distribution function of T. Then

$$H(t) = 1 - \{1 - G(t)\}^N \tag{6.32}$$

and when N is large, one can approximate this by

$$H(t) \approx 1 - e^{-NG(t)} \tag{6.33}$$

Let $\alpha = N/(e^{\lambda D} - 1)$ and $\gamma = \lambda/k$. Then, using (6.30) in (6.33), we have

$$H(t) \approx 1 - e^{-\lambda(e^{\gamma t} - 1)} \tag{6.34}$$

for $t > 0$. Note that this is a form of the extreme value distribution (Equation (4.47)).

Example 6.5 [Lloyd and Lipow (1962)]

In their example of tubing in a liquid rocket engine, Lloyd and Lipow (1962) provide the following values for a typical application: $D = 0.02$ in, $k = 10^7$ sec/in, $\lambda = 400$/in, and $N = 3000$. Thus $\alpha = 1.00673$ and $\gamma = 4.0 \times 10^{-5}$. The reliability of the tube as a function of t is given by $R(t) = 1 - H(t)$. Suppose we wish to find the value of t for which $R(t)$ is 0.90. From (6.34), we obtain

$$\log(0.90) = -\alpha \, (e^{\gamma t} - 1)$$

Solution for t yields $t = 2488$ seconds. Thus the reliability of this component is 0.90 if the burn time of the engine is 2488 seconds, or 41.5 minutes. A reliability of 0.95 is attained at $t = 1242$ seconds, or about 20.7 minutes. ∎

6.7 MULTISTATE MODELS

In the black-box approach to modeling, an item starts in its working state and changes to a failed state after a random time interval. As such, the model involves two states—working and failed. In the white-box approach to wear-out failure, the state of the item at time t is given by $X(t)$. $X(t)$ is a nondecreasing real variable with $X(0) = 0$ and assumes values in the interval $[0, x^*]$, where x^* is the limiting value of $X(t)$ at failure. This can be viewed as a model with an infinite number of states.

As mentioned in the last section, models based on mechanisms of failure are complex and those based on the black-box approach are too simplistic. A compromise between these two approaches is to model item failures using a formulation with a finite number of states. This is achieved by dividing the interval $[0, x^*)$ into M subintervals. Let x_i $(i = 0, 1, 2, \ldots, M)$ be an increasing sequence in the interval $[0, x^*)$ with $x_0 = 0$ and $x_M = x^*$. Then the state $E_{M-(i-1)}$ $(i = 1, 2, \ldots, M)$ corresponds to $X(t)$ being in the ith interval given by $[x_i - x_{(i-1)}]$. The M states correspond to different levels of item degradation while the item remains in its operational state. The lower the index of the state, the greater the degradation until finally, the item reaches state E_0, corresponding to failure.

When the item is new, it starts in state E_M. Suppose that the item is in state i at time t. Then it moves to either state E_0 (failure) or to state $E_{(i-1)}$ (increased degradation). This captures both wearout failure (which corresponds to the item entering state E_0 through state E_1) and sudden chance failure (which corresponds to the item entering state E_0 from state E_i, $i = 2, 3, \ldots, M$).[5]

Model 6.11

Here the transition between states is modeled by a continuous Markov chain (see Section 2 of Appendix B.2) involving the $(M + 1)$ states $E_0, E_1, \ldots, E_M$. Let $X(t)$ denote the state at time

t. If the state at time t is E_i, then it can only move to two other states: to state $E_{(i-1)}$ with a transition rate λ or to state E_0 with a transition rate μ. The probabilities of moving to these two states are given by

$$P_{i,(i-1)} = \frac{\lambda}{\lambda + \mu} \quad \text{and} \quad P_{i,0} = \frac{\mu}{\lambda + \mu}$$

Note that state E_0 is an absorbing state since once $X(t)$ enters this state, it stays there. The time to failure (T) is the time when $X(t)$ enters state E_0 starting from state E_M at $t = 0$. T is a random variable and the distribution function of T, $H(t)$, can be obtained using results from the theory of continuous time Markov chains (see Bhat, 1972 or Cox and Miller, 1965) for the first passage time to state E_0 starting in state E_M. We omit the derivation and present the final result which is as follows:

$$H(t) = 1 - e^{-\mu t}\left[1 - \frac{\int_0^t \lambda^M x^{(M-1)} e^{-\lambda x} dx}{\Gamma(M)} \right] \tag{6.35}$$

The details of the derivation can be found in Lloyd and Lipow (1962).

For more on reliability modeling with several partial states of operation, see Bendell and Humble (1985).

6.8 MODELING FAILURES OF REPAIRED ITEMS

Thus far we have focused our attention on failures of new items. Often when a new item fails, it can be made operational through repair. The failure distribution of a repaired item is, in general, different from that of a new item, and depends on (1) the age at failure (which we shall denote t_1) and (2) the type of repair. In this section we discuss different types of repair and some related topics. Let $F(t)$ and $G(t)$ denote the failure distributions of new and repaired items, respectively.

6.8.1 Types of Repair

One can define five different types of repair actions:

(i) *Repaired Good as New.* Here, after each repair, the condition of the repaired item is assumed to be as good as that of a new item. In other words, the failure distribution of repaired items is the same as that of a new item, so that

$$G(t) = F(t - t_1) \tag{6.36}$$

for $t > t_1$. In real life, this is seldom the case. A situation in which this may hold approximately is that in which item failure is caused almost exclusively by failure of one key component (all others having substantially lower failure rates), and repair consists of replacement of the failed component by a new one. Examples are the motor of a clock and the power unit of a microwave.

(ii) *Minimal Repair.* When a failed item is subjected to a minimal repair (Barlow and Hunter, 1961) , the failure rate of the item after repair is the same as the failure rate of the item immediately before it failed. In this case

$$G(t) = \frac{F(t - t_1) - F(t_1)}{1 - F(t_1)} \qquad (6.37)$$

for $t > t_1$. This model approximates the situation for items having many components that may fail with comparable failure rates and for which repair consists of replacing only the failed component. Since the remaining components are of the same age as before the repair, the item is essentially returned to its state just prior to failure. Automobiles and other complex items are examples.

(iii) *Repaired Items Different from New (I)*. Here, the failed item is subjected to a major overhaul that results in the failure distribution of the repaired items being $F_1(t)$, say, which is different from the failure distribution, $F(t)$, for new items. Since repaired items are assumed to be inferior to new ones, the mean time to failure for a repaired item is taken to be smaller than that for a new item. In this case,

$$G(t) = F_1(t - t_1) \qquad (6.38)$$

for $t > t_1$.

(iv) *Repaired Items Different from New (II)*. In (iii), the failure distribution for repaired item is different from that of a new item but is independent of the number of times the item has been subjected to repair. In some instances, the failure distribution of a repaired item is a function of the number of times the item has been repaired. This can be modeled by assuming that the failure distribution of an item after the jth repair ($j \geq 1$) is given by $F_j(t)$ with mean μ_j. μ_j is assumed to be a decreasing sequence in j, implying that an item repaired j times is inferior to an item repaired ($j - 1$) times.

(v) *Imperfect Repair*. Minimal repair implies no change in the failure rate, whereas a repair action under (iii) results in a predictable failure rate associated with the distribution function $F_1(t)$. Often, however, the failure rate of a repaired item after repair is uncertain. This is called "imperfect repair" and can be modeled in many different ways. Figure 6.4 shows two different imperfect repair actions: (a) corresponds to the failure rate after repair being lower than that before failure and (b) corresponds to the reverse situation. The sudden change in the failure rate is a random variable in both cases.

Another form of imperfect repair is one where the item becomes operational with probability p after it is subjected to a repair action and continues to be in a failed state with probability $(1 - p)$. In the latter case, the item must be repaired more than once before it becomes operational or perhaps would have to be replaced.

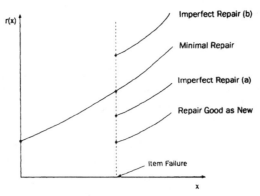

Figure 6.4. Failure rate under various repair options.

6.8.2   Total Repair Time

Repair time consists of several components: investigation time (time needed to locate the fault), the time needed to carry out the actual repair, and testing time after repair. It can also include the waiting times that can result because of lack of spares or because of other failed items awaiting rectification actions. This time is dependent on the inventory of spares and the staffing of the repair facility.

Some of these times can be predicted precisely while others (e.g., time to carry out the actual repair) can be highly variable, depending on the item and the type of failure. The easiest approach is to aggregate all the above-mentioned times into a single repair time $\tilde{X}$ modeled as a random variable with a distribution function $H(x) = P\{\tilde{X} \le x\}$. We assume that $H(x)$ is differentiable and let $h(x) = dH(x)/dx$ denote the density function and $\overline{H}(x) = 1 - H(x)$ the probability that the total repair time will exceed x. Analogous to the concept of the failure rate function, we can define a repair rate function $\rho(x)$ given by

$$\rho(x) = \frac{h(x)}{\overline{H}(x)} \tag{6.39}$$

$\rho(x)\delta x$ is interpreted as the probability that the repair will be completed in $[x, x + \delta x)$, given that it has not been completed in $[0, x)$. In general, $\rho(x)$ would be a decreasing function of x (see Mahon and Bailey, 1975), indicating that the probability of a repair being completed in a short time increases with the duration that the service has been going on. In other words, $\rho(x)$ has a "decreasing repair rate," a concept analogous to that of a decreasing failure rate. Kline (1984) suggests that the lognormal distribution is appropriate for modeling the repair times for many different products.

If the variability in the repair time is small in relation to the mean time for repair, then one can approximate the repair time as being deterministic. If the mean repair time is very small in comparison to the mean time between failures, then we can view service time as being nearly zero. This point of view allows a much simpler characterization of failures over time, as will be demonstrated in the next section.

Example 6.6

Suppose that the time to repair a television set follows a gamma distribution with a mean of 1.2 hours and a shape parameter of 0.75. A plot of $\rho(x)$ is shown in Figure 6.5 and, as can be

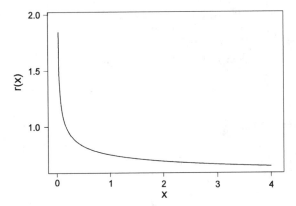

Figure 6.5. Repair rate, gamma distribution with $\alpha = 0.75$, $\beta = 1.60$.

seen, it is a decreasing function. The median (0.5 fractile) of the repair time is 0.73 hours, which means that half the repairs are done in less than this amount of time, and implies that the distribution is highly skewed right (since the median is considerably less than the mean). The 0.95 fractile of the repair time is 3.984. About 5% of the sets require more than 4 hours of repair time. ■

6.8.3 Repair Cost

The cost of repair consists of a variety of cost elements. These include (1) cost of administration, (2) labor cost, (3) material cost, and (4) spare parts inventory costs. Since some of the costs are uncertain (e.g., labor and material costs may be dependent on the type of repair), the total cost of repair (the aggregate of the four items listed) is a random variable and must be modeled by a suitable distribution function. If the variability in the repair cost is small, however, repair cost can be treated approximately as a deterministic quantity.

6.9 MODELING FAILURES OVER TIME

When a repairable item fails, it can either be repaired or replaced by a new item. In the case of a nonrepairable item, the only option is to replace the failed item by a new one. Since failures occur in an uncertain manner, the number of items needed over a time interval is a nonnegative random variable. The distribution of this variable depends on the failure distribution of the item, the actions (repair or replace) taken after each failure, and the type of repair. In this section, we model the number of failures over the interval $[0, t)$, starting with a new item at $t = 0$. We look at several different actions after failure. Let $N(t)$ denote the number of replacements over $[0, t)$. This is a counting process (see Appendix B.3). Let $p_j(t)$ denote the probability that $N(t) = j, j = 1, 2, \ldots$. We begin with the nonrepairable case and then consider repairable items.

6.9.1 Nonrepairable Items

In the case of nonrepairable items, every failure results in the replacement of the failed item by a new item. We assume that all new items are statistically similar and that the failure distribution is given by $F(x)$. Let X_i $(i = 1, 2, \ldots)$ denote the failure time for item i, and Y_i $(i = 1, 2, \ldots)$ denote the time required to replace the ith failed item. We consider two models.

Model 6.12 (Negligible Replacement Time)
In this case, all of the Y_i's are assumed to be zero, so that a failure event coincides with the replacement event. As a result, the time intervals between replacements are identically distributed with distribution $F(t)$ and $N(t)$ is an ordinary renewal process (see Section 2 of Appendix B.3) associated with $F(t)$. Let $p_n(t) = P\{N(t) = n\}, n = 0, 1, 2, \ldots$. Then from (B3.12) we have

$$p_n(t) = F^{(n)}(t) - F^{(n+1)}(t) \tag{6.40}$$

and $M(t)$, the expected number of failures over $[0, t)$, is given by the renewal integral equation (B3.18), i.e.,

$$M(t) = F(t) + \int_0^t M(t - x) dF(x) \tag{6.41}$$

Closed-form expressions exist for a number of distributions—the exponential, Erlang, and normal are examples. In most other cases, it is difficult to obtain an analytical expression for $M(t)$ and computational approaches are used to evaluate it. (See Blischke and Murthy (1994).[6])

Example 6.7

In a chemical plant, the failure of a seal is detected when there is a leak. The time to replace a failed seal is relatively short, so that it can be ignored. The time between failures is given by a Weibull distribution with shape parameter α and scale parameter β. The expected number of failures over a period t is given by the renewal function $M(t)$ obtained from (6.41) with $F(t)$ being a Weibull distribution.

Unfortunately, it is not possible to obtain $M(t)$ analytically. Figure 6.6 shows a plot of $M(t/\beta)$ for a range of α values.

Let $\beta = 1.0$ years. The expected number of replacements over different periods for a range of α values is as shown below.

		α		
t (years)	0.5	1.0	1.5	2.0
1.0	1.3077	1.000	0.8457	0.7538
2.0	2.0478	2.000	1.9456	1.8942
3.0	2.7018	3.000	3.0539	3.0219
4.0	3.3141	4.000	4.1616	4.1503
5.0	3.9010	5.000	5.2693	5.2786

These are obtained from the tables given in Blischke and Murthy (1994). ■

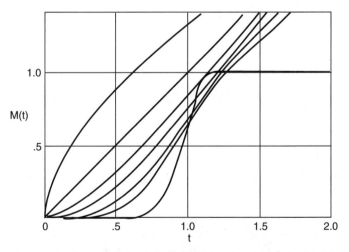

Figure 6.6. Renewal function $M(t)$ for the Weibull distribution. ($\alpha = 0.5$, 1.0, 1.5, 2.0, 3.0, 4.0, 10.0; top to bottom curves at $t = 0.5$.).

Example 6.8

If in Example 6.7, the time between failures is an Erlangian distribution (with k stages and a scale parameter λ) given by

$$F(t) = 1 - e^{-\lambda t}\left\{\sum_{j=0}^{k-1} \frac{(\lambda t)^j}{j!}\right\}$$

then, the renewal function $M(t)$ (see Barlow and Proschan, 1965) is given by

$$M(t) = \frac{\lambda t}{k} + \frac{1}{k}\sum_{j=1}^{k-1} \frac{\theta^j}{1-\theta^j}\{1 - e^{-\lambda t(1-\theta^j)}\}$$

where $\theta = \exp(2\pi i/k)$ with $i = \sqrt{-1}$.
    For the two stage case ($k = 2$), $M(t)$ is given by

$$M(t) = 0.25(2\lambda t - 1 + e^{-2\lambda t})$$

Values of $M(t)$ for $\lambda = 0.5$ are as follows:

t	0.0	1.0	2.0	3.0	4.0	5.0
$M(t)$	0.0000	0.2005	0.7546	1.2506	1.7501	2.2500

∎

Model 6.13 (Nonnegligible Replacement Time)
We assume that the replacement times $\{Y_i\}$ are nonnegligible and are independent random variables with distribution $G(x)$. In this case, the time instants when replacements are completed are renewal points. The time between renewal $(i-1)$ and i (with the item initially put into service (at $t = 0$) being labeled renewal zero) is given by $Z_i = X_i + Y_i$, $(i = 1, 2, \ldots)$. Since X_i and Y_i are independent random variables, $h(x)$, the density function of Z_i, is given by (A3.14), i.e.,

$$h(x) = \int_0^x f(y)g(x-y)dy \qquad (6.42)$$

and the number of renewals $N_r(t)$ in $[0, t)$ is the renewal process associated with the density function $h(x)$. The number of failures $N(t)$ in $[0, t)$ can be either $N_r(t)$ or $N_r(t) + 1$, depending on whether the item is working at time t or in the process of being replaced.

6.9.2 Repairable Items

In this case, the characterization of the number of failures over time depends on the type of repair. We consider the case where the repair times are small so that they can be treated as being negligible. We discuss two models.

Model 6.14 (Minimal Repair)
In this case, the failure rate of an item after a repair is the same as that just before failure. We assume that failures are statistically independent. As a result, $N(t)$ is a nonstationary Poisson process (see Section 2.2 of Appendix B.3) with intensity function $\lambda(t) = r(t)$, the failure rate associated with the failure distribution $F(t)$. (See Nakagawa and Kowada, 1983 and Murthy, 1991a for a proof of this.) From (B3.4) and (B3.6), we have

$$p_n(t) = \frac{e^{-\Lambda(t)}\{\Lambda(t)\}^n}{n!} \qquad (6.43)$$

where

$$\Lambda(t) = \int_0^t r(x)dx \tag{6.44}$$

and

$$E[N(t)] = \Lambda(t) \tag{6.45}$$

Example 6.9
The failure distribution of an item is given by a Weibull distribution with shape parameter α and scale parameter β. Whenever a failure occurs, it is repaired minimally and the time to repair is negligible so that it can be ignored. From (6.45), the expected number of repairs over a period $[0, T)$ is given by

$$E[N(T)] = \int_0^T \frac{\alpha}{\beta}\left(\frac{x}{\beta}\right)^{(\alpha-1)} dx = \left(\frac{T}{\beta}\right)^\alpha$$

$E[N(T)]$ for different combinations of (T/β) and α are as follows:

T/β	$\alpha = 0.5$	$\alpha = 1.0$	$\alpha = 1.5$	$\alpha = 2.0$
1.0	1.000	1.000	1.000	1.000
2.0	1.414	2.000	2.828	4.000
3.0	1.732	3.000	5.196	9.000
4.0	2.000	4.000	8.000	16.000
5.0	2.236	5.000	11.180	25.000

∎

Model 6.15 (Repaired Items Different from New)
We consider the case where the failure distribution of repaired items is given by $G(x)$, which is different from $F(x)$, the failure distribution of a new item. This implies that $N(t)$ is a delayed renewal process (see Section 2.3.2 of Appendix B.3). From (B3.13), we obtain the expected value of $N(t)$ as

$$E[N(t)] = F(t) + \int_0^t M_g(t-x)f(x)dx \tag{6.46}$$

where $M_g(t)$ is the renewal function associated with $G(t)$, given by

$$M_g(t) = G(t) + \int_0^t M_g(t-x)g(x)dx \tag{6.47}$$

Example 6.10
Let the distribution function for the time to failure for a new [repaired] item, $F(t)$ [$G(t)$], be given by an exponential distribution with parameter $\lambda_1[\lambda_2]$ with $\lambda_2 > \lambda_1$. From (6.47), we have $M_g(t) = \lambda_2 t$ and using this in (6.46) we have, after some simplification,

$$E[N(t)] = \lambda_2 t - \frac{\lambda_2 - \lambda_1}{\lambda_1}[1 - e^{-\lambda_1 t}]$$

Suppose that $\lambda_1 = 0.4$ (implying a MTTF of 2.5 years) and $\lambda_2 = 0.5$ (implying a MTTF of 2.0 years), then the expected number of failures over one year is given by $E[N(1)] = 0.5 - (0.1/0.4)(1 - e^{-0.4}) = 0.4176.$ ∎

6.10   MODELING ENVIRONMENTAL EFFECTS

The time to failure (and the failure distribution) are affected by the level of stress on the item. As mentioned earlier, stress is defined in the broad sense—it could be temperature, voltage, force, and so forth, depending on the item. The effect of increasing stress is to accelerate the time to failure. This aspect has been used in accelerated testing for reliability assessment and will be discussed further in Chapter 14. In this section, we focus our attention on modeling the effect of stress on failure time and on the failure distribution.[7]

6.10.1   Arrhenius Life–Temperature Relationship

According to the Arrhenius rate law, the rate at which chemical reactions occur is a function of the absolute temperature T and is given by

$$\text{rate} = Ae^{-(b/T)} \tag{6.48}$$

where A is a constant that is characteristic of the item failure mechanism and $b = Ek$, where E is the activation energy of the reaction and k is Boltzman's constant. As a result, as T increases, the rate of reaction increases and hence hastens the time to failure.

Model 6.16
Let X_1 and X_2 denote the times to failure when the temperatures are T_1 and T_2, respectively. As a first approximation, the time to failure is taken to be inversely related to the rate of reaction, that is $X_1 \propto e^{(b/T_1)}$ and $X_2 \propto e^{(b/T_2)}$, so that we have the ratio

$$\frac{X_1}{X_2} = e^{b\{(1/T_1)-(1/T_2)\}} \tag{6.49}$$

This ratio is less than 1 if $T_1 > T_2$ and greater than 1 if the reverse is true.

Example 6.11
The life of an electronic component is 4000 hours when operating under an operating temperature of 100°C. Suppose the equipment is used under an operating temperature of 150°C. Due to the higher stress, the life of the component under this new operating environment is $X < 4000$ hours.

Suppose that the activation energy is 0.191 eV. The Boltzman constant is 8.623×10^{-5}. Then from (6.49) we have

$$X = 4000 \exp\left(\frac{0.191}{8.2623} \times 10^{-5}\left[\frac{1}{273 + 150} - \frac{1}{273 + 100}\right]\right) = 4000 \exp(-0.7019) = 1982.5 \quad \blacksquare$$

In (6.49), X_2 is a linear function of X_1 since it is based on an approximation. In general, X_2 is a nonlinear function of X_1. This implies that the failure distribution of X_1, $F_1(x)$, will be different from $F_2(x)$, the failure distribution of X_2. From an analytical point of view, it is desirable to have the two belonging to the same class of distribution functions. Various models have been proposed where $F_1(x)$ and $F_2(x)$ satisfy this condition. We discuss a few of these.

Model 6.17 (Arrhenius Exponential Model)
Here $F(x)$ is an exponential distribution with mean time to failure, $\theta(T)$, given by

$$\log\{\theta(T)\} = \gamma_0 + \frac{\gamma_1}{T} \tag{6.50}$$

implying that $\theta(T)$ decreases as T increases. In this case, both $F_1(x)$ and $F_2(x)$ are exponential distributions, with mean time to failure $\theta_1 = \theta(T_1)$ and $\theta_2 = \theta(T_2)$, respectively.

Model 6.18 (Arrhenius Weibull Model)
Here $F(x)$ is a Weibull distribution [Equation (4.44)] with shape parameter α (which does not change with T) and scale parameter $\beta(T)$. The relationship that links $\beta(T)$ with T (measured as absolute temperature) is given by

$$\log\{\beta(T)\} = \gamma_0 + \frac{\gamma_1}{T} \tag{6.51}$$

As a result, $F_1(x)$ and $F_2(x)$ are Weibull distributions with the same shape parameter but with scale parameters given by $\beta_1 = \beta(T_1)$ and $\beta_2 = \beta(T_2)$, respectively.

Example 6.12
The failure distribution of an electronic component is given by a Weibull distribution with shape parameter $\alpha = 1.25$. Suppose that the scale parameter is a function of the operating temperature and is given by (6.51). Extensive testing at $30°$ C indicates that the mean time to failure is 1000 hours at that temperature and at $40°$ C it is 800 hours. The manufacturer is interested in knowing the mean time to failure at $50°$ C and $60°$ C.
   Noting that $30°$ C corresponds to $303°$ $(273 + 30)$ absolute, we have from (4.47),

$$1000 = \Gamma(1.8)\beta(303)$$

This yields $\beta(303) = 1000/0.9314 = 1073.6526$ hours. Similarly, we have $\beta(313) = 800/0.9314 = 858.9221$ hours. Using these in (6.51) we have the following two equations:

$$\log(1073.6526) = 3.03086 = \gamma_0 + \frac{\gamma_1}{303} \quad \text{and} \quad \log(858.9221) = 2.93395 = \gamma_0 + \frac{\gamma_1}{313}$$

Solving these two equations yields $\gamma_1 = 91.9085$ and $\gamma_0 = 2.63758$.
   When the item is used at 50 °C, we have, from (6.51),

$$\log(\beta(323)) = 2.63758 + (91.9085/323) = 2.92210$$

which yields $\beta(323) = 835.795$ and a mean time to failure of $\mu = \Gamma(1.8)\beta(323) = 778$ hours. Similarly, at $60°$ C, the MTTF is 763 hours. ∎

6.10.2   Inverse Power Law

The underlying basis of the relationship called the *inverse power law* is similar to the Arrhenius model, except that here the stress variable, say V, can represent any kind of stress (rather than only temperature). The time to failure is modeled by $X \propto (1/V^\gamma)$, where γ is a constant. Two models based on this type of relationship follow.

Model 6.19 (Power-Exponential Model)
Here $F(x)$ is an exponential distribution, with mean time to failure given by

$$\theta(V) = \frac{e^{\gamma_0}}{V^{\gamma_1}} \tag{6.52}$$

implying that $\theta(V)$ decreases as V increases. In this case, the failure distributions at stresses V_1 and V_2, say $F_1(x)$ and $F_2(x)$, respectively, are both exponential distributions, with respective mean times to failure $\theta_1 = \theta(V_1)$ and $\theta_2 = \theta(V_2)$.

Model 6.20 (Power-Weibull Model)
Here $F(x)$ is a Weibull distribution with shape parameter α (which does not change with V) and scale parameter given by

$$\log\{\beta(V)\} = \frac{e^{\gamma_0}}{V^{\gamma_1}} \tag{6.53}$$

In this case, $F_1(x)$ and $F_2(x)$ are both Weibull distributions with same shape parameter but with scale parameters are given by $\beta_1 = \beta(V_1)$ and $\beta_2 = \beta(V_2)$, respectively.

6.10.3   Other Relationships

Many other relationships have been used to link $E[X]$ (the time to failure) or θ (some parameter of the failure distribution) to the stress level. We list a few of these.

Exponential Relationship:

$$E[X] = e^{(\gamma_0 - \gamma_1 V)} \tag{6.54}$$

Exponential-Power Relationship:

$$E[X] = e^{(\gamma_0 - \gamma_1 V^{\gamma_2})} \tag{6.55}$$

Polynomial Relationship:

$$E[X] = \gamma_0 + \gamma_1 V + \gamma_2 V^2 \tag{6.56}$$

Eyring Relationship (For temperature stress)

$$E[X] = \frac{A e^{(b/T)}}{T} \tag{6.57}$$

where $b = B/k$ is as in the Arrhenius model.
Nelson (1990) discusses these and the earlier relationships in great detail and gives references to their application in modeling many different types of products.

6.10.4   Proportional Hazard Models

Failure times of repairable items are affected by several factors. These can be grouped into the following three categories:

1. operating environment (temperature, pressure, humidity, vibration, dust, etc.)
2. operating history (failure repairs, preventive maintenance)
3. design (selection of material) and manufacturing

To take these factors into account, the failure rate (or hazard rate) of an item can be modeled as

$$\lambda(t; z) = \lambda_0(t)\psi(z; \beta) \tag{6.58}$$

where $\lambda_0(t)$, called the baseline failure rate, is dependent only on time, and $\psi(z; \beta)$ is a functional term which is independent of time but incorporates the effects of the different factors that affect item failure through a row vector z (called the covariates) and a column vector β of regression parameters which characterize the effect of z. Equation (6.58) is the general form of the proportional hazards model.

The covariates influence the failure rate so that the observed failure rate is greater due to factors such as poor maintenance, higher temperatures, etc. (See Figure 6.7.) Various forms of $\psi(z; \beta)$ have been proposed, including

Exponential: $\psi(z; \beta) = e^{z\beta}$

Logistic: $\psi(z; \beta) = \log\{1 + e^{z\beta}\}$

Inverse linear: $\psi(z; \beta) = \dfrac{1}{1 + z\beta}$

Linear: $\psi(z; \beta) = (1 + z\beta)$

For further discussion of proportional hazard Oakes (1983) and Kumar and Klefsjö (1994).

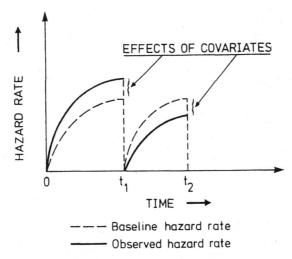

Figure 6.7. Relation between total hazard rate and baseline hazard rate in the presence of influential covariates. Reprinted from *Reliability Engineering and System Safety*, Vol. 44, p. 178, Copyright 1994, with permission from Elsevier Science.

6.11   MODELING FAILURES OF INTERMITTENTLY USED ITEMS

Sometimes an item is used in an intermittent usage mode rather than being used continuously. Murthy (1992) models this using a two-state continuous Markov chain formulation. The two states correspond to the item being in use and idle, respectively. The failure rate depends on the usage history as well as the state (in use or idle). When the periods of usage are relatively small in relation to idle periods, then usage incidents can be treated as points along the time axis. Murthy (1991b) deals with a model formulation for this case. It is assumed that a failure is detected only when an item is put into use. As such, the modeling of failures is done through a point process formulation. An alternate approach is to model such failures by a discrete failure distribution function.

NOTES

1. This section is based on this tutorial paper which is the first in a series of several tutorial papers in the IEEE Transactions on Reliability. Each paper deals with a different mechanism of failure and these are referred to later in the chapter.

2. Our presentation of the physics of failure has been very terse. Material science deals with the study of failures at the microstructural level. One can broadly group materials into the following categories—metals, polymers, and ceramics. There are several books dealing with the failure of these different types of materials. Most of them are qualitative in nature and contain interesting experimental data and real failure case studies. Similarly, the physics of failure for electronic components has received a good deal of attention. Many books, journals and conference proceedings deal with this topic.

3. Kapur and Lamberson (1977) consider many different distributions for both X and Y. In general, the reliability R cannot be expressed analytically and it is necessary to use a computational approach to obtain it. See also Bilikam (1985).

4. Singpurwalla (1995a) provides a taxonomy of the different models of item failure based on wearout. We conclude with the observation that development of models based on the mechanisms of failure is still in its infancy and this is a topic of active research.

5. Misra (1992) deals in detail the analysis of Markov chain models.

6. Renewal theory has been used extensively in the modeling and analysis of unreliable systems. For most distributions, one needs to use computational methods to solve the renewal equation. Blischke and Murthy (1994) provide a FORTRAN program for numerical solution of the renewal equation. The renewal equation has been analyzed extensively. Of particular interest to reliability are the bounds on the renewal function and its asymptotic properties. Some results along these lines can be found in Barlow and Proschan (1965) and Blischke and Murthy (1994).

When the time to replace a failed component cannot be ignored, then the renewal process is not appropriate for modeling failures over time. In this case, one can use an alternate renewal process to model the failures. Let X_1 denote the lifetime of the item put into use at time $t = 0$. Let X_{i+1} and Y_i, $i = 1, 2, \ldots$, denote the failure time of the item used at the ith replacement and the time needed for replacing the item, respectively. Item renewals occur at time instants $(X_i + Y_i)$, $i = 1, 2, \ldots$. If the X_i are independent and identically distributed with distribution function $F(x)$ and the Y_i are independent and identically distributed with distribution function $G(y)$, then the counting process associated with item renewals is an *alternating renewal process*. See Cox (1972) for further details. Note that when the Y_i are all zero, the alternating renewal process reduces to an ordinary renewal process.

7. Often, the deterioration of a component is a function of its age and usage. One approach to modeling this is through a two-dimensional distribution function $F(t, x)$ where

$$F(t, x) = P\{\text{age at failure} < t \text{ and usage at failure} < x\}$$

Johnson and Kotz (1972) discuss several two-dimensional distributions appropriate for the selection of $F(t, x)$. Another approach is to model usage as a function of time, so that the failure rate is a function of age and usage. Usage can be either deterministic or stochastic. In the latter case, one can use Cox's (1972) regression models to model the distribution function for time to failure. Many complex intensity-based models (also known as proportional hazards models) have been developed for modeling item failures. For more on these, see Cox and Isham (1980) and Kumar and Klefsjö (1994).

EXERCISES

1. The distribution of variations in the output voltage from a power pack is found to be approximately an exponential distribution with a mean of 100 volts. The required minimum voltage for the power pack is also exponentially distributed with a mean value of 60 volts. What is the reliability of the power pack? How does this change if the required voltage is single valued at 60 volts?

2. The variations in resistors can be modeled by a uniform distribution over the interval 8 ohms to 12 ohms. The requirements for a specific task are given by a lower limit L and an upper limit U. These are uncertain and can be assumed to have a uniform distributions over the intervals (9, 10) and (11, 12) ohms, respectively. What is the probability that a randomly selected resistor is adequate to the task?

3. The output of diesel engines can be modeled by a Weibull distribution with shape parameter $\alpha = 1.25$ and scale parameter $\beta = 100$ (kW). What is the reliability of an engine selected randomly meeting the requirements for a power output of (a) 80 kW, (b) 90 kW, and (c) 100kW?

4. Reliability testing indicates that the failure rate of a transistor rated at 10 volts is constant over time but depends on the temperature (X_1, in °C) and the ratio of the operating voltage to the rated voltage (X_2) and is given by

$$\lambda(t|x_1, x_2) = e^{-(10.0+0.02x_1+5.5x_2)} \times 10^{-3}$$

The operating temperature and voltage are 50 °C and 12V. Determine the following:
(i) Mean time to failure
(ii) The probability that it will not fail in 1000 hours.
(iii) The reliability of not failing for 1000 hours given that it has not failed for 300 hours.

5. A component has been designed to withstand stresses up to 100 psi. Determine the static reliability of the component if it is subjected to a random load that is exponentially distributed with a mean of 80 psi. How does the reliability change if the load is normally distributed with a mean of 80 psi and a standard deviation of 30 psi?

6. A fixed load of 1000 pounds is applied to a support beam whose strength is uncertain. Determine the probability that the beam will collapse if the strength is

(a) Exponentially distributed with a mean of 7500 lbs
(b) Weibull distributed with scale parameter 7000 lbs and shape parameter 0.80
(c) Lognormal with mean 7500 lbs and shape parameter $\sigma = 10.0$

7. A dam has been designed to withstand a flood level of 30 feet. Floods occur according to a Poisson process with a mean rate of 1 every 2 years. The height of water after a flood is a random variable distributed according to an exponential distribution with a mean of 5 feet. Determine the reliability of the dam for (a) a 15 year period and (b) a 30 year period.

8. Determine the reliability of a component with lognormal strength subject to a lognormal load. Assume that the parameters of the two distributions are the same.

9. The failure rate for ball bearings (according to Ploe and Skewis, 1990) is given by

$$\lambda = \lambda_b \left(\frac{L_a}{L_s}\right)^y \left(\frac{A_e}{0.006}\right)^{2.36} \left(\frac{\nu_0}{\nu_1}\right)^{0.54} \left(\frac{C_1}{60}\right)^{0.67} \left(\frac{M_b}{M_f}\right) C_W$$

where λ_b = base failure rate of a bearing per million hours of operation
L_a = actual radial load in pounds
L_s = specification radial load in pounds
$y = 3.33$ for roller bearings and 3.0 for ball bearings
A_e = alignment error in radians
ν_0 = specification lubricant viscosity
ν_1 = operating lubricant viscosity
C_1 = actual contamination level ($\mu g/m^3$)
M_b = material factor of base material in psi (yield strength)
M_f = material factor of operating material in psi (yield strength)
C_W = water contamination factor (leakage of water into oil lubricant)

$$= \begin{cases} 1 + 460x & \text{for } x < 0.002 \\ 2.036 + 1.0294x - 0.0647x^2 & \text{for } x \geq 0.002 \end{cases}$$

where x is the percentage of water present in the oil

Calculate the mean time to failure for a set of nominal values for the various parameters and study the effect of small changes to each of the parameters.

10. A small AC motor has a failure distribution that can be modeled by a Weibull distribution with shape parameter 1.5. The scale parameter (in hours) depends on the voltage applied (x in volts) and can be modeled as follows:

$$\text{scale parameter} = e^{2.5 - 0.1x}$$

Determine the reliability of the motor for a mission period of 2000 hours when the applied voltage is 100 volts. How does this change when the voltage increases to 110 volts?

11. Suppose that the strength of a component and the stress applied are both random variables distributed according to a lognormal distribution with parameters (m_X, s_X^2) and (m_Y, s_Y^2), respectively. Show that the static reliability R is given by

$$R = \Phi\left(\frac{\ln(m_X/m_Y)}{\sqrt{s_X^2 + s_Y^2}}\right)$$

12. Derive the expression for $E[N(t)]$ in Example 6.10.

CHAPTER 7

Modeling and Analysis of Multi-Component Systems

7.1 INTRODUCTION

The illustrative examples of Section 1.2 reveal that even very simple products can be decomposed into several parts. For complex systems, the number of parts may be very much larger, and in Section 1.2 a multilevel approach to decomposition of such systems was suggested. The number of levels that is appropriate depends on the system (or product) under consideration. The performance of the system depends on the state of the system (working, failed, or in one of several partially failed states) and this in turn depends on the state (working/failed) of the various components. By a component we mean a unit at one of the lower levels—for example, a part, assembly, or subsystem. The focus of this chapter is on the modeling of failures in multicomponent systems.

One can model system failures in several different ways. We first discuss the black-box approach, where the state of the system is described either in terms of two states (working/failed) or more than two states (allowing for one or more partially failed states) without explicitly linking them to the components of the system. Following this, we discuss the white-box approach, where we model the state of the system specifically in terms of the states of the various components of the system. We confine our attention to the case where each component is characterized by two states (working/failed), as discussed in Chapter 4. One can model the state of each component by more than two states or in terms of the mechanism of failure (as discussed in Chapter 6 for failures at the part level). However, these lead to models of sufficient complexity that they are of limited use in the study of failures at the system level.

The outline of the chapter is as follows. In Section 7.2, we commence with a brief discussion of the characterization of a system in terms of its components. This is the starting point for modeling and analysis of system failures. Section 7.3 deals with black-box modeling at the system level. In Section 7.4, we look at white-box modeling and discuss both the qualitative and quantitative approaches. The latter focuses on cause–effect relationships. These are discussed in more detail in Section 7.5, which deals with failure mode, effects, and criticality analysis, and in Section 7.6, which deals with fault tree analysis. Quantitative reliability analysis of systems modeled using reliability block diagrams is discussed in Section 7.7. In Section 7.8 we discuss the reliability importance of a component in a multicomponent system. Section 7.9 of deals with Markovian model formulations. These are appropriate for the case in which all components have constant failure rates. Sections 7.10 and 7.11 deal with model formulations for modeling dependent failures and failure interactions between components.

Some of the models in Sections 7.3 and 7.7 involve complex stochastic processes. Appendix B gives a brief introduction to these processes and presents various results without giving the details of the derivation. The details can be found in the references cited.

7.2 SYSTEM CHARACTERIZATION AND FAILURES

In this section we discuss several different system structures that will be studied later in the chapter and methods for characterization of system failures.

7.2.1 System Characterization

The structure of a system can be represented schematically using special symbols or icons to represent the different components and connecting arcs that show the linkings between the components. The following two examples illustrate the concept.

Example 1.2 (Continued)
In Example 1.2, a pneumatic pump was discussed. A schematic representation of the pump is shown in Figure 1.3, where special symbols (or icons) are used to indicate the different components—e.g., the electromagnet, bellows, valves, and so forth. ■

Example 1.8 (Continued)
A schematic representation of an electrical power system is shown in Figure 1.5(a) and discussed in Example 1.8. Here each power plant (demand center) is treated as a component of the system and indicated by a square (circular) node. The transmission lines are the connecting arcs. This type of characterization is adequate for modeling power flow in the system. The output from each power plant depends on its state. In the simplest characterization, a power plant generates maximum rated output if it is working or zero if it is in a failed state. A more complex characterization involves several intermediate states for each power plant—each state corresponding to a different level of output.

A more detailed characterization of a power plant is shown in Figure 1.5(b). This degree of detail is not needed for power flow studies in the network, but it is needed for the study of power plant failures in the context of production of maximum output. This degree of detail allows one to relate output to failure of different elements of the power plant. For other purposes (e.g., efficiency of a boiler), one needs a still more detailed characterization. ■

Block Diagram Representation
In a block diagram, each component is represented by a block with two end points, as shown in Figure 7.1(a). When the component is in its working state, there is a connection between the two end points and this connection is broken when the component is in a failed state. A multicomponent system can be represented as a network of such blocks with two end points. The system is in working state if there is a connected path between the two end points. If no such path exists, then the system is in a failed state. Two well-known network structures are (1) series and (2) parallel configurations and these are shown in Figures 7.1(b) and 7.1(c), respectively, for an n-component system. In a series structure, the system is functioning only when all of its components are functioning. In contrast, in a parallel

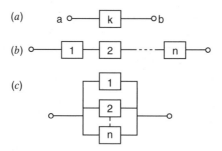

Figure 7.1. Examples of block diagrams.

structure, the system is not working (i.e., is in a failed state) only when all of the components are not working.

In many instances, the system structure is a combination of series and parallel substructures as indicated in the following example.

Example 7.1 (Stereo System)
A schematic representation of a typical stereo system is shown in Figure 7.2. Note that the system can be viewed as a series structure consisting of three subsystems. Subsystems 1 and 3 have parallel structures, whereas Subsystem 2 consists of a single component. We define system failure as inability of the stereo to reproduce sound from a nondefective source such as a CD or cassette or a clear radio signal.

Note that the performance of the system is degraded when one of the speakers fails. There is no sound output when both speakers (Subsystem 3) fail. Similarly, when the amplifier (Subsystem 2) fails, there is no sound output. Finally, Subsystem 1 fails only when all three components fail. ■

It is worth noting that a block diagram representation is not a physical layout; rather, it is a logical relationship between components that indicates the link between system performance and component performance.

Modular and Multilevel Structures. A typical multi-level structure is shown in Figure 7.3, where there are several modules at each level.[1] Here one can number the levels according to a numbering scheme, where Level 1 corresponds to the system level, Level 2 corresponding to the subsystem level, and so on, with Level K corresponding to the lowest level. Failure of a module at level k, $1 < k \leq K$ can affect the performance of the system through its impact on modules at other levels.

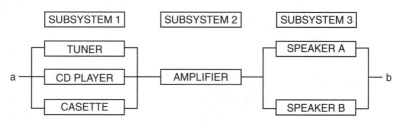

Figure 7.2. Block diagram of a stereo system.

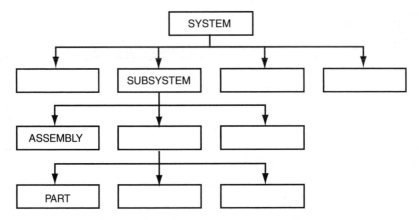

Figure 7.3. Block diagram of a multilevel system.

Example 7.2 (Computer)

The CPU of a personal computer has several printed circuit boards and each board in turn has many discrete components. The PC can be viewed as a system with three levels. Level 1 corresponds to the system level, Level 2 corresponds to the printed circuit board level, and Level 3 corresponds to the component level. Note that there can be duplication at component and/or board levels. ■

The modules at each level can represent different units at that level. If modules are replicated, it is assumed that the replicates are identical. Replication is done to improve system reliability. This will be discussed further in Chapter 15. Another reason for the use of a modular structure is that it facilitates maintenance of the system. Rectification of a failed module requires its replacement by a working module and this can be carried out without the need of a technician.

7.2.2 System Failures

As with modeling at the component level, at the system level it is necessary to differentiate the first failure from subsequent failures, as the latter depend on the type of rectification actions used.

A system failure (either complete or partial) is due to the failure of one or more of its components. Thus the starting point for analysis of the system is analysis of failures at the component level. Henley and Kumamoto (1981) propose the following classification of failures:

1. Primary failure
2. Secondary failure
3. Command fault

A "primary" failure of a component occurs when the component fails due to natural causes (e.g., failing due to natural aging). An action (e.g., repair or replacement by a working unit) is needed to make the component operational. A "secondary" failure is the failure of a component due to excessive stresses resulting from one or more of the following causes: (1)

the (primary) failure of some other component(s) in the system, (2) environmental factors, and/or (3) actions of the user. Murthy and Nguyen (1985) call primary failures "natural failures" and secondary failures "induced failures." Finally, a "command fault" occurs when a component is in a nonworking (rather than a failed) state because of improper control signals or noise (e.g., a faulty action of a logic controller switching off a pump). Often, no corrective action is needed to restore the component to its working state in this case.

In view of the above discussion, the characterization of system failures requires a methodology that incorporates the different modes of component failures and their impact on system failure. The "fault tree" approach Section 7.6 is one such methodology.

Independent vs. Dependent Failures

Often, the failure times of components are influenced by environmental conditions. As indicated in Chapter 6, as the environment gets harsher, the time to failure decreases. Since the components of a system share the same environment, failure times of the components are, in general, statistically dependent. However, if the dependence is weak, one can ignore it and treat the failure times as being statistically independent. The advantage of this is that failure times of the components can be modeled separately using univariate failure distribution functions. If the dependence is significant, however, multivariate failure distributions must be used and the analysis may be much more complicated. This is discussed in Section 7.9.

7.3   BLACK-BOX MODELING

As mentioned earlier, failure of a system is often due to the failure of one or more of its components. At each system failure, the number of failed components that must be restored back to their working state is usually small relative to the total number of components in the system. The system is made operational by either repairing or replacing these failed components. If the time to restore the failed system to its operational state is very small relative to the mean time between failures, then it can effectively be ignored. For practical purposes, this situation is equivalent to minimal repair (see Chapter 6) and system failures can be modeled as follows.

Model 7.1

Here the system failures are modeled by a point process (see Appendix B.3), with intensity function $\Lambda(t; \theta)$, where t represents the age of the system. $\Lambda(t; \theta)$ is an increasing function of t, reflecting the effect of age. In general, it is necessary to specify a form for $\Lambda(t; \theta)$ and estimate its parameters using failure data. One form of $\Lambda(t; \theta)$ is

$$\Lambda(t; \theta) = \alpha\beta(\beta t)^{(\alpha-1)} \tag{7.1}$$

with the parameter set $\theta = \{\alpha, \beta\}$, $\alpha > 1$ and $\beta > 0$ Note that $\Lambda(t; \theta)$ can be viewed as the failure rate of the system at age t. ∎

Systems often undergo a major overhaul (a preventive maintenance action) or a major repair (due to a major failure) that alters the failure rate of the system significantly. The failure rate after such an action is smaller than the failure rate just before failure or overhaul. In the case of an automobile, this would correspond to actions such as the reconditioning of the engine, a new coat of paint, and so on. The time instants at which these actions are carried out can be either deterministic (as in the case of preventive maintenance based on age) or random (as, for example, in the case of a major repair subsequent to an accident). An approach to modeling this situation is as follows.

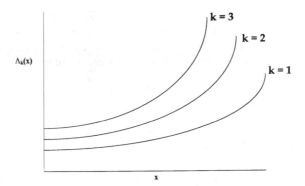

Figure 7.4. System failure rate function after k repairs ($k = 1, 2, 3$).

Model 7.2

Let $t = 0$ correspond to the time when a new system is put into use. Suppose that the system is subjected to $(K - 1)$ overhauls at time instants t_k, $1 \leq k \leq (K - 1)$, and discarded at time t_K. The failure rate of the system after the kth action, $1 \leq k \leq (K - 1)$, is given by $\Lambda_k(t - t_k; \theta)$. The failure rate of the system when it is new is given by $\Lambda_0(t; \theta)$. All failures between t_{k-1} and t_k, $1 \leq k \leq K$ (with $t_0 = 0$), are repaired minimally. The functions $\Lambda_k(x; \theta)$, $0 \leq k \leq (K - 1)$, have the following properties:

(a) For a fixed k, $\Lambda_k(x; \theta)$ is an increasing function in x for $0 \leq k \leq (K - 1)$
(b) For a fixed x, $\Lambda_k(x; \theta) - \Lambda_{k-1}(x; \theta) > 0$ for $1 \leq k \leq (K - 1)$.

(a) implies that the failure rate is always increasing with x (the time lapsed subsequent to an overhaul) and (b) implies that after each overhaul the failure rate decreases and the decrease becomes smaller as k (the number of overhauls) increases. Figure 7.4 shows a typical plot of $\Lambda_k(x; \theta)$ as a function of x and k.

As in Model 7.1, it is necessary to determine the form for $\Lambda_k(x; \theta)$, $0 \leq k \leq (K - 1)$, based on either failure data or some other basis (either theoretical or intuitive) and estimate the parameters using failure data. ■

7.4 WHITE-BOX MODELING

In the white-box approach, system failure is modeled in terms of the failures of the components of the system. We discussed the modeling of component failures in Chapters 4 and 6. The linking of component failures to system failures can be done using two different approaches. The first is called the *forward* (or bottom-up) approach and the second is called the *backward* (top-down) approach.

In the forward approach, one starts with failure events at the part level and then proceeds forward to the system level to evaluate the consequences of such failures on system performance. *Failure mode and effects analysis* (FMEA) uses this approach. In the backward approach, one starts at the system level and then proceeds downward to the part level to link system performance to failures at the part level. *Fault tree analysis* (FTA) uses this approach.

The linking of the system performance to failures at the part level can be done either qualitatively or quantitatively. In the former case, the focus is on the causal relations that link part level failures to system performance (or failure). In the latter case, one obtains various measures of system performance (e.g., system reliability) in terms of the component reliabilities.

The next two sections deal with modeling for qualitative analysis and later sections with modeling for quantitative analysis.

7.5   FAILURE MODE, EFFECTS, AND CRITICALITY ANALYSIS

According to IEEE Standard 352 (IEEE Std 352) the objectives of failure mode and effects analysis (FMEA) are as follows:

1. Assist in selecting design alternatives with high reliability and high safety potential during the early design phase.
2. Ensure that all conceivable failure modes and their effects on operational success of the system have been considered.
3. List potential failures and identify the magnitude of their effects.
4. Develop early criteria for test planning and the design of the test and checkout systems.
5. Provide a basis for quantitative reliability and availability analysis.
6. Provide historical documentation for future reference to aid in the analysis of field failures and consideration of design changes.
7. Provide input data for trade-off studies.
8. Provide a basis for establishing corrective action priorities.
9. Assist in the objective evaluation of design requirements related to redundancy, failure detection systems, fail-safe characteristics, and automatic and manual override.

FMEA involves reviewing a system in terms of its subsystems, assemblies, and so on, down to the part level, to identify failure modes and causes and the effects of such failures. According to IEEE Standard 352, the basic questions to be answered by FMEA are the following:

1. How can each part conceivably fail?
2. What mechanisms might produce these modes of failure?
3. What could the effects be if the failures did occur?
4. How is the failure detected?
5. What inherent provisions are provided in the design to compensate for the failure?

For each component at the part level, the failure modes and their effects are usually documented on worksheets. The documentation involves the following:

A) Description of the different parts. This is done through
    a proper reference number
    the intended function of the part
    the normal operational mode

B) Characterization of failure. This involves

listing the different possible failure modes

failure mechanisms responsible for the different failure modes

the various means of detecting the different failure modes

C) Effect of failure on

other components of the system

system performance

If, in addition to FMEA, a criticality analysis is carried out, the process is called a *failure mode, effects, and criticality analysis* (FMECA).[2] In this case, in addition to (A)–(C) of FMEA, the procedure involves documentation of the following:

D) Severity ranking, which characterizes the degree of the consequence of each failure. As mentioned in Section 1.3.3, a severity ranking scheme (MIL-STD 882) is as follows:

catastrophic

critical

marginal

negligible

The different failure modes for many components have been well documented. The following is a small list from AMPC-706-196 (1976).

Component	Failure modes	Occurrence (%)
Bearings	Lubrication loss	45
	Contamination	30
	Brinelling	5
	Corrosion	5
	Other modes	15
Connectors	Shorts	30
	Solder joint (mechanical)	25
	Insulator resistance	20
	Contact resistance	10
	Miscellaneous mechanical	15
Electrolytic capacitor	Open circuit	35
	Short circuit	35
	leakage	10
	Decrease in capacitance	5
	Others	15

FMECA is usually carried out during the design phase. The objective is to reveal weaknesses and potential failures. This enables the design engineer to make appropriate modifications that may reduce the likelihood of failures and/or the seriousness of their consequences.

Example 7.3 (Differential Pressure Transducer)

A differential pressure transducer measures the differential pressure across the measuring diaphragm. It involves a complicated mechanism comprising of beams, balancing springs, nozzle, compressed air supply, and other parts. A schematic representation of the transducer can be found on page 217 (Figure A.3.1) in Green (1983) along with a description of the mechanism of its operation. The major failure modes for some of the components are as follows:

Component	Failure mode	Effect of failure mode (category)
Measuring Diaphragm	Leaking/ruptured	Beam not actuated (D)
Bellows seal	Leaking/ruptured	Signal in measuring cell reduced (D)
Main beam pivot	Broken/loose	Nozzle opens (S)
Balance spring	Broken	Nozzle opens (S)
Nozzle and flapper	Blockage	Full output (D)
	Breakage	Nozzle leaks (S)
Feedback diaphragm	Leaking/ruptured	Output falls (S)

One can classify the impact of failure in terms of a severity code. For this example, Green suggested two categories—dangerous (D) and not dangerous (S). The categories for the various failure modes are as indicated above. ■

Example 1.2 (Continued) (Hydraulic Valve)
The basic function of the hydraulic valve is to control the flow of fluid in a pipe. The different failure modes for the valve and the effect of the failures are as follows:

Failure mode	Effect
Fails closed	Flow can not be stopped
Fails open	No flow
Does not open fully	Flow less than full
Does not respond to controller	Loss of control over flow
Leaks through valve	Loss of fluid, potential hazard
Leaks around valve	Loss of fluid, potential hazard

The severity ranking would depend on the environment and the type of fluid flowing through the pipe.■

The main drawback of FMECA is that it focuses on hardware failures and does not effectively capture failures resulting from human errors or due to complex combinations of events. Fault tree analysis, which we turn to next, is better suited to dealing these additional factors.

7.6 FAULT TREE ANALYSIS

A fault tree is a logic diagram that displays the relationship between a potential event affecting system performance and the reasons or underlying causes for this event. The reason may be failures (primary or secondary) of one or components of the system, environmental conditions, human errors, and other factors. In this section we focus on qualitative fault tree analysis.[3]

The values of a fault tree (Fussel, 1976) are as follows:

1. Directing the analysis to ferret out failures
2. Pointing out the aspects of the system important to the failure of interest

3. Providing a graphical aid by giving visibility to those in systems management who are removed from design changes

4. Providing options for qualitative and quantitative systems reliability analysis

5. Allowing the analyst to concentrate on one particular system failure at a time

6. Providing an insight into system behavior

A fault tree illustrates the state of the system (denoted the TOP event) in terms of the states (working/failed) of the system's components (denoted as basic events). The connections are done using *gates,* where the output from a gate is determined by the inputs to it. A special set of symbols is used for this purpose; these will be discussed later.

A fault tree analysis involves the following steps:

1. Definition of the TOP event

2. Construction of the fault tree

3. Qualitative and, if desired, quantitative analysis of the fault tree

7.6.1   Definition of the TOP Event

It is important that the TOP event be defined in a clear and unambiguous manner. The description of the TOP event should always answer the following questions:

What: Describes the critical event that is the focus of attention (e.g., release of toxic material into the atmosphere from a chemical plant)

Where: Describes where the critical event occurs (e.g., storage tank Number 2)

When: Describes when the critical event occurs (e.g., during normal operation, under fire)

7.6.2   Fault Tree Construction

The fault tree analysis is carried out by diagrams involving these two types of building blocks: *gate* symbols and *event* symbols. These are used in drawing the fault tree diagrams. Figures 7.5 and 7.6 show the different gate symbols and event symbols used in fault tree analysis along with brief descriptions of each. The fundamental structure of a fault tree is shown in Figure 7.7. Construction of a fault tree begins with the TOP event and proceeds downward to link to basic events through the use of different gates. Completion of the fault tree requires specification of the output of each gate, as determined by the input events to the gate. We illustrate the construction of fault trees by means of a few examples.

Example 1.1 (Continued) (Incandescent Electric Bulb)
The components of a typical incandescent electric bulb are shown in Figure 1.1. We define the TOP event as "no light from the bulb" under normal operation. A fault tree diagram for this event is shown in Figure 7.8. Note that the TOP event occurring due to the output event of the top gate is either a primary or secondary failure, whereas failure due to mechanical impact is a secondary failure. The oxidation of the filament and mechanical impact are two modes of failure under normal operational conditions.

Example 1.3 (Continued) (Pneumatic Pump)
The components of a typical pneumatic pump are shown in Figure 1.3. We define the TOP event as "no air flow into the tank" under normal conditions. The fault tree diagram for this

	Gate Symbol	Gate Name	Causal Relation
1		AND gate	Output event occurs if all input events occur simultaneously.
2		OR gate	Output event occurs if any one of the input events occurs.
3	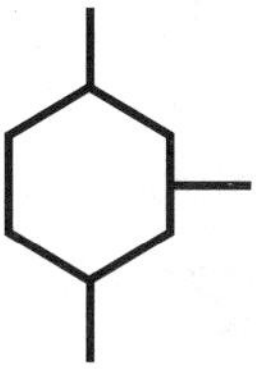	Inhibit gate	Input produces output when conditional event occurs.
4		Priopity AND gate	Output event occurs if all input events occur in the order from left to right.
5		Exclusive OR gate	Output event occurs if one, but not both, of the input events occur.
6	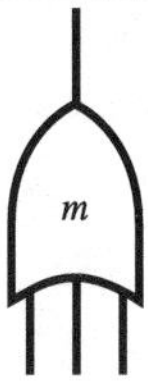	m Out of n gate (voting or sample gate)	Output event occurs if m out of n input events occur.

Figure 7.5. Gate symbols for fault tree analysis.

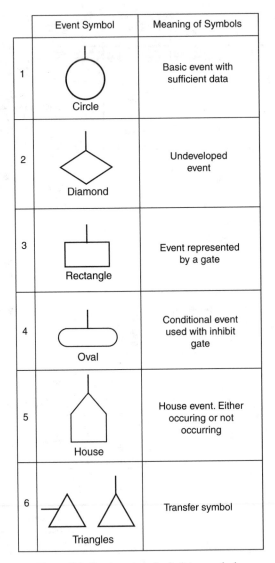

Event Symbol	Meaning of Symbols
1 ⭕ Circle	Basic event with sufficient data
2 ◇ Diamond	Undeveloped event
3 ▭ Rectangle	Event represented by a gate
4 ⬭ Oval	Conditional event used with inhibit gate
5 ⌂ House	House event. Either occuring or not occurring
6 △△ Triangles	Transfer symbol

Figure 7.6. Event symbols for fault tree analysis.

event is shown in Figure 7.9. Note that here there are several modes of failure if we move down to more basic events, such as the reasons for blocked valves or no power to the electromagnet, etc.

Example 7.4 (Pressure Tank System)

A schematic of a pressure tank system is given in Figure 7.10. The operation of the system is as follows. Each cycle consists of two phases—pumping (lasting approximately 20 seconds) and shutdown. Pumping starts with the closing of the reset switch S1 and its immediate re-

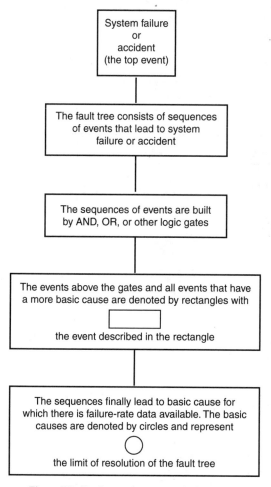

Figure 7.7. Fundamental structure of a fault tree.

opening. This allows for the activation of relays K1 and K2. Pumping action is initiated by the closing of K2. In the shutdown phase, the pressure switch contacts should open (since excess pressure should be detected by the pressure switch). This leads to deenergizing the K2 coil, thereby shutting off the motor. If there is a pressure switch malfunction, the timer relay should open after 60 seconds and deenergize the K1 coil, which in turn should lead to shutting off the pump by deenergizing the K2 coil. The TOP event is rupture of the tank due to the buildup of excessive pressure.

The fault tree for this system was first published by Vesely et al. (1971) and since then it has reappeared in Barlow and Proschan (1975) and Henley and Kumamoto (1981). The fault tree diagram for the system as proposed in Barlow and Proschan is shown in Figure 7.11. The figure given by Henley and Kumamoto is somewhat different from that of Barlow and Proschan. The reason for this is that there is no unique fault tree for a given system. ■

The main drawback of fault trees is that they are tedious and expensive to construct for

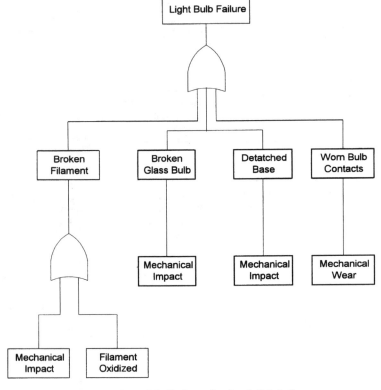

Figure 7.8. Fault tree for electric light bulb.

complex systems. In this context, we are often dealing with situations so complicated that oversights and omissions are all too easy.

7.6.3 Qualitative Analysis

The occurrence of the TOP event (i.e., system failure) is often due to different combinations of basic events (e.g., failures at the part level). A fault tree provides useful information about these different combinations. In this context, the concepts of *cut sets* is very useful.

A cut set in a fault tree is a set of basic events whose (simultaneous) occurrence results in the occurrence of the TOP event. A cut set is said to be minimal if the set cannot be reduced without losing its status as a cut set. The following example of a fault tree, from Barlow and Lambert (1975), will be used to illustrate these concepts:

Example 7.5 (Cut Sets for a Fault Tree)
Consider the fault tree shown in Figure 7.12. The cut sets are obtained by successively replacing each gate with its inputs (events and gates at lower levels) until one reaches the basic events at the lowest level. This is done by proceeding in a systematic manner as follows:

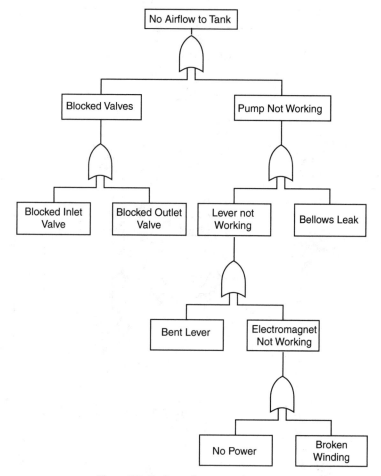

Figure 7.9. Fault tree for pneumatic pump.

Starting at the TOP event (an OR gate), we have cut sets given by

$$\{1\}, \{G1\}, \{2\}$$

If any one of these occurs, the TOP event will occur. Conversely, if all of these fail, the TOP event will not occur. Since G1 is an OR gate, we can replace $\{G1\}$ by the pair $\{G2\}$, $\{G3\}$ and the cut sets become

$$\{1\}, \{G2\}, \{G3\}, \{2\}$$

Since G2 is an AND gate, the output event occurs only when the input events occur. As a result, the cut sets become

$$\{1\}, \{G4, G5\}, \{G3\}, \{2\}$$

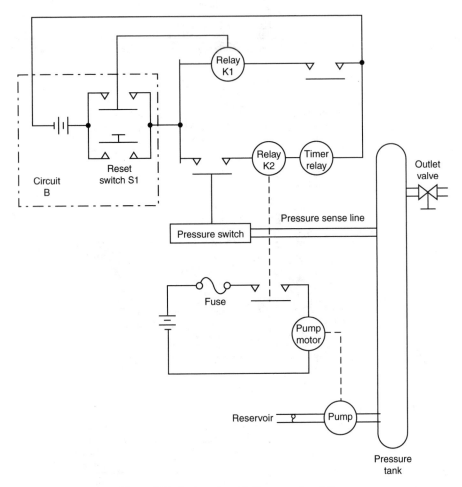

Figure 7.10. System schematic for a pumping station.

and noting that G3 is an OR gate, we have the cut sets

$$\{1\}, \{G4, G5\}, \{3\}, \{G6\}, \{2\}$$

Proceeding further down by noting that G4 is an OR gate, we have the cut sets given by

$$\{1\}, \{4, G5\}, \{5, G5\} \{3\}, \{G6\}, \{2\}$$

Finally, at the lowest level, we have the following nine cut sets:

$$\{1\}, \{2\}, \{3\}, \{6\}, \{8\}, \{4, 6\}, \{4, 7\}, \{5, 6\}, \{5, 7\}$$

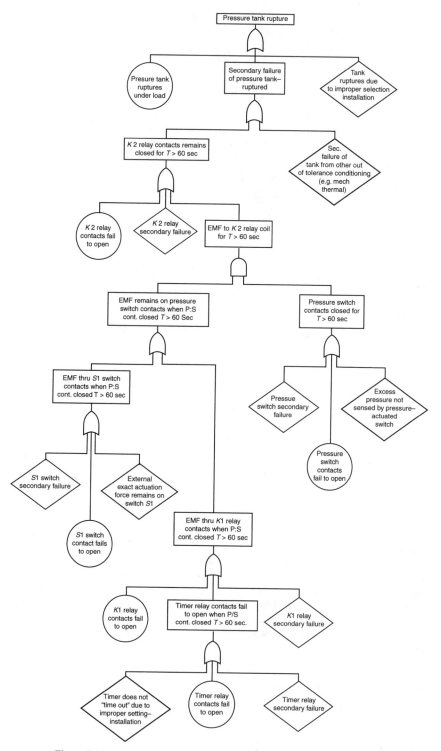

Figure 7.11. Fault tree for pressure tank system (Barlow and Proschan, 1975).

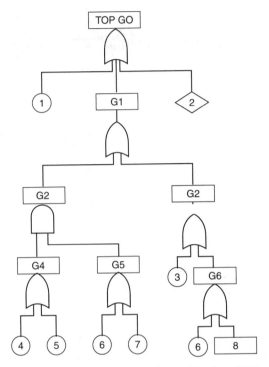

Figure 7.12. Example of fault tree (Barlow and Lambert, 1975).

Since {6} is a cut set, {4, 6} and {5, 6} are not minimal. As a result, by deleting them, we have the following minimal cut sets:

$$\{1\}, \{2\}, \{3\}, \{6\}, \{8\}, \{4, 7\}, \{5, 7\}$$

Note that the TOP event occurs if one of the single events {1}, {2}, {3}, {6}, or {8} occurs, or if one of the joint events {4, 7} or {5, 7} occurs. ■

As can be seen from the above simple example, the number of minimal cut sets and their determination involves considerable analysis. The complexity of this analysis increases as the number of elements in the fault tree diagram increases.

As will be indicated in the next section, fault tree diagrams can be converted into block diagrams and the analysis techniques discussed in the next section can be used for the analysis of fault tree diagrams.

7.6.4 Comparison with FMECA

According to Evans (1997), a fault tree has the following features that a FMECA does not have.

1. Handles multipoint failures (cut sets with more than one element)

2. Allows integration of causes of system *faults*
3. Handles common-cause failures more easily

7.7   RELIABILITY BLOCK DIAGRAM REPRESENTATIONS AND ANALYSIS

In this section we discuss system failure in terms of component failures for a system where the component connections and their interconnections can be represented by a block diagram representation. This type of representation is also referred to as a network representation.

Let n denote the number of components in the system. The way in which the n components are interconnected depends on the system function and can be represented as a network with two end points, (a) and (b). Two special configurations are (1) series and (2) parallel, as shown in Figure 7.1. A more general configuration would involve these two structures as well as other types—e.g., the bridge structure shown in Figure 7.13 (with $n = 5$). The system is in its working state (or fulfilling its function) when there is a connection between the two end points (a) and (b), i.e., when there is a path from (a) to (b) in which all the components are operational. When there is no connection between the two end points, the system is in a failed state.

Example 7.1 (Continued) (Stereo System)
The stereo system as a whole (viewed in terms of its subsystems as in Figure 7.2) has a series structure and Subsystems 1 and 3 have parallel structures. Here there are six components ($n = 6$) and we number them as follows:

Component number	Component description
1	Tuner
2	CD Player
3	Cassette player
4	Amplifier
5	Speaker A
6	Speaker B

First consider the case where system failure is defined as corresponding to no music or other coherent sound output from the stereo. In this case, the network configuration is as shown in Figure 7.2. Note that there is no sound output if there is no connected path between (a) and (b). This implies that as long as one of Components 1 though 3, Component 4, and one of Components 5 and 6 are all functioning, we have output. However, if Component 4 is in a failed state, then there is no connection between (a) and (b) and no output. Similarly if

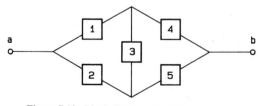

Figure 7.13. Block diagram of a bridge structure.

Components 1 through 3 or Components 5 and 6 are not working, then there is no connection between (a) and (b) and again we have no output.

If we define system failure as corresponding to the system not capable of producing stereophonic sound, then the network configuration changes from that shown in Figure 7.2. The new configuration has Components 5 and 6 connected in series rather than in parallel.

Finally, if one defines system failure corresponding to the owner not being able to listen to music from all three sources (tuner, CD, and cassette players) in stereophonic sound, then for reliability analysis, the block diagram representation of the system is a series configuration with all six components connected in series. ■

7.7.1   Link to Fault Tree Representation

A fault tree representation of a system can be easily converted into a block diagram representation and vice versa. This is illustrated in Figure 7.14, which shows the equivalence between the two for four simple configurations. This implies that a given multicomponent system can be represented either in terms of a fault tree diagram or in terms of a block diagram. The block diagram representation is better suited for certain types of analysis, as will be indicated in the remainder of the section.

7.7.2   Structure Functions

We confine our attention to a binary characterization of the state for each the component. Let $x_i, 1 \leq i \leq n$, denote the state of component i, with

$$x_i = \begin{cases} 1 & \text{if component } i \text{ is in working state} \\ 0 & \text{if component } i \text{ is in a failed state} \end{cases} \tag{7.2}$$

Let $x = (x_1, x_2, \ldots, x_n)$ denote the state of the n components. Note that x can assume one of 2^n values, corresponding to the possible different combinations of the states (working or failed) for the n components.

The state of the system is also characterized by a binary valued function $\phi(x)$ where

$$\phi(x) = \begin{cases} 1 & \text{if the system is in working state} \\ 0 & \text{if the system is in a failed state} \end{cases} \tag{7.3}$$

$\phi(x)$ is called the *structure function of the system.*

Define the operator $\coprod$ as follows:

$$\coprod_{i=1}^{2} x_i = x_1 \coprod x_2 = 1 - (1 - x_1)(1 - x_2) \tag{7.4}$$

The symbol $\coprod$ is read "ip" [the inverse of pi (Π)]. By induction, we have

$$\coprod_{i=1}^{n} x_i = 1 - \prod_{i=1}^{n}(1 - x_i) \tag{7.5}$$

This operator is useful in the reliability analysis of reliability block diagrams.

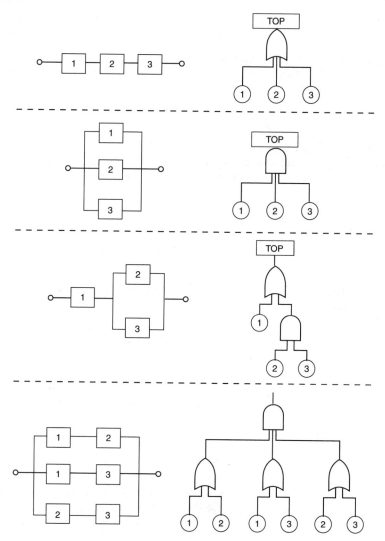

Figure 7.14. Relationship between selected block diagrams and fault trees.

Example 7.6 (Series and Parallel Structures)

A system with series structure is shown in Figure 7.1(b). In this case we have

$$\phi(x) = x_1 \cdot x_2 \cdots x_n = \prod_{i=1}^{n} x_i$$

A system with parallel structure is shown in Figure 7.1 (c). In this case we have

$$\phi(x) = 1 - (1 - x_1)(1 - x_2) \ldots (1 - x_n) = 1 - \prod_{i=1}^{n}(1 - x_i)$$

From (7.5), this result can be expressed as

$$\phi(x) = \coprod_{i=1}^{n} x_i.$$ ■

Example 7.7 (k-out-of-n Structure)
A k-out-of-n system is functioning if at least k of the n components are functioning. For a system of this type, the structure function is

$$\phi(x) = \begin{cases} 1 & \text{if } \sum_{i=1}^{n} x_i \geq k \\ 0 & \text{if } \sum_{i=1}^{n} x_i < k \end{cases}$$

Consider the case $n = 3$ and $k = 2$. In this case, the network structure is as shown in Figure 7.15 and we have, using the results from Example 7.7,

$$\phi(x) = (x_1 x_2) \coprod (x_1 x_3) \coprod (x_2 x_3)$$
$$= 1 - (1 - x_1 x_2)(1 - (x_1 x_3) \coprod (x_2 x_3))$$
$$= 1 - (1 - x_1 x_2)(1 - (1 - (1 - x_1 x_3)(1 - x_2 x_3)))$$
$$= 1 - (1 - x_1 x_2)(1 - x_1 x_3)(1 - x_2 x_3)$$

and, since $x_i^k = x_i$ for all i and k, we have

$$\phi(x) = x_1 x_2 + x_1 x_3 + x_2 x_3 - 2 x_1 x_2 x_3$$ ■

A component is said to be *irrelevant* if the system state is not affected the state of the component. Let $(1_i, \mathbf{x})$ and $(0_i, \mathbf{x})$ denote the state vector where only the state of component i differs, with $x_i = 1$ in the first and 0 in the second. Component i is irrelevant if and only if

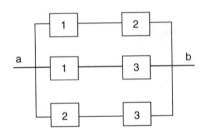

Figure 7.15. Block diagram for 2-out-of-3 system.

$$\phi(1_i, \mathbf{x}) = \phi(0_i, \mathbf{x}) \tag{7.6}$$

for all values of the remaining arguments. An irrelevant component has no effect on the functioning of the system. The components which are not irrelevant are called relevant components and the state of the system is affected only by the state of these components. This is best illustrated by a two-component system with a reliability block diagram shown in Figure 7.16. Here component 2 is irrelevant, as its state (working or failed) has no impact on the state of the system.

7.7.3   Coherent Systems

A system of is said to be coherent if it has no irrelevant components. Let $\phi(\mathbf{x})$ denote the structure function for a coherent system with n components.[4] Some properties of a coherent system are

(i) $\phi(\mathbf{0}) = 0$ and $\phi(\mathbf{1}) = 1$

The first [second] equation implies that the system is in failed [working] state if all the components are in failed [working] states.

(ii) $\displaystyle\prod_{i=1}^{n} x_i \leq \phi(\mathbf{x}) \leq \coprod_{i=1}^{n} x_i$

We omit the proof. The implication of this is that a coherent system functions at least as well as a corresponding series system and at most as well as a corresponding parallel system.

(iii) Let x and y be two state vectors (of same dimension). Define $x \cdot y$ and $x \amalg y$ as follows:

$$x \cdot y = (x_1 y_1, x_2 y_2, \cdots, x_n y_n) \tag{7.7}$$

$$x \amalg y = (x_1 \amalg y_1, x_2 \amalg y_2, \cdots, x_n \amalg y_n) \tag{7.8}$$

Then

$$\phi(x \cdot y) \geq \phi(x) \cdot \phi(y) \tag{7.9}$$

$$\phi(x \amalg y) \geq \phi(x) \amalg \phi(y) \tag{7.10}$$

We omit the proof. The implication of the second inequality will be discussed in Chapter 14 in the context of redundancy.

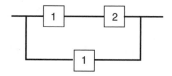

Figure 7.16. Block diagram of system with irrelevant component.

7.7.4   Path and Cut Sets

Path sets are closely related to the cut sets defined in Section 7.6.3. A path set is a set of components of the system which by being in working state ensures that the system is in working state. A path set is said to be minimal path set if it cannot be reduced without losing its status as a path set. This is in contrast to a cut set K, which is a set of components of the system which by failing causes the system to fail. Path and cut sets play an important role in reliability analysis as will be indicated later in the section. We first consider two examples.

Example 7.8 (Bridge Structure)

Consider the bridge structure shown in Figure 7.13. Here $n = 5$. There are four path sets, given by

$$P_1 = \{1, 4\}, P_2 = \{2, 5), P_3 = \{1, 3, 5\} \text{ and } P_4 = \{2, 3, 4\}$$

and each is a minimal path set. It can be seen from Figure 7.17 that for each path set P_i, if all of the components in the set are in working condition, then there is a connection between the end points (a) and (b). The number of minimal cut sets (see Figure 17.9) is also four and these are given by

$$K_1 = \{1, 2\}, K_2 = \{4, 5), K_3 = \{1, 3, 5\} \text{ and } K_4 = \{2, 3, 4\}$$

Note that for each cut set K_i, if all of the components in the set fail, then there is no connection between the end points (a) and (b) and as a result the system is in a failed state. ∎

Example 7.1 (Continued) (Stereo System)

Consider the block diagram shown in Figure 7.2. Here $n = 6$. Under the definition that a reliable system is one that is able to produce acceptable sound, there are six minimal path sets. These given by

$$P_1 = \{1, 4, 5\}, P_2 = \{1, 4, 6), P_3 = \{2, 4, 5\}, P_4 = \{2, 4, 6\}, P_5 = \{3, 4, 5\}, \text{ and } P_6 = \{3, 4, 6\}$$

The number of minimal cut sets is three and these are given by

$$K_1 = \{1, 2, 3\}, K_2 = \{4\}, \text{ and } K_3 = \{5, 6\} \qquad ∎$$

For a coherent system, let p and k denote the number of minimal path and cut sets respectively. The system structure function $\phi(x)$ is

$$\phi(x) = \coprod_{j=1}^{p} \prod_{i \in P_j} x_i = \prod_{j=1}^{k} \coprod_{i \in K_j} x_i \qquad (7.11)$$

where P_j, $1 \leq j \leq p$ are the p minimal path sets and K_j, $1 \leq j \leq k$ are the k minimal cut sets. For a proof of this, see Hoyland and Rausand (1994). Define

$$\rho_j(x) = \prod_{i \in P_j} x_i \quad \text{and} \quad \kappa_j(x) = \coprod_{i \in K_j} x_i \qquad (7.12)$$

Then, we have

$$\phi(x) = \prod_{j=1}^{p} \rho_j(x) = \prod_{j=1}^{k} \kappa_j(x) \tag{7.13}$$

Example 7.8 (Continued) (Bridge Structure)
The minimal path and cut sets were identified in the earlier part of this example. System performance can be represented in terms of a parallel structure with four minimal paths in parallel configuration as shown in Figure 7.18. Note that for each minimal path, the components are in a series configuration. Since there are four minimal path sets we have

$$\rho_1(x) = x_1 x_2, \; \rho_2(x) = x_2 x_5, \; \rho_3(x) = x_1 x_3 x_5, \; \text{and} \; \rho_4(x) = x_2 x_3 x_4,$$

Using this in (7.13), we have, after simplification,

$$\phi(x) = x_1 x_4 + x_2 x_5 + x_1 x_3 x_5 + x_2 x_3 x_4 - x_1 x_3 x_4 x_5 - x_1 x_2 x_3 x_5 - x_1 x_2 x_3 x_4 - x_2 x_3 x_4 x_5$$
$$- x_1 x_2 x_4 x_5 + 2 x_1 x_2 x_3 x_4 x_5$$

System performance can also be represented in terms of a series structure with the four minimal cut sets in a series configuration. This is shown in Figure 7.18. For each cut set, the components are in a parallel configuration. As a result, we have

$$\kappa_1(x) = 1 - (1 - x_1)(1 - x_2)$$
$$\kappa_2(x) = 1 - (1 - x_4)(1 - x_5)$$
$$\kappa_3(x) = 1 - (1 - x_1)(1 - x_3)(1 - x_5)$$
$$\kappa_4(x) = 1 - (1 - x_2)(1 - x_3)(1 - x_4)$$

Using this in (7.13) yields the result for $\phi(x)$ given above. ∎

7.7.5 System Reliability

Let $X_i(t)$, $1 \leq i \leq n$, denote the state of component i, at time t, with

$$X_i(t) = \begin{cases} 1 & \text{if component } i \text{ is in working state at time } t \\ 0 & \text{if component } i \text{ is in failed state at time } t \end{cases} \tag{7.14}$$

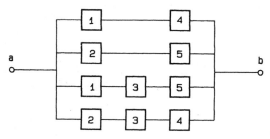

Figure 7.17. Bridge structure as a parallel structure of minimal path series structures.

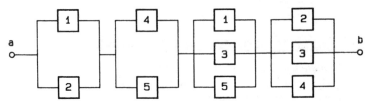

Figure 7.18. Bridge structure as a series structure of minimal cut parallel structures.

Let $X(t) = (X_1(t), X_2(t), \ldots, X_n(t))$ denote the state of the n components at time t. The state of the system at time t is then given by $\phi(X(t))$.

We assume that component failures are statistically independent, that all of the components are in their working state at $t = 0$, and that the components and the system are nonrepairable, so that the system is discarded when it fails. Let

$$p_i(t) = P\{X_i(t) = 1\} \tag{7.15}$$

for $1 \leq i \leq n$ and

$$p_S(t) = P\{\phi(X(t)) = 1\} \tag{7.16}$$

Note that these represent the survivor functions for the n components and for the system. In other words, $p_S(t) = 1 - F_S(t)$ and $p_i(t) = 1 - F_i(t)$ where $F_S(t)$ and $F_i(t)$ are the failure distributions for the system and component i. Since the component and system states are binary valued, we have

$$p_S(t) = P\{\phi(X(t)) = 1\} = E[\phi(X(t))]. \tag{7.17}$$

This is a function of $p_i(t)$, $1 \leq i \leq n$, and can be written as

$$p_S(t) = E[\phi(X(t))] = h(p_1(t), p_2(t), \ldots, p_n(t)) = h(p(t)) \tag{7.18}$$

where $p(t)$ is the vector $(p_1(t), p_2(t), \ldots, p_n(t))$.

When the component failures are dependent, the expressions linking system reliability to component reliabilities are more involved. They can be found in, for example, Hoyland and Rausand (1994).

Example 7.6 (Continued) (Series and Parallel Structure)
For a system with series structure such as that shown in Figure 7.1(b),

$$\phi(X(t)) = X_1(t) \cdot X_2(t) \cdots X_n(t) = \prod_{i=1}^{n} X_i(t)$$

and as result, from (7.18) , we have

$$p_S(t) = E[\phi(X(t))] = E\left[\prod_{i=1}^{n} X_i(t) \right] = \prod_{i=1}^{n} E[X_i(t)] = \prod_{i=1}^{n} p_i(t)$$

or

$$\overline{F}_s(t) = \prod_{i=1}^{n} \overline{F}_i(t)$$

Note that this the same as that for the competing risk model discussed in Section 4.6.2.
   Similarly, for a system with parallel structure as shown in Figure 7.1(c), we have

$$p_s(t) = E[\phi(X(t))] = E\left[\coprod_{i=1}^{n} X_i(t)\right] = \coprod_{i=1}^{n} E[X_i(t)] = \coprod_{i=1}^{n} p_i(t) = 1 - \prod_{i=1}^{n}(1 - p_i(t))$$

and as a result

$$F_s(t) = \prod_{i=1}^{n} F_i(t)$$

This is the same form as the multiplicative model discussed in Section 4.6.3. ∎

Example 7.1 (Continued) (Stereo System)

Suppose that under normal usage the failure distributions $F_1(t)$ and $F_4(t)$ of Components 1
(tuner) and 4 (amplifier) are exponential with means $\mu_1 = \mu_4 = 15$ years, that the two speakers are identical with failure distributions $F_5(t)$ and $F_6(t)$ exponential distributions with
means $\mu_5 = \mu_6 = 20$ years, that the CD player has a Weibull failure distribution $F_2(t)$ with
scale parameter $\beta = 10$ years and shape parameters $\alpha = 1.1$, and that the cassette player has a
Weibull failure distribution with parameters $\beta = 8$ and $\alpha = 1.2$.
   The probability that no component of the system fails for a period T is given by

$$R(T) = \prod_{i=1}^{6} (1 - F_i(T))$$

Values of $R(T)$ for T ranging from 1 to 20 years are as follows:

T	1	2	3	4	5	10	15	20
$R(T)$	0.6735	0.4376	0.2797	0.1767	0.1106	0.0097	0.0008	0.0001	∎

7.8   RELIABILITY AND STRUCTURAL IMPORTANCE OF A COMPONENT

In a multicomponent system, some components are more important than others in determining whether the system is in its working state or not. Also, the reliability of a system is a
function of the reliabilities of the various components of the system. In this section, we look
at some measures which determine the importance of a component to the overall performance of the system.

7.8.1   Structural Importance of Components

For a series system, all the components are equally important, since failure of any results in
the failure of the system. For systems with more complex structures, Birnbaum (1969) sug-

gested a measure to evaluate the structural importance of a component. In order to define this measure, we need the following:

A *critical path vector* for component i is the state vector $(1_i, x)$ such that

$$\phi(1_i, x) = 1, \qquad \phi(0_i, x) = 0$$

This is equivalent to

$$\phi(1_i, x) - \phi(0_i, x) = 1 \tag{7.19}$$

In other words, given the states of the other components $(\cdot_i, x)$, the system is in working state if and only if component i is in working state.

A *critical path set* $C(1_i, x)$ corresponding to a critical path vector $(1_i, x)$ for component i is defined as

$$C(1_i, x) = \{i\} \cup \{j: x_j = 1, \ j \neq i\} \tag{7.20}$$

The total number of critical path sets for component i is

$$\eta_\phi(i) = \sum_{(\cdot_i, x)} [\phi(1_i, x) - \phi(0_i, x)] \tag{7.21}$$

The structural importance of component i , according to Birnbaum (1969), is given by

$$B(i) = \frac{\eta_\phi(i)}{2^{(n-1)}} \tag{7.22}$$

This represents the relative proportion of the $2^{(n-1)}$ possible state vectors which are critical path vectors for component i.

Example 7.7 (Continued) (k-out-of-n structure)
Let $k = 2$ and $n = 3$. For component 1, we have the following:

$(\cdot, 0, 0)$	$\phi(1, x_1, x_2) - \phi(0, x_1, x_2)$	$C(1, x_1, x_2)$
$(\cdot, 0, 1)$	0	
$(\cdot, 0, 1)$	1	$\{1, 3\}$
$(\cdot, 1, 0)$	1	$\{1, 2\}$
$(\cdot, 1, 1)$	0	

As a result, the total number of critical path vectors for component 1 is $\eta_\phi(1) = 2$. Using this in (7.22), we have the structural importance of component 1 equal to 0.5. ∎

7.8.2 Reliability Importance of Components

Different measures for the reliability importance of components have been proposed by several authors. In this section we discuss a measure proposed by Birnbaum (1969).[5] For a system with n independent components, the system reliability in terms of the component reliabilities is given by (7.18). Birnbaum's measure of the reliability importance of component i at time t

is given by

$$I(i|t) = \frac{\partial h(p(t))}{\partial p_i(t)} \tag{7.23}$$

for $i = 1, 2, \ldots, n$. Note that this can be viewed as the sensitivity of the system reliability to the reliability of component i at time t.

Example 7.6 (Continued) (Series and Parallel Structures)
For a system with series structure

$$h(p(t)) = p_1(t) \cdot p_2(t) \cdots p_n(t) = \prod_{k=1}^{n} p_k(t)$$

From (7.23) we have

$$I(i|t) = \prod_{\substack{k=1 \\ k \neq i}}^{n} p_i(t) \qquad\blacksquare$$

7.9   MARKOVIAN MODEL FORMULATIONS

7.9.1   Basic Concepts

Suppose that a system consists of n components. Whenever a component fails, it is repaired minimally and put back into use. We assume that the failure rate for component i, $1 \leq i \leq n$, is λ_i. As indicated in Chapter 4, most components have a bathtub failure rate, where the failure rate is roughly constant over an interval $[t_1, t_2)$ and we confine our attention to this interval. In this case the assumption of constant failure rate is valid. At any given time, since each component can be in either working or failed state, the system can be in one of 2^n states. Let these states be denoted by the set $\mathbf{S} = \{1, 2, \ldots, 2^n\}$. We also assume that the repair times are exponentially distributed (implying constant repair rates) for each component. It follows that the transitions in the system state $X(t)$ can be modeled by a continuous Markov chain formulation (discussed in Appendix B3).[6] Since the performance of the system is determined by the state of the system, various performance measures such as availability, number of failures over a time period, and so on, can be determined given knowledge of the state of the system. We illustrate these concepts by considering the case $n = 2$. The results can easily be extended to arbitrary n.

Example 7.9 (2-Unit Power Plant)
Consider a power station with two generating units, labeled A and B. Unit A [B] generates 200 [100] MW when it is in its working state and 0 MW when it has failed. This results in the following $2^2 = 4$ possibilities with regard to the state of the system:

System state	Characterization	System output
1	A and B both working	300 MW
2	A working and B failed	200 MW
3	A failed and B working	100 MW
4	A and B both failed	0 MW

Note that States 2 and 3 correspond to degraded system performance, where the system cannot generate the rated output. State 4 corresponds to a total system failure, with no output. ■

7.9.2   Special Case ($n = 2$)

Assume that in a small time interval only one event (failure of a unit or completion of repair of a failed unit) can occur. Figure 7.19 is a diagram characterizing the transition between the four states in this case. Let λ_1 and λ_2 denote the failure rates for the two units. Let μ_1 and μ_2 denote the parameters of the exponential distributions for the repair times for the two units, with mean repair times then being $1/\mu_i$, $i = 1, 2$.

We first consider the scenario where repair of a failed unit commences immediately after failure. This implies that both units are simultaneously undergoing repair when the system state is 4. The transition matrix $P = (P_{ij})$ is given by (See Appendix B3)

$$
P = \begin{bmatrix}
0 & \dfrac{\lambda_1}{\lambda_1 + \lambda_2} & \dfrac{\lambda_2}{\lambda_1 + \lambda_2} & 0 \\[2ex]
\dfrac{\mu_1}{\mu_1 + \lambda_2} & 0 & 0 & \dfrac{\lambda_2}{\mu_1 + \lambda_2} \\[2ex]
\dfrac{\mu_2}{\mu_2 + \lambda_1} & 0 & 0 & \dfrac{\lambda_1}{\mu_2 + \lambda_1} \\[2ex]
0 & \dfrac{\mu_2}{\mu_2 + \mu_1} & \dfrac{\mu_1}{\mu_2 + \mu_1} & 0
\end{bmatrix}
\tag{7.24}
$$

This is obtained as follows. Suppose that $X(0) = 1$, i.e., at time $t = 0$, the system is in State 1. If a failure takes place in the interval $[t, t + \delta t)$, the system can move to either State 2 or State 3, depending on whether Unit 1 or Unit 2 fails. (By assumption, it cannot move to State 4.)

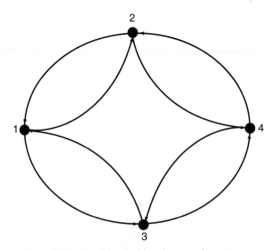

Figure 7.19. Possible transitions between four states.

Since the probability that Unit 1 [2] fails in a small interval δt is given by $\lambda_1 \delta t [\lambda_2 \delta t]$, which does not depend on t, and since the failures are independent, the probability of the event "Change from State 1" is given by $(\lambda_1 + \lambda_2)\delta t$. Given that the event has occurred, the probability that it was due to Unit 1 failing (implying a change from State 1 to State 2) is given by $\{\lambda_1/(\lambda_1 + \lambda_2)\}$. This is the second entry in the first row of the transition matrix (i.e., P_{12}). Similar reasoning gives the remaining entries of the transition matrix. Note that $P_{ij} = 0$ implies that the transition from State i to State j cannot occur.

Once the system enters State i, the time that it stays in that state before making a transition to State j is a random variable with distribution function $F_{ij}(\ )$. Note that this needs to be defined only for transitions for which $P_{ij} \neq 0$. In our case, these distribution functions are as follows:

$$F_{12}(t) = F_{34}(t) = 1 - e^{-\lambda_1 t} \tag{7.25}$$

$$F_{13}(t) = F_{24}(t) = 1 - e^{-\lambda_2 t} \tag{7.26}$$

$$F_{21}(t) = F_{43}(t) = 1 - e^{-\mu_1 t} \tag{7.27}$$

$$F_{31}(t) = F_{42}(t) = 1 - e^{-\mu_2 t} \tag{7.28}$$

Let $p_i(t)$ denote the probability that $X(t)$ is in state i at time t and let $p(t)$ be a column vector given by $p^T(t) = [p_1(t), p_2(t), p_3(t), p_4(t)]$. Then we can derive an expression for $p(t)$ as follows. Note that $X(t + \delta t) = 1$ is a result of one of the following three cases.

Case (i): $X(t) = 2$ and completion of Unit 1 repair took place in $[t, t + \delta t)$. This occurs with probability $\mu_1 \delta t$.

Case (ii): $X(t) = 3$ and completion of Unit 2 repair took place in $[t, t + \delta t)$. This occurs with probability $\mu_2 \delta t$.

Case (iii): $X(t) = 1$ and no failure occurs in $[t, t + \delta t)$. This occurs with probability $\{1 - (\lambda_1 + \lambda_2)\delta t\}$.

This implies that

$$p_1(t + \delta t) = p_1(t)\{1 - (\lambda_1 + \lambda_2)\delta t\} + p_2(t)\mu_1 \delta t + p_3(t)\mu_2 \delta t \tag{7.29}$$

and on taking the limit as $\delta t \to 0$, we have

$$\frac{dp_1(t)}{dt} = -(\lambda_1 + \lambda_2)p_1(t) + \mu_1 p_2(t) + \mu_2 p_3(t) \tag{7.30}$$

Proceeding along similar lines, we can derive the differential equations for $p_i(t)$, $i = 2, 3, 4$. This yields

$$\frac{p_s(t)}{dt} = A p_s(t) \tag{7.31}$$

where A is given by

$$A = \begin{bmatrix} -(\lambda_1 + \lambda_2) & \mu_1 & \mu_2 & 0 \\ \mu_1 & -(\mu_1 + \lambda_2) & 0 & \mu_2 \\ \lambda_2 & 0 & -(\mu_2 + \lambda_1) & \lambda_1 \\ 0 & \lambda_2 & \lambda_1 & -(\mu_1 + \mu_2) \end{bmatrix} \tag{7.32}$$

and

$$p^T(0) = [1, 0, 0, 0]. \tag{7.33}$$

since the system starts in State 1.

Equation (7.31) is an ordinary first-order linear differential equation and has an analytical solution, which can be obtained using one of several methods (See Misra, 1992). Using the Laplace transform approach, (7.31) can be written as

$$s\hat{p}_s(s) - p_s(0) = A\hat{p}_s(s) \tag{7.34}$$

which yields

$$(Is - A)\hat{p}_s(s) = p_s(0) \quad \text{or} \quad \hat{p}_s(s) = (Is - A)^{-1}p_s(0) \tag{7.35}$$

On taking the Laplace inverse, we have

$$p_s(t) = e^{At}p_s(0) \tag{7.36}$$

where e^{At} is a matrix.

Using this approach, we have the solution to (7.31) given by

$$p_1(t) = \frac{\{\mu_1\mu_2 + \lambda_1\mu_2 e^{-(\lambda_1+\mu_1)t} + \lambda_2\mu_1 e^{-(\lambda_2+\mu_2)t} + \lambda_1\lambda_2 e^{-(\lambda_1+\lambda_2+\mu_1+\mu_2)t}\}}{(\lambda_1 + \mu_1)(\lambda_2 + \mu_2)} \tag{7.37}$$

$$p_2(t) = \frac{\{\lambda_1\mu_2 - \lambda_1\mu_2 e^{-(\lambda_1+\mu_1)t} + \lambda_1\lambda_2 e^{-(\lambda_2+\mu_2)t} - \lambda_1\lambda_2 e^{-(\lambda_1+\lambda_2+\mu_1+\mu_2)t}\}}{(\lambda_1 + \mu_1)(\lambda_2 + \mu_2)} \tag{7.38}$$

$$p_3(t) = \frac{\{\lambda_2\mu_1 + \lambda_1\lambda_2 e^{-(\lambda_1+\mu_1)t} - \lambda_2\mu_1 e^{-(\lambda_2+\mu_2)t} - \lambda_1\lambda_2 e^{-(\lambda_1+\lambda_2+\mu_1+\mu_2)t}\}}{(\lambda_1 + \mu_1)(\lambda_2 + \mu_2)} \tag{7.39}$$

$$p_4(t) = \frac{\{\lambda_1\lambda_2 - \lambda_1\lambda_2 e^{-(\lambda_1+\mu_1)t} - \lambda_1\lambda_2 e^{-(\lambda_2+\mu_2)t} + \lambda_1\lambda_2 e^{-(\lambda_1+\lambda_2+\mu_1+\mu_2)t}\}}{(\lambda_1 + \mu_1)(\lambda_2 + \mu_2)} \tag{7.40}$$

From this we can derive various performance measures, as indicated below.

Suppose that we are interested in the fraction of time $\gamma(t)$ that the system is in State 4 (complete failure) over the interval $[0, T)$. $\gamma(T)$ can be obtained as follows. Let

$$Y(t) = \begin{cases} 1 & \text{if } X(t) = 4 \\ 0 & \text{otherwise} \end{cases} \tag{7.41}$$

Then

$$\gamma(T) = \frac{\int_0^T Y(t)dt}{T} \tag{7.42}$$

The expected value of this (also called interval unavailability) is given by

$$E[\gamma(T)] = \frac{\int_0^T p_4(t)dt}{T} \tag{7.43}$$

Suppose that the system performance is deemed to be acceptable if the system is in states 1 or 2 and unacceptable otherwise. Then the fraction of the time $\eta(T)$ that the system performance is acceptable is obtained as a function of the random variable $Z(t)$, defined as

$$Z(t) = \begin{cases} 1 & \text{if } X(t) = 1 \text{ or } 2 \\ 0 & \text{if } X(t) = 3 \text{ or } 4 \end{cases} \tag{7.44}$$

In this case, following a similar approach, we have the expected fraction of the time the system can generate the required output given by

$$E[\eta(T)] = \frac{\int_0^T \{p_1(t) + p_2(t)\}dt}{T}. \tag{7.45}$$

Example 7.9 (Continued) (2-Unit Power Plant)
Let the two components be statistically identical with $\lambda_1 = \lambda_2 = 0.1$ (per year) and $\mu_1 = \mu_2 = 10$ (per year). This implies that the MTTF for the two units is 10 years and the mean time to repair a failure is 0.1 year. From (7.37) we have

$$p_1(t) = 0.0098\{100.00 + 2e^{-10.1t} + 0.01e^{-20.2t}\}$$

From (7.43), we have $E[\gamma(T)]$ given by

$$E[\gamma(T)] = 0.98 + 0.00196\left\{\frac{1 - e^{-10.1T}}{T}\right\} + 0.00000485\left\{\frac{1 - e^{-20.2T}}{T}\right\}$$

The results for various values of T are as shown below.

T Years	0.1	0.2	0.5	1.0	2.0	5.0	10.0
$E[\gamma(T)]$	0.99250	0.98850	0.98389	0.98195	0.98088	0.98039	0.98020

Note that $E[\gamma(T)]$ is a decreasing function of T and it approaches the asymptotic value of 0.98 as T approaches infinity.
    One can easily calculate the corresponding values for $E[\eta(T)]$. ∎
    Suppose that only one failed unit can be repaired at any given time and repair is done under the rule first-failed-first-repaired. In this case, the system model includes five states, as

shown in Figure 7.20. Here State 4 represents the situation where Unit 1 failed first and is undergoing repair and Unit 2 fails before the repair of Unit 1 is completed. Repair of Unit 2 cannot begin until Unit 1 is repaired. State 5 represents the reverse situation—Unit 1 failing while Unit 2 is undergoing repair. Sates 1–3 are as defined earlier. In this case, the matrix A is given by

$$A = \begin{bmatrix} -(\lambda_1 + \lambda_2) & \lambda_1 & \lambda_2 & 0 & 0 \\ \mu_1 & -(\mu_1 + \lambda_2) & 0 & 0 & \lambda_2 \\ \mu_2 & 0 & -(\mu_2 + \lambda_1) & \lambda_1 & 0 \\ 0 & \mu_2 & 0 & -\mu_2 & 0 \\ 0 & 0 & \mu_1 & 0 & -\mu_1 \end{bmatrix} \qquad (7.46)$$

The analysis of this scenario can be carried out in a similar manner to that of the previous case.

7.9.3   General Case

The dimension of the matrix P is 2^n for a system with n components. The number of nonzero entries in P is determined by the connection between the components and the rectification action. These are functions of the n failure rates and n repair rates (one for each component). It is assumed that once the system enters a given state, the duration for which it stays in that state is exponentially distributed with the parameter a function of the failure and repair rates.

The analysis of this model follows along the same lines as that for the special case and can be found in most textbooks on stochastic processes, e.g., Bhat (1972) or Cox and Miller (1965). Estimation of the model parameters is a more involved task as it depends on the type of data available. If the data available are the times at which the transitions take place and the states involved, then one can estimate the parameters of the Markov chain (nonzero entries of P and means of the exponential distributions characterizing the duration for which the system stays in the various states) by standard statistical methods. Results can be found in, for example, Bhat (1972).

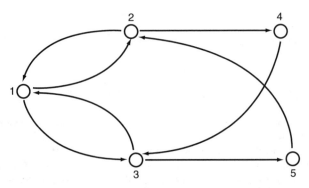

Figure 7.20. Possible transitions between system states.

7.10   DEPENDENT FAILURES

Thus far we have focused our attention on the case where component failures are statistically independent. Failure times are often not independent, for a number of reasons, the most important being environmental effects. Since the components of a system share the same environment, the environment has an impact on the failure of all components. This implies that failure times of components are not statistically independent and it is necessary to consider their interrelationships in assessing system reliability. Here we look at models that take this into account in various ways.

7.10.1   Multivariate Distribution Models

Nonindependent failures of n components of a system may be modeled by a multivariate distribution function, defined as

$$F(t_1, t_2, \ldots, t_n) = P\{T_1 \le t_1, T_2 \le t_2, \ldots, T_n \le t_n\} \tag{7.47}$$

We shall confine our attention to the special case $n = 2$. The extension to the general case follows along similar lines.

In the bivariate case, Equation (7.47) reduces to

$$F(t_1, t_2) = P\{T_1 \le t_1, T_2 \le t_2\} \tag{7.48}$$

The two marginal distributions $F_1(t_1)$ and $F_2(t_2)$ are given by

$$F_1(t_1) = F(t_1, \infty) \quad \text{and} \quad F_2(t_2) = F(\infty, t_2) \tag{7.49}$$

From this, we have

$$P\{T_1 > t_1, T_2 > t_2\} = \overline{F}(t_1, t_2) = 1 - F_1(t_1) - F_2(t_2) + F(t_1, t_2) \tag{7.50}$$

A useful result due to Frechet (1951) is the following:

$$\max[1 - F_1(t_1) - F_2(t_2), 0] \le F(t_1, t_2) \le \min[F_1(t_1), F_2(t_2)] \tag{7.51}$$

The bivariate density function associated with the distribution function given in (7.48) is given by

$$f(t_1, t_2) = \frac{\partial^2 F(t_1, t_2)}{\partial t_1 \partial t_2} \tag{7.52}$$

provided the derivative exists.

A well-known bivariate distribution is the Marshall–Olkin *bivariate exponential distribution*, which is given by

$$\overline{F}(t_1, t_2) = e^{-(\lambda_1 t_1 + \lambda_2 t_2 + \lambda_{12}\max(t_1, t_2))} \tag{7.53}$$

It will be shown later that this distribution arises in the context of modeling the reliability of a two-component system subjected to external shocks. Many other bivariate distributions can be found in Johnson and Kotz (1972).

In contrast to the univariate ($n = 1$) case discussed in Chapter 4, we have four different notions of hazard rates for the bivariate case. When the bivariate distribution function is absolutely continuous, these four hazard rates are as follows.

The conditional hazard rate for components 1 and 2 at time t given that $T_1 > t$, $T_2 > t$ are given by

$$\lambda_1(t) = \lim_{\delta t \to 0} \frac{P\{t < T_1 \le t + \delta t | T_1 > t, T_2 > t\}}{\delta t}, t \ge 0 \qquad (7.54)$$

$$\lambda_2(t) = \lim_{\delta t \to 0} \frac{P\{t < T_2 \le t + \delta t | T_1 > t, T_2 > t\}}{\delta t}, t \ge 0 \qquad (7.55)$$

and the conditional hazard rates of component 1 [2] at time $t > t_2$ [$t > t_1$] given that $T_2 = t_2$ [$T_1 = t_1$] are given by

$$\lambda_1(t|t_2) = \lim_{\delta t \to 0} \frac{P\{t < T_1 \le t + \delta t | T_1 > t, T_2 = t_2\}}{\delta t}, t \ge 0 \qquad (7.56)$$

and

$$\lambda_1(t|t_1) = \lim_{\delta t \to 0} \frac{P\{t < T_2 \le t + \delta t | T_1 = t_1, T_2 > t\}}{\delta t}, t \ge 0 \qquad (7.57)$$

Given a bivariate distribution function, these four hazard functions are defined uniquely. Conversely, this set of four conditional hazard rates determines the bivariate distribution function uniquely. (See, Cox, 1972 or Shaked and Shantikumar, 1986 for further discussion on the topic.)

Example 7.10 (Freund Distribution)

Let $\lambda_1(t) = \alpha$, $\lambda_2(t) = \beta$, $\lambda_1(t|t_2) = \alpha'$, $\lambda_2(t|t_1) = \beta'$ for some positive constants α, β, α', and β'. Then the bivariate distribution is the Freund distribution given by

$$\overline{F}(t_1, t_2) = \begin{cases} \dfrac{\alpha}{\alpha + \beta - \beta'} e^{-(\alpha + \beta - \beta')t_1 - \beta' t_2} + \dfrac{\beta - \beta'}{\alpha + \beta - \beta'} e^{-(\alpha + \beta)t_2} & \text{for } t_1 \le t_2 \\[3mm] \dfrac{\beta}{\alpha + \beta - \alpha'} e^{-(\alpha + \beta - \alpha')t_2 - \alpha' t_1} + \dfrac{\alpha - \alpha'}{\alpha + \beta - \alpha'} e^{-(\alpha + \beta)t_1} & \text{for } t_1 \ge t_2 \end{cases}$$

The marginals are not exponential, but mixtures of exponentials. For further details, see Freund (1961) and Hutchinson and Lai (1990).[7] ∎

7.10.2   Shock Damage Model

The model formulation that follows is due to Marshall and Olkin (1972). The system consists of two components and receives shocks from three sources (denoted Sources 1, 2, and 3). Shocks from Source i arrive according a stationary Poisson process with intensity parameter λ_i. A shock from Source 1 causes the failure of Component 1 with probability q_1 ($0 \le q_1 \le$ 1). A shock from Source 2 causes the failure of Component 2 with probability q_2 ($0 \le q_2 \le$

1). Finally, a shock from Source 3 causes the failure (a) of both components with probability q_{11}, (b) of component 1 only with probability q_{10}, (c) of component 2 only with probability q_{01}, and (d) of neither component with probability q_{00}, with these four probabilities adding to one. Then it can be shown that

$$P\{T_1 > t_1, T_2 > t_2\} = e^{-\{\lambda_1^* t_1 + \lambda_2^* t_2 + \lambda_{12}^* \max(t_1, t_2)\}} \qquad (7.58)$$

where

$$\lambda_1^* = \lambda_1 q_1 + \lambda_3 q_{10}, \qquad \lambda_2^* = \lambda_2 q_2 + \lambda_3 q_{01}, \qquad \lambda_{12}^* = \lambda_3 q_{11} \qquad (7.59)$$

Note that (7.58) is the same as (7.53).

When $q_1 = 1$, $q_2 = 1$, or $q_{11} = 1$, then each shock received is fatal in the sense that it causes the failure of either one or both of the components.

7.10.3   Common Environment Model

Since the various components of a system operate in the same environment, the environment of the system influences the components. Lindley and Singpurwalla (1986) suggested an approach to modeling this through a random parameter that affects the parameters of the failure distributions of the components. They analyzed the following model in detail.

Suppose that the system consists of two components with a parallel structure, so that the system fails only when both components fail. Assume that the failure distributions of the two components are exponential with failure rates λ_1 and λ_2, respectively. The effect of the common environment is modeled by a positive-valued random parameter $\tilde{\eta}$ with distribution function $G(\eta)$. System reliability is modeled as

$$R(t) = P\{T \geq t\} = \int_0^\infty \{e^{-\eta(\lambda_1 t)} + e^{-\eta(\lambda_2 t)} - e^{-\eta(\lambda_1 + \lambda_2)t}\} dG(\eta) \qquad (7.60)$$

Note that η has the effect of scaling the two hazard rates. Equation (7.60) can be rewritten as

$$R(t) = G^*(\lambda_1 t) + G^*(\lambda_2 t) - G^*((\lambda_1 + \lambda_2)t) \qquad (7.61)$$

where

$$G^*(y) = \int_0^\infty e^{-\eta y} dG(\eta) \qquad (7.62)$$

is the Laplace transform of $G(\eta)$.

It is not possible to obtain $R(t)$ analytically for a general $G(\eta)$. However, if $G(\eta)$ is a Gamma distribution with density function given by

$$g(\eta) = \frac{b^{(a+1)} \eta^a e^{-(b\eta)}}{\Gamma(a+1)} \qquad (7.63)$$

then

$$R(t) = \left(\frac{b}{\lambda_1 t + b}\right)^{(a+1)} + \left(\frac{b}{\lambda_2 t + b}\right)^{(a+1)} - \left(\frac{b}{(\lambda_1 + \lambda_2)t + b}\right)^{(a+1)} \tag{7.64}$$

Lindley and Singpurwalla (1986) discuss the relationship of this to other bivariate distributions.

Example 7.12

The lifetimes of electronic components are influenced by the applied voltage. The voltage in turn is dependent on the power pack. Due to variability in manufacturing, the voltage output of power packs deviates from the design specification and this variability can be modeled by $G(\eta)$. As a result, all the components that get their power from a given power pack share the same environment. If there are two components that are connected in parallel and each has constant failure rate, then the system failure can be modeled by (7.60). ■

7.10.4 Other Models

Several other models for dependent failures can be found in the literature. One such is the binomial failure model. Here components can either fail independently or due to a "common cause" shock. When such a shock occurs, the number of components that fail is a random variable. A review of this model can be found in Atwood (1986), where references to other common cause models can also be found.

7.10.5 Other Issues

The use of the models discussed in this section to model real-world problems is very limited. However, they offer potential for new and interesting applications. Analysis and parameter estimation issues have received little attention and the results available are very limited.

7.11 FAILURE INTERACTION MODELS

Often in a multicomponent system, the failure of a component affects one or more of the remaining components of the system. This type of interaction between components was termed "failure interaction" by Murthy and Nguyen (1985) and they defined three types (called Types I–III) of such interactions.

In a Type I failure interaction, whenever a component fails it can induce the failure of one or more of the remaining components of the system. As such, we have two types of failure— natural and induced—with the former being the cause of the latter. We illustrate this situation by considering a two-component system. The distributions of natural failures of the two components are given by $F_1(t, \theta_1)$ and $F_2(t, \theta_2)$, respectively. Whenever Component 1 [2] fails naturally, it can induce a failure of Component 2 [1] with probability p_1 [p_2] or have no affect with probability $(1 - p_1)$ [$(1 - p_2)$]. The failures of the two components are modeled by two interacting point processes.[8]

In Type II failure interactions, whenever a component fails, it acts as a shock to one or more of the remaining components and affects their failure rate. Thus, the failure rates of components are functions not only of their ages but also of the number of shocks received.

Finally, a Type III failure interaction is a combination of Types I and II. In this case, when a component fails, it can induce the failure of one or more components of the system and/or affect the failure rate of one or more of the remaining components of the system.

Example 7.13

Consider a hoist mechanism used in an underground mine operation, where the load is hauled by two cables. When one of the cables fails, it can cause the failure of the second cable if the load being carried is heavy. On the other hand, if the load is light, then it has no affect on the second cable. In this case, the probability of an induced failure occurring corresponds to the probability that the load is so heavy that one cable is not adequate to hoist the load. ∎

NOTES

1. The modular approach to design is used extensively in electronic products such as computers. Murthy (1983) deals with the analysis of an unreliable of multicomponent system with modular structure and some related design issues.

2. For further details of FMECA, see MIL STD-882 and IEEE Standard 352. A number of computer programs are available for carrying out FMEA/FMECA studies. For further details regarding these, see Chapter 20.

3. A more detailed discussion of fault tree analysis can be found in Henley and Kumamoto (1981) and Vesely et al. (1971).

4. For more on coherent systems, see Barlow and Proschan (1975).

5. Other measures for the reliability importance of components can be found in Lambert (1975), Natvig (1979), and Henley and Kumamoto (1981).

6. Markov chain models have been studied extensively. These deal with problems such as the number of visits to a particular state or the duration for which the process is in a particular state over a specified time interval. Bhat (1972) is an introductory-level text where more details may be found. For more on the use of Markov chain models in reliability, see Misra (1992).

7. For Weibull extensions of the Freund (1961) and Marshall–Olkin bivariate exponential models, see Lu (1989).

8. The analysis of Type I failures for a system with two components can be found in Murthy and Nguyen (1985). Some results for a more general case can be found in Murthy and Nguyen (1984). The use of the method of maximum likelihood method to estimate the model parameters can be found in Murthy and Wilson (1994).

EXERCISES

1. Two batteries provide electrical power for operation of a remote transmitter. If the two are operating in parallel, they have an individual constant failure rate of 0.1 per year. If one fails, the other can keep the transmitter operational but its failure rate doubles (due to an increased load). Determine the system reliability for 1, 2, and 3 years. What is the mean time to failure?

2. Consider a domestic hot water system that uses gas for heating. Describe the systems in terms of its components. Construct fault tree diagrams for the following top events.

    (a) "hot water tank ruptures"
    (b) "gas leakage"
    (c) "no hot water"

3. Three identical temperature sensors are connected in parallel. An alarm is sounded if two or more of the sensors record a temperature above a prescribed safe level. The times to failure for the sensors are exponentially distributed with mean values of 5000 hours. The sensors are nonrepairable and inspected regularly every 2000 hours. What is the probability that a situation where the temperature exceeds the safe limit is not detected? How does this value change if the sensors are inspected every 1000 hours?

4. A radio system consists of three major components—a power supply, a receiver, and an amplifier. The static reliabilities are 0.9, 0.7, and 0.95, respectively. Compute the system reliability. If one uses two receivers in parallel, how does this affect the system reliability?

5. An electronic system consists of four major components. At least three of these must be functional in order for the system to operate. If each component has reliability 0.9, determine the reliability of the system.

6. For a simple network consisting of two components (A and B) connected in series and a third component (C) connected in parallel, carry out the following:
   (a) Draw a schematic of the system.
   (b) Write the structure function.
   (c) Find the system reliability in terms of the structure function.

7. An engine and gear box for a truck have failure distributions that can be modeled by Weibull distributions with shape parameters 1.25 and 1.5 and mean life times of 7 and 5 years, respectively. Both units must be operational for the normal operation of the truck. What is the reliability of the system for 3, 4, and 5 years of operation?

8. A string of decorative lights is connected in series. Each bulb has a mean lifetime of 1.8 years. Calculate the probability that the string burns for 100 hours without failure.

9. A system consists of three components, A, B, and C, connected in series and operating independently. The reliabilities of the components are $R(A) = 0.98$, $R(B) = 0.99$, and $R(C) = 0.995$. Calculate the reliability of the system.

10. Suppose that redundancy of all three components in the previous exercise is used in the design, i.e., each component is replicated.
    (a) Draw a block diagram of the system with redundancy.
    (b) Calculate the reliability of the system with redundancy.

11. For which of the components in the previous two exercises is redundancy most important? Calculate the reliability of the system if redundancy is used only for that component.

12. Discuss the importance and use of fault tree analysis and FMEA in reliability analysis. How do these methods differ, and what information do they give the design engineer?

13. Select a simple product with which you are familiar. Prepare a fault tree for the product.

14. Assign probabilities to each of the failure modes identified in the previous exercise and use the results to predict the reliability of the system.

15. Conduct a FMEA for the product used in the previous exercises.

16. A subsystem consists of two assemblies, A and B, connected in series. Assembly B consists of two components, B_1 and B_2, connected in parallel. Assume that components operate independently. Components are purchased from one of three suppliers. Reliabilities of the three components from the suppliers are:

	Supplier		
Component	1	2	3
A	0.99	0.92	0.97
B_1	0.85	0.90	0.93
B_2	0.90	0.99	0.82

(a) Draw a block diagram of the subsystem.

(b) Determine which supplier should be selected as the source for the three components.

(c) If the purchase policy were changed so that components could be bought individually, what would be the reliability of the best subsystem design.

17. An electronic system consists of three components as shown in the following diagram:

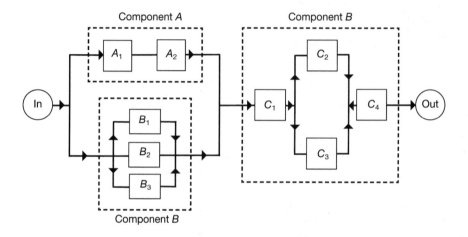

Reliabilities of the parts are as follows:

Part	Reliability	Part	Reliability	Part	Reliability
A_1	0.95	B_2	0.96	C_2	0.92
A_2	0.98	B_3	0.99	C_3	0.91
B_1	0.92	C_1	0.99	C_4	0.98

Assuming that all parts operate independently, calculate the reliability of the system.

CHAPTER 8

Advanced Statistical Methods for Data Analysis

8.1 INTRODUCTION

Chapter 3 introduced the notion of statistical data, discussed various data sources and data structures of importance in reliability applications, and provided some statistical tools for preliminary analysis and description of data. In Chapter 5, basic statistical methods for data analysis were discussed. These included estimation of parameters of the probability functions given in Chapter 4, confidence intervals and tests of hypotheses about these parameters, estimation and tests for functions of parameters, and tolerance intervals.

In this chapter, we extend these results in a number of ways, and show how statistical methods may be applied to provide data-based information concerning the models discussed in Chapters 6 and 7. Included in the discussion are methods for analysis of censored data (see Chapter 3), use of various types of information in statistical analysis, including judgmental or subjective information, and estimation of system reliability using various models, data sources, and methods.

In Section 8.2, we look at analysis of grouped data, that is, data for which we know only frequencies in specified intervals, rather than the individual X values. As discussed in Chapter 3, there are a number of situations in which this occurs, and the calculations must be modified to take the grouping into account. We look at estimation of the mean and variance, and comment on MLE procedures.

Censored data are data for which we do not know exact values for at least some of the observations. This may happen for various reasons and in many ways, as discussed in Chapter 3. Data of this type often occur in reliability studies and analysis of such data can become quite complex. Some techniques for handling censored data are discussed in Section 8.3, and references to additional analytical methods are given.

In Section 8.4, we introduce the concepts of Bayesian statistical methodology. This method of analysis is particularly appropriate in cases where one or more types of information are available to the analyst prior to actual collection of data. The key concept, "Bayes' theorem," enables incorporation of the prior information in analysis of test or other subsequent data. It also provides a natural means of updating reliability estimates and predictions. In Section 8.5, we review data sources and aggregation of data, and discuss the use and importance of this in Bayesian reliability analysis.

Estimation of system reliability is considered in Section 8.6. In this analysis, it is essential to be able to incorporate data at the part, component, etc., levels. Both the classical (i.e., non-Bayesian) and Bayesian approaches are discussed.

In Chapter 5, methods for estimating a number of functions of importance in reliability analysis were discussed. In Section 8.7, we consider several additional such functions, including intensity functions and stress–strength models.

8.2 ANALYSIS OF GROUPED DATA

Grouped data were discussed briefly in Chapter 3. Grouping of observations is, in fact, what is done when a frequency distribution is formed. When calculating summary statistics, the original, ungrouped data should be used if these are available. The reason that the ungrouped data are preferable is that the estimates based on grouped data are inefficient by comparison. This has been known for some time (Lindley, 1950, Fisher, 1922) with regard to estimating moments of the normal distribution and distributions with similar characteristics. On the other hand, in some instances grouped data are all that one has. This is often true when using data reported in the literature; it is always true of data obtained from the U.S. Census Bureau and other agencies where data on individuals is confidential. Census data are not directly relevant to reliability projects, but such data are often important in related marketing studies.

Another problem with grouped data is that bias may be introduced. This comes about because the calculation of sample moments uses the midpoint of each group to represent each observation in the interval. For unimodal distributions that tail off in both extremes, more observations are likely to lie on the side closer to the mean. The commonly used *Sheppard's corrections* were developed to deal with this problem. (See Lindley, 1950 and Stuart and Ord, 1991).

8.2.1   Estimating the Mean and Variance Using Grouped Data

Here we assume that the data are grouped into intervals of equal length h. The lower limit of the initial group is denoted x_0. The lower limit of the ith group is then $x'_i = (x_0 + ih)$, $i = 1, \ldots, k$. The frequency in the ith group is f_i. The formulas for the sample mean and variance [Equations (3.4) and (3.5)] become

$$\bar{x} = x_0 - \frac{h}{2} + \frac{h}{n}\sum_{i=1}^{k} if_i \tag{8.1}$$

and

$$s^2 = \frac{h^2}{n-1}\left[\sum_{i=1}^{k} i^2 f_i - \frac{1}{n}\left(\sum_{i=1}^{k} if_i\right)^2\right] \tag{8.2}$$

The Sheppard's correction for the variance is $h^2/12$. This quantity should be *subtracted* from the variance calculated using Equation (8.2) if it is thought that the population from which the sample was obtained is unimodal and roughly symmetric, with relatively long tails. For other distributions, the use of this and other corrections is questionable. (There is no correction for the mean, but there are corrections for higher moments. In addition, corrections of higher order have been developed. See Stuart and Ord, 1991). When Sheppard's corrections for the sample moments are appropriate, the corrected versions should be used in calculating moment estimates.

Example 8.1 (Case 2.24)

The data of Case 2.24 are grouped data on time to first failure of $n = 241$ airborne radar receivers. Here $h = 40$, $x_0 = 0$, and $k = 25$. We find $\Sigma i f_i = 1(39) + 2(35) + \ldots + 1(25) = 1610.0$ and $\Sigma i^2 f_i = 18674$. From (8.1) and (8.2), we find

$$\bar{x} = 0 - \frac{40}{2} + \frac{40(1610)}{241} = 247.220$$

and

$$s^2 = \frac{40^2}{240}\left(18674 - \frac{1610^2}{241}\right) = 52{,}789.3$$

The standard deviation is $\sqrt{52789.3} = 229.76$. The Sheppard's correction is $40^2/12 = 133.33$, and has very little effect on the result. Note, incidentally, that the frequency distribution of the data is considerably right-skewed, indicative of, for example, the exponential distribution. The fact that the mean and standard are roughly equal would support that conclusion, since for the exponential distribution, the population mean and standard deviation are equal. ∎

8.2.2   Maximum Likelihood Estimation Using Grouped Data

For grouped data, with class boundaries x_i', $i = 0, 1, \ldots, k$, the likelihood function becomes

$$L = \frac{n!}{N_1! \cdots N_k!} \prod_{i=1}^{k} p_i^{N_i} \tag{8.3}$$

where N_i is the (random) number of observations falling in the ith interval and

$$p_i = \int_{x_{i-1}'}^{x_i'} f(x)dx = F(x_0 + ih) - F(x_0 + (i-1)h) \tag{8.4}$$

Here f and F are the density and CDF of X, respectively.

Solution of the likelihood equations (obtained in the usual manner) is difficult, even in cases where the p_i's can be written in closed form. References to solutions for specific distributions can be found in the bibliography of Govindarajulu (1964). For further discussion, see Nelson (1982).[1]

Corrections for grouping are also sometimes appropriate for the MLEs. See Lindley (1950) for additional details.

8.3   ANALYSIS OF CENSORED DATA

In this section, we look at basic inference problems—point estimation, confidence intervals, and hypothesis testing—in the context of censored data. In estimation, we consider primarily the method of maximum likelihood (see Chapter 5). The type of censoring must be taken into account in formulating the likelihood function. Additional inference problems for censored data, e.g., estimation of system reliability, will be discussed later in the chapter. The most often used of the failure distributions given in Chapter 4, including the exponential, Weibull, gamma, lognormal, and extreme value distributions, will be considered.

8.3.1 Basic Concepts

Censored data were discussed in Chapter 3. Analysis of censored data requires some special techniques, which depend on the type of censoring. The two types of censoring that will be considered in this chapter are:

Type I Censoring: n items are put on test; testing is stopped at a predetermined time T.

Type II Censoring: n items are put on test; testing is stopped after a predetermined number r failures have occurred.

Data may be *singly* or *multiply* censored. The definitions given above are those of singly censored Type I and Type II data. Multiply censored Type I data occur when testing is stopped at different times for different subgroups of the sample (e.g., n_1 items removed from testing at time T_1, n_2 removed at time T_2, and so forth). They also occur when items are put on test (or in service) at different times and testing stops at time T. The aircraft windshield data of Case 2.2 are of the latter type. For Type II censoring, data are multiply censored if different subsamples are observed until varying numbers of failures have occurred (e.g., n_1 units observed until r_1 failures occur, n_2 units observed until r_2 occur, etc.). Both singly and multiply censored samples will be considered.

Data also may be left- or right-censored, or both. We look only at right censoring. Results are easily extended to left-censored data and to data that are both left- and right-censored.

Other types of censoring, discussed in Chapter 3 and the references cited, will not be considered. These include random censoring, combinations of Types I and II, and others. In many cases (including some of those listed above), significant mathematical difficulties are encountered

8.3.2 Inference for Censored Data

Estimation of Parameters. We assume a sample of size n, with $X_1, X_2, \ldots, X_n$ being (random) times to failure, and assumed independent. $F(.; \theta)$ and $f(.; \theta)$ are the associated distribution and density functions, respectively, with parameter θ (which may be multidimensional). Note that for censored data, not all of the X_i are observed exactly, though some information (specifically, at least a bound on its value) must be available about each. We wish to estimate one or more of the components of θ by the method of maximum likelihood. Note that the method of moments is not usually appropriate in the case of censored data. Instead, alternatives to MLEs are based on graphical methods. These will be dealt with in Chapter 11.

Recall that for complete data and independent observations the likelihood function $L(.)$ is the product

$$L(x_1, \ldots, x_n; \theta) = \prod_{i=1}^{n} f(x_i; \theta) \tag{8.5}$$

where x_i is the observed value of X_i, and that it is convenient to maximize $\log(L)$ rather than L itself. For censored data, the structure of the likelihood function depends of the type of censoring. The expressions for $\log(L)$ are:

Type I Censoring: We consider singly right-censored data. Thus the value of X_i is observed only if $X_i \leq T$, the censoring time. Suppose that r $(0 \leq r \leq n)$ items have lifetimes less than or equal to T. Reorder the observed values so that the first r values are observed lifetimes. Denote these lifetimes $x_1, \ldots, x_r$. Then $x_i = T$ for $i = r + 1, \ldots, n$. Log L is given by

$$\log L = \sum_{i=1}^{r} \log[f(x_i; \theta)] + (n-r) \log[1 - F(T; \theta)] \qquad (8.6)$$

The result is easily extended to multiply censored data. The MLEs are obtained as usual by minimizing $\log(L)$ with respect to the components of θ.

Type II Censoring: For Type II censoring, r is determined prior to the test. As a result, the data consist of the r smallest lifetimes in a sample of size n. Denote the ordered observed values $x_{(1)}, \ldots, x_{(r)}$. $\text{Log}(L)$ is given by

$$\log L = c + \sum_{i=1}^{r} \log[f(x_{(i)})] + (n-r) \log[1 - F(x_{(r)})] \qquad (8.7)$$

where c is a constant.

For multiply censored data with Type II censoring, further results on Types I and II censoring, and other types of censoring, see Lawless (1982) and Nelson (1982).

Confidence Intervals and Tests of Hypotheses. Derivation of confidence intervals (which can then be used to test hypotheses as indicated in Chapter 5) requires the distribution of the estimator of the parameter(s). An alternative frequently employed is to use asymptotic properties of the MLE to obtain approximate CI's and tests. The use of the normal distribution and asymptotic variances of the MLEs for this purpose is discussed in Chapter 5. Most of the results of the next section are asymptotic results.

For Type II censoring, the situation is essentially that of analysis of complete samples, and exact distributions of the MLE can often be determined. In addition, the usual likelihood theory is straightforward and results are, in principle, easily obtained, though numerical methods may be required for implementation. For Type I censoring, however, the theory is considerably more complex. In this case, r is an observed value of a random variable R, exact distributions are usually impossible to obtain (Lawless, 1982), and asymptotic results are nearly always used. (See Lawless, 1982, Section 1.4, for additional discussion and references.)

8.3.3 Application to Selected Failure Distributions

MLEs and confidence intervals for the exponential, Weibull, extreme value, gamma, normal, and lognormal distributions are given below. We assume singly censored data and consider Type I and Type II censoring. Recall that these results can also be used to test hypotheses about a parameter, say θ. To test H_0: $\theta = \theta_0$ versus H_a: $\theta \neq \theta_0$ at level α, calculate a $100(1 - \alpha)\%$ CI for θ. Reject H_0 if θ_0 is not in the CI; otherwise do not reject. Similarly for one-sided alternatives.

Exponential Distribution [Equation (4.36)]: *Type I censoring.* The MLE of λ is

$$\lambda^* = \frac{r}{\sum_{i=1}^{n} x_i} \qquad (8.8)$$

with asymptotic variance

$$V_a(\lambda^*) = \frac{\lambda^2}{n(1 - e^{-\lambda T})} \qquad (8.9)$$

The MLE of the MTTF μ is $\mu^* = 1/\lambda^*$, with asymptotic variance

$$V_a(\mu^*) = [n(1 - e^{-\lambda T})\lambda^2]^{-1} \tag{8.10}$$

An alternative approach to estimating λ is to calculate a CI for μ and use reciprocals of these limits as a CI for λ. The normal approximations based on (8.8) and (8.9) or (8.10) are good only for fairly large samples. Three alternatives, good for small samples, are given by Lawless (1982, p. 108). These are said to be about equally effective. One of the alternatives is an easily calculated CI based on the fact that the distribution of the transformed estimator $\hat{\phi} = \mu^{*-1/3}$ is better approximated by a normal distribution. The variance of the approximating normal distribution is $[9n\mu^{2/3}(1 - e^{-T/\mu})]^{-1}$. To calculate a CI for μ, substitute $\mu^* = 1/\lambda^*$ into this result, use the normal approximation and cube the reciprocal of the limits. To calculate a CI for λ, take reciprocals for the limits for μ.

Multiply Type I Censored Data. Suppose that $x_1, \ldots, x_r$ are the uncensored observations and that censoring times for censored observations are $T_{r+1}, \ldots, T_n$. For Type I censoring, use the results of (8.8) and (8.9) with the term $(n - r)x_{(r)}$ replaced by $\sum_{i=r+1}^{n} T_i$ and $n(1 - e^{-\lambda^* T})$ replaced by $\sum_{i=1}^{n}(1 - e^{-\lambda^* T_i})$.

Example 8.2 (Case 2.4)

We consider the data on kilometers driven until a warranty claim involving an engine problem. There were $r = 32$ such claims. Suppose we restrict attention to the 329 cars in the data set. Here $T = 40$, the length of the warranty period in thousands of kilometers. For these Type I censored data, $\Sigma x_i = 586.3 + 297(40) = 12466.3$. Thus, from (8.8), $\lambda^* = 32/12466.3 = 0.002567$ per 1000 km. (Note that this translates to 0.1027 per 40000 km, or 0.1027(329) = 34 per 329 cars; 32 were observed.) The estimated MTTF is $\mu^* = 1/\lambda^* = 389.6$ (or 389,600 km).

From (8.9), the estimated variance of λ^* is

$$\hat{V}(\lambda^*) = \frac{0.002567^2}{392[1 - e^{-0.002567(40)}]} = 0.0000002053$$

giving an estimated standard deviation of 0.0004531. A 90% CI is 0.002567 $\pm$ 1.645(0.0004531) or (0.001822, 0.003312). Using the alternate approach, we calculate a confidence interval for $\phi = \mu^{-1/3} = \lambda^{1/3}$. This is estimated as $(0.002567)^{1/3} = 0.1369$, with estimated variance $(0.1369)^2/[9(329)(1 - e^{-0.002567(40)})] = 0.00006489$. The estimated standard deviation is 0.008055, giving a 90% CI for ϕ of (0.1236, 0.1502) This translates to a CI for λ of $(0.1236^3, 0.1502^3) = (0.001888, 0.003389)$, not too different from the original result. These may be expected to differ more in smaller samples. Note that a 90% CI for μ, gotten as reciprocals of these limits, is (295.1, 529.6). ∎

Type II Censoring. The result is basically the same as that for complete samples. The MLE of λ is

$$\lambda^* = \frac{r}{\displaystyle\sum_{1}^{r} x_{(i)} + (n - r)x_{(r)}} \tag{8.11}$$

Again, $\mu^* = 1/\lambda^*$. A $100(1 - \alpha)\%$ confidence interval for λ is

$$\left(\frac{\lambda^* \chi^2_{2r, \alpha/2}}{2r}, \frac{\lambda^* \chi^2_{2r, 1-\alpha/2}}{2r} \right) \tag{8.12}$$

where $\chi^2_{2r,p}$ is the p-fractile of the chi-square distribution with $2r$ df. A $100(1 - \alpha)\%$ confidence interval for μ is gotten as the reciprocals of the values given by (8.12).

Example 8.3 (Case 2.4)

Suppose that Type II censoring had been employed on the usage data of Case 2.4, with "testing" stopped after $r = 25$ failures. In that case, the values 27.8, 29.0, 29.2, 31.6, 32.4, 35.5, and 38.6 would not have been observed. We now have the maximum observation as $x_{(25)} = 27.4$, and $\Sigma x_i = 362.2 + 304(27.4) = 8691.8$, so $\lambda^* = 25/8691.8 = 0.002876$, in reasonable agreement with the previous estimates. To calculate a 90% CI, we need the 0.05 and 0.95 fractiles of the χ^2-distribution with $2r = 50$ df. These are 34.76 and 67.50, respectively. From (8.12), the CI is

$$\left(\frac{0.002876(34.76)}{50}, \frac{0.002876(67.50)}{50} \right) = (0.001999, 0.003883). \qquad \blacksquare$$

Two-Parameter Exponential Distribution [Equation (4.36) with x replaced by $(x - a)$]. Here a is a shift or threshold parameter. The results are:

Type I Censoring: The MLEs are

$$a^* = x_{(1)}, \qquad \lambda^* = \frac{r}{\sum_{i=1}^{n}(x_{(i)} - x_{(1)})} \qquad (8.13)$$

The distribution of the MLEs is complicated and exact results are impossible to obtain. See Lawless (1982) for approaches to confidence intervals and additional discussion. From a practical point of view, one might estimate a by a^* or by $x_{(1)} - 1/n$, subtract this value from all data points, and use the resulting data to estimate λ as in the previous case.

Type II Censoring: The MLEs are

$$a^* = x_{(1)}, \qquad \lambda^* = \frac{r}{\sum_{i=1}^{r}(x_{(i)} - x_{(1)}) + (n - r)(x_{(r)} - x_{(1)})} \qquad (8.14)$$

A confidence interval for λ is obtained from (8.12) using $(2r - 2)$ df rather than $2r$. A $100\gamma\%$ CI for a is

$$a^* - \frac{rc_1}{n\lambda^*}, \qquad a^* - \frac{rc_2}{n\lambda^*} \qquad (8.15)$$

where $c_1 = [(\alpha/2)^{-1/(r-1)} - 1]$, and $c_2 = [(1 - \alpha/2)^{-1/(r-1)} - 1]$, with $\alpha = 1 - \gamma$. Note that this is an unusual result in that the point estimate is not included in the confidence interval, which must be the case since the threshold cannot exceed the minimum observation.

Example 8.4 (Case 2.15)

The estimators of (8.14) and (8.15) are also appropriate for uncensored data. The failure data for Part 1 in Case 2.15 appear to be exponentially distributed with an upward shift (see Exercise 3.18). We have $n = r = 14$. The minimum observation is $a^* = 156.5$. From (8.14), we find $\lambda^* = 14/4340.1 = 0.003226$. Note that the MTTF is estimated as $\mu^* = (1/\lambda^*) + a^* = 310.0 +$

156.5 = 466.5, which is, in fact, the sample mean. The estimate of λ, however, is affected by the shift in origin; if the shift is ignored, λ would be estimated as $1/466.5 = 0.002144$, which is a ⅓ reduction in the estimated failure rate.

A 90% CI for a is obtained from (8.15) with $c_1 = (0.05)^{-1/13} - 1 = 0.2592$ and $c_2 = (0.95)^{-1/13} - 1 = 0.0003953$. The result is (76.2, 156.4). Note that this interval does not include the value $a = 0$, meaning that the hypothesis H_0: $a = 0$ would be rejected at the 10% level, i.e., the shift appears to be real. A CI for λ is easily calculated from (8.12) using the χ^2 distribution with 26 df. ∎

Weibull and Extreme Value Distributions [Equations (4.45) and (4.53), respectively]. To avoid confusion, we replace the parameter α in (4.53) by μ, and β by λ. If X has a Weibull distribution with parameters α and β, then $Y = \log(X)$ has an extreme value distribution with parameters $\mu = \log \beta$ and $\lambda = 1/\alpha$. This relationship can be used to determine estimators of the parameters of either distribution in terms of those of the parameters of the other. Following most of the literature, we consider multiple censoring, with censoring times $T_1, \ldots, T_n$. (Thus the actual observations are $\min\{X_i, T_i\}$.) For singly censored data, we have $T_i = T, i = 1, \ldots, n$. We assume that $r > 0$, and, as before, order the observations so that the first r observations are failure times, ordering the T_i accordingly. (The notation x will be used to indicate an observation from either distribution. Which is meant will be clear from the context.)

Type I Censoring, Extreme Value Distribution. The MLE of λ is obtained as the solution of

$$\frac{\sum_{i=1}^{n} x_i e^{x_i/\lambda^*}}{\sum_{i=1}^{n} e^{x_i/\lambda^*}} - \lambda^* - \frac{1}{r}\sum_{i=1}^{r} x_i = 0 \tag{8.16}$$

The MLE of μ is

$$\mu^* = \lambda^* \log\left(\frac{1}{r}\sum_{i=1}^{n} e^{x_i/\lambda^*}\right) \tag{8.17}$$

(8.16) is easily solved numerically, e.g., by means of a simple search. To calculate confidence intervals, we use the normal approximation. The estimators are asymptotically bivariate normally distributed. To obtain the standard deviations needed for use in the normal approximation, we calculate an estimate I of the 2×2 asymptotic *information matrix* (see Lawless, 1982). The inverse of I is an estimate of the asymptotic *covariance* matrix, the diagonal elements of which are estimates of $V_a(\mu^*)$ and $V_a(\lambda^*)$, respectively, and the off-diagonal elements of the covariance between the estimators, which we denote $\text{Cov}_a(\mu^*, \lambda^*)$. The elements of the estimated information matrix (Lawless, 1982) are

$$I_{11} = r/\lambda^{*2} \tag{8.18}$$

$$I_{12} = \frac{1}{\lambda^{*2}}\sum_{i=1}^{n} z_i e^{z_i} \tag{8.19}$$

$$I_{22} = \frac{r}{\lambda^{*2}} + \frac{1}{\lambda^{*2}}\sum_{i=1}^{n} z_i^2 e^{z_i} \tag{8.20}$$

where $z_i = (x_i - \mu^*)/\lambda^*$. Note that these results are valid for large samples. For alternatives, many of which are suitable for small samples, see Lawless (1982) and Nelson (1982).

Type I Censoring, Weibull Distribution. Transform to $y_i = \log(x_i)$ and $t_i = \log(T_i)$ and analyze the resulting data using (8.16) through (8.20). The MLEs of the Weibull parameters are

$$\alpha^* = 1/\lambda^*, \qquad \beta^* = e^{\mu^*} \tag{8.21}$$

The estimates of the elements of the covariance matrix are calculated by means of the relationships

$$V_a(\alpha^*) \approx (\lambda^*)^{-4}V_a(\lambda^*) \tag{8.22}$$

$$V_a(\beta^*) \approx (e^{\mu^*})^2 V_a(\mu^*) \tag{8.23}$$

$$\mathrm{Cov}_a(\alpha^*, \beta^*) \approx -(\lambda^*)^{-2}e^{\mu^*}\mathrm{Cov}_a(\mu^*, \lambda^*) \tag{8.24}$$

Type II Censoring. For Type II censoring, the likelihood equations are also given by (8.16) and (8.17). For large samples, the results for Type I censoring apply here as well. In this case, r is not a random variable, however, and a number of alternatives are available, including linear estimators and several approaches to obtaining exact confidence intervals. The computations are somewhat complex and the reader is referred to Lawless (1982, Chapter 4), for details. The large-sample maximum likelihood results should not be used for small samples.

Additional results for the Weibull and extreme value distributions, including MLEs and approximate MLEs are given for multiply Type II censored data by Balakrishnan et al. (1995) and Fei et al. (1995).

Example 8.5 (Case 2.17)

Mechanical components such as jet engines may be expected to have an increasing failure rate. The data of Case 2.17 include six failure times and service times for 25 additional engines. This is Type I censoring, and we assume a Weibull distribution. First, transform to logs and fit the extreme value distribution. The solution of (8.16) may be obtained by a simple search or from many statistical program packages (e.g., Minitab). The result is $\lambda^* = 0.6188$. From (8.17), we find $\mu^* = 0.6188 \log(3247593/6) = 8.169$. The corresponding estimates of Weibull parameters are $\alpha^* = 1/0.6188 = 1.616$, and $\beta^* = e^{8.169} = 3530.5$.

To estimate the standard deviations, we calculate the elements of the information matrix from (8.18)–(8.20), obtaining $I_{11} = 15.669$, $I_{12} = I_{21} = -22.909$, and $I_{22} = 53.999$. The inverse of the resulting matrix is an estimate of the covariance matrix $\hat{V}$ of μ^*, λ^*. We find

$$\hat{V} = \begin{pmatrix} 0.16807 & 0.07130 \\ 0.07130 & 0.04877 \end{pmatrix}$$

From (8.22) and (8.23), we find the estimated variances of α^* and β^* to be $\hat{V}(\alpha^*) = 1.616^4(0.04877) = 0.3326$, and $\hat{V}(\beta^*) = (3530.5)^2(0.16807) = 2,094,897$. The corresponding standard deviations are 0.5767 and 1447.4. Both are quite large relative to the estimated values of the parameters, reflecting the relatively small sample size. Note, however, that value of α^* does correspond to an increasing failure rate. ∎

Gamma Distribution [Equation (4.41)]. *Type I and Type II Censoring.* The likelihood equations are the same for both types of censoring. We again denote the observed lifetimes

$x_1, \ldots, x_r$ and the censored observations $x_{r+1}, \ldots, x_n$. The MLEs are expressed in terms of the arithmetic and geometric means of observed lifetimes, given, respectively, by

$$\bar{x}_r = \frac{1}{r}\sum_{i=1}^{r} x_i \qquad \text{and} \qquad \tilde{x}_r = \left(\prod_{i=1}^{r} x_i\right)^{1/r} \tag{8.25}$$

The log likelihood function is

$$\log L = -r\alpha \log \beta - r \log \Gamma(\alpha) + r(\alpha - 1) \log \tilde{x}_r - \frac{r\bar{x}_r}{\beta} + \sum_{i=r+1}^{n} \log[1 - I(\alpha, x_i/\beta)] \tag{8.26}$$

where $I(a, x)$ is the incomplete gamma function, given by

$$I(a, x) = \frac{1}{\Gamma(\alpha)} \int_0^x u^{a-1} e^{-u} du \tag{8.27}$$

Lawless (1982) suggests direct maximization of (8.26) to obtain MLEs. Alternatively, solve the likelihood equations

$$0 = r \log \beta^* - r\psi(\alpha^*) + r \log \tilde{x}_r + \sum_{i=r+1}^{n} \frac{\partial \log[1 - I(\alpha^*, x_i/\beta^*)]}{\partial \alpha^*} \tag{8.28}$$

$$0 = -\frac{r\alpha^*}{\beta^*} + \frac{r\bar{x}_r}{\beta^*} + \sum_{i=r+1}^{n} \frac{\partial \log[1 - I(\alpha^*, x_i/\beta^*)]}{\partial \beta^*} \tag{8.29}$$

where $\psi(x)$ is the digamma function. The incomplete gamma and digamma functions are tabulated by Abramowitz and Stegun (1964) and a number of approaches to evaluation of these functions are given. These functions are also available in some computer packages. A number of approximations and tables for aids in solution to (8.28) and (8.29) are available. The derivative in the last term of (8.28) is difficult to evaluate. Tables for this purpose are given in Bain and Englehardt (1991). See Lawless (1982) for further discussion and alternative approaches.

Elsayed (1996) suggests the following alternative, which results in approximately unbiased estimators. Let $m = \bar{x}_r/x_{(r)}$ and $q = (1 - \tilde{x}_r/\bar{x}_r)^{-1}$. The estimate $\hat{\alpha}$ of α is calculated as follows. If $m < 0.42$, use

$$\hat{\alpha} = 1.061(1 - \sqrt{q}) + 0.2522q[1 + \sqrt{m}(r/n)^4] + 1.953(\sqrt{m} - 1/q)$$
$$- [0.220 - 0.1308q](r/n)^4 + 0.4292(q\sqrt{m})^{-1} \tag{8.30}$$

If $0.42 < m < 0.80$, use

$$\hat{\alpha} = 0.5311q[(r/n)^2 - 1] + 1.436 \log q + 0.7536m(q - 1)$$
$$- 2.040r/n - 0.260qm(r/n)^2 + 2.489\sqrt{n/qr} \tag{8.31}$$

If $0.80 < m$, use

$$\hat{\alpha} = 1.151 + 1.448rq(1 - m)/n - 1.024(q + m) + 0.5311 \log q$$
$$+ 1.541qm - 0.515\sqrt{rq/n} \tag{8.32}$$

The approximately unbiased estimator of β is

$$\hat{\beta} = \frac{r\bar{x}_r + (n-r)x_{(r)}}{n\hat{\alpha}[1 - 1/(r\hat{\alpha})]} \qquad (8.33)$$

For approximate confidence intervals, see Bain and Englehardt (1991) and Elsayed (1996).

Example 8.6 (Case 2.17)
Suppose that a gamma distribution is assumed. We have $r = 6$ and $n = 31$. The required statistics are found to be $\bar{x}_r = 1330.0$, $\tilde{x}_r = 769.7$, and $x_{(6)} = 2050.0$. Thus $m = 1330.0/2050.0 = 0.6488$ and $q = (1 - 769.7/2050.0)^{-1} = 1.6012$. We use (8.31) to obtain $\hat{\alpha} = 4.2175$. From (8.33), $\hat{\beta} = 59230/125.58 = 471.7$. ∎

Normal and Lognormal Distributions. We write $\phi(.)$ and $\Phi(.)$ to indicate the standard normal ($\mu = 0$ and $\sigma = 1$) density and CDF, respectively, with $\bar{\Phi}(x) = 1 - \Phi(x)$. The standard normal hazard function is given by $h(x) = \phi(x)/[1 - \Phi(x)]$; $h'(x)$ is the derivative of $h(x)$, given by

$$h'(x) = h(x)[h(x) - x] \qquad (8.34)$$

As before, we deal with the normal distribution; for lognormal data, transform to $y = \log x$ and use the normal results. The likelihood equations (for either type of censoring) are

$$0 = \frac{1}{\sigma^{*2}} \sum_{i=1}^{r}(x_i - \mu^*) + \frac{1}{\sigma} \sum_{i=r+1}^{n} h\left(\frac{x_i - \mu^*}{\sigma^*}\right) \qquad (8.35)$$

$$0 = -\frac{r}{\sigma^*} + \frac{1}{\sigma^{*3}} \sum_{i=1}^{r}(x_i - \mu^*)^2 + \frac{1}{\sigma^{*2}} \sum_{i=r+1}^{n} (x_i - \mu^*)h\left(\frac{x_i - \mu^*}{\sigma^*}\right) \qquad (8.36)$$

Methods for solution of the ML equations are discussed by Lawless (1982). Approximations, good for $n > 20$, are given by Elsayed (1996), who also gives approximations to the variances of the estimates and discusses smaller samples.

For large samples, we may use the asymptotic normality of the estimators to obtain approximate tests and confidence intervals, as discussed above for the Weibull and extreme value distributions with Type I censoring. Let $z_i = (x_i - \mu^*)/\sigma^*$. The elements of the information matrix are estimated by

$$I_{11} = \frac{r}{\sigma^{*2}} + \frac{1}{\sigma^{*2}} \sum_{i=r+1}^{n} h'(z_i) \qquad (8.37)$$

$$I_{12} = \frac{2}{\sigma^{*2}} \sum_{i=1}^{r} z_i + \frac{1}{\sigma^{*2}} \sum_{i=r+1}^{n} [h(z_i) + z_i h'(z_i)] \qquad (8.38)$$

$$I_{22} = \frac{-r}{\sigma^{*2}} + \frac{3}{\sigma^{*2}} \sum_{i=1}^{r} z_i^2 + \frac{1}{\sigma^{*2}} \sum_{i=r+1}^{n} [2z_i h(z_i) + z_i^2 h'(z_i)] \qquad (8.39)$$

The inverse of the resulting matrix is an estimate of the covariance matrix of the estimators. See Nelson (1982) and Lawless (1982) for further details.

Example 8.7 (Case 2.2)

Suppose that time to failure for the windshield data can be assumed to be approximately normally distributed. The estimates of μ and σ are obtained by numerical solution of (8.35) and (8.36). The result (obtained from Minitab) is $\mu^* = 3.041$ and $\sigma^* = 1.241$ (both in thousands of hours of flight time). The elements of the inverse of the covariance matrix of the estimates are found, after some calculation, to be $I_{11} = 460.31$, $I_{12} = 614.74$, and $I_{22} = 1186.58$. The estimated covariance matrix is

$$\hat{V}(\mu^*, \sigma^*) = \begin{pmatrix} 0.007051 & -0.003653 \\ -.003653 & 0.002735 \end{pmatrix}$$

The estimated standard deviation of μ^* is $\sqrt{0.007051} = 0.0840$; that of σ^* is 0.0523. The relative high precision in the estimates indicated by these standard results is due to the relatively large sample size. ∎

8.3.4 Related Analyses and Results

There is a vast literature on censoring and analysis of censored data. Here we note a few related results regarding the distributions discussed above. Some additional results (e.g., estimation of reliability with censored data) will be given later in this chapter.

Censored data inherently involve incomplete information regarding times to failure. For practical purposes, it is necessary to assure that the data obtained are adequate in this context. This requires a well thought out test plan. Test plans for Type I and Type II censored data, with and without replacement of failed items, as well as sequential test plans are discussed by Lawless (1982, pp. 117–126). See also MIL-STD 781C.

One of the most commonly used alternatives to maximum likelihood estimation has been linear estimators, i.e., estimators that can be expressed as linear functions of the observations. These were developed prior to the advent of high-speed, easily accessible computers. They are still used in some applications, particularly in the case of the Weibull distribution, because of certain desirable properties and in situations where computation is a problem. For details, see Mann et al. (1974), Lawless (1982), and Nelson (1982).

Data that are both censored and grouped pose some special problems. Grouping of exponential data is discussed by Lawless (1982, pp. 125–126). Maximum likelihood estimation for censored, grouped data is discussed by Huet and Kaddour (1994).

Nonparametric approaches to analysis of censored data have been developed for a few situations. See Abdushukurov (1998), Hu et al. (1998), and the referenced cited in these articles.

8.4 BAYESIAN STATISTICAL ANALYSIS

Throughout the life cycle of a product, a good deal of information relevant to reliability is available to the engineer and manager. Even at the conceptual stage, information based on the performance of similar products and/or components, knowledge of material properties, judgmental assessments of reliability, and many other types of information are usually available. As the process proceeds through design, development, testing, and production, additional sources of information may become relevant and additional data—test data, vendor data, and so forth—will become available. This information, called *prior* information, can contribute significantly to reliability assessment, and its inclusion in the analysis and interpretation of data can greatly influence the results.

Bayesian statistical analysis provides a formal methodology for incorporating prior information, including subjective information, into the analysis of data. This is done by means of
a "prior distribution" (prior in the sense of before the acquisition of new data). Assumptions
concerning the form of the distribution of time to failure (exponential, Weibull, etc.) are
made as usual, and uncertainty in the parameters is modeled by the prior distribution. In addition to making use of the prior information, there are two additional distinct advantages to
this approach, if done properly: (1) It enables the analyst to make sensible statements concerning reliability estimation and prediction in cases when few or no failures are observed;
(2) It provides a natural means of updating reliability assessments as additional information
and data are obtained.

The use of a Bayesian approach is especially important in the context of very highly reliable parts or very high reliability requirements. This is the case, for example, in aerospace
applications, the nuclear industry, and certain medical applications. To achieve even reasonably high reliability (say, 0.99) in a space vehicle, very high reliabilities (e.g., 0.99999)
may be required at the part level. Reliabilities at this level are impossible to demonstrate
with any reasonable (i.e., affordable) amount of testing. Bayesian methods provide a mechanism for assessing reliability and providing a nontrivial interval estimate even when no
failures occur.

The Bayesian approach, although broadly applicable, does introduce a new level of difficulty in applications. It is still necessary, in a parametric framework, to select a specific failure distribution (or probability distribution for discrete data). In addition, it is necessary to
select a prior probability distribution to model uncertainty in the parameters, as well as to select parameters for the selected prior distribution In the remainder of Section 8.4, we look at
the basic concepts of Bayesian analysis, Bayesian methods for selected probability distributions, and some applications to reliability.[2]

8.4.1 Concept and Basic Approach

Bayes' Theorem. Bayesian statistical analysis is based on Bayes' theorem, a result involving
conditional probability (see Appendix A) which was first discovered by Thomas Bayes
(1763). Bayes' theorem for k events is as follows: Suppose that $E_1, \ldots, E_k$ form a partition
of the sample space, i.e., the events are mutually exclusive and their union is the entire sample space. Let A be any event with $P(A) > 0$. Then

$$P(E_j|A) = \frac{P(A|E_j)P(E_j)}{\sum_{i=1}^{k} P(A|E_i)P(E_i)} \tag{8.40}$$

for $j = 1, \ldots, k$.

Example 8.8 (Bayes' Theorem)

An implanted medical device is tested on an annual basis for a faulty lead. The test procedure
used is such that it indicates "faulty" 93% of the time when the lead is actually faulty and indicates "not faulty" 98% of the time when the lead is, in fact, not faulty. It is known from
past experience that the reliability of the lead is such that after one year, 1% of the leads are
faulty. A patient is tested after one year and tests positive for a faulty lead. What is the probability that the lead is actually faulty? The solution is obtained from Bayes' theorem. Here k
= 2, E_1 = "faulty lead," E_2 = "nondefective lead," and A = "test indicates faulty." We wish to

determine $P(E_1|A)$, the probability that the lead is faulty given a positive test result. We have $P(E_1) = 0.01$, $P(A|E_1) = 0.93$, and $P(A|E_2) = 0.02$. By Bayes' Theorem,

$$P(E_1|A) = \frac{P(A|E_1)P(E_1)}{P(A|E_1)P(E_1) + P(A|E_2)P(E_2)} = \frac{0.93(0.01)}{0.93(0.01) + 0.02(0.99)} = 0.32.$$

We conclude that only 32% of the time is a lead that tests positive actually faulty. This result reflects the unreliability of the test procedure. ∎

Application to Probability Distributions. In Chapter 4, probability distributions were written as the distribution of a random variable X *given* the parameter set θ. The *classical* approach to estimation of and testing hypotheses about the parameters of the distribution was discussed in Chapter 5. In that approach, the parameters are assumed to be (unknown) constants. In the Bayesian approach, the parameters are effectively assumed to be random variables. The *prior distribution,* which we denote $g(\theta)$, is the probability distribution of θ.

Formally, we have the following: $f(x_1, \ldots, x_n|\theta)$ is the conditional distribution of the data, given θ. We assume that the X_i are independent, in which case this distribution is the product of the individual distributions $f(x_i|\theta)$. Assume for now that θ is one-dimensional and continuous. The joint distribution of θ and the X's is

$$h(x_1, \ldots, x_n, \theta) = \left[\prod_{i=1}^{n} f(x_i|\theta)\right] g(\theta) \tag{8.42}$$

Note that $h(.)$ is the product of the likelihood function and the prior distribution. The marginal distribution of the X's is

$$f(x_1, \ldots, x_n) = \int_{-\infty}^{\infty} h(x_1, \ldots, x_n, \theta) d\theta = \int_{-\infty}^{\infty} f(x_1, \ldots, x_n|\theta) g(\theta) d\theta \tag{8.43}$$

Bayes' theorem is now used to obtain the *posterior distribution* of θ, given the X's, as

$$g(\theta|x_1, \ldots, x_n) = \frac{h(x_1, \ldots, x_n, \theta)}{f(x_1, \ldots, x_n)} \tag{8.44}$$

Thus the posterior distribution is proportional to the product of the likelihood function and the prior distribution, which is another interpretation of Bayes' theorem.

Inference problems regarding θ are now addressed through the agency of the posterior distribution, which encompasses both prior information and current data, i.e., presumably all of the information about θ that we have at hand. We look first at estimation, with additional topics to be taken up later.

Parameter Estimation. The Bayesian point estimator $\hat{\theta}_b$ of a parameter θ is determined as $E(\theta|x_1, \ldots, x_n)$, namely

$$\hat{\theta}_b = E(\theta|x_1, \ldots, x_n) = \int \theta g(\theta|x_1, \ldots, x_n) d\theta \tag{8.45}$$

We illustrate the approach by considering the binomial distribution with parameter $\theta = p$. From (4.24), the distribution is

$$p_x = f(x|p) = \binom{n}{x} p^x (1-p)^{n-x}$$

A natural prior distribution on p is the *beta distribution*, given by

$$g(p) = \frac{\Gamma(\alpha + \beta)}{\Gamma(\alpha)\Gamma(\beta)} p^{\alpha-1}(1-p)^{\beta-1} \tag{8.46}$$

$0 \le p \le 1$. The parameters are $\alpha > 0$ and $\beta > 0$. The mean and variance of the distribution are

$$\mu_p = \frac{\alpha}{\alpha + \beta}, \sigma_p^2 = \frac{\alpha\beta}{(\alpha + \beta)^2(\alpha + \beta + 1)} \tag{8.47}$$

The posterior distribution of p given x is found to be

$$g(p|x) = \frac{\Gamma(\alpha + \beta + n)}{\Gamma(\alpha + x)\Gamma(\beta + n - x)} p^{\alpha+x-1}(1-p)^{\beta+n-x-1} \tag{8.48}$$

For the binomial distribution with beta prior, we note that the posterior distribution is again a beta distribution, with parameters $(\alpha + x)$ and $(\beta + n - x)$. It follows that

$$\hat{p}_b = \frac{\alpha + x}{\alpha + \beta + n} \tag{8.49}$$

In applying these results, an issue that must be dealt with is determination of values for α and β. There are a number of approaches to this problem, some of which will be discussed in the next section. A relatively straightforward approach is to determine values for the mean and standard deviation of the prior distribution and then calculate the corresponding parameter values. The determination may be based on historical information, engineering judgment, and so forth. This requires solution of the expressions for the mean and standard deviation, given in (8.47), for α and β. The result is

$$\alpha = \frac{1}{2 - \mu_p}\left(\frac{(1 - \mu_p)\mu_p}{\sigma_p^2} - 1\right), \beta = \alpha(1 - \mu_p) \tag{8.50}$$

Example 8.9 (Case 2.27)

In Case 2.27, tests of starting capability of a low-pressure coolant pump (Pump A) in a nuclear reactor resulted in 236 successful starts in 240 trials. The maximum likelihood estimate of the proportion p of successes is [from (5.34)] $p^* = x/n = 236/240 = 0.9833$. Suppose a Bayesian approach is taken and it is determined that p is expected to be 0.999 with a standard deviation of 0.0005. From (8.50), we obtain the beta parameters as $\alpha = 3991$ and $\beta = 3.991$. The resulting Bayes estimate is $\hat{p}_b = 4227/4235 = 0.9981$. ∎

Note that in this example, the prior information, represented by a distribution with a high mean and little dispersion, essentially overwhelms the data. There is only a slight movement away from the prior value of $p = 0.999$ toward the sample proportion. This would not be the case if the standard deviation of the prior were larger. Small values of σ_p should be used only if there is great confidence in the prior information. As the sample size increases, however, the influence of the prior distribution can be expected to diminish.

We saw that when the beta distribution is used as a prior for the binomial parameter p, the posterior distribution is also beta. Priors having this property—posterior belonging to the same family of distributions—are called *conjugate* priors. Conjugate priors are often desirable for purposes of mathematical tractability, as is evident in the binomial case. *If* they are also able to represent the prior information adequately, their use in Bayesian analysis of data is justified. We look next at some additional distributions of importance in reliability.

8.4.2 Bayes Estimation for Selected Distributions

In this section, we consider a number of distributions previously discussed that are of importance in reliability applications. In each case, the conjugate prior distribution and the corresponding Bayesian point estimators will be given.

Binomial Distribution. The Bayes estimator with a beta prior is given in (8.49). Another reasonable choice for a prior distribution in some applications is a uniform distribution over the interval [0, 1]. The uniform distribution is given by $f(x) = 1$, $0 \leq x \leq 1$, and represents lack of any information about p. (As such, it is called a *noninformative* prior.) This and other choices are discussed by Martz and Waller (1982) (hereafter referred to as MW) and Sinha (1986).

Poisson Distribution [Equation (4.33)]. The random variable X is the number of failures in a fixed time period; λ is the failure rate per time period. The conjugate prior is a gamma distribution [Equation (4.41)] with parameters α and β. The resulting posterior distribution is

$$g(\lambda|x) = \frac{\lambda^{\alpha+x-1}}{[\beta/(\beta+1)]^{\alpha+x}\Gamma(\alpha+x)} e^{-\beta\lambda/(\beta+1)} \tag{8.51}$$

The Bayes estimator of λ is the mean of this distribution, namely

$$\hat{\lambda}_b = \frac{\beta(\alpha+x)}{\beta+1} \tag{8.52}$$

Exponential Distribution [Equation (4.37)]. We assume a sample of size n, with sample mean $\bar{x}$. The conjugate prior is a gamma distribution. The posterior distribution is

$$g(\lambda|x_1, \ldots, x_n) = \frac{(n\bar{x}+1/\beta)^{n+\alpha}}{\Gamma(n+\alpha)} \lambda^{n+\alpha-1} e^{-\lambda(n\bar{x}+1/\beta)} \tag{8.53}$$

The Bayes estimator of λ is

$$\hat{\lambda}_b = \frac{n+\alpha}{n\bar{x}+1/\beta} \tag{8.54}$$

Note: The Bayes estimator for a censored sample is given by (8.54) with $\bar{x}$ replaced by $\sum_{i=1}^{r} x_i + (n-r)x_{(r)}$, where, as before, $x_1, \ldots, x_r$ are the observed lifetimes and $x_{(r)}$ is the largest observed lifetime.

Weibull Distribution [Equation (4.45)]. For the Weibull distribution, the analysis is considerably more complex. The solution is straightforward if the shape parameter α is known. In this case, $Y = X^{\alpha}$ is exponentially distributed with parameter $\lambda = 1/\beta$. The results of the previous section may be used on the transformed data to estimate β. To avoid confusion, we write a and b for the parameters of the gamma prior in this case. The Bayes estimator of β is

$$\hat{\beta}_b = \left(\frac{n+a-2}{\sum_{i=1}^{n} x_i^{\alpha} + 1/b} \right) \tag{8.55}$$

If α and β are both unknown, it is usually assumed that these are independent random variables, so that their joint distribution is the product of the marginal distributions. For nearly any choice of prior distributions, the posterior is a complex expression and numerical integration is required for calculation of the Bayes estimates. See MW, Sinha (1986), and Sinha and Guttman (1988) for solutions. Choices of priors analyzed include uniform priors for both parameters, uniform distribution for α and gamma distribution for β, and a discrete distribution for α and gamma for β.

Gamma Distribution [Equation (4.41)]. Some work has been done on Bayesian estimation for the gamma distribution, again with complex results requiring numerical methods for solution. Conjugate priors have been considered, as well as a discrete distribution for selected values of the shape parameter α and an inverted gamma distribution for the scale parameter β. See MW (Section 9.5) and Tummala and Sathe (1978).

Inverse Gaussian Distribution [Equation (4.51)]. Bayesian analysis here is also quite complex. For a few results, see MW (Section 9.4) and Banarjee and Bhattacharyya (1979).

Normal Distribution [Equation (4.50)]. We assume that μ and σ are independent. Results are given in terms of μ and $\eta = 1/\sigma$. The joint conjugate distribution for μ and η is the product of a normal distribution with parameters λ, τ, and a gamma distribution with parameters α, β. The resulting posterior distribution is

$$g(\mu, \eta | x_1, \ldots, x_n) = \eta^{\alpha - 0.5 + n/2} e^{-\beta\eta - (\tau+n)\eta(\mu-y)^2/2} \tag{8.56}$$

where

$$y = \frac{\tau\lambda + n\bar{x}}{\tau + n} \tag{8.57}$$

The Bayes estimator of μ is $\hat{\mu}_b = y$. The Bayes estimator of σ^2 requires integrating out μ in (8.57) and then taking expectation with respect to η. For alternative approaches, see Sinha (1986).

Lognormal Distribution [Equation (4.55)]. As usual, analyze $Y = \log(X)$. The results follow from those for the normal distribution. See MW for details.

Example 8.10 (Case 2.10)

Data are time between failures of air conditioning units on a number of aircraft. Suppose that an analyst had data on Aircraft Numbers 7908, 7909, and 7910, and wished to use this information, in addition to some subjective evaluation, to form a prior for analysis of the data on Aircraft 7911. It is decided to assume an exponential distribution for TBF with a gamma prior.

The statistics on the three initial aircraft are

Aircraft	7908	7909	7910
MTBF$_i$	95.7	83.5	121.3
n_i	23	29	15

The grand mean is 96.2 with $n = 67$. Analysis of the estimated failure rates $\hat{\lambda}_i = 1/\text{MTBF}_i$, leads to a mean rate of 0.01022 and a standard deviation of 0.001876. To determine a prior distribution, we proceed as follows: The gamma distribution (See Section 4.4.2) has mean $\alpha\beta$ and variance $\alpha\beta^2$. We equate $\alpha\beta$ to 0.01022 and $\alpha\beta^2$ to $0.001876^2(20)$, the latter to account for the fact that we are dealing with means of about 20 observations. Suppose that after solution for α and β, and a subjective evaluation, the parameter values selected are $\alpha = 1.25$, $\beta = 0.008$.

The data on Aircraft 7911 provide $\bar{x} = 130.9$, with $n = 14$. From (8.54), the estimated failure rate is

$$\hat{\lambda}_b = \frac{14 + 1.25}{14(130.9) + 125} = 0.00779$$

Note that if we used the classical approach and only the data on this aircraft, the estimate would be $\hat{\lambda} = 1/130.9 = 0.00764$. If it were assumed that the four samples were from the same population, the classical approach would pool the data, obtaining a grand mean of 102.2, and $\hat{\lambda} = 1/102.2 = 0.00979$. ∎

8.4.3 Bayesian Probability Intervals

In confidence interval estimation, we determined intervals having the property that in the long run a specified proportion of intervals would contain the true value, which is assumed to be an unknown constant. Thus probability in this context was interpreted as the long-run frequency of occurrence. In Bayesian analysis, the parameter itself is assumed to be a random variable. In this context, we can make probability statements (usually subjective) about the parameter(s) of the distribution. We consider the one-dimensional parameter case. The results extend in a natural way to multidimensional parameters. A *Bayesian interval estimate* (also called a *Bayesian probability interval* or *Bayesian credible interval*) is an interval having the property that the parameter lies in the interval with specified probability $\gamma = 1 - \alpha$. Here the probability is not interpreted as a long-run frequency, but as applicable to the specific interval calculated from a set of data in a Bayesian analysis.

Bayesian probability intervals are calculated from the posterior distribution. Specifically, a Bayesian probability interval for a parameter θ is any interval (θ_1, θ_2) having the property that

$$\int_{-\infty}^{\theta_1} g(\theta | x_1, \ldots, x_n) d\theta = \alpha_1 \tag{8.58}$$

and

$$\int_{\theta_2}^{\infty} g(\theta | x_1, \ldots, x_n) d\theta = \alpha_2 \tag{8.59}$$

where $\alpha_1, \alpha_2 \geq 0$ and $\alpha_1 + \alpha_2 = \alpha$. As in confidence interval estimation, we usually take $\alpha_1 = \alpha_2 = \alpha/2$. These provide the shortest intervals for given α if the posterior distribution is symmetrical. Other choices, which lead to shorter intervals for asymmetrical posteriors, are discussed by Sinha (1986). (Shorter intervals having the same coverage are also available, incidentally, for the classical confidence interval estimators.)

We consider a few specific cases:

Binomial Distribution. With a beta prior with parameters α and β (and hence beta posterior), the $100\gamma\%$ Bayesian interval is given by

$$p_1 = \frac{\alpha + x}{\alpha + x + (\beta + n - x)F_{1-(1-\gamma)/2;2(n+\beta-x),2(\alpha+x)}} \tag{8.60}$$

$$p_2 = \frac{(\alpha + x)F_{1-(1-\gamma)/2;2(\alpha+x),2(\beta+n-x)}}{\beta + n - x + (\alpha + x)F_{1-(1-\gamma)/2;2(\alpha+x),2(\beta+n-x)}} \tag{8.61}$$

where $F_{P;a,b}$ is the P-fractile of the F-distribution with a and b df, given in Appendix C, Table C8. (If α and/or β are not integers, take df to be the closest integers to α and β.)

Example 8.11 (Case 2.27)

In Example 8.9, we calculated a Bayes point estimate of the proportion of times a pump in a nuclear reactor started successfully. The data were 236 successes in 240 tries; the Bayes estimate of p was 0.9981. For a 95% Bayes interval, we require 0.975 fractiles of the F distribution with 8454,16 and 16,8454 df. These values of df are beyond the limits of Table C8. In this case, 8454 df can be reasonably approximated by ∞ df. The resulting tabulated values needed in (8.60) and (8.61) are 1.80 and 2.32, respectively. The Bayes limits are $p_1 = 4227/[4227+8(1.80)] = 0.9966$ and $p_2 = 4227(2.32)/[8 + 4227(2.32)] = 0.9992$.

Note that this interval does not include the MLE $p^* = 236/240 = 0.9833$, again reflecting the precision specified in the selected prior. It is useful at this point for the analyst to question this choice. In general, the prior must be selected carefully and perhaps reevaluated to determine whether or not it expresses the true uncertainty in the prior information on which it is based.

Another point to be made with regard to the Bayes interval is that the interval is meaningful even if $x = 0$ or $n - x = 0$. This is not true of the classical confidence interval based on the MLE; if $x = 0$ or $n - x = 0$, the CI reduces to a single point (namely, 0 or 1). See Gross and Rust (1987) for additional discussion and tables. ■

Poisson Distribution. For a gamma prior with parameters α and β, the $100\gamma\%$ Bayesian interval for λ is

$$\lambda_1 = \frac{\chi^2_{(1-\gamma)/2,2(x+\alpha)}}{2(x + 1/\beta)}, \lambda_2 = \frac{\chi^2_{1-(1-\gamma)/2,2(x+\alpha)}}{2(x + 1/\beta)} \tag{8.62}$$

where $\chi^2_{P,a}$ is the P-fractile of the chi-square distribution with a df, given in Table C3.

Exponential Distribution. For a gamma prior with parameters α and β, the $100\gamma\%$ Bayesian interval is

$$\lambda_1 = \frac{\chi^2_{(1-\gamma)/2,2(n+\alpha)}}{2(n\bar{x} + 1/\beta)}, \lambda_2 = \frac{\chi^2_{1-(1-\gamma)/2,2(n+\alpha)}}{2(n\bar{x} + 1/\beta)} \tag{8.63}$$

Example 8.12 (Case 2.10)

In Example 8.10, data on selected aircraft were analyzed using the Bayesian approach. The Bayes estimate of λ was 0.00779. We calculate a 95% Bayesian probability interval. From the results of Example 8.10, the denominator of the terms in (8.63) is $2(n\bar{x} + 1/\beta) = 3915.2$. The df for the χ^2 are $2(14 + 1.25) = 30.5$, which we take as approximately 30. From Table C3, the required fractiles of the χ^2 distribution with 30 df are 16.79 and 46.98. The 95% Bayesian interval is $(16.79/3915.2, 46.98/3915.2) = (0.00429, 0.01200)$. ■

Weibull Distribution. As before, if the shape parameter α of the Weibull distribution is known, transform to $Y = X^\alpha$ and use the results for the exponential distribution. For unknown α, see Sinha (1986) and Sinha and Guttman (1988).

Inverse Gaussian Distribution. See MW, Section 9.4.

Normal and Lognormal Distributions. For the normal distribution with normal, gamma priors on μ, σ as above and Bayes estimator of the mean $y = (\tau\lambda + n\bar{x})/(\tau + n)$, the $100\gamma\%$ Bayes interval is

$$y \pm \left(\frac{2\beta}{(2\alpha + n)(\tau + n)} \right)^{1/2} t_{1-(1-\gamma)/2,2\alpha+n} \tag{8.64}$$

where $t_{P,a}$ is the P-fractile of the Student-t distribution with a df, obtained from Table C2 of Appendix C. Numerical integration is required to determine Bayesian intervals for σ. See MW for additional results.

For the lognormal distribution, transform to $Y = \log(X)$ and use the above results.

8.5   USE OF AGGREGATED DATA IN RELIABILITY ANALYSIS

A key problem in the analysis of reliability data (and in data analysis generally) is incorporation of various other types of information that might be available into the analysis and interpretation of results. In principle, the Bayesian approach provides a natural framework for doing this, through the prior distribution. It does not, however, provide a general methodology for formulating the prior. How that is done depends, in part, on the nature of the information that is to be utilized for this purpose. There are a number of approaches and some disagreement as to how a prior should be determined. We look briefly at a few approaches.

Note that the lack of a universally (or even widely) accepted formal structure in this context, the subjectivity of the analysis, and the resulting differences of opinion lead to a situation in which different analysts can on occasion arrive at quite different conclusions based on the same set of data. This has led many data analysts to reject the Bayesian approach altogether. The problem with this decision is that it leads to rejection of much valuable and useful information in analysis of the data. This is especially the case in many reliability applications involving complex, evolving systems or product lines. A reasonable approach would appear to be a compromise: Use the classical approach if prior information is not available, if the information that is available is thought to be of questionable relevance, e.g., because of changing circumstances, or if the available information is considered to be highly unreliable. If the available information is relevant and thought to be reliable, formulate a prior distribution that accurately represents this information (along with judgmental information, if appropriate) and use a Bayesian approach.

8.5.1   Data Sources

A number of sources of information that may be relevant to a reliability analysis were discussed in Chapter 3. These included

1. Historical data—test results and other information on previous models of a product, similar parts or components, and so forth
2. Vendor data—test results and other information obtained from a supplier of parts or components
3. Handbook data—information on materials' properties, expected lifetimes, etc.
4. Operational data—data collected on similar products or similar or identical parts or components under operational conditions

5. Warranty data—information from warranty claims and causes of failures leading to the claims

6. Environmental data—information concerning the operating conditions for the product and comparison of these to those of previous products, if any

7. Judgmental information—engineering judgment; a composite of a subjective assessment of reliability based on knowledge and experience of the engineering staff, an understanding of the product and process designs, and a subjective evaluation of some or all of the other information available

In a proper and thorough reliability analysis, all of these types of information (and any relevant information not listed) will be carefully collected at the outset. Whether or not a formal analysis leading to the formulation of a prior distribution is undertaken immediately, much of this information will be valuable in the initial design stages and in preliminary product evaluation. See Section 3.2.3 for additional discussion.

8.5.2   Formulation of Prior Distributions

Conjugate Prior Distributions. In order to use information from the various sources described above in a Bayesian analysis, a means of aggregating the different types of data into a prior distribution must be found. The first step in the process is selection of the form of the prior. In most cases in the previous section, a conjugate prior was used. Justification for the use of conjugate priors is based partly on the fact that they often provide mathematical tractability in that the posterior and its expectation are relatively easily determined and the posterior is a distribution of the same form as the prior.

The use of conjugate priors is further justified if, in addition, the distribution is flexible enough to represent many different types of information and the results are readily interpretable. The aim is to provide a tool for the analyst that will accurately and adequately model prior information and beliefs and at the same time is easy to use.

Martz and Waller (1982) discuss conjugate priors for most of the more important distributions in reliability. A question that must be addressed in the theory of Bayesian analysis is how one determines a conjugate prior for the parameters of a given distribution. Raiffa and Schlaifer (1961) provide a solution based on the theory of sufficient statistics and certain properties of the distribution of the random variable in question. See Raiffa and Schlaifer (1961) and De Groot (1970) for details of the analysis as well as for lists of conjugate priors for many commonly used distributions.

Noninformative Prior Distributions. Conjugate priors are often particularly useful in situations where a considerable amount of prior information is available. Noninformative priors are used in cases where there is little or no prior information. The usual choice for a noninformative prior is a uniform distribution over a finite range of possible parameter values, say $[a, b]$. The use of the uniform distribution implies that the analyst believes that values in any subinterval of $[a, b]$ of equal length are equally likely to occur. The resulting distribution is $f(x) = 1/(b - a)$, $a \leq x \leq b$. There are many other choices for noninformative priors. A formal definition is given by MW, along with a method for deriving a noninformative prior, again based on the theory of sufficient statistics. A thorough development and discussion of the theory of noninformative priors may be found in Box and Tiao (1973). Noninformative priors for many distributions are given by MW.

Other Prior Distributions. It is not necessary to use either conjugate or noninformative priors, and, in fact, they should not be used in situations where they may not represent the prior information adequately or where they cannot be completely specified. One alternative that is sometimes used is to select a finite set of possible parameter values and assign probabilities (usually subjectively) to each member of the set. This approach and a few others are discussed by MW, Chapter 6.

The key objective is to somehow formulate a model for the prior information that represents it adequately, in principle, regardless of the level of mathematical intractability. This is a workable approach in most cases, given the widespread availability of powerful computers.

The philosophy and practice of assigning prior probabilities is discussed in detail by Jaynes (1968), in which the "principle of maximum entropy" and analysis of transformations are proposed as a means of removing the subjectivity in selection of priors.

Selection of Parameter Values. Once the form of the prior distribution has been specified, it is necessary to determine parameter values for the prior. An approach discussed previously involved specification of the mean and standard deviation of the prior distribution, and determination of its parameters accordingly. There are many other versions of this approach, based on specification of characteristics of the prior distribution. In general, it is necessary to specify as many characteristics as there are parameters, express the characteristics in terms of the parameters, and solve for the parameters, as is done in the derivation of moment estimators.

In place of moments, one may use fractiles for this purpose. The median, quartiles, and 5th and 95th percentiles are common choices. A combination of the mean and one or more fractiles is also sometimes employed. Again, there is a vast literature on this subject and we make no attempt to survey it. The following sources and the many references cited by these authors will provide an introduction to certain aspects of the subject and some ideas that will be useful in important applications:

- Martz and Waller (1982)—As noted, considerable discussion on priors is given in Chapter 6 of MW. Included are selection of parameters for a beta prior based on the prior mean and 5th and 95th percentiles of reliability, selecting parameters for a gamma prior based on fractiles of the failure rate, and several other procedures for use of expert opinion (e.g., engineering judgment).

- Kapur and Lamberson (1977)—Selection of a beta prior given a prior value for the proportion of defectives and a gamma prior given a prior value for the failure rate are discussed in Chapter 13.

- Singpurwalla (1988)—A discussion is given of the elicitation and use of engineering judgment or expert opinion in determining a prior for Weibull parameters. An interactive PC program is used for this purpose. A Bayesian analysis of the ball bearing data of Case 2.14 is included.

- Zonnenshain and Haim (1984)—The authors discuss factors that affect reliability and use of questionnaires to obtain expert opinions regarding these factors. The questionnaire results are aggregated with prior experimental and other information to form a beta prior for the reliability of a system.

- Kuo (1985)—The use of engineering judgment in determining the parameters of gamma priors for the parameters of gamma and exponential distributions is discussed. Applications to electronic systems are included.

- Winkler (1967, 1981, 1986)—In these and a number of other papers, Winkler discusses the use of expert information, aggregating prior information from different sources, and the effect of this on data analysis and conclusions.

- Meeker and Hamada (1995)—excellent discussion of tests, data sources, and the use of these in product and process design.

- O'Leary (1998)—Marginal and conditional probabilities are elicited from groups of experts and the quality of the results is assessed. The results of an empirical study of the ability of groups versus individuals to provide consistent subjective probabilities are given.

8.6 ESTIMATION OF SYSTEM RELIABILITY

8.6.1 System Models

In Chapter 5 and in the previous section, we were concerned with estimation of the parameters of life distributions and other distributions used in the analysis of reliability. All of this was oriented to a single level—part, component, system, etc. In this section, we extend these results to estimation of reliability as a function of the life distribution and its parameters. We then extend the results to estimation of system reliability as a function of reliability at lower levels.

In this analysis, we assume that data (which may be test data, prior information, etc.) are available at two or more levels, or that data on two or more elements at a given level are to be used to estimate reliability at a higher level. The foundation for the extension to higher levels is provided by the system models discussed in Chapter 7. Reliability models that relate system to component reliabilities were given for series and parallel systems and for combinations consisting of series and parallel subsystems.

8.6.2 Inferences Based on Life Test Data

Here we consider point and interval estimation of reliability under various distributional assumptions on time to failure of a part, component, system, etc. (i.e., at a single level). It is assumed that data are available at the level in question and that parameters of the assumed distribution have been estimated by one of the methods (usually maximum likelihood) discussed in Chapter 5 or in previous sections of this chapter.

Testing hypotheses about reliability can also be accomplished, based on calculated confidence intervals, as indicated in Chapter 5: If the interval does not include the null-hypothesized value, reject the null hypothesis; if this value is in the interval, do not reject H_0. Note in this context that in reliability applications we are usually interested in testing whether the true reliability R is at least a certain value. (That is, we want to know only that it is lower than required, not that it is too high.) Correspondingly, we are ordinarily interested in a lower confidence interval (or *lower confidence limit*) on $R(t) = 1 - F(t)$, the reliability expressed as a function of time or a related characteristic. In the ensuing, we will look at confidence statements of the form "$100\gamma\%$ confident that R is in the interval $(R_L, 1]$," and tests of $H_0: R \le R_0$ versus $H_a: R > R_0$.

Estimators and confidence intervals for $R(t)$ for various distributions are as follows.

Exponential Distribution [Equation (4.36)]. The reliability function is $R(t) = e^{-\lambda t}$. This is estimated by

$$\hat{R} = \hat{R}(t) = e^{-t/\bar{x}} \tag{8.65}$$

A lower $100\gamma\%$ confidence interval for $R(t)$ can be obtained directly from the CI for λ. The result is

$$R_L(t) = e^{-[t\chi^2_{\gamma,2n}]/2n\bar{x}} \tag{8.66}$$

where $\chi^2_{P,a}$ is the P-fractile of the chi-square distribution with a df, obtained from Table C3.

Under Type II censoring, the procedure is the same, except that nx is replaced by $\Sigma_{i=1}^{r} x_{(i)} +$ $(n-r)x_{(r)}$, as usual in this case, and df$=2r$ rather than $2n$. If two-sided limits are desired, R_L and R_U are calculated using the $(1 + \gamma)/2$ and $(1 - \gamma)/2$ fractiles of the χ^2 distribution, respectively.

For approaches under Type I censoring, see Lawless (1982). Lawless also considers the two-parameter exponential distribution.

Bayes Intervals. In Bayesian analysis generally, two approaches to obtaining interval estimators for R have been considered. These are: (1) Choose a prior for λ, then transform to obtain a prior for R and analyze the data based on this prior distribution; (2) Select a prior for $R(t)$ and analyze reliability directly based on this prior. In either case, a Bayesian analysis is carried out as indicated above. See MW for details. We give the result for the exponential distribution based on approach (1) and assuming a gamma prior on λ with parameters α and β as in the previous section. The resulting lower confidence limit $R_{b,L}(t)$ is

$$R_{b,L}(t) = e^{-[t\beta\chi^2_{\gamma,2(\alpha+n)}]/[2(1+\beta n\bar{x})]} \tag{8.67}$$

Weibull and Extreme Value Distributions [Equations (4.44) and (4.53)]. The analysis in this case is quite complex. Exact solutions based on the MLEs are not possible, though exact methods based on linear estimation are. Linear estimators of parameters and reliability, as well as other exact and approximate methods and some required tables are given by Englehardt (1975), Englehardt and Bain (1974, 1977), and Mann and Fertig (1975). For a thorough discussion of these results for complete and censored samples, see Lawless (1982). An exact, asymptotically efficient confidence interval estimator for reliability is also given by Johns and Lieberman (1966), though the tables provided are somewhat limited.

For the Weibull distribution, $R(t)$ is given by

$$R(t) = R(t; \alpha, \beta) = e^{-(t/\beta)^\alpha} \tag{8.68}$$

For the extreme value distribution with parameters $\mu = \log \beta$ and $\lambda = 1/\beta$, the reliability function is

$$R(t; \mu, \lambda) = e^{-e^{(t-\mu)/\lambda}} \tag{8.69}$$

These can be estimated by substituting MLEs or other estimators into these expressions. A straightforward approach, good for moderate to large samples, is to use the results of Section 5.4 and Appendix A, Sections A5.4 and A5.5, to obtain an estimate of the asymptotic variance of the resulting estimators and use the normal approximation in the usual way. We pursue this approach for the Weibull distribution. The asymptotic variance of $R^* = R(t; \alpha^*, \beta^*)$, where α^*, β^* are the MLEs given in (5.53) and (5.54), is estimated by

$$\hat{V}_a(R^*) = \hat{V}_a(\alpha^*)\left(\frac{\partial R}{\partial \alpha}\right)^2 + \hat{V}_a(\beta^*)\left(\frac{\partial R}{\partial \beta}\right)^2 + 2\hat{\text{Cov}}_a(\alpha^*, \beta^*)\left(\frac{\partial R}{\partial \alpha}\right)\left(\frac{\partial R}{\partial \beta}\right) \tag{8.70}$$

where the derivatives are evaluated at α^*, β^*. The asymptotic variances of α^* and β^* are given in (5.55) and (5.56). The covariance is

$$\text{Cov}_a(\alpha^*, \beta^*) = \frac{0.254}{n\beta} \tag{8.71}$$

The required derivatives are

$$\frac{\partial R}{\partial \alpha} = \left(\frac{t}{\beta}\right)^\alpha \log(\beta/t)e^{-(t/\beta)^\alpha}, \quad \frac{\partial R}{\partial \beta} = \frac{\alpha t^\alpha}{\beta^{\alpha+1}} e^{-(t/\beta)^\alpha} \tag{8.72}$$

Example 8.13 (Case 2.20)
In Case 2.20, we consider 30 mm individual fibers and assume a Weibull distribution for breaking strength. Here $n = 70$. The MLEs were calculated in Example 5.7. The results are $\alpha^* = 5.524$ and $\beta^* = 2.648$. Suppose that in a certain application, fiber strength of $t > 1$ pound is required. From (8.68), we find $\hat{R}(1) = 0.9954$. The partial derivatives are estimated by substitution of α^* and β^* into (8.72), resulting in 0.00447 and 0.009619, respectively. The estimated variances were calculated in Example 5.7 as $\hat{V}_a(\alpha^*) = 0.00007257$ and $\hat{V}_a(\beta^*) = 0.2647$. The estimated covariance is found to be 0.001370. Thus, from (8.70),

$$\hat{V}_a(R) = 0.00007257(0.00447)^2 + .02647(0.009663)^2 + 2(0.001370)(0.00447)(0.009663)$$

$$= 0.000024834.$$

The resulting estimated standard deviation is 0.004983. A 95% lower confidence limit for $R(1)$ is $0.9954 - 1.645(0.004983) = 0.9872$. Suppose that 99% reliability is required with 95% confidence. To determine if this requirement is met, we test $H_0: R \leq 0.99$ versus $H_a: R > .99$. Since 0.99 is included in the confidence interval, we conclude that the null hypothesis cannot be rejected, i.e., that there is not statistical evidence that the requirement is met.

If the application required $t = 1.5$ pounds, the corresponding values are $R^*(1.5) = 0.9576$ with an estimated standard deviation of 0.04557, giving a 95% lower confidence limit of $0.9576 - 1.645(0.04557) = 0.8843$. ■

The Bayesian approach is again quite complicated if both α and β are unknown. See MW for a few results.

Gamma Distribution (Equation (4.41)). For the gamma distribution with shape parameter α and scale parameter β, the reliability function is given by

$$R(t) = R(t;\ \alpha,\ \beta) = 1 - \int_0^{t/\beta} \frac{x^{\alpha-1}e^{-x}}{\Gamma(\alpha)}\,dx \qquad (8.73)$$

The integral in (8.73) is the incomplete gamma function and is evaluated as indicated previously [e.g., see Abramowitz and Stegun (1964)]. $R = R(t)$ is estimated by substituting the MLEs α^* and β^* into (8.73). Since the integral is not expressible in closed form (unless α is an integer), the approach used for the Weibull distribution is messy. Lloyd and Lipow (1962) suggest the following alternative: Based on the *Edgeworth expansion* of the gamma distribution, $R(t)$ can be approximated by

$$R(t) \approx 1 - \Phi(z) + \frac{z^2 - 1}{3\sqrt{2\pi\alpha}}e^{-(1/2)z^2} \qquad (8.74)$$

where $\Phi(.)$ is the standard normal CDF and $z = (t/\beta - \alpha)/\sqrt{\alpha}$. The corresponding estimate $\hat{R}$ of R is calculated by substitution of α^*, β^* into (8.74). The asymptotic variance of $\hat{R}$ is approximated by

$$\hat{V}_a(\hat{R}) = \frac{e^{-z^2}}{2\pi n}\left[\frac{2t^2}{\alpha^*\beta^{*2}} - (2\alpha^* - 1)\left(1 + \frac{z}{2\sqrt{\alpha^*}}\right)\left(1 + \frac{3z}{2\sqrt{\alpha^*}}\right)\right] \qquad (8.75)$$

Interval estimates and tests are based on the normal approximation.

Because of analytical difficulties, only a few Bayesian results are available. See MW for some approaches.

Normal and Lognormal Distributions. We give the results for the normal distribution with parameters μ and σ. $\bar{x}$ and s are the sample mean and standard deviation, and we write $z = (t - \bar{x})/s$. The estimated reliability is $R^*(t) = 1 - \Phi(z)$. An approximate $100\gamma\%$ lower confidence bound for R, good for $n \geq 20$, is

$$R_L(t) = 1 - \Phi(z_{L,\gamma}) \tag{8.76}$$

where

$$z_{L,\gamma} = z + \left(\frac{z_\gamma}{\sqrt{n}} \right) \left[1 + \frac{nz^2}{2(n-1)} \right]^{1/2} \tag{8.77}$$

For the lognormal distribution, transform to $y = \log x$ and apply (8.76) and (8.77) to the resulting data.

The exact result, good for any n, is a function of the noncentral t-distribution. See Lawless (1982, p. 230). Bayesian point and interval estimation are dealt with in MW, Section 9.2 and Padgett and Wei (1978).

Inverse Gaussian Distribution. A thorough analysis of this distribution has not been done. The MLE of $R(t)$ may be obtained by substitution of MLEs of the parameters in the usual way. Some Bayesian results are given in MW, p. 448.

Nonparametric Estimation of Reliability. For complete data, reliability as a function of t may be estimated without explicit use of the distribution function F, as follows. Let $X =$ number of items in the sample that survive to at least time t. X is a binomial random variable with parameters n and $p = 1 - F(t)$. Calculate a lower confidence limit for p as indicated in Chapter 5.

Example 8.14 (Case 2.20)

Estimates of reliability based on these data were given in Example 8.12 under the assumption of a Weibull distribution. Take $t = 1.5$. We have three failures (i.e., three observed lifetimes less than 1.5) and $n = 70$, so $\hat{R}(1.5) = 67/70 = 0.9571$, which is very close to the Weibull estimate of 0.9576. The nonparametric confidence limit based on the normal approximation differs somewhat, however. The lower 95% confidence limit is found to be $0.9571 - 1.645[0.9571(0.0429)/70]^{1/2} = 0.9173$, a somewhat better result than that based on the Weibull assumption. (The approximation may not be appropriate, however, since the condition $n\hat{p} > 5$ and $n(1 - \hat{p}) > 5$ is not satisfied.) ∎

Updating Reliability Estimates. As testing continues, it is useful to update reliability estimates, incorporating the new data into the estimation process. There are many ways of doing this. The two basic principles that are usually followed are the classical and Bayesian approaches.

Classical Approach. If the new data are collected under the same conditions as the previous data set and all data are considered to be of equal relevance and importance, form a composite sample and analyze this as a single data set. If some parts of the composite data set are considered to be more important than others (more precise, more relevant, etc.), use a weighted analysis (calculating weighted means, etc.), weighting the more important elements more heavily. The problem, of course, is determining the weights; here engineering judgment (based on the context, *not* on the data) may play an important role.

Bayesian Approach. The Bayesian framework provides a natural methodology for incorporation of new information, since that is the very essence of the Bayesian approach—begin with prior information and then form the posterior distribution on which inferences are based. The posterior distribution thereby becomes a prior for further test data and the procedure continues sequentially as additional data become available.

Example 8.15 (Case 2.27)
Our previous Bayesian analysis of this case resulted in a beta posterior distribution with $\alpha = 4227$ and $\beta = 4235$. Suppose that a second test consisting of 400 attempted starts of the pump is done, resulting in 390 successes. The updated Bayesian estimate of p is

$$\hat{p}_b = \frac{4227 + 390}{4235 + 400} = 0.9961.$$

The result is a lower value for the estimated proportion, but still quite high relative to the sample proportion. ∎

8.6.3   Estimation of System Reliability from Component Reliabilities

We consider systems consisting of components connected in series, in parallel, or in combinations of series and parallel components. Several approaches to estimating system reliability as a function of component reliabilities will be considered.[3] Note that we are in actuality considering reliability analysis at a given level as a function of reliability assessments at the next lower level. "System" may actually be a component or subsystem, etc., and "component" may by a subsystem, component, part, etc. In fact, the analysis that will be discussed can be built up through more than two levels in a somewhat obvious fashion. Thus we use "system" and "component" in a generic sense.

As in calculating reliabilities, the basis of estimation of system reliability is a model of the relationship between reliability at the system level and reliability of the components. This is discussed in Chapter 7 for a number of important system structures. Many additional system models may be found in Kececioglu (1994) and Hoyland and Rausand (1994).

We look first at basic estimation for various system configurations and then at methods for determining confidence intervals.

Basic Point Estimation
Series Systems of Independent Components. We consider k independent components, with respective reliabilities $R_1, \ldots, R_k$, connected in series. (We suppress the dependence of reliability on time in most of this section.) System reliability is given by $R = R_1 R_2 \cdots R_k$, and estimated by $\hat{R} = \hat{R}_1 \hat{R}_2 \cdots \hat{R}_k$, the product of the estimated reliabilities of the components. By independence,

$$E(\hat{R}) = E(\hat{R}_1) \cdots E(\hat{R}_k) \tag{8.78}$$

If the $\hat{R}_i$ are unbiased, $\hat{R}$ is an unbiased estimator of R.

The variance of $\hat{R}$ can be determined by the results of Appendix A5. For $k = 2$, we have

$$V(\hat{R}) = V(\hat{R}_1 \hat{R}_2) = [E(\hat{R}_1)]^2 V(\hat{R}_2) + [E(\hat{R}_2)]^2 V(\hat{R}_1) + V(\hat{R}_1)V(\hat{R}_2) \tag{8.79}$$

For $k = 3$, the result is given by Equation (A5.9), with $\mu_i = E(R_i)$ and $\sigma_i^2 = V(\hat{R}_i)$. The result extends to $k > 3$ in a straightforward manner. If the estimators of the component reliabilities are unbiased, the expectations in (8.79) are simply the true values R_i. In practice, we assume that the estimators are at least approximately unbiased and use the true values (estimated or hypothesized) in our computations.

Parallel Systems of Independent Components. For parallel systems, it is convenient to look at $\overline{R} = 1 - R$. Since the system fails only if all components fail and failures occur independently,

$$\overline{R} = P(\text{All fail}) = (1 - R_1)(1 - R_2) \cdots (1 - R_k) = \overline{R}_1 \overline{R}_2 \cdots \overline{R}_k \tag{8.80}$$

R is estimated by

$$\hat{R} = 1 - \hat{\overline{R}} = 1 - \hat{\overline{R}}_1 \hat{\overline{R}}_2 \cdots \hat{\overline{R}}_k \tag{8.81}$$

Since for any random variable X, $V(1 - X) = V(X)$, (8.78) and (8.79) and the extensions of these to $k > 2$ apply directly to $\hat{R}$ in this case as well. Thus the estimator is unbiased if the $\hat{R}_i$ are, and the variance of $\hat{R}$ is calculated from (8.79) and its generalization for $k > 2$, with $\overline{R}$'s replaced by R's in the expectations.

Independent Components in Other Configurations. For systems consisting of combinations of components in series and components in parallel, the analysis is done as either a series or parallel system of subsystems, with each subsystem analyzed separately and the results then aggregated to the system level. The methodology used is that given above. The analysis is straightforward (though often tedious) as long as all elements are independent. A block diagram, as discussed in Chapter 7, is essential for the analysis of even relatively simple systems.

If the system cannot be modeled in this way (e.g., as a series system of series and parallel subsystems), use a block diagram to represent the system and use this to express the reliability R as a function of component reliabilities. Use the methods of Chapter 5 and Appendix A5 to determine the asymptotic variance of the estimator of R in terms of the variances and covariances of the estimators of component reliabilities. For large, or even moderately large, systems, computer analysis is essential.

Confidence Interval Estimation
Methodology. In all of the above results, estimation of reliabilities and their variances (exact or asymptotic) may be based on any of the models discussed in the previous sections—binomial data, models based on specific life distributions, or combinations of these. Crucial assumptions are independence and the form of the distribution.

Equations (8.78) and (8.79) are exact results for the mean and variance of $\hat{R}$. To determine exact confidence intervals, the distribution of $\hat{R}$ is required, and this in not uniquely determined by its mean and variance. Except for a few cases, the mathematics involved is rather complicated, and in almost no case provides a tractable result. A simple case where an exact result is easily obtained is that in which the k components are connected in series and are independent with identical exponential distributions with parameter λ. In this case, $\hat{R}(t) = e^{-k\lambda t}$, the estimate of λ is based on the composite of failure data from all components, and the methods for the exponential distribution apply directly. (If the λ's are different, the reliability function is easily estimated as above, but the distribution problems are more difficult.)

Most results provided in the considerable literature on this subject assume binomial (pass/fail) data, and some of the results given below are based on that approach. Others are based on asymptotic results, and require moderate to large sample sizes. We look at several approximations.

As before, attention will be focused on a lower confidence limit for R. If a two-sided confidence interval is desired, the results can be modified as indicated above and in Chapter 5.

The approximations to be considered are based on asymptotic normality, a binomial approximation, and a bound on the lower confidence limit. The results are as follows:

Asymptotic Normality Approach. Calculate point estimates as indicated. Substitute these into (8.79) or its equivalent for other models to estimate $V(\hat{R})$. (Estimated variances for the components may be binomial variances, exact results based on a specific distributional assumption, asymptotic variances, or other approximations.) Once the estimate, $\hat{V}(\hat{R})$, is calculated, the lower $100\gamma\%$ confidence limit is calculated as

$$\hat{R} - z_\gamma \sqrt{\hat{V}(\hat{R})} \tag{8.82}$$

where z_γ is the γ-fractile of the standard normal distribution. Moderate to large sample sizes are required for validity of this approach.

__Easterling Approach.__ Easterling (1972) notes that the normal distribution is symmetrical, whereas that of $\hat{R}$ is almost always skewed, which can lead (in the two-sided case) to CIs that include values outside the interval [0, 1]. In any case, approximating a skewed distribution by the normal distribution will lead to imprecision in the result. Easterling proposed instead the use of a binomial distribution with variance equal to the calculated variance of $\hat{R}$. The approximating binomial has parameters $p = \hat{R}$ and an effective sample size n', calculated by equating variances, of

$$n' = \frac{\hat{R}(1 - \hat{R})}{\hat{V}(\hat{R})} \tag{8.83}$$

The confidence limit is calculated by determining the p-value for which the lower-tail binomial probabilities add to exactly $1 - \gamma$. This is done by use of the relationship between the binomial CDF and the incomplete beta function. The latter is available in many computer packages (e.g., Minitab). The lower confidence limit is

$$R_L = B(1 - \gamma; x', n' - x' + 1) \tag{8.84}$$

where $x' = n'\hat{R}$ and $B(p; a, b)$ is the p-fractile of the beta distribution with parameters a and b. (Note that a and b are not required to be integers in this calculation.)

__Spencer–Easterling Bound.__ A lower bound on R_L is given by Spencer and Easterling (1986). The bound is calculated on the basis of a reduction of component data to effective binomial system data. Binomial test data are assumed, with n_i being the number of items tested for the ith component and x_i the number of successes. The bound depends on the structure of the system.

For a series system, the estimate of system reliability is calculated as

$$\hat{R} = \prod_{i=1}^{k} \left(\frac{x_i}{n_i} \right) \tag{8.85}$$

The effective sample size is taken to be $n_s = \min(n_i)$, and the effective number of successes is taken to be $x_s = n_s \hat{R}$. The bound on the lower confidence limit on R is calculated as $B(1 - \gamma; x_s, n_s - x_s + 1)$.

For a parallel system, calculate

$$q_0 = \prod_{i=1}^{k} \left(\frac{n_i - x_i}{n_i} \right) \qquad q_1 = \prod_{i=1}^{k} \left(\frac{n_i - x_i + 1}{n_i + 1} \right) \tag{8.86}$$

The effective n is calculated as

$$n_s = \frac{1 - q_1}{q_1 - q_0} \tag{8.87}$$

The effective number of successes is $x_s = (1 - q_0)n_s$. The lower confidence bound on reliability is calculated as $B(1 - \gamma; x_s, n_s - x_s + 1)$.

For other systems of series/parallel components, including those with repeated components, and many examples, see Spencer and Easterling (1986).

Example 8.16 (Case 2.27)

Consider Pump A in Case 2.27. Suppose that a downstream valve is required to open within two seconds of activation of the pump and that in 400 tests, the valve met this requirement 392 times. The pump and valve must both operate properly for this system to be in perfect operational condition. The system at this level consists of Pump A as Component 1 and the valve as Component 2, connected in series. We have $\hat{R}_1 = 236/240 = 0.983333$, $\hat{V}(\hat{R}_1) = 0.9833(0.0167)/240 = 0.00006842$, $\hat{R}_2 = 392/400 = 0.980000$, and $\hat{V}(\hat{R}_2) = 0.00004900$. The estimated reliability of the system is $\hat{R} = (0.983333)(0.9800) = 0.9637$. From (8.79), the estimated variance is

$$\hat{V}(\hat{R}) = 0.9800^2(0.00006842) + 0.9833^2(0.00004900)$$
$$+ (0.00006842)(0.00004900) = 1.13094 \times 10^{-4}$$

We calculate 95% lower confidence limits by the three methods. The asymptotic normal approach gives

$$R_L = 0.9637 - 1.645(0.01063) = 0.9462$$

To calculate the Easterling limit, we use (8.83) and (8.84). For these data, we obtain

$$n' = 0.9637(0.0363)/0.000113094 = 309.3$$

and $x' = 0.9637(309.3) = 298.1$. From Minitab, $B(0.05; 298.1, 12.2)$ is found to be 0.9410.

The Spencer–Easterling bound is obtained using $n_s = 240$, giving $x_s = \hat{R}n_s = 231.3$. The bound on the lower limit is $B(0.05; 231.3, 9.7) = 0.9370$. The two approximations are in reasonably close agreement and both are less than the calculated lower bound. ■

Example 8.17 (Case 2.27)

Suppose that Pump C is a backup to Pump A. The system is then a parallel system in that the system fails only if both Pumps A and C fail. For C, the data are 238 successes out of 240 trials, giving $\hat{R}_2 = 0.9917$, with an estimated variance of 0.000034433. The estimated reliability of the system is

$$\hat{R} = 1 - (1 - 0.9833)(1 - 0.9917) = 0.999861$$

The estimated variance of $\hat{R}$ is

$$\hat{V}(\hat{R}) = 0.0083^2(0.00006842) + 0.0167^2(0.00003443) + (0.00006842)(0.00003443)$$
$$= 1.6633 \times 10^{-8}$$

The estimated standard deviation is 0.0001291, giving a 95% lower confidence limit based on normality of $0.999861 - 0.000212 = 0.9996$. For the Easterling limit, $n' = 8355.8$, giving a limit of 0.9994. The Spencer–Easterling bound is also 0.9994. ∎

Comment: Note that the estimated reliability for a parallel system is much higher than that of a series system, as expected. Here the variance of a parallel system is much lower that than of a series system. This will always be the case for relatively high reliability components. The opposite would be true of highly unreliable components.

8.6.4 Bayesian Analysis of Multicomponent Systems

The Bayesian approach provides a natural method of aggregating data at different levels of a system to obtain an estimate and interval for system reliability. In the classical approach discussed in the previous sections, confidence intervals for system reliability are obtained by analysis of the system structure and use of asymptotic or other approximations. This is done by expressing system reliability as a function of the reliabilities at all lower levels, then using this relationship to determine the asymptotic variance of the estimate, and, finally, using asymptotic normality to obtain the confidence interval.

In contrast, the Bayesian approach uses information obtained at a given level to determine a prior distribution for reliability (or a related characteristic) at the next higher level. This also, of course, requires an understanding of the system structure. It does provide the advantage of a logical method of incorporating not only information from previous levels, but also both prior and test information at the current level. The difficulties are that, as always in the Bayesian approach, quantifying the prior information may be highly subjective and therefore subject to criticism, and that the mathematics may become quite intractable.

A schematic representation of the process is given in Figure 8.1. The figure shows the path from part to system reliability calculation, with details shown at an intermediate component level. It is assumed that prior distributions on part reliability have been obtained by methods such as those discussed in Section 8.4. These are combined by means of the reliabil-

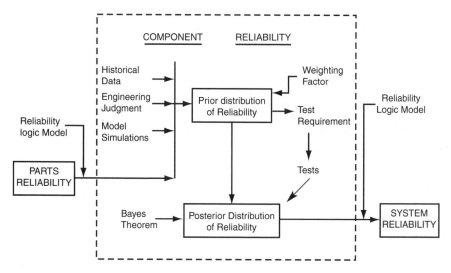

Figure 8.1. Bayesian system reliability analysis

ity logic model to form an induced prior distribution on component reliability. In addition, other information may be available at this level as well, including relevant historical data, subjective evaluations, results of simulations, and so forth. This information is used to form an independent assessment of the prior distribution (a "natural" prior). Weights are assigned to these two priors and they are combined to form a composite prior distribution of component reliability. Tests are performed and the posterior distribution of component reliability is determined. This is done for each component, and the process proceeds through the various levels of the system, again using the system logic model, up to the formulation of the posterior distribution of system reliability.

We illustrate the Bayesian approach in the context on multilevel data and priors with pass/fail (binomial) data at each level.

The Martz–Waller–Fickas Approach. A detailed description of an approach such as that discussed above is given by Martz et al. (1988) for binomial components in a series system. The system structure is given in Figure 8.2. It consists of m subsystems linked in series, with the ith subsystem consisting of k_i components, also in a series configuration. Prior and test results are denoted by n and s (nominally numbers of tests and successes, but may be any positive values for priors, determined as indicated in Section 8.4). The superscript "0" indicates prior parameters; lack of a superscript indicates test results. Double subscripting indicates component level values, single subscripts indicate subsystem values, and lack of a subscript indicates system level values. Prior distributions are taken to be beta distributions with parameters $s + 1$ and $n - s + 1$, with sub- and superscripts on s and n as appropriate.

The steps in determining the subsystem and system posterior distributions are given in

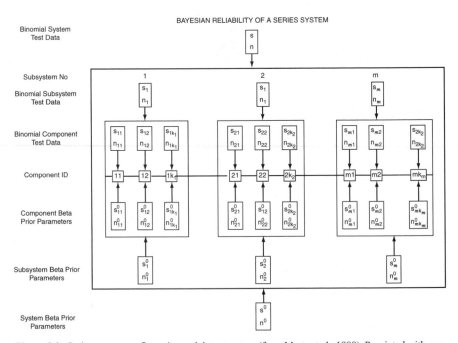

Figure 8.2. Series system configuration and data structure (from Martz et al., 1988). Reprinted with permission from *Technometrics,* Copyright 1988 by the American Statistical Association. All rights reserved.

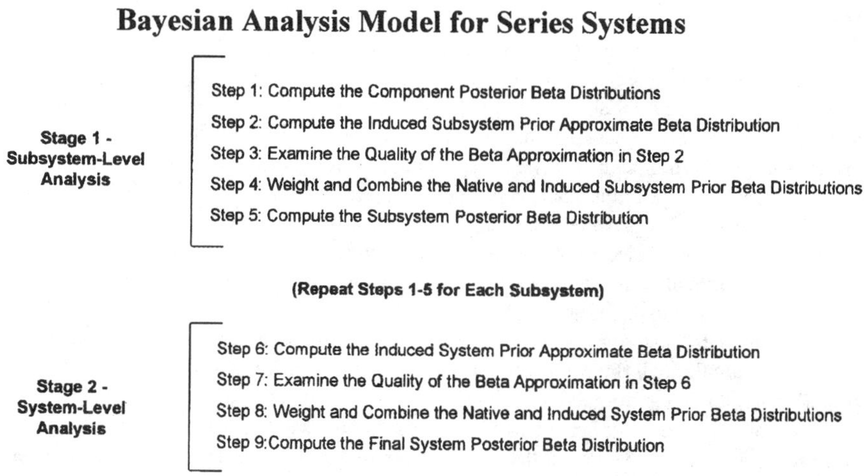

Bayesian Analysis Model for Series Systems

Stage 1 - Subsystem-Level Analysis

Step 1: Compute the Component Posterior Beta Distributions

Step 2: Compute the Induced Subsystem Prior Approximate Beta Distribution

Step 3: Examine the Quality of the Beta Approximation in Step 2

Step 4: Weight and Combine the Native and Induced Subsystem Prior Beta Distributions

Step 5: Compute the Subsystem Posterior Beta Distribution

(Repeat Steps 1-5 for Each Subsystem)

Stage 2 - System-Level Analysis

Step 6: Compute the Induced System Prior Approximate Beta Distribution

Step 7: Examine the Quality of the Beta Approximation in Step 6

Step 8: Weight and Combine the Native and Induced System Prior Beta Distributions

Step 9:Compute the Final System Posterior Beta Distribution

Figure 8.3. Steps in a two-stage reliability analysis (from Martz et al., 1988). Reprinted with permission from *Technometrics,* Copyright 1988 by the American Statistical Association. All rights reserved.

Figure 8.3. At the component level, the parameters of the beta prior are $(s_{ij}^0 + 1)$ and $(n_{ij}^0 - s_{ij}^0 + 1)$. The corresponding posterior distribution is a beta with parameters $(s_{ij}^0 + s_{ij} + 1)$ and $(n_{ij}^0 + n_{ij} - s_{ij}^0 - s_{ij} + 1)$. (If no testing is done for a particular component, the posterior is taken to be the same as the prior.)

To carry this to the next level, we use the fact that the reliability of a subsystem is the product of the component reliabilities. Thus the subsystem prior distribution *induced* from the component priors is the distribution of a product of beta variates. If the parameters of the component priors are nonnegative integers, Mellin transforms can be used to obtain an exact solution (see Springer and Thompson, 1966), but the computations are complex. Martz et al. (1988) suggest instead that a beta distribution be used to approximate the induced subsystem prior. For the ith component, the parameters of the approximating beta distribution, denoted a_i and b_i, are given by

$$a_i = [M_i^2(1 - M_i) - V_iM_i]/V_i \tag{8.88}$$

and

$$b_i = [M_i(1 - M_i)^2 - V_i(1 - M_i)]/V_i \tag{8.89}$$

where

$$M_i = \prod_{j=1}^{k_i} \frac{s_{ij}^0 + s_{ij} + 1}{n_{ij}^0 + n_{ij} + 2} \tag{8.90}$$

and

$$V_i = \prod_{j=1}^{k_i} \frac{(s_{ij}^0 + s_{ij} + 1)(s_{ij}^0 + s_{ij} + 2)}{(n_{ij}^0 + n_{ij} + 2)(n_{ij}^0 + n_{ij} + 3)} - M_i^2 \tag{8.91}$$

Martz et al. (1988) suggest that in applying the method, the adequacy of the approximation be investigated. A simulation is suggested for this purpose.

The final step in preparation of the subsystem prior distributions is to combine the predetermined ("native") priors at this level with the induced priors. This is done by assigning weights to the respective beta parameters that express the relative importance of the induced and native prior information. The weights, denoted w_{i1} and w_{i2}, then determine the parameters of the combined prior distribution for the ith subsystem. These are $w_{i1}a_i + w_{i2}s_i^0 + w_{i2}$ and $w_{i1}b_i + w_{i2}(n_i^0 - s_i^0) + w_{i2}$. A problem that must be addressed is selection of the weights. It is suggested that they be selected so that their sum is between 1 and 2 and that this sum be close to two if the information in the induced and native priors can reasonably be assumed to be independent and close to one if not.

The subsystem posterior distributions are next determined by combination of prior and test information through Bayes' theorem in the usual way. The process used to determine the posterior distribution at the subsystem level is now repeated at the system level. The induced prior is approximated in the same way since the subsystems are also assumed to be connected in series and weighted sums of the resulting parameters and those of the native priors are formed. Finally, the system posterior distribution is determined in the usual way. A thorough discussion of this process and a worked example can be found in Martz et al. (1988).

This type of analysis can be extended in an obvious way to additional levels—parts, assemblies, components, etc., connected in series. Martz and Waller (1990) give extensions of the analysis to complex systems of series and parallel subsystems with binomial data. Hulting and Robinson (1994) extended the results for series systems to lifetime data and combinations of lifetime and pass/fail data. Weibull distributions are assumed for the lifetime data. Repairable items and censoring are included in the analysis.[4]

8.7 OTHER INFERENCE PROBLEMS IN RELIABILITY

8.7.1 Estimation in Stress–Strength Analysis

In stress–strength models, discussed in Section 6.5, failure is characterized by means of two random variables: X, the strength of the item, and Y, the stress to which it is subjected. The item fails if $X < Y$, and item reliability is calculated as $R = P(X > Y)$. R is given for X and Y both normally distributed and both exponentially distributed in Section 6.5, and we look later at these cases. A number of other models are given by Kapur and Lamberson (1977, Chapter 6). In this section, we discuss estimation of R based on data on X or on both X and Y.[5]

Note that the very simplest situation of this type is that in which Y is a fixed value, say y, rather than a random variable, and we simply observe whether or not each item in a sample of size n fails on test under stress y. Thus we have a binomial situation with n the sample size and $p = 1 - F_X(y)$, and a confidence interval for R ($= p$) is obtained by the methods of Chapter 5.

A somewhat more complicated situation is that in which y is fixed and actual stress at failure is recorded for each item (as in Cases 2.6, 2.20, 2.21, and 2.22 discussed in Chapter 2.) In this case, one can estimate parameters of the assumed strength distribution and use these to estimate $F_X(y)$ by the methods for estimating fractiles of a distribution. Nonparametric methods (also discussed in Chapter 5) may also be used for this purpose. Nonparametric confidence intervals when both X and Y are random variables and independent samples of each are obtained are given by Ury (1972).

We turn next to parametric analyses in which both X and Y are random variables and specific distributional assumptions are made. Note that the usual situation is that in which the distribution of stress Y is known or assumed, and a sample of strengths $X_1, \ldots, X_n$ is observed, with the X_i independent and identically distributed.

Normally Distributed Stress and Strength. We assume that X and Y are independent normal variates with respective parameters μ_X, σ_X, and μ_Y, σ_Y. This is a very reasonable assumption in many problems of this type since stress and strength are much more likely to have symmetrical or nearly symmetrical distributions than are, for example, item lifetimes. This is apparent in the histograms of the strength data in the several cases discussed in Chapter 2. μ_Y and σ_Y are assumed known.

From Equation (6.8), item reliability is given by

$$R = \Phi\left(\frac{\mu_X - \mu_Y}{\sqrt{\sigma_X^2 + \sigma_Y^2}} \right) \tag{8.92}$$

where $\Phi(.)$ is the CDF of the standard normal distribution. This is estimated by

$$\hat{R} = \Phi\left(\frac{\bar{x} - \mu_Y}{\sqrt{s_x^2 + \sigma_Y^2}} \right) \tag{8.93}$$

where $\bar{x}$ and s_x^2 are the sample mean and variance of observed strengths. To obtain an approximate confidence interval for R, we write

$$W = \frac{\bar{X} - \mu_Y}{\sqrt{S_x^2 + \sigma_Y^2}} \tag{8.94}$$

W is asymptotically normally distributed (Church and Harris, 1970) with mean $(\mu_X - \mu_Y)/\sqrt{\sigma_X^2 + \sigma_Y^2}$ and variance

$$\sigma_W^2 = \frac{\sigma_X^2}{\sigma_X^2 + \sigma_Y^2}\left(\frac{1}{n} + \frac{(\mu_X - \mu_Y)^2}{2(n-1)(\sigma_X^2 + \sigma_Y^2)^2} \right) \tag{8.95}$$

This variance is estimated by substituting the sample mean and variance of X into (8.95). The approximate (two-sided) CI for R is given by

$$\Phi(W - z_{1-\alpha/2}\hat{\sigma}_W), \; \Phi(W + z_{1-\alpha/2}\hat{\sigma}_W) \tag{8.96}$$

where $\hat{\sigma}_W$ is the estimated standard deviation of W. A one-sided CI is obtained by the usual modification of (8.96).

Example 8.18 (Case 2.6)
Suppose that fibers of length 12.7 mm are to be used in an application in which stress is normally distributed with mean $\mu_Y = 1.50$ and standard deviation $\sigma_Y = 0.50$. The mean and standard deviation of strength are estimated from the data to be $\bar{x} = 3.091$ and $s_X = 0.6353$. The sample value of W is found, from (8.94), to be $W = (3.091 - 1.5)/(0.6353^2 + 0.5^2)^{1/2} = 1.968$. The estimated reliability is $\Phi(1.968) = 0.9755$. From (8.95), the estimated variance of W is found to be $\hat{\sigma}_W^2 = 0.0497$, giving an estimated standard deviation of 0.2229. An approximate 95% two-sided confidence interval for R is

$$\Phi(1.968 - 1.96(0.2229)), \; \Phi(1.968 + 1.96(0.2229)) = 0.9371, \, 0.9919$$

Note that if a lower one-sided CI is desired, this can be obtained from the first term in this expression with (for 95% confidence) 1.96 replaced by 1.645. The resulting lower confidence limit is 0.9453. ∎

Stress and Strength Exponentially Distributed. We assume independent samples from exponential distributions with respective parameters λ_X and λ_Y. From Equation (6.9), the reliability of an item is

$$R = \frac{\lambda_Y}{\lambda_X + \lambda_Y} = \frac{\mu_X}{\mu_X + \mu_Y} \tag{8.97}$$

R may be estimated by substituting sample means into (8.97). By the methods of Appendix A5, the approximate (asymptotic) variance of the estimator is

$$V(\hat{R}) \approx \left(\frac{\mu_X}{(\mu_X + \mu_Y)^2} \right)^2 \sigma_{\bar{X}}^2 + \left(\frac{\mu_Y}{(\mu_X + \mu_Y)^2} \right)^2 \sigma_{\bar{Y}}^2 = \frac{\mu_X^2 \sigma_X^2 + \mu_Y^2 \sigma_Y^2}{(\mu_X + \mu_Y)^2} \tag{8.98}$$

Since the standard deviation is equal to the mean for the exponential distribution (see Section 4.4.1), this variance may be estimated simply by substitution of the sample means into (8.98) and an asymptotic confidence interval calculated in the usual way.

If the mean stress μ_Y is known, the estimated reliability is simply a function of x, and the analysis is straightforward.

8.7.2   Comparison of Parameters

In many applications it is desired to compare one or more characteristics of different populations from which samples have been drawn. Most commonly, a comparison of means is of interest. In reliability applications, many other characteristics, e.g., failure rates, fractiles, parameters of specific life distributions, and so forth, may also be of interest.

Comparison of population means will be discussed in detail in Chapter 10, where methods based on the assumption of normality (also appropriate for nonnormal cases if sample sizes are adequate) will be given. Comparison of population variances will also be discussed. Here we look at a few examples of comparison of parameters of life distributions.

Comparison of Means of Exponential Distributions. Assume samples from k exponential populations. The ith population has parameter λ_i, n_i is the sample size from this population, and x_{ij} is the jth observation in the sample from population i. We wish to test H_0: $\mu_1 = \mu_2 = \cdots = \mu_k$ against the alternative that not all μ_i's are equal. Note that this is equivalent to testing that the λ_i are all equal. The test statistic is

$$X^2 = \frac{2}{C} \left(n \log \bar{x} - \sum_{i=1}^{k} n_i \log \bar{x}_i \right) \tag{8.99}$$

where $n = \Sigma_{i=1}^{k} n_i$, $\bar{x}_i$ is the ith sample mean, $\bar{x}$ is the grand mean, and

$$C = 1 + \frac{1}{6(k-1)} \left(\sum_{i=1}^{k} \frac{1}{n_i} - \frac{1}{n} \right) \tag{8.100}$$

X^2 is approximately distributed as χ^2 with $(k-1)$ df. Note that the test is an upper-tail test, i.e., large values of X^2 lead to rejection of H_0.

Example 8.19 (Case 2.10)

In Example 8.9, Aircraft 7908, 7909, and 7910 were used in a Bayesian analysis to develop a prior distribution. A common λ was assumed. The MTBF's of the air conditioning units for these three aircraft were 95.7, 83.5, and 121.3, with n's of 23, 29, and 15, respectively. We test the hypothesis that the corresponding population means are equal, testing at the 1% level. Here $k = 3$ and $n = 67$. From (8.100), $C = 1.011$. The calculated value of the test statistic, from (8.99), is $X^2 = 1.48$. With 2 df, the critical value of χ^2, gotten from Table C3, is 9.210. There is no evidence, at the 1% level, that the means are different. ∎

Comparison of Weibull Parameters. Lawless (1982) discusses comparison of shape parameters of two Weibull distributions with complete or censored samples. Comparisons of scale parameters and fractiles of two or more Weibull populations are also covered.

Other Comparisons. For comparison of parameters or other characteristics of two populations under specific distributional assumptions and independent samples, when exact tests are not available (very often the case), tests based on asymptotic normality are usually appropriate unless sample sizes are small. The procedure is to estimate the parameters of each population and the variance of the estimate, as indicated in Chapter 5, use the fact that under independence the variance of the difference is the sum of the variances, and apply the z-test in the usual manner.

8.7.3   Estimation of the Parameters of Intensity Function

In Section 7.3, we discussed the black-box approach to modeling failures at the system level. In Model 7.1, the system failures occur according to a point process with intensity function $\Lambda(t; \theta)$ and it is assumed that the failures are independent. Note that in this model formulation the time intervals between failures are not identically distributed. The distribution at any point in time depends on the age of the system.

The problem is to estimate the parameter θ based on failure data over the interval $[0, T]$. We consider the following two cases.

Case (i): Actual failure times known. Let $T_j, j = 1, 2, \ldots$ denote the time of the jth failure. This is an increasing sequence of random variables. Let t_j denote the observed value of T_j and n denote the number of failures over the time interval $[0, T]$. The likelihood function can be obtained as follows. Note that

- $P\{\text{no failure in } [0, t_1)\} = e^{-\int_0^{t_1} \Lambda(t; \theta)dt}$
- $P\{t_1 \leq T_1 < t_1 + \delta t\} = \Lambda(t_1; \theta)\, \delta t + 0(\delta t^2)$
- $P\{\text{no failure in } [t_1 + \delta t, t_2)\} = e^{-\int_{t_1}^{t_2} \Lambda(t; \theta)dt} + 0(\delta t^2)$
- $\ldots$

and so on. This follows from the assumption that failures occur independently. This yields

$$P\{t_j \leq T_j < t_j + \delta t_j;\ 1 \leq j \leq n\} = \left[\prod_{j=1}^{n} \Lambda(t_j;\ \theta)\delta t_j \right] e^{-\int_0^T \Lambda(t;\theta)dt} \qquad (8.101)$$

As a result, the likelihood function is given by

$$L(\theta;\ T_j = t_j;\ 1 \leq j \leq n) = \left[\prod_{j=1}^{n} \Lambda(t_j;\ \theta) \right] e^{-\int_0^T \Lambda(t;\theta)dt} \qquad (8.102)$$

The maximum likelihood estimator of θ is the value which maximizes $L(\ )$ or $\log L(\ )$.

The approach is easily extended to the case of K identical systems. Let T_{jk} denote the jth failure for system k and let n_k denote the number of failures for system k over the time interval $[0, L_k)$. Then the likelihood function is given by

$$L(\theta; T_{jk} = t_{jk}; 1 \le j \le n, 1 \le k \le K) = \prod_{k=1}^{K}\left\{\left[\prod_{j=1}^{n_k}\Lambda(t_{jk}: \theta)\right]e^{-\int_0^{L_k}\Lambda(t;\theta)dt}\right\} \quad (8.103)$$

Let the intensity function be given by

$$\Lambda(t) = \alpha\beta(\beta t)^{(\alpha-1)} \quad (8.104)$$

This is the Weibull intensity function with parameter $\theta = \{\beta, \alpha\}$. It is easily seen that

$$\int_0^{L_k}\Lambda(t; \theta)dt = (\beta L_k)^{\alpha} \quad (8.105)$$

The likelihood function is given by

$$L(\theta; T_{jk} = t_{jk}; 1 \le j \le n, 1 \le k \le K) = \prod_{k=1}^{K}\left\{\left[(\alpha\beta)^{L_k}\prod_{j=1}^{n_k}(\beta t_{jk})^{(\alpha-1)}\right]e^{-(\beta L_k)^{\alpha}}\right\} \quad (8.106)$$

Define $n = \Sigma_{k=1}^{K}N_k$. Then from (8.106) the log likelihood function is given by

$$\log L(\theta) = n\alpha\log(\beta) + n\log(\alpha) + (\alpha-1)\left\{\sum_{k=1}^{K}\sum_{j=1}^{N_k}\log(t_{jk})\right\} - \beta^{\alpha}\left\{\sum_{k=1}^{K}\{L_k\}^{\alpha}\right\} \quad (8.107)$$

The maximum likelihood estimators α^* and β^* are obtained by solution of

$$\frac{\partial \log L(\theta)}{\partial \alpha} = 0 \quad \text{and} \quad \frac{\partial \log L(\theta)}{\partial \beta} = 0 \quad (8.108)$$

This yields

$$\beta^* = \left\{\frac{n}{\sum_{k=1}^{K}\{L_k\}^{\alpha^*}}\right\}^{1/\alpha^*} \quad (8.109)$$

and

$$\frac{n}{\alpha^*} + \sum_{k=1}^{K}\sum_{j=1}^{N_k}\log\{t_{jk}\} = \frac{n\sum_{k=1}^{K}\{L_k\}^{\alpha^*}\log\{L_k\}}{\sum_{k=1}^{K}\{L_k\}^{\alpha^*}} \quad (8.110)$$

Example 8. 20 (Case 2.10)

The failure data on air conditioners were analyzed in Examples 8.10 and 8.19 for Planes 7908–7910, assuming an exponential distribution of time to failure. Suppose instead that we look at this as a point process, assuming a Weibull intensity function. The data required for the analysis are partial sums of failure times (e.g., 413, 413 + 14, 413 + 14 + 58, etc., for Plane 7908). A problem in completing the analysis is that it is not known when the observation period stopped for each plane. As an approximation, we take the L_k to be the time of the last observed failure. The resulting values for the three planes are 2201, 2422, and 1819, respectively. Here $n = 67$, and the double summation on the LHS of (8.110) is found to be 454.113. A simple search (e.g., using Minitab) yields $\alpha^* = 1.11$. From (8.109), we find $\beta^* = 0.007639$.

Note that $\alpha^* = 1.11$ indicates a distribution close to the exponential and that the value of β^* is reasonably close to 0.01022, the value of the parameter obtained in Example 8.10 under the exponential assumption. Here the intensity function, given in (8.104), is estimated to be $0.004960 t^{0.11}$. ∎

Case (ii): Failures over subintervals grouped. Suppose that data are grouped into J subintervals, with the jth subinterval given by $[\tau_{j-1}, \tau_j)$, where $\tau_0 = 0$ and $\tau_J = T$. Note that τ_j is an increasing sequence in j. Let N_j denote the number of failures in the jth subinterval. Then we have

$$P(N_j = n_j) = p_j(n_j;\ \theta) = \frac{\left\{ \int_{\tau_{j-1}}^{\tau_j} \Lambda(t;\ \theta)dt \right\}^{n_j} e^{-\int_{t_{j-1}}^{t_j} \Lambda(t;\theta)dt}}{n_j!} \tag{8.111}$$

If the failure data consist of the set $\{n_j, j = 1, \ldots, k\}$, then the likelihood function is given by

$$L(\theta;\ n_j,\ 1 \le j \le k) = \prod_{j=1}^{k} p_j(n_j;\ \theta) \tag{8.112}$$

The maximum likelihood estimator is the value of θ that maximizes $L(\ )$ or $\log L(\ )$.

NOTES

1. A detailed discussion of maximum likelihood estimation for grouped data in the case of the three-parameter Weibull distribution, along with many references to the three-parameter estimation problem generally, is given by Hirose and Lai (1997).

2. There is a vast literature on the subject of Bayesian statistical analysis. For a good introduction, see Press (1989). For a very thorough treatment of Bayesian analysis in reliability, see Martz and Waller (1982). A good summary of Bayesian estimation in reliability is given by Sinha (1986).

Dey and Lee (1992) discuss the use of Gibbs sampling to provide computationally feasible procedures for complex life distributions.

3. There are many other approaches to dealing with estimation of system reliability based on information on component reliabilities. The following are approaches based on classical statistical methodology:

- Mann and Grubbs (1974)—exponential data with Type I and Type II censoring and binomial data; approximations to optimal bounds
- Winterbottom (1974)—series systems with binomial component data; comparison of methods
- Myhre (1978) and Myhre et al. (1978)—confidence limits based on data with few or no failures.

- Tillman et al. (1982)—review of a large segment of the literature on Bayesian reliability to that time
- Rice and Moore (1983), Moore et al. (1985), and Chao and Huwang (1987)—use of Monte Carlo methods in simulation studies to determine lower confidence bounds
- Winterbottom (1984)—approximations to exact binomial analysis for single items, series systems, coherent systems; exponential time to failure; some Bayesian analysis
- Murthy and Wilson (1994)—ML estimation for systems with dependent component failures
- Mhyre and Rennie (1986)—coherent systems; binomial data; empirical Bayes methods.
- Soms (1989)—CI for series system based on exact distribution of a product of binomial random variables
- Rekab (1993)—sequential sampling plan for estimating component reliabilities, series system
- Baxter (1993)—confidence intervals for system reliability for coherent systems based on component lifetime data

4. Some additional results on Bayesian analysis of system reliability as a function of component reliabilities are:

- Thompson and Chang (1975)—Bayesian analysis of reliability of redundant systems assuming exponentially distributed lifetimes
- Mastran and Singpurwalla (1978)—Bayesian analysis of reliability of coherent structures
- Thompson and Hayes (1980)—computational approaches to evaluation of posterior distributions for multicomponent systems
- Martz and Duran (1985)—use of simulation to estimate system reliability with binomial component data
- Blischke (1994)—use of the Martz–Waller–Fickas method in reliability analysis of rocket propulsion systems
- Tang et al. (1994, 1997)—asymptotic form of posterior distribution of reliability; further results on Bayesian analysis of system reliability based on component data
- Phillipson (1996)—discussion of some logical difficulties with the Bayesian approach

5. Some additional results on statistical analysis of stress–strength models are:

- Draper and Guttman (1978)—Bayesian analysis of stress–strength models for multicomponent, s-out-of-k systems with identical components. See also MW, Section 11.5.3
- Weerahandi and Johnson (1992)—Tests of hypotheses about R, X and Y normal, distribution of X known or unknown; comparison with Bayesian results. See also Pham-Gia and Turkkan (1995)
- Kunchur and Munoli (1993)—Multicomponent parallel system with nonindependent failures, exponential distribution
- Hanagal (1996)—Two components in parallel, several bivariate exponential models
- Gupta and Subramanian (1998)—Bivariate normal with same coefficient of variation; confidence interval for R

- English et al. (1996)—discrete approximations for complex stress and strength functions; analysis of three systems

EXERCISES

1. For Case 2.6, fiber lengths 265 and 5, calculate the sample mean and standard deviations. Group the data to form a frequency distribution using the procedure given in Chapter 3. Calculate the mean and standard deviation of the grouped data, with and without the Sheppard's correction and compare the results with those obtained for the raw data.

2. Repeat Exercise 1 for Machines LHD1 and LHD3 of Case 2.7, and for the combined data for these two machines.

3. Repeat Exercise 1 for the combined data over all 6 machines in Case 2.7.

4. Repeat Exercise 1 for each of the clerks in Case 1.13, and for the combined data over all five clerks.

5. Summarize the results of the previous exercises. Does grouping appear to affect the results? Is the effect, if any, consistent over the data sets? For which data sets is Sheppard's correction appropriate? (You may wish to analyze other of the data sets in Chapter 2 to test the generalizability of your conclusions.)

6. Suppose that TTF for Part 3 in Case 2.15 can be modeled by a two-parameter exponential distribution. Calculate point estimates and 90% confidence intervals for the shift and scale parameters.

7. Recalculate the results of Example 8.2 using $n = 5000$ (the total number of vehicles sold) instead of 329. Comment on the validity of this result.

8. Analyze the data of Case 2.2 assuming exponential lifetimes.

9. Use the results of the previous exercise to test the null hypothesis H_0: $\mu \leq 3.0$ versus H_a: $\mu > 3$, where μ is the MTTF in thousands of flight hours.

10. Analyze the data of Case 2.16 assuming a Weibull distribution of time to failure. Include asymptotic confidence intervals for the parameters in your analysis.

11. For Case 2.17, plot the gamma and Weibull distributions obtained using the results of Examples 8.5 and 8.6, and compare the results.

12. For the GIDEP data of Case 2.26, assume exponential distributions and compare the MTTFs for the different entries for amplifier circuits. (Note that entries with zero failures cannot be included in the analysis.) Test at the 1% level.

13. Repeat the previous exercise for logic circuits.

14. In the previous two exercises, combine the data used in your analysis for each manufacturer and repeat the analyses using the combined data.

15. Repeat the previous exercise including data with zero failures, where possible. Compare the results.

16. Use a statistical test to compare the MTTFs (times between errors in this context) of the five clerks in Case 2.13. Assume exponential times to failure and test at the 1% level.

17. A certain type of failure in a system can be caused by failure of one of three modules, A, B, and C. The reliabilities of these modules are 0.998, 0.992, and 0.995, respectively. The probabilities of system failure given module failure are 0.105, 0.005, and 0.007 for A, B, and C, respectively. Suppose that the system has failed. What is the probability that this is due to failure of Module A? Module B? Module C?

18. Example 8.10 involved the use of failure data of air conditioners from three aircraft (Case 2.13). Assuming exponentially distributed times to failure, compare the failure rates of air conditioners in all aircraft. Test at the 5% level.

19. Calculate a 95% Bayes interval for λ in Example 8.10.

20. Calculate Bayes point estimates of p for each of the components in Case 2.27. Assume a beta prior with parameters corresponding to a prior expectation of 0.99 and a standard deviation of 0.006.

21. Repeat the previous exercise with a prior standard deviation of 0.01.

22. Calculate 90% Bayesian probability intervals for each of the last two exercises.

23. Compare the results of the previous three exercises with the classical solution.

24. The number of data entry errors per 1000 processed checks in a bank is a Poisson random variable. Prior to a study of the operation, it is decided that the mean number should be about 8.0, with a standard deviation of 1.0. If a gamma distribution is to be used to express this prior information, what are the appropriate values of α and β?

25. (Continuation) If a sample of 1000 checks is taken and four errors are found, what is the Bayes estimate of λ? Calculate a 95% Bayes probability interval. Calculate the classical estimate and a 95% confidence interval and compare the results.

26. For Example 8.12, calculate the 95% confidence interval and compare the result with the Bayesian probability interval. Which of these is "correct?"

27. Calculate 95% Bayesian probability intervals for $R(t)$ using the results of Example 8.9 and $t = 50, 75, 100$, and 150 hours.

28. Assume that the data of Case 2.20 used in Example 8.13 are gamma distributed rather than Weibull. Analyze the data accordingly and give an approximate 95% confidence interval for $R(1)$. Compare the result with that of Example 8.13.

29. A system consists of three components, A, B, and C, connected in series. Components are assumed to operate independently. In a reliability study, 500 components of each type are tested under extreme conditions, resulting in three failures of Component A, two of Component B, and eight of Component C. Estimate the reliability of the system under these conditions based on this data, and calculate a 95% lower confidence limit for system reliability, using the asymptotic normality approach.

30. Calculate a 95% lower confidence limit for system reliability in the previous exercise if the components are connected in parallel.

31. Suppose that a system consists of two subsystems connected in series. The first subsystem consists of two components connected in parallel. The second consists of a single component. Draw a block diagram of this system and derive a formula to express the reliability of the system in terms of the reliability of the components. Suppose that tests are done on each component, resulting in x_i failures in n_i tests, $i = 1, 2, 3$. Give a formula for the estimated reliability of the system based on these results and derive a formula for the asymptotic variance of this estimator.

32. Suppose that in Exercise 29 Components A, B, and C are configured as in Exercise 30. Use the test results to calculate a 95% one-sided confidence interval for system reliability.

33. Use the results of the previous exercise to test the null hypothesis that system reliability is less than or equal to 0.995 against the alternative that it is greater than 0.995

34. Use the Easterling approach to calculate confidence limits in Exercises 29 through 31. Compare the results with those obtained in the previous exercises.

35. Compute the Spencer–Easterling lower bound for the reliabilities calculated in Exercises 29 through 34 and compare the results obtained with this bound.

36. In Example 8.18, reliability was estimated for fibers of length 12.7 mm. Calculate a lower 95% confidence limit for the reliability of each of the four fiber lengths for which data are given in Case 2.6. Comment on the reliability of fibers as a function of length.

37. Test results on bond strengths in audio components under four environmental conditions are given in Table 2.1(b). Suppose that in a certain application stresses are normally distributed with mean $\mu = 5.0$, and standard deviation $\sigma = 1.1$. Estimate the reliability of the bond for each environmental condition and calculate the 95% lower confidence limit for each situation. Discuss reliability as a function of environment based on these results.

38. Repeat the analysis of the previous exercise assuming exponential distributions and using only the information concerning the means. Compare the results with those of the previous exercise and comment.

39. In example 8.20, set the end points of the intervals of observations for the three planes to be 2500, 2500, and 2000, respectively, and reanalyze the data under the assumption of a Weibull intensity function. Compare the results with those of Example 8.20.

40. Estimate the parameters of a Weibull intensity function using the data for all planes in Case 2.10.

41. Assume that the data of Case 2.7 are independent observations from exponential distributions. Test the null hypothesis that the exponential parameters are identical against the alternative that they are different.

42. Analyze the data of Case 2.7 assuming a point process with Weibull intensity function. Does it appear that the data may be exponentially distributed?

CHAPTER 9

Software Reliability

9.1 INTRODUCTION

Up to this point, we have been concerned with hardware reliability—time to failure, modeling at the component and system levels, calculation of reliabilities, description and analysis of failure data, estimation, and related topics. These issues are also basic concerns in dealing with the reliability of software, and many of the same models are used in its analysis. There are some important differences between hardware and software, however, some obvious and some rather subtle, and these lead to some significant differences in definitions, modeling, data, and interpretation of results. In this chapter, we highlight some of these differences, and discuss reliability analysis in the context of software.

9.1.1 The Need for Reliable Software

Society as a whole has become crucially dependent on computers in production, commerce, finance, aerospace, medicine, and in our daily lives, and there is no doubt that this dependency will continue to increase. It is self-evident that a computer system, to operate properly, requires both reliable hardware and reliable software. In fact, the greatest difficulties in designing a reliable computer system are often in the software area. There are long-standing principles and procedures for dealing with hardware design, development and testing, some dating back to the Guilds in the Middle Ages. Software, on the other hand, is a relatively recent development, and its development is a purely intellectual activity as opposed to the "hammer and tongs" endeavor of developing hardware in earlier times or a highly sophisticated piece of machinery used in modern manufacturing.

The criticality of reliable software is dramatically apparent in complex systems such as the space shuttle. The propulsion system and the shuttle itself contain some tens of thousands of parts. The on-board software, however, includes over 500,000 lines of code, and the ground control system about 3.5 million lines of code. Other space systems include comparable amounts. Vast software systems are used in communications, government and military operations, avionics, airline and other scheduling operations, control systems in industry, financial analysis, and in many other areas. Even in our now essential PC's, operating systems and software packages can and do run from one to five million lines of code, or more. As software becomes more and more complex, it becomes the critical cause of system failure.

Software defects have led to many unwanted consequences, ranging from inconvenient to disastrous. Phone and power outages, delays in space missions and flights of experimental aircraft, disruption of communications, and many similar events, some of which have led to very significant loss of revenue or very significant cost, have been reported. An example of the latter is the 1979 Venus probe that cost several hundred million dollars and missed its target because of a period instead of a comma in a FORTRAN statement. Disasters include aircraft accidents with significant loss of life, patients killed by radiation machines having faulty software, and nuclear and chemical accidents. Near disasters include emergencies in aircraft (including a military aircraft that did a 180° roll on crossing the equator in 1983 because of a program bug), spacecraft, and ground transportation systems, and numerous other incidents that were traced to software defects.

Software reliability is affected by three basic factors (Ireson and Coombs, 1988):

1. Defects in code (design errors, omissions, incorrect usage of language, files, subroutines, subsystems, data, typographical errors, simple coding errors)
2. Defects in interfaces with other code (linkages, data transmission, etc.)
3. Operational defects that cause changes to code (improper editing, indexing, and so forth).

In this chapter, we shall elaborate on these elements and provide more precise definitions of the terms in the context of software reliability.

It has been estimated (Raheja, 1991) that 60% of software errors are logic and design errors; the remaining are coding and operations-related errors. Möller (1991) estimates that 10% occur in requirements specifications, 40% in design, and 50% in implementation. Lloyd and Lipow (1962) claim that 60% are in requirements and preliminary and detailed design, and 40% in coding. Other analysts provide different breakdowns, and the numbers will vary, depending on the type of software and the context. (More detailed breakdowns will be given later in the chapter.) In any case, except for very small programs, it is unlikely that error-free software will be produced, and for very complex systems and systems with millions of lines of code, it is virtually certain that errors will occur.

The effect of software errors is felt in other ways as well. It has been estimated (Wallmüller, 1994) that only 5% of software projects are completed on time, 60% have time overruns of 20% or more, and many projects are terminated because of delays. Lalli, et al. (1998) claim that only 1% of major projects are finished on time, and that 25% are never finished. Furthermore, costs tend to increase exponentially with increasing software complexity, with 60% often spent on maintainability (i.e., after-installation error correction). Delays such as these and maintainability problems can lead to significant loss of market and loss of profitability. Pullman and Doyle (1998) estimate that in the United States 25% of software projects are cancelled, that up to 15% of defects remain in software upon its release, and that software failures exceed hardware failures by an order of magnitude.

A primary objective in software design is to minimize the number of software errors. The has led to the development of *software engineering,* defined (Priest, 1988) as "application of engineering principles . . . to the development, implementation and maintenance of computer software over its life cycle." From a user perspective, software reliability is the "applied science of predicting, measuring and managing the reliability of software-based systems to maximize customer satisfaction." (Vouk, 1998). In this chapter, we look at the concept of software life cycle and the application of engineering and management principles and procedures in software design, development, testing, and use.

9.1.2   Chapter Outline

The outline of the chapter is as follows: In Section 9.2, we look at basic definitions of software quality and reliability terms, the different interpretations of these terms in analysis of software and hardware, and the broader context of software quality. Section 9.3 is concerned with issues and procedures in the design of reliable software. Modeling and measurement of software reliability is discussed in Section 9.4. Many of the models used in analyzing hardware are appropriate here as well; a number of these as well as models specific to software will be listed and their interpretation in the context of software discussed. Section 9.5 deals with testing of software. Section 9.6 is a brief introduction to software management. This is discussed in more detail in Chapter 12.

9.2   SOFTWARE QUALITY AND RELIABILITY ISSUES

In this section, we look first at definitions of reliability terms in the context of software. An initial notion of software reliability was that it was either zero (if it contained one or more errors of any kind) or one (if it was perfect). Many more elaborate definitions have been proposed. We then look at severity and risk issues, contrast software and hardware reliability concepts, and discuss other dimensions of software quality.

9.2.1   Definitions

The basic definitions of reliability and of many of the terms used in reliability analysis arose in the context of hardware. Many of the terms have been carried over to the study of software reliability, but the concepts are often interpreted differently and there is not complete unanimity with regard to their meaning in the context of software. In fact, opinions vary from that held by many authors that most of the definitions, models, and other concepts used in analyzing hardware reliability carry over directly (or with minor modifications) to software, to statements to the effect that the failure rate of software "does not exist" (Singpurwalla, 1995b).

The following list of definitions is based on our ideas plus a number of other sources (Ireson and Coombs, 1988; Lloyd and Lipow, 1962; Lyu, 1995; Musa, et al., 1990; Raheja, 1991; Singpurwalla, 1995b; Vouk, 1998).

Defect: a generic term used to describe either a fault (cause) or failure (effect)

Error: (1) a discrepancy between a computed, measured, or observed value or condition and the true, specified, or theoretically correct value or condition

(2) a programming action that results in software containing a fault

Failure: (1) inability of a system to produce an expected output or service; incorrect result that is different from specifications

(2) any action or error that causes discontinuance of operations

(3) departure of an observed result from requirements or user expectations

Note that a failure may be caused by either software, hardware, or their interface.

Failure Intensity: number of failures per unit of time

Fault: internal program error that causes a failure to occur (also called a **bug**); any error in coding, logic or structure. Faults may be *inherent faults,* which are present in the original system, or *modification faults,* namely, those introduced during correction or as a result of design or specification changes.

Fault Density: number of faults per thousand lines of code (KLOC)

Nonconformance: failure to meet specifications; failure to perform as intended or to perform as intended consistently

Operational Profile: a listing of the set of operations that a software entity can perform along with their probability of occurrence

Reliability Allocation: as in hardware (See Chapter 13), a specification of reliability requirements for lower-level units based on an overall reliability goal

Reliability Growth: increase in reliability through time

Software Reliability: $R = P$(failure-free operation of a software entity for a specified period of time in a specified environment). Example: $R = 0.95$ in 4 hours means that on average 95 out of 100 such runs will be failure free.

System Reliability Model: an expression that relates system reliability to component (module) reliability

Time: *calendar time:* elapsed time as ordinarily measured (minutes, days, etc.)

clock time: elapsed time from start to end of a computer run (not counting downtime) = actual run time plus waiting time

execution time: CPU time that is spent in executing a specified software system or subsystem

Note that these are related (in ways that depend on the particular application), and that a manager may wish to translate all times to calendar time.

Two of the most important definitions is the above list are those of fault and failure; these are worth exploring further. The IEEE definitions of these terms (Raheja, 1991) are:

Failure: 1. Termination of a functional unit to perform its required function

2. Inability of a component or system to perform its required function within specified limits (e.g., when a fault is encountered)

3. Departure of operations from requirements

4. Inoperable state

Fault: 1. Accidental condition that causes a functional unit to fail to perform its required function

2. An error (bug) in software that may lead to software failure

3. Immediate cause of failure (e.g., maladjustment)

4. Any condition, action, or element of software that, when encountered in operations, results in a failure

The key notion here is that, as in hardware, a failure is malfunction of the software that leads to incorrect conditions and a fault is the cause of a failure. An error could be either of these.

9.2.2   Types and Severity of Errors in Software

Whether or not a failure occurs is a function of usage. If a particular module of a program is never used in certain applications, then faults in that module will not lead to failures. Furthermore, some failures are tolerable. This occurs, for example, when a user knows that a failure has occurred but is able to use the output after a slight or modest modification. Thus software errors, as in the case of other errors and in the case of hardware, differ in their impact and severity.

Failure Taxonomies

Many types and causes of errors have been identified. In a sense, all are basically "design" errors, since, unlike hardware, software does not wear out or age. There are a number of taxonomies of errors (bugs). The following is due to Lloyd and Lipow (1962):

- Computational (incorrect variables, parentheses, truncation, etc.)
- Logic (sequencing, calling wrong variables, incorrect loops, etc.)
- Input (invalid, incorrect format, incorrect file)
- Data manipulation (initialization, sort error, problems with variables)
- Output (wrong files, incomplete, missing, garbled, misleading)
- Interface (user, subroutines, other software)
- Data definition (incorrect, wrong dimension)
- Data base (units, initialization)
- Operations (hardware, operator, user)
- Other (time limits, storage, compatibility, etc.)

Another classification is given by Beizer (1990), who also lists the percent of each type in a sample of 6,877 KLOC in which 16,209 bugs were discovered. The taxonomy is:

- Requirements errors (incorrect, logic error, incomplete, documentation, changes)   8.1%
- Features and functions (incorrect, incomplete, domain bugs, user messages, diagnostics, etc.)   16.2%
- Structural (control, sequencing processing)   25.2%
- Data (definition, structure, access, handling, etc.)   22.4%
- Implementation and coding (coding, typographical, standards violations, documentation, etc.)   9.9%
- Integration (internal interfaces, external interfaces, timing, throughput, etc.)   9.0%
- System, software architecture (O/S call and use, recovery, accountability, performance, etc.)   1.7%
- Testing (test design bugs, execution, documentation, completeness, etc.)   2.8%
- Other or unspecified   4.7%

Many studies have reported that the most significant sources of error are logic, data handling, and interfaces, accounting for over 50% of errors. A study done by Rome Air Development Center, USAF, (Raheja, 1991, p. 274) arrived at the following breakdown of software errors:

Logic	21.3%
I/O	14.7%
Data handling	14.5%

Computational	8.3%
Preset database	7.8%
Documentation	6.3%
User interface	5.7%
Software interface	5.6%

In summary, software errors occur for many reasons—inadequate requirements specifications (incomplete, inconsistent, unclear, incorrect); coding errors (misinterpretation or inadequate understanding of the programming language, typographic errors, improper data handling, faulty logic, incorrect or misunderstood assumptions, inappropriate or inaccurate approximations); data problems (I/O, definitions, alteration of stored data, etc.); and interface problems (of modules within a program, with user, with hardware, with other software). Other sources of error include incomplete design, unreliable tools (e.g., compilers), unanticipated conditions (use in applications not originally planned), and inadequate controls, testing, and validation.

Risk Factors

Risk factors that determine software quality and reliability basically involve size and complexity of the software item, but a number of other features of both product and process may have an impact as well. These include (Wallmüller, 1994):

- Size of the project (person years, cost)
- Competency of staff
- Complexity of the software
- Degree of novelty
- Number of requirements
- Level of uncertainty of requirements
- Amount of in-house labor versus outsourcing
- Time constraints (e.g., for product release)

Failure/Bug Severity

A number of scales of severity or criticality of bugs have been proposed. A relatively simple classification of failure severity, due to Musa et al. (1990) is

A. Basic service interruption
B. Basic service degradation
C. Inconvenience, immediate correction necessary
D. Minor defect, correction deferrable

A more complex ten-point scale, including consequences of bugs, is (Beizer, 1990; Pullman and Doyle, 1998):

1. Mild—aesthetics (e.g., spelling error)
2. Moderate— misleading or redundant output
3. Annoying—e.g., bills for $0.00 sent out; names misspelled

4. Disturbing—software refuses to handle legitimate transactions (e.g., rejection of a valid credit card)
5. Serious—information/transactions lost
6. Very serious—performs wrong transactions
7. Extreme—frequent and arbitrary errors
8. Intolerable—unrecoverable, hard to detect corruption of data base
9. Catastrophic—system failure
10. Infectious— corrupts other systems

It is essential that bugs in at least categories 6 through 10 be removed from software to the greatest extent possible. This is typically the biggest single cost factor in software development. The cost is in detection, designing and running tests, and correction.

9.2.3 Hardware Versus Software Reliability

A fundamental difference in dealing with software versus hardware reliability is that hardware reliability tends to decrease in time, due to aging, wear-out, and so forth, whereas software reliability tends to increase through time, due to the removal of bugs, or will remain the same if no action is taken. There are exceptions to both, of course. Hardware reliability may increase in time as a result of a reliability improvement program, and software reliability may decrease as a result of the introduction of new bugs.

Some other differences are:

- Hardware: a small anomaly may lead to a predictable failure or have little or no effect
  Software: one incorrect bit can lead to disaster
- Hardware: Design and production predominate
  Software: Nearly 100% design (production is trivial—copy to diskette or CD ROM)
- Hardware: Can test "all" events (general physical principles, mathematical models apply)
  Software: The number of events is huge and events tend to be unique to the software
- Hardware: Causes of failure are design, manufacture, maintenance, misuse
  Software: Failures are due to design defects
- Hardware: Redundancy can be used to increase reliability
  Software: Redundancy does not necessarily lead to improvement. (Simply repeating a module, as is done in hardware, is of no use whatsoever, because the duplicate module contains the same bugs. Having separate versions prepared by different programmers or teams introduces another set of bugs.)
- Hardware: Maintenance improves reliability
  Software: Reprogramming may introduce new errors
- Hardware: Failures may be described by physical laws
  Software: Comparable laws do not exist
- Hardware: Interfaces are physical structures
  Software: Interfaces are conceptual
- Hardware: Standard parts are commonly used
  Software: "Standard" parts are seldom used

- Hardware: Safety margins can be designed into hardware
  Software: There are no "safety margins" in software

Other characteristics that differentiate software from hardware are that in software

- There are many more "paths" than in a physical system
- There are many more "parts"
- Failure modes are different
- Repair is not "good-as-new"

Because of the large number of differences, one might conclude that methods developed for hardware would not be appropriate for analysis of software reliability. As will be seen, however, many of the techniques used in analysis of hardware reliability are useful here as well, but care must be taken in selecting models and interpreting results in the software context. For further discussion, see Lloyd and Lipow (1962), Ireson and Coombs (1988), Raheja (1991), and Lalli et al. (1998).

9.2.4   Dimensions of Software Quality

Our primary concern in this book is reliability. As noted previously, this is only one of the many features that comprise product quality. Here we look briefly at the dimensions of software quality. Again, many attempts have been made to define software quality. The list of desirable characteristics that follows is a composite of several of these. High quality software has the following features:

- Performance—achieves system requirements and goals
- Reliability—consistently performs all expected functions
- Usability—designed to permit easy use by personnel with minimal training
- Manageability—easily integrated, tested, enhanced, adapted, and maintained
- Portability—easily transported and modified for use in other systems
- Testability—structured so that general algorithms that facilitate step-by-step testing may be used
- Efficiency—conserves computer resources by use of high-performance algorithms and efficient computer resource management techniques
- Maintainability—program is easy to read, well documented, modularized, and easily modified
- Accuracy—calculated values are sufficiently close to the true values to satisfy their intended use
- Completeness—all parts of the code are present and each is fully developed.
- Conciseness—excessive and/or unnecessary information is not present
- Consistency—the code possesses both internal consistency (uniform notation and terminology) and external consistency (names, units, definitions, etc., consistent with standard usage)
- Robustness—ability to perform correctly in the presence of some violation of assumptions and/or specifications

- Modifiability—facilitates incorporation of changes
- Supportability—can be installed, serviced, reconfigured, and upgraded at reasonable cost
- Correctness—extent to which specifications are satisfied
- Interoperability—ease with which one subsystem or module can be coupled to another
- Flexibility—ease with which a program can perform or be modified to perform tasks outside the scope of the original requirements
- Validity—ability to perform adequately in the intended user environment
- Generality—ability to perform over a wide range of usage modes and inputs, even beyond those specified as requirements
- Understandability—purpose of the code is clear, modules are self-descriptive, control is simple
- Reusability—program or modules can be used or easily adapted for use in other programs or software systems

Additional discussion of these and many other features can be found in Lloyd and Lipow (1962), Ireson and Coombs (1988), Raheja (1991), and in many of the other references cited in this chapter.

9.2.5   Software Quality and Reliability Standards

Standards for software quality have been established by many organizations, including, among others, the U.S. Department of Defense (DoD), NASA, ISO, and IEEE. Relevant documents available from the latter four organizations are:

DoD: DoD-STD-2167A, "Defense System Software Development"

  MIL-STD-498, "Software Development and Documentation"

  MIL-STD-882C, "System Safety Program Requirements"

IEEE: IEEE Standard 730-1989, "Standard for Software Quality Assurance Plans"

  IEEE Standard 829-1983, "Standard for Software Test Documentation"

  IEEE Standard 1008-1987, "Standard for Software Unit Testing"

  IEEE Standard 1028-1988, "Standard for Software Reviews and Audits"

  IEEE Standard 1044-1993, "Standard for Classification of Software Anomalies"

  IEEE Standard 1059-1993, "Guide for Software Verification and Validation Plans"

  IEEE Standard 1061-1992, "Standard for a Software Quality Metrics Methodology"

ISO: ISO 9000-3, "Software Guidelines"

NASA: NASA Management Instruction 2410.10, "NASA Software Management Assurance and Engineering Policy"

  NSS 1740.13. "NASA Software Safety Standards"

For further information, see Lalli (1998) and Pullum and Doyle (1998).

9.3   DESIGN OF RELIABILE SOFTWARE

9.3.1   Software Life Cycle

As is the case with hardware, software reliability is a function of design, use, environment, maintenance, and many other factors, and these may change or assume different levels of importance in different stages of the software life cycle. The life cycle of a software product will vary with the type of software and application, but generally proceeds through a sequence of stages similar to those in the life cycle of hardware. For software, these stages typically include conceptualization, specifications and requirements, initial design, detailed design, coding, testing, debugging, release, support, and obsolescence. Definitions of software product life cycles can be found in many sources including Evans and Marciniak (1987), Priest (1988), Vincent et al. (1988), and Ireson and Coombs (1988).

Figure 9.1 (from Priest, 1988) is a description of a software life cycle through the maintenance phase. Included in the figure are estimates of the amount of effort typically devoted to each phase from inception to deployment, as well as suggested periodic reviews and the documentation that should be produced at each stage. The following are brief descriptions of the phases and activities:

- Requirements   Mission statement; system   [Requirements review]
  definition   requirements; system software
                     requirements
- Design   Conceptual/top level design   [Preliminary design review]
                     Detailed design   [Critical design review]
- Programming   Coding and unit testing; debugging

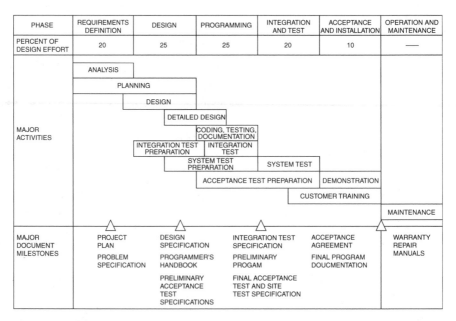

PHASE	REQUIREMENTS DEFINITION	DESIGN	PROGRAMMING	INTEGRATION AND TEST	ACCEPTANCE AND INSTALLATION	OPERATION AND MAINTENANCE
PERCENT OF DESIGN EFFORT	20	25	25	20	10	——

ANALYSIS
PLANNING
DESIGN
DETAILED DESIGN
CODING, TESTING, DOCUMENTATION
INTEGRATION TEST PREPARATION | INTEGRATION TEST
SYSTEM TEST PREPARATION | SYSTEM TEST
ACCEPTANCE TEST PREPARATION | DEMONSTRATION
CUSTOMER TRAINING
MAINTENANCE

MAJOR ACTIVITIES

MAJOR DOCUMENT MILESTONES	PROJECT PLAN	DESIGN SPECIFICATION	INTEGRATION TEST SPECIFICATION	ACCEPTANCE AGREEMENT	WARRANTY REPAIR MANUALS
	PROBLEM SPECIFICATION	PROGRAMMER'S HANDBOOK	PRELIMINARY PROGAM	FINAL PROGRAM DOUCMENTATION	
		PRELIMINARY ACCEPTANCE TEST SPECIFICATIONS	FINAL ACCEPTANCE TEST AND SITE TEST SPECIFICATION		

Figure 9.1. Software life cycle (Reprinted from Priest, 1988, p. 144 by courtesy of Marcel Dekker, Inc.).

- Integration   Software integration and testing   [Test review]
                       System integration and testing
- Installation   Operational testing; production
                       and deployment
- Maintenance   Postsale service; error correction;
                       warranty service
- Obsolescence   Discontinuation of support

Additional discussion regarding the infrastructure and organization necessary for effective management of software development—project management, personnel, resource management and control, data management, technology and quality management— can be found in Evans and Marciniak (1987).

9.3.2   Software Reliability Requirements and Specifications

It is useful to elaborate on software requirements and specifications, since failure to delineate these adequately is one of the leading causes of software defects. A fundamental principle (see Leveson, 1995) is that specifications must be sufficiently complete so that desirable and undesirable behaviors can be distinguished. Three components of requirements specifications are (1) the basic function or objective; (2) constraints on operating conditions; and (3) quality goals, prioritized to aid in trade-off decisions.

Ireson and Coombs (1988) discuss reliability objectives in the various phases of the life cycle (which are slightly different from those of the previous section). These are summarized in Table 9.1 (based on their Table 16.5).

9.3.3   Software Reliability Engineering

Good engineering practice is essential in the design of reliable software. The process is inherently more difficult in dealing with software than in the case of hardware. Flaws in hardware are often easily seen, whereas in software they are essentially invisible except to the programming team. Furthermore, software does not experience random failures (e.g., due to aging), has infinite life, and, as previously noted, is such that nearly 100% of errors are human errors (in specifications, coding, modifications, etc.) As a result, according to Raheja (1991, p. 271), "The vast majority of [software] systems are deeply flawed from the viewpoint of reliability."

Software reliability engineering (SRE) addresses these problems. SRE is the "quantitative study of the operational behavior of software-based systems" (Vouk, 1998). It includes

1. Software reliability measurement, estimation and prediction
2. Analysis of the effects of design and development activities on operational software behavior
3. Application of this knowledge in software development, testing, acquisition, use, and maintenance

In software engineering, basic principles of design, many of which are also applicable to hardware (See Chapter 13), must be followed. For software, these include the following:

- Requirements—adequate definition of requirements and specifications
- Preliminary evaluation— feasibility studies

Table 9.1. Software reliability objectives

Phase	Objectives
Requirements	Stated reliability objectives; prioritization of conflicting objectives
Specification	Specification of recovery modes, error checking and handling List of elements to receive detailed inspection
Design	Inspections to be done List of design problems uncovered Design elements to receive detailed inspection Module interfaces to be inspected; interfaces to be prototyped Tools for design analysis
Code	Inspection schedule List of problems uncovered Identification of code to receive detailed inspection Style and complexity analysis
Testing	Amount of coverage at unit level Definition of defect levels acceptable User acceptance test plan
Release	Determination of defect level remaining Test coverage achieved List of test sites; results of tests MTTF, MTTR at test sites
Support	Reported defects from users Additional testing and analysis Defect levels of updated software

- Modular design— separate modules for each function
- Robust design—make the software fault tolerant
- Design standards —standardize specifications; control complexity
- Design reviews— include independent review teams
- Method—use structured programming
- Technique— use a top-down approach to design (system, subsystem, program, module, routine, instruction) with traceability of steps
- Process standards—standardize the process; comply with existing standards
- Configuration—rigorous control
- Coding standards—standardize the coding process
- Inspection/reviews—of requirements, specifications, flow charts, code, etc.
- Inspection standards—standardize the inspection process
- Testing—use independent testing teams
- Local data definition—assure that data cannot be changed by another module
- Data editing routines— verify form and limits for incoming data
- Soft failure—program recognizes and corrects errors and stores the correct results

- Redundancy— for critical components, have two or more design teams
- Documentation—complete, understandable, logical, organized

Coding for reliability requires that the specifications and design be complete and accurate, reliability objectives be clearly understood, coding standards be adhered to, and test tools and strategies be properly formulated and understood.

For further information on engineering aspects of software reliability, see Myers (1976), Ireson and Coombs (1988), Priest (1988), Raheja (1991), Lalli et al. (1998), and Vouk (1998).

9.3.4 Software Quality Assurance

The many dimensions of software quality were discussed in Section 9.2.4. In the previous section, we dealt with one component of software quality—software reliability. Here we look briefly at the broader picture. There are many excellent sources of further information in this area, including excellent texts on the subject. Some of these are Evans and Marciniak (1987), Vincent et al. (1988), Raheja (1991), Wahlmüller (1994), and Oskarsson and Glass (1996).

Again, planning is the key to software quality. This includes program planning, project planning, organization, personnel, planning for assessment of quality characteristics, determining data needs, planning for tests, and so forth. Two essential ingredients in all of this are involvement of management at all levels, including the top levels, and teamwork.

A detailed plan for software quality assurance (SQA), including quality criteria, metrics, principles, and procedures, is given by Wahlmüller (1994). The author also provides constructive quality assurance measures (involving program features and related issues), analytical measures for use in design and testing, and a discussion of the systems approach and organization in this context.

Vincent et al.(1988) give a detailed description of SQA through the product life cycle, providing a detailed list of activities at each stage. Also included are data collection efforts, required documentation, and reviews.

Standards organizations have also provided plans and guidelines for SQA. The most widely used procedure of this type (Raheja, 1991, Figure 9.1) is DoD-STD 2167A, referred to previously. The ISO 9000 series also addresses the issue, though most of the standards were written for hardware.

A detailed discussion of each of the twenty quality elements of ISO 9000 in the context of software is given by Oskarsson and Glass (1996). Specific issues in the standards that are interpreted in the context of software include concurrent engineering, the research organization, prototypes ("throw-away" programs that constitute attempts at the easiest possible solution), use of consultants, incorporation of prior software products, and training. Issues of standards and certification are also addressed, and ISO standards are compared with those of other organizations. Additional information and guidelines for implementation of ISO 9000-3, particularly from a management perspective, may be found in Kehoe and Jarvis (1996).

9.4 MODELING AND MEASUREMENT OF SOFTWARE RELIABILITY

9.4.1 Metrics and Data

Many types of data may be collected in a software project—number of defects found, types of defects, criticality, CPU time utilized in various activities, analysis hours, and many other qualitative and quantitative measures. Here we deal with problems of modeling, assessing, and predicting software reliability. How this is accomplished depends on the type of data planned for and available. Characterization of the data involves definition of variables, the metrics used, model requirements, interactions between these, and related issues.

Measurement problems involve not only how to measure reliability, but how to measure factors that affect reliability, such as complexity, program size, and so forth. Both discrete and continuous variables are used. Some typical discrete variables are

- Total number of defects
- Program length in thousands of lines of code (KLOC)
- Thousands of noncomment source statements (KNCSS)
- Number of defects per KLOC
- Number of failures per unit of time
- Number of distinct operators (e.g., keywords)
- Number of distinct operands (e.g., data inputs)

A number of these are discussed by Ireson and Coombs (1966) and Beizer (1990) in the context of assessing quality, comparing programs, comparing a program to specifications, assessing customer satisfaction, and so forth.

Metrics such as KLOC, statement counts, and related measures are called *linguistic* metrics. Beizer also discusses *Halstead's metrics*, which are based on n_1 (n_2) = number of distinct operators (operands). In these terms, program length is defined to be (Halstead, 1977)

$$H = n_1 \log_2(n_1) + n_2 \log_2(n_2) \tag{9.1}$$

Note that this is not the same as the number of statements in the program, but is a function of the number of distinct elements in the programming language that will be used in the program and the number of variables that will be used. This number is useful in estimating program length (and hence cost and reliability) prior to writing the program.

See the references cited as well for discussions of *structural metrics,* which are measures of complexity based on the numbers of links, nodes, and disconnected parts of a program.

Example 9.1
The number of keywords used in programming a medical device is 86 and the number of distinct data elements used is 180. The Halstead metric for this application is

$$H = 86 \log_2(86) + 180 \log_2(180) = 1901$$

If half as many keywords and half as many variables are used, the corresponding value is $H = 818$, somewhat less than one-half the previous value. ■

Continuous variables (or variables that are typically modeled as continuous) used in analysis of software include:

- Time to detection of a failure (TTF)
- Time between failures (TBF)
- Time to repair (removal of a defect) (TTR)

- Fault density = number of faults per KLOC
- Usage (total time)

Here time may be execution (CPU) time, clock time, or calendar time. (See Section 9.2.)

Many other variables should be tracked in software development and operation. These include

- Date and time of failure
- System on which software was run
- Which version of software
- Apparent cause of failure
- Consequences of failure

Many additional software quality and reliability measures are discussed by Jones and Vouk (1996), with particular emphasis on field data and its analysis. Metrics that are especially useful in tracking software reliability throughout the life cycle of the product are discussed by Walters and McColl (1979) and Iyer and Lee (1996). The use of metrics for identification of failure-prone software, with application to the space shuttle avionics software, is discussed by Munson (1996).

Many of the variables listed above are used in assessment of software reliability and related quantities. How this is done depends on the type of data and the models employed. As usual, basic data analysis begins with summarization of the data, as indicated in Chapter 3, and proceeds to estimation and testing of parameters and models, as indicated in Chapters 5 and 8. The probability models discussed in Chapter 4 play an important role in this analysis.

Example 9.2 (Case 2.23)
The data of Case 2.23 are times between failures (detection of a bug) in a program. Thirty-four bugs were ultimately found. Summary statistics for this sample are $\bar{x} = 24.97$, $x_{.50}$ (the median) $= 7.00$, $s = 50.87$, $x_{(1)} = 1.00$, $Q_1 = 3.00$, $Q_3 = 13.00$, and $x_{(34)} = 258.00$. [See Hossain and Dahiya (1993) for the original data set.] By any measure of symmetry, these data are seen to be highly skewed right. Distributions such as the Weibull, lognormal, or gamma would appear to be reasonable candidate models. (This will be discussed further in Chapter 11.) ∎

9.4.2   Software Reliability Models

As in modeling hardware failures (see Chapter 7), white-box and black-box models are used to model software failures (Friedman and Voas, 1995). The white-box approach involves an assessment of software reliability based on analysis of the complexity of the program and structure of the code. The black-box approach is an external assessment, involving frequency of execution, time between failures, etc., and, as in hardware reliability analysis, is based on certain assumptions regarding the associated probability distributions. (There is also a "gray-box" approach to reliability assessment; for example, estimating MTTF simply as number of errors per KLOC. This neither looks at the structure of the program nor makes any probabilistic assumptions regarding TTF.) In this section, we consider primarily black-box models. For further discussion, see Musa et al. (1990), Farr (1995), Singpurwalla and Wilson (1992/1993), and the texts on software reliability cited previously.

Models may be either continuous (e.g., modeling TTF) or discrete (e.g., modeling number of failures per unit of time). Singpurwalla and Wilson (1992/1993) call these Type I (or Type Ia, if the failure rate changes through time) and Type II models. The distinction is also discussed by Farr (1995), who notes that there are some models that can be used for both types of data. They may be further classified in a number of ways. Let $N(t)$ = number of failures occurring in the interval $[0, t)$. Farr (1995) classifies models as *finite* if $\lim_{t \to \infty} E[N(t)] < \infty$, and *infinite* otherwise. A more detailed classification of models is given by Musa and Okumoto (1983).

The models given below require a number of assumptions. These may vary from model to model. Here we list the more common assumptions, with additional information given briefly below. The following notation is used:

N = number of faults in the software item (may be assumed random or deterministic)

$N(x)$ = number of failures in the interval $[0, x)$ (a random variable)

T_i = time until ith failure (with $T_0 = 0$), $i = 1, \ldots, N$

$X_i = T_i - T_{i-1}$ = time between failures

$\mu(t) = E[N(t)]$ = expected number of failures in $[0, t)$

$\lambda(t) = \mu'(t)$ = failure intensity at time t

$t_0 = 0 < t_1 < \ldots < t_n = t$ provide a partition of the time interval $[0, t]$

$F_i(x)$ = CDF of X_i, $x > 0$

$f_i(x)$ = density function of X_i

Most models are based on the assumptions that

1. Failures are independent
2. Every failure within a failure class has the same chance of being discovered
3. Faults are corrected instantaneously
4. The software is operated in the same environment as that under true operational conditions
5. Removal of a fault does not introduce new defects
6. Each time a failure is detected, one fault is corrected

For many models, it is assumed that the time to failure follows an exponential distribution, with the mean time to failure possibly varying from fault to fault. Thus the number of faults may follow a Poisson process or a nonhomogeneous Poisson process (NHPP). (See Chapter 7 and Appendix A.)

Complete details regarding model assumptions may be found in the references cited for specific models and in Farr (1995) and Xie (1991, 1993), which provide thorough discussions of the models given below and of many other software reliability models, giving, in addition to assumptions, motivation for the model, data needs, estimation, comments on implementation, and a number of examples. Much of the remainder of this section is based on Farr (1995) and Singpurwalla and Wilson (1982/1983). Additional information on modeling software reliability may be found Sagola and Albin (1984) and in the sources cited below, the references cited by these authors, and in the several hundred references given by Xie (1991, 1993).

Reliability Models 1. Modeling Time to Failure

Jelinski–Moranda Model.   Jelinski and Moranda (1972) developed the first software reliability model. It assumes that the X_i's are independent, exponentially distributed random
variables, with respective parameters $\lambda[N - (i - 1)]$, i.e.,

$$f_i(x) = \lambda[N - i + 1]e^{-\lambda[N-i+1]x} \tag{9.2}$$

for $x > 0$. Thus the expected time between failures, $\mu = 1/\lambda[N - (i - 1)]$, increases with each
successive failure found. The mean and intensity functions are, respectively, $\mu(t) = N(1 -
e^{-\lambda t})$ and $\lambda(t) = N\lambda e^{-\lambda t}$. The failure-rate function (See Chapter 4) for x in the ith subinterval is
$r_{X_i}(x) = \lambda(N - i + 1)$, which is constant in each interval and decreases to a new constant value
after each failure. (Thus these are *conditional* failure rate or hazard functions given that
$(i - 1)$ failures have occurred.)

Schick–Wolverton Model.   Schick and Wolverton (1978) introduced a time element as
well as number of failures previously found into the failure-rate function, resulting in $r_{X_i}(x) =
\lambda(N - i + 1)x$. The corresponding density function is

$$f_i(x) = \lambda(N - i + 1)xe^{-\lambda(N-i+1)x^2/2} \tag{9.3}$$

$x > 0$, which is a Weibull distribution with shape parameter 2 (see Section 4.4.3). Here the
failure-rate function jumps to zero at each failure instant and then increases linearly until the
next failure instant.

Weibull Model.   Here X_i is assumed to follow a Weibull distribution with shape parameter
α and failure rate given by

$$r_{X_i}(x) = \lambda\alpha(N - i + 1)x^{\alpha-1} \tag{9.4}$$

This is a generalization of the Jelinski–Moranda and Schick–Wolverton models. The corresponding density function of the ith time to failure X_i is

$$f_i(x) = \lambda\alpha(N - i + 1)x^{\alpha-1}e^{-\lambda\alpha(N-i+1)x^\alpha} \tag{9.5}$$

for $x > 0$. For this model, the risk function decreases to zero at each failure instant and increases nonlinearly until the next failure if $\alpha > 1$ (except for $\alpha = 2$). If $\alpha < 1$, the failure rate
goes to ∞ at each failure instant and decreases until the next failure instant, which is an unlikely model for software reliability.

Musa Basic Execution Time Model.   This is one of the more widely used models. Time
is defined to be execution time until failure. The failure rate function after $(i - 1)$ failures
have occurred (the last at time t_{i-1}) is

$$r_{X_i}(x) = \lambda_0 e^{-\beta t_{i-1}}e^{-\beta x} \tag{9.6}$$

where λ_0 is the initial failure rate and β is the decrease in failure rate per failure. For this
model, the mean function is

$$\mu(x) = \frac{\lambda_0}{\beta}(1 - e^{-\beta t}) \tag{9.7}$$

Musa et al. (1990) then relate execution time to calendar time by means of a model for usage of each computer resource, the argument for this being that for planning purposes, a manager needs to know calendar time, not execution time. They recommend this model as one that generally predicts well, is simple and easily understood, is thoroughly developed and widely used, has parameters having a physical interpretation (related to characteristics of the software, such as size), and can handle evolving systems.

Imperfect Debugging Model.   Goel and Okomoto (1979) proposed a generalization of the Jelinski–Moranda model under which it is assumed that a fault is not fixed with certainty. It is assumed that there is a constant probability p of fixing a bug when it is found. In this case, the failure rate after failure $(i - 1)$ is

$$r_{X_i}(x) = \lambda[N - p(i - 1)]$$ (9.8)

$0 \leq p \leq 1$.

Hyperexponential Model.   Here it is assumed that there are k classes of faults (for example, in different sections of software), with different failure rates within each class. This type of model was first considered by Ohba (1984). It is assumed that in the ith class, which occurs with probability p_i, the time to failure is exponentially distributed with parameter λ_i (with declining rates of failure within each class as in the Jelinski–Moranda and other models). The mean function for this model is

$$\mu(x) = N \sum_{i=1}^{k} p_i(1 - e^{-\lambda_i x})$$ (9.9)

where $0 < p_i < 1$ and $\Sigma_{i=1}^{k} p_i = 1$.

Ohba S-Shaped Model.   Ohba (1984) developed an S-shaped reliability model to account for learning in the early phase of debugging, more rapid detection as familiarity with the software increases, and a tailing off as the remaining faults become more difficult to detect. The mean function that provides this shape is

$$\mu(x) = N[1 - (1 + \beta x)e^{-\beta x}]$$ (9.10)

Models of this type are discussed further in Yamada et al. (1984). Extensions of this model to multiple, dependent versions of a software system are discussed by Lin and Chen (1993).

Reliability Models 2. Bayesian Models of Time to Failure

Bayesian models assume that the model parameters are random, and various assumptions are made with regard to the probability distributions of these random quantities (see Chapter 8). We briefly look at a few such models.

Littlewood–Verrall Model.   Littlewood and Verall (1973) proposed a model in which the random variable X_i is assumed to be exponentially distributed with parameter Λ_i, which is assumed to be a random variable having a Gamma distribution (see Section 4.4.2) with shape parameter α and scale parameter $\psi(i)$, so that the distribution of Λ_i is given by

$$f_{\Lambda_i}(\lambda) = \frac{[\psi(i)]^\alpha \lambda^{\alpha-1}}{\Gamma(\alpha)} e^{-\psi(i)\lambda}$$ (9.11)

and the posterior distribution of X_i is (Farr (1995))

$$f_{X_i}(x) = \frac{\alpha[\psi(i)]^\alpha}{[x + \psi(i)]^{\alpha-1}}$$

(9.12)

for $x > 0$. This is a Pareto distribution (see Johnson and Kotz, 1970a).

A common approach is to assume that the scale parameter is a linear function of the TBF index, i.e., $\psi(i) = \beta_0 + \beta_1 i$. In this case the intensity function becomes $\lambda(x) = (\alpha - 1)[\beta_0^2 + 2\beta_1 t(\alpha - 1)]^{-1/2}$.

Littlewood Differential Debugging Model. Littlewood (1980) extended the previous model to the case where the Λ_i are not identically distributed. It is assumed under this model that Λ_i has a gamma distribution with shape parameter $\alpha(N - i)$.

Empirical Bayes Model. Mazzuchi and Soyer (1988) developed two empirical Bayes models as extensions of the Littlewood–Verrall model. In the first, is assumed that α and Λ are both random variables, that they are independent and that α is uniformly distributed on an interval $[0, v)$ and Λ_i is gamma distributed. In the second, Λ_i is distributed as in the original Littlewood–Verrall model, with $\psi(i) = \beta_0 + \beta_1 i$ and β_0 and β_1 dependent and having gamma distributions. See Mazzuchi and Soyer (1988) and Singpurwalla and Wilson (1983/1984) for details.

Reliability Models 3. Models Based on Number of Failures

Goel–Okumoto Model. Goel and Okomoto (1979) proposed a model for $N = N(x)$, the number of failures in the interval $[0, x)$ under which it is assumed that N follows a nonhomogeneous Poisson process with intensity function $\lambda(x) = \alpha\beta e^{-\beta x}$. (This model is also called the NHPP model.) It follows that the mean function is

$$\mu(x) = \alpha(1 - e^{-\beta x})$$

(9.13)

Thus the distribution of $N(x)$ is given by

$$P(N(x) = n) = \frac{[\alpha(1 - e^{-\beta x})]^n}{n!} e^{-\alpha\{1 - e^{-\beta x}\}}$$

(9.14)

Musa–Okumoto Log Poisson Model. This model assumes exponential decay in expected number of failures, and is due to Musa and Okomoto (1984). The mean function is taken to be

$$\mu(x) = \frac{1}{\beta} \log(\lambda_0 \beta x + 1)$$

(9.15)

As a result, the distribution of $N(x)$ is given by

$$P[N(x) = n] = \frac{[\log(\lambda_0 \beta x + 1)]^n}{\beta^n (\lambda_0 \beta x + 1)^{1/\beta} n!}$$

(9.16)

Note that the Poisson assumption is equivalent to the assumption that time between failures is exponential. This model is an alternative to the basic execution time model discussed previously. Musa et al. (1990) suggest the following uses of these two models:

Use the basic model for: 1. Analysis and prediction prior to execution of the program

                                2. Analysis of the effect of engineering/technology (via faults)

                                3. Analysis of the effect of significant changes in program size

Use the log Poisson model for: 1. Early predictive validity

                                2. Highly nonuniform operations environments

Reliability Models 4. Other Reliability Growth Models

Here we look at a few additional reliability growth models that have been used in the analysis of software. We simply characterize the models by means of their associated mean and/or intensity functions. Details of the model assumptions, applications and parameter estimation may be found in Friedman and Voas (1995), Farr (1995), and the sources cited below.

Duane Power Model. One of the earliest reliability growth models is that of Duane (1964). It is called a power model because the mean function is expressed as a power of x, namely $\mu(x) = \alpha x^\beta$. The result is an NHPP with failure intensity function having the same form as the hazard function corresponding to a Weibull distribution. Duane arrived at the model by empirical means, plotting data on various scales.

Geometric Model. This is a generalization of the Jelinski–Moranda model proposed by Moranda (1975). It assumes a Poisson process with mean decreasing geometrically as failures are discovered. The mean function is given by

$$\mu(x) = \frac{1}{\beta} \log(\lambda \beta x e^\beta + 1) \tag{9.17}$$

Log Power Model. The log power model proposed by Zhao and Xie (1992) has intensity function

$$\lambda(x) = \frac{\alpha\beta[\log(1 + x)]^{\beta-1}}{1 + x}, \tag{9.18}$$

so that the mean function is

$$\mu(x) = \alpha[\log(1 + x)]^\beta \tag{9.19}$$

Comment. It is apparent from the above that many models for representing, estimating and predicting software reliability have been developed. Still others are given in the references cited. At present, few programmers, analysts, or reliability engineers use *any* models of this type. Instead, good (usually) engineering principles and practice are used in developing software; reliability and maintainability are only vaguely addressed, and management by crisis is used in supporting the software.

One of the problems, of course, is choosing a particular model that is appropriate for a specific application. Many of the models appear to be of an ad hoc nature, but all are based on assumptions (explicit or implicit) about the nature of the failures and probability of their occurrence. It is suggested that these assumptions (only very briefly discussed above) be looked at carefully in attempting to select an appropriate model. They are usually discussed in some detail in the references cited for the specific models. Goodness-of-fit techniques (discussed in Chapter 11) can be used to select among competing models. An excellent summary of assumptions for many of the models discussed in this section is given by Farr (1995). Additional guidelines for applications are given by Musa et al. (1990)

Applications to data sets from JPL (14,000 lines of code used on a spacecraft) and Bellcore (telecommunications software system) are discussed by Nikora and Lyu (1995). In application to several data sets, it was found that no single model was consistently best. Model adequacy is tested using chi-square and Kolmogorov–Smirnov goodness-of-fit tests. (See Chapter 11.)

9.4.3   Estimation of Model Parameters and Software Reliability

We now look briefly at estimation of parameter for selected models. Most of the procedures given below are based on the method of maximum likelihood (see Chapter 5). ML estimators for parameters of many of the remaining models may be found in Farr (1995) and, in many cases, in the original references cited. In other cases, it is necessary to formulate the likelihood function based on the assumptions of the model, differentiate to find the maximum, and solve the resulting equations (often by numerical methods). If this is not possible, the method of moments (see Chapter 5) may provide a solution to the basic estimation problem. Note that application of the estimation procedures presumes the existence of an appropriate database, which we describe below as well. Careful attention must be paid to this, as it is essential that proper data be collected, particularly during the test, debugging, and operational phases of any software project.

Estimation of reliability at any given stage as well as prediction of future reliability can usually be fairly easily accomplished for the models discussed. For most of these models, the reliability function is a one-to-one function of the parameters and the MLE of reliability is that function of the MLEs of the parameters. Asymptotic confidence intervals can also be obtained for the parameters and for reliability by use of standard asymptotic maximum likelihood results.

Data Needs and Parameter Estimation for Selected Models

Jelinski–Moranda Model.   We wish to estimate the number of defects N and the scale parameter λ. Data required are the x_i, the times between failures, $i = 1, \ldots, n$. The MLE N^* of N is obtained as the solution of

$$\sum_{i=1}^{n} \frac{1}{N^* - i + 1} = \frac{n}{N^* - (1/\Sigma x_i)[\Sigma(i-1)x_i]} \tag{9.20}$$

This equation is solved by numerical methods. (See Farr, 1995 for a computer algorithm.) Once N^* is obtained, the MLE of λ is calculated as

$$\lambda^* = \frac{n}{N^* \Sigma x_i - \Sigma(i-1)x_i} \tag{9.21}$$

From this, it is possible to estimate many other characteristics. For example, the MTTF of the $(n + 1)$st fault, given that n faults have been detected and corrected, is $[\lambda^*(N^* - n)]$ (Farr, 1995).

Example 9.3 (Case 2.23)

Suppose we use just the production phase of the project (errors 1 through 26) and estimate the parameters of the Jelinski–Moranda model. From the data in Hossain and Dahiya (1993), we obtain $\Sigma x_i = 250$, and $\Sigma(i - 1)x_i = 4008$. A simple search yields $N^* = 31$. From (9.21), λ^* is found to be 0.006948. If the entire data set is used, we find that $\Sigma x_i = 849$, and $\Sigma(i - 1)x_i = 21851$, giving $N^* = 34$ and $\lambda^* = 0.004847$. Note that, in fact, 34 errors have already been discovered, indicating that we estimate that no bugs remain in the program. ■

Weibull and Schick–Wolverton Models. Moment estimators, MLEs, and graphical estimators are given by Coutinho (1973). The MLEs require solution of a complex set of equations. Farr (1995) suggests instead the following least squares fit. Partition the observation period $[0, t)$ into k time intervals, $[0, t_1), [t_1, t_2), \ldots, [t_{k-1}, t_k)$. Let n_i be the number of failures in period i, $i = 1, \ldots, k$, and $n = \Sigma_{i=1}^k n_i$. Calculate $F(i) = (1/n)\Sigma_{j=1}^i n_j$ and let $x_i = \log(t_i)$ and $y_i = \log\{\log[1/(1 - F(i))]\}$. Perform a regression analysis (see Section 10.9) with x as the predictor and y the response variable, obtaining the sample regression equation $\hat{y} = a + bx$. The estimates of α and λ are $\hat{\alpha} = a$ and $\hat{\lambda} = e^b$.

Musa Basic Execution Time Model. The data required are $x_i =$ time of occurrence of the ith failure, $i = 1, \ldots, n$, and $x =$ remaining execution time after detection of the last failure (at time x_n). The likelihood equations are

$$n - \frac{n\beta^*(x_n + x)}{e^{\beta^*(x_n+x)} - 1} - \sum_{i=1}^n x_i = 0 \tag{9.22}$$

and

$$\lambda_0^* = \frac{n\beta^*}{1 - e^{\beta^*(x_n+x)}} \tag{9.23}$$

Obha S-Shaped Reliability Growth Model. The data required are the partition of $[0, t)$ and $n_i =$ number of failures in subinterval i, as in the Schick–Wolverton model. The MLE's are obtained as the solution of

$$n = N^*[1 - (1 + \beta^* t_k)e^{-\beta^* t_k}] \tag{9.24}$$

and

$$N^* t_k^2 e^{-\beta^* t_k} = \sum_{i=1}^k \frac{n_i t_i^2 e^{-\beta^* t_i} - t_{i-1}^2 e^{-\beta^* t_{i-1}}}{(1 - \beta^* t_{i-1})e^{-\beta^* t_{i-1}} - (1 - \beta^* t_i)e^{-\beta^* t_i}} \tag{9.25}$$

See Farr (1995) for further discussion and a computer algorithm for solution of the ML equations.

Goel–Okumoto NHPP Model. Goel and Okumoto (1979) derive MLEs for N and β based on observed times to failure, $t_1, \ldots, t_n$. MLEs are obtained as the solution of

$$\frac{n}{\beta^*} = \sum_{i=1}^{n} t_i + \frac{n t_n}{e^{\beta^* t_n} - 1}$$

(9.26)

and

$$N^* = \frac{n}{1 - e^{-\beta^* t_n}}$$

(9.27)

MLEs for grouped data are given by Farr (1995).

Example 9.4 (Case 2.23)

Suppose that the Goel–Okumoto NHPP model is used to model the debugging process. We begin again with the production phase data only ($n = 26$). The t_i can be calculated from the original data set. The results are $t_{26} = 250$ and $\Sigma t_i = 2492$. Equation (9.26) is easily solved by means of a simple search. The result is $\beta^* = 0.00579$. From (9.27), we obtain $N^* = 34$. Using all of the 34 data points, the results are $\beta^* = 0.004312$ and $N^* = 34$, again indicating that all bugs have been found. ■

Hyperexponential Model.   MLEs for β_i and $N_i = N p_i$ are calculated for each category using the formulas for estimating β and N in the Goel–Okumoto NHPP model.

Littlewood–Verrall Model.   For the linear form of $\psi(i)$, the ML equations, based on the times between failures x_i are

$$\frac{n}{\alpha^*} + \sum_{i=1}^{n} \log\left(\frac{\beta_0^* + \beta_1^* i}{x_i + \beta_0^* + \beta_1^* i}\right) = 0$$

(9.28)

$$\alpha^* \sum_{i=1}^{n} \frac{1}{\beta_0^* + \beta_1^* i} - (\alpha^* + 1) \sum_{i=1}^{n} \frac{1}{x_i + \beta_0^* + \beta_1^* i} = 0$$

(9.29)

and

$$\alpha^* \sum_{i=1}^{n} \frac{i}{\beta_0^* + \beta_1^* i} - (\alpha^* + 1) \sum_{i=1}^{n} \frac{i}{x_i + \beta_0^* + \beta_1^* i} = 0$$

(9.30)

Duane Reliability Growth Model.   The MLEs are

$$\alpha^* = \frac{n}{t_n^{\beta^*}}$$

(9.31)

and

$$\beta^* = \frac{n}{\sum_{i=1}^{n-1} \log(t_n/t_i)}$$

(9.32)

9.4.4   Estimation of the Number of Remaining Defects

For many of the models considered above, the number of remaining faults in the software, given that n faults have been detected and fixed, can be estimated in a straightforward manner based on estimates of the model parameters. We consider a few of these.

Jelinski–Moranda Model.   The model provides a direct estimate N^* of the total number N of faults. The MLE of the number remaining is $N^* - n$.

Obha S-Shaped Model.   As in the previous case, the estimate of the number remaining is calculated as $N^* - n$.

Goel–Okumoto NHPP Model.   The estimate of the number of faults remaining is $N^* - n$.

Hyperexponential Model.   N is estimated as $\Sigma_{i=1}^{k} N_i^*$, where N_i^* is the MLE of the number of faults remaining in the ith category.

Example 9.5 (Case 2.23)

With $n = 26$, the Jelinski–Moranda model estimates N to be 31, so the estimate of the number of remaining defects is $31 - 26 = 5$. The corresponding value for the Goel–Okumoto model is $34 - 26 = 8$. In fact, 8 were ultimately found. Using all data, we estimate that there are no remaining defects with both models. ∎

A different approach to estimating the total number of defects in a program is suggested by McConnell (1997). It is based on setting up two independent test groups, say A and B, each of which tests the complete product. We denote by N_A, N_B, and N_{AB} the number of defects found by Group A, Group B, and by both groups, respectively. The number of unique defects discovered after completion of testing is then $N_A + N_B - N_{AB}$. The total number of defects is estimated to be $N_A N_B / N_{AB}$. The difference between these two quantities is an estimate of the number of defects remaining.

Additional sources of information relevant to the problem of estimating N are Blumenthal and Marcus (1975), Joe and Ried (1985), and Nayak (1986).

9.5   SOFTWARE TESTING PROCEDURES

The point has been made that it is very important to make provisions for obtaining relevant, valid, and accurate data in any attempt at analysis of software reliability. An important consideration in assuring that this takes place is the use of proper software testing procedures. There is a very substantial literature on this subject as well. Here we provide some basic guidelines and refer the reader to some of the key works in this area for further information.

Prior to testing, careful inspection of the code is essential. The objective is to remove defects as soon as possible in the development process. (Lalli et al. (1998) estimate that it takes, on average, 0.7 hours to fix a defect found in inspection, and 5 to 18 hours to fix one found in testing.) Inspection should be done by means of a structured review process—preferably a small group with assigned roles, a vested interest in the product, and shared responsibility. A discussion of the structured approach and rules for inspection are given by Lalli et al. (1998).

Plans for testing begin at inception of a project and continue throughout the product life cycle. The point has also been made that it is critical to detect defects prior to product release. The cost of correcting a defect is 80 to 100 times as much after a product is released than if it is found during development, and testing typically accounts for 30 to 50% of a software project budget (Pullum and Doyle, 1998).

In this section, we look at some basic principles of software testing, types and levels of tests, and tools and procedures for testing. IEEE standards for testing are listed in Section 9.2.5.

9.5.1   Basic Principles of Software Testing

Well-planned and carefully executed testing is essential to the production of reliabile software. Testing is *not* the same as debugging. (See Beizer, 1990 for a thorough discussion of the distinction between the two.) The purpose of testing is to discover program defects and to provide a basis for estimation of software reliability. The purpose of debugging is to identify faults—errors in logic, coding, etc., that caused the failure—and correct them.

Some basic rules for testing have been given by Myers (1976) (who called them "axioms"), Priest (1988), and Pullum and Doyle (1998). Although not all of these are applicable in any given project and not all programmers/analysts would agree with all of them, these rules provide a sound philosophy and foundation for testing. The following list is a compilation based on the several sources.

- Testing begins with a clear statement of objectives.
- A good test case is one that leads to the discovery of a previously undiscovered defect.
- One of the most difficult problems in testing is knowing when to stop.
- A program cannot be adequately tested by the person who wrote it.
- The expected results of every test must be specified prior to testing.
- Nonreproducible or ad hoc testing is of little or no use.
- Write test cases for invalid as well as valid input conditions.
- Carefully analyze all test results.
- As the number of defects discovered increases, the probability that more defects exist also increases.
- Test teams should include the organization's most competent programmers.
- Ensuring testability must be a key objective in software design.
- Never alter a program to make testing easier unless the alteration is a permanent change.
- Design a system so that each module is integrated only once.

Test plans should be developed for each phase of a program and should include objectives, a plan to accomplish the objectives, and a method of verification that the objectives have been met. A test schedule should be prepared and expected milestones indicated. These must be updated regularly as the project proceeds. It is essential that complete records be kept of all tests and their outcomes.

A simplified overview of the test process is given in Figure 9.2 (from Ireson and Coombs (1988)). This schematic emphasizes the two major phases of testing—(1) unit testing, in which modules are tested in isolation from the rest of the system; and (2) integration testing, where the complete system is tested. Both show the test–fix–test cycle, which will lead to increased reliability unless new defects are introduced in the "fix" process. Preparation of an extended flow diagram of the process, including many of the elements discussed here and in

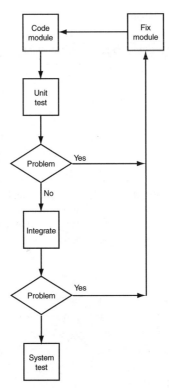

Figure 9.2. Software test–fix–test process (from Ireson and Coombs, Copyright 1988; reproduced with permission of the McGraw-Hill Companies).

the references cited, would be a valuable aid in formulating and structuring a software testing strategy.

9.5.2   Testing Methodology

As previously noted, two basic types of testing may be undertaken:

- Black-box testing, also called *functional testing* or *static testing,* in which the system is treated as a black box and tested by providing sets of inputs and verifying that the resulting outputs conform to specifications
- White-box testing, or *structural testing* or *dynamic testing,* which is an analysis of the details of the structure of the program, coding, language, data base design, and so forth

The use of black- versus white-box testing in different phases of a product development cycle is discussed by Ireson and Coombs (1988). Specific features of the software that must be tested include:

- Functionality—a check that all required features are provided
- Reliability—the program operates without failure over time

- Regression—the program does not operate less reliably than previous versions
- System tests—the completely integrated product is tested
- Stress tests—use of the product in multiuser, multiprogram environments

Automation of the process, so that repeated testing is easily done, is suggested as a very desirable feature of a test program.

Note that testing takes place after design and coding, that is, relatively late in the development cycle. The most complete description of the testing process is given by Beizer (1990). Beizer describes in great detail the following key elements of software testing, with excellent summaries of foundations, methodology, and implementation:

- Path testing—Select a set of test paths through the program that thoroughly tests the program's capabilities and correctness. A key technique is preparation of a "flowgraph," which is an upper-level flow chart, showing only key elements ("blocks") and omitting the details shown in ordinary flow charts of a process. This is basically a structural model of the software system. Beizer provides guidelines for selection of appropriate paths, instrumentation of the paths, and other details, and discusses implementation and applications.

- Transaction Flow Testing—A transactional flowgraph is prepared to represent the functionality of the system. Here transactions are tasks such as accept input, validate input, acknowledge input receipt, process input, etc. The purpose of the testing is to validate functional operation of the program. Testing techniques, implementation and examples are discussed.

- Domain Testing—The domain of a variable in this context is the set of acceptable values for an input variable. Domain testing attempts to determine whether or not a set of inputs is acceptable, usually by checking one variable at a time, without regard to program calculations. Domain errors can be the result of many bugs. These include domains that are contradictory, ambiguous, or overspecified, representations of zero (negative or positive), boundary and logic errors, etc. Detailed descriptions of these and methods for avoiding bugs of this type are discussed.

- Syntax Testing—The objective here is testing of conformance of internal and external inputs to the system. This involves understanding the program language and its format. Syntax errors can have disastrous results, and methods for designing and implementing tests and selection of test cases are discussed.

- Logic Testing—Logic testing in hardware is highly automated, but this is not true in software design. Methods for test design and data selection based on Boolean algebra and decision tables are presented.

- State Testing—A table of states for the software system is prepared and allowable transitions between the states are defined. Input ordinarily passes through a number of states until output is produced. State testing tracks the individual states rather than just input and output. Methods for identifying states and testing for validity of paths are discussed.

Beizer provides much additional information on bugs and their frequency of occurrence, implementation of the techniques and applications, as well as numerous references. Other valuable sources are Kaner (1993), Woodward (1991), Friedman and Voas (1995), and Pullum and Doyle (1998).

Comment. Nothing has been said in this section concerning the *number* of tests that should be run (except for Myer's comment that knowing when to stop is one of the most difficult problems in testing!). Estimates of the remaining number of defects, based on the models of Section 9.4 and discussed in Section 9.4.4, will shed some light on the subject, and this may provide a strong motivation for undertaking a modeling effort.

9.5.3 Tools for Testing Software

Testing of software begins in the design stage, as noted, and continues throughout the life cycle. This is also the case in testing hardware, and many of the same techniques and tools used in testing hardware may be used here. A significant difference is that, except under rare circumstances, far less testing and redesign or correction of errors is done on hardware after product release than is the case with software.

Two of the key design review and defect detection techniques used in analysis of hardware, fault tree analysis (FTA) and failure modes and effects (and criticality) analysis (FMEA and FMECA, see Chapter 7) have been used in analysis of software as well. These

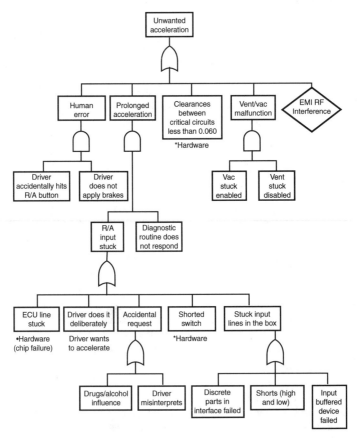

Figure 9.3. Fault tree of human/hardware/software system (from Raheja, 1991; reprinted with permission of the author).

are appropriate in the early stages of software development and are discussed in this context in Raheja (1991, Chapter 9). An example of a fault tree, from Raheja, is shown in Figure 9.3. Here the undesired outcome is unwanted acceleration of an automobile. The failure could be due to software, hardware, or human faults, as indicated in the chart. Examples of FMEAs and their application are also discussed by Raheja.

There are many other tools for analysis of software. Raheja reports the results of a survey of users. Although admittedly a small sample, it was found that FTA was the most used technique. Other techniques reported—33 in all—included software hazard analysis, FMEA, traceability matrices, and so forth, ranging down to one honest soul who claimed to use black magic!

A very detailed list of tools is given by Pullum and Doyle (1998). These are classified by purpose of the tool, as follows:

- Software standards—code auditor, test auditor
- Requirements analysis—test requirements list, verification matrix
- Product management—compare sets of code, software patch manager, configuration manager, software library manager
- Error detection—editor, flow chart generator, data logging, simulator, statistical tools, emulator, hardware monitor, interface checker, and many others
- Test design—test data generation, job control language data
- Test stop decision—economic trade-off model

Pullum and Doyle also deal with another important aspect of testing, report preparation. Included are recommended versions of

1. Test Incident Form—documents computer system, program, problem description, action taken, and disposition
2. Bug Report Form—documents description and identification, program version, correction needed, priority, programmer assigned to fix, list of items changed, and a number of related items

Detailed reports of this type are essential to effective software program management.

Software is available for implementing many of the tools listed above and for preparation of reports in various formats.

9.6   MANAGEMENT OF SOFTWARE RELIABILITY

Many of the issues addressed in this chapter require significant management involvement, and this has been pointed out in the discussion of these issues. As is true in dealing with quality in general, a successful program requires the participation of management at the highest level. Software is, in many respects, more difficult to deal with than hardware. It is often much more complex, more difficult to test, and lacks the structure (e.g., blueprints) and formal theory of physical processes (e.g., material's properties, metal fatigue, stress strength models). Because of this complexity, many software projects are never completed.

A second aspect of software management that is essential is communication, that is, documentation that is thoroughly done and understandable, from the onset of a program until its completion. This must include the program requirements and objectives, unambiguously stated, test reports, reports of problems found, and so forth.

Note that few of these issues are solely managerial or completely technical. It is self-evident that managers must be aware of the process being managed. To be effective, managers must be aware of the technical issues and involved in the resolution of any problems that arise.

The point is that in software management, the problems are somewhat different from those encountered in dealing with hardware. In order for a software project to be successful, complexity must be controlled, to the extent possible, personnel must be assigned to the areas in which they will be most effective (design, coding, testing, evaluation, etc.), and responsibilities delegated accordingly. This is not as easily done in managing software as it is in managing hardware.

A very important tool for management that is often overlooked or underutilized is modeling. The systems approach is to develop overall system models, models of individual components, and integration of these models, and involves many other facets. Models not only represent the system, they may enable the manager to assess progress, evaluate the quality of a product at a given point in time, and determine when it is ready to be released into the marketplace. System models and reliability allocation in the context of software are discussed by Friedman and Voas (1995). See this and the many other references cited in this chapter for flow diagrams and models for software design and predictions of software system reliability.

Tools to improve software reliability are discussed by Lalli et al.(1998). In order to accomplish this, organizations must provide for improvement in communications, concurrent engineering, review teams and joint training. Documentation, standardization, and personnel management are essential.

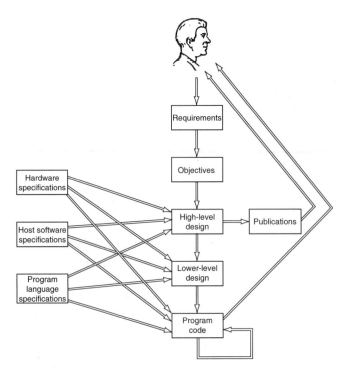

Figure 9.4. Software reliability management system (from Genadis, 1988).

A discussion of the management process is given by Genadis (1988). Figure 9.4 is a model of the management of software. Note the many feedback loops in the process. Each of the twelve points labeled is a possible source of error. Software management issues are discussed in more detail in Chapter 12. Other aspects of software that require significant management involvement are software safety considerations and software maintainability. These are also discussed in Chapter 12.

For additional information on software management, see Lloyd and Lipow (1962), Myers (1976), Vincent et al. (1988), Ireson and Coombs (1988), and Wallmüller (1994). The ISO 9000 approach to software reliability management is discussed in detail by Kehoe and Jarvis (1995) and Oskarson and Glass (1996).

EXERCISES

1. Contrast hardware and software faults.

2. Contrast hardware and software failures.

3. List cases where it may be difficult to determine whether a failure is due to hardware or due to software.

4. Software errors have been found to occur in clumps. Discuss the possible reasons and the effect this has on software reliability.

5. Give some examples of the sequence fault → error → failure in the context of software.

6. Two programmers independently debug a program. The first finds 81 bugs and the second finds 63. Fifty bugs are found by both. Calculate the number of unique bugs found and estimate the number of bugs remaining in the program.

7. Give examples of severity of bugs using each of the two severity scales.

8. Discuss the types of problems that one might encounter in collecting data on both counts per unit of time and time between software failures.

9. Discuss the concept of mean time to repair (MTTR), contrasting hardware and software.

10. Software and hardware failures may not be independent. Discuss the reasons for this and gives examples where this may be the case.

11. Contrast reliability modeling for hardware and software.

    Note: Problems 12 through 19 refer to the data of Case 2.23, summarized in Table 2.23. To complete the exercises, you will need to obtain the data from the original source.

12. Prepare a histogram and boxplot of the data of Case 2.23 and interpret the results.

13. Assume that the data comprise a sample of iid observations from a Weibull distribution. Estimate the Weibull parameters using only the production phase data (i.e., the first 26 observations).

14. Use the production phase data to estimate the parameters of a lognormal distribution. Calculate estimates of $P(X < x)$ for $x = 1, 2, 5, 10, 50, 100,$ and 200 using both the Weibull and lognormal distributions and compare the results.

15. Repeat Exercises 13 and 14 using all 34 data points.

16. Estimate the parameters of the NHPP model assuming a Weibull distribution.

17. Estimate the parameters of the Jelinski–Moranda model using only the first 15 observations (i.e., the first three month's data). Compare the results with those of Example 9.2.

18. Estimate the parameters of the Goel–Okumoto model using only the first three month's data. Compare the results with those of Example 9.3.

19. Estimate the parameters of the Duane reliability growth model (using all data) and interpret the results.

20. The following are times between failures (hours) for a statistical analysis program:

Failure No.	1	2	3	4	5	6	7	8	9	10	11	12	13	14	15	16	17	18
Time	0.5	1.4	1.9	2.8	1.6	2.9	3.5	0.9	1.2	3.4	1.6	2.8	4.5	5.7	1.0	4.1	2.0	7.2

   Prepare graphical and numerical summaries of the data and discuss the results.

21. Fit exponential, Weibull, and lognormal distributions to the data of Exercise 20 and compare the results.

22. Estimate the parameters of appropriate software reliability models using the data of Exercise 20 and compare the results.

23. Estimate the number of remaining bugs in the program using each of the models of the previous exercise and compare the results.

24. The following is a frequency distribution of data on time to failure (days) for a software product for a computer device:

Days	Frequency	Days	Frequency
1	32	11	2
2	12	12	0
3	8	13	1
4	6	14	1
5	5	15	1
6	4	16	0
7	2	17	0
8	3	18	0
9	1	19	0
10	1	20	1

   Analyze the data.

CHAPTER 10

Design of Experiments and Analysis of Variance

10.1 INTRODUCTION

In Chapter 3, types and sources of data were discussed and basic methods of summarization and description were given. Sources of data include historical data, test data, vendor data, and so forth. In subsequent chapters, we looked at basic statistical inference regarding the parameter(s) of specified probability distributions: estimation (method of moments and MLE), confidence and tolerance intervals, and hypothesis testing. We looked at continuous and discrete distributions, complete and incomplete data, and grouped data.

In this chapter, we are concerned with the collection of *test* data, specifically in the context of structured experiments. The basic objective will be comparisons of two or more populations. The emphasis will be on comparison of population means, although comparisons of other characteristics (e.g., variances or fractiles) will be considered to some extent as well. For example, an adhesive may be bought from several suppliers and we may wish to compare average bond strengths of the different products. Problems of this type are also encountered in analysis of dynamic reliability. For example, we may wish to compare MTTFs for several product designs under consideration.

By a structured experiment is meant a plan for data collection under which the experimenter has control over the important experimental conditions and other factors that could affect the results. Design of experiments (DOE) is the statistical discipline that deals with the development of such test plans. We will discuss complete data, excluding grouped and censored data sets, and consider standard research designs, including complete randomization as well as designs that feature blocking, multifactor experiments (called factorial designs), incomplete designs, and look briefly at accelerated testing.

The importance of the notion of structured data is that it is intended to preclude haphazard data collection or uncontrolled factors. Haphazard data collection is unlikely to produce test results that provide a proper basis (such as a true random sample) for inference about the population actually of interest. If important factors are left uncontrolled, the experimenter can never have confidence that the results obtained are reflecting the factors included in the design or are due to those not included. (Of course, we can never be certain that this is not the case, even in the most carefully designed and run experiments, because there are always a vast number of factors remaining that we do *not* control, but that is the nature of scientific inquiry.)

The situation just described is an ideal and is often not attainable. When that is the case, all is not lost and, in fact, the basic methods of data analysis discussed in this chapter are often employed in cases that are less than ideal as well. A few cases in point are:

- Missing data—Even carefully designed experiments can result in less than a complete data set. Equipment fails; resources are inadequate, information is found to have been recorded incorrectly or is lost, and so forth. In most cases, the analyses given in this chapter can be modified to take this into account. When there are many missing observations in complex experiments, however, the conclusions may be very tenuous.

- Uncontrolled factors—There are sometimes factors that are thought to be important but cannot be controlled but can be measured (e.g., ambient temperature, impurities in raw materials). In these cases, it is important to make such measurements. The analyses can be modified to make use of this information.

- Observational data—There are many instances where the only data that can be obtained are gotten under conditions under which the observer has little or no control. This is particularly true in the social sciences (e.g., one cannot experiment to any great extent with societal factors; mainly, one observes outcomes), where data of this type are called "observational data." It also occasionally occurs in engineering studies, where we have called such data "haphazard." Typical cases are when data are collected with no particular purpose in mind other than, say, occasional monitoring, e.g., of raw materials, metal parts, or simple components received from suppliers. An approach to dealing with observational data is to structure the data (i.e., design an experiment) after the fact and analyze the data as if obtained in a true experiment. This is legitimate *if* it can be assumed that the observational data are a true random sample from the population of interest. This is seldom the case and caution must be exercised in interpreting the results, since significant bias can be present when it is not.

The previous discussion illustrates a situation in which data collected for other purposes are to be used in a "scientific" study. Data of this type—uncontrolled data, observational data, and so forth—should, of course, be analyzed, because they almost always contain *some* relevant information, but the results must be judiciously interpreted and limitations of the study and the data carefully noted. A good maxim to keep in mind in this regard is, "A little knowledge is *not* a dangerous thing, *as long as you know it's a little!*"

In analyzing the data collected in structured experiments, the key tool that will be used is *analysis of variance* (ANOVA). This is the most widely used tool in statistical analysis for comparison of two or more means. It requires certain assumptions regarding the data, which will be discussed in the chapter, but is quite robust in that it is not sensitive to small or even modest violations of the assumptions. Note, incidentally, that for some distributions (e.g., the exponential) comparison of means is equivalent to comparison of reliabilities. For others, this may be true under certain constraints (e.g., Weibull distributions with the same shape parameter).

For unstructured data, a commonly used method of analysis is regression analysis. In cases where some factors are uncontrolled but are measured, an appropriate analysis is *analysis of covariance* (ANCOVA), which is basically a combination of ANOVA and regression. We briefly discuss both of these topics. (Note: ANOVA, regression, and ANCOVA can all be subsumed into a structure called the general linear model, and, in a general sense, are mathematically identical.)

The outline of the chapter is as follows. In Section 10.2, we begin with some important basic concepts of DOE and scientific experimentation in general. Some standard experimental designs are given in Section 10.3. The basic principles and assumptions of ANOVA, and the relationship of ANOVA to the structure of the experiment, are discussed in Section 10.4. The ANOVA for the most basic design and multiple comparison of treatments are also given in this section. In Sections 10.5 and 10.6, analyses of variance for more complex designs are given. In section 10.7, we look briefly at incomplete designs. In Section 10.8, we present some basic results in regression analysis and ANCOVA. Accelerated tests designs are discussed briefly in Section 10.9. These tests and analysis of the resulting data will be discussed in more detail in Chapter 14.

10.2   BASIC CONCEPTS OF EXPERIMENTATION

10.2.1   Experimentation and the Scientific Method

The scientific method is a formal methodology for investigation of natural phenomena, dating to Sir Francis Bacon in 1620. As it has evolved over the past several centuries, the scientific method may be represented as consisting of four essential steps, which occur in sequence. These may be succinctly stated as follows:

$$\text{Hypothesis} \rightarrow \text{Method} \rightarrow \text{Results} \rightarrow \text{Conclusion}$$

The hypothesis essentially delineates the area of investigation and the specific problems to be addressed. These issues are dealt with by the experimenter (engineer, scientist, manager), and are obviously very applications-specific. The hypotheses are translated to statistical hypotheses as discussed in Section 5.6. For example, an engineer and production manager may be interested in determining whether or not two different processes have the same yield under specified operating conditions. Appropriate hypotheses would be H_0: $\mu_1 = \mu_2$ versus H_a: $\mu_1 \neq \mu_2$, where μ_i is the mean yield for process i.

By "method" is meant experimentation. It follows that DOE is an essential component of the scientific method. An experiment must be both effective and efficient. An effective experiment is one that is feasible and that enables the experiment to draw inferences to the population of interest; an efficient experiment is one that provides the most information at a given cost or the required information at minimum cost. In order to effect this, the investigator must address a number of important issues prior to setting up the experiment. These include:

- Population(s) to which inferences are to be drawn
- Variables to be measured and units of measurement
- Factors to be varied in the experiment (e.g., temperature, amount of an additive)
- Choice of levels of these factors (e.g., 70°, 100°, 130°; 0.5%, 1.0%, 1.5%, 2.0%)
- Conditions under which the experiment is to be run
- Preparation of as complete a list as possible of other factors that could affect the results
- Preparation of a list of these factors that cannot be controlled
- Limitations of the experiment
- Feasibility, cost issues, facilities, and so forth

Results include the data obtained along with any other information concerning the experiment, especially including any deviation of the actual experiment from the designed experiment. Summary statistics, data plots, and the results of all appropriate statistical analyses are also included here.

Conclusions include the statistical conclusions (e.g., which null hypotheses are rejected and which are not) as well as the interpretation of these in the context of the original experimental objectives and in the associated managerial context. Conclusions should include recommendations for actions to be taken as well as suggestions for future research and the reasons for it. A discussion of the limitations of the experiment and conclusions should also be given.

Note that in practice, experiments are seldom conducted in isolation. Typically, there is some prior information available, perhaps from previous similar or related experiments. This is often useful and may even be essential in designing the current experiment. Furthermore, it often happens that results and conclusions of an experiment suggest other issues to be addressed, beginning a new cycle of the four steps. Thus the linear sequence given above is usually, in fact, a cyclical process.

10.2.2 Definitions

In this section, we discuss some basic concepts and approaches that are common to all true scientific experimentation. We begin with some basic definitions.

- *Treatment*—In DOE, the term *treatment* is used in a very generic sense. It may be a treatment in the usual sense of the word (such as exposure of a part to a temperature of 300 °C for 2 hours) or a qualitative characteristic such as laboratory in which the test was conducted, gender of a production worker, etc. In this general sense, a treatment is a method of classification of observations. Related terms, often used synonymously, are *factor, group, population, class,* and *category.*

- *Experimental Unit*—The basic unit of experimental material to which a treatment is applied or assigned is called an *experimental unit.* In the 747 windshield study (Case 2.2), experimental units are the individual windshields. In the fiber strength tests of several of the cases, experimental units are the individual fiber or fiber bundles. In cases where several units are treated in one operation, the collection of units is the experimental unit. For example, if five parts are exposed to a temperature of 300 °C for 2 hours, the group of five is the experimental unit and the observation on that experimental unit is the mean (or sum) of the five individual measurements. If, however, five separate runs had been made (i.e., each part exposed separately), then the individual parts would be the experimental units and we would have data on five experimental units. *Note:* In order to analyze data properly, it is essential that the experimental units be properly identified

- *Experimental Error*—The variability of responses among experimental units treated alike is called *experimental error.* Experimental error, usually denoted σ^2, is estimated by (directly or indirectly) calculating a variance of the measurements recorded on experimental units that received the same treatment. It is a reference measure that is fundamental to the analysis of experimental data. It is used in calculating confidence intervals, testing hypotheses, comparing treatments, and in solving many other statistical inference problems. It is important to note that if experimental units are not properly identified, it is unlikely that experimental error will be properly estimated.

Experimental error reflects the inherent variability in the measurements that cannot be controlled (or, at least, were not controlled) during the course of the experiment. Many elements of the experimental procedure contribute to experimental error, including

1. inherent variability in the experimental material (experimental units will not be perfectly homogenous)
2. variability in the treatment (a test chamber set to 300°will not be exactly that temperature, and the temperature on the next run will also not be 300°, nor will it be the same as the temperature on the first run)
3. uncertainty in the measurement process itself (with the amount depending on the precision of the measuring instrument)
4. variability caused by lack of control of many other factors that could affect response

Note that there may be more than one level of experimental error. For example, if batches of five items are subjected to three different temperatures, with one item of each batch removed and tested at the end of each day in a five-day period, the batches of five would be the experimental unit for temperature, but the individual items would be the experimental unit for time. This notion extends in an obvious way to more than two factors.

It is apparent that experimental error cannot be eliminated. In a well-designed and run experiment, however, we can attempt to control it, i.e., make it as small as possible. The features of an experiment that provide this control include

- careful selection of experimental material in order to obtain homogeneous experimental units (but not to the extent that we limit the population to which we can legitimately draw inferences)
- blocking (grouping experimental units into relatively homogeneous subgroups)
- improvement of the measurement process
- measurement and use of information on other variables (called *covariates*) that may be related to the response of interest

We next look briefly at some additional features that are essential to proper experimentation.

10.2.3 Randomization

As noted previously, a true experiment, as opposed to an observational study, always involves random sampling. Randomization in some form is a key requirement in the experimental designs we will consider. The reason this is important is that, if done properly, it avoids bias, i.e., provides the experimenter with the legitimacy of objectivity. In DOE, randomization is done by randomly assigning experimental units to treatments. The procedure for doing this depends on the structure of the experiment.

The randomization process itself may be carried out by use of random number tables, chance devices (e.g., dice, drawing numbers from a hat, etc.) or by use of a computer. The last is the more common practice today, and statistical program packages typically include random number generators and various options for sampling. In fact, many include programs for design of experiments.

10.2.4 Replication

In principle, replication means repetition, in the sense that each treatment is applied to more than one experimental unit. If this is, in fact, done, we have a direct estimate of experimental error, since this is defined to be the variability among experimental units treated alike. As will be seen, however, there are many experiments in which experimental error is estimated even when there are no two experimental units treated exactly alike, all of which require some special assumptions.

Another sense in which replication is used in DOE is forming groups of like experimental units (called "replicates") and then replicating over groups. This is a special feature of many designs and will be discussed in the next section.

10.2.5 Mathematical Models in DOE

One of the key tools used in data analysis in the context of designed experiments is representation of the data by a mathematical model. The model is a representation that describes both the structure of the experiment and the structure of the analysis of the resulting data. An excellent description of this relationship is given by Lorenzen and Anderson (1993).

As an example, the following is a model that describes the data (X) from a simply structured experiment in terms of that structure:

$$X = \mu + \tau + \varepsilon \tag{10.1}$$

Here μ represents and overall true mean, and τ is the "treatment effect," i.e., the change in μ as a result of the treatment. ε is the "error" term, also called "experimental error" (another use of this term) or "residual." It is not an error in the sense of a mistake, although if that occurs it will be a part of ε (unless the mistake is systematic, in which case it is a part of τ). Instead, it represents the inherent variability in the data that cannot be accounted for by the treatments.

In the models we will consider, ε is assumed to be a random variable with mean zero, and μ, as usual, an unknown constant. Typically (though not always), τ is considered to be a constant as well, in which case $E(X) = \mu + \tau$, i.e., the overall mean plus the change in mean associated with this particular treatment.

10.2.6 Approach to Designing an Experiment

As noted previously, any scientific investigation begins with a description of the context and a problem statement, usually translated into statistical terms (hypotheses to be tested, and so forth). This was discussed in Section 10.2.1, and a list of activities in this phase of a study was given. To actually implement this in designing the experiment itself, the following issues must be addressed:

- The number of observations to be obtained for each treatment
- The structure of the experiment (order, grouping of experimental units, etc.)
- Randomization to be used
- Mathematical models to describe the data and assumptions made

Since, as we shall see, a proper analysis of experimental data depends crucially on a proper understanding of the structure of the experiment, data analytical issues should also be addressed at this point. These include

- Choice of appropriate descriptive statistics
- Basic analysis of the data

- Tests of assumptions
- Detailed analyses to be done

Each of these issues will be discussed in the following sections as we look at a number of experimental designs. Note that experimental design and data analysis are very much applications specific, and not all of the issues raised will necessarily be applicable. Further, there may be issues not on the list that may be of importance and care should be taken to address these as completely as feasible.

10.3 SOME BASIC EXPERIMENTAL DESIGNS

Here we look at several basic designs that are widely used and form the basis for many additional experimental procedures. Various treatment structures and some more complex designs will also be discussed. In each case, we describe the experimental objectives, layout, method of randomization, and appropriate mathematical models.

10.3.1 The Completely Randomized Design

The *Completely Randomized Design* (CRD) is characterized, as the name implies, by complete randomization in the assignment of experimental units, the only constraint being that each treatment is assigned a predetermined number of experimental units. Suppose that the experiment involves k treatments and that we wish to assign n_i experimental units to the ith treatment, $i = 1, \ldots, k$. The total sample size is then $n = \Sigma_{i=1}^{k} n_i$. Each experimental unit is assigned to a treatment by selecting a random integer between 1 and k, stopping in each case when n_i units have been assigned to treatment i, and continuing the process until all n units have been assigned.

We assume, as in all experiments, that experimental units have been selected at random from some population (usually considered conceptually infinite in size). The treatments, if they have different effects on the response, partition this into k different (or, at least, potentially different) populations. As a result, the CRD may be thought of as an experiment in which simple random samples are drawn from each of k populations.

The CRD is also called a *one-way classification,* a *single-factor experiment,* a *one-factor design,* and so forth. In looking at other texts or computer program manuals, it is necessary to read the description of the design carefully in order to properly identify it. The principal objective of the experiment is to make inferences about the treatment means. Other objectives may be to test assumptions, calculate tolerance intervals, estimate reliabilities, compare variances or other population characteristics, and so forth. We will concentrate on the comparison of means, but will briefly consider some other of these inference problems as well.

For the CRD, we use the notation X_{ij} to identify the jth observation on treatment i. The model is that given in (10.1), written in terms of the data as

$$X_{ij} = \mu + \tau_i + \varepsilon_{ij} \tag{10.2}$$

were τ_i is the effect (change in μ) of the ith treatment, and ε_{ij} is the error associated with the jth observation on the ith treatment. In terms of the model, inferences about means are equivalent to inferences about the τ_i's.

The advantages of the CRD are that it is flexible, allowing for unequal replication, easily analyzed, with the analysis not complicated by missing observations in the single-factor case, and efficient in cases where experimental units are relatively homogeneous. If the experimental units are heterogeneous, however, experimental error will be large, and alternative designs should be considered. The CRD was used in several of the cases of Chapter 2, where experimental material could be expected to be relatively homogeneous. These include Case 2.6, with fiber length as treatment and strength as response; Case 2.9, with week as treatment and lifetime of light bulbs as response; Case 2.12, with machine setting as treatment and strength of laser weld as response; and Case 2.22, with stress rate as treatment and strength as response.

10.3.2 Randomized Block and Latin Square Designs

When experimental units are heterogeneous, one might be tempted to reduce heterogeneity by using units that are as nearly alike as possible. This is usually not a good idea, because it can greatly restrict the population to which inferences can legitimately be drawn. One approach to dealing with heterogeneous experimental units is to group the units into relatively homogeneous sets (called *blocks* or *replicates*) and then assign treatments to units within each block. We look at two such designs.

The Randomized Complete Block Design (RCBD)

The RCBD is the most simply structured design that features blocking. Experimental units are grouped into blocks of k relatively homogeneous units. Suppose there are r such blocks. Then the total sample size is $n = rk$. All k treatments appear in each block (hence the name "complete" block), with experimental units randomly assigned to treatments. A separate randomization is done for each block. The intent of blocking is to structure the experiment so that experimental units are homogeneous within blocks but differ from block to block.

The data are X_{ij} = response for ith treatment in jth block. The model is

$$X_{ij} = \mu + \tau_i + \beta_j + \varepsilon_{ij} \tag{10.3}$$

where β_j is the jth block effect and the remaining symbols are as previously defined. This is the usual model given for the RCBD and it will suffice for our purposes under the assumptions to be stated later, but it has been recognized as incomplete. The problem is that lack of complete randomization, because of the constraint that each treatment appear in each block, introduces an additional term, a "restriction error," into the model. See Lorenzen and Anderson (1993, Chapter 5); Anderson (1970); and Anderson and McLean (1974) for detailed discussion and analysis of this issue.

The RCBD is clearly not as flexible as the CRD, requiring equal replication. This is a disadvantage in certain applications, for example, where experimental material is limited for some treatments or where there is much more interest in certain treatments than in others. The advantage of the RCBD is that some of the variability in the data can be accounted for by differences among blocks, thus reducing experimental error, and, as we shall see, leading to more powerful tests.

Case 2.21 is an example of an RCBD, with the 26 fibers being blocks and the four fiber segment lengths being treatments. In other applications, blocks are often test chambers (with each treatment run in each chamber), days on which the tests were run (each treatment run on each day), laboratories, factories, technicians, and any of many other such factors.

The Latin Square Design (LSD)

The LSD features blocking in two dimensions (called *rows* and *columns*). Thus blocks are formed according to two characteristics and the variability due to both of these is removed from experimental error. The structure of the design is such that each treatment appears once in each row and once in each column. As a result, column and row lengths are both k, and the design may be represented schematically as a square. (The "Latin Square" terminology dates to antiquity; this pattern forms the basis of a number of games and puzzles, and of some challenging mathematical problems.) An example of a 4×4 Latin Square (with A, B, C, and D as treatments) is given in Table 10.1.

An application in which a design such as that of Table 10.1 might be used is in comparing wear for four brands of automobile tires (treatments). Four different cars are used as rows and columns are location of the tire (right-front, etc.) It is assumed that the cars are each driven a fixed distance under as nearly identical conditions as possible (for example, with professional drivers on a test track). There are many other such applications. The design is used in marketing studies as well. An example would be sales for five displays of a product (treatments) for five weeks in five stores (rows and columns).

The proper randomization of the LSD becomes complex. The problem is that one would like to select a square at random from all of the possible squares of size k. This cannot be done simply by randomization of the rows, columns, and treatments in a given square such as that of Table 10.1, because that process will not generate all of the possible squares. The simplest approach is to select a basic square from the tables of Latin squares given by Fisher and Yates (1957), then randomize the order of rows, randomize the order of columns, and assign letters to treatments at random. (This still does not make all possible squares equally likely, but it is adequate for our purposes.)

A difficulty in using the LSD in many applications is its lack of flexibility because the number of rows and number of columns must both be k. This is a problem if k is either small or large; 2×2 and 3×3 squares give very little information, whereas a 20×20 square constitutes a large and possibly expensive experiment. For this reason, the LSD is usually used, when appropriate, with something on the order of 4 to 8 or so treatments.

The notation for the data in an LSD is $X_{ij(h)}$, where i indicates the row, j the column, and h the treatment. The treatment index is put in parentheses to indicate that the i, j combination uniquely identifies the observation, but the treatment must be identified as well for purposes of analysis. The model for this design, in an obvious notation, is

$$X_{ij(h)} = \mu + \rho_i + \gamma_j + \tau_h + \varepsilon_{ij(h)} \tag{10.4}$$

10.3.3 Factorial Experiments

The designs discussed in the previous sections involved single-factor treatments—length of fiber, setting of machine, stress rate, etc. In many studies, particularly in engineering applica-

Table 10.1. Example of a 4×4 Latin square.

Row	Column			
	1	2	3	4
1	B	C	D	A
2	D	A	C	B
3	C	B	A	D
4	A	D	B	C

tions (but also in agricultural, biological, and many other applications), the investigator wishes to vary more than one treatment factor in an experiment. A simple example is Case 1b, where temperature and humidity are both varied. Very often, many factors (sometimes as many as 15 to 20) are involved.

The treatment factors (e.g., temperature and humidity) are called simply *factors*. The amounts of a factor present in an experiment (e.g., 100 °C and 200 °C; 70% and 90% relative humidity) are called *levels* of the factor. In this context, a *treatment* is a combination of one level of each factor, and a complete experiment is one that includes all possible treatments, at least once each. Experiments of this type are called *factorial* experiments. Note that "factorial" refers to the treatment design, not the experimental design. Factorial treatment structures may be used in any of the experimental designs described above and in many other designs.

Note that in the example of the previous paragraph, we have two factors, each at two levels. This is often referred to as a 2×2 or 2^2 experiment. The result is a total of 4 treatments. With 15 factors, each at two levels, we would have a 2^{15} experiment, with a total of 32,768 treatments—a very large experiment, indeed! This extends in an obvious way to any factorial structure. For example, a $2^2 \times 3 \times 5$ experiment is one with four factors, two at two levels each, one at three levels, and one at five levels, with a total of 60 treatments.

In a complete factorial experiment, not only is it possible to assess the affect of each factor, it is also possible to determine whether or not there is evidence of *interactions* among the factors. Two factors, say A and B, do *not* interact if the differences between average responses for levels of A are the same at all levels of B. An interaction exists if the differences between average responses at the different levels of A are different at different levels of B. We illustrate this in the case of a 2×2 experiment, with factors temperature (T) at levels t_1 and t_2, say, and humidity (H) at levels h_1 and h_2. The treatments are $h_1 t_1$, etc. Suppose that response is pull strength required to break an adhesive bond after exposure to treatment for a period of 96 hours. A numerical example, showing different levels of interaction, is given in Table 10.2.

In Table 10.2(a), there is no interaction, since the differences in average responses at the two temperatures is two at each level of H (equivalently differences for H are the same at each level of T). Note that here the overall difference between levels of T, called the *main effect* of T, is $15.5 - 13.5 = 2.0$. The main effect of H is $16.0 - 13.0 = 3.0$. In Table 10.2(b), the difference between temperatures is $17 - 13 = 4$ at h_1, but $14 - 12 = 2$ at h_2, so there is a temperature–humidity (TH) interaction. The main effects here are 2.0 for H and 3.0 for T. In Table 10.2(c), the interaction is extreme, the mean difference between temperatures being 4.0 at h_1 and -4.0 at h_2. Note that both main effects in this case are zero!

This illustrates an important point in analysis and interpretation of data in a factorial experiment: *test for interaction first, then look at main effects*. The fact that main effects are

Table 10.2. Examples of data from a two-factor experiment.

	(a) No interaction			(b) Slight interaction			(c) Extreme interaction		
	T			T			T		
H	t_1	t_2	Average	t_1	t_2	Average	t_1	t_2	Average
h_1	17	15	16.0	17	13	15.0	17	13	15.0
h_2	14	12	13.0	14	12	13.0	13	17	15.0
Average	15.5	13.5	14.5	15.5	12.5	14.0	15.0	15.0	16.0

zero does not necessarily indicate that there are no treatment differences. When interaction is extreme, main effects may make no sense at all, and levels of a given factor must be compared separately within each level of the other factor in order to properly interpret the data.

It is usually useful to look at interaction plots in interpreting the results in a factorial experiment. The results of Table 10.2 are plotted in Figure 10.1. The nature of interaction is apparent in these plots. In Figure 10.1(a), we see that parallel lines (or, in general, parallel curves) are indicative of no interaction. In Figure 10.1(b), we see that the main effects are somewhat misleading, while in Table 10.1(c) they are totally misleading, since both are zero. In the last case, if interaction were ignored, one would conclude that neither main effect affected response. Nothing could be further from the truth! Incidentally, many statistical program packages will prepare interaction plots. With more than two levels per factor, we again look for parallel response curves as an indication of no interaction. Anything else indicates that interaction is present.

Extension of the results to more than two factors leads to the possibility of many more interactions. For example, with three factors, say A, B, and C, the possible interactions are AB, AC, BC, and ABC. The interpretation of a three-factor interaction is that each two-factor interaction differs over levels of the third factor. This can be seen in Table 10.2 and Figure 10.1. Suppose that (a), (b), and (c) in the table and figure represent three levels of a third factor C. At the first level of C, there is no AB interaction; at the second level of C, there is a small AB interaction; and at the third level of C, there is an extreme AB interaction. Thus the nature of the AB interaction depends on the level of C, indicating an ABC interaction. In this case, one has to interpret the data by looking at comparisons of means for a given factor separately for each combination of levels of both other factors. These notions extend in a straightforward manner to higher-order factorials.

The model for a factorial experiment is a function of both the treatment design and the experimental design, with terms that identify the characteristics of each. We give two illustrations. The first is a two-factor factorial in a CRD. Denote the factors A and B and suppose that the experiment includes a levels of A and b levels of B, with r experimental units assigned to each treatment, for a total of ab treatments and $n = rab$ observations. The model is constructed by modifying (10.2), replacing τ_i by terms that reflect the factorial nature of the treatments. This requires three subscripts, two of which identify the treatment. We write X_{ijh} = observation h on the treatment consisting of the ith level of A and the jth level of B. The model is

$$X_{ijh} = \mu + \alpha_i + \beta_j + (\alpha\beta)_{ij} + \varepsilon_{ijh} \tag{10.5}$$

where α_i is the effect on response of the ith level of factor A (representing the main effect of A) β_j is the main effect of B, and $(\alpha\beta)_{ij}$ is the interaction term. The main effects are measured as deviations from μ, while the interaction term (which is interpreted as a single entity, not the produce of α and β) is measured as a deviation from $\mu + \alpha_i + \beta_j$.

As a second illustration, we look at the model for a three-factor factorial (factors A, B, and C) in a RCBD. This requires four subscripts—three to identify the treatment and one to identify the block. The model is

$$X_{ijhg} = \mu + \alpha_i + \beta_j + \gamma_h + (\alpha\beta)_{ij} + (\alpha\gamma)_{ih} + (\beta\gamma)_{jh} + (\alpha\beta\gamma)_{ijh} + \rho_g + \varepsilon_{ijhg} \tag{10.6}$$

The model contains the overall mean, three main effect terms, three two-factor interaction terms, a three-factor interaction term, a replicate (block) term, and an error term, in that order. We have again omitted the restriction error.

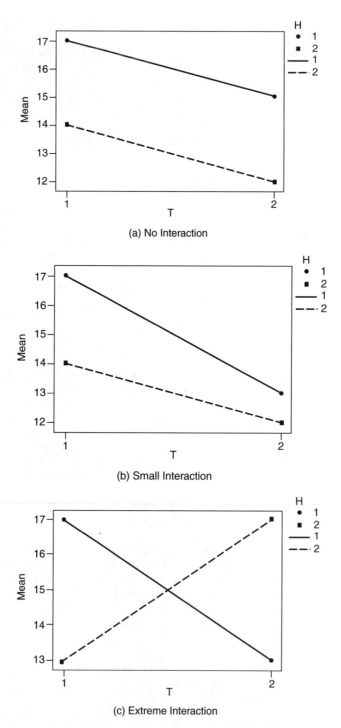

Figure 10.1. Examples of interaction (from Table 10.2).

Three important comments on the use and analysis of factorial designs are worthy of note. (1) For ease of analysis and interpretation, it is highly desirable that equal replication be used in the experimental design. The analysis can become quite messy if unequal replication is used, although most computer program packages can handle it. Furthermore, the interpretation is not nearly as straightforward as in the above examples, especially if there is evidence of interaction. These problems can become much worse yet if there are missing cells (treatment combinations that were not run at all) in the experiment. In extreme cases, it may be necessary to select subsets of the experiment that are complete, or nearly so, and analyze these separately.

(2) If the number of factors in an experiment increases, the number of possible treatments increases geometrically. This leads to experiments that are impossible to conduct and that no one could afford to do, even if it were possible. The 2^{15} experiment with 32,768 treatments illustrates the problem. In analyzing this experiment, we would deal with 15 main effects, $_{15}C_2 = 105$ two-factor interactions, $_{15}C_3 = 455$ three-factor interactions, and so forth, most of which would hopefully be negligible. (If not, much smaller experiments must be done first. It is impossible to envision a response surface in 15-dimensional space!) An alternative in such experiments is to run only a subset of the treatments. This subset must be selected very carefully, so that effects of interest are estimable. Experiments of this type are called *fractional factorial* experiments. We discuss this topic briefly later in the chapter.

(3) In experiments with many factors (or even a few), the approach of varying one factor at a time (say to a higher value than nominal and then to a lower), while holding all other factors at nominal, is almost always a bad idea. This experiment gives the analyst no chance at all to investigate interactions, and these might be the most important effects to identify.

10.3.4   Other Experimental Designs

In this section, we briefly describe a few additional experimental designs that are used in engineering and many other studies. Details and still other designs can be found in the many DOE texts available, e.g., Hicks (1982), Lorenzen and Anderson (1993), Montgomery (1997), and Winer (1971). The last provides extensive coverage of DOE, including the theory underlying the analysis of experimental data.

Methods for analysis of the designs listed below can also be found in the references. Documentation for most of the statistical program packages will also include descriptions of the designs and their analysis along with instructions for setting up the program to perform the analysis. A few of the designs included in this section will be discussed in some detail. Others will simply be illustrated by a simple example to aid in recognizing the structure of the design.

Split-Plot Experiments. In the discussion on experimental units and experimental error, it was noted that there may be more than one level of experimental unit in an experiment. An example is an experiment in which groups of five items are exposed to different temperatures in a chamber, with one item removed and tested each day for a five-day period. In this case, the batches of five, called *whole plots,* are the experimental unit for temperature. Temperature is called the *whole plot treatment,* and variability among whole plots treated alike is called *whole plot error.* Similarly, days are called *subplots,* time is the *subplot treatment,* and variability among subplots within whole plots is the *subplot error.*

The split-plot structure may be employed in many types of experimental designs. For example, the experiment just described may be run as a RCBD, using several temperature chambers as blocks. In this experiment, temperature levels are run in random order, with a new randomization done for each chamber and the items to be tested each day are selected at random.

This notion extends to a third level—split-split-plot experiments—and to many other configurations involving plots and associated treatments at various levels, arranged in blocks in an RCBD, or rows and columns in an LSD, and so forth. These experiments are widely used in certain applications in engineering and agricultural experimentation, where this type of experiment originated. [In agriculture, "plots" were, indeed, plots in a field (hence the name), with, for example, whole plots being harvest time and subplots variety of a crop, etc., with many variations and factors.]

The model for a split-plot experiment, with factor A as whole-plot treatment and B as sub-plot treatment, run in a RCBD, is

$$X_{ijh} = \mu + \rho_h + \alpha_i + \delta_{ih} + \beta_j + (\alpha\beta)_{ij} + \varepsilon_{ijh} \qquad (10.7)$$

where X_{ijh} is the observation on the ith level of A, jth level of B in block h, δ_{ih} is the whole-plot error term, ε_{ijh} is the subplot error, and the remaining terms are defined as before.

Note: The model of (10.7) assumes no block–treatment interactions. If interactions exist, then δ_{ih} represents the interaction between blocks and whole-plot treatments, and ε_{ijh} is replaced by the interactions terms $(\rho\beta)_{jh} + (\rho\alpha\beta)_{ijh}$.

Extensions of the Latin Square Design. It was noted that the LSD poses certain problems when the number of treatments in either small or large. In fact, 2×2 squares are of no use whatever, since they provide very little information and no way of estimating experimental error. 3×3 squares are of limited use because they are quite small experiments. To increase the size of the experiment, the squares can be repeated any number of times (with a new randomization for each repetition). The model and analysis are similar to those of a split-plot experiment, and depend on the exact structure (e.g., whether or not rows and/or columns are the same in the different squares).

For larger numbers of treatments, partial squares (called "Youden Squares" and others) can be used. These are formed by omitting one or more rows or columns of the Latin Square. Another modification is the Greaco–Latin Square. This is an experiment in which a second treatment factor is used, also appearing once in every row and column and once with each level of the first factor.

Nested Designs. In factorial designs, the structure may be described as a cross-classification—each level of factor A occurs with each level of factor B, and they are the same levels in each case. In some experiments, the levels of B may be different in the different levels of A. For example, we may have four machines, each of which has three dies that operate in tandem and punch out parts simultaneously. Thus we have a total of 12 different dies. Since the dies in the different machines are different, this factor is *nested* rather than crossed with the other factor. Nesting can occur at more than one level. For example, samples of batches of raw materials may be sent to several labs for analysis, and each lab may split each sample and have it analyzed by two different technicians, who may, in turn, split each subsample into two parts for separate analysis. Thus we have sub-subsamples within technicians within labs within batches.

Nesting can occur within cells of the various experimental designs and in factorial experiments as well. Inferences of interest with regard to the nesting factors in these experiments usually relate to the relative amounts of variability associated with these factors rather than comparisons of means. In the machine experiment, for example, the interest would be in determining how much of the variability in parts is due to machines and how much is due to dies.

Models for these experiments must, as always, incorporate all features of the design, including crossed factors (if any) as well as nested factors, and will often include more than one error term. As an example, the model for a two-factor experiment with B nested in A is

$$X_{ijh} = \mu + \alpha_i + \beta_{j(i)} + \varepsilon_{h(ij)} \tag{10.8}$$

where the parentheses indicate the nesting structure (e.g., j within i).

Repeated Measures Experiments. In repeated measures experiments, as the name implies, experimental units are measured more than once, usually before and after some event or in a longer such sequence. For example, glucose content of samples of a cola product may be measured before and after storage for 4 months. Although often treated separately in texts and manuals, these can be viewed simply as nested experiments of a particular type (see Hicks, 1982). Many examples of repeated measures experiments, the corresponding models, and analyses are given by Winer (1971).

Incomplete Block Designs. The RCBD, as defined, requires that each treatment appear once in each block. In fact, each can appear two or more times, as well. This provides for a larger experiment, with the repetitions being a form of nesting. Often, however, a smaller, rather than larger, block size is desired. This occurs when blocks of size k with homogeneous experimental units cannot be formed or when blocks of size k are not possible. Suppose, for example, that the experimenter wishes to compare mean lifetimes of like components from five potential suppliers, but the test chamber is only able to accommodate three items at a time. In this case, it is necessary to run incomplete blocks, each containing only three of the treatments.

Care must be taken in setting up incomplete block experiments because if not done properly, the analysis can become complex and information on comparisons of interest may be impossible to obtain. The most straightforward approach is to use a *balanced incomplete block design* (BIBD). In a BIBD, each treatment appears in the same block with each other treatment the same number of times. An example of a BIBD with five treatments in blocks of three, as in our illustration, is given in Table 10.3. Here each treatment appears with each other in the same block three times. Note that it requires 10 blocks of size three to accomplish this.

BIBD's can also be used with factorial treatments (see Winer, 1971). There are a number

Table 10.3. Example of a balanced incomplete block design—five treatments in blocks of size three

Block	Treatments		
1	1	2	3
2	1	2	4
3	1	2	5
4	1	3	4
5	1	3	5
6	1	4	5
7	2	3	4
8	2	3	5
9	2	4	5
10	3	4	5

of other options, including partially balanced incomplete block designs, in which treatments appear with some other treatments a certain number of times and with other treatments another number of times. All such designs have a well-defined structure that admits of a relatively easy analysis and interpretation. Randomly constituted incomplete blocks will not do. It is a bit of an art to produce BIBD's and other such designs. See the DOE texts for further information.

Comment on Data Analysis: We have noted that in order to analyze experimental data properly, it is necessary to determine the exact structure of the experiment and structure the analysis accordingly. The purpose of the lists of experimental designs given elsewhere in this chapter, in the various texts cited, and in statistical program documentation, is to assist in this process. If a given experiment does not fit into any of the categories listed, it is not appropriate to ignore certain features of the experiment, make unwarranted assumptions, or otherwise modify the structure to make it fit. The resulting analysis is unlikely to be correct. Instead, write out the proper model for the data and determine how to use the program package to do the corresponding analysis. See Lorenzen and Anderson (1993) for an excellent approach to this process. Most statistical packages include a program, usually called "general linear model" or "GLM," that will allow the user to specify a model and will structure the analysis accordingly.

10.4   ANALYSIS OF VARIANCE I. THE COMPLETELY RANDOMIZED DESIGN

In this section, basic concepts of analysis of variance (ANOVA), a fundamental tool in the analysis of experimental data, are introduced. We look at the basic assumptions, concepts, and structure of ANOVA and apply these to the analysis of data in a CRD. Some related aspects of data analysis, including confidence intervals, additional tests of hypotheses, and tests of assumptions, will also be addressed. These analyses will be extended to other designs in ensuing sections.

10.4.1   Objectives, Models, and Assumptions of ANOVA

In discussing DOE, the point has been made that the primary objective of the experiment was to provide a basis for inference about treatments. In the CRD with a single-factor treatment, this is the only factor involved in the experiment. In ANOVA, the test done is that of

$$H_0: \mu_1 = \mu_2 = \ldots = \mu_k \text{ versus } H_a: \mu_1 \neq \mu_2 \text{ and/or } \mu_1 \neq \mu_3 \text{ and/or } \ldots \text{and/or } \mu_{k-1} \neq \mu_k \quad (10.9)$$

where μ_i is the ith treatment mean. Note that the alternative is simply that at least two means differ. The test of the null hypothesis of (10.9) is called the *overall* test of significance. Unless $k = 2$, it is seldom satisfactory as a complete analysis of the data. Rather than simply knowing that differences between treatments exist, the experimenter usually wants to know what is different from what. Later we will look at some methods of addressing this issue.

In the context of the model of (10.2), we have $\mu_i = \mu + \tau_i$, i.e., the treatment effect τ_i is expressed as a deviation from the overall mean μ. It follows that the hypotheses of (10.9) are equivalent to

$$H_0: \tau_1 = \tau_2 = \ldots = \tau_k = 0 \text{ versus } H_a: \tau_i \neq \tau_j \text{ for at least one } i, j \text{ combination } (i \neq j) \quad (10.10)$$

The objectives in factorial and more complex designs will be more extensive. In factorial designs, we will wish to investigate main effects as well as interactions at various levels. In experiments with blocks (e.g., the RCBD and LSD), we may wish to compare these as well. In other experiments, still other inference problems may be of interest. With quantitative factors (e.g., temperature, fiber length), we may wish to look at polynomial fits or response surfaces. Factorials will be addressed later in the chapter; other issues will be mentioned with references given for details.

Assumptions made in ANOVA for the single-factor CRD are as follows:

- μ is an unknown constant.
- The ε_{ij} are independent random variables.
- The ε_{ij} are normally distributed with mean zero.
- The variance of the ε_{ij} is σ^2, which is the same for all treatments. (We express the assumptions on ε_{ij} symbolically as $\varepsilon_{ij} \sim N(0, \sigma^2)$.)
- τ_i may be either deterministic (the *fixed effects model* or *model I*), in which case we assume that $\Sigma_{i=1}^k \tau_i = 0$, or a random variable (the *random effects model* or *model II*), in which case we assume that $\tau_i \sim N(0, \sigma_\tau^2)$.

We will primarily be interested in the fixed effects model, but will mention random effects occasionally. The distinction is as follows: Treatment effects are "fixed" if the treatments are selected by the experimenter and inferences of interest are only to those included in the experiment. Treatment effects are random if the treatments were selected at random from some population of treatments and inferences of interest are to this larger population. In the fixed effects case, we are primarily interested in inferences about means; in the random effects case, variances are usually of interest.

The following is an example of a random effects factorial experiment. The two major components of a liquid rocket engine are an injector and a combustion chamber. These come off separate assembly lines and are paired up and tested as a unit before final installation into a propulsion system. Suppose an experiment is done in which three combustion chambers and three injectors are paired up in all 9 possible combinations and each configuration is tested four times, with measurements taken of thrust coefficient (and many other performance characteristics) on each run. If runs are done in a completely random order, we have a 3×3 factorial experiment in a CRD. The appropriate model is (10.5), but all of the elements of the model (except for μ) will be considered random variables. The reasoning is that we assume that the injectors and combustion chambers used are random samples from the populations of all items that could be produced by each assembly line and we wish to draw inferences to these populations. In particular, we would like to know how much of the variability in thrust is due to injectors, how much is due to combustion chambers and interaction of the two factors, and how much is simply run-to-run variability (experimental error).

In relating the experiment just described to the model, our assumptions would be that $\alpha_i \sim N(0, \sigma_A^2)$, $\beta_j \sim N(0, \sigma_B^2)$, $(\alpha\beta)_{ij} \sim N(0, \sigma_{AB}^2)$, and $\varepsilon_{ijh} \sim N(0, \sigma^2)$. It follows that the variance of an observation is $V(X_{ijh}) = \sigma^2 + \sigma_A^2 + \sigma_B^2 + \sigma_{AB}^2$. As a result, these variances are called *variance components*—each is a component of the variance of the observation. Variance component analysis, which involves inferences about the components of this and other random models, can be found in the texts cited and in many other texts on DOE.

10.4.2 ANOVA for the CRD, Test Statistic and Interpretation

By analysis is meant breaking an entity down into its component parts. In analysis of variance, the "entity" is the total amount of variability in a set of data (as measured, for example, by the sample variance of the totality of observations). In breaking this into its component parts, we must identify the parts and then determine how much of the total variability is associated with each.

The component parts are determined by the structure of the design. *Every* factor in an experiment could account for some of the variability observed in the data, so, again, it is necessary to understand the structure of the experiment in order to identify the factors involved. In a CRD, the only factor determined by the experimenter is the treatment itself. Whatever cannot be accounted for by treatments is unexplained variability, i.e., experimental error. In the ANOVA, we therefore have three sources of variability—total, treatments, and experimental error (usually called simply "error").

The key concept is ANOVA is to look at variability due to treatments, measured conceptually as variability among individual measurements from different treatment groups, and compare this with experimental error, which is a measure of variability among individuals treated alike, i.e., individuals from the *same* treatment group. If these measures of variability appear to be the same, we conclude that there are no differences among treatment means. As we shall see, the "measures of variability" are sample variances. Thus in ANOVA, we use sample *variances* to test hypotheses about population *means*. Specifically, the test is based on the *ratio* of sample variances, called an F ratio or F test. For the CRD, this ratio is of the form

$$F = \frac{\text{measure of variability among treatments}}{\text{measure of variability within treatments}} \tag{10.11}$$

How this works may be seen from Figure 10.2, which shows the distributions of the data under H_0 and H_a in a CRD with $k = 4$. In Figure 10.2(a), H_0 is true and $\mu_1 = \mu_2 = \mu_3 = \mu_4 = 40$, with $\sigma^2 = 36$. (By definition, this value is the true experimental error.) In Figure 10.2(b), $\mu_1 = 30$, $\mu_2 = 40$, $\mu_3 = 48$, and $\mu_4 = 52$, again with $\sigma^2 = 36$. The denominator of the F test is an estimate of the variability among individuals treated alike, i.e., experimental error, whether or not H_0 is true. The numerator of the F test may be thought of as a measure of variability among individuals treated differently, i.e., among individuals in the composite population. If H_a is true, this variability will be much larger than that in the denominator. On the other hand, if H_0 is true, we are sampling from the same population whether "within" or "among" and so the measures of variability should be about the same. We conclude that if H_0 is true, the F ratio should be about 1; if H_a is true, it should be greater than 1. Thus the test is an upper-tail test, with large values of F leading to rejection of H_0.

The computations of the ANOVA and the F test are arranged in tabular form in an *ANOVA table*. Table 10.4 shows the structure of the table, and indicates the entries for the CRD. In the ANOVA table, we indicate the sources of variability, and the computations leading to the sample variances, called *mean squares* (MS) in ANOVA. As in the case of an ordinary sample variance, a MS is calculated as a sum of squares of deviations from a mean (SS) divided by degrees of freedom (df). F is calculated as the ratio of two mean squares. The final entry in the table is a p value that is used in determining whether or not to reject H_0. The p value is usually not determined in hand calculations, but is given in most computer output.

To complete an ANOVA, then, we need to determine the sources of variability, df, and SS for each source and, in general, the appropriate denominator for the F test. The sources of variability are determined by the structure of the design, as reflected by the terms in the associated model. Generally, there will be a line in the ANOVA table for each term in the model.

For the CRD, the sources of variability are as we have indicated. The rationale for the df listed in Table 10.4 is as follows: There are a total of n observations, so df for total is $(n - 1)$;

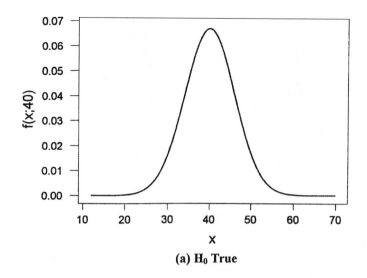

(a) H_0 True

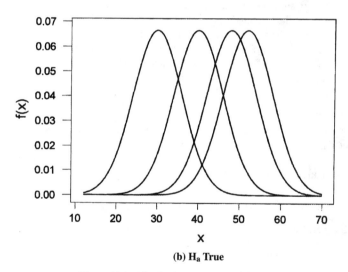

(b) H_a True

Figure 10.2. Distributions for four treatments.

Table 10.4. ANOVA for the CRD

Source	df	SS	MS	F	p
Treatments	$k-1$	SST	$\text{SST}/(k-1)$	MST/MSE	
Error	$n-k$	SSE	$\text{SSE}/(n-k)$		
Total	$n-1$	SSTot			

there are k treatment means, so $(k-1)$ df. Error is a pooled within-treatments variance over the k treatments; variability within the ith treatment is measured with $(n_i - 1)$ df; pooling gives $\sum_{i=1}^{k}(n_i - 1) = n - k$ df for Error. Note that treatment and error df add to total df. This additivity is also true of the SS column, and is a feature of ANOVA generally, at least for balanced designs. This suggests that in our analysis, we have, indeed, identified all of the parts and they add to the whole.

Sums of squares for the CRD are calculated as

$$\text{SSTot} = \sum_{i=1}^{k} \sum_{j=1}^{n_i} x_{ij} - \frac{T^2}{n} \tag{10.12}$$

$$\text{SST} = \sum_{i=1}^{k} \frac{T_i^2}{n_i} - \frac{T^2}{n} \tag{10.13}$$

and

$$\text{SSE} = \text{SSTot} - \text{SST} \tag{10.14}$$

where T_i is the ith treatment total and T is the grand total.

Mean squares and the F ratio are calculated as indicated in Table 10.4. The p value, if given, is used to test H_0 at level of significance α as follows: Reject H_0 if $p < \alpha$; do not reject if $p > \alpha$. If hand calculations are used or a p value is not given in the computer output, a tabulated $F_{1-\alpha}$ with $k-1$, $n-k$ df is obtained from Table C8 and the decision rule is: Reject H_0 if the calculated F is greater than $F_{1-\alpha}$; otherwise, do not reject.

Example 10.1 (Case 2.6)

Data on tensile strength for fibers of four lengths are given in Table 2.6. Sample sizes are $n_1 = 50$, $n_2 = 64$, $n_3 = 50$, and $n_4 = 50$. Suppose we wish to test H_0: $\mu_1 = \mu_2 = \mu_3 = \mu_4$ (or equivalently, H_0: $\tau_1 = \tau_2 = \tau_3 = \tau_4 = 0$), at the 1% level ($\alpha = 0.01$). Minitab was used to perform the ANOVA calculations, resulting in the following output:

One-Way Analysis of Variance

```
Analysis of Variance for Strength
Source      DF         SS         MS         F          P
Length       3      88.000     29.333      78.58      0.000
Error      210      78.393      0.373
Total      213     166.393
```

```
                                    Individual 95% CIs For Mean
                                    Based on Pooled StDev
Level     N      Mean      StDev    --+----------+---------+---------+-
1        50    1.7128     0.5318    (-*-)
2        64    2.8494     0.6687                    (-*-)
3        50    3.0908     0.6353                         (-*-)
4        50    3.5004     0.5818                               (-*-)
                                    --+----------+---------+---------+-
Pooled StDev =    0.6110            1.80      2.40      3.00      3.60
```

The first part of the output is the ANOVA table, including the calculated p value. A p value of 0.000 indicates that H_0 is rejected at any level of significance of 0.001 or greater. Testing at the 1% level, we reject H_0 and conclude that there is evidence that the average strengths of fibers of the different lengths are different. Equivalently, we find the tabulated value of F with 3, 120 df[1] to be 3.95. Since the calculated value is 78.58 > 3.95, this again leads to rejection of H_0. [The two procedures are exactly equivalent (when exact tabulated values are used) and will always lead to the same conclusion.]

The remainder of the output includes sample means and standard deviations for each treatment and 95% CIs for the treatment means. We will look at these in more detail later. Note at this point, however, that the standard deviations are remarkably similar, lending credence to the ANOVA assumption of equal population variances. ∎

10.4.3 Confidence Intervals

In calculating confidence intervals (and in addressing other inference problems) in the context of ANOVA, the most efficient approach is to use all of the information we have about σ^2, i.e., to use the pooled estimate, given by MSE, the mean square for error. This is used in the calculations of CIs shown graphically in the Minitab output in Example 10.1. Note that the pooled estimate of σ is also listed in the output. The value is $s = 0.6110$. This value is calculated as $\sqrt{\text{MSE}}$.

The two-sided $100\gamma\%$ CIs for the means are calculated as

$$\bar{x}_i \pm t_{1-\alpha/2} \frac{s}{\sqrt{n_i}} \tag{10.15}$$

where $\bar{x}_i$ is the ith treatment mean, $\alpha = 1 - \gamma$, and $t_{1-\alpha/2}$ is the $(1 - \alpha/2)$-fractile of the Student-t distribution with $(n - k)$ df, gotten from Table C2. One-sided CIs are calculated in the usual way by modification of the tabulated value.

Confidence intervals for mean differences are also sometimes of interest. A $100\gamma\%$ CI for $\mu_i - \mu_j$ is calculated as

$$\bar{x}_i - \bar{x}_j \pm t_{1-\alpha/2} s \sqrt{\frac{1}{n_i} + \frac{1}{n_j}} \tag{10.16}$$

Example 10.2 (Case 2.6, Continued)
From the computer output, we have $s = 0.6110$. To calculate a 95% CI for $\mu_1 = \mu + \tau_1$, we use $t_{.975} = 1.9713$. The result is $1.7128 \pm 1.9713(0.6110)/\sqrt{50}$, giving $(1.5425, 1.8831)$. This is the CI plotted for Treatment 1 in the Minitab output.

We illustrate (10.16) by calculation of a 95% CI for $\tau_1 - \tau_2$. The result is

$$1.7128 - 2.8494 \pm 1.9713(0.6110)\sqrt{\frac{1}{50} + \frac{1}{64}} = -1.1366 \pm 0.2273$$

giving $(-1.3639, -0.9093)$. Note that if we wish to test $H_0: \mu_1 - \mu_2 = 0$ at the 5% level, we reject H_0 because the hypothesized value of zero in not included the 95% CI. We conclude that the means are not equal. ∎

10.4.4 Multiple Comparisons

One of the most common objectives in analysis of data is comparison of treatment means. For example, we may wish to compare treatments with a control, compare the highest or lowest mean with each other, or, in the absence of any more specific objective, simply compare each mean with each other. The F test is an overall comparison; we have concluded in Example 10.1 that means differ. This does not address the (usually more interesting) issue of more specific comparisons such as those just listed.

Here we consider comparison of each treatment mean with each other. This is a special case of the general problem of *multiple comparisons;* many other multiple comparison procedures can be found in the DOE texts and statistical program manuals. We look at several approaches to the problem.

Least Significant Difference. The simplest (and earliest) approach to making all possible comparisons among means is to apply (10.16) to each pair of treatments. This procedure is called the *least significant difference* or *lsd* procedure. (It is called the Fisher procedure in Minitab.) The term lsd is used because, in the equal numbers case, the procedure is equivalent to calculating the "± factor" in (10.16) and comparing each mean difference with this quantity, which is the lsd in the sense that any difference found to be this large in absolute value is significant. In general, one simply calculates the CIs of (10.16) for each pair of treatments, concluding that the treatment means are different if the interval does not include zero, and that there is no evidence of a difference if the interval includes zero.

A problem with the lsd is that the Type I error rate is on a per-comparison basis. As a result, the overall Type I error rate increases as the number of comparisons made increases. To illustrate this, suppose we consider c independent comparisons, testing at level of significance α, with H_0 true in each case. The overall Type I error rate, defined as the probability of making at least one Type I error is then $1 - (1 - \alpha)^c$. For $\alpha = 0.05$ and $c = 1$, Type I error rate is, of course, 0.05. For $c = 2$, the overall Type I error rate is 0.0975; for $c = 10$, it is 0.40; for $c = 40$, it is 0.87! The lsd has the highest overall Type I error rate. A number of procedures for reducing this rate have been proposed. We look at three of them.

The Bonferroni Procedure. This procedure is based on a Bonferroni inequality and gives a maximum overall Type I error rate. It is the most conservative procedure with regard to Type I error (and thus has the highest Type II error rate). The procedure is to use the lsd, but replace α by α/c, where c is the number of comparisons to be make. For example, if all possible comparisons among k treatments are to be make, $c = k(k - 1)/2$, and α in (10.16) is replaced by α/c. With k treatments, we determine the $(1 - \alpha/2c)$-fractile of the Student-t distribution and proceed as in the lsd.

Honestly Significant Difference. Tukey's *honestly significant difference,* or *hsd,* is a procedure for making all possible comparisons for which the *overall* Type I error rate is α. For equal sample sizes, the hsd is calculated as

$$w(k) = q_\alpha(k, \text{df})s/\sqrt{r} \tag{10.17}$$

where r is the common sample size and $q_\alpha(k, \text{df})$ is obtained from Table C9 with df = df for error. Values for $\alpha = 0.05$ and 0.01 are given in Table C9. If the sample sizes are unequal, the hsd for comparison of treatment i with treatment j is given by

$$w_{ij}(k) = q_\alpha(k, \text{df})s\sqrt{\frac{1}{2}\left(\frac{1}{n_i} + \frac{1}{n_j}\right)} \tag{10.18}$$

The procedure is to conclude that the treatment means are different if their difference exceeds the hsd, and not different otherwise. Alternatively, calculate a CI for the mean difference as the difference between the sample means ± the hsd and conclude that the means are different if the CI does not include zero. (Minitab uses the latter approach.)

The Newman–Keuls Procedure. We have noted that the lsd procedure with $\alpha = 0.05$ leads to a high overall Type I error rate. A consequence of this is that the Type II error rate is low, which is desirable in exploratory research. The Bonferroni and hsd procedures are the most conservative with respect to Type I error rate and consequently have relatively high Type II error rates. This is appropriate in later stages of research when definitive conclusions regarding treatment differences are desired. There are a number of procedures that are a compromise between these two extremes. The Newman–Keuls multiple range test is one such procedure; for others, see the texts cited.

The procedure is as follows:

1. Order the treatment means from smallest to largest. The ordered array is denoted $\bar{x}_{(1)}, \ldots, \bar{x}_{(k)}$.
2. Select a level of significance ($\alpha = 0.05$ or 0.01 in the table given).
3. For equal sample sizes, calculate $w(h)$ from (10.17) for $h = 2, \ldots, k$. For unequal sample sizes, calculate $w_{ij}(h)$ from (10.18) for all h and each i, j for which $i \neq j$.
4. Compare $\bar{x}_{(k)} - \bar{x}_{(l)}$ with $w(k)$ or with the modified value if sample sizes are unequal. If the difference exceeds the appropriate w, conclude that the corresponding μ's are different and proceed to the next step. If not, conclude that there are no differences among the treatment means.
5. Compare $\bar{x}_{(k)} - \bar{x}_{(2)}$ with $w(k-1)$, modified if necessary for unequal sample sizes. If the difference exceeds the w value, conclude that these means are different and proceed to the next step. If not, make no further comparisons among the means $\bar{x}_{(2)}, \ldots, \bar{x}_{(k)}$, and conclude that these are all equal. Apply the same test to $\bar{x}_{(k-1)} - \bar{x}_{(1)}$.
6. Compare the differences $\bar{x}_{(k)} - \bar{x}_{(3)}$, $\bar{x}_{(k-1)} - \bar{x}_{(2)}$, and $\bar{x}_{(k-2)} - \bar{x}_{(1)}$, as appropriate, to $w(k-2)$, again modified as appropriate. Any differences that exceed the appropriate w value are significant; those that do not are not significant, and no further comparisons are made within that group of treatments.
7. Continue in this fashion until all allowable comparisons down to the level of adjacent treatments in the ranked array are made.

Recommendation: In practice, one must choose one of the above procedures (or some other, as mentioned). The following approach is suggested:

- If all comparisons are to be made and Type I error is not a major concern but Type II error is, use the lsd, with $\alpha = 0.05$ or some other reasonable choice.
- If it is very important to guard against Type I error, use the hsd.
- If only a few out of all possible comparisons are of interest and Type I error is a concern, use the Bonferroni procedure.
- For a balance between Type I and Type II errors, use the Newman–Keuls procedure.

Comments: (1) Applications: These procedures may by modified for use in most experimental designs, including factorial experiments. The modification is to use the MS for the appropriate error for testing the factor in question and use the corresponding df in obtaining tabulated values.

(2) Presentation of Results: When many comparisons among means are made, it quickly becomes awkward to describe the outcome verbally. A convenient method of succinctly expressing the results is to present the ordered set of sample means or symbols that identify them, and then underline all sets in which no differences are statistically significant. For example, if comparisons were made among three means, with the ordered sample means labeled 1, 2, 3, and only 1 and 3 were found to be significantly different, the result would be written as

$$\underline{1 \qquad 2} \qquad 3$$
$$ \qquad \underline{}$$

Note that there appears to be a logical contradiction here: 1 is not different from 2 and 2 is not different from 3, but 1 is different from 3. The interpretation is that there is statistical evidence that 1 is different from 3, but there is no such evidence regarding 1 versus 2 or 2 versus 3; we cannot tell if 2 is the same as 1, or the same as 3, or is in a class by itself.

Example 10.3 (Case 2.6, Continued)

The lsd with $\alpha = 0.05$ gives the following CIs, here presented as Minitab output. (The pairs of numbers are the upper and lower confidence limits and the treatments are numbered in the original order, which, by chance, is also the rank order.)

```
Fisher's pairwise comparisons

    Family error rate = 0.202
Individual error rate = 0.0500

Critical value = 1.971

Intervals for (column level mean) - (row level mean)

                    1              2              3

    2         -1.3639
              -0.9093

    3         -1.6188       -0.4687
              -1.1372       -0.0141

    4         -2.0284       -0.8783       -0.6504
              -1.5468       -0.4237       -0.1688
```

We conclude on this basis that all treatment means are different. Note that the overall (called "family") Type I error rate is given as 20.2%

The hsd, with overall error rate at 5%, gives the following results:

```
Tukey's pairwise comparisons

    Family error rate = 0.0500
Individual error rate = 0.0103

Critical value = 3.66
```

```
Intervals for (column level mean) - (row level mean)

                    1                  2                 3

        2        -1.4351
                 -0.8382

        3        -1.6942           -0.5398
                 -1.0618            0.0571

        4        -2.1038           -0.9494          -0.7258
                 -1.4714           -0.3525          -0.0934
```

Here all treatment means except 2 and 3 differ (the CI for 2 versus 3 including zero) and the individual error rate is given as 0.0103. The conclusion is expressed as:

$$1 \quad\quad \underline{2 \quad\quad 3} \quad\quad 4.$$

The total number of comparisons in this example is $c = 6$. The Bonferroni procedure is done using the lsd with $\alpha = 0.05/6 = 0.08333$. The result is the same as that for the hsd.

The Newman–Keuls procedure (with $\alpha = 0.05$) is applied as follows:

The q values for $\alpha = 0.05$ are obtained from Table C9 (by interpolation) as:

j	2	3	4
$Q_{0.05}(j, 210)$	2.79	3.34	3.66

The treatment means are:

Treatment (i)	1	2	3	4
Fiber length (mm)	265	25.4	12.7	5.0
n_i	50	64	50	50
$\bar{x}_{(i)}$	1.7128	2.8494	3.0908	3.5004

Comparisons:

- 1 versus 4: $w = 3.66(0.6110)/\sqrt{50} = 0.3163$; observed difference is $3.5004 - 1.7128 = 1.7876$; significant. We can proceed to compare 1 with 3 and 2 with 4. (Note that these comparisons would not be made.)
- 1 versus 3: $w = 3.34(0.6110)/\sqrt{50} = 0.2886$. The difference is 1.3880, which is significant. We can compare 1 with 2 and 2 with 3.
- 2 versus 4: $w = 3.34(.6110)\sqrt{\frac{1}{2}\left(\frac{1}{50} + \frac{1}{64}\right)} = 0.2724$. The difference is significant, and we can compare 3 with 4.
- 1 versus 2: $w = 0.2275$; significant.
- 2 versus 3: $w = 0.2275$; significant.
- 3 versus 4: $w = 0.2411$; significant.

Conclusion: each treatment differs from each other. ∎

10.4.5 Tests of Assumptions

The assumptions of ANOVA were stated in Section 10.4.1. The assumption of independence is not easily tested except for a few special situations (e.g., if the data were obtained sequentially in time and this sequence is recorded). Normality and equality of variance are easily tested in the CRD, and most statistical packages have options for doing so. We illustrate this using Minitab. It is suggested that tests of assumptions be done at the 1% level. The reason is that the tests given are quite robust (i.e., not greatly affected by small to modest departures from assumed conditions) and the alternatives are sometimes not nearly as powerful.

Test for Normality. The test of normality can be done for each treatment or on the entire set of *residuals* e_{ij}, defined to be

$$e_{ij} = x_{ij} - \hat{\mu} - \hat{\tau}_i = x_{ij} - \bar{x}_i \qquad (10.19)$$

where $\hat{\mu}$ and $\hat{\tau}_i$ are estimated effects and $\bar{x}$ is the overall mean. A straightforward approach is to plot the data or residuals on normal probability paper. If the normal distribution is appropriate, the data will plot as a straight line. For a more precise test, we may use any of several statistical tests.

Details of some test statistics that can be used to test for "fit" to any specified distribution will be given in the next chapter. For the time being, we will use the Anderson–Darling test without definition. The Minitab output will give a p value for this test. As usual, if the p value is less than the chosen level of significance α, the null hypothesis (normality in this case) is rejected.

Test for Homogeneity of Variance. Two common tests of equal population variances are the Bartlett test for normal data, and the Levine test, which makes no specific distributional assumption. It is suggested that both tests be made (Minitab does both, as well as providing plots of confidence intervals for the individual treatment variances) unless there is evidence on nonnormality, in which case only the Levine test should be used. (There are other tests, as well. See Lorenzen and Anderson, 1993.)

The test statistic for Bartlett's test is

$$X^2 = \frac{1}{C}\left\{\left[\sum_{i=1}^{k}(n_i - 1)\right]\log(s^2) - \sum_{i=1}^{k}(n_i - 1)\log(s_i^2)\right\} \qquad (10.20)$$

where

$$C = 1 + \frac{1}{3(k-1)}\left[\sum_{i=1}^{k}\frac{1}{n_i - 1} - \frac{1}{n - k}\right] \qquad (10.21)$$

Under the null hypothesis of equal variance, X^2 is distributed as χ^2 with $(k-1)$ df. H_0 is rejected if the calculated value exceeds the tabulated value for an upper-tail test at level α.

Levine's test is based on the absolute differences of the observations from the sample medians, $|x_{ij} - \tilde{x}_i|$, where $\tilde{x}_i$ is the ith treatment median. Perform an ANOVA on these quantities. If $p > 0.01$, do not reject equal variances.

Notes on Assumptions:

1. Plots of residuals were mentioned above. Residuals may be plotted in many ways, e.g., against treatments, versus time if the data were taken sequentially, against fitted values, and so forth. In more complex designs, they may be plotted against other factors as well. In fact, any plot that makes sense should be done and this may be the only way

to check some of the assumptions in some designs. In all cases, a random, horizontal plot is indicative of no violation of assumptions.

2. If either normality or homogeneity of variance is rejected at $\alpha = 0.01$, it is suggested that the data be transformed to a scale on which the assumptions are more nearly met. See the cited texts for suggested transformations.

3. If the ANOVA assumptions are grossly violated and no suitable transformation can be found, alternative methods of analysis should be used. There are nonparametric procedures available for some of the designs discussed. When all else fails, do the ANOVA but be cautious in interpretation of the results. (The logic is still there, but the p values can be very misleading.)

Example 10.4 (Case 2.6, Continued)
The normal probability plot of residuals is given in Figure 10.3. The p value for the Anderson–Darling test is 0.290, indicating no evidence of nonnormality. The relative linearity of the plot supports this conclusion. The p values for the individual treatments are 0.353, 0.927, 0.091, and 0.039, and these also plot as straight lines, further supporting the normality assumption.

The Minitab output for testing homogeneity of variance is given in Figure 10.4. The p values for the Barlett and Levine tests are 0.366 and 0.323, respectively, indicating no evidence of unequal variances. ∎

10.5 ANALYSIS OF VARIANCE II. FACTORIAL EXPERIMENTS

10.5.1 The Completely Randomized Design

As noted, factorial experiments may be run in any of many experimental designs. To begin with, we illustrate the analysis by considering a two-factor factorial in a CRD. The model for

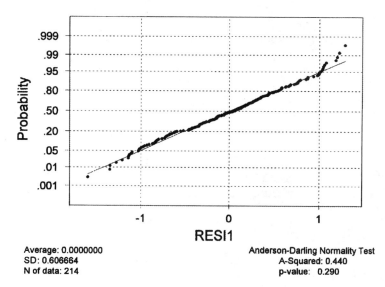

Average: 0.0000000
SD: 0.606664
N of data: 214

Anderson-Darling Normality Test
A-Squared: 0.440
p-value: 0.290

Figure 10.3. Normal plot of residuals, Case 2.6.

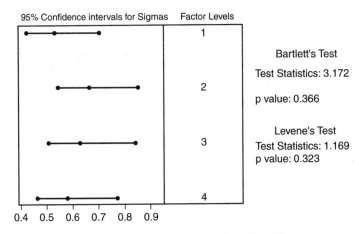

Figure 10.4. Test for equality of variance, Case 2.6.

this experiment is given in (10.5). Note that the treatment effect in this model consists of three terms, the A and B main effects and an AB interaction term. Accordingly, the treatment source in the ANOVA will be broken down into the corresponding three sources. Thus the ANOVA may be considered a two-stage analysis: Total variability is broken into its component parts (treatments and error) and then treatment is broken into its component parts (main effects and interaction).

The ANOVA for this experiment, with a levels of A, b levels of B and r experimental units on each treatment combination, is given in Table 10.5. Note that in the table, we indent the entries under treatment to indicate the second-level analysis. Sums of squares for treatments, error, and total are calculated as for the CRD, remembering that Treatments here are all combinations of a level of A with a level of B. The SS for A is calculated using the totals for levels of this factor, dividing by rb after squaring. Similarly, totals for levels of B are used to calculated SSB. Finally, $SSAB = SST - SSA - SSB$.

These computations are usually done by computer. Hand calculations are easy, however, using the following rules of thumb:

- For any main effect (A, B, blocks, etc.), calculate the SS using totals for levels of that factor.
- Whenever a total is squared, divide by the number of numbers in that total.

Table 10.5. ANOVA for a two-factor factorial in a CRD.

Source	df	SS	MS	F	p
Treatments	$ab - 1$	SST			
A	$a - 1$	SSA	$SSA/(a - 1)$		
B	$b - 1$	SSB	$SSB/(b - 1)$		
AB	$(a - 1)(b - 1)$	$SSAB$	$SSAB/(a - 1)(b - 1)$		
Error	$ab(r - 1)$	SSE	$SSE/[ab(r - 1)]$		
Total	$n - 1$	$SSTot$			

- For two-factor interactions, prepare two-way tables (*AB*, *AC*, etc.), summing over all levels of all other factors. Calculate a SS using the table entries and the above rules and then subtract the SSs for the corresponding main effects.
- For three-factor interactions, prepare a three-way table, calculate a SS based on the table entries and then subtract the corresponding main effects and two-factor interactions.
- Etc.

Example 10.5 (Two-Factor Factorial in a CRD)
Shrinkage of fabrics is an important characteristic to a clothing manufacturer. Excessive shrinkage ("failure" in this context) can lead to product returns and high cost. In a comparative study of four suppliers, batches of material were cleaned at four relatively high temperatures (degrees Farenheit). Eight batches of material from each supplier were used, with the 32 runs done in a completely random order, and average percent shrinkage measured in each case. The following results were obtained:

	Temperature							
Material	170°		180°		190°		200°	
A	1.8	3.2	3.5	2.9	2.9	3.6	9.6	8.4
B	4.8	5.1	5.6	4.2	6.2	5.5	10.3	9.9
C	4.2	3.7	5.7	4.6	10.2	7.9	12.9	12.3
D	3.6	4.1	3.1	4.7	5.1	4.4	9.8	11.8

The ANOVA for this experiment is:

Source	DF	SS	MS	F	P
Temp	3	233.068	77.689	116.39	0.000
Material	3	42.542	14.181	21.24	0.000
Temp*Material	9	18.425	2.047	3.07	0.024
Error	16	10.680	0.668		
Total	31	304.715			

The main effects of temperature and material are significant at any level down to 0.0005, since the p values are less than that amount. Interaction is significant at $\alpha = 0.05$, but not at $\alpha = 0.01$. It is instructive to look at a plot of the interaction. This is given in Figure 10.5. The lack of parallelism of the response curves shows the nature of the interaction, which is slight in comparison with the main effects. It would not be misleading to say that material 1 exhibits the lowest shrinkage and fabric 3 the highest, and that temperature increases shrinkage at an increasing rate.

The average shrinkage at the four increasing temperatures are 3.81, 4.29, 5.73, and 10.63. the averages for materials are 4.49, 6.45, 7.69, and 5.83. Multiple comparisons may be made among either set of means using the lsd, hsd, or Newman–Keuls procedure. The appropriate MSE in either case is 0.668, with 16 df.

As a quick check on assumptions, we look at residual plots. These are given in Figure 10.6 and show no particular unusual patterns, though the histogram of residuals is a bit ragged. This is due, at least in part, to the relatively few significant digits in the data. The p values (not shown) for the Anderson–Darling and Bartlett tests are 0.48 and 0.95, indicating no evidence of nonnormality or unequal variances. ■

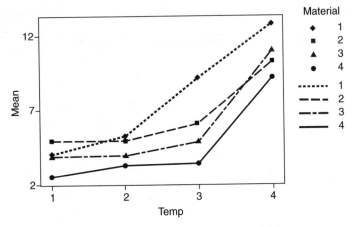

Figure 10.5. Interaction plot for temperature–material data.

10.5.2   Other Experimental Designs

For designs other than the CRD, do the following:

- Set up the basic ANOVA corresponding to the *experimental design*. (See the following sections and the DOE texts for details.)
- Partition the treatment df and SS to reflect the factorial structure as indicated above.

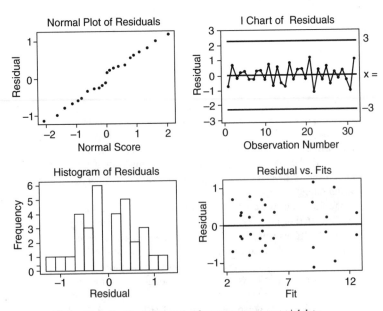

Figure 10.6. Residual diagnostics for temperature–material data.

10.6 ANALYSIS OF VARIANCE III. COMPLETE BLOCK DESIGNS

As we have noted several times, the ANOVA for a given design follows the structure of the design and is reflected in the model for that design. The analyses for the RCBD, LSD, and split plot are given in this section. There are many designs related to these—multiple squares, nesting, various types of split plots, etc., and combinations of these concepts in great profusion. We give only the basic analyses for the three designs; see the texts cited for additional analyses.

10.6.1 ANOVA for the Randomized Complete Block Design

The basic ANOVA for the RCBD is given in Table 10.6. The calculation of sums of squares proceeds along the lines indicated in the previous section (treatment totals used to calculate SST, dividing by the number of blocks r, and so on). The error SS is calculated as SSE = SSTot − SSB − SST. If blocks are a random effect, the error MS, which is actually an estimate of block by treatment interaction, is the appropriate denominator for the F test of treatments. This is also true in the fixed effects case if we can assume that the interaction is zero. Error can also be used to test blocks *if* there is no interaction and no restriction error. *Note:* Most computer packages will give the F for blocks, whether or not it is meaningful.

Example 10.6 (Case 2.21)
The data of Case 2.21 consist of breaking strength of segments of four different lengths cut from each of 26 fibers embedded in resin. The fibers may be considered blocks, and the lengths treatments. For ease of analysis, we omit the two fibers with missing observations, resulting in four treatments in 24 blocks. (A GLM program could handle the missing observations.) The resulting ANOVA is

```
Source    DF       SS        MS       F       P
Fiber     23    4.98430   0.21671   10.19   0.000
Length     3    3.57836   1.19279   56.08   0.000
Error     69    1.46747   0.02127
Total     95   10.03012
```

We find that the mean strength of fibers of different lengths are significantly different at any reasonable level ($p = 0.000$). These could again be compared using any of the multiple comparison procedures, with MSE = 0.02127 and 69 df. The means for length are 3.64, 3.55, 3.38, and 3.14.

Note also that fibers (blocks) are a significant source of variability in the results; whether or not the F test for fibers is valid, fiber differences account for 50% of the total variability.

Table 10.6. ANOVA for a Randomized Complete Block experiment

Source	df	SS	MS	F
Blocks	$r-1$	SSB	SSB/(r − 1)	
Treatments	$k-1$	SST	SST/(k − 1)	MST/MSE
Error	$(r-1)(k-1)$	SSE	SSE/(r − 1)(k − 1)	
Total	$rk-1$	SSTot		

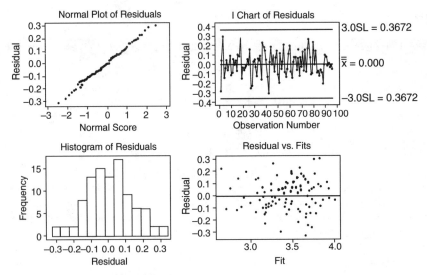

Figure 10.7. Residual plots for Case 2.21.

Residual plots for this experiment as given in Figure 10.7. The p-value for the Anderson–Darling test is 0.90. There are no indications of violations of assumptions. ■

10.6.2 ANOVA for the Latin Square Design

The breakdown for df and SS in the ANOVA for the LSD is given is Table 10.7. The sums of squares are calculated in the usual way. As in the case of the RCBD, the tests for rows and columns (the two types of blocks in this experiment) may or may not be possible, depending on the assumptions one is willing to make. The sums of squares are calculated using row, column, and treatment totals, with error gotten by subtraction. Under the assumptions of no interactions among rows, columns, and treatments and no restriction errors, rows and columns may also be tested with MSE as denominator of the F tests. The table entries are quite easily gotten by hand calculation; most computer packages use GLM with a main-effects-only model.

Example 10.7
Four brands of tires (treatments A, B, C, and D) were tested for wear. The design was a Latin square, using four cars and mounting a different brand in a different position on each car.

Table 10.7. ANOVA for the Latin square design.

Source	df	SS	MS	F
Rows	$k-1$	SSR	SS$R/(k-1)$	
Columns	$k-1$	SSC	SS$C/(k-1)$	
Treatments	$k-1$	SST	SS$T/(k-1)$	MST/MSE
Error	$(k-1)(k-2)$	SSE	SSE$/(k-1)(k-1)$	
Total	k^2-1	SSTot		

Cars were driven over a fixed course by professional drivers and wear (mm) was measured after a predetermined number of miles driven. The design and resulting data were as follows:

Position	Car 1	Car 2	Car 3	Car 4
1	A 2.7	B 2.4	C 2.0	D 1.9
2	C 2.2	D 2.1	A 2.3	B 1.8
3	B 2.4	C 2.2	D 2.1	A 2.3
4	D 2.3	A 2.4	B 2.3	C 1.9

The ANOVA for this experiment is:

```
Source      DF    Seq SS     Adj SS     Adj MS       F      P
Car          3   0.386875   0.386875   0.128958    14.40   0.004
Position     3   0.061875   0.061875   0.020625     2.30   0.177
Brand        3   0.306875   0.306875   0.102292    11.42   0.007
Error        6   0.053750   0.053750   0.008958
Total       15   0.809375
```

We conclude that brands differ with regard to mean wear. ∎

10.6.3   ANOVA for the Split-Plot Design

The key element of a split-plot experiment is that the subplots are arranged within whole plots. The whole plots themselves may be arranged in any of many experimental designs. The key to analysis is to set up the whole plot analysis first, based on the experimental design for whole plots, and then do a "within whole plots" analysis to deal with the subplots and the one or more error terms associated with these. We illustrate the procedure with a RCBD, assuming only the two error terms of the model of (10.7). The whole-plot and split-plot errors will be denoted E_a and E_b, respectively. A is the whole-plot treatment; B the subplot treatment. The ANOVA is given in Table 10.8. Computations are as before for the RCBD and factorial treatments. The sum of squares for E_a is calculated as block by A interaction; E_b is gotten by subtraction of the sums of squares above this line from SSTot.

An alternative analysis, appropriate if E_b is decomposed into two errors, is to calculate block $\times B$ and block $\times A \times B$ interactions. The first is used to test B and the second to test AB.

Table 10.8. ANOVA for a split-plot experiment with whole plots in a RBCD.

Source	df	SS	MS	F
Blocks	$r-1$	SSR	$SSR/(k-1)$	
A	$a-1$	SSA	$SSC/(k-1)$	MSA/MSE_a
E_a	$(r-1)(a-1)$	SSE_a	$SSE_a/(r-1)(a-1)$	
B	$b-1$	SSB	$SSB/(b-1)$	MSB/MSE_b
AB	$(a-1)(b-1)$	SSAB	$SSAB/(a-1)(b-1)$	$MSAB/MSE_b$
E_b	$a(r-1)(b-2)$	SSE_b	$SSE_b/a(r-1)(b-1)$	
Total	$rab-1$	SSTot		

Example 10.8 [Split-Plot Design; Hicks (1982), p. 267]

Life testing of an electrical component was done after exposure to four different temperatures (A; 580, 600, 620, and 640 °F) for three periods of time (B; 5, 10 and 15 minutes). The
experiment was done as a split plot in three blocks, with A as whole-plot treatment. The response was time to failure (TTF) and the results obtained were as follows:

Block	Time	Temperature			
		580°	600°	620°	640°
1	5	217	158	229	223
	10	233	138	186	227
	15	175	152	155	156
2	5	188	126	160	201
	10	201	130	170	181
	15	195	147	161	172
3	5	162	122	167	182
	10	170	185	181	201
	15	213	180	182	199

The ANOVA for this experiment is

Source	df	SS	MS	F	p
Blocks	2	1962.7	981.4		
A	3	12494.3	4164.8	14.09	0.004
E_a	6	1773.9	295.7		
B	2	566.2	283.1	0.46	0.639
AB	6	2640.4	433.4	0.70	0.566
E_b	16	9333.3	620.8		
Total	35	29331.0			

As a check on the computations, we calculate SSA. The totals for temperature levels are
1754, 1338, 1591, and 1742, and the grand total is 6425, so the sum of squares is

$$SSA = (1754^2 + 1338^2 + 1591^2 + 1742^2)/9 - 6425^2/36 = 12494.3$$

If we test at the 1% level, the only significant effect is temperature. Note, however, that
the effect of temperature is a bit unusual. In most cases, one would expect increasing temperature to degrade the performance of an item, but in this case the average lifetime falls and
then rises again as temperature increases. The engineer interpreting the analysis would have
to explain this phenomenon.

Another somewhat unusual feature of this experiment is the relative magnitudes of E_a and
E_b. Ordinarily we expect E_a to be the larger of the two, because it is associated with larger
plots, which are expected to be more variable.

A thorough analysis of the data would require plots and tests of assumptions. This is left
as an exercise for the reader. This exercise will also provide an opportunity to set up whatever program package is in use for analysis of a split-plot experiment. ■

10.7   ANALYSIS OF VARIANCE IV. BIBD AND FRACTIONAL FACTORIAL DESIGNS

In the previous sections, we considered designs that were complete in the sense that all treatment combinations were run and, in blocking designs, all treatments were included in each block. In incomplete designs, one or both of these conditions are not met. The two main reasons for running incomplete experiments are (1) It is impossible or prohibitively expensive to run a complete experiment, and (2) blocks of the required size are not available.

In this section, we look at incomplete block designs, i.e., designs in which the block size is smaller than the number of treatments in the experiment, and fractional factorial designs, in which only a fraction of a complete factorial set of treatments is included in the experiment. In both cases, we give only a brief introduction to the topic, confining attention to a few simple designs. Some comments on a special type of fractional factorial called a Taguchi design will also be included. There is an extensive literature on all of these topics, and some references from which further information may be obtained will be provided.

10.7.1   Balanced Incomplete Block Designs

An example of a balanced incomplete block design (BIBD) is given in Table 10.3. The design has five treatments in blocks of three. It is balanced because each treatment appears with each other in the same block the same number of times (three in this case). This feature makes the analysis quite straightforward and it is easily done with most standard statistical computer packages.

There are many reliability applications in which the use of such designs is either necessary or highly desirable. Examples are lack of space in test chambers (so that not all treatments can be included in a run), lack of an adequate amount of test materials, inability to run all treatments in a given period of time, and so forth.

The structure of the ANOVA for a BIBD is conceptually relatively simple. For a design with b blocks, t treatments and a total of n observations, the breakdown of the ANOVA is

Source	df
Blocks	$b - 1$
Treatments	$t - 1$
Error	$n - b - t + 1$
Total	$n - 1$

Note, however, that the sums of squares are not calculated in the usual way. The block SS must be adjusted for the fact that each block contains different treatments and treatment SS must be adjusted for blocks. Formulas for these adjusted sums of squares are not given. Rather, it is suggested that the analysis be done on computer (using the GLM program).

Example 10.9

Components are subjected to a temperature of 300° and then tested until failure. Treatments are length of time in the temperature chamber (1, 2, 4, 7, and 10 hours). The chamber can hold only three items at a time, and one run per day is done. The BIBD of Table 10.3 is used for this experiment. The resulting data are (with treatment and TTF given in each cell):

Block

1	1	26.6	2	21.7	3	16.9
2	1	25.4	2	18.8	4	10.4
3	1	26.9	2	19.8	5	8.3
4	1	26.9	3	17.3	4	11.5
5	1	26.3	3	17.3	5	7.3
6	1	26.6	4	12.8	5	9.8
7	2	20.7	3	17.2	4	11.6
8	2	21.5	3	16.0	5	12.7
9	2	22.0	4	11.0	5	8.2
10	3	17.1	4	11.9	5	9.8

The ANOVA for this experiment (from GLM in Minitab) is:

Source	DF	Seq SS	Adj SS	Adj MS	F	P
Block	9	167.900	13.933	1.548	1.36	0.282
Time	4	996.668	996.668	249.167	219.30	0.000
Error	16	18.179	18.179	1.136		.
Total	29	1182.747				

There is strong evidence of differences in MTTF for the five treatments. The means follow an expected pattern and are shown in the following table:

Hours	Mean
1	26.710
2	20.863
4	16.683
7	11.583
10	9.210

■

10.7.2   Fractional Factorial Designs

As the number of factors in an experiment increases, the number of treatments increases exponentially. In a 2^{10} experiment (ten factors at two levels each), there are 1024 treatments; a 2^{20} experiment involves 1,048,576 treatments. Complete experiments of this size are not feasible (except perhaps in computer simulations). Alternatives that are feasible are *fractional factorials,* that is, designs in which only a fraction of the complete factorial set of treatments are run. Note, however, that not any fraction will do; the fraction to be used must be carefully chosen.

The idea is as follows. Consider, as an example, a 2^{10} factorial arrangement. In the ANOVA for this experiment, there are 1023 df for treatments, including 10 df for main effects, 45 for two-factor interactions, etc., up to 10 df for nine-factor interactions, and 1 df for the 10-factor interaction. In virtually all applications, high-order interactions do not exist or, at least, are very small compared to main effects and low-order interactions, so that one may safely assume these to be zero. In fact, most experimenters would be willing to assume all interactions involving three or more factors are zero, being content to draw inferences regarding main effects and some or all of the two-factor interactions. In some cases, particularly in preliminary or screening experiments, all interactions may be assumed to be zero. Thus, in the 2^{10} example, we need to estimate only the 10 main effects in the most extreme case, or 10 main effects plus up to 45 two-factor interactions in most applications. This can be done with far fewer than 1024 treatments, provided that an appropriate subset of treatments is employed.

In order to estimate the desired effects, it is necessary to select a fraction of the complete set having the property that all higher-order interactions that are assumed to be zero are confounded with main effects and the interactions not assumed to be zero, and such that all effects not assumed to be zero are estimable. A completely general solution to the problem of constructing such designs is an unsolved problem. Solutions are available, however, for many factorials, notably the 2^k and 3^k series, and some mixed factorials.

There are a number of approaches to construction of fractional factorial designs. We refer the reader to Raghavarao (1971), Box et al. (1978), Dey (1985), Lorenzen and Anderson (1993), and Montgomery (1997) for details.[2] Several of these sources give tables of fractional factorial designs. Blocking may also be used in fractional factorials. This is also discussed in the references. Many computer packages include specialized programs for construction and analysis of fractional factorial experiments. The ANOVA may also be done by use of the GLM program available in most packages.

The basic approach in ANOVA of fractional factorials is to calculate sums of squares for main effects and interactions that are included in the model and to pool all of the remaining SSs into a "Residual SS" and use the corresponding mean square for testing the model effects.

Here we simply illustrate the concept by considering one experiment of this type and its analysis. Suppose we consider five factors $(A, \ldots, E)$, each at two levels (1 and 2), for a total of 32 treatments. A one-half fraction would give a total of 15 df, which would provide a means of estimating the five main effects and the ten two-factor interactions, but not experimental error, since zero df would remain. Suppose that it is possible to assume that some of these interactions are zero. In that case, the data would be analyzed by pooling these interactions and using the result as an estimate of experimental error.

A fractional factorial experiment of the structure just described is given in Table 10.9.

Table 10.9. Example of a one-half fraction of a 2^5 factorial experiment

Treatment	A	B	C	D	E
1	2	1	1	1	1
2	1	1	1	2	1
3	1	1	1	1	2
4	2	1	1	2	2
5	1	2	1	1	1
6	2	2	1	2	1
7	2	2	1	1	2
8	1	2	1	2	2
9	1	1	2	1	1
10	2	1	2	2	1
11	2	1	2	1	2
12	1	1	2	2	2
13	2	2	2	1	1
14	1	2	2	2	1
15	1	2	2	1	2
16	2	2	2	2	2

This experiment has a special structure in that each main effect is confounded with a four-factor interaction and each two-factor interaction is confounded with a three-factor interaction. As noted, in order to analyze the data to the point of testing statistical hypotheses, it would be necessary to assume that all three- and four-factor and at least some two-factor interactions were negligible.

Example 10.10
An experiment on bond strength is conducted with factors A = Alloy, B = Temperature, C = Technician, D = Relative Humidity (RH), and E = Time in oven. The design is the fractional factorial given in Table 10.9. The responses are:

Treatment	1	2	3	4	5	6	7	8	9	10	11	12	13	14	15	16
Break Strength	19	20	42	58	28	32	53	66	16	20	41	60	25	31	54	73

Suppose that it is known that A, B, C, and D do not interact, but information concerning interactions with E is not available. Then we may pool all interactions among the first four factors and use this result as an estimate of experimental error.

The ANOVA (from Minitab), without pooling, is

General Linear Model

```
Factor    Levels Values
A              2    0    1
B              2    0    1
C              2    0    1
D              2    0    1
E              2    0    1
```

Analysis of Variance for Strength

Source	DF	Seq SS	Adj SS	Adj MS	F	P
A	1	1.00	1.00	1.00	**	
B	1	462.25	462.25	462.25	**	
C	1	0.25	0.25	0.25	**	
D	1	420.25	420.25	420.25	**	
E	1	4096.00	4096.00	4096.00	**	
A*E	1	0.25	0.25	0.25	**	
B*E	1	1.00	1.00	1.00	**	
C*E	1	16.00	16.00	16.00	**	
D*E	1	169.00	169.00	169.00	**	
A*B	1	1.00	1.00	1.00	**	
A*C	1	4.00	4.00	4.00	**	
A*D	1	4.00	4.00	4.00	**	
B*C	1	2.25	2.25	2.25	**	
B*D	1	0.25	0.25	0.25	**	
C*D	1	12.25	12.25	12.25	**	
Error	0	0.00	0.00	0.00		
Total	15	5189.75				

** Denominator of F-test is zero.

Pooling interactions except those with E gives the ANOVA

```
Analysis of Variance for Strength
```

Source	DF	Seq SS	Adj SS	Adj MS	F	P
A	1	1.00	1.00	1.00	0.25	0.633
B	1	462.25	462.25	462.25	116.78	0.000
C	1	0.25	0.25	0.25	0.06	0.810
D	1	420.25	420.25	420.25	106.17	0.000
E	1	4096.00	4096.00	4096.00	1034.78	0.000
A*E	1	0.25	0.25	0.25	0.06	0.810
B*E	1	1.00	1.00	1.00	0.25	0.633
C*E	1	16.00	16.00	16.00	4.04	0.091
D*E	1	169.00	169.00	169.00	42.69	0.001
Error	6	23.75	23.75	3.96		
Total	15	5189.75				

It is apparent that temperature, RH, and time significantly affect breaking strength, and that there is a strong RH–time interaction. ■

10.7.3   Taguchi Designs

A class of experimental designs studied extensively by Taguchi includes fractional factorial designs of a particular structure, with a minimal number of treatment combinations. They are often main-effects-only designs or designs that allow for estimation of main effects and a few interactions. The goal is to minimize the number of runs made and thereby minimize cost. Experiments of this type are most useful in screening experiments, where the main objective is to identify factors having a large effect on response, and in experiments where it is known a priori that most interactions are negligible. They are also used widely in quality applications, including quality control and process design and monitoring (see Ryan, 1989).

Taguchi designs are based on *orthogonal arrays,* i.e., variables that are perpendicular to one another when plotted two at a time. An example of this is the following 1/2 fraction of a 2^3 factorial:

	Factor		
Treatment	A	B	C
1	1	1	1
2	1	2	2
3	2	2	1
4	2	1	2

This is known as a Taguchi L_4 array. Methods for constructing such arrays and many resulting designs are given by Taguchi and Wu (1979), Ross (1988), Ryan (1989), and Condra (1993).

Taguchi introduced a number of additional innovations in the analysis of quality, including the use of loss functions in decision making with regard to quality actions and the use of signal-to-noise ratio in analysis of quality and process data. These are also discussed in the references cited.

10.8   REGRESSION, CORRELATION, AND COVARIANCE ANALYSIS

10.8.1   Introduction

Regression and correlation analyses are statistical techniques for studying relationships between two or more variables. Regression analysis is concerned with the form of the relationship; correlation measures strength of relationship. Analysis of Covariance (ANCOVA) is a combination of regression and ANOVA of the type described in this chapter for designed experiments. All of these—regression, ANOVA, and ANCOVA—are special cases of what is called the *general linear model,* and the mathematics underlying the analyses, as well as the computations, can be expressed in a straightforward manner in matrix notation. For details, see Searle (1971), Neter, et al. (1996), and Draper and Smith (1998).

In this section, we look at basic regression models, comment on correlation, and give the ANCOVA for a few basic designs. In this discussion, we present the standard linear regression models. address inference problems regarding the model parameters, and deal with computational aspects by means of computer programs.

10.8.2   The Linear Regression Model and Assumptions

In regression analysis, we investigate the relationship between *k predictor variables,* $x_1, \ldots,$ x_k (also called *explanatory* or *independent* variables) and a *response variable, Y* (also called the *dependent* variable). Specifically, the *multiple linear regression model* expresses the relationship between the predictor variables and the expected value of the response variable as a linear function, namely

$$E(Y) = \beta_0 + \beta_1 x_1 + \beta_2 x_2 + \cdots + \beta_k x_k \tag{10.22}$$

In this model, called the *regression of Y on* $x_1, \ldots, x_k,$ β_0 is the *intercept,* and the β_i ($i =$ $1, \ldots, k$) are the *partial regression coefficients.* This plots as a hyperplane in $k + 1$ dimensions. (If $k = 1$, the plot reduces to a line in two-dimensional space, and the analysis is called *simple linear regression.*)

In the multiple linear regression model, β_0 is the height of the plane where it passes through the *y*-axis and β_i measures the change in $E(Y)$ per unit of change in x_i, holding all other *x*'s constant. Thus β_i is a measure of the influence of the *i*th predictor variable on the average response, exclusive of the influence of the other predictor variables (given, of course, that the model is correct). For example, if we were investigating the effect of $x_1 =$ temperature and $x_2 =$ humidity on the strength, *Y,* of a material and the linear model were appropriate, then β_1 would indicate the change in $E(Y)$ per unit of increase in temperature, holding humidity fixed, and β_2 the change in average strength per unit increase in humidity at constant temperature.

In the regression model, the *x*'s may be deterministic (with values selected by the experimenter) or they may be random. In the latter case, the analysis done is a conditional analysis, given the values of the (random) *X*'s, i.e., we replace $E(Y)$ of the LHS of (10.22) by $E(Y|X_i =$ $x_i, i = 1, \ldots, k$). Note that in either case, in regression analysis we are investigating the relationship between $E(Y)$, the expected value of the random variable *Y,* and a set of observed values of the predictor variables.

Data

Although a regression relationship expresses $E(Y)$ as a function of the observed values of the predictor variables, we, in fact, observe values of Y, not the mean value. We assume a sample of size n, with X_{ij} = observed value of ith predictor variable for the jth individual in the sample, for $i = 1, \ldots, k, j = 1, \ldots, n$, and y_j = corresponding value of the response variable. In terms of the data, the model is expressed as

$$y_j = \beta_0 + \beta_1 x_{1j} + \beta_2 x_{2j} + \cdots + \beta_k x_{kj} + \varepsilon_j \tag{10.23}$$

In this model ε_j is the "error" term and plays the same role as in the ANOVA models given earlier in the chapter.

Qualitative Predictors

In the applications discussed to this point, the predictors were quantitative variables—temperature, time, etc. It is also possible to include qualitative predictors, e.g., supplier, laboratory, temperature chamber used, and so forth. These categorical variables are equivalent to qualitative treatments in ANOVA models. In order to analyze such data by regression methods, we use *indicator variables* (sometimes called "dummy vairables").

Indicator variables are defined as follows. Suppose that a levels of a categorical variable A are included in a study. These are represented by $(a - 1)$ predictor variables in the regression model. The value of the ith predictor for the jth individual in the sample is given by

$$x_{ij} = \begin{cases} 1 & \text{if in category } i \\ 0 & \text{otherwise} \end{cases} \tag{10.24}$$

for $i = 1, \ldots, a - 1, j = 1, \ldots, n$. For example, if the categories are male and female, one indicator is required and it takes on the value 0 for one gender and 1 for the other. If we wish to include materials from four suppliers, three indicator variables are required, the first indicating supplier 1, etc.

The category omitted in setting up the indicators is called the *reference group*. The indicator variables are included in the regression analysis along with any other predictor variables in the model. The partial regression coefficient associated with an indicator variable is a measure of the difference between the means of the group identified by that indicator and the reference group. (Thus the partial regression coefficient associated with gender would be a measure of the difference in response between males and females.)

Nonlinear Regression Models

Note that regression analysis is not restricted to linear relationships. Linear methods may also be used for transformations of the predictors as well as for nonlinear models that can be transformed to linearity. This will work as long as the model is linear in the β's. Thus polynomial models such as

$$E(Y) = \beta_0 + \beta_1 x + \beta_2 x^2 + \cdots + \beta_k x^k \tag{10.25}$$

and

$$E(Y) = \beta_0 + \beta_1 x_1 + \beta_2 x_2 + \beta_3 x_1^2 + \beta_4 x_2^2 + \beta_5 x_1 x_2 \tag{10.26}$$

may be treated by linear regression methods. (Simply rename the variables $x_1, \ldots, x_k$ and

$x_1, \ldots, x_5$ in models (10.25) and (10.26) and set up the data accordingly.) Note that the cross-product term in (10.26) is a special form of interaction between the two predictor variables x_1 and x_2, similar (but not identical) to interaction in an ANOVA model.

An exponential model of the form

$$E(Y) = \alpha x_1^{\beta_1} e^{\beta_2 x_2} \tag{10.27}$$

may be linearized by a logarithmic transformation. [Thus we regress $\log(Y)$ on $\log(x_1)$ and x_2.] In all of these models, we assume an additive error term on the transformed scale, as in the multiple linear regression model.

Models that can be linearized in this way are called *intrinsically linear*. Intrinsically nonlinear models, i.e., those that cannot be linearized, must be analyzed by nonlinear regression methods. An example would be the model of (10.27) with an additive error term. For these, the least squares equations are nonlinear in the estimators, and numerical methods are usually required for solution. Many computer packages include routines for solution of selected nonlinear regression models. A thorough treatment of nonlinear regression may be found in Bates and Watts (1988).

Assumptions
In the usual multiple regression analysis, it is assumed that:

- The ε_j are independent.
- The ε_j are normally distributed with mean zero.
- The ε_j have constant variance σ^2 (which does not depend on the predictor variables).
- The predictor variables are measured without error.

These assumptions are equivalent to those made in ANOVA models. The assumptions regarding the ε_j imply that the Y_j are normally distributed about the model (line, plane, or hyperplane in linear regression), and have constant variance.

Applications in Reliability
Standard regression models are used in analyzing both static and dynamic reliability. Examples of the former are breaking strengths of materials as functions of temperature, storage time, geometrical configuration, and so forth; thickness of an insulation layer as a function of raw materials, amount of an additive, etc. Examples of dynamic reliability include numerous applications in which MTTF, MTBF, or other characteristics of life distributions, as well as MTTR, computer execution times, etc., are investigated as a function of various predictor variables,

In many applications of this type, it is necessary to transform data to a scale on which the assumptions given above are at least approximately satisfied. In cases where this is not possible, alternative assumptions (commonly based on the Weibull distribution) are made and the analysis is modified accordingly. In addition, certain reliability models, notably the proportional hazards model (see Chapter 6), can be expressed and analyzed in a regression formulation. A good introduction to the topic is given by Meeker and Escobar (1998). The use of Weibull, gamma, lognormal, nonparametric, and other regression models is discussed in detail by Lawless (1982). Censored and grouped data are also considered. The use of regression models in analysis of field failure data, particularly in the context of warranty, is discussed by Kalbfleisch and Lawless (1988).

Regression methods are also used for parameter estimation in graphical approaches to data analysis. This will be discussed briefly in Chapter 11.

10.8.3 Inferences in Linear Regression

The primary inferences of interest in regression analysis are

- Estimation of model parameters (regression coefficients and σ), including confidence intervals
- Testing hypothesis about the model parameters
- Calculation of point and interval estimates of $E(Y)$ at a selected set of x-values
- Calculation of a predicted Y-value at a selected set of x-values along with a confidence interval for the prediction

We briefly discuss each of these, and then look at an example. Additional details may be found in Draper and Smith (1996), Neter et al. (1996), or any of the many other texts on regression analysis.

Estimation of Model Parameters in Multiple Linear Regression

Regression coefficients in multiple linear regression are estimated by the method of least squares. Under the stated assumptions, these are equivalent to the MLEs. The LS estimators of $\beta_0, \beta_1, \ldots, \beta_k$ are denoted $b_0, b_1, \ldots, b_k$ The LS approach leads to a system of $k + 1$ linear equations in the $k + 1$ unknown b's. These are easily solved on a computer and all statistical packages include routines for calculation of the estimates.

The sample variance about the regression model (also called the residual or error variance) is calculated as

$$s^2 = \frac{1}{n-k-1} \sum_{j=1}^{n} (y_j - b_0 - b_1 x_{1j} - \ldots - b_k x_{kj})^2$$

$$= \frac{1}{n-k-1} \left\{ \sum_{j=1}^{n} (y_j - \bar{y})^2 - \sum_{i=1}^{k} b_i \left[\sum_{j=1}^{n} (x_{ij} - \bar{x}_i)(y_j - \bar{y}) \right] \right\}$$

(10.28)

where $\bar{x}_i$ is the sample mean of the ith predictor variable. Note that s^2 is expressed in the first instance as a function of the sum of squares of deviations from the estimated regression relationship; it is an estimate of σ^2, the variance of the ε_j, and has $(n - k - 1)$ df. The second expression gives s^2 in the form of the total sum of squares of the response variable [with $(n - 1)$ df] minus the reduction in this sum of squares attributable to the predictor variables (with k df).

Hypothesis Testing in Regression

The two sets of hypotheses of most interest in multiple linear regression are usually

1. $H_0: \beta_1 = \ldots = \beta_k = 0$ versus $H_a: \beta_i \neq 0$ for at least one i
2. $H_0: \beta_i = 0$ versus $H_a: \beta_i \neq 0$ for $i = 1, 2, \ldots$, and/or k

The test of the first is called the *overall test* and tests whether or not there is any (linear) relationship between $E(Y)$ and the predictor variables. If there is not, we conclude that $E(Y)$ is not linearly related to the x's, i.e., the linear model with these predictors has no predictive value. The second set of hypotheses comprises individual tests of the predictor variables in the linear model. Essentially, each such test is a test of whether or not the variable in question contributes to predictive power above and beyond that of the remaining variables in the set of predictors.

The overall test is an F test with $n - k - 1$, k df. The computations leading to the test are usually presented in an ANOVA table, as shown in Table 10.10. In this ANOVA, the Total SS is given by the first term in braces in the last expression in (10.28); the second term in this expression is SSR; and SSE is calculated as $SSE = SST - SSR$. The rationale of the F test is that if there is no (linear) relationship between $E(Y)$ and the predictors, then the amount of reduction due to regression will be zero except for sampling error, i.e., the reduction will not be statistically significant. Numerical results may be obtained from any regression program.

The individual tests are t-tests of the form

$$t = \frac{b_i}{s_{b_i}} \tag{10.29}$$

Under H_0, this statistic is distributed as Student's t with $(n - k - 1)$ df. Tests of null hypotheses other than $\beta_i = 0$ may be performed as well. To test H_0: $\beta_i = \beta^*$, say, simply replace b_i by $(b_i - \beta^*)$ in the numerator of (10.29). *Note:* If $k = 1$, the t and F tests yield identical results. If $k > 1$, the tests are not the same, nor is the F test equivalent to the set of t tests.

Estimation and Prediction in Regression

For estimating or predicting $E(Y)$ or a single Y value at a set of values of the predictor variables, we simply use a point, $\hat{y}$, on the sample regression function, given by

$$\hat{y} = b_0 + b_1 x_1 + \ldots + b_k x_k \tag{10.30}$$

where $x_1, \ldots, x_k$ are the values at which the prediction is to be made. The variance of the estimator, however, is not the same in these two applications. It is substantially larger when predicting a single outcome than when estimating or predicting a mean. It follows that the confidence interval for a single outcome will be much wider than that for a mean.

Notes: (1) The CI for $E(Y)$ is often called simply a confidence interval. That for Y is called a *prediction interval.*

(2) It is essential that a proper distinction be made between these two inference problems; otherwise, the results can be very misleading. Regression programs will usually provide these estimates and confidence intervals as optional output.

Table 10.10. Analysis of variance in multiple regression.

Source	df	SS	MS	F	p
Regression	k	SSR	MSR	MSR/MSE	
Error	$n - k - 1$	SSE	MSE		
Total	$n - 1$	SST			

Analysis of Residuals

In regression, the residuals are simply $e_j = y_j - \hat{y}_j$, where $\hat{y}_j$ is given by (10.30) using the x values of the data. Most regression programs provide options for calculation, plotting, and analysis of residuals. These may be used for tests of assumptions as in ANOVA and other data analyses, as well as tests of the model. Some of the many tests and analyses of interest in regression analysis are

- tests of normality of the ε's
- tests of independence of the ε's
- plots of residuals against all variables in the model, against $\hat{y}$, and against any other variables of interest (e.g., time of observation, variables not in the regression model, and so forth)

In the residual plots, the normal pattern is a horizontal pattern with random scatter about the center. Abnormal patterns may indicate skewed distributions, unequal variance, or an incorrect model (either nonlinearities, interactions among predictors, or incorrect choices of predictor variables). See the regression texts for details.

Example 10.11 (Case 2.1, Part a)

Data on bond strength versus length of storage are given in Table 2.1a. A plot of the data is given in Figure 10.8. Note that no particular pattern is apparent in the plot. We use regression analysis to analyze the data, with y = bond strength and the single predictor x = storage time (days). Included in the analysis will be predictions of average and individual bond strengths at $x = 40$ and $x = 100$ days. A Minitab run resulted in the following output:

Regression Analysis
```
The regression equation is
Strength = 291 - 0.414 Day
```

Predictor	Coef	StDev	T	P
Constant	291.290	6.275	46.42	0.000
Day	-0.4140	0.1503	-2.75	0.007

```
S = 40.40       R-Sq = 8.4%       R-Sq(adj) = 7.3%
```

Analysis of Variance

Source	DF	SS	MS	F	P
Regression	1	12386	12386	7.59	0.007
Error	83	135478	1632		
Total	84	147864			

Unusual Observations

Obs	Day	Strength	Fit	StDev Fit	Residual	St Resid
6	4.0	197.00	289.63	5.86	-92.63	-2.32R
69	68.0	132.00	263.14	7.21	-131.14	-3.30R
78	80.0	360.00	258.17	8.71	101.83	2.58R
82	82.0	356.00	257.34	8.98	98.66	2.50R

```
R denotes an observation with a large standardized residual
```

```
     Fit  StDev Fit        95.0% CI              95.0% PI
   274.73     4.64   (  265.50,  283.96)   (  193.83,  355.63)
   249.89    11.41   (  227.18,  272.59)   (  166.37,  333.41) X
X  denotes a row with X values away from the center
```

The interpretation of these results is as follows:

- The sample regression equation is the first element of the output. The result is $\hat{y} = 291 - 0.414x$. We estimate that on average bond strength decreases 0.414 pounds per day of storage.

- The t test of H_0: $\beta_1 = 0$ is the same as the F test in the ANOVA, and both give a p value of 0.007, which would lead to rejection at the 1% level. Note, however, that $b_1 = -0.414$, indicating a very small, though significant, decrease in strength with increasing storage time. As noted, this change is not apparent in the plot of y versus x.

- The estimate of σ is $s = 40.40$, which appears in the printout. It may also be gotten as the square root of $s^2 = $ MSE in the ANOVA (1632 in this case).

- 95% confidence intervals for $E(Y)$ and Y at $x = 40$ and $x = 100$ (denoted CI and PI, respectively) are given at the end of the output. Note that the prediction intervals in each case are much wider than the CIs, and are, in fact, so wide as to be almost useless. Note that Minitab flags the second set of intervals as being at an x value outside the range of the data. (Extrapolation in regression, as in any other analysis, can be dangerous.)

- Minitab has also flagged four observations as "unusual." All have standardized residuals in excess of 2.0, meaning that the values are more than two standard deviations from the mean, and one observation is a bit more than three standard deviations from the mean. This is about what one would expect with $n = 85$, except for observation 69, which should be considered a possible outlier.

- Residual plots are shown in Figure 10.9. The patterns do not indicate a perfect fit to the model and assumptions, but they are not indicative of serious departures either. The

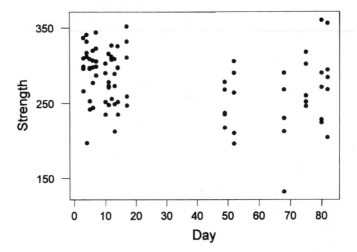

Figure 10.8. Plot of data, Case 2.1a.

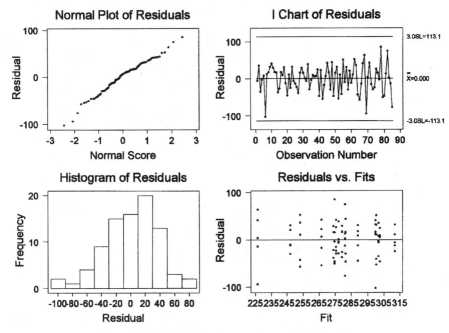

Figure 10.9. Residual plots for regression analysis, Case 2.1a.

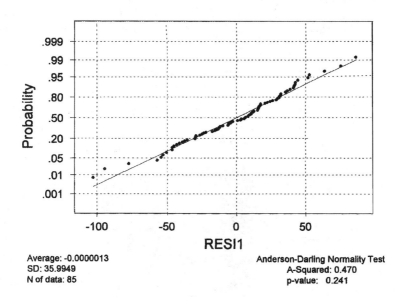

Figure 10.10. Normal probability plot of residuals, Case 2.1a.

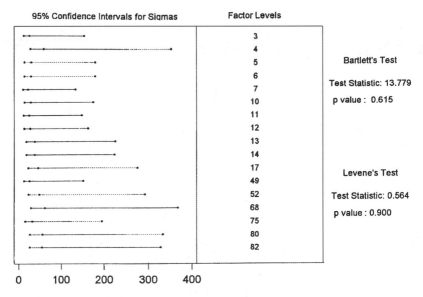

Figure 10.11. Test of homogeneity of variance, Case 2.1a.

Anderson–Darling test of normality is given in Figure 10.10. With a p value of 0.241, normality is not rejected. Finally, Figure 10.11 gives the results of the tests of homogeneity of variance for the 17 different storage times. Again, the p-values indicate that there is no evidence of a violation of this assumption. ■

Example 10.12 (Case 2.1, Part b)

As an example of multiple predictors, we consider the data of Table 2.1b, with x_1 = temperature and x_2 = humidity. A regression analysis of the data provides the following Minitab output:

Regression Analysis

```
The regression equation is
Data = 420 + 4.16 Temp - 4.97 Humidity

Predictor          Coef        StDev            T          P
Constant          419.8        110.0         3.82      0.000
Temp              4.160        3.745         1.11      0.269
Humidity        -4.9723       0.6243        -7.96      0.000

S = 87.17        R-Sq = 38.8%      R-Sq(adj) = 37.6%

Analysis of Variance

Source         DF            SS           MS          F          P
Regression      2        482039       241020      31.72      0.000
Error         100        759915         7599
Total         102       1241954
```

The p value of 0.000 for the F test indicates that H_0: $\beta_1 = \beta_2 = 0$ is rejected, i.e., we conclude that bond strength is related to the two predictor variables. The t tests of the individual coefficients indicate that humidity is a significant explanatory variable, while temperature is not. The estimated standard deviation of 87.17 indicates that there is a great deal of variability in bond strength. ∎

10.8.4   Linear Correlation

A *simple correlation coefficient r* is a measure of strength of linear relationship between two variables, say x and y. r is an index number in the sense that it is unit-free, with possible values in the interval $[-1, 1]$, A correlation of 1 (-1) is called a perfect positive (negative) correlation; in these cases, the two variables are perfectly colinear. A correlation of zero indicates no linear relationship. (In that case the slopes of the two regression relationships—y on x and x on y—are both also zero.) The closer r is to either extreme, the stronger the linear relationship.

A related quantity that is important in regression analysis is the *coefficient of determination*, r^2 (usually expressed in percent and calculated as $100r^2$). In fact, $100r^2$ is the percent reduction in the total variance of y that can be attributed to x (often called the "percent of 'explained' variability").

For two variables, the correlation coefficient can easily be calculated directly. Based on a sample of size n, with data (x_i, y_i), $i = 1, \ldots, n$, r is given by

$$r = \frac{\sum_{i=1}^{n}(x_i - \bar{x})(y_i - \bar{y})}{\sqrt{\left(\sum_{i=1}^{n}(x_i - \bar{x})^2\right)\left(\sum_{i=1}^{n}(y_i - \bar{y})^2\right)}} \tag{10.31}$$

r^2 can also be calculated directly from the ANOVA results, namely

$$r^2 = \frac{SSR}{SST} \tag{10.32}$$

which shows this quantity as the proportion of "explained" variability. (r may be calculated as the square root of r^2, with the sign taken to agree with that of b_1.)

In multiple linear regression, the *multiple coefficient of determination, R^2*, is given by (10.32), where SSR and SST are taken from the ANOVA for multiple regression analysis. This is standard output for most statistical regression programs, where R^2 is ordinarily expressed as a percent (i.e., is calculated as $100SSR/SST$). It is interpreted as the percent of variability in y "explained" by the set of predictors $x_1, \ldots, x_k$.

The *multiple correlation coefficient R* is gotten as the positive root of R^2. It can also be calculated as the simple correlation between y and $\hat{y}$ (for the sample data), and is thus seen as a measure of closeness of fit of the model to the data.

The F test in the ANOVA in regression analysis is also a test of the null hypothesis that the true R is zero against the alternative that it is positive.

Example 10.13 (Case 2.1)

We consider both 2.1a and 2.1b. In simple linear regression (2.1a), R^2 in the output is $100r^2$. The values of R^2 are given (in percent) in the Minitab output. The values can also be calculated directly from (10.32). For the storage data, we find $R^2 = 100(12386)/147964 = 8.38\%$. Storage time accounts for only 8.4% of the variability in bond strength, indicating a weak relationship, as previously observed.

For the temperature-humidity data analyzed in Example 10.12, $R^2 = 38.8\%$, indicating a modest linear relationship between bond strength and the two predictor variables. ■

10.8.5 Analysis of Covariance

In the designed experiments considered in this chapter, treatments play the role of predictor variables in regression. Designed experiments are structured so that factors are balanced, equal replication is used when necessary (i.e., usually), and experimental conditions are controlled. In regression studies, we usually deal with situations where this is not the case. Here the experimenter may have little or no control over the values of the predictor variables included in the experiment. Although regression analysis provides a method of analysis of such data, it is not possible to obtain the detailed types of information on, for example, interactions between factors that can be gotten from ANOVA of a structured experiment. Thus both approaches to data analysis are needed.

Analysis of covariance (ANCOVA) employs both techniques and is used in designed experiments in which information on one or more predictor variables is recorded along with the response. In Case 2.15, for example, a comparison of TTF for three helicopter parts is of interest. Also included in the data are flight hours for the helicopters on which these parts were installed. Since flight time may affect the failure rate, it is desirable to include this factor in the analysis. (The design is thus a completely random design with a covariate.)

The role of covariates in designed experiments is twofold. First, we can thereby account for uncontrolled factors. Secondly, we may thereby substantially reduce experimental error, if there is, in fact, a relationship between response and the predictor variable(s) and that relationship is properly modeled. This occurs as in regression analysis when the regression is significant.

The structure of ANCOVA is analogous to that of ANOVA. A table is set up, indicating sources of variability, df, and so forth, as appropriate for the basic experimental design, except that an additional source for covariates is included. The df for this source are the number of covariates in the study; df for error are reduced accordingly. The computations involve estimating regression coefficients for error and each source in the ANCOVA and adjusting SSs for the covariates. Statistical computer packages will handle most designs.

Example 10.14 (Case 2.15)

In an ANCOVA for these data, the structure is that of a CRD with part as treatment, time to failure as response, and flight hours as covariate. The Minitab output for the analysis is as follows:

Analysis of Covariance (Orthogonal Designs)

```
Factor          Levels  Values
Part            3       1    2    3

Analysis of Covariance for PartHrs

Source        DF    Adj SS        MS       F       P
Covariates    1        20         20    0.00   0.994
```

Part	2	3184772	1592386	4.64	0.016
Error	36	12360597	343350		
Total	39	15780724			

Covariate	Coef	StDev	T	P
AC Hours	0.000216	0.0281	0.007668	0.994

Note that part is significant at the 5% but not at the 1% level ($p = 0.016$) and that the covariate is not significant. Average times to failure are different for the different parts, but there is no evidence that they are affected by time in flight. ∎

10.9   ACCELERATED TEST DESIGNS

In many reliability applications, parts, components, or even the final products are very highly reliable. This is typically true in, for example, space applications, nuclear reactor components, and so forth, but is often true as well of many consumer products, e.g., small appliances, tools, etc. In such cases, it may be impossible or extremely costly and time-consuming to test a batch of items until failure. An alternative that is sometimes feasible is to use test plans that feature either Type I or Type II censoring. In many cases, these are not an acceptable solution either, because testing for an acceptably short time period may lead to few or no failures and insufficient information, whereas testing until a fixed number of failures are observed may require an unacceptably long time period.

An alternative that may provide useful information at reasonable cost and within an acceptable time frame is to test under conditions of stress. Such tests are called *accelerated tests*. Depending on the item being tested, stress factors may be temperature, humidity, vibration, electrical, radiation, light, and so forth. Usually, only one stress factor is employed in a given test, but two or more may be varied simultaneously as well. In addition, stresses may cycle through two or more levels during the test (e.g., high temperature, then low, return to high, etc.) Another method of applying stress to a product is to apply a higher usage rate (e.g., constant rather than intermittent use or frequent start-ups).

Accelerated tests are also used in studies of product characteristics other than failure. Alternatives are typically measures of item degradation, e.g., changes in electrical conductivity, bond strength of an adhesive, and so forth.

Tests of this type are usually run on materials, parts, and small components rather than on more complex subsystems or systems. The assumption on which accelerated tests are based is that items will behave in basically the same way in a short period of time under high stress as they do in a longer period of time under low stress. This is much more likely to be the case for simple parts and components than for an entire system.

In representing and analyzing accelerated test data, models are used to relate item characteristics at the various stress levels. Many mathematical models have been developed for this purpose and empirical models, for example, graphical methods or regression models, may be used as well. Selection of an appropriate mathematical model requires some understanding of the failure mechanisms involved. Graphical methods are discussed in Chapter 12. Modeling and data analysis issues will be discussed in Chapter 14.

In order to design an experiment for accelerated testing, an investigator must have adequate knowledge about the item, its normal operating environment, including the stresses to which it might be subjected, its failure modes and/or degradation patterns, and the form of the relationship between these under normal and stress conditions. This information is used to select appropriate stress levels for testing and appropriate models for analysis. Samples sizes to be used in the experiment should ideally be determined based on the required precision of the results, but in practice are often determined on the basis of cost, materials, facilities, and time limitations (e.g., limited availability of items for test, test chambers for conducting tests, etc.)

An example of a small-scale accelerated test is given in Case 2.1. The part being tested is an adhesive that was found to degrade at high relative humidity. In this relatively small preliminary experiment, groups of items were placed in a test chamber for a fixed period of time at 32 °C, 70% RH; 27 °C, 50% RH; 27 °C, 70% RH; and 27 °C, 100% RH. On removal of the items from the test chamber, bond strength was determined.

Again, there is a vast literature on accelerated testing. Good introductions to tests, modeling, and data analysis, along with many examples, are given by Condra (1993, Chapters 17–20) and Meeker and Escobar (1998, Chapters 18–20). A very thorough treatment of all aspects of the subject is given by Nelson (1990). All three sources discuss many applications as well as some of the difficulties and pitfalls that might be encountered.

NOTES

1. Here we actually need 3, 210 df, which is not in the table. In such cases, we use a conservative approach and take the value for the highest value of df that does not exceed the desired value. An alternative is to use a computer program to obtain the value. In this case, Minitab calculates the value for 3, 210 df as 3.88.

2. There is also a vast literature on fractional factorial designs, much of which is highly technical. Some important references are given in the texts cited.

EXERCISES

1. For each of the cases of Chapter 2, identify the experimental units. State your answer as precisely as possible based on the information given.

2. Identify the treatments for each case in Chapter 2

3. Discuss the concept of randomization in the context of each of the cases of Chapter 2. Which of these are true experiments and which are observational studies? What assumptions must be made in order to interpret the observational studies as true experiments?

4. (Based on Lorenzen and Anderson, 1993, Chapter 1.) Four experimental units are used in an experiment with two treatments, A and B. List the outcomes if (a) treatments are assigned to experimental units, and (b) experimental units are assigned to treatments. Calculate the probabilities of each of the six outcomes for each of these schemes. What does this tell you about the proper method of randomization? (Is it (a) or (b) or does it matter?) *Note:* Your answer applies to DOE generally.

5. For each of the cases in Chapter 2, determine which, if any, of the experimental designs discussed in this chapter was used in obtaining the data.

6. Describe an experimental situation in your own field, addressing the issues discussed in Section 10.2.

7. Based on your analysis in Exercise 6, design an experiment that will achieve your experimental objectives.

8. A two-factor experiment in which main effects are both zero but there is a large interaction is described in Section 10.3. Extend this notion to a three-factor experiment by set-

ting up a three-dimensional table in which all main effects and two-factor interactions are zero, but there is a large three-factor interaction.

9. Construct a balanced incomplete block design with five treatments in blocks of four. What is the minimum number of blocks required for a BIBD with five treatments in blocks of two? Can you generalize this result for number of blocks required?

10. Construct a BIBD with six treatments in blocks of three.

11. Table 2.7 lists data on time to failure for hydraulic systems of six machines. Use ANO-VA to compare MTTFs for the six machines. Make all possible comparisons using the lsd with $\alpha = 0.05$.

12. Reanalyze the data of Table 2.7 assuming a lognormal distribution for TTF. Compare the results with those of the previous exercise.

13. Analyze residuals and perform tests of assumptions in each of the previous exercises. Which model (normal or lognormal) appears to more nearly satisfy the ANOVA assumptions.

14. In Case 2.7, the machines are classified into three age groups of size two each. Suppose that the machines can be assumed to be a random sample of size two from each group and it is desired to analyze the age effect. What is the experimental design in this case and how should the ANOVA be structured?

15. Perform an ANOVA to compare days in Case 2.9. Use the lsd with $\alpha = 0.01$. How many significant differences are found? If the tests were independent (which they are not), and H_0 were true in each test, how many significant differences would you expect to find?

16. Use ANOVA to analyze the data of Case 2.12. Prepare plots as appropriate and analyze residuals. Are there any problems with the data or assumptions?

17. Analyze the data on Part Hours of Case 2.15, comparing the three treatments. Are there any problems with the data or assumptions? If so, make any necessary adjustments and reanalyze the data.

18. Use ANOVA to analyze the data on single carbon fibers in Case 2.20. Check assumptions and comment. Reanalyze, if necessary.

19. Repeat the pervious exercise on tows of 1000 carbon fibers in Case 2.20. Compare the results with those of the previous exercise.

20. Analyze the data of Table 1(b). Comment on the structure of the experiment, the results obtained, and the assumptions.

21. An experiment was done to investigate the effects of three factors on production—experience (low/high), gender (M/F), and shift (day/night). The response was number of defectives per 1000 items produced. Three workers in each of the eight treatment groups were randomly selected. The results were as follows:

			Shift				
Experience	Gender	Day			Night		
Low	M	24	29	25	18	19	22
	F	20	22	18	15	10	11
High	M	15	15	12	15	20	13
	F	16	9	11	10	14	6

Analyze the data, testing at the 5% level. State your conclusions with regard to main effects and interactions. What would be your conclusions if the lsd were used to test main effects and interaction at $\alpha = 0.05$? What would they be if you used the hsd?

22. Apply the Newman–Keuls procedure to compare mean responses to temperature in Example 10.5. Do the same for mean responses to materials.

23. Analyze the data of Case 2.11 as a RCBD with four treatments in a 2×2 factorial arrangement and 30 blocks.

24. Reanalyze the data of Case 2.11 as a split plot with trial as whole plot and insertion/withdrawal as subplot. Which is the correct analysis? Are your conclusions different? If not, why not?

25. Use a multiple-comparisons procedure to compare mean strengths for fiber lengths in Example 10.6.

26. In a study of productivity in a factory, five types of background music were tested. These were

A   Slow tempo, instrumental and vocal

B   Medium tempo, instrumental and vocal

C   Fast tempo, instrumental and vocal

D   Medium tempo, instrumental only

E   Fast tempo, instrumental only

The experiment was run over a five-week period in a Latin Square, with weeks as rows and Days to the week as columns. The results (coded values for productivity) were as follows:

Day

Week	M	Tu	W	Th	F
1	D 18	C 17	A 14	B 21	E 17
2	C 13	B 34	E 21	A 16	D 15
3	A 7	D 29	B 32	E 27	C 13
4	E 17	A 13	C 24	D 31	B 25
5	B 21	E 26	D 26	C 31	A 7

Analyze the data. Include in your analysis the Newman–Keuls procedure for comparing each treatment mean with each other. Use the underlining method to display your conclusions.

27. Use a multiple comparisons procedure to compare the mean wear for the four brands of tires in Example 10.8. Do a further analysis of the data of Example 10.8, as follows:

(a) Prepare residual plots.

(b) Test for normality and, if possible, equality of variance.

(c) Redo the analysis using blocks $\times B$ to test B and blocks $\times A \times B$ to test AB. (See the comments on the alternative model for this experiment.)

(d) Test for homogeneity of the three variances, E_a and the two error terms of Part (c). Comment on the implications of your result.

28. Perform a residual analysis on the data of Example 10.7.

29. Analyze the data of Example 10.10 using a statistical program of your choice and compare the output with results given in the example.

30. Reanalyze the data of Example 10.10 assuming a main-effects-only model and compare the results with those given in the example.

31. Look up a fractional factorial design with at least eight factors in one of the references cited. Give a list of the treatment combinations to be used and state what effects are estimable in the selected design.

32. Use ANOVA (with storage time as treatment) to analyze the data of Table 1a, and compare the results with those of the regression analysis of Example 10.11. Include residual plots in your analysis.

33. Repeat the regression analysis of Example 10.12, including in your analysis confidence and prediction intervals for 30°, 90% RH, and 35°, 60% RH.

34. Analyze residuals in the previous example and test for homogeneity of variance.

35. Using the data of Table 2.4, regress warranty cost (y) on km driven (x) for autos for which claims were made. Does there appear to be a relationship between x and y? Include in your analysis a confidence interval for the mean warranty cost of autos driven 30,000 km for which claims are made and a prediction interval for a single such automobile. Interpret the results.

36. Use regression analysis with x = fiber length to analyze the data of Case 2.6 and compare the results with those of Examples 10.1 and 10.4.

37. Analyze the data of Case 2.6 using indicator variables for the treatments. Compare the results with those of Example 10.1.

38. State and interpret the R^2 values in Exercises 33, 36, and 37.

39. Analyze the data of Case 2.15 by ANOVA, ignoring aircraft hours, and compare the results with the ANCOVA given in Example 10.14.

40. Analyze residuals in the ANCOVA of Example 10.14.

41. Prepare a list of five applications in which accelerating testing would be highly desirable or essential in analysis of item lifetimes.

CHAPTER 11

Model Selection and Validation

11.1 INTRODUCTION

As is apparent in the material covered in the previous chapters, modeling plays a very important role in reliability analysis. In Chapter 1, we began with the system characterization and corresponding mathematical models. Subsequently, many additional models were discussed, including models of failure mechanisms, time to failure, system failure as a function of component failures, and models representing data, e.g., those used in regression and ANOVA. In this chapter, we are concerned mainly with probabilistic models, and are concerned with selection and validation of an appropriate model. By this is meant choosing a "best" model from a set of candidate models and then validating this choice, i.e., providing evidence that the chosen model is, in fact, adequate and valid for the intended reliability objective.

In Chapter 4, we dealt with probability distributions used in reliability. These are basically models representing the probabilistic structure of time to failure and related random quantities. In Chapters 6 and 7, reliability models for components and multicomponent systems were developed and analyzed. These were based on the system structure and, where possible, on the underlying failure mechanisms of the item.

In cases where the underlying failure mechanism is well understood, the appropriate probability distribution can be deduced, and white-box models are used in the analysis. In situations where this is not the case, black-box models, employing probability distributions that are thought to be reasonable in the particular application, are used. In black-box analyses, the choice of models is rarely completely arbitrary. Choices are made on the basis of whatever limited knowledge is available and with the use of well-considered judgment. It is important in these cases to select models that are flexible enough to represent a fairly broad spectrum of phenomena, i.e., to make a reasonable compromise between simplicity and ease of analysis and complexity and wide applicability. For example, use the Weibull distribution rather than the exponential; use system models that allow for the possibility of interactions.

Chapters 3, 5, and 8 through 10 were devoted to various aspects of data collection, description and analysis. Included were graphical and numerical methods for description of data, methods for estimation and hypothesis testing, analysis of other problems in statistical inference in reliability applications, and design of experiments. In nearly all of these analyses, the form of the underlying distribution and/or the structural reliability model were assumed to be known and inferences regarding the parameters of the models or related reliability quantities were the objectives of the analyses.

All of these analyses, and hence the validity of the conclusions drawn, rely, to a greater or lesser extent, on the validity of the assumptions made. Many analyses are not sensitive to small or even modest departures from assumed conditions, especially if sample sizes are modest to large. This is true, for example, of ANOVA and of other inference procedures that are based on sample means. Procedures that are found to be relatively insensitive in this way are called *robust*.

In other reliability applications, the analysis and conclusions may be very sensitive to the assumptions made. This is particularly true when dealing with the tails of a probability distribution, as, for example, in analyses of situations requiring high reliability. The amount of probability in the tails (say beyond three or more standard deviations from the mean) can be very different for the normal, exponential, Weibull, and other distributions.

Because of the importance of assumptions and the uncertainties associated with their selection, it is very useful to be able to test their validity, especially in data-based reliability analysis. In previous chapters, we have looked at some approaches to checking assumptions. For example, graphical procedures such as plots of residuals, and statistical tests such as the Anderson–Darling test of normality and Bartlett's and Cochran's tests of homogeneity of variance. In this chapter we concentrate on tests of distributional assumptions, and look at the Anderson–Darling and other tests more generally and more carefully, elaborating on some techniques previously introduced and presenting some important additional methods.

In applications, we usually want to do more than simply test a given model. We, in fact, want to choose one of two or more competing models. In this, we are concerned with the two types of problems discussed above—selection of a probability model and selection of a system model. Here we deal primarily with the first of these. The procedures to be discussed are both graphical and analytical, the latter involving statistical hypothesis testing.

Graphical methods of model selection involve preparation of probability plots or other graphs, such as hazard plots, and checking visually to determine whether or not the plots appear to show certain characteristics indicative of a particular model. A side benefit of this analysis is that one can often also estimate parameters from the plots. Graphical estimation is sometimes used as an alternative to MLE and other analytical methods, either as a quick initial analysis or in cases where the analytical approach is somewhat intractable.

The second tool for testing assumptions is a "goodness-of-fit" test. Goodness-of-fit tests are statistical procedures for testing hypothesized distributions and/or models. The Anderson–Darling test is one such test; we will look at several others as well. Note, incidentally, that a poor fit (either graphical or analytical) may occur for two reasons: (1) the model is incorrect, or (2) the model is correct but the parameter values specified or estimated may differ from the true values by too great an amount.

Finally, in practice, the modeling problem is not resolved by selection of a model by one of the methods to be discussed (or any other). Once a model is selected, validation of the model is necessary to complete the analysis. This generally requires further testing, additional data and other information, and careful evaluation of the results.

Note that in both the selection and validation processes, particularly when graphical methods are employed, both art and science are involved. The art part is especially apparent in visual fitting of data to a model. Further, there are many at least partly subjective decisions that must be made in the analysis. This includes selection of models to be considered, selection of a level of significance in statistical testing, which statistical procedures are to be used, and so forth.

The outline of the chapter is as follows. Graphical methods are discussed in Section 11.2. Section 11.3 deals with goodness-of-fit tests. In both sections, complete and censored data are considered. Some procedures and suggestions for model selection and validation will be given in Sections 11.4 and 11.5.

11.2 GRAPHICAL METHODS

As noted in Chapter 3, the principal role of graphical techniques in data analysis is in the description and presentation of data. In Chapter 4, many probability distributions were presented as models of randomness. Here we look at graphical analysis in the context of these models. The main objective is to "fit" data to one or more of these probability distributions (or, conversely, to "fit" the distributions to data). Visual fits provided by graphical methods are quite useful for this purpose, and these will be the main topic of this section. In Section 11.3, we will look at statistical tests of how well a candidate distribution fits the data.

There are many plotting papers, computer algorithms, and other aids in plotting data under specific distributional assumptions. Many of these also provide graphical techniques for estimating the parameters of the assumed distribution. In addition, some nonparametric techniques are available. These will also be discussed briefly.[1]

The benefits of the graphical approach are the visual insights obtained and (usually) ease of application, particularly in the case of a number of distributions, including the exponential, Weibull, normal, extreme value, and others, for which plotting and computer aids are available. The problems with this approach are (1) it is, to some extent, subjective; and (2) there is no well-developed statistical theory for determining the small sample or asymptotic properties of the procedures. The first problem is not serious; it means that different analysts are likely to arrive at somewhat different results if the plotting is done by hand (though not if it is done by computer), but usually will arrive at the same conclusions. The second problem is more serious; it means that standard errors and distributions of estimators are not known, even asymptotically, and that test statistics based on these cannot be obtained. As a result, a proper analysis of a set of data cannot be based exclusively on graphical procedures; the analytical methods of the previous sections are essential as well.

In the following sections, we look at the sample CDF, various probability plots, hazard plots, and, where available, graphical estimates. In many cases, both complete and censored data will be considered.

11.2.1 Plotting the Sample CDF

The *sample cumulative distribution function* (also called the *empirical distribution function or empirical* CDF) is a plot of the cumulative proportion of the data that lie below x, against x ($-\infty < x < \infty$). The result is a *step function,* in this case a function having steps of height $1/n$ at each data point x_i. This is most easily done by use of the order statistics $x_{(1)}, \ldots, x_{(n)}$: plot i/n versus $x_{(i)}$; then draw a horizontal line of height i/n from $x_{(i)}$ to $x_{(i+1)}$. Alternatively, the ordinates can simply be connected to form a *cumulative frequency polygon.* Both of these plots are nonparametric estimators of the population CDF, $F(x) = P(X < x)$.

There are many other graphical estimators of these types. For example, another commonly used plot of the sample CDF is formed by plotting steps of height $i/(n + 1)$, called the "mean plotting position." The idea is to plot at the center of the steps rather than at an end, which distorts the estimate of the intercept. Another alternative that accomplishes this is plotting $(i - 0.5)/n$. For further discussion and other alternatives, see King (1971), Nelson (1982), Lawless (1982), and Meeker and Escobar (1998). Computer packages vary with regard to the type of plot provided, and some have options for various plots.

Example 11.1 (Case 2.1)
Figure 11.1 is a plot of the sample CDF as a step function for the bond strength data of Table 2.1a. Note that with a relatively large sample size, most of the steps are obscured in the plot.

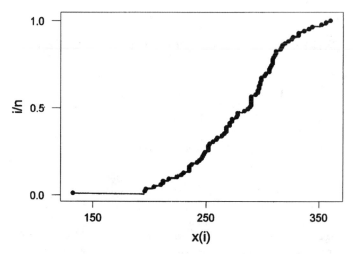

Figure 11.1. Sample CDF for bond strength data.

On the other hand, plots of this amount of data tend to give a fair impression of the general shape of the CDF. In this case, the shape exhibited is that of a sigmoid curve that is characteristic of the CDF of the normal distribution. For small samples, it is often not possible to discern a pattern, and care must be taken in interpreting the plot, as it is easy to reach unwarranted conclusions. ■

11.2.2 Probability Plots, Complete Data

In Example 11.1, it was noted that a sigmoid shape indicates normality. Other probability distributions lead to data that plot as different characteristic shapes. These may or may not be easily recognizable or distinguishable in a plot of a given set of data. *Probability plots* have been developed as an alternative method of plotting data. As opposed to the sample CDF, which is a nonparametric procedure, probability plots assume a particular underlying distribution. The idea is to transform the data and/or probability scales so that the plot on the transformed scale is linear (within chance fluctuations). Equivalently, we plot sample fractiles against the fractiles of a specified distribution. As a result, the plots are sometimes referred to as "P-P plots."

Probability plotting papers have been developed for a number of distributions, including the normal and most of the important distributions used in reliability applications: exponential, lognormal, Weibull, gamma, and extreme value. Many statistical computer packages (e.g., Minitab, SPSS, STAT, S, etc.) include some or all of these plotting options as well.

A consequence of the transformed scales is that data that do not follow the distribution specified will usually (but not always) plot as a nonlinear function on that paper. Thus a nonlinear plot can be taken as an indication that the assumption regarding the distribution underlying the data is incorrect. This will be apparent in some of the examples given in this section. It is important to note, however, that the converse is not true, that is, the fact that a linear plot is obtained does *not* necessarily mean that the correct distribution has been found. The notion of selecting an appropriate distribution will be pursued more precisely in the next section, which is devoted to statistical tests of distributional assumptions.

In interpreting the plots, we look for linearity as indicative of an adequate fit, as noted. In addition, the plots may give an indication of potential outliers in the data, particularly if the overall pattern appears to be linear but there appear to be a few points that lie outside this pattern. (These may occur in one or both tails of the distribution.) Plots that appear to be linear for the bulk of the data but veer off at one or both ends may indicate multiple outliers or longer-tailed distributions than the one plotted. Another aspect that must be considered in interpretation is that plots of this type tend to be inherently more variable in the tails than in the center of the distribution. Thus more points that appear to be outside the pattern can be expected in the tails and caution must be exercised in interpreting these as outliers.

Other anomalies that may be suggestive of specific model violations can also be found in the plots. For example, plots for which segmented lines (i.e., linear segments with different slopes) appear to fit the data suggest that a mixture model, which may result, for example, when multiple failure modes are present, may be appropriate. [Note, however, that mixture models can lead to many different shapes; see Jiang and Murthy (1995a,b, and 1997a–d.] Graphical analysis involves a good deal of judgment, especially in cases such as these. In the process, one often plots data on the various plotting papers available and then applies rough guidelines such as those given here in an attempt to select one or more reasonable candidate models for the failure distribution.

Transformations for Probability Plots

The theory of probability plots involves the relationship between the $x_{(i)}$ and the corresponding fractiles of the selected distribution. This relationship is discussed by Lawless (1982), Nelson (1982), and Meeker and Escobar (1998) and given for the exponential, normal, lognormal, gamma, Weibull, and extreme-value distributions. Additional discussion of plotting for these distributions, along with many additional examples and applications, is given by King (1971).

The plots are constructed by plotting the sample CDF corresponding to a given distribution against the $x_{(i)}$ after suitable transformation of one or both variables. Transformations for some distributions that are frequently used in reliability are as follows.

Exponential Distribution. For the exponential distribution with parameter λ, we have

$$p = F(x) = 1 - e^{-\lambda x} \tag{11.1}$$

so that the p-fractile x_p is

$$x_p = \lambda^{-1}\log[1/(1 - p)] \tag{11.2}$$

As a result, $\log[1/(1 - p)]$ plotted against x_p will result in a linear relationship with intercept 0 and slope $\lambda^{-1} = \mu$. A probability plot is a plot of the transformed sample CDF versus x.

Weibull Distribution [Equation (4.44)]. An analysis similar to that for the exponential distribution results in

$$\log[\log(1/(1 - p))] = \alpha \log(x_p) - \alpha \log(\beta) \tag{11.3}$$

Thus a plot of $\log[\log(1/(1 - p))]$ versus $\log(x_p)$ is linear with slope α and intercept $-\alpha \log(\beta)$.

Extreme Value Distribution [Equation (4.60)]. For the distribution of smallest extreme value, use relation (11.3) with $\log(x_p)$ replaced by x_p.

Gamma Distribution [Equation (4.41)].   For the gamma distribution,

$$x_p = \beta \Gamma^{-1}(p; \alpha) \tag{11.4}$$

where $\Gamma^{-1}(\ )$ is the inverse of the incomplete gamma function, given by

$$I(p; \alpha) = \int_0^p f(x)dx \tag{11.5}$$

with $f(\ )$ given by (4.41). Thus x_p is a linear function of β for fixed α. For unknown α, plots can be generated for various values and α selected as that value (if any) for which the plot is most nearly linear.

Normal Distribution [Equation (4.50)].   The relevant relationship is

$$x_p = \mu + z_p \sigma \tag{11.6}$$

where z_p is the p-fractile of the standard normal distribution. Thus x_p is linearly related to z_p.

Lognormal Distribution [Equation (4.55)].   As usual, the lognormal distribution is analyzed by transforming to $\log(x)$ and using the procedure for the normal distribution.

     Transformations for a number of other distributions, including the two-parameter exponential and three-parameter Weibull and gamma distributions, are given by King (1971), Meeker and Escobar (1998), and others. A very good discussion of the relationships between plots for different distributions, and suggestions for alternatives when the plot appears to be nonlinear, are given by King (1971). Also see Meeker and Escobar (1998).

Examples of Probability Plots

We look at probability plots for a few distributions. Note that data plots are based on estimates of the CDF. For purposes of estimation (to be discussed below), data plots using relationships such as those given above are usually centered, with the plot based on $(i + 0.5)/n = 0.5[\hat{F}(x_{(i)}) + \hat{F}(x_{(i+1)})]$, suitably transformed. The scales of probability plots in reliability applications are adjusted to show probability versus time, though other scales (e.g., untransformed values) are sometimes shown as well. On some plotting papers, the coordinates are reversed. (This is sometimes useful for interpretation and comparison of slopes.) Plots for many distributions are easily generated by computer. Computer output often includes confidence bands for the CDF as well as the estimated CDF. Minitab will also provide a "four-way probability plot," which includes the normal, lognormal, exponential, and Weibull distributions.

     Plotting paper is available from TEAM, Inc., Keuffel & Esser Co., and other sources. The examples given use Minitab or the TEAM plotting paper for preparation of the plot.

Example 11.2 (Case 2.1; Normal Probability Plot)

As noted in the discussion of Example 11.1, the plot of the CDF for the bond strength data of Table 2.1a appears to exhibit the features of a normal CDF. A plot of these data on normal probability paper as calculated by Minitab is given in Figure 11.2(a). Note three additional features in the output—a line fit through the data, bands about the fitted line, and the sample mean and standard deviation. The line is a least-squares fit (using linear regression analysis; see Chapter 10) to the sample CDF on the transformed scale. Here it indicates a reasonable fit of the data to a normal distribution and indicates the presence of one outlier and a few oth-

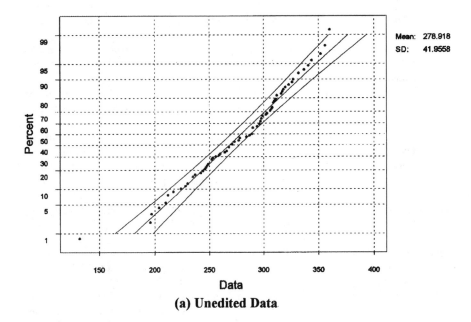

(a) Unedited Data

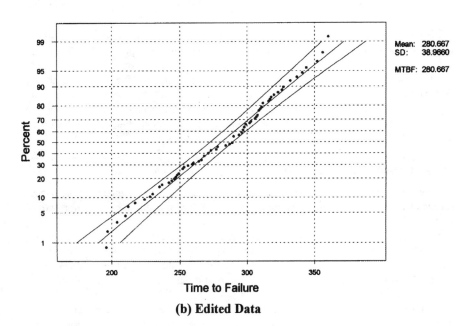

(b) Edited Data

Figure 11.2. Normal probability plots, bond strength data.

er potential outliers. The mean and standard deviation given are calculated on the basis of the original data and not from the plot.

Figure 11.2(b) is a normal probability plot of the bond strength data omitting the apparent outlier. Note that the fit to the remaining data is good, as expected, and that the mean has been increased slightly and the standard deviation reduced by about 7%.

In analyzing the data, other plots may be of interest as well. Figure 11.3(a) is a plot of the bond strength data on normal, lognormal, exponential, and Weibull probability paper[2]; Figure 11.3(b) shows the same plots, omitting the apparent outlier. We look at the latter, since the outlier leads to some distortion in all of the plots. Under normality, one would not expect a linear relationship in the nonnormal plots, and this is apparent, particularly in the lognormal and exponential plots. Note that in both the normal and Weibull plots, the possibility of at least one additional outlier is indicated. Aside from this, the plots are nearly linear in both cases. A more detailed analysis of the Weibull fit (not given here) would provide an estimate of the shape parameter of about $\beta = 8.0$. Weibull distributions with $\beta > 3$ are very nearly normal, so the result is not surprising. ■

Example 11.3 (Case 2.21; Weibull Probability Plot)
We consider the data on fiber strength for 12 mm lengths. The four-way probability plot of the data is given in Figure 11.4(a). Note that an excellent fit to the Weibull distribution is apparent, with the normal plot nearly linear, but exhibiting some concavity. The lognormal and exponential fits are very bad. Figure 11.4(b) shows the Weibull fit in more detail and includes MLEs of the parameters and confidence bands. The estimated shape parameter of 16.3 indicates a distribution very close to the normal. ■

Example 11.4 (Case 2.13; Exponential Probability Plot)
If clerical errors occur randomly, the exponential distribution should provide a good fit to the data of Case 2.13. We check this for Clerks 1 and 5. Exponential plots for these two clerks are given in Figure 11.5. Note that in both cases, the fits appear to be reasonable, though there is some curvature in both plots and a possible outlier in the data for Clerk 5. As a further check, the four-way plots for these two clerks are given in Figure 11.6. Note that the Weibull gives approximately the same fit as the exponential (taking into account the differences in scale in the plots). ■

11.2.3 Hazard Plots, Complete Data

As noted in Chapters 1 and 4, the hazard function $h(x)$ gives the instantaneous failure rate associated with a given distribution function. [It is also called the failure rate function and denoted $r(x)$.] The *cumulative hazard function* $H(x)$ is the integral of $h(x)$, which yields $H(x) = -\log(1 - F(x))$. Hazard plots are plots of the sample (or empirical) hazard function or cumulative hazard function. Traditionally, cumulative hazard is plotted as the abscissa and time to failure as the ordinate (both transformed as appropriate so that the plot under the assumed distribution is linear), but the scales may be reversed if desired and the untransformed data may be plotted instead.

Plotting papers for the cumulative hazard function are available for the distributions mentioned in the previous section, and many computer packages include hazard plots. Those available from TEAM are for plotting the cumulative hazard function on the transformed scale. Again, the plot should be linear on suitably transformed paper if the assumed distribution provides a good fit to the data. The information obtained from and interpretation of a hazard plot is basically the same as for a probability plot. (In fact, hazard plotting paper often includes the probability scale as well.) In particular, from a hazard plot we can estimate the

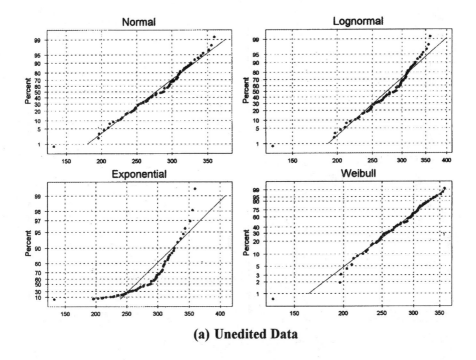

(a) Unedited Data

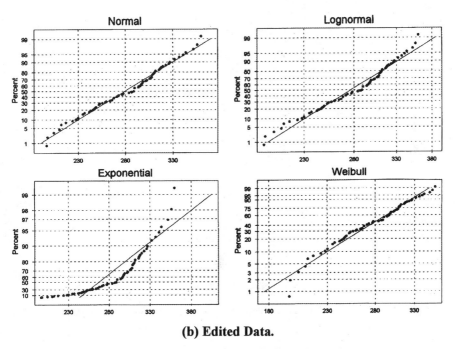

(b) Edited Data.

Figure 11.3. Four-way probability plot for bond strength.

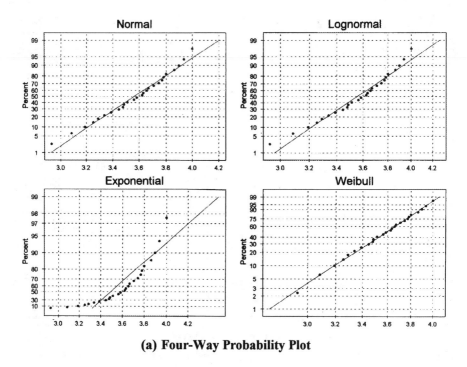

(a) Four-Way Probability Plot

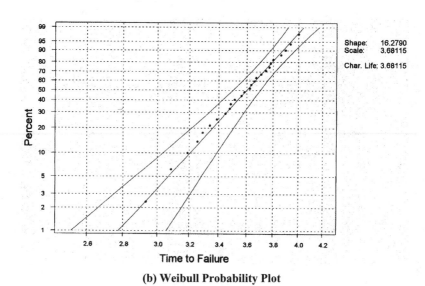

(b) Weibull Probability Plot

Figure 11.4. Probability plots for 12 mm fibers.

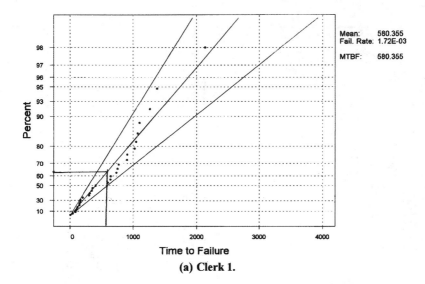

(a) Clerk 1.

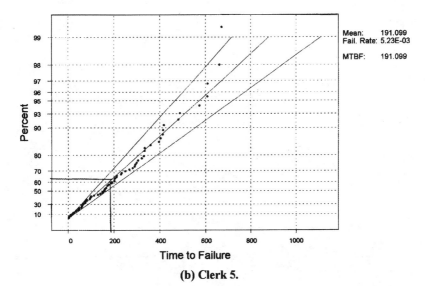

(b) Clerk 5.

Figure 11.5. Exponential probability plots for time between errors.

fractiles of the distribution, the failure rate as a function of age, and a number of related quantities. (See Nelson, 1982.) More on interpretation of the hazard and cumulative hazard functions may be found in the references cited above.

Hazard plots for complete data are constructed as follows:

1. Form the order statistics $x_{(1)}, \ldots, x_{(n)}$.
2. Calculate the hazard, $h_i = 100/(n - i + 1)$, $i = 1, \ldots, n$.

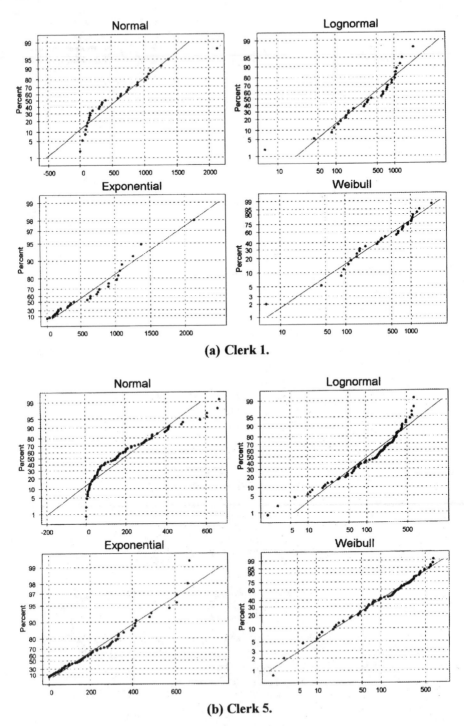

Figure 11.6. Four-way probability plots for time between errors.

3. Calculate the cumulative hazard, $H_i = \sum_{j=1}^{i} h_j$
4. Plot $x_{(i)}$ versus H_i (or H_i versus $x_{(i)}$) on either ordinary graph paper or cumulative hazard paper. Alternatively, the sample hazard function itself may be plotted by plotting h_i versus $x_{(i)}$ on ordinary graph paper.

One or more of these options are available in computer packages. For example, Minitab will plot the hazard and cumulative hazard functions on the untransformed scale.

Example 11.5 (Case 2.1; Normal Hazard Plot)
Figures 11.7, 11.8, and 11.9 illustrate some of the types of plots available as applied to the bond strength data of Table 2.1a, edited by removal of the apparent outlier. Figure 11.7 is a Minitab plot of the hazard function on the untransformed scale under the assumption of normality. Figure 11.8 is a plot of the cumulative hazard function (with axes reversed) on TEAM normal hazard paper. The probability scale, based on the normal distribution, is given at the top of the graph. Note that the data appear to plot reasonably linearly. Figure 11.9, also produced by Minitab, is an "overview" plot that includes both hazard and cumulative normal hazard functions as well as the normal probability plot and a plot of the survival function, which is an estimate of $S(x) = 1 - F(x)$. ■

Example 11.6 (Case 2.21; Weibull Hazard Plot)
The Weibull hazard plot and the set of overview plots from Minitab are given in Figures 11.10(a) and (b). The plot on TEAM Weibull hazard paper (not shown) would again show a very pronounced linear pattern, again indicating a good fit to the Weibull distribution. ■

11.2.4   Graphical Estimation of Parameters

There are a number of approaches to obtaining parameter estimates based on the empirical CDF or the empirical cumulative hazard function. The methods basically depend on the rela-

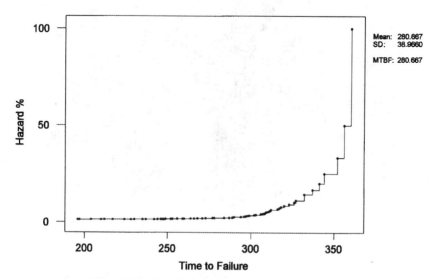

Figure 11.7. Hazard function for bond strength, edited data.

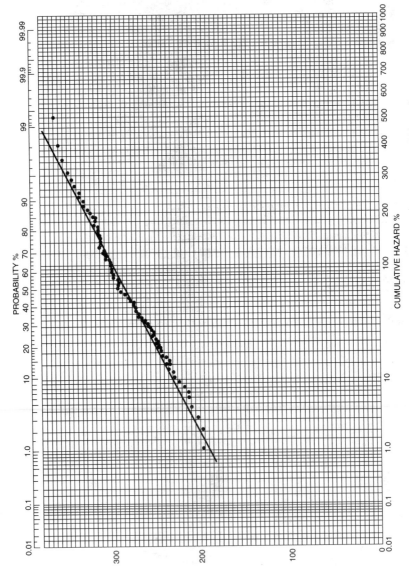

Figure 11.8. Cumulative normal hazard plot of edited bond strength data on TEAM hazard paper.

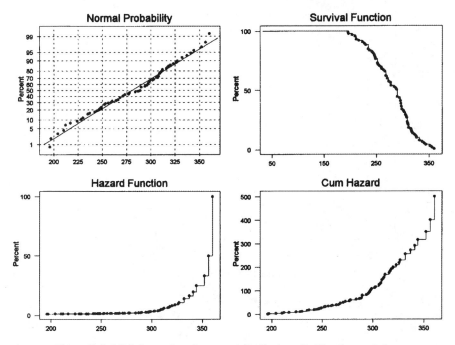

Figure 11.9. Minitab overview plot, normal distribution, edited bond strength data.

tionship between the fractiles of the distribution and its parameters, with the fractiles being estimated from the probability or hazard plots. Note again that it is suggested that estimates of this type be used either as rough initial estimates or as starting points in numerical solutions of maximum likelihood equations (or both).

Fractiles may be read from graphs with free-hand lines or curves drawn or may be calculated from least squares fits to the plotted data. They may also be calculated directly by use of the analytical relationship between the CDF or hazard function and transformed time to failure that is used to obtain the linearized plot. See King (1971), Nelson (1982), Lawless (1982), and Kececioglu (1991) for details. We look at a few illustrations of graphical estimation for some of these approaches and for selected distributions. See the references for other methods and many additional examples.

Normal Distribution
Estimated fractiles may be used in many ways to estimate the parameters of a normal distribution. The following is a relatively simple approach based on the normal probability plot. Since the 0.5-fractile is both the mean and the median, μ is easily estimated by determining the x-value corresponding to 0.5 on the probability plot. Since about 68% of the normal distribution lies within one standard deviation of the mean, the standard deviation can be estimated as one-half the distance between the graphical estimates of a $x_{.84}$ and $x_{.16}$. Similarly, the difference $x_{.975} - x_{.025}$ is equivalent to about 4σ (more precisely, 3.92σ), so one-quarter of the estimated difference estimates σ. Any other convenient fractiles can be used in this way. A refinement, particularly for small samples, would be to use fractiles of the Student-t distribution instead of the standard normal.

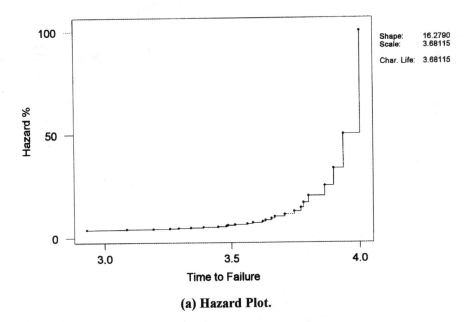

(a) Hazard Plot.

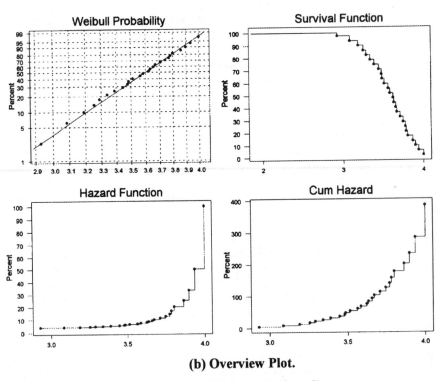

(b) Overview Plot.

Figure 11.10. Minitab Weibull plots for 12 mm fibers.

Example 11.7 (Case 2.1; Normal Distribution)
We use the data omitting the suspected outlier. Fractiles may be read from either Figure 11.2(b) or 11.8. We find $x_{.50}$ to be about 280. To estimate σ, we may use $x_{.975} - x_{.025} \cong 360 - 205 = 155$; equating this to $4\hat{\sigma}$ yields $\hat{\sigma} = 39.5$. Also $x_{.84} - x_{.16} \cong 320 - 240 = 80$, giving $\hat{\sigma} = 40$. Note that these values are quite close to the calculated values of $x = 280.67$ and $s = 39.00$, which are the MLEs. This will often be the case for moderate to large samples and a good fit of the distribution to the data. ■

Exponential Distribution.
The MTTF μ can be estimated directly from the probability plot as $x_{.63}$. In a hazard plot, this corresponds to $H = 100$. Thus μ is estimated by reading from 100 on the cumulative hazard scale to the corresponding value on the data scale. λ is estimated as $\hat{\lambda} = 1/\hat{\mu}$.

Example 11.8 (Case 2.13; Exponential Distribution)
The data for Clerks 1 and 5 are plotted in Figure 11.5. The estimated MTBF may be read by locating $x_{.63}$ as indicated on the charts. For Clerk 1, the approximate value is 570; for Clerk 5, it is about 185. The observed sample means (which are unbiased estimates of the MTBF) are 580 and 191, respectively, in relatively close agreement with the graphical results. ■

Weibull Distribution
We consider first the two-parameter Weibull distribution. Parameters may be estimated using either cumulative hazard paper or a probability plot. In the former case, plot the data as usual and sketch a line through the data. Locate the dot in the upper left quadrant of the paper and draw a line through this point parallel to the sketched hazard line. The point at which this intersects the "shape parameter" scale at the top of the page is the estimate of α. The estimate of the scale parameter β is the x-value corresponding to a cumulative hazard of $H = 100$. Figure 11.11 is a reproduction of the TEAM Weibull hazard paper with the dot circled.

To use a probability plot for estimation of α and β, the linearizing transformation on which the plot is based may be used. This is

$$\log\{-\log[1 - F(x)]\} = -\alpha \log \beta + \alpha \log x \qquad (11.7)$$

Thus the slope of the line on the probability plot estimates α, while the intercept estimates $\alpha \log(1/\beta)$.

Regression analysis may be used to provide a "best fit" estimator. The variables in the regression analysis are $x_i' = \log x_i$ and y_i defined as the left-hand side of Equation (11.7) with $F(.)$ replaced by the empirical CDF. The parameters are estimated from the slope and intercept of the resulting regression as in the previous paragraph.

For the three-parameter Weibull distribution, use the minimum observation $x_{(1)}$ to estimate the location parameter, plot $x_{(i)} - x_{(1)}$, $i = 2, \ldots, n$ on cumulative hazard paper and proceed as above.

Example 11.9 (Case 2.21; Weibull Distribution)
We use the regression approach, with plotting positions $\hat{F}(x_{(i)}) = (i - 0.5)/n$. Thus we regress $y_i = \log[-\log(1 - \hat{F}(x_{(i)}))]$ against $\log(x_i)$. The results of the regression analysis (from Minitab) follow.

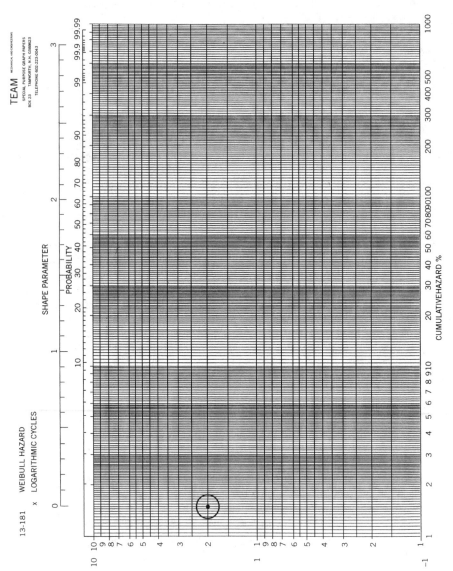

Figure 11.11. Aid for graphical estimation of Weibull parameters.

Regression Analysis. The regression equation is

$$y = -21.0 + 16.1 \log(x)$$

Predictor	Coef	Stdev	t-ratio	p
Constant	-20.9955	0.2769	-75.82	0.000
$\log(x)$	16.1140	0.2180	73.90	0.000

$$s = 0.08442 \qquad R\text{-sq} = 99.6\% \qquad R\text{-sq(adj)} = 99.5\%$$

Analysis of Variance

SOURCE	DF	SS	MS	F	p
Regression	1	38.925	38.925	5461.85	0.000
Error	24	0.171	0.007		
Total	25	39.096			

The estimate of α is the slope of the line, i.e., $\hat{\alpha} = 16.11$. β is estimated as $e^{21.00/16.11} = 3.68$. Note that these are in very close agreement with the MLEs, given in the annotation of Figure 11.4b as $\alpha^* = 16.58$ and $\beta^* = 3.68$.

Note that because of the nature of the data, the remaining output of the regression analysis can only loosely be interpreted in the usual way. The high R^2 is, however, indicative of a very good linear fit. ■

Lognormal Distribution. Here one may plot $\log x$ on normal plotting paper and proceed as for the normal distribution or the original data may be plotted on lognormal paper. In the latter case, μ is estimated as the 0.50-fractile and the slope of the line is an estimate of $\log(\sigma)$.

Other Distributions. Plotting procedures and methods of parameter estimation for the extreme value and gamma distributions are given by King (1971). Plotting papers are not available for other distributions of interest in reliability. SPSS and other programs give P-P plots for many distributions.

11.2.5 Plots for Censored Data

In reliability applications, data are often censored on the right; that is, failure times are known for some items, but only service times (which provide lower bounds on the time to item failure) are known for others. Probability and hazard plots are easily modified to deal with right-censored data. The procedure is to merge the failure and service times, calculate H in the usual way, but only for failure times (with n the total sample size), and plot the results on ordinary graph paper (which provides a nonparametric estimate of the CDF), probability paper, or hazard paper. In the latter two cases, linearity is again an indication of a good fit to the distribution selected for plotting. Parameters are estimated as indicated for complete data.

The survival function is also plotted in this way and various plotting positions are employed. These are also offered as options in several of the computer packages. See the references cited for details.

Example 11.10 (Case 2.2; Hazard Plots)
Case 2.2 deals with 747 windshields, with $n = 153$, of which 88 are failure times and 65 service times. Figure 11.12a gives plots of the data for four distributions. The plots were pre-

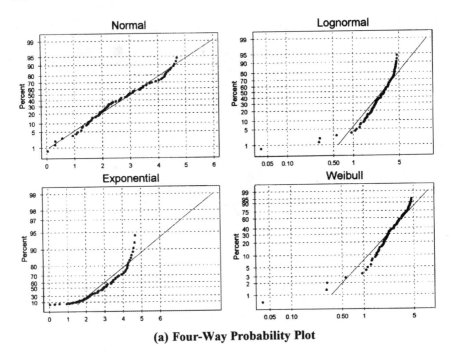

(a) Four-Way Probability Plot

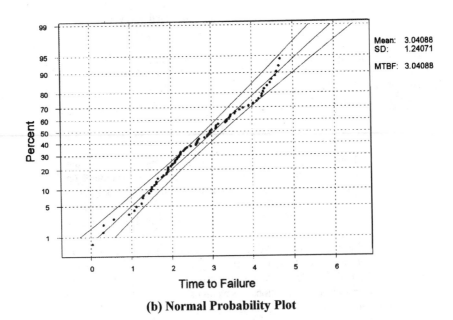

(b) Normal Probability Plot

Figure 11.12. Probability plots for TTF of windshields, incomplete data.

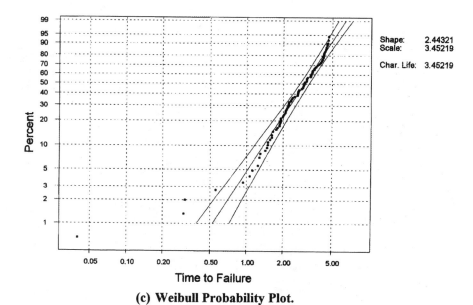

(c) **Weibull Probability Plot.**

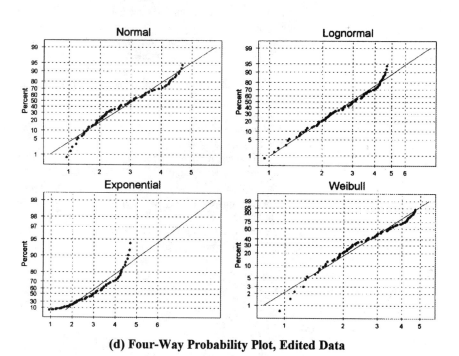

(d) **Four-Way Probability Plot, Edited Data**

Figure 11.12. *Continued.*

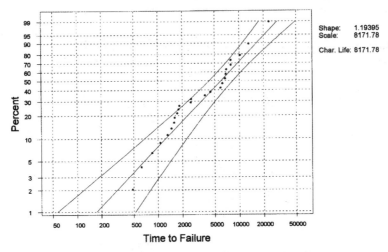

Figure 11.13. Weibull probability plot for throttle failure data.

pared by Minitab. Note that the fit appears fairly reasonable on the normal plot, with some tailing off in the right-hand tail. On the Weibull plot, the fit appears to be reasonable, except that there appear to be four or five outliers on the left-hand side, and again a tailing off on the right-hand side. This would suggest either a mixture of two or three Weibull distributions, or true outliers, or simply a poor fit. The lognormal and exponential are very poor fits. Figures 11.12b and 11.12c are the normal and Weibull plots showing more detail and giving parameter estimates, which are MLEs based on the incomplete data (see Chapter 8). Figure 11.12d is the four-way plot omitting four of the suspected outliers. The fits are not much improved. ■

Example 11.11 (Case 2.19; Weibull Plot)
In attempting to model the throttle failure data, which is a mechanical component, a Weibull distribution may be expected to provide a good fit. A plot of the data, which include 25 failure times and 25 censored observations, is given in Figure 11.13. As can be seen, the fit does not appear to be good. The points appear to cluster about two lines of the Weibull plot, suggesting a mixture of two Weibull distributions. This would result if two different failure modes were operative. Note that the second line (which appears to fit the points from about 2000 onwards) has a smaller slope, indicating a larger value of the shape parameter α than that of the first component of the mixture. The structure and analysis of these data are discussed in detail in Jiang and Murthy (1995a,b). ■

11.3   GOODNESS-OF-FIT TESTS

11.3.1   Basic Concepts

In the previous section, data were "fit" to specified distributions by plotting and observing the shape of the resulting plot. The analyst may look at the empirical CDF itself or determine whether the plot appears to be linear on the probability plotting papers for the several distributions for which these are available. In this section, we look at formal statistical tests of goodness of fit. The classical test of this type is a chi-square test, which is appropriate for discrete data or grouped continuous data. In the continuous case, much more powerful tests are available for a few specific problems. Included are procedures that test specifically for the normal, exponential, Weibull, and a few other distributions; and nonparametric tests that are used to test for any completely specified continuous distribution. We look at several of these procedures.[3]

In practice, in applying goodness-of-fit tests we encounter two types of problems: (1) those in which a specific distribution is suggested by theoretical considerations, past experience, and so forth, and we wish to determine whether or not a set of data follow this distribution; and (2) those in which we have only a vague idea of the correct distribution (e.g., that it is skewed right) and want to do some screening of possible candidate distributions. In the former case, it is usually appropriate to test at a low level of significance, say $\alpha = 0.01$, the idea being that we do not want to reject the theoretical distribution unless there is strong evidence against it. In the second case, it may be more appropriate to test at a much higher level of significance, say $\alpha = 0.1$ or 0.2, in order to narrow the list of candidates somewhat. As always, the choice is to a great extent subjective and depends on the particular application.

Actual selection of a distribution based on a sample of data is often a mixture of art and science. In most of the situations to be discussed below, several approaches are presented (with many more given in the references). Some of the suggested approaches are more powerful against specific alternatives and some are more powerful against others. If *any* of the tests rejects a particular distribution, particularly at $\alpha = 0.01$ or less, that distribution should be rejected as a candidate life distribution. The only exception might be if it appears to provide a good fit in a graphical analysis, but there is evidence of possible outliers. In that case, a distribution should not be rejected if a good fit is obtained after removal of the outliers.

A good rule of thumb is that results should be used cautiously unless a good fit is found, and perhaps even then unless it is validated (e.g., by analysis of one or more additional independent sets of data). This is discussed in Section 11.5. Another good rule of thumb whenever there is a fair amount of doubt, is to complete subsequent reliability analyses using whatever distributions appear to provide a reasonable fit, compare the results, and then use judgment as well the results of this analysis to arrive at an appropriate (tentative) decision.

11.3.2   Fitting Discrete Distributions; the Chi-Square Test

The classic chi-square test, due to Karl Pearson, may be applied to test the fit of discrete data to any specified distribution. The test statistic is

$$\chi^2 = \sum_{i=1}^{k} \frac{(O_i - E_i)^2}{E_i} = \sum_{i=1}^{k} \frac{O_i^2}{E_i} - n \qquad (11.8)$$

In the goodness-of-fit application of this test, we assume a sample of size n, with each observation falling into one of k possible classes. O_i is the observed frequency in class i; E_i is the frequency that would be expected if the specified distribution were the correct one. The classes are usually a set of integers, e.g., the possible values of the random variable involved, truncated or grouped at one or both ends if necessary (for example, in cases where the set of possible values is infinite). The E_i are calculated as np_i, where p_i is the probability assigned to Class i by the hypothesized distribution. Ordinarily $p_i = P(X = i)$, or the sum of such probabilities. In calculating χ^2, it is necessary to assure that expected frequencies are not too small; if they are, this may greatly distort the result. A rule of thumb is that if $E_i < 1$ for some i, combine that class with either the previous or the succeeding class, repeating the process until $E_i \geq 1$ for all i. A more conservative rule is to combine classes if $E_i < 5$.

Under the null hypothesis of a completely specified distribution (including parameter values), the test statistic has approximately a chi-square distribution with $(k-1)$ df. If r parameters are estimated from the data and the expected frequencies are calculated on this basis, the distribution of the test statistic is approximately a chi-square with $(k-r-1)$ df. The test is an upper-tail test; large values of χ^2 result when observed and expected frequencies differ significantly, so these lead to rejection of the hypothesized distribution.

We illustrate the technique by application to the binomial and Poisson distributions. Other discrete distributions are handled similarly.

Test of Fit to the Binomial Distribution

It is assumed that $x_1, x_2, \ldots, x_n$ is a sample of size n from a discrete distribution on $\{0, 1, \ldots, N\}$ and that we wish to test the null hypothesis that this distribution is a binomial with parameters N and p. The expected frequencies are $E_i = Np_i$ $(i = 1, \ldots, N+1)$, where

$$p_i = \binom{N}{i-1} p^{i-1}(1-p)^{N-i+1} \tag{11.9}$$

Here $k = N+1$ unless the E_i are too small at either or both ends of the range, in which case classes are combined. The O_i are the observed frequencies; these are calculated as the number of x's taking on the value $(i-1)$.

Example 11.12 (Goodness-of-Fit Test for the Binomial Distribution)

A machine punching out metal cabinet parts is intended to produce parts 20 cm long. As a quick check, four parts are selected from each hour's production and manually checked for length against a gauge. X_i is the number of parts that are identified as over 20 cm in the ith sample. If the machine is adjusted properly, half the parts should be over 20 cm and half should be under. Data for a three week period (120 samples) are summarized in Table 11.1. We wish to test the hypothesis that the data follow a binomial distribution with $p = 0.5$. Here $N = 4$ and no classes have to be combined, so $k = 5$. Table 11.1 also includes the calculated expected values. In this calculation, p_i is the binomial probability with $p = 0.5$ and $N = 4$.

The calculated value of χ^2 is

$$\chi^2 = \frac{14^2}{7.5} + \frac{38^2}{30} + \cdots + \frac{3^2}{7.5} - 120 = 11.86$$

with 4 df. From Table C3, the tabulated value for testing at the 5% level is is found to be 9.49. There is evidence, testing at the 5% level, that the data do not fit a binomial with $p = 0.5$.

Table 11.1. Observed and expected values, binomial example

i	$p_i = P(X = i - 1)$	$E_i = 120\,p_i$	O_i	$120\hat{p}_i$
1	0.0625	7.5	14	12.93
2	0.2500	30.0	38	38.55
3	0.3750	45.0	40	43.11
4	0.2500	30.0	25	21.42
5	0.0625	7.5	3	3.99

To fit a binomial with p not specified, we calculate the overall sample proportion, namely $\hat{p} = [38 + 2(40) + 3(25) + 4(3)]/480 = 0.4271$. The resulting expected numbers are given in the last column of Table 11.1, and the calculated χ^2 is 1.16 with 3 df (since one parameter was estimated from the data). In this case, the tabulated value at the 5% level is 7.81 and we conclude that there is no evidence of a poor fit. We may conclude that the data follow a binomial distribution, but not with $p = 0.5$.

Example 11.13 (Case 2.13; Goodness-of-Fit Test for the Poisson Distribution)
If data entry errors occur randomly and independently, then the number of errors per unit of time will follow a Poisson distribution. We check this for Clerk 4 with the "time unit" being 500 data entries. The frequencies of numbers of occurrences per 500 entries (O_i) for Clerk 4, along with the corresponding expected frequencies (E_i) under the Poisson assumption with λ estimated from the data are

Number of errors:	0	1	2	3	4	5
Frequency:	7	14	8	6	0	1
Expected frequency:	8.26	12.16	8.95	4.39	1.62	0.62

Here $n = 36$ and we estimate the mean by $\bar{x} = 1.4722$. The expected frequencies are calculated as $E_i = 36(1.4722)^{i-1}e^{-1.4722}/(i - 1)!$, $i = 1, \ldots, 5$, and $E_6 = 36 - (8.26 + 12.16 + \cdots + 1.62)$. Since the last expected number is less than one, we combine the last two groups, getting $E_5 = 2.24$ and $O_5 = 1$ for the combined group. The calculated value of χ^2 is

$$\chi^2 = \frac{7^2}{8.26} + \cdots + \frac{1^2}{2.24} - 36 = 1.848$$

with $(k - 1) - 1 = 3$ df. We test at the 1% level. From Table C3, the tabulated value is found to be 11.34, and the hypothesis of a Poisson distribution is not rejected. ∎

11.3.3 Fitting Continuous Distributions, General Approaches

Chi-Square Tests, Continuous Distributions
For large samples, the approach of the previous section may be used as well. In applying the χ^2 analysis, the data are grouped into k intervals as in forming a frequency distribution and the expected frequencies are calculated as np_i, where

$$p_i = \int_{L_{i-1}}^{L_i} f(x)dx \tag{11.10}$$

where $[L_{i-1}, L_i)$ is the ith interval, with $L_0 = -\infty$ and $L_k = \infty$. As before, the intervals must be chosen so that $E_i \geq 1$ (or 5), $i = 1, \ldots, k$. If the distribution $f(.)$ is completely specified, the resulting χ^2 has $(k - 1)$ df. As before, if r parameters are estimated from the data, df are $k - r - 1$.

This χ^2 test has the advantages of being easy to apply and being applicable even when parameters are unknown. However, it is not a very powerful test and is not of much use in small or sometimes even modest size samples.

Much more powerful tests are available for completely specified distributions F. We look at two of these, the *Kolmogorov–Smirnov test* and the *Anderson–Darling test*. Both of these tests are nonparametric in the sense that the distribution of the test statistic does not depend on F. Modifications of these tests when parameters are unknown have been developed for a few distributions. In this case the nonparametric property no longer holds.

The Kolmogorov–Smirnov (K-S) Test
The null hypothesis being tested is H_0: $F = F_0$. The test statistic D_n is simply the maximum distance between the hypothesized CDF and the empirical CDF. This is most easily calculated as $D_n = \max\{D_n^+, D_n^-\}$, where

$$D_n^+ = \max_{i=1,\ldots,n}\left[\frac{i}{n} - F_0(x_{(i)})\right], \quad D_n^- = \max_{i=1,\ldots,n}\left[F_0(x_{(i)}) - \frac{i-1}{n}\right] \tag{11.11}$$

Fractiles of the distribution of D_n were given by Massey (1951). A very close approximation to these fractiles that is a function only of n and tabulated constants d_α is given by Stephens (1974). The constants d_α are given in Table 11.2 for $\alpha = 0.15, 0.10, 0.05$, and 0.01. The critical value of D_n is calculated as $d_\alpha/(n^{1/2} + 0.11n^{-1/2} + 0.12)$. Values of D_n in excess of the critical value lead to rejection of the hypothesized distribution.

Example 11.14 (Case 2.7; K-S Test of Normality)
Suppose that it is hypothesized that time to failure is normally distributed with $\mu = 150$ and $\sigma = 50$. We calculate the K-S test of this hypothesis for Vehicle LHD-1, testing at the 1% level. The calculations require the evaluation of the hypothesized distribution $F_0(\ )$ for each of the ordered data points. The value of the test statistic is gotten as the maximum of the two quantities in (11.11). For these data, we find $D_n^+ = 0.5308$ and $D_n^- = 0.1684$, so $D_n = 0.5308$. The critical value at $\alpha = 0.01$ and $n = 23$ is gotten from Table 11.2 as $1.628[23^{1/2} + 0.11(23)^{-1/2} + 0.12] = 0.3296$. Since D_n exceeds this value, the hypothesized distribution is rejected at the 1% level. ∎

The Anderson–Darling (A-D) Test
The Anderson–Darling test is also based on the difference between the hypothesized and empirical CDF's. The test statistic is

$$A^2 = A_n^2 = \frac{-1}{n}\sum_{i=1}^{n}(2i - 1)\{\log F_0(x_{(i)}) + \log[1 - F_0(x_{(n-i+1)})]\} - n \tag{11.12}$$

Fractiles a_α of the distribution of A_n^2 are also given in Table 11.2. If the calculated value of A_n^2 exceeds a_α, the hypothesized distribution is rejected at level of significance α. Note that the critical value a_α does not depend on n. Although this is an asymptotic result, it has been found to be a very good approximation in samples as small as $n = 3$.

Table 11.2. Factors for calculating critical values for the Kolmogorov–Smirov and Anderson–Darling tests

α	0.15	0.10	0.05	0.01
d_α	1.138	1.224	1.358	1.628
a_α	1.610	1.933	2.492	3.857

Example 11.15 (Case 2.17; A-D Test of Normality)
We analyze the data discussed in Example 11.13 using the Anderson–Darling test, again testing at the 1% level. The summation in (11.12) is found to be -1197, giving the value of the test statistic as $A_n^2 = 1197/23 - 23 = 29.04$. The critical value, from Table 11.2 with $\alpha = 0.01$, is $a_n = 3.857$. The hypothesized distribution is rejected at the 1% level. ■

Many additional details regarding these tests, a number of additional tests, and modifications of these tests for censored data are discussed in Lawless (1982) and D'Agostino and Stephens (1986). Additional results regarding incomplete data are given by Guilbaud (1988), which deals with the Kolmogorov–Smirnov test for censored data, and Hollander and Peña (1992), which deals with chi-square tests with random censoring.

11.3.4   Fitting Specific Continuous Distributions, Parameters Unknown

The goodness-of-fit problem is much more difficult when the parameters of the distribution are not specified in the null hypothesis, unless the sample size is large and the chi-square test is used. In this case, incidentally, *minimum chi-square* estimators, that is, those that minimize the joint distribution expressed in terms of the p_i, given in Equation (11.10), should be used rather than the moment or maximum likelihood estimators. If the MLEs are used, the resulting statistic is not distributed as χ^2.

A natural approach when H_0 specifies only the form of the distribution would be to use some method of estimating the parameters (e.g., maximum likelihood) and then apply the test statistics of the previous section (or others given in the references). The problem is that the tests are no longer nonparametric in this case, and the distribution of the test statistic is very complicated, depending on the statistic, the estimation procedures used, the hypothesized distribution, and the true distribution.

Three basic approaches to the problem have been proposed. These are:

1. Embed the distribution in question in a larger family of distributions and test the null hypothesis that the parameter values are those that reduce the distribution to the one in question. For example, the Weibull distribution reduces to the exponential when the shape parameter $\alpha = 1$, so a test for the exponential can be done by estimating the Weibull parameters and testing H_0: $\alpha = 1$. A difficulty that has been encountered in using this approach is that the test may be powerful only against alternatives in the same family (e.g., other Weibull distributions). This approach is discussed further in Section 11.4.

2. Determine the asymptotic distribution of the test statistic and use these results to obtain critical points. This has been accomplished for only a few distributions and has the added disadvantage that the results may not be appropriate for small samples.

3. Use the tests discussed in the previous section, but use estimates of the unknown parameters and modify the critical values accordingly. This leads to some very difficult analytical problems, and the modified critical values are usually determined by means of simulation studies. A number of published tables of the results are given or cited in the references previously cited. Quite simple modifications are provided for a number of distributions and these have been found in many cases to be very good approximations, even for quite small samples.

In the remainder of this section, we look at some specific distributions, giving tests and either critical values or references to sources where they may be found. See Lawless (1982) and D'Agostino and Stephens (1986) for additional details and references. For each distribution considered, we give results for the case when all parameters are unknown, which is the usual situation in most applications. For test procedures and tables of critical values in cases where some parameters may be known, see the references cited.

One or more of the three approaches discussed above will be used, with emphasis on the modified K-S and A-D tests. We begin with tests for normality since this is the most widely studied distribution.

Test for Normality

Many tests for normality have been developed. Here we look at modifications to the K-S and A-D tests when μ and σ are unknown and are estimated by $\bar{x}$ and s.

K-S Test. Modified critical values of D_n for small n were first obtained through simulation studies by Lilliefors (1967 and 1969). Improved values, based on more extensive simulations, are given by Mason and Bell (1986). The authors also give critical values for $n \geq 30$ as $d_\alpha (n^{1/2} - 0.01 + 0.83n^{-1/2})^{-1}$, where d_α is given in Table 11.3. (In fact, the values are a good approximation (within 0.001) to the tabulated critical values if $n \geq 10$.) The hypothesis of normality is rejected if D_n exceeds the critical value.

A-D Test. Critical values for the A-D test are calculated as $a_\alpha(1 + 4n^{-1} - 25n^{-2})^{-1}$, where a_α is given in Table 11.3 (see Stephens, 1974).

These are but a few of the many tests for normality that have been devised. Critical values for several additional tests are given by D'Agostino and Stephens (1986, Chapter 4). See Lawless (1982) and the references cited therein for further information and many additional tests.

Example 11.16 (Case 2.1; Tests of Normality)

The bond strength data of Table 2.1a have been extensively analyzed. In examples 11.1 and 11.2, it was noted that the graphical analysis strongly suggested a normal distribution, except for one or more possible outliers. Here we test this assumption, estimating the parameters from the data, with and without the minimum observation, which was identified as suspect. For the complete data set, we have $n = 85$, $\bar{x} = 278.9$, and $s = 41.96$. Omitting the minimum, we have $n = 84$, $\bar{x} = 280.7$, and $s = 38.97$. The estimated CDF is evaluated at each data point using these sets of values. The resulting values of the test statistics for the modified K-S and A-D tests (along with critical values for testing at the 5% level) are as follows:

	$n = 85$	$n = 84$
K-S	0.101 (0.0962)	0.0965 (0.0968)
A-D	0.548 (0.754)	0.495 (0.754)

The critical values are calculated using the values given in Table 11.3. For example, the critical value at the 5% level for the modified K-S test with $n = 85$ is $0.895/(85^{1/2} - 0.01 + 0.83(85)^{-1/2}) = 0.0962$. Note that the K-S test rejects normality at the 5% level for the com-

Table 11.3. Factors for calculating critical values for fitting the normal distribution with unknown parameters

α	0.20	0.15	0.10	0.05	0.01
d_α	0.741	0.775	0.819	0.895	1.035
a_α	—	0.576	0.656	0.787	1.092

plete sample, but not for the edited data. Normality would not be rejected at the 1% level by any of these tests, and this suggests that normality is a reasonable assumption for both cases.

Test of Fit to the Exponential Distribution

To test for exponentiality in the single-parameter case, the approach based on embedding in a larger class of distributions is particularly straightforward. Either the Weibull or gamma distributions can be used. In either case, the test is of H_0: $\alpha = 1$. MLEs of the shape and scale parameters may be used and the test performed as indicated in Section 5.6.

A more general approach is to use the modified K-S or A-D test, with the CDF estimated using the MLE λ^*. These are applied as follows.

K-S Test. The critical value for the K-S statistic is given by Mason and Bell (1986) as $0.2n^{-1} + d_\alpha(n^{1/2} + 0.26 + 0.5n^{-1/2})^{-1}$, where d_α is given in Table 11.4.

A-D Test. For the modified A-D test, the critical values, obtained by Stephens (1974), are calculated as $a_\alpha(1 + 0.6n^{-1})^{-1}$, with a_α given in Table 11.4. Again, the hypothesized distribution is rejected if the calculated value of the test statistic exceeds the critical value.

Goodness-of-fit tests for the two-parameter case are discussed by D'Agostino and Stephens (1986).

Example 11.17 (Case 2.1; Test for the Exponential Distribution)

To illustrate the procedure, we test for goodness-of-fit of the data of Table 2.1a to the exponential distribution, using the modified K-S test and the complete data set, and testing at the 1% level. The test is based on calculating the CDF using the MLE of λ, namely $\lambda^* = 1/\bar{x} = 0.0035855$. Thus the estimated CDF is $\hat{F}(x_{(i)}) = 1 - e^{-0.0035855x_{(i)}}$. The value of the test statistic is found to be $D_n = \max(0.2751, 0.5048) = 0.5048$. The critical value is $10308/9.5338 = 0.1372$. As expected, the exponential distribution is rejected at the 1% level. The normal distribution is a reasonable fit to these data and the exponential is a very bad fit, as can be seen in Figure 11.3(a). ■

Test of Fit to the Extreme Value Distribution

We consider the extreme value distribution with CDF given in Equation (4.53) (with, for clarity, α and β replaced by a and b). To estimate the parameters, use the fact that if X has this distribution, then $Y = e^X$ has a Weibull distribution with parameters $\alpha = 1/b$ and $\beta = e^a$. Calculate MLEs α^*, β^* on the transformed (Y) scale from Equations (5.53) and (5.54), transform to $b^* = 1/\alpha^*$ and $a^* = \log \beta^*$, and calculate the empirical CDF based on these estimates. The two modified tests are:

K-S Test. To use asymptotic results, large samples ($n > 100$) are required. In this case, calculate the critical value as $d_\alpha n^{-1/2}$, where d_α is given in Table 11.5. For small n, the critical value depends on the value of the shape parameter. Tables of critical values as functions of the MLE of this parameter are given by Chandra et al. (1981) and Woodruff et al. (1983). These should be used for tests when $n < 100$.

Table 11.4. Factors for calculating critical values for fitting the exponential distribution with unknown parameter

α	0.20	0.15	0.10	0.05	0.01
d_α	0.861	0.926	0.990	1.094	1.308
a_α	—	0.922	1.078	1.341	1.957

Table 11.5. Factors for calculating critical values for fitting the extreme value and Weibull distributions with unknown parameters

α	0.25	0.10	0.05	0.01
d_α	—	0.803	0.874	1.007
a_α	0.474	0.637	0.757	1.038

A-D Test. The critical values, apparently good for any n, are given by $a_\alpha (1 + 0.2n^{-1/2})^{-1}$, where a_α is given in Table 11.5 (Stephens (1977)).

Test of Fit to the Weibull Distribution

For the two-parameter Weibull distribution, calculate the MLEs α^* and β^*, then transform to the log scale to obtain the extreme value distribution and fit using the K-S or A-D statistic as above.

An alternative approach, found to be more powerful in many cases, is given by Mann et al. (1973). The test statistic is based on the normalized sample spacings (calculated as differences between successive order statistics) and is given by

$$M = \frac{\left[\frac{n}{2}\right] \sum_{i=[n/2]+1}^{n-1} L_i}{\left[\frac{n-1}{2}\right] \sum_{i=1}^{[n/2]} L_i} \tag{11.13}$$

where $[x]$ is the smallest integer less than or equal to x, and

$$L_i = (X_{(i+1)} - X_{(i)})/(M_{i+1} - M_i) \tag{11.14}$$

The M_i are expected sample spacings and have been extensively tabulated by Mann et al. (1973). An approximation, good for $n \geq 10$, is given by

$$M_i \approx \log\left[\log\left(1 + \frac{i - 0.5}{n + 0.25}\right)^{-1}\right] \tag{11.15}$$

[See also Lawless (1982).] For small n and for censored samples, critical values are given by Mann et al. For $n \geq 20$, the distribution of M is approximately an F distribution with numerator and denominator df given by $\nu_1 = 2[(n-1)/2]$ and $\nu_2 = 2[n/2]$, respectively (where "[ ]" indicates integer part, as before). Fractiles of the F distribution are given in Table 9 of Appendix C.

Many additional results, including tests for the three-parameter Weibull, analysis of censored data, and many other tests are given by D'Agostino and Stephens (1986) and Lawless (1982). For the three-parameter Weibull, also see Lockhart and Stephens (1994).

Example 11.18 (Case 2.21; Fit to Weibull and Exponential Distributions)

In Example 11.3, the Weibull distribution appeared, from the graphical analysis done, to provide a good fit to the breaking strengths of 12 mm fibers. Here we perform statistical tests of fit to the Weibull distribution and the use of the Weibull to test for exponentiality. Specifically, we look at (1) the K-S test, (2) the A-D test, (3) the Mann et al. test, and (4) a test of H_0: $\alpha = 1$ versus H_a: $\alpha \neq 1$. The results are

1. The analysis is performed on e^x, where the x-values are the data for the 12 mm fibers, given in Table 2.21. The MLEs for the parameters of the Weibull distribution (gotten from Minitab) are $\alpha^* = 16.28$ and $\beta^* = 3.681$. The corresponding parameter estimates for the extreme value distribution are $a^* = \log(\beta^*) = 1.303$ and $b^* = 1/\alpha^* = 0.06143$. Thus $\hat{F}(x_{(i)}) = 1 - \exp(-e^{(x_{(i)}-1.303)/0.06143})$. The value of the test statistic is found to be 0.04865. Note that for this sample, $n = 26$, so that the approximate critical value (calculated to be $1.007(26)^{-1/2} = 0.198$ for testing at the 1% level) may not be applicable. The observed value of the test statistic is very small, however, and would not lead to rejection of the hypothesis of an extreme value distribution for the transformed data or, equivalently, of a Weibull distribution for the original data.

2. The A-D test is performed similarly. The value of the test statistic is found to be $A^2 = 0.0772$. The critical value (which is appropriate in this case) at $\alpha = 0.01$ is $1.038/[1 + 0.2(26)^{-1/2}] = 0.5139$. The Weibull distribution is not rejected.

3. To calculate the test statistic of Mann et al. (1973), we require the M_i of (11.15) and the L_i of (11.14). These are easily calculated on a computer, and lead to the value of the test statistic of $M = 0.3069$. The corresponding critical value is obtained from the F distribution with 24, 26 df. By use of Minitab, the critical value at the 1% level is found to be 1.95. Again, the hypothesis of a Weibull distribution is not rejected, testing at the 1% level. (The p value for the test, also gotten by use of Minitab, is 0.998.)

4. A test of exponentially distributed data within the Weibull family is performed by testing H_0: $\alpha = 1$, as noted. We use the asymptotic normality of the estimator to perform the test. The point estimate of α is 16.28. The approximate variance of the estimate is found from (5.55) to be

$$\hat{V}(\alpha^*) \approx \frac{1.1087\beta^{*2}}{n\alpha^{*2}} = \frac{1.1087(3.681)^2}{26(16.28)^2} = 0.002180$$

giving an estimated standard deviation of $\sqrt{0.002180} = 0.04669$. A confidence interval at any level would not contain the hypothesized value of $\alpha = 1$, and the exponential distribution is rejected. ∎

Test of Fit to the Lognormal Distribution

Transform to $Y = \log X$ and use any of the tests for normality.

Example 11.19 (Case 2.7; Fit to the Lognormal Distribution)

In Examples 11.14 and 11.15, the data for Vehicle LHD1 were fit to a normal distribution with parameters specified as $\mu = 150$ and $\sigma = 50$. The fit was found to be very poor. For these data, the sample mean and standard deviation are $\bar{x} = 108.5$ and $s = 133.0$, which are not only quite different from those of the hypothesized distribution, they are not tenable values for a normal distribution as an approximate distribution of lifetimes, since a significant segment of the distribution would correspond to negative values. In fact, the modified K-S and A-D tests of normality (obtained from Minitab) lead to p values of $p < 0.01$ and $p = 0.000$, respectively.

As an alternative, we fit a lognormal distribution. Transforming to $y = \log(x)$, we find the sample mean and standard deviation to be 3.908 and 1.369, respectively. The test of normality on the transformed scale results in values of the test statistics of $D_n = 0.103$ and $A^2 =$

0.318, with p values (again from Minitab) of $p > 0.15$ and $p = 0.514$, respectively. The log-normal distribution is not rejected.

Test of Fit to the Gamma Distribution
Use MLE to estimate the parameters. Critical values for the A-D and other tests depend on the value of the shape parameter. For tables of critical values as functions of level of significance and the value of the shape parameter, see D'Agostino and Stephens (1986) and Woodruff et al. (1984).

Test of Fit to the Inverse Gaussian Distribution
Modified procedures for use when MLEs are used to estimate the parameters have been developed for the K-S test by Edgeman et al. (1988) and for the A-D test by Pavur et al. (1992). The critical values, obtained by simulation, are expressed as functions of n and the estimated scale parameter. Tables are given in the references.

11.4   MODEL SELECTION

The selection of a model in the context of reliability is the same as in any other application: it is a difficult problem and it involves art as well as science. If the modeler (analyst, engineer, manager, or other expert) truly understands the underlying structure, a model may be built on basic principles. More often than not, however, this is not the case, and the modeling process proceeds iteratively through many steps, a number of which involve data of many types. Each step in the process leads to an incremental improvement in the model and its relation to the real world.

In reliability applications, modeling involves representation of two phenomena. The first is the basic reliability structure itself—series, parallel, combination systems, and so forth; independent or interacting components. This is usually reasonably well understood, with the possible exception of the existence and nature of interactions. The second is the life distribution of the item (part, component, etc.). This, too, may be determined on the basis of an understanding of the underlying process, i.e., the physical basis of failure. Again, more often than not, data-based decisions are involved. In this section, we look at some procedures for making such decisions.[4]

11.4.1   Selection of a Life Distribution

In the previous two sections of this chapter, we have looked at various graphical and statistical methods of modeling and fitting life distributions, and have discussed various data-based techniques for *rejecting* a particular model. In practice, a problem is that failure to reject a hypothesized distribution does not imply its acceptance. (This is the nature of statistical hypothesis testing.) In fact, it frequently happens that any of a number of distributions will provide an adequate fit to the data. The question is, then, which shall we use?

In many cases in classical statistical inference, e.g., in the use of asymptotic results for confidence intervals and other inferences, it does not matter much. The procedures are quite robust, so that the actual confidence level is quite close to the nominal, and whether the nominal 95% confidence interval is, in fact, 94%, 96%, or even 91% is often not important from a practical point of view. In reliability studies, however, the difference between 95% and 94% is often critical—0.94 reliability means 1.2 times as many failures than does a reliability of 0.95. If this means 20% more warranty claims, the impact on profitability can be very substantial. The difference between 0.995 and 0.990 is a doubling of the failure rate that could be disastrous in a space application or nuclear power plant.

As is apparent, in many reliability applications we are primarily concerned with the tails of the distribution, wishing to know, for example, the number that will survive beyond a certain extreme point or the point beyond which at least 99.9% will survive. Unfortunately, the tails of the distribution are most sensitive to distributional assumptions.

There is no definitive answer as to how a life distribution should be selected. A reasonable approach is the following:

- Perform good-of-fit tests to all feasible life distributions, estimating parameters from the data, testing for fit at the 1% level (or some alternative low level of significance). Any distribution rejected at this level should be considered an unacceptable model.

- Any distribution not rejected should be considered to be a reasonable candidate model, and should be analyzed further. If a physical model that accounts for the failure mechanism underlying a candidate distribution can be deduced and the parameters can be assigned a physical meaning, this model should be accepted.

- Use Bayesian methods where appropriate to incorporate all relevant information into the decision process.

- If two or more candidate models fit the data and no other information is available, accept that all of these are possible models and check on the sensitivity of any estimators (e.g., of model parameters, reliability, etc.) to the distributions.

- Calculate confidence limits for each parameter or other characteristic of interest under each distributional assumption.

- Compare the results and do a worst-case analysis. This involves using the maximum upper confidence bound and minimum lower confidence bound for each parameter and considering these extremes as real possibilities. Make reliability decisions accordingly.

- If these extremes are not acceptable, accept the fact that more data and/or other information will be required to arrive at a rational decision concerning the reliability of the item.

- Design and conduct an experiment that will provide the additional data needed, analyze the data, and continue the modeling process.

In the remainder of this section, we elaborate on some of these points, revisit some of the statistical techniques, and illustrate the process by use of some of the data sets analyzed previously. The well-known observation that there are no correct models, but there are some useful ones, certainly holds here. Another basic principle that should be applied is that of parsimony. There is no point in formulating a complex model involving many variables and parameters if, for all practical purposes, a much simpler, more understandable, and more easily statistically estimable model will do as well.

In some cases, the analysis is relatively straightforward. These "easy" problems have some or all of the following characteristics:

- The theoretical structure underlying the failure mechanism is well understood. This includes materials properties, physics of failure, geometries and their impact on failure, if relevant, and so forth.

- In addition to this body of knowledge, there is a substantial body of empirical evidence supporting the theoretical structure and conclusions. This includes a history of careful experimentation and measurement as well as compilation and proper statistical analysis of data.

- Component, subsystem, and system models have been carefully developed and confirmed.
- The models and experimental data have been validated under many conditions and the conclusions have been replicated and confirmed.
- Conclusions other than those at which the investigators have arrived have been ruled out as extremely less likely explanations for the phenomena observed.

"Hard" problems in reliability modeling have some or all of the following characteristics:

- The underlying theory of failure is poorly understood.
- The technology is new and background and experience in dealing with it are not well developed.
- The data for testing most aspects of the theory are not available or are of doubtful quality.
- Experimentation has been done haphazardly, if at all.
- The measurement process, including selection of variables and instrumentation to measure and record them, is not at the level required to provide the necessary information.

Without doubt, these lists could be expanded and depend on the nature of the application. For example, software considerations are quite different in many respects; simple products may be much easier to analyze (but the analysis may not have been considered important and therefore not have been done); and so forth. Furthermore, there are the usual two types of modeling errors that can be made here—selecting a model that is too simple for the purpose intended, or selecting one that is too complex. There are also the usual Type I and Type II statistical errors (see Chapter 5), which are the two problems of "bad" data—data that fail to reject a model that is incorrect and those that lead to rejection of a correct model.

Again, the modeling problem is a difficult one. By the very nature of what we are doing in reliability, statistics, engineering, management, and many other disciplines, there will always be uncertainty, variability, and randomness. This has been a recurring theme of this book. The two key issues that must be dealt with are (1) how much uncertainty can we tolerate, and (2) how much can we afford to spend to reduce uncertainty, realizing that it can never be eliminated entirely and that there are definitely diminishing returns for efforts expended.

We look at several examples of modeling failure times, including an "easy" problem and a "hard" problem. A number of additional examples are included in the exercises. Attention is restricted to complete data. Incomplete data can also be analyzed, but the problems are more difficult. Some results are given in the references cited previously. We note that these examples are illustrative of the types of data and situations that are encountered in practice, but no attempt is made to present a complete and thorough analysis along the lines suggested above.

Example 11.20 (Case 2.13)
In previous analyses of the data on clerical errors, in Examples 11.4 and 11.8, it was noted that errors should occur randomly, so that an exponential distribution of time to failure was expected. This expectation is based on the interpretation of "random" to mean independent occurrences and a constant failure rate. The probability plots for the exponential distribution

given in Figures 11.5a and 11.5b for Clerks 1 and 5, respectively, appear to indicate a good fit, except for a possible outlier. The four-way plots in Figure 11.6 tend to confirm this. The normal and lognormal distributions give a bad fit, and the Weibull fit is very similar to that of the exponential. Weibull plots showing more detail and providing estimates of the Weibull parameters are given in Figure 11.14. The estimated values of the shape parameter of 1.10 and 1.04 are very close to the exponential value of $\alpha = 1$. In fact, $\alpha = 1$ would not be rejected by a statistical test in either case. Goodness-of-fit tests also do not reject the exponential.

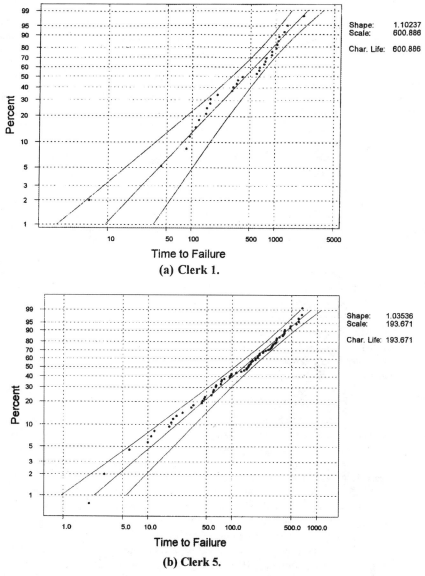

Figure 11.14. Weibull probability plots for time between errors.

In this example, we have an "easy" problem. There is a logical basis for the assumption of exponentiality, the data support this, and other conclusions are not suggested by the results and would violate the principle of parsimony. The point in this regard is that the Weibull, gamma, three-parameter Weibull and gamma, two-parameter exponential, generalized gamma, and many other distributions would also fit the data, but would be unnecessarily complicated models. There are additional considerations that might be pursued, however. For example, Clerks 1 and 5 appear to have quite different failure rates, and the remaining clerks may be compared as well. The question is that of whether these clerks (the systems in this case) are to be treated separately or jointly. The modeling problem becomes more complex in the latter case.

Finally, note that this case may be taken as a prototypical example of the exponential. The conclusions are applicable to a broad range of electronic and a few electro-mechanical items that have constant failure rates. ■

Example 11.21 (Case 2.14)

The ball bearing failure data given in Table 2.14, originally given by Leiblein and Zelen (1956), have been analyzed by many authors, e.g., Lawless (1982) and Richards and McDonald (1987). An immediate difficulty in analysis of the data is that several types of failure modes are indicated. This suggests the possibility of a mixture of distributions as an appropriate model. A second problem is that the sample size is relatively small, making it very difficult to estimate the parameters of a mixture. Instead, in previous analyses fits to single-component distributions were studied, with the Weibull and lognormal distributions providing good fits to the data.

In analysis of these data, both x and $\log(x)$ have been considered. Four-way plots of both are given in Figure 11.15. The plots support the claimed fits. Note that the Weibull distribution appears to fit $\log(x)$ as well. The A-D statistic for testing fit to the Weibull (on the original scale) is $A^2 = 0.329$, which is not significant at the 5% level. Fitting $\log(x)$ to the normal distribution gives $A^2 = 0.186$, also not significant. Beyond this, Richards and McDonald (1987) found that *ten additional* failure distributions out of a total of 16 tried, including the gamma and generalized gamma and beta distributions, fit the data equally well. In addition, a number of distributions fit $\log(x)$ as well as does the Weibull. Because of the nature of the application, there may be some physical justification for the Weibull distribution, but here many others must be considered as possible models as well. The real problems are as indicated previously—multiple failure modes and a small sample. Here the modeling issue may well be unresolvable. ■

Example 11.22 (Case 2.9)

This case involves recording lifetimes of light bulbs for samples of size 10 over a period of 42 weeks (with three missing observations), for a total n of 417. If the process is stable, a sample of this size should give a good indication of the failure distribution. A four-way plot of the complete data set is given in Figure 11.16a. The normal distribution appears to provide the best fit, but note the presence of an apparent outlier in all of the plots. The sample mean is 1045.85; $s = 190.585$. The outlying value is 225, which is 4.31 standard deviations below the mean. This is a very unlikely value ($p = 0.003$), even in samples of this size and on this basis, we eliminate the outlier from the data set. Normal and Weibull probability plots of the remaining 416 observations are given in Figures 11.16b and c, respectively. Inspection of the plots indicates that the normal provides a much better fit. Neither the K-S nor the A-D tests reject normality. Here there is no apparent physical explanation of normality, but in view of the size of the data set, the use of this distribution as a good working model is justified. ■

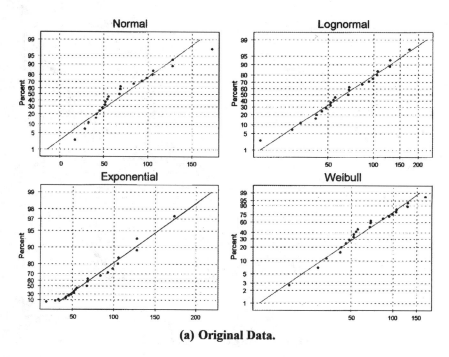

(a) Original Data.

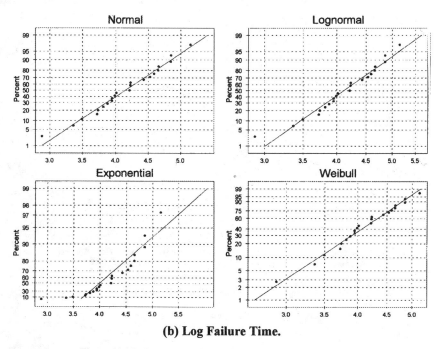

(b) Log Failure Time.

Figure 11.15. Four-way probability plots, ball bearing failure data.

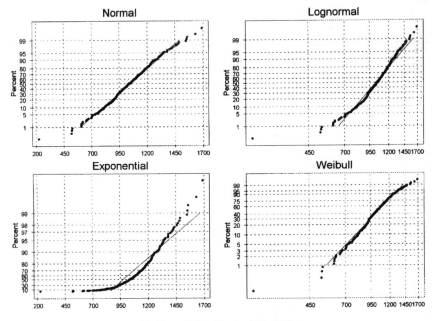

(a) Four-Way Probability Plot.

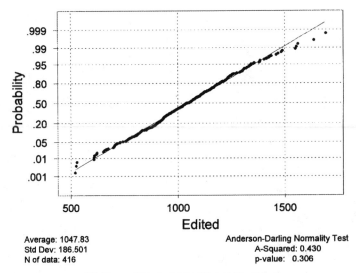

Average: 1047.83
Std Dev: 186.501
N of data: 416

Anderson-Darling Normality Test
A-Squared: 0.430
p-value: 0.306

(b) Normal Probability Plot, Edited Data.

Figure 11.16. Probability plots of TTF for light bulbs.

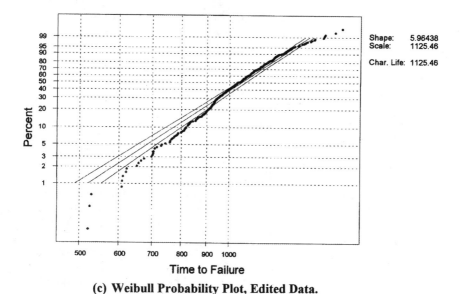

Time to Failure

(c) Weibull Probability Plot, Edited Data.

Figure 11.16. *Continued.*

11.4.2 Reliability Models

Selecting structural models of reliability—series, mix of series and parallel components, etc.—presents somewhat different challenges from those relating to selection of life distributions. Reliability modeling in this sense has been discussed in detail in earlier chapters. It involves more in the way of engineering analysis, whereas the issues addressed in the previous section involved more in the way of statistical anlysis. Engineering input is particularly important in system design and in understanding operational aspects of the system, for example, whether or not components can be assumed to operate independently, and, if not, the nature of any dependencies. In addressing these types of problems, both analytical and computer anayses are often required, and sensitivity studies (to be discussed briefly below) are very important.

In collecting and analyzing data in this context, many of the statistical approaches discussed in this chapter and previously play an important role. For example, after product design and modeling, testing is required. This may include testing of parts, components, etc., and the methods of statistical modeling, including selection of life distributions and estimation of parameters are employed. These provide the essential information for decisions regarding product evaluation, design changes, cost and risk analysis, and related issues.

Other statistical methods discussed previously—analysis of residuals, tests of independence, Bayesian methods, etc., are also relevant. The approach to be taken is to aggregate all relevant data and other information, complete the engineering analysis, design appropriate experiments, and do a careful and thorough analysis of the resulting data.

11.4.3 Interpretation of Lack of Fit: Alternative Models

Earlier in this chapter, we discussed the interpretation of graphs and the results of related goodness-of-fit tests. If the plot is linear on a scale designed for a given distribution and the test statistic is not significant at an appropriate level, the fit to that distribution is good. If the plot is highly nonlinear and the test statistic is significant, the fit is not acceptable. The problem is to determine the implication of plots that fall into neither of these categories; for example, a plot that is nearly linear, but not quite, and the test statistic is not significant.

For purposes of model selection, all such models must be considered as reasonable candidates and analyzed further. One possible explanation for nonlinearity is the presence of outliers. Statistical tests for this may be found in many introductory and intermediate statistical tests. As noted previously, other possible explanations include mixed distributions, indicating multiple failure modes, and simply the fact that, in data analysis, the outlying regions of the plot are the least well determined. In these cases, further engineering and statistical analysis is required.

Another possible explanation that has been suggested for the Weibull and gamma for which the plot exhibits a curve at the bottom is the possibility of a threshold effect. Often a third parameter (or second, in the case of the exponential) is included to account for this (see Wolstenholme, 1996 and Abernethy et al., 1983).

Here and in the previous section, much of the concern has been with situations in which too many models fit, or nearly fit. The other important concern, underlying some of the comments above, is what to do if none fit. Many additional models are given in Chapter 4 that have not been dealt with explicitly here. Some comments and references with regard to these and others are given in the Notes at the end of the chapter.

11.4.4 Sensitivity Studies

In cases where a number of candidate models appear to fit the data, and these cannot be distinguished on other bases (e.g., a physical theory), a careful analysis of the implications of the various alternatives in the reliability context is required. This may be accomplished, at least in part, by means of sensitivity studies. Here we look at a few of the many approaches to this type of analysis.

In setting up a sensitivity study, one or more key reliability objectives and/or assumptions are selected for analysis. These may be assumptions of independence of the elements of a system, life distributions, parameters of the life distribution, or any of many other characteristics. Objectives may be estimation of reliability to within a specified level of error, estimation of a MTTF, comparison of several engineering designs, and so forth. The objective of the sensitivity study is to determine which, if any, of the variables, assumptions, or other factors affects the outcome and to what extent. The study itself may involve mathematical analysis or numerical methods. Most often, it is the latter or a combination of the two. Some typical situations and approaches are the following:

- In looking at the sensitivity of reliability models, e.g., to the assumption of independence, computer simulations are often found to be the most useful approach. Simulation in this context requires the construction of models to represent the interactions between components.
- Simulation may also be the only feasible approach to analysis of very highly reliable items with complex geometries, sophisticated materials, and new technologies. When very high reliability requirements are demanded, an adequate level of testing, even under accelerated conditions, is not possible. In the space program, huge simulation studies were undertaken to determine the sensitivity of even individual parts such as blades of a turbopump to environmental factors and various stress variables.

- In comparing various life distributions, use MLEs for each that provide a good fit, and calculate the reliability related quantity in question. The analysis may also include distributions for which the fit cannot be assessed. Calculate confidence intervals if possible. Take the extremes of the results as bounds on the quantity.
- Another approach to comparing the effect of distributional assumptions is to perform sampling studies in which samples are drawn from each candidate distribution using the parameter estimates from the original data and compare the results for the characteristic being studied.
- A related approach is to use bootstrapping or resampling techniques (see Efron and Tibshirani, 1993 for general methodology and Meeker and Escobar, 1998 for some reliability applications). In these studies, repeated subsamples are selected from the original data and the characteristic is calculated for each subsample. The methodology provides estimates of variances. Although this approach to testing goodness of fit is not well developed, it may provide some valuable insights.

We conclude the discussion of sensitivity testing by looking at the data sets considered earlier. Much more can be done in this area of reliability analysis. The current state of the art is, again, a mixture of art, science, and good judgment.

Example 11.23 (Case 2.13)

The exponential distribution has been assumed in previous analyses and has been found to fit the data. Since this is a special case of the Weibull distribution, this will fit as well. For Clerks 1 and 5, the sample mean and estimated parameters for these two distributions are:

	Clerk 1	Clerk 5
Mean	580.36	191.10
λ^*	0.001723	0.005233
α^*	1.10237	1.03536
β^*	600.89	193.67

Suppose that reliability is calculated as $R(x) = P(X > x)$. For selected x-values, the results are:

	Clerk 1		Clerk 5	
x	Exponential	Weibull	Exponential	Weibull
1	0.9983	0.9991	0.9948	0.9957
5	0.9914	0.9949	0.9742	0.9776
10	0.9829	0.9891	0.9490	0.9546
50	0.9175	0.9375	0.7698	0.7818
100	0.8417	0.8707	0.5926	0.6039

Note that in spite of the closeness of the estimated Weibull shape parameter to 1 (the exponential distribution), the probabilities indicate that the exponential distribution would predict significantly more failures (nearly twice as many for Clerk 1 and small x-values) than would the best fitting Weibull distribution. ■

Example 11.24 (Case 2.14)

In Example 11.21, it was noted that Richards and McDonald (1987) found that at least 12 distributions fit the ball bearing failure data. Although some of these are special cases of others, many distinct shapes are represented. Here we look at distributional sensitivity by considering the Weibull, gamma, lognormal, and Raleigh distributions. (The Raleigh distribution is a Weibull with $\alpha = 2.0$.) The results for selected x-values are:

		$P(X > x)$		
x	Weibull	Gamma	Lognorml	Raleigh
5	0.997198	0.999812	1.00000	0.996198
10	0.988029	0.997547	0.99980	0.984878
15	0.972156	0.989913	0.99715	0.966296
20	0.949619	0.974114	0.98656	0.940870
25	0.920693	0.948638	0.96291	0.909159
50	0.701397	0.699689	0.67614	0.683217
75	0.435325	0.404096	0.37438	0.424382
100	0.218177	0.196915	0.19168	0.217889

The results are again found to be quite sensitive to the distributional assumption. ∎

These examples illustrate the importance of goodness-of-fit tests and sensitivity studies. Although the samples sizes in both cases are relatively small, even for large samples, more than one distribution may provide a good fit and sensitivity to the assumed distribution becomes essential in assessment of reliability.

In applications, sensitivity results can be used to perform a worst-case analysis. For example, in Example 11.24, if $R(10) = P(X > 10)$ were of interest, a point estimate of this reliability would be taken to be 0.9849, the lowest of the calculated values. Another approach would be to calculate a lower confidence bound under each assumption (as well as those not listed in the example) and take the lowest of these as a working value.

11.5 MODEL VALIDATION

Up to this point, we have been concerned primarily with model selection. In many situations, in order for a model to be used, some confirmation of its validity is required. In this section, we look at a few approaches to validation criteria and methods.

11.5.1 Criteria for Validation

Two basic principles in validating a model are replicability of the results and assessment of the predictive power of the model under various conditions. Replication of results generally requires additional experimentation. To be effective as validity checks, the experiments must include the same conditions used in the original studies as well as broader conditions, all under the control of the experimenter. The first assures true repeatability; the second provides a basis for generalization. Both of these are essential for validation to be successfully achieved. Engineering knowledge and judgment and statistical principles of experimental design and data analysis are employed.

Assessment of the predictive power of a model is basically a statistical problem. The idea is to use an estimated model to predict outcomes of other observations, which may be data set aside for this purpose or future observations, and to evaluate the closeness of the predictions to observed values. A key requirement for credibility is that data used for validation not be a part of the data base used in model selection and estimation.

We will look at two statistical procedures that may be used for this purpose. These are the coefficient of determination R^2, discussed in Section 10.9, and the goodness-of-fit approach discussed in Section 11.3. There are many other approaches to this problem.[5] These include:

- Mean squared error (MSE), defined as the average squared deviation of the observed from predicted values. This may also be taken as a measure of closeness and is related to R^2.
- Chi-square tests of various forms.
- Subjective evaluations of model adequacy based on validity data.
- Bayesian analysis with the estimated model as a prior for the validity data.

R^2 Criterion. The correlation coefficient r is a measure of the strength of relationship between two variables. Its square is a measure of the reduction in variation in a regression relationship. With more than two variables, usually a set of predictors and a response, R^2 measures the strength of the relationship between the response and the group of predictors. R^2 is also an important measure in the comparison of models. In the evaluation of predictive power in this context, adjusted R^2 (see Section 10.9) is particularly useful, because the adjustment made accounts for differences in the numbers of variables and parameters in the competing models. This is important, since a higher parameterization can be expected to lead to a better fit even if the model is, in fact, not a better predictor. Two other advantages of the correlation coefficient are: (1) it is an index number, i.e., is unit-free, so scale does not enter into any assessment or comparison of models; and (2) there are some relatively easy and robust tests for these inference problems.

In Section 10.9, we looked at a test of H_0: $\rho = 0$. The test is a t test and is equivalent, in regression analysis with one predictor variable, to the test of H_0: $\beta_1 = 0$, where β_1 is the slope. This is not an appropriate test in the context of model validation because rejecting H_0: $\rho = 0$ simply means that there is some nonzero relationship between observed and predicted. To validate a model, we look for evidence of a strong relationship. Thus a more appropriate test is one of the null hypothesis H_0: $\rho \le \rho_0$ versus H_a: $\rho > \rho_0$, where ρ_0 is some minimal value that will be accepted as evidence of model validation, say 0.9 or 0.98. The required test is not equivalent to any in regression. Instead, the test statistic is

$$z = \frac{z_r - \zeta_0}{\sqrt{\dfrac{1}{n-3}}} \tag{11.16}$$

where $z_r = \tanh^{-1}(r)$ and $\zeta_0 = \tanh^{-1}(\rho_0)$. Under H_0, z is approximately normally distributed and the usual upper-tail z test is used.

For comparing models using R^2, an F test is used. In regression, the test basically addresses the question of whether or not inclusion of additional predictors improves predictive power, or, equivalently, whether or not unexplained variability is thereby significantly reduced. [See any regression text, e.g., Neter et al. (1996) for details.] The analysis is straightforward in ordinary multiple regression, but may not be in reliability model selection and validation.

Goodness-of-Fit Criteria. A straightforward approach to validation using goodness-of-fit techniques is to select a model using the methods of the Section 11.4 and estimate parameters from the original data, then use K-S or A-D or some other procedure with this model as the null hypothesized distribution. The test is applied to the validation data. Since the hypothesized distribution is completely specified, conditional on the independently estimated parameters, the test is distribution-free. This avoids the use of the special tables for each distribution, as required when the parameters are estimated from the same data that are used in fitting. Another advantage is that the procedure is easily applied if the validation data are incomplete.

11.5.2 Data-Based Model Validation

As noted, validation procedures are conceptually straightforward if independent samples are available for estimation and modeling and for use in validation. An alternative is to use split samples, that is, randomly divide the original sample into subsamples, with each subsample used for one of the purposes. Ideally, we would form three subsamples, using one for modeling, the second for estimation of the parameters of the selected model, and the third for validation. This requires a large initial data set and one that was collected under all conditions of interest. It also requires that a decision be made regarding the size of each of the subsamples. In most cases, 40% for modeling, 40% for estimation, and the remaining 20% for validation may be a reasonable allocation. If the sample size is not large, splitting into two subsets, one for both modeling and estimation and the second for validation, is an alternative. In this case, 70–30, 80–20, or 90–10 splits are suggested.

Some modifications of the subsampling approach that may be useful and informative are:

- Even if an independent validation sample is available, combine all of the data and then subsample.
- Analyze the separate samples first, then combine them and subsample, and compare the results.
- Repeat the subsampling process several times and compare the results.
- Repeat the process with different sampling rates and compare the results.

Note that for any of these methods of validation to work, the data must be valid and reliable, i.e., "clean." This requires proper random sampling, careful experimentation and measurement, removal of obvious outliers, identification of possible outliers, etc. If some data are doubtful, the analysis should be done with and without the questionable observations. If the results do not differ materially, this can be considered further evidence of validation. If they do differ, judgement and/or further validation with additional data are required in the decision.

In the best of worlds, modeling and validation proceed smoothly and a correct model results. In the real world, the conclusion may finally be that the model is invalid and the process must continue. Reasons for lack of validation include:

- The distributions from which the samples were drawn are different
- The distributions are of the same form but the parameters are different
- Both of the above plus differences in other factors/variables
- Change in conditions since the original study
- Changes in materials and/or procedures and/or personnel
- A Type I or Type II statistical error was made
- Many other possible reasons

In some cases, if reasons for the problems can be identified, adjustments in the analysis can be made to account for this and a valid model found. Regression methods are sometimes useful in this regard; more general models subsuming a broader set of conditions may be employed; and so forth.

If a reason for lack of validation cannot be found, assume that the model has not been validated. Finding a reason is also not enough, unless it can be demonstrated that the corresponding problem has been fixed. This usually requires further study and experimentation.

We conclude the chapter with analyses of two validation problems. These are not intended to illustrate all of the problems that may be encountered, nor to be complete solutions to those found, but will illustrate the techniques. This is again a highly applications-specific exercise. Some additional examples, with more complete solutions, will be discussed in Chapter 19.

Example 11.25 (Case 2.11)

We illustrate the correlation test by considering the data on pull strength for leads on successive trials. An important aspect of reliability is repeatability. Suppose that a process is acceptable only if the correlation between trials is at least 0.7. To check this, we test $H_0: \rho < 0.7$ versus $H_a: \rho \geq 0.7$. We test at $\alpha = 0.05$; the critical value of the test statistic is $z_{0.95} = 1.645$. Here $n = 30$ and the sample correlation coefficient is $r = 0.884$. We find $\tanh^{-1}(0.7) = 0.867$ and $\tanh^{-1}(0.884) = 1.394$. From (11.16),

$$z = \frac{1.394 - 0.867}{\sqrt{1/27}} = 2.738$$

H_0 is rejected and we conclude that the device/procedure passes the repeatability test. ∎

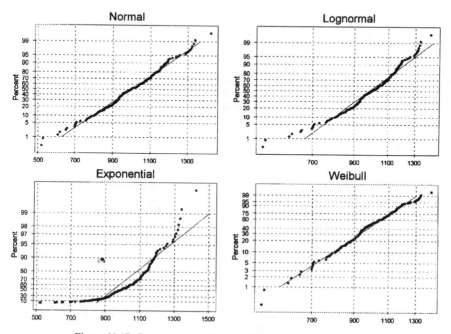

Figure 11.17. Four-way probability plot, TTF for light bulbs, data set 1.

Example 11.26 (Case 2.9)

The data on failure times of light bulbs consist of 10 observations in each of 42 weeks, with three missing observations. In Example 11.22, goodness-of-fit tests were applies to the entire data set and the normal distribution was found to provide a good fit. Suppose we perform a split-sample analysis with the following arbitrary data sets: Set 1—Weeks 1–20; Set 2—Weeks 21–32; Set 3—Weeks 33–42. The four-way probability plot of the data of Set 1, given in Figure 11.17, indicates a good fit to both the Weibull and normal distributions. (The estimate of the shape parameter of the fitted Weibull distribution is $\alpha^* = 7.1$, indicating a distribution close to the normal.) We illustrate the validation process using the normal distribution. The parameter estimates from Set 2 are $\bar{x} = 1096.6$ and $s = 200.1$. The corresponding normal distribution is tested for goodness-of-fit using Set 3. The K-S test gives $D = 0.05435$. With $n = 99$, the critical value for testing at the 5% level is $D_{0.05} = 1.358/[\sqrt{99} + 0.11/\sqrt{99} + 0.12] = 0.1347$ Normality is not rejected.

We note that the split selected was a fortuitous choice. The means of the three sets are significantly different, with that of Set 1 being smaller than those of the other two sets. A more thorough analysis of the data would consider other splits as well as other distributions and would include a sensitivity analysis. ■

NOTES

1. There are many good sources for more information on graphical methods. King (1971) is one of the first and still one of the best on the techniques discussed in this section. Nelson (1982) and Meeker and Escobar (1998) discuss plotting techniques in detail. Tufte (1983, 1990, and 1997) deals with graphical techniques for the presentation of information more generally.

The classical paper on hazard plots for incomplete data is Nelson (1969). A review of graphical methods for censored data is given by Sengupta (1995).

2. This is a standard Minitab output and will henceforth be referred to as a four-way probability plot.

3. As noted in Section 11.3, detailed discussions of goodness-of-fit techniques in the context of reliability, including censored data, are given by Lawless (1982), Woodruff and Moore (1988), and Meeker and Escobar (1998). There is a very large literature on goodness-of-fit. The following is a small sampling of this, covering several aspects of this type of analysis. Many more can be found in the references cited.

Specific models and distributions:

- K-S and modified K-S tests for the inverse Gaussian distribution based on transformation to normality; Edgeman (1990)
- Modified K-S, A-D, and Cramer–von Mises tests for the Pareto distribution; Porter et al. (1992)
- Goodness-of-fit to a mixed-hazard model in the context of auto warranty data; Majeske and Herrin (1995)

Additional tables for the A-D test:

- Modification giving more weight to the tails of the distribution, often important in reliability applications; Sinclair et al. (1990)
- Improved accuracy of tables for the original test; Moss et al. (1990)

Statistical tests and visual assessment of fit based on graphical methods:

- Visual assessment of fit based on probability and quantile–quantile plots; correlation test of goodness-of-fit; Gan et al. (1991) and Gan and Kohler (1990)
- Graphical representations other than probability plots for testing fit; Shirhata (1987)
- Use of total time on test plots and fit to a power law process; Barlow and Campo (1975), Klefsjö and Kumar (1992)

4. Many reliability models and failure distributions were introduced in earlier chapters. As a result, the list of possible models is endless. In spite of this, finding a specific model that will fit a set of data is often quite difficult. Many additional models of relevance in reliability applications can be found in the literature. Richards and McDonald (1987) and McDonald and Richards (1987b) provide many generalized distributions and analyze the air conditioning data of Case 2.10 and the ball bearing data of Case 2.14. Wolstenholme (1996) discusses an alternative to the Weibull distribution. A huge number of additional sources can be found in the *Current Index to Statistics* and in many other sources and on the Internet.

5. There are many other criteria for selecting and validating models. A common approach to discriminating between competing models is the likelihood ratio test (see Lawless, 1982, Chapter 9). This is based on the ratio of the likelihood functions under the competing failure distributions. A difficulty is that the distribution of the test statistic is very complicated and procedures have been developed for only a few cases.

Some recent results in the context of multiple regression models are given by Rahman and King (1999).

EXERCISES

1. Plot the sample CDF for each of the data sets in Table 2.1(b). Do the distributions appear to be the same? Compare the plots with plots of CDF's for various distributions given in Chapter 4. Can you tell anything from this?
2. Repeat Exercise 1 using the data sets of Case 2.6.
3. Plot the sample CDF for the data of Case 2.9, combining data over all 42 days. What distribution does the plot suggest?
4. Repeat Exercise 1 using the data sets of Case 2.21.
5. A Weibull probability plot is given for 12 mm fibers using the data of Case 2.21 in Example 11.3. Prepare Weibull probability plots for the remaining data sets and compare the results.
6. Prepare normal probability plots for the data sets of Case 2.21. Compare the results with each other and with the results of Exercise 5.
7. Prepare probability plots for various distributions for the data of each fiber length of Case 2.6. Which distribution(s) appear to provide a good fit to the data.
8. Repeat Exercise 7 for each truck for which data are given in Case 2.7.
9. Plot the data of Case 2.9 on the distribution plot suggested by you in Exercise 2.3. Do you concur with your original conclusion?
10. Repeat Exercise 7 for each of the data sets in Cases 2.11 and 2.12.
11. Do a graphical analysis of all of the data of Case 2.13.
12. Repeat Exercise 11 for all of the remaining data sets for which complete (uncensored) data are given.

13. Prepare Weibull and normal hazard plots for each of the data sets of Case 2.21. Interpret the results and compare your conclusions with those of Exercises 5 and 6.

14. Repeat Exercise 7 using hazard plots. Interpret the results and compare your conclusions with those made in Exercise 7.

15. Repeat Exercise 8 using hazard plots. Interpret the results and compare your conclusions with those made in Exercise 8.

16. Prepare hazard plots for the data of Case 2.13 and interpret the results.

17. Repeat Exercise 16 for all of the remaining data sets for which complete (uncensored) data are given.

18. Repeat the regression analysis of Example 11.9 using plotting positions $i/(n + 1)$ and $(i - 0.5)/(n + 1)$. Calculate the resulting Weibull parameter estimates and compare the results with those of the example.

19. Use regression analysis with all three plotting positions to calculate graphical estimates of Weibull parameters for the remaining three data sets of Case 2.21.

20. Prepare probability plots for the air conditioning data of Case 2.10. Use planes 7908 through 7913. Prepare lognormal and Weibull plots and compare the results.

21. Prepare a Weibull probability plot of the (incomplete) data of Case 17. Calculate maximum likelihood and graphical estimates of the parameters and compare the results.

22. Perform a graphical analysis of the incomplete data of Case 2.19, fitting various distributions. Which, if any, appear to provide a good fit. Calculate graphical estimates of the parameters of the selected distribution(s).

23. Fit a Poisson distribution to the data of Clerk 5 in Case 2.13. Use "time units" of both 500 and 1000 entries and compare the results.

24. Fit the inverse Gaussian distribution to the data sets of Case 2.20. Does this distribution appear to provide a good fit to the data?

25. Perform tests of goodness-of-fit to the normal, exponential, Weibull and lognormal distributions for the air conditioner data of Case 2.10. Use the data on each of the aircraft and also analyze the composite data. Use both the K-S and A-D tests. State your conclusions.

26. Perform tests of goodness-of-fit to the normal, exponential, Weibull and lognormal distributions for each of the six vehicles in Case 2.7 and for the composite set of data. State your conclusions.

27. Fit distributions to each of the data sets of Case 2.6.

28. Fit distributions to time to first failure, second failure, etc., in Case 2.8.

29. Do a goodness-of-fit analysis for the data of Case 2.2.

30. Fit distributions to the four data sets of Case 2.11.

31. Fit distributions to the three data sets of Case 2.12.

32. Fit distributions to the entire data set of Case 2.9, calculating both the K-S and A-D statistics.

33. Group the data and use χ^2 to test for goodness-of-fit to each of the distributions for each data set in Exercises 27, 29, and 32. Compare the results with those of the K-S and A-D tests.

34. Use a chi-square test to perform tests of fit of various distributions using the grouped data of Case 2.24.

35. Perform tests of goodness-of-fit to the normal, exponential, Weibull, and lognormal distributions for any remaining cases in Chapter 2, as appropriate.

36. Select a set of appropriate life distributions to represent the data of Case 2.20. Perform a sensitivity analysis including all selected life distributions.

37. Repeat Exercise 36 using the data of Case 2.23.

38. Apply the K-S and A-D tests of goodness-of-fit to the Weibull distribution to the data of Case 2.9, omitting the outlying observation. Do the results support the conclusion of Example 11.22?

39. Perform a sensitivity analysis on the data of Case 2.9, using the Weibull and normal distributions and any other distribution that may appear to provide a reasonable fit.

40. Test for repeatability using the withdrawal data of Case 2.11 (see Example 11.25).

41. Repeat the analysis of Example 11.26 using several splits of the data, including weeks 1–17 for model selection, 18–34 for estimation, and 35–42 for validation and at least two other splits. Do the results change the conclusions of Example 11.26? If so, in what way?

42. *Class project:* Summarize the results of the graphical analyses and goodness-of-fit tests for all of the cases in Chapter 2, as appropriate. In addition, aggregate all of the results of statistical analyses previously performed on these data, including estimates of parameters (especially for the distributions that appear to fit the data), confidence intervals, comparisons of means, and any other analyses discussed in Chapters 3, 5, and 8–11. Include these results in the chart begun for each case in Chapter 2.

PART D

Reliability Management, Improvement, and Optimization

CHAPTER 12

Reliability Management

12.1 INTRODUCTION

As indicated in Chapter 1, buyers of products (or systems) can be grouped into three categories:

1. Individuals (buyers of consumer durables and nondurables)
2. Businesses (buyers of commercial and industrial products as well as durables and nondurables)
3. Governments agencies (buyers of advanced high-tech, defense products as well as all of the previous types of products)

The reasons for buying a product are different for each group. Individual buyers purchase products to derive some satisfaction or pleasure (e.g., a stereo system to listen to music) or benefits (e.g., a washing machine to wash clothes or refrigerator to store food to prevent spoilage) or both (e.g., a new car). Businesses buy products to produce other products and/or services. An important goal in this context is making a profit and achieving other business goals such as market share, growth, etc. Finally, governments buy high-tech defense products to ensure national security and protection from unfriendly governments.

As discussed in Parts A–C, all products are unreliable in the sense that they deteriorate over time and ultimately fail. As a result, the performance of the product is affected and this in turn has a significant impact on the buyer. For individual buyers, it is usually the loss of satisfaction or benefits, but in some cases there is also an associated economic loss (e.g., the contents of a freezer spoiling due to failure of the compressor). The cost of rectification, either through repair or replacement of the failed item, can also have a significant impact on the satisfaction derived (or lack of it). In the case of businesses, the deterioration and failure of an item can have a major impact on overall business performance. This can include greater operating costs, delays in meeting orders, loss of sales, and so on. Finally, in the case of high-tech defense products and systems, deterioration and failure can have a significant impact on national security and this in turn can have political and social consequences or a serious economic impact.

Failure to meet buyer satisfaction has a major implication for manufacturers. The paramount driving force for manufacturers is to ensure that the products they produce meet (or still better, exceed) the needs and expectations of the buyers. This is especially important as new products are appearing on the market due to technological innovations that are occurring at an ever-increasing pace. In addition, competition is becoming more global and fierce, and buyer expectations are continually increasing. As a result, manufacturers have to manage the reliability of the products they produce in an effective manner. Failure to do so can have serious economic consequences and in the worst case can lead to the manufacturing businesses going bankrupt. Similarly, buyers need to understand the reliability of products in terms of their implications for achieving the desired goals or satisfaction. This is of special importance when the buyer is a business or a government.

In Chapter 1 we briefly mentioned *reliability management* as a discipline that deals with various reliability-related management issues in the context of the manufacture and/or operation of products (systems) that are unreliable. Reliability management deals with various reliability-related decisions regarding the product (system) from the overall business viewpoint. These are different for manufacturer and buyer. The reliability-related decisions must be made in the context of the overall product life cycle and this, again, is different for manufacturer and buyer. In this chapter, we deal with reliability management from both manufacturer and buyer perspectives. More specifically, we discuss the need to develop optimal strategies for various reliability-related issues over the product life cycle and from an overall business viewpoint. This is best done using the strategic management paradigm.

In this chapter we focus our attention on the management of hardware reliability. Several of the issues involved are also applicable to management of software reliability. Two sections at the end of the chapter will focus on software reliability management. The outline of the chapter is as follows. We start with a discussion of the different notions of product life cycle in Section 12.2 and indicate the notion most appropriate for reliability management. We then give a brief introduction to strategic management in Section 12.3. Following this, we discuss the relationship between reliability and quality in Section 12.4. This is done by first discussing the different notions of quality and then highlighting some of the issues that are of particular relevance in the context of reliability management. Section 12.5 deals with the total quality management paradigm. The material of Sections 12.3 and 12.5 sets the scene against which we discuss reliability management strategies. These are done in Sections 12.6 and 12.7 from the manufacturer and buyer perspectives, respectively. In Section 12.6, we highlight the interaction between technology and commercial issues. Product reliability is affected by technology-related decisions and these in turn have an impact on commercial decisions. We develop a framework for reliability strategy in the context of new product development. In Section 12.8 we discuss various reliability programs that have been used by different organizations for reliability management. Risk is an important factor that needs to be addressed in effective reliability management. This issue is discussed in Section 12.9 from both manufacturer and buyer perspectives. An important element of reliability management is assuring the buyer that the product delivered is reliable. Section 12.10 deals with this topic. In Section 12.11 we give a simple illustrative example of reliability management. This also highlights the use of quantitative models for effective reliability management and this is pursued further in Section 12.12. Section 12.13 deals with software safety issues and Section 12.14 deals with software maintainability. The maintainability issue in the context of hardware reliability is discussed in Chapter 13. Finally, we conclude with some general comments in Section 12.15.

12.2   PRODUCT LIFE CYCLE

There are a number of approaches to the concept of a product life cycle. The concept is quite different in meaning, intent, and importance for buyer and manufacturer. In addition, for each, there are different life cycles that may be of interest. Note that the product life cycle can be viewed in a larger overall context, with important strategic implications (Betz, 1993, p. 272). In this structure, the product life cycle is seen as embedded in the product line life cycle, which, in turn, is embedded in the technology life cycle.

12.2.1   Manufacturer's versus Buyer's Points of View

We first consider the product life cycle from the manufacturer's viewpoint. Here, we have two different approaches, the first based on a marketing and the second on a production perspective. From the marketing perspective, the product life cycle is defined as the curve that represents the unit sales for some product extending from the time it is first entered into the marketplace until it is removed (Rink and Swan, 1979). In this traditional form, the product life cycle describes the evolution of product sales and market conditions over the life of the product from its introduction into the market through to its ultimate withdrawal. Usually, the life cycle is characterized in terms of the following four phases:

1. Introduction phase (with low sales)
2. Growth phase (with rapid increase in sales)
3. Maturity phase (with near constant sales)
4. Decline phase (with decreasing sales)

For further details on the marketing perspective, see Kotler and Armstrong (1996) or Steffens (1991).

From a production perspective, the product life cycle is the time from the initial conception of the product to the final withdrawal of the product from the marketplace. It can be broken into two stages—prelaunch and postlaunch. As the name implies, the prelaunch stage deals with activities undertaken by the manufacturer prior to the release of the product in the marketplace. It consists of following six phases:

1. Product concept (initial idea for the product)
2. Product evaluation (target characteristics and pricing)
3. Research and development
4. Product design
5. Prototype development and testing
6. Manufacturing

Once items are manufactured, they are launched in the market; the postlaunch stage consists of two phases:

7. Marketing
8. Postsale servicing

The marketing phase can be divided into several subphases, as in case of the product life cycle from the marketing perspective.

From the buyer's viewpoint, the product life cycle is the time from the purchase of an item to its being discarded at the end of its useful life or its becoming obsolete due to technological changes. The life cycle involves three phases, namely

1. Acquisition
2. Operation and maintenance
3. Discard and replacement

12.2.2   Product Life Cycle for Reliability Management

For reliability management purposes, the life cycle of a product as proposed in the document *Reliability and Maintainability Guideline for Manufacturing Machinery and Equipment* (SAE M-110, 1993) is more appropriate and consists of the following five phases:

1. Concept
2. Design and development
3. Manufacture (and installation, if relevant)
4. Operation and maintenance
5. Conversion (or upgrade) or decommission (scrap)

The reliability-related decisions in the first three phases are made by the manufacturer and these need to be done in a manner so as to ensure that buyer's expectations are satisfied. The reliability-related decisions (e.g., maintenance, upgrades) in the last two phases are made by buyers. However, the manufacturer needs to take these two phases into account in making decisions in the first three phases (e.g., ensuring maintainability requirements). The specifics of these reliability-related issues will be discussed further in the later sections of the chapter.

12.3   STRATEGIC MANAGEMENT

In this section, we give a brief introduction to strategic management as a starting point for the discussion of reliability management strategies of Sections 12.5 and 12.6.

12.3.1   Basic Concepts

Strategic management is the process for the long-term planning of a business firm and involves several steps, as shown in Figure 12.1 The process begins with formulation of *mission* and *vision* statements. The former defines the reason for the existence of the firm and address the question "Why are we in business?" The mission statement is usually in the form of a definition of products and/or services and can include other issues such as technologies used, focus on customer needs, distinctive competencies, and so on. The vision statement describes where the firm is heading and what it wants to be in the future. The next step is formulation of (strategic) *goals*. These are broad statements that set the direction that the firm must take in order to realize its mission. *Strategies* are key actions taken towards achieving the goals.

For a business firm with several diversified business units, the strategies have a nested hierarchical structure, as shown in Figure 12.2. At the top is the *corporate strategy*, which is the overall managerial plan for the future. A *business strategy* is the managerial plan for a business unit. Responsibility for achieving the goals set out in the strategy is shared among the business units. A business can be broken down into functional areas (or departments), each responsible for its own performance and strategies. These are called *functional strategies*. The functional activities for a firm (manufacturing or service oriented) can be broadly grouped into three categories—technical, commercial, and support. For a manufacturing

Figure 12.1. Strategic planning process.

firm, the technical activities are design, research and development, manufacture, quality control, and so on. The commercial activities are marketing, postsale service, finance, and accounting. Support activities relate to legal issues, human resources, etc. The functional strategies are supported by *operational strategies.* Responsibility for operational strategies lies with the lower-level managers within functional areas, although approval for these lies with those responsible for functional strategies.

The different strategies must address both the long-term *strategic objectives* and the shorter-term *operational goals.* Strategic management aims to integrate these various strategies into a consistent overall business strategy that outlines the future direction of the company (medium- to long-term), while operational management is responsible for achieving the day-to-day intermediate steps needed in order to reach those strategic objectives (short- to

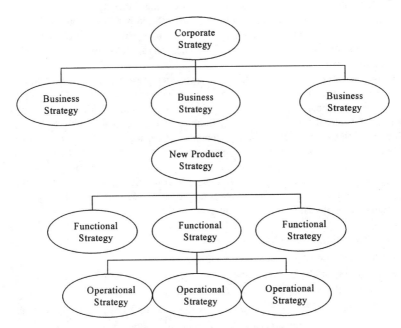

Figure 12.2. Hierarchy of strategies for a business.

medium-term). Effective strategic management requires that strategies at each level support those at higher levels in the hierarchy. The hierarchy of strategies is often called the "business plan" for the firm.

12.3.2   Planning and Implementation

Many factors, both internal and external to the company, influence the firm's strategies. Internal factors include the strengths, weaknesses, and competitive market position of the organization, as well as the ambitions, philosophies and principles of key executives. External factors include those of society, including regulatory bodies, political climate, changes in the industry, and various external opportunities and threats. All of these should be considered when formulating the hierarchy of strategies needed by a business firm. Strategies must be formulated and implemented and progress monitored through the use of feedback. The need for the organization to learn and adapt is important, as the strategic plan cannot anticipate every eventuality, and if new circumstances arise that fit with the vision but not with the plan, then the plan may need to be reevaluated. The organization must have the flexibility to experiment with new ideas and processes, and provide feedback to other areas in order to help the business improve and achieve its goals.

Strategic planning involves several stages. These are:

Setting Objectives. Setting objectives allows for the mission to become specific and quantifiable. Managers should endeavor to make objectives challenging in order to inspire employees, but achievable to avoid frustration and complacency. Long-range objectives are necessary to provide a focus for the future, and to make managers think about what can be done *now* to obtain those objectives. Short-term objectives provide stepping stones to reach the long-term goals, and can range from a number of days to a number of years, depending on the industry.

Crafting Plans. The crafting of a plan involves translating the mission and objectives into a hierarchy of strategies. This involves breaking the corporation or business strategy into functional strategies, with corresponding functional objectives. These need to be further broken down into operating strategies, responsibilities, and tasks assigned, and specifying how the objectives will be achieved.

Implementation. Strategic management does not end with the formulation of a strategic plan. Even the best strategy will not produce the intended results unless it is implemented successfully. Implementation of the strategic plan requires a set of skills different from those necessary for formulation. While strategy formulation requires vision, contemplation, and inspiration, implementation of the strategy requires action, administration and continued inspiration. Managers responsible for implementation in their areas must deal with many factors from the launch of new initiatives in line with the strategy to overcoming the resistance to change that is often found. Although each business firm is in a different situation, there are a number of administrative tasks that have been identified as significant. The key administrative components of strategy implementation are as follows:

1. Establishing strategy-supportive budget
2. Installing internal administrative support systems
3. Exercising strategic leadership
4. Shaping the corporate culture to fit the strategy
5. Devising rewards and incentives that are linked to achievements of objectives and goals
6. Building an organization capable of executing the strategy

It is difficult to generalize, since each firm is unique, and different strategies require different implementation approaches. The level of importance of each of the tasks will differ from firm to firm. For example, implementation for a company making minor changes to an existing strategy will be different from that for one making major changes to its strategic goals. The success of the implementation process is measured by the ability to produce the intended results of the strategy. Further details of strategic management can be found in most textbooks on the subject, for example, Thompson and Strickland (1990), and Betz (1993).

12.3.3   Comment

Traditionally, manufacturing firms competed on either cost or product differentiation. As a result, strategic business planning centered on financial and marketing goals and strategies relating to issues such as product development, technology, manufacturing, etc., were delegated to lower levels of operational management. This has changed over the last decade, primarily as a result of the influence of total quality management. This is discussed in Section 12.5.

12.4   RELIABILITY AND QUALITY

There is no universally agreed definition of quality; it is a concept that takes on different meanings depending on the perspective. Garvin (1988) proposes five different criteria for defining the notion of quality. These are as follows:

1. *Judgmental Criteria:* This is also called the *transcendent* definition and here quality is associated with something universally recognizable as a mark of high standard, achievement, or degree of excellence.
2. *Product-Based Criteria:* Here quality is defined in terms of some measurable variable such as the acceleration of a car, efficiency of an engine, and so on.
3. *User-Based Criteria:* These are determined by what the user wants and is best defined through "fitness for intended use." In the case of a car, this would be smoothness of the ride, ease of steering, etc.
4. *Value-Based Criteria:* These link the usefulness of, or the satisfaction derived from, a product to the price of the product. This involves the notion of "value for money" in defining quality.
5. *Manufacturing-Based Criteria:* Here quality is defined in terms of manufactured items conforming to the design specification. Items that do not conform either need some rectification action to make them conform or need to be scrapped.

As can be seen from the above list, the "quality of a product" is a concept that involves many dimensions. Garvin (1988) suggests the following eight quality dimensions for a product:

1. *Performance:* This characterizes the primary operating characteristics of the product. In the case of a car, it can include acceleration, braking distance, efficiency of engine, emissive pollution generated, and so on.
2. *Features:* These are the "extras" or the "bells and whistles" of a product. In the case of a new automobile, these include air conditioner, cruise control, CD deck, leather upholstery, and so on.

3. *Reliability:* This is a measure of the product performing satisfactorily over a specified time under stated conditions of use. In the case of a car, failure of the engine to start in bad weather, the windshield wiper not lasting beyond one rainy season, or brakes failing to perform satisfactorily are some of the many examples of lack of reliability.

4. *Conformance:* This indicates the degree to which the physical and performance characteristics meet some preestablished standards. The fuel consumption of a car or the wear on its tires being significantly higher than specifications would imply that the item is a nonconforming item.

5. *Durability:* This is an indicator of the time interval after which the product has deteriorated sufficiently so that it is unacceptable for use and needs to be scrapped or replaced. In the case of a car, it might correspond to corrosion affecting the frame and body to such a level that it is no longer safe to drive.

6. *Serviceability:* This deals with all of the postsale issues. These include frequency and cost of maintenance, ease of repair, availability of spares, and so on.

7. *Aesthetics:* This deals with issues such as appearance, feel, sound, etc. The body design and interior layout of a car would be good indicators of quality in this sense.

8. *Perceived quality:* This refers to the perceptions of the buyers or potential buyers. This impression is shaped by several factors such as advertising, the firm's reputation, consumer reports, etc.

As can be seen, reliability can be viewed as one of the many dimensions of quality. Yang and Kapur (1997) define reliability as "quality over time." It is important to note that there is a strong interlinking between several of these dimensions. For example, the link between reliability and performance can be seen through the concept of *availability* and between reliability and serviceability through the concept of *maintainability.*

Availability is the probability that an item will be operational at any given point in time or the proportion of time that it is operational. Availability is of considerable concern in some applications, often depending on the context. For example, storing a few spare engines in strategic places may be considered a part of the normal cost of operations for a large airline, but would be prohibitively expensive for a small airline in a remote location. For the small airline, ability to repair the engine quickly and easily (thereby attaining high availability) may be crucial to its commercial success. Similarly, breakdown of a computer will ordinarily not be critical to a large firm with many such systems, but for a small firm with only one such system, breakdown may mean shutdown of the operation.

Maintainability is an intuitively obvious concept. Technically, this term is defined to be the probability that a repairable item can be made operable within a specified period following a failure. Repairability considers only the probability that the time required for repair is within a specified interval, not including idle time while waiting for repair to begin.

In some applications, other performance measures may need to be guaranteed by the manufacturer. For example, jet engine manufacturers may guarantee thrust, fuel efficiency and low noise levels of their product. For complex systems, particularly in military applications, the list of performance characteristics may be quite extensive.

The link between reliability and conformance is due to variability in the manufacturing process. As a result of this, not all items produced meet the design specification. Such items are called nonconforming and their performance is, in general, inferior to those that conform to the design specification. Nonconforming items are less reliable than conforming items in terms of measures such as mean time to failure (MTTF), reliability $R(T)$, and so on.

Over the last few decades, a customer-driven concept of quality has become more widely accepted and it is defined as follows: *Quality is meeting or exceeding customer expectations.* In this thinking, a reliable product is one that performs in a manner and for a time period that exceed customer expectations.

12.5   TOTAL QUALITY MANAGEMENT (TQM)

12.5.1   The TQM Concept

Total quality is a system concept that has evolved over the past few decades. The following definition from Procter and Gamble (1992) describes the concept succinctly: "Total quality is the unyielding and continually improving efforts by everyone in the organization to understand, meet and exceed expectations of customers."

Total quality management (TQM) is used to denote a system for managing total quality. The core principles of TQM are

- focus on achieving customer satisfaction
- strive for continuous improvement
- involvement of the entire work force

A variety of practices along with many different tools and techniques have been developed for effective TQM. These include quality systems, quality assurance, quality function deployment, process control, quality circles, and so forth. Details of these can be found in most texts on TQM, for example, Evans and Lindsay (1996).

Customer-driven quality requires a feedback cycle approach, as shown in Figure 12.3. Note how the different notions and dimensions of quality are interlinked to achieve continuous improvements in meeting customer needs and expectations.

As discussed in the previous section, reliability is a very important dimension of quality that is closely linked to several other dimensions of quality. The composite of these is very critical for ensuring customer satisfaction, as illustrated by the discussion on the implications of the lack of reliability in Chapter 1.

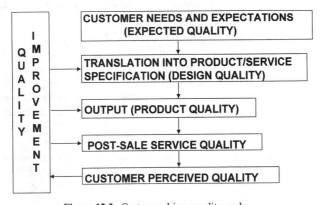

Figure 12.3. Customer-driven quality cycle.

The need for continuous improvement and the participation of the entire work force is essential for increasing the reliability of products. This must be done in a manner that fits within the strategic objectives of the business. Effective reliability management requires the blending of TQM and strategic management principles.

Because TQM takes a holistic approach to business, strategies are no longer based solely on the traditional marketing and financial strategic goals. Technology, new product development, and manufacturing assume the same importance as marketing and finance in formulating strategic goals. As a result, new product strategies and other related strategies (manufacturing, technology, quality, reliability, etc.) are no longer viewed as elements of operational strategies. Rather, they have become part of the higher-level functional strategies. This issue is discussed further in the next section where we discuss the linking of commercial and technology issues in the strategy formulation process so that the hierarchical distinction has become slightly blurred

12.5.2   Customer Satisfaction

Buyers (individuals, businesses, and governments) are becoming more aware that the products they buy must meet their needs and expectations. According to Knowles et al. (1995), the manufacturer must answer the following questions in order to produce a product that will satisfy the buyer's needs:

1. Does the manufacturer understand the buyer's needs?
2. Can the manufacturer develop the product to meet the buyer's needs?
3. Can the manufacturer assure the buyer that the needs will be met?

Buyer satisfaction is assured if the manufacturer ensures the following:

1. *Ascertaining the needs of the buyer.* This involves defining product requirements in terms of buyer's needs and manufacturer's capabilities to meet those needs. To do this requires understanding the buyer's needs (through customer surveys) and translating them into product development constraints, goals, and requirements.
2. *Developing a product that meets the buyer's needs.* This involves designing a product to meet the requirements defined in (1). Determining the reliability requirements requires identifying potential problems, their impact on product performance, and approaches to overcoming the problems.
3. *Assuring the buyer that the needs are met.* This is achieved through the use of TQM practices that ensure that the items produced always meet the buyer's needs. One of these is quality assurance. Use of warranties and other postsale services to assure buyers of appropriate actions when an item fails to perform as expected is important in this context. In addition, setting up procedures to collect feedback from buyers to determine root causes and initiating corrective actions also leads to increased buyer assurance.

Brunelle and Kapur (1997) propose several customer-satisfaction-oriented reliability measures based on a multistate reliability characterization for the product. Here product reliability is viewed as a time-oriented quality characteristic and the measures are based on disutility and continuous degradation of the state of the product.

12.6 STRATEGIC RELIABILITY MANAGEMENT (MANUFACTURER'S PERSPECTIVE)

Strategic reliability management for consumer durables and commercial and industrial products are, to a large extent, similar to that for specialized defense products. However, there are some differences and we discuss them separately.

12.6.1 Consumer Durables and Commercial and Industrial Products

Figure 12.4 is a schematic showing a hierarchy of strategies for a manufacturing firm, beginning with the overall business strategy. Typically, the firm in its long-range planning will be considering a number of possible new products or product lines as a part of its overall business strategy. From this will emerge a set of goals with regard to new product development and a strategy for achieving these goals. Goals may be quite broad, but will ordinarily include market share, profit objectives, per-unit production cost, and so forth.

In planning to achieve these goals, strategies for addressing the technological issues, as shown of the left side of Figure 12.4, and the commercial issues, shown on the right side, must be developed. (Strategies for other issues, not shown in the figure, such as human resource, finances, etc., must also be defined.) Technical issues include the engineering aspects of product design and manufacturing. Commercial issues in this context involve marketing and servicing aspects. To be effective, these strategies must be coherent and integrated. Failure to address both sets of issues adequately can seriously affect profitability of the product. In particular, lack of attention to technological issues can lead to inadequate product quality (in the broad sense of the usage as discussed earlier) and hence to excessive warranty claims and customer dissatisfaction. Excessive claims, in turn, will lead to increased warranty costs. The consequence of customer dissatisfaction is loss of sales for both this and future products. Undue delays, hidden costs to the customer, improper repairs, and inconveniences in obtaining service also adversely affect both costs and level of dissatisfaction. The end result of all of this is declining profits. As a result, the first step in strategic management of a new product is an understanding of the technical and commercial issues and their interaction.

Technical Issues

The principal focus is on the quality of the items produced. Product quality is determined primarily by design and manufacturing choices. Technical issues in this context are those concerned with quality in all of its dimensions of performance and conformance. Technical issues also involve methods for testing and for prediction of expected numbers and types of product failures and the probable cost of repair or replacement associated with each type of failure.

The technical issues for product and process design are listed in broad terms in the first two columns of Figure 12.4. The research and development strategy is developed in response to the basic design and reliability goals specified in the new product strategy. From this, the strategy for building and testing prototypes follows. This, in turn, leads to development of the final product design. A manufacturing strategy also needs to be developed at the outset. This includes selection of raw materials, selection of suppliers of raw materials and outsourced parts and components, and a process strategy. Finally, a quality control strategy, including sampling procedures, testing schemes, acceptance standards, and so forth, for proper monitoring of the manufacturing process also needs to be developed.

To assure high quality in the long run, attention must be paid to this product attribute be-

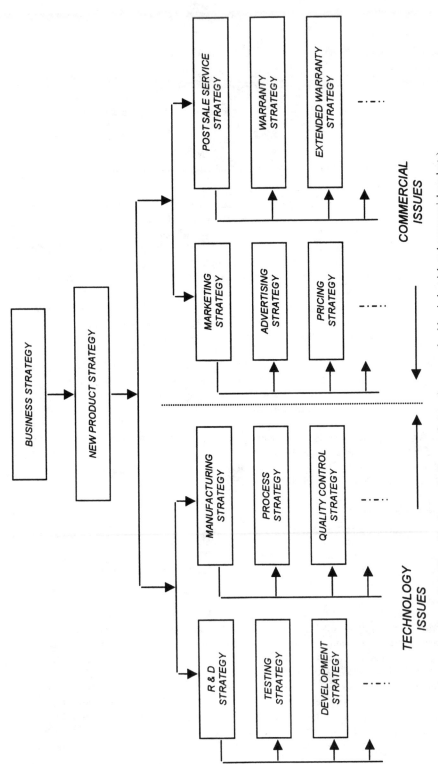

Figure 12.4. Hierarchy of strategies for manufacturer (consumer durables, industrial, and commercial products).

438

ginning early in the product design process. Methods for reliability prediction play an important role here. Predictions are based on past experience with similar products, analysis of various design trade-offs, use of engineering judgment, and so forth. These predictions are refined throughout the research and development phase, as prototypes are built, testing is done, redesign is undertaken, and the final version of the product evolves. The extent and cost of this process depends on the type of product and the context. It may be a relatively straightforward process for a simple product based on existing technology, or quite costly and time-consuming if new technologies are involved.

Development of an effective and efficient manufacturing process deals with the issue of producibility—process designs that minimize unit production cost while maximizing the fraction of nondefective units produced. Feedback in both directions between product design and manufacturing greatly enhances the chances of achieving the goals of producibility and design effectiveness.

Ideally, the analysis and testing of both product and process should lead to a reasonable prediction of product failure rates as a function of time. In some cases, reliability prediction may be relatively easy. This will usually be true, for example, for simple products of the type that the company has produced for a relatively long period of time and for which the newest models do not involve a significant technological leap. In other cases, reliability prediction can be extremely complex. [See Blischke (1994) for an approach to reliability prediction for rocket propulsion systems.] Maintainability and availability are almost always even more difficult to predict.

Commercial Issues

Commercial issues include marketing, postsale support, accounting and related areas, and deal with pricing, selection of warranty terms, promotion, and warranty service. Actions of competitors and the marketplace into which the product is to be introduced are also important factors. Strategies in this area that should be developed at the same time as the technology strategies are listed in the right-hand two columns of Figure 12.4.

The two major areas that directly or indirectly involve warranty are marketing and postsale service. Until relatively recently, the latter have received far too little attention, especially in the early planning stages. For the most part, warranty remains in this category. To be effective, strategies in these areas need to be developed concurrently, with frequent feedback from one to the other.

In developing a new product strategy, it is essential to assess the marketplace at an early stage. From this, a marketing strategy for the new product will evolve. A proposed advertising strategy will be a part of the marketing strategy. Assessment of the market and the overall marketing objectives will also lead to setting of goals with regard to pricing of the product and warranty costs.

Prediction of warranty costs is an important consideration in product pricing, particularly if warranty is to be used as a marketing variable. If the strategy is to convince the potential customer that the product is highly reliable by offering a longer warranty than that of the competition, this will almost always have the effect of increasing warranty costs. Unless the product has very high reliability, this increase could be substantial.

Interaction between Technological and Commercial Issues

Commercial and technical issues tend to interact strongly in determining the profitability of a product. This needs to be taken into account in selecting the various strategies. For example, in the case of the warranties, longer warranties require higher quality (performance, conformance, and reliability) in order to control warranty costs. There is a trade-off between design and production costs to improve reliability and the cost of servicing a warranty. As one increases, the other decreases. The difficult problem is to find the optimal cost allocation, i.e., the balance between design and production cost (to improve reliability) and future warranty costs that minimizes the total cost per item. In seeking such an optimum, close cooperation among engineering, manufacturing, marketing, cost analysis, and management functions is essential.

Recognition of the importance of the linkage between technological and commercial issues has developed slowly over the past few decades. For effective strategic management of new products, R&D, manufacturing, marketing and postsale service sections of the business must interact. Figure 12.5 shows schematically the linking needed between the different functional units of the business to formulate functional strategies that are coherent and well integrated.

Reliability-Related Issues and Strategies

For product reliability management, a life cycle approach is necessary, and in Section 12.2 a five-phase characterization for the product life cycle was suggested. The manufacturer must make decisions with regard to various reliability issues during the first three phases—(1) product concept, (2) design and development, and (3) manufacturing. The decisions include defining various reliability goals and crafting strategies for the different reliability-related activities. In the three phases, activities may include any or all of the following.

Product Concept Phase. In this phase, the reliability and maintainability goals for the product are defined by the manufacturer based on market analysis and buyer surveys. On this basis, preliminary design concepts are considered and evaluated.

Design and Development Phase. In this phase, the details of the preliminary design selected are sorted out. Reliability allocation to various subsystems, assemblies, etc., down to part level are finalized and techniques to achieve these reliabilities are identified. This may involve reliability-development strategies. The reliability-testing strategies for the prototype must also be defined. Based on the test results, the reliabilities (at different levels) are assessed and the development and testing strategies modified accordingly.

For certain products, a limited number of items are released to selected buyers for their evaluation and feedback from them is used in the development process to improve product performance. Again, one needs to define strategies for this type of testing by buyers.

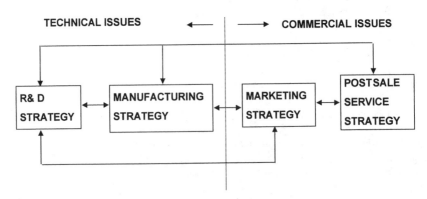

Figure 12.5. Linkages between technological and commercial aspects.

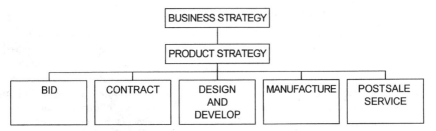

Figure 12.6. Hierarchy of strategies for the manufacturer (defense products).

Manufacturing Phase. Here the focus is on quality of conformance. This is affected by input material and process control. Quality can be controlled through proper process control and appropriate inspection and testing strategies.

Note that the decisions during the last two phases—(4) operation and maintenance and (5) conversion or decommission—are made by the buyer but the manufacturer needs to take these into account in making the decisions during the first three phases.

12.6.2 Specialized Defense Products

As mentioned earlier, here the request for the product (system) is initiated by the government through a *request for proposal* (RFP). The manufacturer has to decide on whether to respond to this request through a bid proposal. If the proposal is successful, then contracts are signed that clearly define the technical issues (product quality and other performance measures), economic issues, delivery date, and so on. This stage is followed by development of the product, leading to manufacture and finally delivery. The contract usually specifies actions by the manufacturer subsequent to delivery. These postdelivery actions can involve reliability improvement if the reliability measures agreed to in the contract are not met. An example of this is the *reliability improvement warranty* (RIW) which most government contracts require in the case of complex systems as part of any new defense acquisition programs. The manufacturer must craft various strategies as shown in Figure 12.6.

Reliability-Related Issues and Strategies

The implications for reliability-related issues and strategies are similar to those for consumer durables and commercial and industrial products. Reliability-related strategies must be defined for activities in the first three phases of the product life cycle. However, there are some differences. In the product concept phase, the reliability and maintainability goals are defined initially by the buyer and mutually agreed to in the contract stage and hence market focus is of no relevance. In the design, development, and manufacturing phases, the testing strategies are also often determined by the buyer or negotiated and included in the contract. This implies that the manufacturer has a greater obligation to assure the buyer that all of the reliability-related measures in the contract are met. It may also require monitoring in the field, data collection, report preparation, field repairs, and many other activities subsequent to the sale.

12.7 STRATEGIC RELIABILITY MANAGEMENT (BUYER'S PERSPECTIVE)

We first consider the case of a buyer of industrial or commercial products and then look at government as a buyer of special defense products.

12.7.1 Commercial and Industrial Products

Figure 12.7 is a schematic showing the hierarchy of strategies for the buyer from a business viewpoint. The buyer uses the product (usually machinery or equipment such as computers) to either produce other products and/or services for delivery to customers or to assist in the process (for example, airconditioners to provide a more pleasant environment). Again, we deal with both technology and commercial issues. The technology issues are relevant in the context of acquisition, operation, and maintenance. The factors that influence the buyer's acquisition strategy are many. They include the ability of the product to operate satisfactorily, cost of operation, maintenance requirements, and so on. The buyer's maintenance strategy involves choosing between in-house maintenance versus outsourcing. In the former case, it is necessary to define strategies for other issues, such as preventive maintenance, corrective maintenance, spares, etc. Similarly, the operations strategy is closely linked to commercial strategies—economic quantities for production in the case of batch production, strategies with regard to meeting variable demand, and so on.

Technical Issues

The reliability of a product has a significant impact on operation and maintenance requirements. A product with low reliability has a smaller acquisition cost but the operating and maintenance costs can be high. On the other hand, a more reliable product will cost more but have smaller operating and maintenance cost. This implies that the reliability of the product is a very important factor in choosing between different options.

One approach to deciding on the strategies for acquisition, operation and maintenance is the *life cycle cost* (LCC) approach. The LCC is the total cost of owning, operating, maintaining, and finally discarding the product. Figure 12.8 shows the different elements of the life cycle cost model for defense contracts. Maintenance costs are influenced by product reliability and the maintenance strategies (for corrective and preventive maintenance) used. For further discussion on LCC, see Blanchard (1981).

Commercial Issues

The impact of product failure on the buyer's business performance is significant in terms of customer satisfaction. The costs of dissatisfaction can be assessed in terms of the penalty incurred, loss of customers, damage to the reputation of the company, and so on. This implies

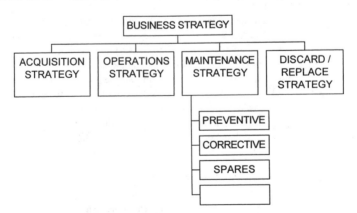

Figure 12.7. Hierarchy of strategies for the buyer.

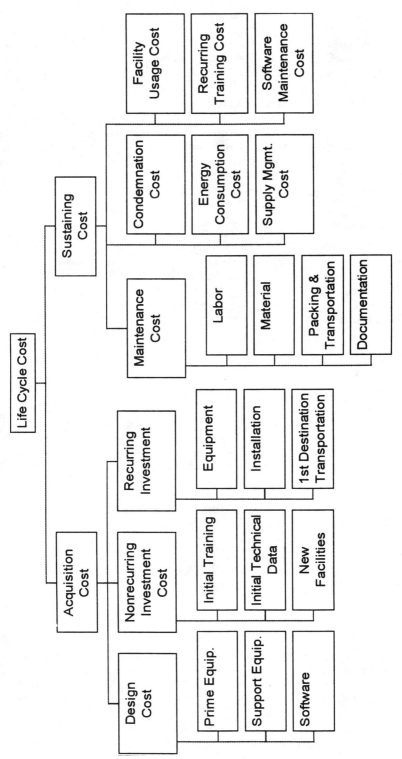

Figure 12.8. Life cycle costs (from Design to Cost Directive 5000.28, DoD, Washington, May 1997).

that the buyer should include these costs into the overall model in working out an acquisition strategy. This again illustrates the point that the technology and commercial issues must be integrated into the strategic management of the business.

Interaction between Buyer and Manufacturer

For standard products, the buyer has little influence over the manufacturer, since the manufacturer is producing the product to sell to many buyers. However, for products built specially to a buyer's request (in other words, custom-built products), the buyer specifies the performance requirements. In this case, the manufacturer has to assure the buyer that the product delivered meets the performance requirements. In a sense, this scenario is similar to the case of specialized defense products bought by governments.

12.7.2 Defense Products

Here the buyer is the government. The buyer defines the performance requirements for the product and these can include several reliability-related performance measures. The buyer is involved in reliability issues in the different phases of the product life cycle discussed in Section 12.2. To assist in this, various governments have reliability and maintainability standards that provide the basis for effective reliability programs that ensure that the products have the desired reliability.

12.8   RELIABILITY PROGRAMS

The reliability of a product is influenced by activities such as design, material selection, manufacturing, quality control and testing, etc. A *reliability program* provides a framework for a systematic approach for definition and management of the various reliability-related tasks. It includes a comprehensive list of activities that are considered to be essential to the success of the product. It further contains a description of each task and an assignment of responsibility and accountability. Reliability programs deal with reliability strategies at the functional as well as the operational levels.

The departments of defense (DOD) in several countries and a few professional organizations have produced standards for reliability programs. Reliability programs are influenced by the policies of the organizations involved, the product being developed, and by the unique practices of the organizations. In this section we briefly discuss some of these standards.

We first consider defense standards. Following this we look at standards developed by professional societies and standards organizations. Finally, we look at a few standards for commercial products.

12.8.1 MIL-STD-785

This program originated in the United States for use in government and military acquisitions of high-tech products. The request for proposal (RFP) usually contains the reliability specifications, reliability demonstration requirements, and the requirement for a reliability program plan developed in accordance with MIL-STD 785. The standard consists of the following four phases.

Phase 1: Concept. This phase deals with the identification and exploration of alternative solutions or solution concepts to satisfy a validated need as stated in the RFP.

Phase 2: Demonstration/Validation. As the name suggests, this phase requires the manufacturer to define the procedures for demonstrating the reliability performance of the product and its validation in terms of meeting the reliability measures stated in the RFP. The demonstration requires that a test or a series of tests be performed at component and other levels in usage-simulated environments. Based on the test data and root cause analysis, the component reliabilities can then be determined with confidence. Validation involves techniques to assure that the estimated component reliabilities will ensure that the reliability-related performance goals are met.

Phase 3: Full-Scale Engineering Development. This phase involves formulation of the detailed engineering design and construction of a prototype. The prototype can be viewed as an engineering model of the product. The first model is the "breadboard model". It is not intended to resemble the final product, but instead, is intended to demonstrate that the technology needed to complete the development program actually exists. The models evolve over time and come closer and closer to the final product in all forms. Testing of the final model can demonstrate the reliability of the product and validate that it meets the specifications of the RFP.

The reliability issues in the design stage are reliability allocation and reliability apportionment, which define target values for component reliabilities. Product reliability is predicted in terms of the reliability of the various components and the reliability analysis carried out using such tools as fault tree analysis (FTA), and failure mode, effects, and criticality analysis (FMECA). Development programs to achieve the target values are initiated through reliability improvement programs involving test–fix–test cycles.

Concurrent with this, process engineering looks at the production of the product. Among issues that need to be addressed are the impact of the production process on the reliability of the product.

Phase 4: Production. This is the final phase, during which products are produced and finally delivered to the buyer. The important factor is quality control that ensures that the items produced conform to the design specification.

The IEEE Reliability Program Standard (RPS) will serve as a commercial replacement for MIL-STD-875. It is being developed by representatives from the U.S. Army and Air Force, U.K. Ministry of Defense, electronic firms (Westinghouse, Motorola, Texas Instruments), and several professional societies. It will replace MIL-STD-785 in due course for defense contracts. The IEEE RPS is a performance-based approach that standardizes fundamental reliability program objectives:

- Ascertaining the needs of the buyer
- Developing a product to satisfy the needs
- Assuring the buyer that the needs have been met

Utilizing this approach, the buyer does not specify the tasks that must be performed. The manufacturer then has the freedom to use innovative means to develop the product, but also has the responsibility to provide the buyer with a product that meets the specified needs.

12.8.2 Other Defense Standards

The defense departments of several countries have their own standards. The R&M Standard for the Royal Australian Air Force is very similar to MIL-STD 785 but it involves three phases rather than four:

1. Concept (overlapping with Phases 1 and 2 of MIL-STD 785): This involves two sub-phases—concept development and project approval
2. Acquisition (Overlapping with Phases 2, 3, and 4 of MIL-STD 785): This involves three subphases—contracting, design and development, and production
3. In-Service (which corresponds to operations and support in the U.S. system)

Further details of this can be found in Bayley and Tabbagh (1995).

The NATO Standards evolved from the need to unify the various standards across the different members of NATO. This led to a series of Applied Reliability and Maintainability Publications (ARMP 1–8). For further details of these documents and a comparison with the standards in the United Kingdom, see Hockley and Comer (1993).

12.8.3   ISO Standards

The International Standards Organization (ISO) 9000 series deals with standards for quality. Of relevance to reliability and maintainability is ISO 9000-4—Guide to Dependability Program. It defines dependability as the collective term used to describe availability performance and its influencing factors: reliability performance, maintainability performance, and maintenance support performance. It also covers the essential features of a comprehensive dependability program for planning, organization, direction, and control of resources to produce products that will be reliable and maintainable. The following statements from the standard indicate the requirements of the supplier.

1. The supplier should identify and perform dependability analysis, prediction and formal design review activities (program tasks) adequate for the product or project.
2. The supplier should establish and maintain procedures for effective and adequate verification and validation of dependability requirements.

12.8.4   British Standard

Figure 12.9 shows schematically the reliability program according to British Standard BS 5760 (Part 1). As can be seen, it involves five phases—definition, design and development, production, install and commission, and function. The execution of the program requires many different tools and techniques. Some of these (e.g., FMEA, stress–strength analysis, system reliability in terms of component reliabilities) have been discussed in earlier chapters. Others (such as accelerated tests, reliability assessment, reliability growth, maintainability, etc.) will be discussed in later chapters.

12.8.5   IEC Standards

The International Electrotechnical Commission (IEC) Technical Committee 56 (IEC/TC 56) deals with dependability issues. The IEC 300 series deals with dependability issues and has links with the ISO 9000 series. The four documents in this series are the following:

IEC 300-1 (1993): Dependability Management—Part 1. Dependability Program Management

IEC 300-2 (yet to be issued): Dependability Management—Part 2 Dependability Program Elements and Tasks

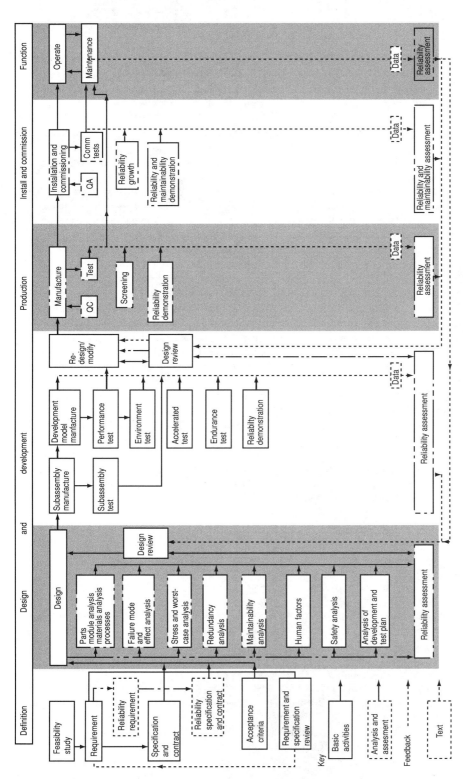

Figure 12.9. Reliability Program. (Extract from BS5760: Part 1: 1985 is reproduced with the permission of BSI under license number PD/1999 1421. This standard has been withdrawn and superceded by BS5760: Part I: 1996. Readers should ensure they have current information.)

447

IEC 300-3 (1991): Dependability Management—Part 3. Application Guide—Section 1: Analysis Techniques and Dependability: Guide to Methodology

IEC 300-4 (1993): Dependability Management—Part 3. Application Guide—Section 2: Collection of Dependability Data from the Field

O'Leary et al. (1996) discuss the IEC and ISO series and list many other IEC documents dealing with standards for other reliability-related activities such as equipment reliability testing, program for reliability growth, etc. Contact addresses for obtaining copies of the various ISO, IEC, and other standards are also given.

Comment: The BS5760 and IEC-300 documents are essentially the same as the ISO 9000 documents. As a consequence, there is a convergence in the approach to reliability management taken by the different organizations, and in due course the ISO series will become a true universal standard for reliability management.

12.8.6   SAE M-110

SAE M-110 (1993) is a guideline (rather than a standard) and evolved under the auspices of the Society of Automotive Engineers, Inc., with the aim of improving the reliability and quality of automotive products and manufacturing processes associated with the automotive industry in the United States. It consists of five phases as follows.

Phase 1: Concept. In this phase, the user specifies the reliability and maintainability requirements. The manufacturer (or supplier) should be prepared to address how the reliability and maintainability (R&M) qualitative and quantitative requirements dictated by the user will be met. Preliminary design concepts are considered and proposals are evaluated.

Phase 2: Development Design. In this phase, the user should verify the supplier's capabilities to undertake the actions specified in the R&M requirement and monitor the supplier's progress through scheduled design reviews. The supplier, in turn, utilizes various R&M techniques to achieve the R&M requirements.

Phase 3: Build and Install. During this phase, process variables affecting R&M should be identified and targeted for control during the manufacture, assembly, and installation of the equipment. Information collected during testing should be used to establish an equipment R&M baseline.

Phase 4: Operation/Support. This phase deals with user and supplier R&M activities during the operation of the equipment and the support needed. The user is expected to implement a system of R&M data collection, analysis, and feedback. The supplier uses this feedback to improve the R&M of existing equipment designs.

Phase 5: Conversion and/or Decommission. This phase involves either an upgrade (using later-generation technologies) or the scrapping of the system.

Figure 12.10 (based on Kindree et al., 1994) shows the sequential linking between these phases and highlights some of the critical R&M elements for each phase.

12.8.7   Norwegian Petroleum Industry

The Norwegian Petroleum Industry has developed a series of standards relating to reliability and risk management. The NORSOK Z-016 (Regularity Management and Reliability Technology) deals with reliability management. It is closely related to the A-SG-015 Technical Standard (Reliability Management in the Early Phases of Project Development) of STATOIL. These standards are for equipment used in offshore oil industry. The key element of NORSOK Z-016 is the concept of "regularity," which is defined as follows: "Regularity: A

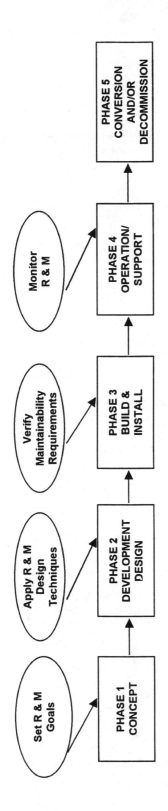

Figure 12.10. Elements of the SAE M-110 reliability program.

term used to describe how a system is capable of meeting the demand for deliveries or performance. Production availability, deliverability or other appropriate measures can be used to express regularity." Figure 12.11 (from NORSOK Z-016) shows the relationship between these measures. Also, the figure shows the impact of reliability and maintainability on regularity.

A regularity management program serves as a management tool in achieving regularity objectives by cost-effective means. This involves appropriate actions over the different phases of a life cycle, which are as follows:

- Feasibility
- Concept definition
- Engineering
- Procurement
- Fabrication/Construction
- Commissioning
- Operation
- Modifications

The Standard discusses the different reliability issues that must be addressed at each of these phases.

12.8.8 Other Standards for Commercial Products

Hough (1997) discusses a reliability program for commercial products that is very similar to the MIL-STD 875 standard. It has one additional phase, which deals with the manufacturer formulating the reliability specifications based on market survey data. (In MIL-STD 875, this

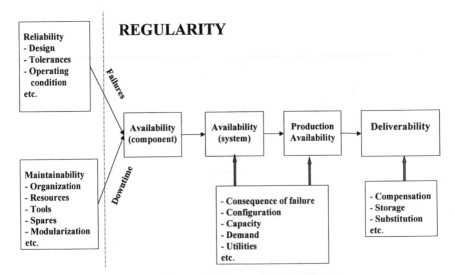

Figure 12.11. Regularity concept (NORSOK Z-016).

specification is provided to the manufacturer by the buyer through the RFP.) Hough compares and contrasts the commercial environment with that for the military.

12.8.9   Some Other Issues

It is important to realize that reliability management involves many other issues. Three of these are discussed here.

Reliability Team
Successful implementation of a reliability program requires a reliability team consisting of people from different sections (marketing, accounting, engineering, manufacturing, etc.) working with reliability specialists. This cross-functional team follows the reliability of the product over its life cycle and provides a link between upper management and functional managers (making strategic decisions) and lower-level managers (making operational decisions).

Implementation
Successful implementation of an R&M program requires the integration of various activities and proper planning as shown in Figure 12.12. When implemented in a team atmosphere involving all functions of the business, successful attainment of the R&M quantitative and qualitative objectives can be obtained. This is consistent with the TQM approach discussed earlier.

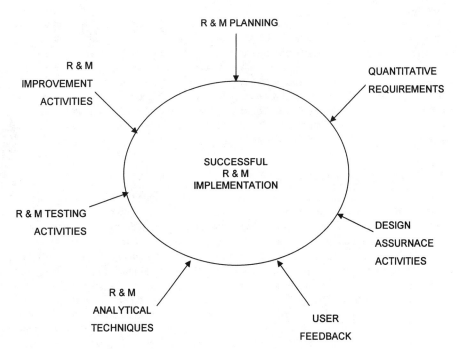

Figure 12.12. Different activities in a R&M program.

Need for Effective Data Collection Systems

Effective reliability management requires proper data collection systems—especially for failure reporting. Data must be collected over the product life cycle and all standards deal with this topic. A closed loop failure reporting system allows for failures during the operational phase to be communicated by the buyer to the manufacturer. This information is used by the manufacturer to implement corrective actions leading to continuous improvement in reliability and maintainability. See Brall (1994) for the closed loop failure reporting system used at Landis, a manufacturing enterprise that produces machinery.

There are several databases that contain reliability data for a range of components that are useful for evaluating product reliability in terms of component reliability at the design stage. A list of these can be found in Chapter 20.

12.9   RELIABILITY, RISK, AND LIABILITY

12.9.1   New Product Risk

In the case of new products, the cost of development can be very high. These costs are recovered if the product is a success. Failure of the product can be due to technical and/or market-related factors. Such failures can be costly and it is necessary for manufacturing firms to evaluate the associated risk before beginning any new product venture.

From a reliability point of view, an important risk element is the research and development needed to improve reliability, since an R&D venture is a highly risky proposition. Market risk can be due to liability claims that can follow because of the failure of a product to perform satisfactorily. A classic example of this is the silicon breast implant that proved to be not very reliable.

The notion of *vulnerability* is very relevant in this context. A product is vulnerable to a cause, either external or internal to the product, if it leads to a failure with serious consequences. A spectacular example of this is the space shuttle disaster, where the failure of the O-ring led to loss of lives and the vehicle. All products are vulnerable to some cause or combination of causes. A product can be made less vulnerable through increased investment in design and manufacturing. Reliability management examines the trade-off between the increase in the cost of making the product less vulnerable and the benefits derived from reducing its vulnerability. In many cases, the benefits are mainly economic, but often they are noneconomic (e.g., customer good will) as well.

12.9.2   Product Liability

When product reliability has an impact on safety, lack of adequate reliability of a component (for example, an unreliable brake system in a car) may result in the manufacturer incurring high legal costs. In such cases, the manufacturer may be forced to recall the product to repair or replace the unreliable component. This can be very costly, not only in terms of the direct costs involved, but also as a result of damage to the reputation of the product and/or of the manufacturer. There are several examples that highlight this. The recall of the Ford Pinto some years ago because of the danger of fire in a rear-end collision is perhaps the best known of these.

Evans and Lindsay (1996, p. 234) note that many businesses do not introduce new products into the marketplace because of liability suits resulting from inadequate product reliability. Because reliability has serious implications in terms of product liability, businesses must ensure that they have effective liability prevention mechanisms to minimize liability-related claims. Gooden (1996) suggests that businesses must establish a product safety and reliability review board with the responsibility of implementing such mechanisms and formulating and promoting a product reliability policy statement within the firm. He suggests that businesses adopt the following policy statement:

PRODUCT RELIABILITY POLICY STATEMENT

(Company or firm) is committed to ensuring that the products we design, engineer, and manufacture will be safe and reliable for our customers and the end users in the application that the product was intended. This policy statement issued as part of the overall Quality Program and the objective will be accomplished through the following means:

- All newly designed and manufactured products will be reviewed by the Product Safety and Reliability Review board to ensure that they are safe and reliable and will function as required in their intended application.

- All newly designed products will be reviewed to ensure that we as a manufacturer, or any outside subcontractor we select, have the ability and provisions in place to produce the product in a consistent manner.

- We will ensure that all new products, components, materials and processes, receive the necessary testing, prior to their use in production.

- We will concentrate our design and engineering process on "designing out" potential risks and hazards, and adequate provision for warnings and instructions concerning any other "residual" hazards, as well as to warn against any "reasonable foreseeable" misuse of the product.

- We will ensure through routine testing and inspection during the course of manufacturing that the products being produced conform to the standards, specification, and performance factors that apply.

- We will maintain a record that proves that the above safeguards were properly addressed, and retain these records for a reasonable period.

The policy statement will serve as our guideline in ensuring safety and reliability in the products we manufacture, and will be incorporated into thew appropriate procedures of our Quality System. Compliance to these statements and guidelines will be the responsibility of every employee of our organization. Assistance in the interpretation and application of these policy statements and guidelines will be given by the Product Safety and Reliability Review Board. © 1996 American Society for Quality. Reprinted with permission.

12.10   CUSTOMER ASSURANCE

Effective reliability management requires that the manufacturer assure buyers that the products being delivered are reliable. There are many devices that a manufacturer may use to achieve this. One such is the product reliability policy statement mentioned in the previous section. In this section we discuss two other ways of assuring customers of reliable products: reliability contracts and warranties.

12.10.1   Reliability Contract

A proper legal contract between buyer and manufacturer can ensure that the manufacturer delivers a reliable product. Contracts of this type are generally negotiated in commercial and government transactions; seldom, if ever, in consumer purchases. Such contracts may apply to both purchases of equipment and to detailed engineering, for example, in the case of custom-built and new equipment.

To prevent possible misunderstanding between the two parties or inadequately specified reliability requirements, it is important that the legal contract deals with the following issues:

- Definitions
- Documentation requirements
- Quality control requirements
- Work schedules
- Testing for reliability assessment and verification
- Data collection and analysis

These should be done in an unambiguous manner and should be complete. A proper drafting of such a contract requires a team comprised of lawyers, engineers, reliability experts, and managers.

12.10.2   Warranties

A warranty is a manufacturer's assurance to a buyer that a product is or shall be as represented. It may be considered to be a contractual agreement between the buyer and manufacturer entered into upon the sale of the product or service. In broad terms, the purpose of a warranty is to establish liability among the two parties (manufacturer and buyer) in the event that an item fails. An item is said to fail when it is unable to perform its intended function satisfactorily when properly used. The contract specifies both the performance that is to be expected and the redress available to the buyer if a failure occurs.

The terms warranty and guarantee are often used synonymously. The distinction is that a guarantee is defined to be a pledge or assurance of something; a warranty is a particular type of guarantee, namely a guarantee concerning goods or services provided by a seller to a buyer.

At present, nearly everything purchased or leased, whether by an individual, a corporation, or a government agency, is covered by warranty, either express or implied. An express warranty is one whose terms are explicitly stated in writing, while an implied warranty is a contract that is automatically in force upon purchase of an item because of statutes. Warranties are an integral part of nearly all consumer and commercial and many government transactions that involve product purchases.

In the procurement of complex military equipment, warranties of a certain type play a very different and important role, that of giving the seller an incentive to increase the reliability of the items after they are put into service. This is accomplished by requiring that the contractor service the items in the field and make design changes as failures are observed and analyzed. The incentive is an increased fee paid to the contractor if it can be demonstrated that the reliability of the item has, in fact, been increased. Warranties of this type are called reliability improvement warranties (RIW).

We describe some illustrative warranty policies for consumer durables and commercial products (sold in lots) along with a RIW policy. Warranties will be discussed in more detail in Chapter 17.

Warranty Policies for Consumer Durables

Free Replacement Policy (FRW). The manufacturer agrees to repair or provide replacements for failed items free of charge up to a time W (the warranty period) from the time of the initial purchase. The warranty expires at time W after purchase.

Pro-rata Rebate Policy (PRW). The manufacturer agrees to refund a fraction of the purchase price should the item fail before time W from the time of the initial purchase. The buyer is not constrained to buy a replacement item. The refund depends on the age of the item at failure (X) and can be either linear or a nonlinear function of ($W - X$), the remaining time in the warranty period.

Warranty Policy for a Commercial Product

Cumulative FRW. A lot of n items is warranted for a total (aggregate) period of nW. The n items in the lot are used one at a time. If S_n (the sum of the lifetimes for the n units) $< nW$, free replacement items are supplied, also one at a time, until the first instant when the total lifetimes of all failed items plus the service time of the item then in use is at least nW.

RIW Policy

Reliability Improvement Warranty. Under this policy, the manufacturer agrees to repair or provide replacements free of charge for any failed parts or units until time W after purchase. In addition, the manufacturer guarantees the MTBF of the purchased equipment to be at least M. If the computed MTBF is less than M, the manufacturer will provide, at no cost to the buyer, (1) engineering analysis to determine the cause of failure to meet the guaranteed MTBF requirement, (2) engineering change proposals, (3) modification of all existing units in accordance with approved engineering changes, and (4) consignment spares for buyer use until such time as it is shown that the MTBF is at least M.

12.11 AN ILLUSTRATIVE EXAMPLE OF AN INTEGRATED MODEL

In this section we give an illustrative example from the oil industry to highlight some of the salient features of reliability management. Figure 12.13 indicates the different elements (technical, operational, and commercial) that need to be taken into account in making reliability decisions from the overall business perspective.

Once crude oil (or natural gas) is extracted from the ground, it often must be pumped over long distances through a network of pipes. The pipes deteriorate over time as a result of aging and corrosion. The effect of this is a reduction in the thickness.

Consider a new pipe put into use at time $t = 0$ and let x_0 denote the strength of the new pipe. Strength is defined in terms of the pressure that the pipe can withstand before it ruptures. Let $x(t)$ denote the strength of the pipe at time t. $x(t)$ is related to the thickness of the pipe, with a larger value of $x(t)$ corresponding to a stronger pipe. Let $y(t)$ denote the pressure at which the oil is being pumped. This can be viewed as the stress on the pipe.

Safe operation of the pipe requires that

$$y(t) \leq x(t) - \delta \tag{12.1}$$

TECHNICAL

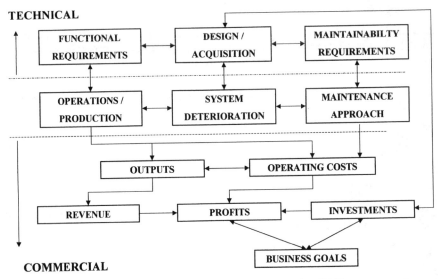

Figure 12.13. Technical, operational, and commercial elements in reliability management.

where $\delta\ (> 0)$ is the safety factor. A functional requirement is that

$$y(t) > y_m \qquad\qquad (12.2)$$

implying that effective pumping requires that the pressure be above a specified minimum value.

Deterioration (leading to the reduction in the strength) of the pipe is modeled by a first-order deterministic linear equation given by

$$\frac{dx(t)}{dt} = -\theta x(t) - \varphi y(t) \qquad\qquad (12.3)$$

The parameter θ characterizes the aging effect and the parameter φ characterizes the effect of the stress (or load) on the pipe. Note that the rate of deterioration increases as the pressure at which oil is pumped increases.

The design variable is x_0 and this has a cost implication since the acquisition cost, $C(x_0)$, is a nonlinear function of x_0. The only relevant maintenance action is the replacement of the pipe when its strength reaches $y_m + \delta$. Operating the pipe at below this strength implies either lowering the pressure below y_m and/or having a safety factor of less than δ, neither of which is acceptable for functional and safety reasons.

The output rate (volume of oil pumped) is proportional to the pressure $y(t)$. It follows that the revenue rate is given by $\gamma y(t)$ with $\gamma > 0$. The revenue generated over a period T is given by

$$R(T) = \int_0^T \gamma y(t)\,dt \qquad\qquad (12.4)$$

The business goal is to maximize the profit over the life of the pipe. Profit is given by

$$J(T; y(t), 0 \leq t \leq T; x_0) = \gamma \int_0^T y(t)dt - C(x_0) \qquad (12.5)$$

where the decision variables to be selected optimally are

1. x_0: Thickness of the pipe (design variable)
2. $y(t)$: Pumping pressure (operations variable)
3. T: Time to replacement of pipe (maintenance variable)

Note that the decision variables are constrained by the constraints given in (12.1) and (12.2) and the deterioration process is given by (12.3). We shall discuss the optimal choices of these quantities in Chapter 18.

In the above example, we assume that the deterioration process is deterministic. In real life, it is stochastic and is influenced by other environmental factors (e.g., chemical constituents of the crude oil). This requires modifying (12.3) to take these into account.

Note that the above model allows one to determine the optimal strategies for acquisition, operation, and replacement of the item. If the investment is constrained, then we have an additional inequality constraint, which x_0 must satisfy.

In the above model formulation, the safety and risk issues are accounted for through the parameter δ. One can model the effect of this and T on a catastrophic failure (due to unforeseen causes, such as an earthquake of magnitude exceeding some specified level) by a reliability function $\Gamma(\delta, T)$ which increases with increasing δ and decreases with increasing T. If the consequence of such a catastrophe is a loss worth Z, then (12.5) becomes

$$J(T; y(t), 0 \leq t \leq T; x_0) = \gamma \int_0^T y(t)dt - C(x_0) - Z[1 - \Gamma(\delta, T)] \qquad (12.6)$$

Note that in this case, δ becomes an additional decision variable.

Our discussion has been at the planning and strategic management level. The various operational issues, such as drafting a contract, tests to ensure that the pipe has the required strength, and so forth, have not been addressed.

12.12   QUANTITATIVE APPROACH TO RELIABILITY MANAGEMENT

As can be seen from the discussion thus far, product reliability is influenced by, and in turn influences, various commercial and technology factors in a business. From the simple illustrative example of the previous section, one sees that quantitative models play a very important and useful role in reliability management. This implies that effective reliability management requires building quantitative models for the different elements involved (technical, commercial, and operational) and effectively linking these to capture their interactions. Figure 12.14 (from Lyons, 2000) shows the models needed and the interactions between them in the context of a consumer durable or commercial product. The design and development models and the manufacturing models allow one to study the effect of decisions during these stages on the reliability of the product. The postsale and marketing models allow one to study the impact of product reliability on commercial factors. An integrat-

R&D Models   Manufacturing Models

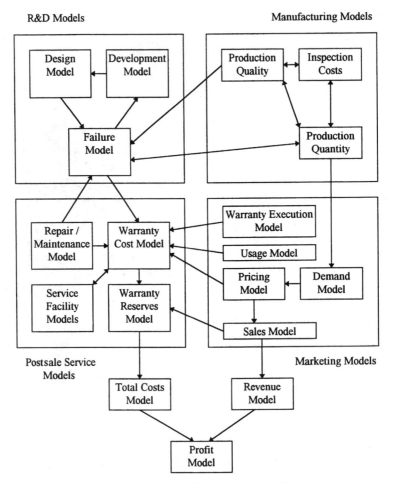

Figure 12.14. Reliability management model.

ed model, built using these different models as building blocks, would allow one to evaluate different decisions and to arrive at the optimal decision choice from an overall business perspective.

The complexity of the final model will depend on the degree of detail required and is problem-dependent. Figure 12.15 illustrates the details needed in the reliability management process for a custom-built product under an RIW policy. Here the warranty terms define the reliability requirements. Reliability issues dominate in the prototype stage and are also important during the manufacturing and postdelivery phases.

The success of a quantitative model depends on the validity and accuracy of the various submodels involved. Building and validation of the model and its components requires considerable data. In addition, effective decision making requires a considerable exchange of information between different models. Lyons (2000) deals with the information needs of the different submodels and the associated data collection issues.

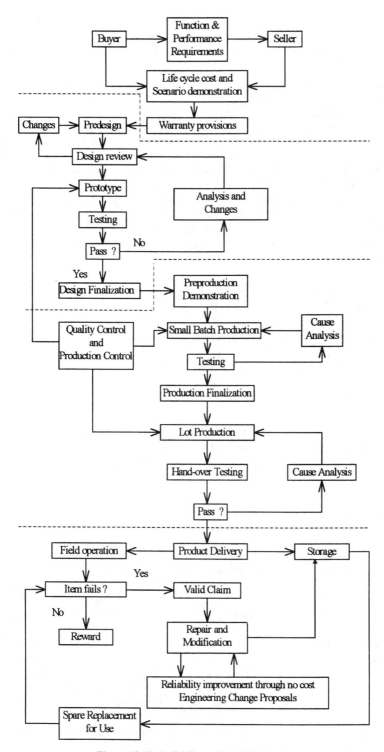

Figure 12.15. Reliability and the RIW process.

12.13 SOFTWARE SAFETY ISSUES

That software errors can seriously impact system safety is apparent in many of the illustrations cited in Chapter 9—aircraft control, operation of nuclear power plants, automotive controls, heart pacemakers. Defects can lead to failures in the software system, the hardware, and larger systems of which these are a part. They can lead to serious economic consequences and to loss of life. Here we look briefly at some of the safety issues that must be addressed in analysis of software and at some of the tools that are used to increase software safety. Safety issues are particularly important in process control, and much of the discussion has this orientation.

The vast majority of accidents involving software are due to requirements flaws (Leveson, 1995). Coding errors tend to effect reliability rather than safety. The primary problem is incompleteness of the specifications. We discuss this briefly below, along with design considerations, and tools to improve software safety. We begin with some basic concepts and definitions.

12.13.1 Basic Concepts

It is apparent that software safety must be considered in the context of system safety. The system perspective is discussed in Chapter 6 of Friedman and Voas (1995) and by Leveson (1995, Chapter 6 and elsewhere).
Some of the key concepts in safety analysis are

- Risk—the possibility of undesirable outcomes
- Safety—freedom from risk
- Mishap—unintended events that result in loss (also called "accidents")
- Hazard—state of a system that could lead to a mishap
- Software hazard—a software condition that could lead to an unsafe condition in hardware

In a qualitative assessment of risk, possible outcomes are ranked in terms of severity (e.g., catastrophic, critical, marginal, negligible) and hazard level (e.g., frequent, probable, occasional, remote, improbable, impossible). Risk may be expressed in a two-dimensional matrix based on these two characteristics. This enables the analyst to identify high-risk conditions and focus attention on avoidance of the most undesirable outcomes.

Quantitatively, risk is defined in terms of the probability and severity of a loss, with loss expressed in economic terms. Specifically, risk is calculated as expected loss, i.e., as a sum of loss times the probability of occurrence of the loss (summing over all possible outcomes) or by the corresponding integral, if the loss function is continuous. The quantitative approach requires that a realistic loss function be developed and that, in applications, realistic numerical values for the parameters of this function be provided. If this can be done, some of the probabilistic models of Section 9.4 may be of use in a quantitative assessment of risk. The quantitative approach also allows for the possibility of more sophisticated analyses, for example, optimization, e.g., minimizing risk, maximizing profits, etc. See Raheja (1991) for additional discussion of risk assessment.

Basic principles of software safety also include guidelines such as those of Myers (1976) for testing. Some of the "Axioms" for software safety listed by Lalli et al. (1998), are

1. Software, in and of itself, is neither safe nor unsafe.
2. Accidents are caused by wrong assumptions about the system or process.
3. It is impossible to build a perfect software system.
4. Lack of up-to-date standards and lack of the use of standards is the cause of many software problems.
5. Software quality, reliability and safety are designed in, not tested in.
6. Many software bugs are timing bugs that are difficult to test for.
7. Safety-critical elements should be kept as small as possible.
8. Determine what must be avoided at all costs and make certain that the program cannot allow that condition to occur.
9. Prepare the program so that it is fault-tolerant.
10. Never treat a program solely as a black box; understand what is inside the box.

How one accomplishes all of this is another matter, but at least being aware of the importance of software where safety is an issue and keeping in mind the above axioms will help to facilitate the development of "safe" software.

12.13.2 Completeness

Completeness in the context of software safety basically means that the specification of software requirements must be absolutely unambiguous and must provide sufficient detail so that proper inputs and outputs can be distinguished from improper ones. There are many more facets of completeness and this is discussed in some detail by Leveson (1995, Chapter 15). Some of the important issues, particularly appropriate in process control, are the following:

- Human–computer interface—requirements specification should include events to be queued, type and number of queues, priorities, operator notification, review and disposal commands.
- State completeness—operation states should be identified as normal and nonnormal states and transitions between them controlled. Initialization of variables, timing, before start-up and after shut-down behavior, off-line actions, fail-safe states, and a number of other factors play a role here.
- Variables completeness—input and output variables must be rigorously specified.
- Trigger event completeness—events that trigger state changes must be robust and unambiguous and must satisfy a number of other criteria with regard to timing, bounds, and so forth. Relation to output and basic feedback loops must be specified.
- Safety-critical outputs—these must be checked for reasonableness, timing, hazards, etc., and contingency actions specified.
- Transitions between states—all states must be reachable from the initial state; cancellation or reversal of a control should usually be possible; special features must be included to monitor and control paths that can lead to hazardous conditions.

12.13.3 Design for Safety

As noted previously, safety must be designed into a system. A thorough discussion of the design process is given in Leveson (1995, Chapter 16). Also see Raheja (1991, Chapter 9). As a rule, software is more difficult to design for safety than is hardware, but some of the same principles apply. Many of the design considerations for creating reliable software discussed in Section 9.3 apply here as well. The two key approaches to designing for safety are (1) design from experience, so as to avoid accidents previously encountered, and (2) use hazard analysis to guide the design process. Good design practice involves both approaches.

The first approach, basically benefiting from "lessons learned," involves the use of standards such as those listed in Section 9.2.5. The second important tool is preparation of a detailed design checklist.

Hazard analysis of the system as a whole is concerned with

1. Hazard elimination—This involves eliminating either the hazardous state or the undesirable consequences of that state. Some possible solutions (substitution, simplification, decoupling, elimination of hazardous conditions and materials) are discussed by Leveson (1995).
2. Hazard reduction—design for control, inclusion of barriers, safety factors, safety margins, redundancy
3. Hazard control—reduction of exposure, isolation, containment, protection, fail-safe design
4. Damage reduction—warning system, protection, etc.

12.13.4   Tools to Improve Safety

Many of the tools for improvement of software safety have been discussed or implied in the previous sections. These include FTA, FMECA, and other approaches to hazard analysis, analysis of code, flow charts, and so forth. Other tools are:

- Use of cut sets (see Chapter 7)
- Common cause analysis
- Sneak circuit analysis—identification of latent paths that lead to unintended outcomes
- Use of Petri nets—computer analysis of graphical models (see Leveson, 1995)
- Use of MIL-STD-882 (see Section 9.2.5)

See Raheja (1991) and Leveson (1995) for details. More information on safety standards, including those in use in avionics, automotive, DoD, and nuclear power plants, and on software safety considerations in general, may be found in Friedman and Voas (1995).

12.14   SOFTWARE MAINTAINABILITY

12.14.1   Basic Concepts

Maintainability is formally defined as the probability that a failed item is restored to working order in a specified period of time. Qualitatively, maintainability deals with the ease with which an item can be repaired, modified, tested and improved. The following are the dimensions of maintainability in the context of software (Raheja, 1991):

- Modularity—the number of independent parts within a software system
- Modifiability—the extent to which modules within a program are isolated from one another

- Availability—the proportion of time that a system is operational
- Testability—properties of a program that allow its functions to be separated and tested in a logical sequence
- Portability—ability to be used in an environment other than that for which it was programmed

Measures of maintainability include the following time elements: problem identification time, delay time, action correction and review, testing time, and total recovery time.

12.14.2 Design Issues.

Unlike most hardware, software can easily be made worse by "fixing" it. This is particularly true of large programs with many modules and interfaces. Effective program maintenance requires configuration control, accurate and complete documentation, adequate staffing and effective management. Important tools are (Raheja, 1991) modularity, top-down structured programming, fault-tolerant design, and "sequential programming as opposed to spaghetti coding." Top-down design assures that faults can be detected by a direct path rather than having to branch into other segments of the program. Modularity and testability are also important in fault isolation. Tools discussed previously, such as FTA and FMECA, are important tools in analysis of designs.

An excellent, very thorough checklist for software design review is given by Raheja (1991), in his Figure 9.8.

12.15 CONCLUDING COMMENTS

In this chapter, we have discussed briefly the issues involved in effective reliability management. The basic principles of reliability management have been illustrated through a very simple example. The example shows how technical, operational, and commercial issues must be integrated to determine an optimal strategy. It also highlights how reliability and risk aspects affect the various strategies and in turn are affected by them. The specific details of this integration vary depending on the firm and product, but the principles are essentially the same.

Reliability management is very broad and involves many different disciplines. Most reliability decision problems involve multicriterion optimization, since the goal is usually not a single economic performance measure. For more on multicriterion decision problems, see Keeney and Raiffa (1993).

Another important factor that must be addressed is the information needed to develop appropriate models and to validate them. The complexity of the model needed depends on the level of detail desired. This in turn depends on the goal of the model. Models for feasibility studies do not require detailed characterizations of reliability, but this changes at the design and manufacturing stages. At these stages, fairly detailed reliability-related information is required.

In the remaining chapters of Part D of the book, we discuss various topics of relevance to reliability management, including reliability assessment, reliability prediction, and maintenance, in more detail and develop models useful for analysis and optimization.

EXERCISES

1. Carry out a life cycle cost analysis for a consumer durable—for example, a refrigerator, a washer, a television set—from the buyer's perspective.

2. Carry out a life cycle cost analysis for an industrial product—for example, a truck, an aircraft, a computer system—from the buyer's perspective.

3. Repeat Exercise 1 from the manufacturer's perspective.

4. Repeat Exercise 2 from the manufacturer's perspective.

5. The plan for strategic management is often referred to as a business plan. Obtain business plans for several businesses (manufacturing and service) and carry out a critical evaluation of these in terms of their missions, goals, and strategies.

6. Suppose that you have been appointed reliability manager for an electronics firm that has a contract to supply antenna dishes for use by the Army in its new microwave communications system. Develop a reliability program that conforms to MIL-STD-785.

7. Repeat Exercise 3 using the ISO 9000 standard. Compare and contrast the two programs.

8. The "regularity" concept in NORSOK Z-016 captures the notion of availability and deliverability. In this chapter, we have defined measures for availability. Define appropriate measures for deliverability.

9. An appliance manufacturer is considering introduction of a new washing machine to replace a current model. A market survey reveals that users can be grouped into two categories—(1) light users and (2) heavy users. For a given design, the wear of the machine depends on the usage rate. Should the manufacturer have two different designs or only one? Develop a conceptual model that highlights the different issues that must be taken into account in making an optimal choice.

10. A mining company is planning to buy a new fleet of heavy haul trucks. Draft a reliability contract that the company should have with the manufacturer of such trucks.

11. How would you define reliability in the context of an umbrella?

12. A manufacturer of umbrellas has two options—build a cheap and unreliable product or an expensive and costly product. How should he choose between the two options?

13. A manufacturer has signed a contract to build a specialized piece of equipment with strict reliability performance measures specified in a reliability improvement warranty (RIW) contract. How should he set up an effective reliability management program to ensure success?

14. (Continuation.) Suppose that the manufacturer is unable to deliver a product that meets the reliability requirement by the date specified in the RIW contract. What options are available to the manufacturer? How should he decide on the best course of action?

15. Lack of reliability in a product has a direct cost (cost of rectification) and an indirect cost (consequential costs resulting from product failure). Make a list of the indirect costs when the product is a bus operated by a travel company.

16. Discuss the risks associated with a heart pacemaker. How should the manufacturer of such devices incorporate risk into reliability management?

17. For more detailed modeling, the elements of Figure 12.14 need to be broken into additional elements. Discuss the degree of detail needed for building a model to predict sales over the life cycle of a product.

18. A large metropolitan area is planning a new electric train system. As a part of the planning process, a team has been set up to analyze the reliability of the system. Discuss the following:

   - The experts from the different disciplines needed for such a team
   - Defining and quantifying the reliability measures for the system

- Drafting of a reliability document to assist in manufacturer's bid preparation
- Verification that the achieved reliability equals or exceeds the desired reliability

19. Often, the performance measures used in reliability management are based on average values of some variables. What are the shortcomings of using average values? Indicate alternative measures, e.g., measures involving higher moments, and comment on their desirability.

20. Ineffective reliability management can be due several reasons. Make a list of such reasons. Rank them in terms of their importance for reliability management.

21. In many cases it is not possible to use a quantitative approach to reliability management for lack of suitable models and/or data. In this case, one has to use a qualitative approach. Discuss this, using an illustrative example.

22. The TQM approach stresses continuous improvement in products and services. Define and discuss different notions of reliability improvement in the context of a product.

23. Repeat Exercise 22 in the context of a service.

24. Product reliability is influenced by design and manufacturing decisions. These are under the control of the manufacturer. However, reliability is also influenced by maintenance and care exercised by the buyer. This can vary significantly across the population. How should this be accounted for in the design of a product?

CHAPTER 13

Reliability Engineering

13.1 INTRODUCTION

In Chapter 1 we defined failure of a product (or system) as the inability of the product to perform its stated function and discussed the consequential customer dissatisfaction resulting from such failures. The customer's view of whether or not product performance is satisfactory depends on their needs and expectations. Product performance is influenced by the following two sets of factors:

1. *Factors prior to the sale of the product:* These are primarily technical and engineering factors related to the design, development, and manufacturing of the product. The manufacturer has reasonable control over these factors and in some cases (for example, defense acquisitions) the customer may have a significant influence.

2. *Factors during use:* These relate to the environment and the mode of usage. The latter includes factors such as duty cycle, intensity of usage, etc. The performance of a product ordinarily degrades as the environment becomes harsher and/or the usage intensity increases. These factors are, to a significant degree, under the control of the customer, and the manufacturer has very little (and often no) control over them.

For most products, the characterization of product performance requires a multidimensional vector. Several components of this vector are related to product reliability. These include various notions of reliability, dependability, availability, and so on. Some of the reliability-related performance measures, such as mean time to failure (MTTF) or mean time to repair (MTTR) were discussed in earlier chapters. Many other measures relating to different notions of availability will be discussed later in this chapter. It is through such measures that one can evaluate product performance and, in turn, customer satisfaction.

Reliability engineering is important in this context. Reliability engineering deals with the design and manufacture (or construction) of products, taking into account the unreliability of its parts and components. It also includes testing and programs to assess and improve reliability. These must be done in such a manner that they are consistent with the overall business context and must take into account technological and commercial aspects and their interactions, as discussed in Chapter 12. Good engineering of a product requires that the manufacturer take these into account during the design stage to ensure that the product produced is reliable. Failure to do so can lead to additional costs resulting from excess warranty and liability claims. Indirectly, it can affect the reputation of the firm and its products and this can have a very significant impact on total sales.

This chapter deals with the engineering of reliable products (systems). The engineering process is a multiphase process with several stages in each phase. We focus our attention on the reliability process needed in this engineering activity and develop a framework that encompasses the various issues that must be addressed in order to produce a reliable product. We briefly discuss the different phases of the process, highlighting the reliability decisions that need to be made at each stage of each phase and the models and tools needed for carrying this out. Some of these tools have been discussed in earlier chapters and others will be discussed in detail in the chapters to follow.

The outline of the chapter is as follows. In Section 13.2 we discuss the three different phases in the engineering of a product. These are design, manufacturing, and presale assurance. Section 13.3 deals with the concept of design for reliability (DFR). In Section 13.4, we discuss the different stages in the design phase. The first stage deals with customer needs and expectations and how they relate to the performance requirement. The subsequent stages deal with conceptual and detailed design of the product. Various reliability related issues in the context of product design are discussed. One of these—reliability allocation and apportionment—is discussed in detail in Section 13.5. Several others are discussed in later chapters. The manufacturing phase is discussed in Section 13.6, where the focus is on modeling the degradation of product reliability during manufacturing. Different approaches for control of reliability degradation are discussed in Section 13.7. Section 13.8 deals, in a general context, with various issues in testing. Testing is carried out for different purposes in the design, manufacturing, and presale assurance phases. One particular issue in testing is burn-in and this is discussed in Section 13.9. We conclude the chapter with notes and references giving further details on topics covered.

13.2   THE ENGINEERING PROCESS

Engineering can be defined as the profession of applying science to the conversion of raw materials and resources into products and energy for use. It requires judgement in order to adapt scientific knowledge to produce new products or systems that meet stated requirements in terms of technical performance and commercial constraints (e.g., unit manufacturing cost, target date for release, and so on). The engineering process to achieve this involves the following three phases:

- Design and development
- Manufacturing (construction)
- Presale (predelivery) assurance

The three phases are sequentially linked as shown in Figure 13.1 and there are several stages in each phase. We discuss each of these briefly in this section. Further details are given later in the chapter. It is important to note that in each stage it is necessary to assess reliability and to address other reliability-related issues, and that this may sometimes be a very difficult problem. Basic functions in the three phases are:

Design and Development Phase. The design and development phase deals with the process that leads to a prototype version of the product that meets the stated technical requirements and can be manufactured in a manner that meets the stated constraints. The design process can be viewed as the consolidation of resources (e.g., human resources, materi-

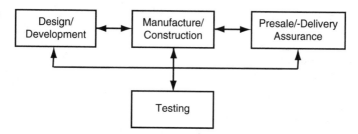

Figure 13.1. Elements of the engineering process.

al, equipment, financial resources, and technologies) and information so as to create a product that meets customer needs and achieves the business goals of the manufacturer. It a systematic procedure involving the following four stages:

- Requirement definition
- Conceptual design
- Detailed design
- Testing and evaluation

The first three are sequentially linked. There are several engineering tasks involved at each stage. These will be discussed in later sections.

Manufacturing Phase. The manufacturing phase deals with the process of producing items in bulk in such a way that all items conform to the stated design performance specifications and doing so in the most economical manner. Due to variability in manufacturing, however, not all items will meet the design specification. In the quality paradigm, the focus in this stage is on quality control in order to assure product conformance.

Presale Assurance Phase. The last phase deals with presale assurance. As the name indicates, the aim of this phase is to assure product performance prior to its being released to customers. For complex, expensive products involving new technologies and custom-built products, this phase is very important and often explicitly addressed in a negotiated contract between the manufacturer and buyer. For such products, each item produced is subjected to a well-defined testing procedure to evaluate its performance. For consumer durables and many other standard industrial or commercial products, not every item produced goes through this phase. A more common practice is to subject a fraction of the items produced (picked either randomly or according to some specified rule that is equivalent to random sampling) to some testing procedure to evaluate product performance.

Testing Phase. Testing is very important in all three phases. The reasons for testing are different in each phase. In the design and development phase, testing is carried out at part, module, and subsystem levels to assess the capabilities of the design to meet the stated goals. During the manufacturing/construction phase, the primary role of testing is to detect item nonconformance. In the final phase, the testing is done at the product level for product assurance.[1]

13.2.1 The Reliability Process

In the context of the engineering process, reliability engineering focuses on the prevention, detection, and correction of reliability-related design deficiencies, weak parts, workmanship defects, and so forth. It is an integral part of the product design process, including design changes. The process through which reliability engineering contributes to product design, manufacture, and assurance is called the *reliability process. Design for Reliability* (DFR) is one such process that evolved in the early 1980s in the United States in response to competition from overseas manufacturers who were producing more reliable products. We will discuss this issue further in the next section. The reliability program discussed in Chapter 12 can be viewed as a plan for implementation of the reliability process. An effective reliability program stresses early investment in reliability engineering tasks to avoid subsequent costs and schedule delays and possible litigation resulting from the sale of unreliable products.

13.2.2 Reliability Engineer and Associated Tasks

New product development requires an interdisciplinary team approach, involving specialists from different engineering disciplines. Reliability engineers are specialists who provide the skills to ensure that reliability issues are tackled at the design stage and that reliability is controlled during the manufacturing stage. In order to do this effectively, reliability engineers are trained to carry out many different reliability-related tasks. Some of these tasks are as follows:

1. Participate in all design reviews and modifications
2. Decide on reliability allocation and apportionment
3. Carry out reliability estimation, prediction, and growth plans
4. Plan and conduct reliability tests
5. Perform statistical analysis of test data
6. Maintain a reliability data system
7. Provide reliability-related assistance to other sections such as (i) manufacturing, (ii) purchasing, and (iii) postsale servicing
8. Write reliability specifications for parts to be purchased
9. Identify causes of reliability degradation
10. Continuously improve the reliability of the product

We will discuss some of these in this chapter and others in later chapters.

13.3 DESIGN FOR RELIABILITY (DFR)

Design for reliability (DFR) arose as a result of market pressure and a challenge (see Moss, 1996 for a more detailed description of the material presented in this section). In the early 1980s, the CEO of a large electronics firm decided to define a goal that required a ten-fold reduction in the failure rates for its products over the period of a decade. This decision was made in response to concerns about a steady increase in warranty costs and an erosion of market share.

About halfway through the decade, it was noted that the rate of improvement was not the same across the product lines manufactured by different divisions of the firm. A survey involving the evaluation of 37 DFR activities (shown in Table 13.1) that could be used for reliability improvement was carried out across the 12 divisions.

The engineering and quality staff were asked to rate the usage of each activity using a nu-

Table 13.1. DFR survey checklist

Management

1. Goal setting for division
2. Priority of quality and reliability improvement
3. Management attention and follow-up (ownership of goals)

Engineering

4. Documented hardware design cycle
5. Reliability goal setting by product or module
6. Priority of reliability-improvement goals
7. Ownership of reliability goals
8. Design for reliability (DFR) training
9. Preferred technology selection (standardization)
10. Component qualification testing
11. Original equipment manufacturer (OEM) selection and qualification testing
12. Physical failure analysis of testing failures
13. Failure and root-cause analysis
14. Statistically designed engineering experiments
15. Design and stress derating rules
16. Design review and design rule checking
17. Failure-rate estimation (prediction)
18. Thermal design and measurements
19. Worst-case analysis
20. Failure modes and effects analysis (FMEA)
21. Environmental (margin) testing
22. STRIFE (cyclical, multistress) testing
23. Design defect tracking (DDT)
24. Lessons-learned databases

Manufacturing

25. Design for manufacturability (DFM)
26. Priority of quality and reliability goals
27. Ownership of quality and reliability goals
28. Quality-training programs
29. Statistical process control (SPC/SQC)
30. Internal process audits
31. Supplier process audits
32. Incoming inspection (1005 or sampling)
33. Component-level burn-in
34. Assembly-level burn-in
35. Product-level burn-in
36. Manufacturing defect tracking
37. Corrective action reports

merical scale. A composite DFR score was computed based on the survey data, using eight of the key DFR activities (shown in Table 13.2). Data regarding reliability improvement over the three previous years were also gathered and correlated with the composite DFR score. It produced some interesting results:

(a) Products from divisions that used the eight key DFR activities 30% of the time had three times higher failure rates than those that used them 90% of the time.

(b) The 30% users were also roughly about three times further behind the 90% users in reaching the improvement-rate goal.

(c) The pretax operating profit for the 90% users was roughly +15% as opposed to +5% for the 30% users.

This was the beginning of DFR. DFR can be defined as the actions to be taken during the design phase of new product development to anticipate and manage product failures. It is through this management that one arrives at a more reliable product.

According to Moss (1996), DFR involves three phases. The key activities in each phase (for an electronic product) are:

1. Definition phase
   - Set goals
   - Select technology
   - Carry out DFR training
2. Design phase
   - Thermal design
   - Worst-case analysis
   - FMEA
3. Transfer to manufacturing
   - Component qualification
   - Supplier process evaluation

(*Note:* These key DFR activities would be different for products from other industry sectors.)
   In addition to the above, some other key DFR activities are

1. Design qualification testing
2. Failure root-cause analysis

Table 13.2. The eight key DFR activities

1. Thermal design and measurement
2. Worst-case analysis
3. Supplier process audits
4. Goals high priority
5. Supplier qualification testing
6. DFR training for engineers
7. Component stress derating
8. Failure modes and effects analysis

3. Statistical data analysis
4. Design reviews
5. Retrospective product reviews

Some of these have been discussed in earlier chapters. Others will be discussed later in the chapter.

13.4 DESIGN AND DEVELOPMENT

As indicated in Section 13.2, design and development involves three phases. In this section, we discuss each of these in detail.

13.4.1 Requirement Definition

The importance of customer satisfaction is illustrated by the following statement:

> Customer satisfaction is definitely essential to survival in today's global dynamic competition and everybody knows that the ultimate proof of a product design is the acceptance by the customer. As a result of open market place, only those companies that listen to what the customer wants and provide high-quality and reliable products, which meet customer expectations, over the product useful life period with minimum cost in a timely fashion will eventually survive. (Wang, 1990)

The requirement definition identifies the overall needs of the user and defines the objectives for the product. For certain types of products (e.g., specialized defense and industrial products) the requirement definition is supplied by the buyer. For other types (e.g., consumer durables, commercial products, and some industrial products) the manufacturer must formulate the requirement definition based on information obtained through market analysis, customer surveys, and so on. A product that meets the requirement definition will satisfy or exceed the customer needs and expectation.[2]

Consumer Durables, Commercial and Industrial Products
The satisfaction parameters for consumer durables and commercial and industrial products are many and typically include one or more of the following:

- Product performance specifications (e.g., efficiency, output level, etc.)
- Product conformance to specifications
- Product reliability (mean time to failure, mean time to repair, durability, longevity, etc.)
- Technical assistance (e.g., manuals, training programs, hot lines, etc.)
- Price and value for money
- Postsale service (warranties, spares, repair service)
- Complaint resolution

A 1989 market survey of early model new car buyers (MMR, 1989)) ranked the top five reasons for buying a new vehicle as:

1. Well made car (quality of workmanship)
2. Previous experience
3. Price or deal offered
4. Value for money
5. Reliability and durability

Wang (1990) reports that a quality function deployment (QFD) study of customer comments on a reliable/durable vehicle showed that the customer perception is usually very generic and vague. Some of the comments are listed below.

- Last for a long time
- Starts every morning
- A well-made car
- No breakdowns
- Consistent performance
- Hassle-free during ownership
- Dependable
- Maintenance-free

The manufacturer must translate these vague notions into precise technical requirements that assure customer satisfaction.

Defense Products

For defense products (e.g., aircraft, radars, ships, etc.) the buyer usually states the need in terms of performance measures expected of the product or system in operation, for example:

> ... the Air Force measures and tracks the availability, mission reliability, sortie generation rates and other figures of merit (FOMs) of its fighter aircraft fleets. These measures account for all factors that affect performance. (Born and Criscimagna, 1995)

The requirements for a new system are stated in the operational requirement document through various measures with which they manage the systems.

Reliability Specification and Goal Setting

Reliability specification and goal setting is closely linked to the requirement definition. In this context, Kohoutek (1982) suggests three types of reliability goals. We discuss these briefly before dealing with the issue of reliability specification.

The three types of reliability goals are as follows:

1. *Company-Level Reliability Goal:* This is critical for firms that manufacture products (such as aircraft, medical equipment, power plants, and communication equipment) where high reliability is very critical. This would be part of the mission statement stated in a succinct manner, such as, "Every new product will have a reliability at least 10 percent higher than that of the product it is replacing." Such a statement signals to customers and employees that top management views product reliability as an important element of its strategy for survival and growth.

2. *Reliability Goals for Product Family:* Here the focus is on reliability-growth strategies to exploit new technologies for producing more reliable products with successive generations. In other words, the reliability-improvement strategies become linked with technology development and acquisition strategies for achieving business goals. Changing technology has implications for the manner in which reliability issues are tackled in the design stage—for example, the derating rule (to be discussed later) might change as a result of new technology.

3. *Reliability Goals for Individual Products:* In contrast to the two previous goals, which are more strategic in nature and focus on technology and market trends, the reliability goal for an individual product is more operational in its focus. As such, setting reliability goals for individual products requires skills that are more analytical in character and very product specific.

For consumer durables and many commercial and industrial products, reliability goals must compete with other performance and commercial goals during the formulation of a new product strategy. As such, for the reliability goals to receive proper management attention, they must be realistic, understandable, measurable, and verifiable throughout the design and development stage. In addition, these must fit into the higher-level goals of the firm.

Reliability goal setting at the product level, in a generic context, requires taking into account the following factors:

- Customers' expectations
- Competitive pressures (industry reliability levels, overseas competition)
- Quality and reliability standards
- Past experience (based on customer feedback and in-house quality and reliability information)
- New technology impact (learning curve, new failure mechanisms)
- New end product requirements
- Company goals

Following Kohoutek (1996), one can define several different bases for setting reliability goals as indicated below. These in turn lead to reliability specifications for the design stage.

1. *Arbitrary Goals:* These can be either qualitative or quantitative, as for example
   (i) Failure rate less than 0.1 per year over the warranty period
   (ii) Expected warranty cost less than 0.5% of the sale price
   (iii) Mean time to repair less than 3 hours

2. *Goals Based on Market Sensitivity Assessment:* Goals of this type are based on fairly vague in-house opinion with regard to market sensitivity to certain reliability-related issues. This vague opinion is translated into measurable quantities by defining different levels of "discomfort" to the business. As an example, every second item failing within the first six months might be deemed as being a "major disaster" whereas one in five hundred items sold causing minor trouble would be "ideal." These two would form the extremes on the "discomfort" axis, with one or more intermediate levels of discomfort—e.g., "difficult" might correspond to profits (sale price – manufacturing cost – warranty servicing cost) being 5% of the sale price.

3. *End Product and Company Requirements:* These deal with long-term reliability goals and warranty performance of the end product. This involves warranty-trend analysis and the linkage of the reliability goal to the end product requirements. For example, the current warranty cost may be 5% of the sale price and the reliability goal in the long term is to reduce this to 3% in two years and to less than 1% in five years.

4. *Goals Based on Past Performance:* This is appropriate when an existing product is replaced by a new product that does essentially the same function, but involves new technology. The reliability goals are set to achieve a higher reliability based on the benefits of the new technology. As an example, reducing the expected failure rate in the first twelve months from 0.05% to 0.03%.

5. *Goals Based on Reliability Cost Optimization:* The total cost of manufacturing a product is the sum of the following three costs—cost of reliability design (a function of the number of cycles needed for achieving the desired reliability), cost of reliability production (fixed and variable costs of controlling reliability during manufacturing), and warranty cost (which depends on the environment and usage intensity). The reliability goal is selected to minimize this total cost.

Reliability Measures

Reliability specification involves defining appropriate reliability measures that can be evaluated objectively. These measures should capture the needs and expectations of the customers. We have defined several of these in earlier chapters. They include measures such as reliability, mean time to failure (MTTF), mean time between failures (MTBF), mean time to repair (MTTR), and so on. There are many other reliability-related measures of potentially great importance to customers that have not been discussed earlier. A number of these involve the concept of *availability*.

Let $X(t) = 1$ if the system is operating at time t, and 0 if it is not. There are several different notions of availability. Four of them are as follows:

(i) The *availability A(t) at time t* (also called *point availability*) is given by

$$A(t) = P\{X(t) = 1\} = E[X(t)] \tag{13.1}$$

(ii) The *limiting availability* is given by

$$\overline{A} = \lim_{t \to \infty} A(t) \tag{13.2}$$

when the limit exists.

(iii) The *average (mean) availability* in $[0, T]$ (also called *mission availability*) is given by

$$\overline{A}_T = (1/T) \int_0^T A(t)dt \tag{13.3}$$

(iv) The *limiting average availability* (also called *steady-state availability*) is given by

$$\overline{A}_\infty = \lim_{T \to \infty} (1/T) \int_0^T A(t)dt \tag{13.4}$$

For products used continuously (e.g., a pump in a chemical plant), the reliability goal might be defined in terms of steady-state availability. For products where usage is determined by a duty cycle (e.g., an airliner on a 12-hour flight or a satellite required to transmit data for two years), the goal might be in terms of mission availability. For products that are required to perform a function at any random time (e.g., a missile to intercept an incoming enemy aircraft or a back-up generator to supply power when the regular power supply is interrupted), the reliability goal might be defined in terms of instantaneous availability.[3]

Example 13.1 (Exponential Distribution)

We consider a nonrepairable system. Whenever a failure occurs, the failed item is replaced by a new item that is statistically similar to the failed item. Failure times are independent and identically distributed with distribution function $F(x)$. The time to replace a failed item by a new one is also uncertain. Replacement times are independent and identically distributed with distribution function $G(x)$. As a result, the failures and renewals (through replacement) occur according to an alternating renewal process. We consider the special case where $F(x)$ and $G(x)$ are exponential distributions means $1/\lambda$ and $1/\mu$, respectively. In this case, $X(t)$ can be viewed as a continuous time Markov chain with two states (0 for failed and 1 for working). Let $p_0(t)$ [$p_1(t)$] denote the probability that $X(t) = 0$ [1] at time t. Note that the two probabilities sum to one for all t and that $p_0(0) = 1$, implying that the system is new and in working state at time $t = 0$.

These probabilities are obtained as the solution of the following set of ordinary linear differential equations:

$$\frac{dp_1(t)}{dt} = -\lambda p_1(t) + \mu p_0(t)$$

$$\frac{dp_0(t)}{dt} = \lambda p_1(t) - \mu p_0(t)$$

with $p_1(0) = 1$ and $p_0(0) = 0$. These equations can be solved by any of a number of methods. The solution is

$$p_1(t) = \frac{\mu}{\lambda + \mu} + \frac{\lambda}{\lambda + \mu} e^{-(\lambda+\mu)t}$$

$$p_0(t) = \frac{\lambda}{\lambda + \mu} - \frac{\lambda}{\lambda + \mu} e^{-(\lambda+\mu)t}$$

From (13.1) we have

$$A(t) = E[X(t)] = p_1(t) = \frac{\mu}{\lambda + \mu} + \frac{\lambda}{\lambda + \mu} e^{-(\lambda+\mu)t}$$

Using this in (13.2) yields

$$\bar{A} = p_1(\infty) = \frac{\mu}{\lambda + \mu}$$

This implies that the limiting availability increases as λ decreases (or mean time to failure increases) and as μ increases (or mean time to repair decreases).

Using $p_1(t)$ in (13.3) and carrying out the integration yields

$$\overline{A}_T = \frac{1}{T}\int_0^T p_1(t)dt = \frac{\mu}{\lambda + \mu} + \frac{\lambda}{(\lambda + \mu)^2 T}[1 - e^{-(\lambda+\mu)T}]$$

In the limit as $T \to \infty$, we have

$$\overline{A}_\infty = \frac{\mu}{\lambda + \mu}$$

This implies that the limiting availability and the limiting average availability are the same.

■

13.4.2   Conceptual Design

During this stage, the capabilities of different design approaches and technologies are evaluated. The goal is to ensure that a product that embodies these design features is capable of meeting all the requirements of (1) performance as stated in the requirement definition, (2) manufacturability, and (3) unit manufacturing cost.

At this stage, the objective is to decide on the best design approach in fairly general terms—the structure for the product, the materials and technologies to be used, and so on. The following activities are involved in analysis of the various design options:

- Trade-off studies
- Mathematical and simulation models
- Cost analysis

Design trade-off studies examine alternative designs in an attempt to optimize the overall performance and at the same time reduce the technical risk. The latter is especially important when the design involves new and untested technologies.

There are many different types of cost analysis. Two commonly used analyses are *design to cost* (DTC) and *life cycle cost* (LCC). In the design to cost methodology, the aim is to minimize the unit manufacturing cost. This cost includes the cost of design and development, testing, and manufacturing. DTC is used to achieve the business strategy of a higher market share through increased sales. It is used for most consumer durables and many industrial and commercial products. As discussed in Chapter 12, in the life cycle cost methodology, the cost under consideration includes the total cost of acquisition, operation, and maintenance over the life of the item as well as the cost associated with discarding the item at the end of its useful life. LCC is used for expensive defense and industrial products. Buyers of such products often require a cost analysis from the manufacturer as a part of the acquisition process.

At the conceptual design stage, there are several reliability-related issues that are to be addressed. These are as follows:

Reliability Allocation
The process of determining subsystem and component reliability goals based on the system reliability goal is called reliability allocation or reliability apportionment. In a sense, this is the reverse of what was done in Chapter 7, where we obtained system reliability in terms of the reliabilities of its subsystems and components. The reliability allocation to each subsystem must

be compatible with its current state of development, expected improvement, and the amount of testing effort (in terms of time and money) needed for the development process. Various approaches for reliability allocation have been proposed. These are discussed in Section 13.5.

Reliability Growth
Achieving a desired reliability at the component or subsystem level often requires further development to improve reliability at that level. This is also known as reliability growth. Development involves time and money and analysis of the process requires models that relate reliability improvement to the time and effort expended. A complicating factor is the fact that the outcome of the development process is often uncertain. This leads to the need for a risk analysis. The topic of reliability improvement is discussed in detail in Chapter 15.

Proof of Concept
Proof of concept is an essential activity during the conceptual design phase of new product development. Basically, it involves verification that the targets for the various product attributes (performance, reliability, maintainability, and so on) can be achieved within the target levels for cost and time.

Output of Conceptual Design
The output of the conceptual design stage includes the following:

- Guidelines
- Design requirements
- Program plans
- Other documentation

These provide a baseline for the activities at the detailed designing stage, the next stage of the design process.

13.4.3   Detailed Design and Development

The detailed design stage begins with an initial design that is subjected to detailed analysis. Based on this analysis, the design is improved and the process is repeated until analysis of the design indicates that the performance requirements are met.

Figure 13.2, from Priest (1988), shows some of many design elements and analyses and their relationship in trade-off studies. The final design must achieve a proper balance between performance, technical risk, cost, manufacturability, and other relevant factors.

Many different types of analysis must be performed by specialists from different engineering disciplines. For example, in the case of an aircraft, the stress analysis of the frame is carried out by a specialist in solid mechanics using finite-elements methods, the lift performance of the wing in a wind tunnel using a scaled model by a specialist in fluid mechanics, and so on. Thus the detailed design analysis must be carried out by a multidisciplinary team as a project under a project leader. Project management and configuration management is used to control the evolution of the detailed design activities .

The engineering knowledge necessary for carrying out such studies and analyses includes rules of thumb, published standards by professional agencies, handbooks, and specialized textbooks. A factor that makes this difficult is that technology is changing very rapidly, and engineering knowledge and expertise often lags behind the appearance of new technologies.

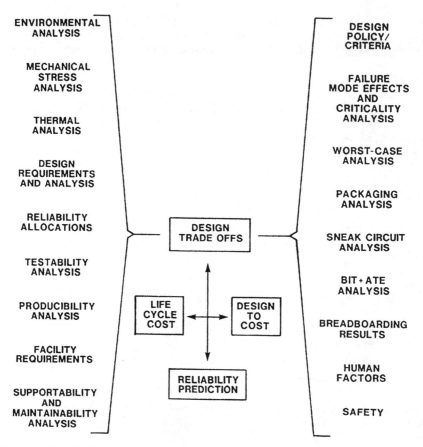

Figure 13.2. Relationship of design elements in trade-off studies. Reprinted from Priest (1988), p. 105, by courtesy of Marcel Dekker, Inc.

This requires that the design go through a rigorous review process involving testing and continuous development to overcome problems and shortcomings.

It is important to discover all potential problems with the product and fix them during this stage of the design process. If this is not done, the cost of changes to the design increases by a significant factor (roughly in the range of 2–5) if a problem requiring a design change is detected during the manufacturing phase, and by a still bigger factor (roughly in the range of 10–50) if a problem requiring a design change is detected after the product is released to buyers.

Reliability issues and techniques used during the detailed design and development phase are as follows.

Reliability Prediction and Assessment

Reliability prediction is a process used for estimating the reliability of a design prior to manufacturing and testing of produced items. Reliability predictions provide a basis for evaluating reliability growth during the development stage

It is worth noting that a reliability prediction is the reliability potential of a product based upon available design information. During the development stage, one can assess the actual reliability achieved by testing. This process is called reliability assessment. Chapter 14 deals with the topics of reliability prediction and assessment and discusses several different methods for carrying these out.

FMEA and FTA Analysis

A major part of designing for reliability is to minimize the potential effects of failures. Two design techniques for analyzing the effects of failures on product performance are

1. fault tree analysis
2. failure mode, effects, and criticality analysis

Both of these are discussed in Chapter 7.

Testing to Failure

This test procedure involves testing the prototype until it fails. The testing is done by increasing the stress level in steps at predetermined time intervals until the item fails. Such tests are called STRIFE (stress life) or HALT (highly accelerated life test). Based on the failure results, the following activities (discussed further in Chapter 15) are carried out.

Root-Cause Analysis. The aim here is to find the fundamental source of failure. Any cause higher up is a symptom or a result of the root cause. For example, the root cause of failure of a slender beam might be material property that in turn caused the failure through fatigue. In a sense, root-cause analysis can be viewed as carrying out an "autopsy" of failed item(s). The root-cause analysis requires asking the question "why?" several times until a satisfactorily explanation is found. Once the root cause is identified, the problem can be fixed by taking appropriate corrective actions by way of changes to the design or the material selection. A good discussion of this type of failure analysis in mechanical components can be found in Nishida (1992).

Example 13.2

Failure of a heart pacemaker can be due to one or more of several causes. Case 2.4 presents data on causes of failure for samples of products manufactured by two companies. A Pareto analysis of the data for the product from Company A is given in Example 3.3. As can be seen from the chart, the first six causes account for roughly 67% of the failures. Hence design changes to reduce the occurrence of these faults will have a significant impact on the reliability of the product. ■

Worst-Case Analysis. Products are often used in environments that vary significantly. The reliability of an item is influenced by the mode of usage and the operational environment. As a result, engineers must consider operational and environmental envelopes (rather than a single value) in determining reliability requirements. This implies that the product must perform satisfactorily over a wide operating environment. The range of this variation is specified as part of the requirement definition. Worst-case analysis deals with testing under the worst scenarios to evaluate product performance and reliability.

Most products go through several stages. These include production, shipping and storage, installation, and field operation. At each stage, an item is subjected to various environments. The following is an illustrative example of a product that is subjected to varying temperatures during the different stages:

Stage	Maximum temperature	Minimum temperature
Production	40 °C	10 °C
Shipping and storage	85 °C	−20 °C
Installation	70 °C	0 °C
Field operation	70 °C	0 °C

A product can undergo degradation while it is in storage and when it is being installed. As a result, the reliability of an item put into use can be lower than the designed reliability. Reliability degradation during storage is dependent on the temperature and humidity of the storage room, the length of storage, and so forth. Similarly, the reliability of the product in the field is dependent on the mode of usage of the product, the temperature profile encountered, and other environmental factors.

A worst-case analysis would involve computing reliability measures based on the extreme temperatures and extremes of other factors at each stage. This yields a very conservative design. From a practical point of view, it is necessary to evaluate the reliability measures at different values of the environmental factors at each stage and then do trade-off analyses to decide on the final design.

Example 13.3 (Case 2.1)
Case 2.1 is an example in which the reliability of the product (an audio component) during storage is affected by temperature and humidity in the storage room. The performance of the product is related to the bond that fastens three different parts, as discussed in Section 2.1. The effect of different storage environments on bond strength is shown in Table 2.1(b). Table 13.3 gives the mean, median, and standard deviation of bond strength for four different environmental conditions. It appears that an increase in temperature greatly reduces bond strength. ∎

Warranty. Warranty is a contractual obligation that requires the manufacturer to correct, or compensate for, all failures occurring within a specified warranty period subsequent to the sale of the product. Warranty is an integral part of the sale process and the servicing of warranty results in additional costs to the manufacturer. The reliability of the product has a very significant impact on warranty cost. Warranty cost analysis is discussed in Chapter 17.

Derating of Components. Derating implies operating the item below its rated stress level and is a means of improving the reliability of the item. Derating criteria and safety margins are used to provide extra insurance that the product can adequately withstand the varying operational levels and environmental stresses.

Table 13.3. Effect of storage environment on product performance

Test	Environmental condition		Statistics			
	Temperature	Humidity	n	$\bar{x}$	$x_{0.50}$	s
1	27 °C	50%	32	204.9	233.5	104.0
2	27 °C	70%	32	191.2	230.0	98.9
3	27 °C	100%	29	278.8	277.5	60.6
4	32 °C	70%	10	25.8	5.7	41.2

Derating is used extensively in the application of electrical and electronic components. The *derating factor* is defined as the ratio of maximum allowable stress to rated working stress. A component designed for operation at 440 volts being used at levels where the maximum voltage is 220 volts would have a derating factor of 0.5.

These procedures are fairly ad-hoc and are often stated as rules of thumb based on practical experience. MIL-HDBK-217 discusses this topic in depth for components subjected to electrical and thermal stresses.

Redundancy. Here one or more components (or subassemblies, etc.) are replicated. This device is used to improve reliability. Many different types of redundancies have been used in the design of reliable products. Chapter 15 deals with topic in more detail.

Reliability Design Review. As the name implies, this is a review of the reliability of the final design to ensure that the reliability goals have been achieved. This must performed independently. Lloyd and Lipow (1962) give a checklist for effective reliability design review. Two important checks in the context of new product design are:

- What are the critical weaknesses? What provision has been made in the design so that modifications can be made at the earliest possible time if these and other weaknesses show up in testing?
- Has the product been designed as simply as possible? Have human factors been considered to prevent errors such as reversed wiring for an electronic product or misassembly for a mechanical product?

Critical for carrying out an effective reliability design review is knowledge of item reliability. This is obtained by development of a proper reliability and failure reporting system.

Design for Qualification. For electronic equipment, Pecht (1993) advocated an approach called "design for qualification," which enables rigorous analysis in order to meet reliability goals. It involves the following steps:

1. Define the system usage environment
3. Identify potential failure sites and failure mechanisms
4. Characterize the materials and the manufacturing and assembly processes
5. Design reliable products within the capabilities of the materials and manufacturing processes used
6. Qualify the manufacturing and assembly processes
7. Control the manufacturing and assembly processes
8. Manage the life cycle usage of the product

In the design for qualification approach, the design tools are the physics-of-failure model and analytical techniques. Pecht illustrates this approach by use of some interesting real-life cases.

Outputs of the Detailed Design and Development Stage

The output of the detailed design stage consists of design documents for use in the manufacturing (construction) phase of the overall engineering process. This documentation deals with various issues such as

- Product (system) characterization in terms of subsystems, modules, etc., down to the part level
- Performance specification at different levels
- Material selection at the part level
- Assembly of parts into modules, subsystems, and so forth, leading to the final system
- Testing at different levels to evaluate performance

Often, the manufacturing processes to be used and the design of such processes are also determined during the detailed design stage.

13.4.4 Key Design Practices

Figure 13.3 shows in schematic form the different elements that have an impact on the design process. The figure highlights the need for a very systematic and team-oriented approach. Good design must take into account manufacturability and postsale service issues.

Priest (1988) lists key design practices at the requirements-definition, conceptual, and detailed design stages and the program organization needed to ensure the effectiveness of the overall process. At the requirements-definition stage, key design practices involve the following:

1. User needs are evaluated and documented
2. Realistic operational requirements are developed into product use profiles. These deal with the environments in which products must perform.

At the conceptual design stage, key design practices provide the following:

1. Design requirements based on user needs, operational requirements, and product use profiles
2. Design goals stated in easily understood, design-oriented terms
3. Design requirements frozen at the end of conceptual design

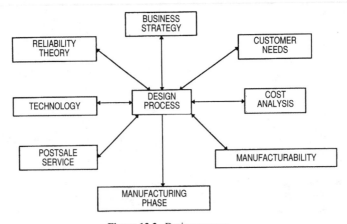

Figure 13.3. Design process.

At the detailed design stage, key design practices provide the following:

1. Design analyses and trade-off studies that take into account manufacturability and postsale issues
2. Stress analysis to evaluate product performance under different environments

13.5   RELIABILITY ALLOCATION AND APPORTIONMENT

Once the overall reliability goals for a system are specified, these must be translated into reliability goals for the subsystems, lower-level assemblies, etc., down to the part level. How this is done depends on the design of the system, but it must be done in a manner that ensures design feasibility and is consistent with current technology. In addition, the allocation must be done in such a way that other constraints are not violated. Kapur and Lamberson (1977) give several reasons for the use of reliability allocation in designing the reliability of a system. Two of these are:

1. It forces the designer to understand and develop the relationship between component, subsystem, and system reliabilities.
2. The designer can consider reliability in a framework that incorporates other issues such as cost, physical dimensions, weight, etc., in the design process.

Suppose that the system is composed on n subsystems, indexed by a subscript i, $1 \leq i \leq n$. The focus is on satisfactory performance of the system for a period of time T. This is given by a reliability goal that $R(T) \geq R^*$, where R^* is the specified minimum target value. Let $R_i(T)$ denote the reliability allocated to component i. Then depending on the configuration of the system (i.e., the linking between the different subsystems), the state of the system (working or failed) can be described in terms of the states of the n subsystems by a structure function. From this, an expression for the system reliability $R(T)$ in terms of the allocated subsystem reliabilities can be obtained as indicated in Chapter 7. The result is

$$R(T) = \phi[R_1(T), R_2(T), \ldots, R_n(T)] \tag{13.5}$$

where $\phi(\ )$ is the structure function for the system.

If the reliability goal is specified in terms of the MTTF, namely that the MTTF for the system be greater than some specified target value T^*, then the reliability of the components must be related to this MTTF. This relationship is given by

$$\text{MTTF} = \int_0^\infty R(t)dt \tag{13.6}$$

where $R(t)$ is given by (13.5) with t replacing T.

13.5.1   Approaches to Reliability Allocation

Many different approaches have been proposed for reliability allocation. The bulk of them deal with the case where the system is a series system comprised of n subsystems and the failure distributions of the subsystems are exponential with failure rate λ_i for subsystem i, $1 \leq i \leq n$. In this case, the system failure rate is given by

$$\Lambda = \sum_{i=1}^{n} \lambda_i \qquad (13.7)$$

Note that both a reliability target R^* [$R(T) \geq R^*$] and an MTTF target T^* (MTTF $\geq T^*$) can be expressed in terms of $\Lambda \leq \Lambda^*$.

Some of the approaches used (mainly in the electronics industry) are as follows:

Approach 1: Equal Apportionment. Here, one selects λ_i as

$$\lambda_i = \frac{\Lambda}{n} \qquad (13.8)$$

for $1 \leq i \leq n$. This implies that all subsystems have the same reliability. However, this might not be possible or may be uneconomical, depending on the component technology.

Approach 2: Feasibility-of-Objectives Technique. Here the subsystem failure rates are weighted by allocation factors a_i, $1 \leq i \leq n$, so that the system reliability is given by

$$\Lambda = \sum_{i=1}^{n} a_i \lambda_i \qquad (13.9)$$

The allocation factors are based on expert opinion and experience and are functions of factors such as complexity, criticality of the subsystem, technological state of the art, and so on. Larger a_i implies that subsystem i is more critical. The reliability allocation is given by

$$\lambda_i = \frac{\Lambda}{a_i n} \qquad (13.10)$$

Approach 3: AGREE Technique. This is a technique proposed by the Advisory Group on Reliability of Electronic Equipment, Office of the Assistant Secretary of Defense, USA. It takes into consideration the complexity (number of modules in each subsystem) and importance of each subsystem.

We use the following notation:

θ_i: MTTF for subsystem i

k_i: Number of modules in subsystem i

w_i: Importance factor for subsystem i

T_i: Mission time for subsystem i

$R_i(T_i)$: Reliability assigned to subsystem i

n: Number of subsystems

T: Mission time for the system

$R(T)$:   System reliability goal

The number of modules, N, is given by $N = \sum_{i=1}^{n} k_i$. The AGREE allocation results in

$$\theta_i = \frac{N w_i T_i}{k_i \{-\log R(T)\}} \qquad (13.11)$$

and subsystem reliability goals are given by

$$R_i(T_i) = e^{-(T_i/\theta_i)} \qquad (13.12)$$

Approach 4: ARINC Technique. This is a technique proposed by Aeronautical Research Inc. The method involves the use of statistical estimates of failure rates. Let λ_{ip} denote the predicted failure rate for subsystem i based on historical data. Then the failure rate, λ_{ia}, allocated to subsystem i is given by

$$\lambda_{ia} = \Lambda\left(\frac{\lambda_{ip}}{\Lambda_p}\right) \qquad (13.13)$$

where Λ_p is the predicted system failure rate based on historical data (Kapur and Lamberson, 1977).

Note that in this case, it may be determined that reliability growth through development is required. This will usually lead to significant additional design and development cost and will impact the time schedule as well.

Example 13.4

Consider an electronic system consisting of four subsystems. The number of modules (k_i), importance factor (w_i) and mission time (T_i) for the four subsystems are as follows:

Subsystem	k_i	w_i	T_i (days)
1	4	1.0	300
2	8	0.8	100
3	1	1.0	300
4	6	0.9	100

The reliability allocation is to be determined based on the AGREE method to ensure a system reliability level of 0.95 for 300 days of continuous operation.

Note that $N = k_1 + k_2 + k_3 + k_4 = 19$. From (13.12) we have

$$\theta_1 = \frac{19 \times 1.0 \times 300}{4 \times \{-\log(0.95)\}} = 2.7781 \times 10^4$$

Similarly,

$$\theta_2 = 3.7042 \times 10^3, \qquad \theta_3 = 2.7779 \times 10^4, \qquad \theta_4 = 5.5575 \times 10^3$$

These yield

$$R_1(300) = 0.9893, \qquad R_2(300) = 0.9734, \qquad R_3(300) = 0.9893, \qquad R_4(300) = 0.9822$$

This yields an overall system reliability $R = 0.9357$. This is somewhat below the desired value of 0.95. The reason for this is that the importance factors for subsystems 2 and 4 are less than 1. A redesign may be necessary to achieve the desired reliability. ∎

13.5.2 Optimal Reliability Allocation

In some applications, it may be desirable to allocate reliability in a manner that optimizes a given criterion related to performance or cost (e.g., system reliability, MTTF, or total cost) subject to constraints on one or both of these. This issue will be discussed in Chapter 18.

13.6 DESIGNING AT THE COMPONENT LEVEL

Designing at the component level is discipline-specific in the sense that designing mechanical components is different from designing electrical or electronic ones. The basic methodology, however, is the same. We focus on this methodology and illustrate the technique by application to the design of mechanical components.[4]

Design for static reliability (where failure occurs due to overstress) involves modeling the strength of, and the stress on, the component. Design for dynamic reliability (where failure occurs due to degradation and wear) involves modeling the degradation process to obtain the distribution function for time to failure. The design parameters are then selected to ensure that the component meets the desired reliability specification. This is done using the stress–strength models discussed in Section 6.6.1 (for static reliability) and Section 6.6.2 (for dynamic reliability).

There are several factors that affect the strength of a component. Some of the important ones are as follows:

1. *Material used:* The strength of the material varies due to variations in the raw material and the processes used. Table 13.4 gives the mean and standard deviations for the yield strength and ultimate yield strength (in 10^3 lbs per square inch) for a sample of materials. More comprehensive tables can be found in books on mechanical design, for example, Haugen (1980).

2. *Manufacturing process:* Due to uncontrollable variability in the manufacturing process, the actual dimensions of a component will differ from the desired. This is captured through the concept of tolerance. If a particular desired dimension (for example, diameter of a shaft) is to be μ_D, then the actual dimension, D, is a random variable which may be assumed to be normally distributed with mean μ_D and standard deviation σ_D. The tolerance is $3\sigma_D$, implying that with probability 0.997 (assuming normality) the dimensions of the component are within $\mu_D \pm 3\sigma_D$. The tolerances for a variety

Table 13.4. Statistical properties of common materials used in design

Material	Yield strength, k (lb/in^2)		Ultimate strength, k (lb/in^2)	
	Mean (μ)	SD (σ)	Mean (μ)	SD (σ)
2014 Aluminum alloy (forging)	63.0	2.23	70.0	1.89
2024-T6 Aluminum alloy (sheet)	50.1	2.85	63.6	2.51
C1006 Carbon steel (hot-rolled sheet)	35.7	0.80	48.3	0.52
Type 202 Stainless steel	49.9	1.32	99.7	2.71
Type 301 Stainless steel	166.8	9.37	191.2	5.82

Table 13.5. Tolerances obtainable in common manufacturing operations

Manufacturing process	Tolerance obtainable (inch)
Turning, shaping, milling	±0.001
Drilling	±0.002
Grinding	+0.002
Sawing	±0.005
Forging	±0.03
Hot extrusion	±0.005
Die casting	±0.002
Spot welding, fusion welding	±0.01

of machining operations are given in Table 13.5. More comprehensive lists can be found in design handbooks such as that by Shigley and Mischke (1986).

3. *Usage mode:* This characterizes the mode of failure—fatigue failure, tensile fracture, etc. Usage mode plays a critical role in the relationship that links the design parameters to the probability of failure, as will be seen in Example 13.5.

4. *Geometry of the component:* This has a bearing on the tolerance and as such its effect is incorporated in the tolerance calculation.

13.6.1 Designing for Static Reliability

Let X denote the strength of a component. Variability in X is due to several factors, as indicated earlier. Let X_i, $i = 1, \ldots, m$, be design parameters that are to be selected properly to ensure a certain level of reliability when the component is put into operation. If the variability in X_i is small and can be ignored, then the corresponding design parameter can be treated as deterministic. If not, it is characterized by its mean μ_i and variance σ_i^2 (which is related to tolerance in the case of a machining operation). In this case, the design decision variable is the mean.

Let X_i, $i = m + 1, \ldots, m + n$, with means μ_i and variances σ_i^2, denote the set of factors that are not under the control of the designer, but affect the strength of the component. The strength X is a function of the X_i. The precise nature of this relationship is component-specific. For example, the relationship for a component failing under axial compressive force would be different from that for a component failing under torsional loading. Let this relationship be given by

$$X = g(X_1, \ldots, X_{m+n}) \tag{13.14}$$

The stress on the component, when put into operation, is a random variable. Let Y denote this stress. It is a function of several factors such as environment, nature of loading (static or dynamic), and so on. We assume that Y is normally distributed with mean μ_Y and variance σ_Y^2. As discussed in Section 6.6.1, we have a component failure when $X < Y$. The design parameters are to be selected so as to ensure that the reliability of the component is R_0. This translates into the requirement that

$$P\{X - Y > 0\} = R_0 \tag{13.15}$$

In order to calculate this probability, we require the distribution of $X - Y$, which may be obtained from the distributions of X and Y. The distribution of X may be quite complex, depending on the distributions of the X_i and the complexity of the function g(.). To overcome this, we use the normal approximation outlined in Section A5 of Appendix A. By linearizing the equation, expressions for the mean and variance of X can be obtained from Equations (A5.4) and (A5.5) with $\tau(\ ) = g(\ )$ and $\sigma_{ij} = 0$ for $i, j = 1, \ldots, m + n$ and $i \neq j$. These are given by

$$\mu_X = g(\mu_1, \ldots, \mu_{m+n}) \tag{13.16}$$

and

$$V_a(X) = \sigma_X^2 = \sum_{i=1}^{m+n} \left(\frac{\partial g}{\partial X_i} \bigg|_{\mu_1, \ldots, \mu_{m+n}} \right)^2 \sigma_i^2 \tag{13.17}$$

As a result of this approximation and the assumption on Y, $\{X - Y\}$ is approximately a normal random variable with mean $\{\mu_X - \mu_Y\}$, and variance $\{\sigma_X^2 + \sigma_Y^2\}$. This is used in (13.15) to select design parameters to ensure the desired reliability.

Example 13.5

The component under consideration is a connecting rod of length L and diameter D that is subjected to an axial compressive force Y. The strength of the component is the Euler buckling load (see Fitzgerald, 1982) of the connecting rod, given by

$$X = \frac{\pi^2 EI}{L^2}$$

where E is the Young's modulus and I is the moment of inertia of the cross section. I is given by

$$I = \frac{\pi D^4}{64}$$

As a result, the strength X is given by

$$X = \frac{\pi^3 ED^4}{64L^2}$$

The mean diameter μ_D is to be selected to ensure a reliability of 0.99 against buckling under the stress Y. We make the following assumptions:

- Y is normally distributed with mean $\mu_Y = 2000$ lbs and standard deviation $\sigma_Y = 200$ lbs
- L is normally distributed with mean $\mu_L = 20$ in and standard deviation $\sigma_L = 0.5$ in
- E is assumed to be deterministic with a value 30×10^6 (lb/in²)
- D is also normally distributed with mean μ_D and standard deviation $\sigma_D = 0.1 \mu_D$

From (13.16) and (13.17), we have

$$\mu_X = \frac{\pi^3 E \mu_D^4}{64 \mu_L^2} = \frac{\pi^3 (30 \times 10^6) \mu_D^4}{64(20)^2} = 36331 \mu_D^4$$

and

$$\sigma_X^2 = \left[\left(\frac{\partial X}{\partial L} \right)^2 \sigma_L^2 + \left(\frac{\partial X}{\partial D} \right)^2 \sigma_D^2 \right]$$

Note that

$$\frac{\partial X}{\partial L} = -2 \frac{\mu_X}{\mu_L} \quad \text{and} \quad \frac{\partial X}{\partial D} = 4 \frac{\mu_X}{\mu_D}$$

and as a result, we have

$$\sigma_X = 15856 \mu_D^4$$

By the approximation, $(X - Y)$ is taken to be normally distributed with mean $(36331 \mu_D^4 - 2000)$ and variance $(200)^2 + (15856 \mu_D^4)^2$. From (13.16) the specified reliability of 0.99 corresponds to a lower limit of -2.3264 for the standard normal distribution and this implies that

$$-2.3264 = -\left(\frac{36331 \mu_D^4 - 2000}{\sqrt{(15856 \mu_D^4)^2 + (200)^2}} \right)$$

This can be rewritten (after squaring and simplifying) as

$$A_2 z^2 + A_1 z + A_0 = 0$$

where $z = \mu_D^4$, $A_2 = 4.6312$, $A_1 = -1.45324$ and $A_0 = 0.037839$. This is a quadratic equation in z, the solutions of which are $z = 0.30829$ and $z = 0.004674$. This yields $\mu_D = 0.7451$ for the first case. The second case yields a value that does not ensure that the reliability is 0.99 and hence is of no interest. ∎

13.6.2 Designing for Dynamic Reliability

In Section 6.6.1, we discussed the effect of component degradation on its strength and developed relationships that link time to failure with the parameters of the model. Some of these parameters are design parameters that must be properly chosen so as to ensure that the component meets its dynamic reliability specification. The reliability specification is usually in the form of mission reliability. This requires that the probability of the component not failing over some specified mission time is greater than some prespecified value. We illustrate this by the following example.

Example 13.6
Consider Model 6.3 of Section 6.6.1. The strength of the component degrades over time and as a result the time to failure T is given by

$$T = \frac{1}{\theta} \log(A/Y)$$

Here Y is the stress on the component, A is its initial strength, and θ is a parameter that characterizes the degradation process. This can be viewed as a model of a pipe in a chemical plant, where A corresponds to the initial thickness of the pipe that decreases with time due to corrosion. Y is the pressure at which the fluid is pumped. This represents the stress on the pipe and is assumed to be constant. The parameter θ is a random variable with a uniform distribution over the interval $[a, b]$. This implies that the density function for θ is given by

$$f_\theta(x) = \frac{1}{b-a}$$

for $a \leq x \leq b$ and zero elsewhere. As a result, the time to failure, T, is a random variable. The reliability requirement for the component is that $P\{T > t_d\} \geq \alpha$ where t_d and α are specified. This is to be achieved through a proper choice of the design parameter A.

Let $F_T(t)$ denote the distribution function for T and $f_T(t)$ be the associated density function. Note that T is a nonlinear function of θ. Let $k = \log(A/Y)$. From Appendix A, we have

$$f_T(t) = \frac{k}{(b-a)t^2}$$

or

$$F_T(t) = \frac{1}{(b-a)}\left[\frac{b}{k} - \frac{1}{t}\right]$$

The reliability requirement then becomes

$$P\{T > t_d\} = 1 - F_T(t_d) = 1 - \frac{1}{(b-a)}\left[\frac{b}{k} - \frac{1}{t_d}\right] = \alpha$$

Solving for k, we have, after some simplification,

$$k = \log(A/Y) = t_d\{\alpha b + (1-\alpha)a\}$$

which yields

$$A = Ye^{t_d\{\alpha b + (1-\alpha)a\}}$$

Note that A increases as

1. stress level Y increases
3. mission time t_d increases
4. variability in θ [or the length of the interval $(b-a)$] increases
4. mission time reliability α increases

These are as to be expected. ∎

13.7   MANUFACTURING AND RELIABILITY DEGRADATION

As mentioned earlier, manufacturing is the process of transforming inputs (raw materials, components) into finished products. The process is often complex and can involve several stages with several different operations at each stage, depending on the product. The reliability of the products produced will, in general, differ from that predicted by the design or the prototype produced under strict laboratory conditions and, in fact, is almost always lower. Degradation in reliability during manufacturing is due to a variety of causes. As first proposed by Shewhart (1931), these can be categorized into two groups:

1. *Variations due to uncontrollable causes.* This is often called *common cause variability.* By and large, nothing can be done about this source of variation, except to modify the process.
2. *Variations due to assignable causes* in the production process, input material and/or human operators. Often called *special cause variability,* this can be controlled through effective quality control schemes and process modifications (e.g., machine adjustment, replacement of worn-out parts, additional training of operators, etc.).

In practice, it is usually found that 80 to 95% of the variability is common-cause variability and the remaining 5 to 20% special-cause.

Items that conform to the design specifications are called "conforming" (or nondefective) and those that do not, are called "nonconforming" (or defective) items. In the remainder of the section, we discuss the characterization of nonconforming items and their occurrence during manufacturing.[5]

13.7.1   Reliability Characterization of Nonconforming Items

For a conforming item, the failure rate is roughly constant until time t_f, beyond which it is increasing due to wear, as indicated in Figure 13.4. Any failure up to t_f is a purely random event. Let $F(t)$ denote the failure distribution associated with this failure rate.

The failure rate of a nonconforming item can be modeled in several different ways, depending on the context. We look at three of these below. Let $H(t)$ denote the failure distribution function for a nonconforming item.

Case (i)
The failure rate of a nonconforming item is as shown in Figure 13.5(a). This differs from that in Figure 13.4 (the failure rate for a conforming item) over the interval $[0, t_i)$. The failure rate starts with a high initial value and is continuously decreasing over the interval.

An explanation for this is as follows. In general, most products are composed of several

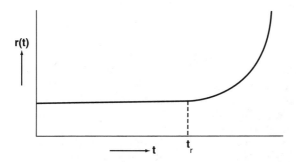

Figure 13.4. Failure rate (design specification).

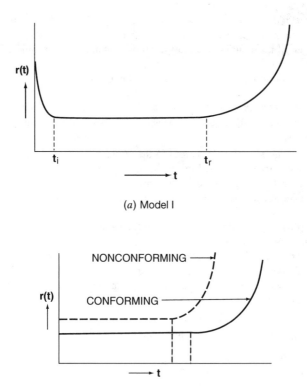

(a) Model I

(b) Model II

Figure 13.5. Failure rate of manufactured items.

parts. Even with all the parts conforming to design specification, an item can fail early due to chance errors in the assembly process (e.g., misalignment, dry solder joint, and so on). This increases the failure rate during the initial life of items. These types of failures are also termed "teething problems" or "infant mortality" and are detected over the interval $[0, t_i)$. Once these problems have been sorted out, the failure rate is the same as that of conforming items.

Case (ii)
The failure rate of a nonconforming item is as shown in Figure 13.5(b). This differs from that shown in Figure 13.4 over the entire time interval $[0, \infty)$. Note that the interval over which the failure rate is constant is smaller, and the failure rate higher, than that for a conforming item. In addition, the rate of increase over the remaining period is larger. In other words, the failure rate for a nonconforming item is larger than that for a conforming item over the interval $[0, \infty)$. This implies that $H(t) > F(t)$ for all t.

An explanation for this is as follows. Due to variability in manufacturing, some of the components have reliability characteristics that are inferior to those of the rest. When such a component is used in the manufacture of an item, the failure rate of the item increases.

Case (iii)

This is a combination of the two earlier models. In this case, a nonconforming item has a decreasing failure rate until time t_1, beyond which it has a failure rate similar to that for a conforming item, but with a higher value.

Comment: The discussion so far has dealt with dynamic reliability. In this case, whether or not an item is conforming can be assessed only in a statistical sense through testing (or operating) the item to failure. Often, degradation results in a nonoperational item. This can be modeled by $H(t) = 1$, for $t \geq 0$. Such nonconforming items can be detected through testing for a very short time interval.

Example 13. 7

Let the failure rate of a conforming item be given by

$$r_F(t) = \begin{cases} a_0 & 0 \leq t < t_f \\ a_0 + \beta(t - t_f) & t_f \leq t < \infty \end{cases}$$

This implies that $F(t)$, the failure distribution for a conforming item, is $F(t) = 1 - e^{-R_F(t)}$, with $R_F(t)$ given by

$$R_F(t) = \begin{cases} a_0 t & 0 \leq t < t_f \\ a_0 t + (\beta/2)(t - t_f)^2 & t_f \leq t < \infty \end{cases}$$

Let the failure rate of a nonconforming item be given by

$$r_H(t) = \begin{cases} a_i + \alpha(t_i - t) & 0 \leq t < t_i \\ a_i & t_i \leq t < t_f \\ a_i t + \beta(t - t_f) & t_f \leq t < \infty \end{cases}$$

This implies that $H(t)$, the failure distribution for a nonconforming item, is $H(t) = 1 - e^{-R_H(t)}$, with $R_H(t)$ given by

$$R_H(t) = \begin{cases} a_i t + (\alpha t/2)(2t_i - t) & 0 \leq t < t_i \\ a_i t + (\alpha/2)t_i^2 & t_i \leq t < t_f \\ a_i t + (\alpha/2)t_i^2 + (B/2)(t - t_f)^2 & t_f \leq t < \infty \end{cases}$$

Suppose that the parameter values are as follows:

$$a_0 = 0.2, \qquad a_i = 0.22, \qquad \alpha = 10.0, \qquad \beta = 0.1, \qquad t_i = 0.5, \qquad t_f = 5.0$$

Figure 13.6 is a plot of $F(t)$ and $H(t)$. As can be seen from these, nonconformance substantially increases the probability of early failures. ■

13.7.2 Modeling Occurrence of Nonconforming Items

Modeling of the occurrence of nonconforming items depends on the type of manufacturing process used. The process to be used depends on the demand for the product and is determined by economic considerations. If the demand is high, then it is economical to use a continuous production process. If the demand is low to medium, then it is more economical to use a batch production process, where items are produced in lots (or batches).

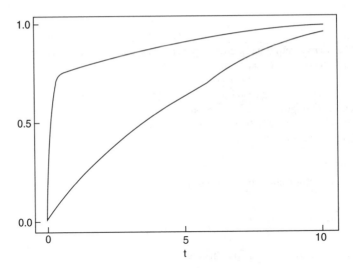

Figure 13.6. Failure distributions $F(t)$ (lower curve) and $H(t)$ (upper curve).

In either case, the state of the manufacturing process has a significant impact on the occurrence of nonconforming items. In the simplest characterization, the process state can be modeled as being in one of two possible states—(1) in control and, (2) out of control. When the process state is in control, all the assignable causes are under control and, although nonconformance cannot be avoided entirely, the probability that an item produced is nonconforming is very small. For a process that has been designed properly, this probability can be as small as 10^{-3} to 10^{-6}. The change from in control to out of control is due to one or more of the process parameters no longer being at required target values. This increases the probability that an item is nonconforming. Let p_{in} and p_{out} denote the probability that an item produced is conforming when the process is in control and out of control, respectively. In general, $p_{in} \gg p_{out}$. In the extreme cases, $p_{in} = 1$, implying that all items produced are conforming when the state is in control, and $p_{out} = 0$, implying that all items produced are nonconforming when the process is out of control.

We discuss two models for the occurrence of nonconforming items. Both will be discussed further later in the chapter.

Model 1 (Continuous Production)

The manufacturing process starts in control and after a random length of time it changes to out of control. When the process is in control, the probability that an item produced is conforming is p_{in} and that it is nonconforming is $(1 - p_{in})$. Since the failure distributions of the two types of items are $F(t)$ and $H(t)$, respectively, the failure distribution of an item, $G(t)$, given that the process is in control, is a mixture of these two distributions, namely

$$G_{in}(t) = p_{in}F(t) + (1 - p_{in})H(t) \tag{13.18}$$

Similarly, given that the process is out of control, the item failure distribution is given by the mixture

$$G_{out}(t) = p_{out}F(t) + (1 - p_{out})H(t) \tag{13.19}$$

Once the process state changes from in control to out of control, it remains in that state until it is brought back to in control through some corrective action.

Example 13.8

Let $p_{in} = 1$ and $p_{out} = p$. Suppose that the failure distribution $F(t)$ is a Weibull distribution with scale parameter β_1 and shape parameter α_1 and that $H(t)$ is also a Weibull distribution with scale parameter β_2 and shape parameter α_2. The failure distribution of items produced, $G(t)$, is given by $F(t)$ when the process is in control and by a mixture of two Weibull distributions when the process is out of control. This implies that all items produced are conforming when the process is in control and the characteristics of items produced when the state is out of control is a function of p, namely

$$G(t) = pF(t) + (1 - p)H(t)$$

We use the same parameter values as in Example 4.17, so that the parameters of $F_1(x)$ are $\alpha_1 = 1.25$ and $\beta_1 = 5.3659$ (implying an MTTF of 5 years) and the parameters of $F_2(x)$ are $\alpha_2 = 1.25$ and $\beta_2 = 1.0732$ (implying an MTTF of 1 year). Then from (4.81), the MTTF for the item is $[5p + (1 - p)]$ years and is a function of p. As the manufacturing process deteriorates with age and usage, p decreases implying that the quality (in terms of conformance) is going down. The effect of this on the component reliability is shown in Figure 4.7 for p varying from 0 to 1 in steps of 0.2. Not surprisingly, the effect is substantial; the reliability for a period of 3 years changes from 0.617 for $p = 1$ to 0.027 for $p = 0$. ∎

Model 2 (Batch Production)

Here the items are produced in lots of size L. At the start of each lot production, the process state is checked to ensure that it is in control. If the process state is in-control at the start of the production of an item, it can change to out of control with probability $(1 - q)$, or continue to be in control with probability q. Once the state changes to out of control, it remains there until completion of the lot. As mentioned previously, an item produced with the state in control [out of control], is conforming with probability p_{in} [p_{out}] with $p_{out} \ll p_{in}$.

Let N_c denote the number of conforming items in a lot. Note that this is a random variable. We obtain the distribution of N_c using a conditional approach. Let N denote the number of items produced before the process changes from in control to out of control. This is a random variable that can assume integer values in the interval $[0, L]$. The probability distribution of N is given by

$$P\{N = n\} = \begin{cases} q^n(1 - q) & 0 \le n < L \\ q^L & n = L \end{cases} \tag{13.20}$$

This follows since the process starts in control and during the production of an item it can change to out of control with probability $(1 - q)$ or remain in control with probability q. Note that $n = 0$ corresponds to the process going out of control during the production of the first item and as a result, all items produced subsequently are produced with the process out of control. Similarly, $n = L$ implies that the process state never changes during the production of the lot and as a result, all items produced are with the process in control. For $0 < n < L$, the first n items are produced with the process in control and the remaining $(L - n)$ with the process out of control.

Let the number of conforming items in the first N items of the lot be N_1 and the number in the remaining $(L - N)$ items be N_2. Conditional on $N = n$, N_1, and N_2 are independent random variables with probability functions given by binomial distributions with parameters (p_{in}, n) and $(p_{out}, L - n)$, respectively. The number of conforming items in the batch, K, is given by $N_c = N_1 + N_2$. Using a straightforward probabilistic argument, we have

$$P\{N_c = k\} = \sum_{n=0}^{L} \sum_{m=\max\{0,k-L+n\}}^{\min\{n,k\}} P\{N_1 = m|N = n\}P\{N_2 = k - m|N = n\}P\{N = n\}$$

$$= \sum_{n=0}^{L} \sum_{m=\max\{0,k-L+n\}}^{\min\{n,k\}} P\{N = n\}\binom{n}{m}p_{in}^m(1 - p_{in})^{n-m}\binom{L-n}{k-m}p_{out}^{k-m}(1 - p_{out})^{L-n-k+m} \quad (13.21)$$

with $P\{N = n\}$ given by (13.20). From this, we have the expected number of conforming items in the lot given by

$$E[N_c] = \sum_{k=1}^{L} kP\{K = k\} = \frac{q(p_{in} - p_{out})(1 - q^L)}{(1 - q)} + p_{out}L \quad (13.22)$$

The expected fraction of conforming items in a lot of size L, $\phi(L)$, is given by

$$\phi(L) = \frac{q(p_{in} - p_{out})(1 - q^L)}{(1 - q)L} + p_{out} \quad (13.23)$$

It is easily seen that $\phi(L)$ is a decreasing sequence in L since

$$\phi(L + 1) - \phi(L) = \left\{\frac{q(p_{in} - p_{out})}{1 - q}\right\}\left[\frac{(1 - q^{L+1})}{(L + 1)} - \frac{(1 - q^L)}{L}\right] < 0 \quad (13.24)$$

This implies that the expected fraction of conforming items in a batch decreases as the batch size (L) increases. Two special cases are:

$$L = 1: \phi(L) = qp_{in} + (1 - q)p_{out}$$
$$L = \infty: \phi(L) = p_{out}$$

Table 13.6. Average outgoing quality $\phi(L)$ versus lot size (L).

L	$\phi(L)$
10	0.9683
20	0.9516
30	0.9350
40	0.9195
50	0.9050
60	0.8914
70	0.8787
80	0.8666
90	0.8544
100	0.8448

In the latter case, only a finite number of items are produced with the process being in control and an infinite number produced with the state being out of control.

Example 13.9

Let $q = 0.99$, $p_{in} = 0.99$ and $p_{out} = 0.60$. Table 13.6 gives the values of $\phi(L)$ for L varying from 10 to 100. Note that $\phi(L)$ decreases as L increases, as expected. The optimal choice of L is discussed in Chapter 18. ■

13.8 CONTROL OF RELIABILITY DEGRADATION

A nonconforming item is less reliable than a conforming item. This has serious implications for both buyers and manufacturers. From a buyer's perspective, it leads to increased maintenance cost as well as affecting the operational cost because of increased down time. This in turn affects the buyer's satisfaction. From the manufacturer's point of view, it leads to increased warranty servicing cost when items are sold with a warranty. In addition, items that fail early in their life can seriously affect the reputation of the product and its subsequent sales. This implies that the manufacture must control reliability degradation during manufacturing.

Reliability degradation during manufacturing can be controlled using any of a number of different approaches. These can be broadly grouped into the following three categories:

1. Weeding: To detect and remove nonconforming items through inspection and testing
2. Prevention: To reduce the occurrence of nonconforming items through control actions
3. Process improvement: To reduce the occurrence of nonconforming items through changes in process design

The first two are "on-line" approaches and the third one is an "off-line" approach to controlling reliability degradation. In this section we briefly discuss some approaches in each of the three categories.

13.8.1 Weeding Out Nonconforming Items

The aim of weeding is essentially to detect nonconforming items before they are released for sale. This can only be done by inspection and testing of each item. For static reliability, testing takes very little time, since a nonconforming item is detected immediately after it is put into operation. In contrast, the detection of nonconforming items for dynamic reliability may involve testing for a significant length of time.

Testing can be done either at the end of the final stage and/or at one or more of the intermediate stages of the manufacturing process. Detection of nonconformance at the earliest possible stage is desirable, as this allows for immediate corrective action. In some cases, the nonconforming item can be transformed into a conforming item by reworking it, and in other cases, the item must be scrapped. In either case, if a nonconformance is not detected at the earliest possible instant, the effort involved until it is detected is either wasted (if the item has to be scrapped) or the amount of rework required to make it conforming increases (if the nonconforming item can be fixed). Both result in extra cost. On the other hand, testing and inspection also cost money and it is necessary to achieve a suitable balance between these two costs. This implies that the location of inspection and testing stations in a multistage manufacturing process must be optimally selected.

Another issue that must be addressed is the level of inspection and testing effort necessary. One can either carry out 100% inspection or less than 100% inspection. Again, cost becomes an important factor in deciding on the level of inspection. The quality of inspection and testing is still another variable that must be considered. If inspection and testing are perfect, then every nonconforming item tested is detected. With imperfect testing and inspection, not only may a nonconforming item not be detected, but a conforming item may be classified as nonconforming. As a result, the outgoing quality (the fraction of conforming items) depends on the level of testing and the quality of testing. It is 100% if testing is both 100% and perfect.

Example 13.10

A mechanical component is required to have a breaking strength of at least 500 lbs. Components are manufactured using casting and are produced in lots of 100. Due to variability in the material and the casting process, p, the probability that an item is not conforming (has strength less than 500 lbs), is uncertain.

The manufacturer uses the following testing scheme. For each batch, one item is selected randomly and tested. If it is nonconforming, the batch is released with no further testing. If not, all of the remaining items are tested.

The probability that a batch is released with no further testing is $(1—p)$. In this case, the average number of nonconforming items in a lot released is $99p$.

When the batch is subjected to 100% testing (i.e., when the first item is found to be nonconforming), then there is no nonconforming item in the lot released. The average number of items rejected in a batch is $1 + 99p$. The quality improvement is achieved at the expense of 100% testing.

One can look at schemes where a sample of n items are tested and the batch released only if there are no nonconforming items in the sample. ■

In the case of dynamic reliability, testing involves putting items on a test bed and operating them (either under normal or accelerated test conditions) for a certain length of time. This is also called "burn-in" and is discussed in Section 13.10.

13.8.2   Prevention of the Occurrence of Nonconforming Items

When the process is in control, the occurrence of nonconforming items is due to uncontrollable factors. When the process is out of control, the occurrence of nonconforming items increases and the reason for the change in the process state is due to one or more of the controllable factors deviating significantly from their set values.

In this subsection we focus on prevention of the occurrence of nonconforming items through actions that attempt to ensure that the process is in control during the production run. The next subsection deals with prevention through process redesign and improvement. We consider both continuous and batch production.

Continuous Production

In continuos production, the process begins in control and changes to out of control with the passage of time. If the change is detected immediately, then it can be brought back into control at once and the high occurrence of nonconforming items when the state is out of control would thereby be avoided.

Control charts are used for the purpose of detecting out of control conditions. These charts can be grouped into two broad categories—(1) variables charts and (2) attribute charts. Variables charts are based on continuous-valued measurements (e.g., physical dimension, hardness), whereas attribute charts are based on integer-valued measurements (e.g., counts of flaws such as the number of dry solder joints, number of items out of spec, etc.). The basic principle of a control chart is to sample periodically and to plot sample statistics (e.g., sample mean, range, or standard deviation in the case of continuous variables, or number or fraction of defectives in the case of attribute variables).

In the case of a variables chart, suppose the variable being observed is normally distributed with mean μ and variance σ^2 when the process is in control. When the process goes out of control, either the mean changes and/or the variance increases. Each inspection involves taking a sample of size n at regular intervals. Let x_{ji} denote the observed value for the ith item at time instant t_j. Note that $j = 1, 2 \ldots$ and $i = 1, \ldots, n$ for each j. The sample mean for the observations at time t_j is given by

$$\bar{x}_j = \frac{\displaystyle\sum_{i=1}^{n} x_{ji}}{n}$$

This statistic is plotted on the control chart. The chart also has a center line and two control lines. The center line is a horizontal line corresponding to the nominal mean μ or the overall sample mean (the average of the $\bar{x}_j$). The two control lines (or control limits) are parallel to the center line and at a distance $(3\sigma/\sqrt{n})$ on either side of the center line. (*Note:* For other charts, including range charts, s-charts, and attribute charts, the center line and control lines are based on the mean and standard deviation of the appropriate distribution.)

If the sample plots lie within the control limits, we conclude that the process is in control with high probability and no action is needed. When one or more sample plots fall outside the limits, this is taken as an indicator of a change in the process state from in control to out of control. In fact, many rules have been developed to determine when a process is out of control and action is to be taken. Some of these are:

1. A single point falling outside the control limits
2. Six points in a row all increasing or all decreasing
3. Two out of three points in a row more than 2σ above or below the center line

These are a few of many different rules that can be found in the quality-control literature. For more on this, see Evans and Lindsey (1996) and Ryan (1989). Lists of rules may also be found in statistical packages (e.g., Minitab lists eight rules).

When it is concluded that the process is out of control, some action is required. This involves stopping the process and checking to see if a change has indeed occurred. If so, corrective actions to restore it back to its in control state are initiated. If not, no corrective action is needed and production resumes.

For any given stopping rule, the following two issues are of importance:

1. A false alarm that leads to a stoppage of the production when the process is in control
2. Corrective action not being initiated for a certain length of time subsequent to the process changing from in control to out of control. When this occurs, one or more batches are produced with the process out of control

One would ideally like to have the probabilities of both of these events to be zero. Unfortunately, this is not possible and the probabilities depend on the rules used for initiating corrective actions. There is a large literature on the determination of optimal rules that take into account the economic consequences of these two types of errors.

Finally, it is worth noting that the control chart does not indicate the cause for the change in the state of the process. To determine the cause requires other tools, such as root-cause analysis.

Batch Production

In batch production, the process starts in control and can go to out of control during the production of a lot. This affects the number of nonconforming items in the lot. As shown in Section 13.7.2 (Model 2), the expected fraction of nonconforming items in a lot increases with the lot size L. This implies that the smaller the lot size, the better the outgoing quality. On the other hand, the size of the lot has implications with regard to unit manufacturing cost, since each batch production results in a fixed set-up cost. This implies that it is necessary to determine the optimal lot size by a proper trade-off between this cost and the benefits derived through better outgoing quality. This issue is discussed further in Chapter 18.

13.8.3 Process Improvement

In the cases of both continuous and batch manufacturing, the design of the manufacturing process has a significant impact on p_{in}, the probability that an item is conforming when the process is in control. Ideally, one would like to have this probability be one, so that no item produced is nonconforming.

A manufacturing process is affected by several factors—some controllable and others, not. Taguchi (1981) proposed a method for determining optimal settings for the controllable factors, taking into account the influence of the uncontrollable factors. The method uses well-known concepts from design of experiments (discussed in Chapter 10) combined with the concept of "signal-to-noise" ratio from electrical communication engineering. Since the pioneering work of Taguchi, there has been considerable development in the design of optimal and robust manufacturing processes. (See Notes at the end of the chapter for some references.)

13.9 TESTING

Testing can be defined as the application of some form of stimulation to a system (or subsystem, module or part) so that the resulting performance can be measured and compared to design requirements. In the context of new products, one can group testing into three categories:

1. Developmental testing
2. Manufacturing testing
3. Field and operational testing

13.9.1 Developmental Testing

As the name suggests, developmental testing is the testing carried out during the development phase to assess and improve product reliability. Some of the tests used follow.

Testing to Failure: Tests to failure are usually performed at module and subsystem levels.

The test involves subjecting the item to increasing levels of stress until a failure occurs. Each failure is analyzed and fixed. This is part of the test, analyze, and fix methodology that is discussed in more detail in Chapter 15.

Environmental and Design Limit Testing: This is done at part, subsystem, and system levels and should include worst-case operating conditions, including operations at the maximum and minimum specified limits. Test conditions can include temperature, shock, vibration, and so forth. Any failures resulting from the test should be analyzed (through root-cause analysis) and fixed through design changes. These tests are to assure that the product performs at the extreme conditions of its operating envelope.

Accelerated Life Testing: When a product is very reliable, it is necessary to use accelerated life tests to reduce the time required for testing. This involves putting items on test under environmental conditions that are far more severe than those normally encountered. Such tests are used to evaluate the useful life of systems-critical parts for problem identification and improvement.

In order to conduct and analyze an accelerated life test, an understanding the relationship between environmental conditions and their impact on product failure is necessary. Chapter 6 discussed some models that link failure times to environmental stress levels, and some designs are discussed in Chapters 10 and 14.

Critical Item Evaluation and Part Qualification Testing: The purpose of these tests is to verify that a part is suitable under the most severe conditions encountered under normal use. The tests to be performed depend on the product. For example, in the case of computer chips, test conditions might involve vibration and temperature; for a mechanical component used to pump chemicals, they might be resistance to solvents and seal tests. Other test factors include strength, thermal shock, and humidity, to name a few.

13.9.2 Testing during Manufacturing

The purpose of testing during manufacturing is to eliminate manufacturing defects and early part failures. The type of testing to be done depends on the product (electrical, mechanical, or electronic). For very expensive products (e.g., defense systems or commercial satellites), where a high level of reliability is absolutely essential, 100% testing would be employed, whereas for most other products (particularly consumer durables), testing can sometimes be substantially less than 100%. In either case, testing may be done under various environmental conditions.

In addition, testing requirements can change over the period of production. For a new product, in the early stages of production, considerable testing is required to establish the process characteristics and the effect of process parameters on the reliability of the product. As the product matures, the testing requirements are reduced. Two types of testing used are

1. Environmental stress screening [ESS]
2. Burn-in

ESS is the process of subjecting a part or assembly to various environmental extremes to identify and eliminate manufacturing defects prior to customer use. Typical methods used are temperature cycling, random vibrations, electrical stress, thermal stress, etc.

Burn-in is a process used to eliminate the high initial failure rate due to manufacturing defects. It involves putting items on a test bed to detect early failures, so that they can be weeded out before the item is released for sale. This topic is discussed in Section 13.10.

13.9.3   Operational Testing

Testing during development and manufacturing is often done under conditions that tend to simulate the real-world environment. Often, the simulated condition is a poor representation of the real-world environment and usage. One way of overcoming this limitation is for the manufacturer to have a small number of users test the product by using it in realistic applications. Such tests provide useful data relating to product reliability and performance in the real world. Based on the resulting data, changes can be made to the product design and/or the manufacturing process to improve the reliability of a product so that it meets the needs and expectations of the customer.

Operational testing is a joint effort involving the manufacturer and buyers. Such tests allow the manufacturer to evaluate all characteristics by utilizing actual users, maintenance procedures, and support equipment in an operational environment.

13.9.4   Testability

Testability is a concept closely related to testing. It is a process through which a failure in a system can be detected and identified so that corrective actions can be initiated to rectify the failure. If the failure is due to an external factor (e.g., loss of power supply to a computer), then testability allows for its detection and identification.

Testability can either be built into the product (called BIT—built-in test) or it can be carried out by equipment external to the product. For complex electronic systems, testability can be done at different levels, ranging from the system level down to the part level. Testing often involves the processing of measurements made by sensors, so that testability involves both hardware and software. As such, both of these are important in the context of design for testability.

13.10   BURN-IN

We first consider the use of burn-in in the context of weeding out nonconforming items and then discuss its use in presale testing.[6]

13.10.1   Weeding Out Nonconforming Items

In Section 13.7.1, we discussed different characterizations of nonconforming items. Here we consider Case (ii), where a nonconforming item has a failure rate that is considerably higher than that of a conforming item. Burn-in is used to weed out the nonconforming items.

We consider continuous production manufacturing, where the process has been stabilized and continuously monitored to ensure that it is always in control. In this case, the probability distribution of item lifetimes is given by

$$G(t) = pF(t) + (1 - p)H(t) \qquad (13.25)$$

where $F(t)$ and $H(t)$ are the failure distributions of conforming and nonconforming items and p is the probability that an item is conforming. The mean time to failure for a nonconforming item is $\mu_F \gg \mu_H$, the mean time to failure for a conforming item.

Burn-in involves testing all items for a period τ. Those that fail during testing are scrapped. The rationale for this is that nonconforming items are more likely to fail than conforming items and hence are weeded out.

The probability that a conforming [nonconforming] item will fail during testing for a period τ is given by $F(\tau)$ [$H(\tau)$]. As a result, the probability that an item that survives the test is conforming is given by

$$p_1 = \frac{p\overline{F}(\tau)}{p\overline{F}(\tau) + (1-p)\overline{H}(\tau)} \tag{13.26}$$

Since $\overline{F}(\tau) > \overline{H}(\tau)$, we have $p_1 > p$. The failure distribution of an item that survives the test is given by

$$G_1(t) = p_1 F_1(t) + (1-p_1)H_1(t) \tag{13.27}$$

where $F_1(t)$ and $H_1(t)$ are given by

$$F_1(t) = \frac{F(t+\tau) - F(\tau)}{1 - F(\tau)} \tag{13.28}$$

and

$$H_1(t) = \frac{H(t+\tau) - H(\tau)}{1 - H(\tau)} \tag{13.29}$$

for $t \geq 0$. Note that as τ increases, p_1 (the probability that an item released is conforming) increases, and hence the outgoing quality is improved. However, this is achieved at the expense of the useful life of conforming items released being reduced by an amount τ. As a result, it is necessary to determine an optimal τ so that a sensible tradeoff is achieved between the cost of testing and the benefits derived through improvements in the quality of outgoing products. This issue is discussed further in Chapter 18.

Example 13.11
Let $F(t)$ be a Weibull distribution with scale parameter β_1 and shape parameter α_1 and $H(t)$ be a Weibull distribution with scale parameter β_2 and shape parameter α_2. Then the failure distribution $G(t)$ is a mixture of two Weibull distributions, as discussed in Section 4.6.5 and Example 4.17. We use the same parameter values as in Example 4.17, namely $\alpha_1 = \alpha_2 = 1.25$, $\beta_1 = 5.3659$, and $\beta_2 = 1.0732$, and let $p = 0.8$. Figure 13.7(a) is a plot of p_1 [given by (13.26)] as a function of τ. Note that as τ increases, p_1 increases (implying that the probability that the item is nonconforming decreases). Figure 13.7(b) is a plot of the probability that an item fails during testing. This is given by $pF(\tau) + (1-p) H(\tau)$. As can be seen, this an increasing function of τ, as expected. ∎

13.10.2 Presale Testing

For products with bathtub failure rate, the failure rate is high in the initial period. As a result, if items are released without burn-in, a high fraction would fail in the early period, leading to high warranty costs and loss of customer goodwill. In this case, burn-in can be used to improve product reliability by consuming a part of the lifetime. The approach is to test each item for a period τ prior to its sale. Any items that fail within this period are minimally repaired. If the time to repair is small (in relation to τ), so that it can be ignored, then the failure distribution of the item at the end of the test, $F_1(t)$, is given by (13.28).

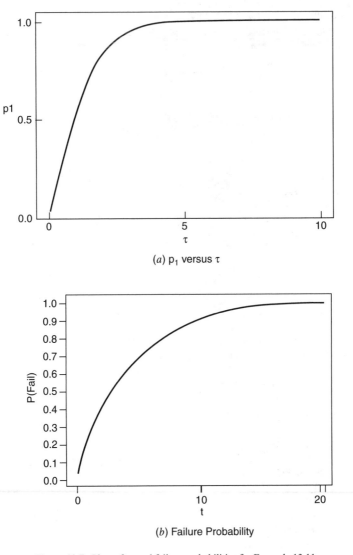

(a) p_1 versus τ

(b) Failure Probability

Figure 13.7. Plots of p_1 and failure probabilities for Example 13.11.

For burn-in to be effective, $F_1(t)$ must be superior to $F(t)$, where $F(t)$ is the failure distribution before testing, so that the reliability is improved. There are several ways of quantifying this improvement, e.g., $F_1(t)$ assumed to have a higher mean time to failure than $F(t)$ or $F_1(t)$ $< F(t)$ for all t, implying higher reliability.

Burn-in results in additional costs due to (1) fixed set up cost of the burn-in facility, (2) variable cost (which increases with τ) for testing each item, and (3) rectification cost for failures during burn-in. Hence, burn-in is worthwhile only if its benefits (measured in terms of the improvements in reliability) exceed the cost of burn-in. Optimal burn-in decisions are discussed in Chapter 18.

Table 13.7. $R_1(T; \tau)$ for Example 13.12

	$\tau = 0.00$	$\tau = 0.05$	$\tau = 0.10$	$\tau = 0.15$	$\tau = 0.20$
$T = 0.1$	0.8420	0.8851	0.9305	0.9661	0.9782
$T = 0.5$	0.7334	0.8005	0.8521	0.8847	0.8958
$T = 1.0$	0.6570	0.7171	0.7634	0.7926	0.8025
$T = 2.0$	0.5273	0.5755	0.6126	0.6360	0.6440
$T = 5.0$	0.2725	0.2974	0.3163	0.3280	0.3315

Example 13.12

Let $F(t)$ be given by

$$F(t) = 1 - e^{-\Lambda(t)}$$

with $\Lambda(t)$ given by

$$\Lambda(t) = \begin{cases} a_i t + (\alpha t/2)(2t_i - t) & 0 \le t < t_i \\ a_i t + (\alpha/2)t_i^2 & t_i \le t < t_f \\ a_i t + (\alpha/2)t_i^2 + (\beta/2)(t - t_f)^2 & t_f \le t < \infty \end{cases}$$

with the following parameter values:

$$a_0 = 0.2, \qquad a_1 = 0.22, \qquad \alpha = 10.0, \qquad \beta = 0.1, \qquad t_i = 0.2, \qquad t_f = 5.0$$

This implies that $F(t)$ has a bathtub shape.

The item is sold with a free replacement warranty with warranty period T. The manufacturer is interested in studying the effect of burn-in time τ on the probability that an item sold does not fail in the warranty period. This probability is simply the reliability of the item after burn-in and is given by

$$R_1(T; \tau) = 1 - F_1(T) = 1 - \frac{F(T + \tau) - F(\tau)}{1 - F(\tau)}$$

Table 13.7 shows $R_1(T; \tau)$ for $T = 0.2, 0.5, 1.0, 2.0$, and 5.0 and $\tau = 0.00, 0.05, 0.10, 0.15$, and 0.20. Note that, as expected, burn-in increases the reliability in every case. ∎

NOTES

1. The handbook by Ireson, Coombs, and Moss (1996) deals with various reliability related issues in engineering design. In particular, the chapter by Rohrbach (1996) deals with management of the reliability process.

2. For more on customer satisfaction, see Rooney (1994). There are many books which deal with reliability design. The Handbook of the Reliability Analysis Centre (RDH376, 1975), Ang (1984), Blanks (1992), and Haviland (1964) are examples of these. Risk is an important factor in reliability design and Henley and Kumamoto (1981) deal with this topic. Meeker and Hamada (1995) and Ueno (1995) deal with two different aspects of product development.

3. Schroeder and Johnson (1990) define more complex availability measures that might be used to define reliability goals. Born and Criscimagna (1995) discuss how to relate user needs for reliability and maintainability into reliability goals for the manufacturer in the context of defense product acquisition. An issue they highlight, based on a study of eleven acquisitions, is the lack of consistency in this context.

4. Design at the component level is discipline-specific. For more on the design of mechanical components, see Carter (1997), Cruse (1997), Dieter (1991), Rao (1992), and Vinogradov (1991). For the design of electronic components, see Brombacher (1992) and Christou (1994). Harr (1987) deals with the design of structures in civil engineering.

5. There are a number of introductory books on quality control that deal with topics such as control charts, acceptance sampling, design of experiments, reliability management, etc. A few of these are Evans and Lindsay (1996), Montgomery (1985), Ryan (1989), Sinha and Willborn (1985), and Taguchi et al. (1989). There are several specialized books that deal with a specific topic in greater depth. For example, Schilling (1982) deals with different acceptance sampling schemes. A list of journals on quality is given in Chapter 20.

There are several papers in the Proceedings of Annual Reliability and Maintainability Symposium on total quality in the context of reliability and maintainability; for example, see Yoo and Smith (1991).

6. For more on burn-in, see Jensen and Petersen (1982) and Leemis and Benke (1990).

EXERCISES

1. In Section 13.5.1, we introduced the notion of reliability allocation in a multicomponent system. Can one define a similar notion for maintainability allocation? If so, develop your notion through some precise concepts.

2. How does reliability stress–strength impact maintainability (if at all)?

3. What is meant by design specification? How are reliability-related issues incorporated into a design specification?

4. A small manufacturing company produces brake components for a large car manufacturer. The company is interested in adopting the DFR philosophy. Indicate how the company should proceed to achieve this.

5. In Exercise 4, what kind of data are needed for improving brake reliability? How can this data be collected?

6. Case 2.3 of Chapter 2 deals with heart pacemakers. Consider the failure data for Company A given in Table 2.3. How should the company proceed in setting up reliability goals for a new generation of pacemakers?

7. A manufacturer of kitchen appliances is considering launching a new product in its electric toaster line. Discuss the reliability specification for the product.

8. How would the reliability specifications for an electric toaster differ from those for a heart pacemaker? Compare and contrast the severity and criticality of failures for the two products. Contrast both products with a jet engine.

9. A pump is used in a continuous mode for pumping the coolant fluid in a nuclear reactor. What type of reliability measure is appropriate for such a pump? Discuss the role and use of preventive maintenance in this context.

10. What type of reliability measure is appropriate for a pump used as a backup for the one discussed in Exercise 7?

11. The failure and repair times for a welding robot are exponentially distributed with MTTF = 6 hours and MTTR = 0.3 hour. The robot is in its working state at the start of each day. Compute the availability of the robot if the shift is 8 hours. How does this change when the shift increases to 10 hours. If the MTTR can be reduced to 0.2 hour, how will this affect the availability?

12. A mechanical component is to be designed to withstand a normally distributed load (or stress) with mean 1000 and standard deviation 500. The strength of the component is also normally distributed with mean μ (a decision variable under the control of the designer) and standard deviation 600. The designer is to select μ to ensure a reliability of 0.95. What is the value of μ that will ensure this? Suppose that the designer wants to build in a safety factor. He does this by assuming that the mean load is 1200 as opposed to 1000. How does this affect the value of μ (assuming reliability of 0.95)? What is the reliability of the item when put into use?

13. Compare and contrast the four approaches to reliability allocation discussed in Section 13.5.1.

14. Consider a system composed of four subsystems with failure rates 0.005, 0.004, 0.003, and 0.006 failures per hour. The system is to be designed so that it survives a 15 hours mission time with probability 0.95. Determine the reliabilities needed for each of the subsystems based on the ARINC apportionment technique.

15. Consider an electronic system consisting of four subsystems. The number of modules (k_i), importance factor (w_i), and mission time (T_i) for the four subsystems are as follows:

Subsystem	k_i	w_i	T_i (hours)
1	6	0.9	30
2	8	0.8	10
3	1	1.0	40
4	4	0.9	20

Determine the reliability allocation based on the AGREE method to ensure a system reliability level of 0.95 for 20 hours of continuous operation.

16. How will the results of Exercise 13 change if the weights are equal?

17. Consider a cantilever beam subjected to a lateral concentrated load P at the free end. The deflection at the tip is given by

$$X = \frac{PL^3}{3EI}$$

where L is the length of the beam, E is Young's modulus, and I is the moment of inertia. E and I are random variables distributed normally. The beam is deemed to have failed if the deflection exceeds a specified value Y. The design variable L is to be selected so that the probability of failure is less than or equal to some prespecified value α. Determine L.

18. The amount of wear (over a specified time interval) for a sliding component is given by

$$X = \frac{KLD}{3p}$$

where L is the force normal to the sliding surface, D is the sliding distance, p is hardness of the wearing surface and K is a dimensionless empirical constant. D and p are normally distributed random variables. The item fails when X exceeds the design limit for volumetric adhesive wear Y, which is also normally distributed. The design problem is to determine the maximum force L that will ensure that the component reliability is greater than $1 - \alpha$. Determine this value.

19. The corrosion rate to exposure to chemicals can be modeled as

$$X = \frac{543\,W}{rAt}$$

where W is the weight loss in milligrams, r is the density of the specimen (in grams per cc), A is the exposed area and t is the exposure time. The item fails when X exceeds the design corrosion penetration rate Y. W and r are random variables. Determine A so that the component can withstand exposure for T units of time with a prespecified probability α.

20. In Case 2.5 of Chapter 2, items were produced in batches of different sizes. Define a conforming item as one that does not fail over the warranty period. How would you use the data to determine the quality (in terms of conformance) of items produced?

21. Prove the inequality in equation (13.24).

22. The failure of a component can be due to one of two causes and can be modeled by a competing risk model. Suppose that it is given by equation (4.83) with the following parameter values: $\alpha_1 = 1.25$, $\beta_1 = 5$, $\alpha_2 = 0.5$, and $\beta_2 = 1.0$. The items are subjected to burn-in (for a length T). Items that survive the burn-in are sold with a warranty period W. The probability that an item released for sale does not fail over the warranty period depends on both T and W. Compute this probability for $T = 0.01$ to 0.05 (in steps 0.01) for $W = 1$, 2, and 3.

23. Repeat Exercise 22 with component failure given by a mixture model [equation (4.79)] with $p = 0.9$.

24. Repeat Exercise 22 with component failure given by a multiplicative model [equation (4.86)].

25. Compare and contrast the results of Exercises 22–24.

26. The MTTF and standard deviation for a mechanical component are 1000 and 600 hours, respectively. The manufacturer initiated a reliability improvement program. The outcome of the program has resulted in the MTTF increasing to 1200 hours but the standard deviation also increasing to 800 hours. Suppose that the product is designed to have a mission life of T. Compare the mission reliability for $T = 600, 800, 1000$, and 1200 hours for the component before and after the reliability improvement program, assuming that the item failure distribution (before and after reliability improvement) is Weibull. Based on this, comment on whether or not the reliability improvement has been worthwhile?

27. Repeat Exercise 26 assuming that item failure distribution (before and after improvement) is gamma.

28. The market for washing machines can be divided into two broad groups—light and heavy duty. The manufacturer has the choice of either designing two different products to meet the two different load and usage conditions or designing one product for both groups. Discuss the reliability specifications for the two cases.

CHAPTER 14

Reliability Prediction and Assessment

14.1 INTRODUCTION

Determination of product reliability is important in all phases of the life cycle of a product. The process begins at the design stage, where predictions of reliability are made based on whatever information is available at that point, continues through the development and testing stages as additional information becomes available, and proceeds through the production and operational stages of the life cycle. At each of these stages, still more reliability-related information may be obtained, and updated reliability assessments are made periodically as appropriate.

Reliability prediction may be interpreted in several ways. First, it may mean prediction of system reliability in the conceptual stage of product development, where little or no "hard" information is available for many of the parts and components of the system. Second, predictions may be made later in the development cycle in situations where test data and other information are available on components and on the system as a whole. Third, predictions may by made regarding future reliabilities in situations involving reliability growth. In all of these cases, prediction from a statistical point of view is basically equivalent to use of statistical estimates for prediction purposes, with confidence intervals calculated as appropriate in this context.

Reliability assessment is basically concerned with evaluation of the current reliability of a product or system. From a statistical point of view, this involves estimation based on available information and data at the point in time when the assessment is being made.

Thus prediction is concerned with systems not yet built or partially built or with the use of systems in new environments. Prediction is heavily model-oriented and often uses nominal values of parameters as inputs. Assessment, on the other hand, is data based and may or may not use models. Assessment may also be done at the component or system levels.

Whether we are assessing the current state of product reliability or attempting to predict a future state, it is important to use all available information. Furthermore, the methodologies used in both endeavors are basically the same in many instances. In referring to this process, we use the term "reliability evaluation" to describe either assessment or prediction.

In order to predict future reliability or assess current reliability of a system effectively, many inputs are required. These include

- Component, subsystem and system models, e.g., block diagrams
- An understanding of modes of failure and the effects of various failures

511

- Test and other data, appropriately summarized, analyzed, and interpreted
- Information regarding failure rates, distributions of time to failure of parts, components, etc., and parameters of these models
- A compilation of any additional information regarding the product, including historical data, vendor data, judgmental data, and so forth
- Aggregation of the various types of information, e.g., via Bayesian analysis
- Intended product use, including life cycle of the item, operating conditions, mission, and so forth

Thus all of the previous chapters of the book provide input to reliability prediction and assessment. In this chapter, we illustrate the integration and application of many of the techniques in addressing the reliability issues in question.

The outline of the chapter is as follows. Predictions are made in many contexts. To set the scene, in Section 14.2, we consider various ideas and definitions of a product life cycle, elaborating on some of the issues introduced in previous chapters. We then first consider reliability prediction, since this often takes place (or should take place) early in the product life cycle. This is discussed in Section 14.3. The next activity in most applications is acquisition and analysis of additional data. Accordingly, Sections 14.4 and 14.5 deal with test procedures and data analysis. In Section 14.4, we discuss the role of testing and the use of the methods of experimental design given in Chapter 10 and the data analysis techniques given in Chapters 3, 5, 8, 10, and 11. Accelerated testing, important in certain applications and introduced in Chapter 10, will be discussed in more detail in Section 14.5. Finally, in Section 14.6 we look at the integration and use of these results in reliability assessment.

14.2 CONTEXT OF PRODUCT RELIABILITY PREDICTION AND ASSESSMENT

As noted previously, there are many instances where prediction and/or assessment of product reliability is desirable, and, in fact, may be essential. These may occur in the various stages of product design, development and testing, production, and operations, and continue nearly until product obsolescence. Here we look at various concepts of product life cycles, noting the role of reliability evaluation in each stage.

14.2.1 Product Life Cycle

Product life cycle concepts are many and varied, depending on the product, the marketplace, technological developments, the point of view (e.g., consumer or manufacturer), and the analysis under consideration (engineering, marketing, finance, and so forth). Concepts range from a top view that may include product evolution through many generations and technological changes, to a more narrow view of a product line or a specific product, to a very limited view, such as a specific purchase of an item for short-term use. Reliability issues may be very different in these different contexts, and we make no attempt to discuss these fully. Rather, we look at a few specific situations that are illustrative of the types of problems encountered in practice.

Two notions of a product life cycle are discussed in Sections 1.5 and 12.2, and many other formulations have been considered. The discussion in Section 1.5 centers on a specific product from its inception to its ultimate replacement, as shown in Figure 1.9. For the manufacturer, the life cycle begins with a product concept based on perceived customer requirements and proceeds though product design, development, prototype construction and testing, production, marketing, sales, postsale service, repeat sales, and so forth, until production is discontinued. From the buyer's point of view, the life cycle begins with product selection and purchase. If this purchase alone is considered, the life cycle ends when the item is sold, traded in, or discarded. If replacements are considered as well, the life cycle continues through replacement purchases as well, until the product is no longer used.

In this context, the manufacturer is interested in evaluating reliability at many stages in the product life cycle and views reliability in an overall framework in which what happens on the average is the major concern. The buyer, however, is concerned almost exclusively with the reliability of the single item (or the few such items) purchased, and then only during the tenure of his or her ownership. Furthermore, even if the owner were interested in evaluating reliability from the broader perspective, the necessary data and other information needed would almost never be available. Thus reliability assessment and prediction is mainly the purview of the manufacturer and we look at it from this point of view.

As discussed in Chapter 12, the manufacturer may also consider the life cycle from several points of view, e.g., from a marketing perspective, a financial/accounting perspective, a production perspective, or a strategic management perspective. The definition of the life cycle will depend on the perspective. For example, in the last case, a reasonable structure consists of three phases: (1) prelaunch of a new product, including concept design and so forth, (2) launch window, and (3) post-launch. (For additional discussion, see Murthy and Blischke, 1999.)

Reliability issues are best highlighted in a composite of the various approaches to the definition of a product life cycle (see Chapter 1), consisting of:

- Initial product concept
- Preliminary design studies
- Detailed design phase
- Testing and development
- Manufacturing
- Product introduction
- Product operation and use
- Postsale service

Note that this structure highlights both the significant milestones where reliability evaluation is needed and the windows of opportunity during which relevant information is typically available.

14.2.2   Illustrative Cases of Products and Systems

In the following sections and later in the chapter, we look at several products, ranging from quite simple—a pneumatic pump—to very complex—a liquid rocket engine and a space vehicle. In each case, we focus on reliability evaluation during selected phases of the life cycle, addressing various issues.

To illustrate the various levels of complexity, consider the following products:

1. *An ordinary electric lamp.* This is a relatively simple product, consisting of the basic structure, base, socket, bracket and shade, plus switch, cord, light bulb, and fasteners.

The switch is the only mechanical part. Operationally, the most complicated and least reliable part is the bulb (see Figure 1.1 for a schematic of an incandescent bulb). Since the bulb is not ordinarily supplied by the manufacturer of the lamp, the manufacturer is mainly concerned with the remainder of the product. Factors affecting quality and reliability are the switch, cord, and other materials, and the assembly of the lamp. Other than as it relates to the life of a single item, the life cycle concept is of little relevance, since styles change rapidly and there are virtually no repeat purchases of the same item.

2. *A pneumatic pump.* This is also a relatively simple product, but there are several electrical, mechanical, and electromechanical parts, plus fasteners (adhesives, etc.) that could fail. A schematic of a pump used to aerate a small aquarium is given in Figure 1.3. For this application, the life cycle cost is not of interest. For larger pumps in industrial and commercial applications, e.g., in spray painting or similar uses, medium to long-term reliability and repeat purchases may be of interest, and a life cycle accounting in this context would be appropriate.

3. *A loudspeaker.* A third example that is a relatively simple product is a loudspeaker. A loudspeaker is a major component of any audio system and is simple in concept, having relatively few parts, but complex from a design point of view because of the many forms and uses of the product, including equipment for rock concerts, large public address systems, home stereos, car audio systems, and small radios. Schematics of loudspeakers can be found in most books on audio equipment, e.g., Kelly (1998). Parts that can fail range from adhesive bonds (see Case 2.1) to mechanical and electromechanical parts.

4. *A stereo system.* A complete audio system is a much more complex piece of equipment, consisting of two or more speakers, an amplifier, and one or more devices that generate signals that are amplified and otherwise transformed to produce sound. These may include a CD player, a cassette player, AM and FM receivers, a microphone, a television receiver, and various musical instruments. Each of these components may, itself, be a complex piece of equipment. In addition, the speaker subsystem ordinarily consists of much more than a single loudspeaker and will include several loudspeakers, each covering a different range of sound. Finally, the system includes many connectors, both within and between components. Thus there are a large number of possible causes and modes of failure. A block diagram of a relatively simple stereo system, consisting of a tuner, CD and cassette players, amplifier, and two speakers, is given in Figure 7.2.

5. *A liquid rocket engine.* A schematic of the space shuttle main engine, indicating propellant flow but also showing the main components of the engine, is given in Figure 1.5(a). This is a very complex piece of equipment and is designed for very high performance. It consists of many complex subsystems and components. A schematic of one of these, the high-pressure fuel turbopump, is shown in Figure 1.5(b). Reliability analysis of this system and it components, down to the part level, has been very extensive and detailed.

6. *The Space Shuttle.* The liquid rocket engine described very briefly in the previous example is, in fact, a subsystem of the shuttle vehicle as a whole, which consists of liquid engines, fuel and oxidizer tanks, two solid rocket motors, the shuttle itself, other structural components, all of the controls, computers and software, and the ground support system. As a whole, this is an extremely complex system, with very high performance and reliability requirements. Design, development, and testing of such a system is a large, complex, and costly endeavor. An overview of the basic activities in the process, indicating major reliability efforts, is shown in Figure 14.1 (from Blischke, 1994).

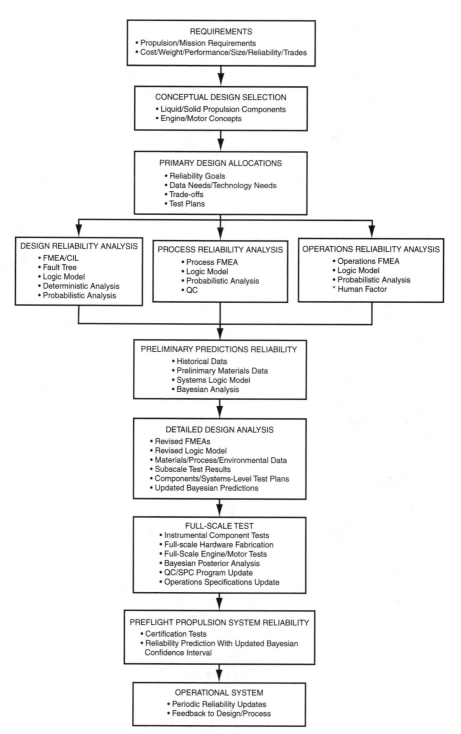

Figure 14.1. Reliability program, liquid rocket engine.

14.2.3   Target Values for Reliability

To be meaningful, reliability evaluation must be done in the context of specified goals. Thus target values for reliability must be set at the outset, used as benchmarks, and modified as necessary as further information is developed, cost factors are analyzed, and realistic, achievable goals evolve.

Because of their interrelationships with numerous other factors, target values for reliability should be set at the product concept stage. They must be known in order to evaluate various design concepts, select appropriate materials and suppliers, estimate costs, consider trade-offs between development costs, reliability, and downstream costs, such as warranty and postsale service demands. In practice, target values may of necessity be modified as further information is developed, goals are modified, and refinements are made to the design.

14.2.4   Reliability Evaluation over the Product Life Cycle

Reliability analyses of products, whether new or existing, range from very cursory or not at all to very complete and detailed. A minimal analysis is ordinarily done when

1. A product is very simple (e.g., a hammer) and obviously reliable
2. A new product is basically the same as an existing product, with perhaps cosmetic changes, and the existing product has a history of high or at least acceptable reliability
3. A product can be overdesigned, so that it is highly unlikely to fail, and it can be produced at an acceptable cost
4. Aong-term cost considerations are not important, e.g., no substantial warranty is to be offered or the product will be discontinued

At the other extreme, reliabilities are predicted very early in the life cycle and assessed frequently thereafter, resulting in updated predictions. This would be the case, for example, in analysis of a rocket propulsion system, a space vehicle, advanced weapons system, or other highly sophisticated systems, where

1. The product is complex, often involving new technology
2. Reliability is critical, with lack of reliability being very costly and possibly resulting in loss of life
3. Overdesign is highly undesirable, as it results in increased weight and hence highly inflated operating costs

In cases such as these, a thorough reliability analysis is undertaken.

Many of the steps involved in such an analysis for a rocket propulsion system are indicated in Figure 14.1. In this formulation, the time frame considered is nearly the complete life cycle of the system, beginning with the conceptual stage, selection and analysis of the preliminary and detailed designs, and proceeding through the operational stage. The key activities and milestones are:

- Specification of reliability and other requirements at the outset
- A reliability prediction made at the conceptual design stage
- Allocation of reliability requirements to subsystems, etc., based on the system reliability requirement

- A thorough reliability analysis of the selected (preliminary) design
- A thorough reliability analysis of the production process
- A thorough reliability analysis of the proposed operations
- Reliability prediction based on the preliminary design
- Formulation and analysis of the detailed design plan
- Full-scale testing
- Evaluation of preflight system reliability
- Operations

Many additional details of the process are indicated in Figure 14.1 and discussed in Blischke (1994).

Most consumer products and many commercial products fall in between the two extremes of little or no reliability analysis and a thorough analysis of the type described. The level of reliability analysis done in these cases is that considered appropriate to the specific situation, and depends in great part on the consequences, financial and otherwise, of unreliability.

14.3 RELIABILITY PREDICTION

Prediction deals with evaluation of a design prior to actual construction of the system. It is an attempt to evaluate the consequences of decisions made before the system is built and/or put into use. Prediction deals with analysis using models rather than actual systems and provides a basis for test planning, manufacturing, evaluation of reliability growth, maintenance, and other management activities. For further discussion, see Priest (1985).

Typical examples of predictions are:

1. Prediction of the reliability of a system for a given design and selected set of components (see Chapter 7)
2. Prediction of the reliability of a system in a different environment from those for which data are available (see Chapter 6, especially environmental factors, physics of failure)
3. Prediction of the reliability of the system at the end of the development program (to be discussed in the next several chapters)

In this section, we look at prediction in the various stages of the product life cycle, some of the important tools used for prediction, and some applications.[1]

14.3.1 Importance of Reliability Prediction

As noted, for the most part, reliability predictions are made in the early stages of the design and development of an item, i.e., prior to its actual operation. Reliability predictions are important in many situations and used for many purposes. These include

- Setting generic requirements for materials, parts, components, and so forth
- Performing trade-off studies
- Setting plans for developmental testing

- Planning for design improvements
- Planning for manufacture
- Setting of factory standards for accept/reject decisions
- Cost analyses, including life cycle cost studies
- Providing a basis for evaluation of reliability growth
- Studies of maintenance requirements and logistics

See Healy et al. (1997) for a discussion of a number of these issues. Preliminary reliability analysis is especially important for large, complex systems.

Predictions made early in the design and development of a product are ordinarily modified and refined in later stages of its life cycle, and, as testing is done and other information is obtained, prediction progresses to assessment of actual reliability. In either case, reliability evaluation becomes more and more precise as more information is entered into the process.

In this section, we look at some of the key engineering, analytical, and statistical tools used in reliability prediction, including many that were introduced previously, indicating their role in the prediction process. The emphasis is on the design stage of a product, which itself includes several stages, as noted, depending on the complexity of the item. For example, Figure 14.1 indicates three major stages in the design of a rocket propulsion system—conceptual design, preliminary design, and detailed design. There are a number of tools that may be used for prediction. Some are appropriate at different stages of product development, some for particular products, and so forth. What should be used in any particular situation is a matter of preference, experience, and the context of the prediction.

Reliability predictions are important for both hardware and software. When a complex system involves both, prediction of total system reliability becomes both more important and more difficult. This is certainly the case in analysis of a rocket propulsion system, and no attempt is made here to present a thorough analysis.

14.3.2   Prediction at the Conceptual Stage of Design

At the conceptual stage of the design of a new product, requirements and objectives are determined, and many alternatives may be considered. The initial stage involves many business decisions regarding the product—its marketability, where it fits in the overall company strategy, and so forth. Once it is determined that a product should be seriously considered, the conceptual design stage is entered. At this point, design *concepts* are evaluated and compared. Considerations include alternatives regarding the basic structure of the item, raw materials to be used, outsourcing and potential suppliers, process implications, and many other factors.

For a simple product, such as an electric lamp, this may be a straightforward process. (Nonetheless, it is an important activity, because if a product fails due to unreliability, as for any other reason, this can have a significant financial impact.) For a very complex item, such as a rocket propulsion system, the conceptual design phase requires a significant effort and capital expense, and is anything but straightforward, involving a complex analysis and many important decisions.

14.3.3   Tools for Failure Analysis

In order to formulate predictions at early stages of development, it is necessary to undertake a careful analysis of potential failures and their underlying causes. Two of the principal tools used for this purpose, failure modes and effects analysis and fault tree analysis, were discussed in Chapter 7. These can be quite simple or they can be extremely complex, depending on the nature of the product, but even for simple products, they can lead to avoidance of future problems and thus be well worth the effort.

A prediction of system reliability is usually determined as a function of the reliability of its individual parts and components. In this analysis, other important tools are block diagrams and other schematics that show major components of a system and their interrelationships. These are also discussed in Chapter 7 and are essential in analysis of the system and in preparation of FMEAs and FTAs.

FMEA

A FMEA typically includes a listing of failure modes, possible causes for each failure, effects of the failure and their seriousness, and corrective actions that might be taken. An example of part of a FMEA for a simple product, an electric lamp, is given in Table 14.1 (from

Table 14.1. Partial FMEA for an electric lamp

Component name	Failure mode	Cause of failure	Effect of failure on system	Correction of problem	Comments
Plug	Loose wiring	Use vibration, handling	Will not conduct current; may generate heat	Molded plug and wire	Uncorrected, could cause fire
	Not a failure of plug per se	User contacts prongs when plugging or unplugging	May cause severe shock or death	Enlarged safety plug	Children
Metal base and stem	Bent or nicked	Dropping, bumping, shipping	Degrades looks	Distress finish, improved packaging	Cosmetic
Lamp socket	Cracked	Excessive heat, bumping, forcing	May cause shock if contacts metal base or stem; may cause shock upon bulb replacement	Improve material used for socket	Dangerous
Wiring	Broken, frayed from lamp to plug	Fatigue, heat, carelessness, dog bite, etc.	Will not conduct current; may generate heat, blow breakers, or cause shock	Use of wire suitable for long life in extreme environment anticipated	Dangerous; warning on instructions
	Internal short circuit	Heat, brittle insulation	May cause electrical shock or render lamp useless	Use of wire suitable for long life in extreme environment anticipated	
	Internal wire broken	Socket slipping and twisting wires	May cause electrical shock or render lamp useless	Use of indent or notch to prevent socket from turning	

Case and Jones, 1978). The complete FMEA extends for a few pages. Note that issues of safety and product liability are involved here as well as product performance. The FMEA for a loudspeaker would also be relatively straightforward. The significant value of these would be to highlight parts that are potential sources of trouble at an early stage so that designs can be evaluated and improved as necessary when this can be done most effectively and at least cost.

For complex products, the process is even more important, as costs can escalate rapidly and safety issues can become critical. The FMEA for a liquid rocket engine, for example, can cover thousands of pages. A tiny fraction of such a document is given in Table 14.2. Here the part in question is the main oxygen valve in a liquid oxygen/liquid hydrogen engine. This valve controls the flow of oxygen to the engine during start-up and main-stage firing and terminates the flow at cutoff of the engine.

FTA

A fault tree analysis begins with a set of unsafe or undesired states of a system and proceeds backward to a list of possible causes of the failure. This provides additional valuable information for evaluation of designs and reliability improvement. Again, the complexity of the fault tree depends on the complexity of the product and its possible modes of failure. A fault tree for a solid rocket motor (SRM) of a rocket propulsion system is given in Figure 14.2. An SRM is not nearly as complex as a liquid rocket engine, but is by no means a simple product. Figure 14.2 shows the major elements of an FTA of the fault involving a burst case.

Many rocket propulsion systems consist of both liquid and solid propulsion components. The conceptual analysis and further analyses of preliminary designs is a very complex undertaking, of which the examples discussed here are a very small part.

14.3.4   Methods of Reliability Prediction

A quantitative prediction of reliability is usually determined by combining the reliabilities of individual parts. In the very earliest stages of conceptual design of a new product, this may not be possible, and qualitative predictions (high, moderately high, etc.) may be made. These are of some use, at least for purposes of comparison of competing design concepts. As the design activity proceeds and specific parts and components are analyzed, quantitative measures are obtained, but in the early stages even these may be useful only in relative rather than absolute terms.

As the early design evolves, more specific and precise information can usually be included in the prediction. Some parts may be the same as previously used and test and operational data can be obtained. Data on prior similar products may also be available, and so forth. Data may be of varying degrees of relevance. Some items may be completely new designs, new materials, etc. Many theoretical and statistical procedures are used, and engineering judgment will play a role as well.

Three commonly used prediction techniques are

- The *part count method:* base the prediction on the number of parts in the proposed design. This approach may be used in purely qualitative evaluations, where, among designs that are otherwise comparable, the one having the fewest parts (or fewest of a particular type, e.g., moving parts, electrical parts, etc.) is likely to be the most reliable. For quantitative measures, the system structure must be considered (e.g., by use of a block diagram) and individual reliabilities must be provided. The most common approach here is to formulate the system as a series system, determine failure rates for

Table 14.2. Partial FMEA for one part of a liquid rocket engine

Mode no.	Failure mode	Detection method	Effect description	Effect on engine
1	Internal leak; premature opening	CC injector pressure and temperature monitors; MOV D/S skin temp.; MOV position monitor	Accumulation of oxidizer in main combustion chamber prior to start signal; possible hard start	Hard start; may damage engine
2	Fail closed	MOV position monitor	Engine fails to start	None
3	Restricted flow; partially opened valve	MOV position monitor; MR monitor	Reduced LOX flow to MCC, MR upset, reduced performance	May cause premature cutoff
4	External leak	Engine LOX flow monitor: MOV D/S skin temperature	Loss of oxidizer to main combustion chamber, injector and ASI.	LOX impinging on adjacent hardware may cause secondary failure
5	Fails open/leaks	CC injector pressure and temperature monitors; MOV position monitor	LOX accumulation in injector; hard start and off-MR operation. High MR transient at shutdown, oxidizer flow continues until vehicle prevalve is closed	Possible combustion chamber/injector burnout

Mode no.	Effect on mission/vehicle	Criticality	Possible causes	Preventive action
1	Possible launch delay	3	Contamination on seat; damaged seat seal; premature open signal	Leak and functional check during prelaunch preparations; Prelaunch purges; controller
2	Possible launch delay	3	Loss of pneumatics; mechanical restriction to actuator/valve motion	Functional checks during prelaunch preparations
3	Possible loss of mission. Engine-out capability may save mission during mainstage	2	Contamination; mechanical restriction to actuator/valve motion	Functional checks during prelaunch preparations
4	Possible loss of mission. Engine-out capability may save mission during mainstage	2	Material or manufacturing defect; seal damage or loss of seal retention	Proof and leak test at build plus leak test prelaunch
5	Possible loss of vehicle/mission	1	Internal leak from contamination on seat or damaged seat; loss of pneumatic pressure to close; mechanical restriction	Close vehicle prevalve. Fail-safe closing spring

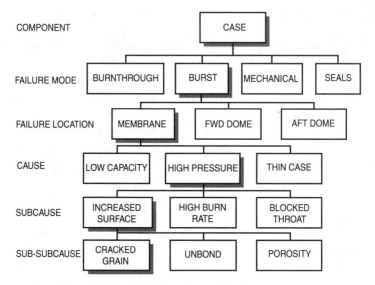

Figure 14.2. Partial fault tree for a solid rocket motor case.

each element, assume independence and exponential failure distributions, and obtain the predicted system failure rate by adding individual failure rates.

- The *stress analysis method:* evaluate designs by comparing predicted strengths with anticipated stresses. This may be done for each key component or for the system as a whole. The stress–strength models discussed in Chapter 6 are used in the analysis. This method of prediction is particularly useful for items where failure is a result of breakage, rupture, etc. An example is the analysis of a case-burst failure of a solid rocket motor, as indicated in Figure 14.2. Probabilities may be assigned or calculated at each level of the fault tree and used to predict system reliability.

- The *design similarity method:* analyze similar systems currently in operation and use the results to predict the reliability of a proposed design. This requires a careful comparison of components to determine which are truly comparable in the new design and a separate evaluation for those that are not. It also requires valid and reliable data on the performance of the similar components.

For simple items, such as the hydraulic gate valve or the audio speaker, any of these approaches may be used, depending on past experience, data, and assumptions about the operating environment. For systems of even modest complexity, most reliability predictions made in any of the design stages are done using a combination of two or more of these three methods. As prototypes are built and test data become available on parts and components in the "final" design, refinements are made, as will be discussed in later sections.

14.3.5 Environmental Factors

Since new products are often employed under conditions that are also new and many items are intended for uses different from typical applications, it is common in many applications to modify predicted values by application of environmental and other factors. The intent is to account for different conditions, such as temperature, voltage stress, and so forth. The adjustment is done by multiplication of the predicted failure rate by appropriate constants (often called "K-factors"). There are many sources—handbooks, MIL-STDs, and so forth—from which such constants may be obtained. Many of these are listed in Chapter 20. Many factors, including temperature, electrical stress, first-year factor to account for infant mortality, and so forth, are discussed by Healy et al. (1997); see also MIL-HDBK-217.

A difficulty that is encountered in practice is that the sources may not agree on the constants or on how to apply them. The result is that very different final values for failures can be obtained. This is discussed in some detail for microelectronic devices by Boyle (1992), who also lists sources for multiplicative factors for applications of this type. A possible solution is to do a worst-case analysis, calculating predictions by all of the suggested methods, and then using the worst of these in arriving at a decision.

Whatever method might be used to arrive at a prediction in cases such as those just described, a fact that must be kept in mind is that predictions in the absence of hard data are, by their very nature, fraught with uncertainty. They may be highly unreliable, and must be interpreted accordingly.

Example 14.1 (Liquid Rocket Engine)

A number of studies have been conducted to analyze designs of liquid rocket engines as possible successors to the space shuttle. In some of these, in the conceptual and preliminary design stages, a gas generator tripropellant engine was considered. A parts/components list for a candidate design is given in Table 14.3. Also included are number required (n_i) and generic failure rates (λ_i) for each item. The failure rates were obtained from various sources, includ-

Table 14.3. Liquid rocket engine reliability prediction

Item no.	Component	n_i	λ_i
1	Injector	1	1.10
2	Combustion chamber	1	0.60
3	Nozzle	1	0.20
4	Nozzle coolant manifold	1	0.20
5	Nozzle coolant assembly	2	2.10
6	Gimbal bearing assembly	1	0.40
7	Propellant shut-off valve	2	0.70
8	Pressurant shut-off valve	2	0.30
9	Low pressure turbopump	1	0.30
10	Bypass valve	1	0.50
11	Turbopump assembly	3	3.50
12	Oxidizer/fuel valves	2	0.40
13	Gas generator	1	0.60
14	Igniter chamber	1	0.20
15	Igniter	2	0.10
16	Igniter valves	4	0.20
17	Lines	20	0.01
18	Orifice	4	0.10
19	Filter	2	0.10
20	Transducer	9	0.06
21	Harness set	1	1.50

ing handbooks, company technical reports, and government and other data banks (e.g., GIDEP; see Case 2.26).

The failure rates given are for a specific mission and are per 10^6 cycles. The environmental factor used in the reliability calculation is $K = 40$. The predicted total failure rate is

$$\lambda = \sum_{i=1}^{21} n_i \lambda_i = 25.44 \text{ per } 10^6 \text{ cycles}$$

The predicted reliability of the engine per mission is $R = e^{-40(25.44)(10^{-6})} = 0.998983.$ ∎

Example 14.2   (Solid Rocket Motor)

A manufacturer of solid rocket motors for space propulsion systems used stress–strength analysis to predict the reliability of the case for which a partial fault tree is given in Figure 14.2. Measurements of strength had been gotten from actual field test, pressurizing until burst, of cases. Stresses were calculated on the basis of knowledge and experience with the propellant. Both were assumed to be normally distributed. The stress–strength analyses (See Chapter 7) resulted in probabilities of failure due to burst assigned to membrane, forward dome, aft dome, and other components (not shown in Figure 14.2) of 0.9999958, 0.9999996, 0.9999996, and 0.9999995, respectively. (These results were obtained from the manufacturer of one of the components.) As a result, the predicted case burst reliability is 0.9999945. Analysis of the remaining identified failure modes led to a predicted case reliability of 0.999932.

Note that very high reliabilities are predicted for each of the components. These result because failure takes place at a value of (stress – strength) many standard deviations from the mean. The precise numbers are highly dependent on the normality assumption. ∎

Note that here we have only discussed reliability predictions for a small portion of a space vehicle such as the shuttle. Reliability evaluation branches in both directions from the examples given. To analyze the reliability of a liquid engine, for example, the analysis has to be carried down to lower and lower levels. Thus each turbopump must be analyzed, and this carries down to the part level, where individual blades are analyzed, and so forth. Proceeding upward, the analyst must consider the overall vehicle structure, the structure of the propulsion system (which may consist of both liquid engines and SRMs), mission, environment, etc. Thus implications for system reliability of uncertainties in predicted reliabilities of major components such as those discussed in Examples 14.1 and 14.2 must be addressed.

14.3.6   Prediction at Later Stages

As noted previously, as designs are refined and development proceeds, reliability predictions are also refined. Materials are tested, parts and components are built and tested, and further information of many types is obtained. This information is used to update predictions of item reliability, and to make additional design changes as necessary when the predicted values do not meet reliability goals.

A key statistical tool in the updating process is Bayesian statistical analysis. This is discussed in Chapter 8 and in the references cited. The process involves formulating a probabilistic model based on realistic assumptions and the information available. This should be done at the earliest stage possible. Bayesian methodology is then used to update the model each time additional results are gotten and introduced into the analysis. Bayesian predictions intervals, which provide a confidence level (in the Bayesian sense) on the prediction are also calculated.

The problem of formulating a prior distribution may be approached at various levels, e.g., it may done at the system level or done for each component, etc. Thus in Example 14.1, a prior distribution on system reliability may be devised in a number of ways, e.g., (1) using the predicted value of 0.9990 and a beta distribution, say, with parameter values selected to give 0.9990 as the mean value; (2) selecting an appropriate prior distribution for the parameter of the exponential distribution; or (3) formulating priors for each part or component; and so forth.

In this context, the approach of Martz et al. (1988), as further developed by Martz and Waller (1990) and Hulting and Robinson (1994), is very useful. See Chapter 8 and the references for details.

14.4 THE ROLE OF TESTING

Testing of items is of obvious importance in reliability evaluation. It is the main source of information leading from reliability prediction to reliability assessment and verification of reliability objectives. A goal of every test and evaluation program should be to improve reliability and producibilty by

- Improving understanding of the product and process
- Improving understanding of failure modes and mechanisms
- Identifying deficiencies
- Identifying areas for design improvement

Data sources during product development include many forms of testing. A number of these are discussed by Meeker and Hamada (1995). These include:

- Laboratory test for materials evaluations
- Laboratory life tests of parts, components, etc.
- Environmental stress screening
- Tests of prototypes
- Degradation tests of materials, parts, etc.
- Test results from suppliers
- Qualification testing
- Stress life tests
- Reliability demonstration tests

To ensure data validity and reliability, carefully designed experiments are necessary. Basic principles of experimentation and a number of experimental designs useful for such studies are discussed in Chapter 10. Additional information, particularly on factorial and fractional factorial designs, is given by Hamada (1995), including models, experimental data, and analyses.[2]

An important additional experimental technique is accelerated testing, that is, testing under conditions involving stresses somewhat or far in excess of those encountered in normal operation. This allows the analyst to obtain data in a shorter time frame but requires a technique for translating results to normal operating conditions. This will be discussed in the next section.

In later stages of the product life cycle, additional studies may be made and information from other sources may become available (and will, if properly prepared for). These potential sources include

- Monitoring of early production models after sale
- Warranty data
- Product service data
- Follow-up studies of consumers
- Long-term tracking of items after sale

Note that these sources of information often do not lend themselves well to controlled exper-imentation and may have other limitations as well. For example, warranty data are subject to bias because of false reporting, failure of some owners to report failures, and so forth, and, in any case, are effectively censored at the end of the warranty period. Efforts can be made to obtain valid data from these sources through random sampling, but nonresponse and other problems are encountered in studies of this type as well, and such data must be interpreted carefully in light of these shortcomings.

14.5   ACCELERATED TESTS

In Chapter 10, we discussed many classical experimental designs, including those with qual-itative factors, quantitative factors, or both. The major purposes of such experiments are to obtain reliability or reliability-related data under controlled conditions and then estimate or test hypotheses about characteristics of interest, e.g., MTTF, material strength, and so forth. In many cases, particularly where time is a primary factor of interest, obtaining data under re-alistic operating conditions can be very costly or, in fact, impossible. Under these circum-stances, testing is often done under conditions of much higher stress than would ordinarily be encountered in actual applications. Such tests are called *accelerated tests,* and were dis-cussed briefly in Chapter 10. Here we look in more detail at such tests, including test design, models for accelerated tests, and data analysis.[3]

Accelerated testing is especially important in the context of highly reliable items, where it is often virtually impossible to demonstrate that a reliability goal has been attained. For ex-ample (Meeker and Hamada, 1995), to demonstrate statistically, with 95% confidence, that the probability that an item lasts at least 5 years is at least 0.999, requires that 2995 items be tested for 5 years with no failures occurring. Much higher sample sizes are needed if one or more failures occur. To demonstrate that a component of a solid rocket motor case has relia-bility 0.9999996 would be astronomically expensive. In these examples, accelerated testing is necessary, and in many less-extreme applications, it is desirable for reasons of cost effec-tiveness and timeliness.

The major difficulty in the use of accelerated testing is that it is necessary to relate the re-sults obtained under conditions of higher stress to those that would result under normal con-ditions. This requires an adequate understanding of failure mechanisms and appropriate mod-els that express the relationship. There is a great deal of literature dealing with these topics. We present an introduction to some of the basic concepts. All aspects of accelerated testing are discussed in great detail by Nelson (1990).

14.5.1   Design of Accelerated Tests

Stress that accelerates the failure process may be applied in many forms: high or low temper-atures, humidity, cycling between excessively high and low conditions, excess usage, electri-cal stress, vibration, and so forth. Test designs follow the basic principles of DOE. Particularly appropriate are factorial experiments, with the factor or factors involved usually being quantitative. Split-plot experiments of various types are often appropriate if two or more factors are involved, particularly when time is a factor.

An additional concern that must be addressed in designing experiments of this type is the possibility of the excess stress causing failures that would not occur in normal operations, for example, melting of materials, weakening of bonds, or expansion of materials at high temperatures. Furthermore, if an item can fail in several ways, acceleration may affect failure rates for the different modes differently. For this and other reasons (e.g., cost and test equipment requirements), accelerated tests are most often done at the part or small component level. Furthermore, because of the complexity of relating failures (or failure-related characteristics) to more than one stress factor, accelerated tests most often involve only a single accelerating factor, e.g., temperature alone rather than temperature and humidity or temperature, humidity, and time. Two-factor experiments are not uncommon, but many-factor experiments are not usually used because of modeling and other analytical difficulties.

For basically the same reasons, relatively straightforward experimental designs are used for accelerated tests. The completely random design (CRD) provides the simplest analysis. With two or more factors, the CRD may be used as well. A split-plot design with the basic CRD structure for whole plots may also be used for two or more factors, but this somewhat complicates the analysis, particularly if interactions among the factors exists and/or confidence intervals are desired.

Selection of levels of a factor to use in testing depends on the context (materials, stresses, and so forth). Important considerations are that

1. Levels are not so extreme that the failure mechanism changes
2. Levels are within the range over which the selected model is appropriate
3. Excessive extrapolation is not required

In the first two cases, the model may be invalidated and the test data not relevant to normal conditions. Excessive extrapolation has not only these difficulties, but the added problem that confidence intervals, even if valid, will be so wide as to be useless.

14.5.2   Models for Data from Accelerated Tests

In accelerated testing, the key to analyzing and interpreting data is the formulation of appropriate models. A number of models have been developed for specific stress factors, e.g., temperature, electrical stress, etc. The most commonly used stress factor is temperature, which tends to accelerate physicochemical processes, and many models have been developed for representing failure and other characteristics in this context. Extensions of these to other factors and to two or more simultaneous factors have also been developed. Here we look at some of the more common models used in accelerated testing. For a detailed discussion and extensive lists of applications, see Nelson (1990).

Arrhenius Model
The most commonly used model relating time to failure to thermal stress is the Arrhenius model. As discussed in Section 6.10.1, the Arrhenius model expresses the MTTF or other characteristics of failure in terms of a rate function, given by

$$r = r_0 e^{-E/kT} \tag{14.1}$$

where r is the reaction time, r_0 is a constant that depends on the part geometry, size, etc., E is the activation energy of the reaction (in electron-volts), k is Boltzmann's constant (8.6171×10^{-5} electron-volts per °C), and T is temperature in degrees Kelvin (°C + 273).

Note that thermal stress may be due to high temperatures or excessively low temperatures. In the latter case, failure may be caused, for example, by increased brittleness of the materials. The Arrhenius equation may be used to model either situation. Applications include electrical insulation, electronic devices, adhesive bonds, batteries, fibers, lubricants, filaments, and plastics.

The Arrhenius model is used to calculate an acceleration factor A relating mean lifetimes, μ_1 and μ_2, at two temperatures, T_1 and T_2. The result is

$$A = \frac{\mu_1}{\mu_2} = e^{(E/k)[(1/T_1)-(1/T_2)]} \tag{14.2}$$

Thus, for example, if T_1 is the normal operating temperature and T_2 is the temperature at which the test was run, the mean time under normal conditions is calculated as $A\mu_2$. If a sample at temperature T_2 is obtained, with sample mean $\bar{x}$, μ_1 is estimated as $A\bar{x}$, with estimated standard deviation $As/\sqrt{n}$, where s is the sample standard of the test data. Unless n is small, this may be used to obtain a confidence interval for μ_1 based on the usual normal approximation.

Example 14.3

The lifetimes of a sample of 50 PC components operated at a temperature of 100 °C are recorded. The sample mean is found to be $\bar{x} = 9.3$ days; the sample standard deviation is $s = 3.45$ days. In this application, $E = 0.85$. Normal operation is at 27 °C. These temperatures correspond to 373 °K and 300 °K, respectively. Thus the acceleration factor is

$$A = e^{(.85/8.617)(100,000)[(1/300)-(1/373)]} = 623.35$$

giving an estimated MTTF for the component of $9.3(623.35) = 5797$ days, or just under 16 years. A lower 95% confidence limit for the MTTF is obtained as $5797 - 1.645(623.35)(3.45)/\sqrt{50} = 5297$ days, or 14.5 years. We are 95% confident that the component has a mean lifetime of at least 14.5 years. ∎

Example 14.3 used the basic Arrhenius relationship and asymptotic normality of the estimated accelerated mean to estimate the MTTF. For small sample sizes, the actual life distribution of the item is used in the analysis, with the Arrhenius equation expressing the relationship between the MTTFs. Methods of analysis for the exponential, lognormal, and Weibull distributions are given by Nelson (1990).

Inverse Power Law Model

Another widely used relationship, with many applications, including electrical devices, metal products such as ball bearings, metal fatigue, fiber, filaments, and so forth, is the inverse power model, discussed in Section 6.10.2. The basic model is

$$\tau = \frac{A}{S^\beta} \tag{14.3}$$

where τ is the MTTF or a related quantity, S is the stress applied, and A and β are constants specific to the item being tested and the test procedure. Here stress is usually voltage. Values of τ, say τ_1 and τ_2, at stress levels S_1 and S_2 are related by the expression

$$\tau_1 = \tau_2 \left(\frac{S_2}{S_1} \right)^{\beta} \tag{14.4}$$

The parameter β is determined by the context. If τ_2 is estimated from data obtained from an accelerated test, τ_1 can be estimated, and a confidence interval calculated, by the methods used for the Arrhenius model.

A number of versions of the power law are discussed by Nelson (1990) and Condra (1993). A detailed analysis of data on ball bearing failures based on the inverse power law is given by Leiblien and Zelen (1956).

Eyring Model

The Eyring model is similar to the Arrhenius model, but is more general in the sense that is may be used to represent a second stress in combination with temperature, for example, an electrical stress or relative humidity. It is based on the reaction rate for chemical degradation derived from quantum mechanics. The basic relationship is

$$\tau = \frac{B}{S} e^{E/kT} \tag{14.5}$$

where τ is the MTTF or a related quantity, S is the applied stress at temperature T, B is a constant, and the exponential term is as in the Arrhenius model.

For this relationship, the acceleration factor becomes

$$A = \frac{\tau_1}{\tau_2} = \frac{S_2}{S_1} e^{(E/k)[(1/T_2)-(1/T_1)]} \tag{14.6}$$

Where τ_1 and τ_2 are the means at stress levels S_1 and S_2, respectively.

The model has applications in many of the same areas as the Arrhenius model, with the added feature of the second stress factor. An example of a model of this type is the Peck (1986) model for failure times of electronic microcircuits, in which stress is expressed as $S = h^3$, where h is percent relative humidity, and E is taken to be 0.9. A more general form is $S = h^{\beta}$, where β is estimated from test data.

Again, if τ_1 is estimated from data, an approximate confidence interval for τ_2 may be calculated by use of the asymptotic normality of the estimator of τ_1.

Example 14.4 (Case 2.1)

Table 2.1b contains data on bond strength at several levels of temperature and humidity. Although this is not the application discussed by Peck (1986), the Peck model may provide a reasonable approximation. We consider Test 1, which is very moderately accelerated from normal conditions. The test was run at 32 °C, 70% RH. Suppose that normal conditions for operation are 27 °C, 50% RH. From the data of Table 2.1b, the mean and standard deviation for Test 1 are calculated to be $\bar{x} = 204.9$ and $s = 18.4$, with $n = 32$. A 95% confidence interval for μ_1 is $204.9 \pm 2.04(18.4)/\sqrt{32} = 167.4, 242.5$, where 2.04 is the tabulated value of t with 31 df. The acceleration factor is

$$A = \left(\frac{50}{70} \right)^3 e^{[0.9(10^5)/8.617][(1/300)-(1/305)]} = 0.64488$$

Since time to failure of the bond may reasonably be assumed to be inversely related to stress, we estimate the mean bond strength at 27 °C, 50% RH, based on the data at 32 °C, 70% RH to be $204.9/0.645 = 317.7$, with the corresponding 95% confidence interval of $(259.5, 375.9)$. Note that Test 3 was run at the stated nominal conditions, resulting in a sample mean of 278.8, which is included in the calculated confidence interval for the mean under these conditions. ■

Multifactor Models

As noted, in accelerated tests it is desirable to consider a single acceleration factor, because of the added complexity of multifactor experiments and the additional assumptions required in the analysis. On the other hand, operational conditions often involve multiple stress factors, which may interact in their effect on item failure, so that testing under realistic conditions must include two or more types of stress. The Eyring model was developed for the situation in which temperature is combined with a second stress factor, such as relative humidity, voltage, etc. More generally, it may be used in other two-variable situations as well, if an understanding of the physics of failure suggests such a model, or as a first approximation in less well understood cases.

A number of additional physical models involving two or more factors have been developed. For example, for vibration stress, e.g., in metal ducts, pump blades, and so forth, Miner's rule (Dowling, 1999, p. 402) expresses remaining life as a function of various stress levels. The model is

$$U = \sum_{i=1}^{k} \frac{c_i}{C_i} \qquad (14.7)$$

where U is the unused lifetime of the item, c_i is the number of cycles experienced under the ith stress level at a particular point in time, and C_i is the average number of cycles to failure at that stress level. For an application involving liquid rocket engines, see Moore et al. (1990). Miner's rule and other aspects of modeling for many types of stresses, including environmental factors, are discussed by Nelson (1990) and Condra (1993).

A commonly used statistical model for representing multiple stress factors and analyzing test data is the proportional hazards model, discussed in Section 6.10.4. The basic model expresses the failure rate $\lambda(.)$ as a function of k predictor variables (e.g., stresses). The most commonly used form of the model, due to Cox (1972), is the exponential relationship

$$\lambda(t) = \lambda_0(t)e^{\sum_{i=1}^{k}\beta_i x_i} \qquad (14.8)$$

where $\lambda_0(t)$ is the baseline failure rate, x_i is the ith predictor variable, and β_i is the corresponding regression coefficient. Transforming to logarithms in (14.8) results in the usual multiple linear regression model relating log failure rate to the predictor variables. Interaction terms can be introduced into the relationship by defining some of the predictor variables as functions of others (e.g., $x_3 = x_1 x_2$).

14.5.3 Analysis of Data from Accelerated Tests

If a single stress level is used and n items are tested at this level, data analysis is relatively straightforward. An estimate of and confidence interval for average lifetime at nominal stress levels based on the accelerated test data can be obtained as in Example 14.3 for the Arrhenius model and Example 14.4 for the Eyring model. These results are based on asymptotic normality of the estimators and may not be appropriate for small samples unless it is reasonable to assume approximate normality of the data. Alternatives based on the exponential, Weibull, and other life distributions are discussed by Nelson (1990) and Condra (1993).

For multiple stress levels with a single accelerating variable, a number of approaches have been developed. A common approach using graphical methods is discussed by Nelson (1990), Condra (1993), and Meeker and Escobar (1998). Included are the exponential, Weibull, and lognormal distributions. Plots of failure data at the different stress levels are done on an appropriate plotting paper, and, assuming that the slopes are the same, differences between the resulting lines are analyzed. (If the life distribution is not known or deduced from physical models, various distributions should be investigated and goodness-of-fit procedures applied as in Chapter 11 to select an appropriate distribution.)

Many of the models for accelerated test data are exponential or multiplicative. These are easily linearized by the logarithmic transformation. Other forms may be linearized as well. In these cases, data may be analyzed by ANOVA or regression methods, as indicated in Chapter 10. This is especially useful in dealing with multivariate models, and is the basis of analysis for the Cox proportional hazards model. Since estimation of characteristics at normal operating conditions using accelerated test data always involves extrapolation, it is especially important in applications to calculate confidence intervals for the characteristic in question. These may become very wide if the extrapolation is extreme and/or the variability of the test results is not small, and are indications that caution must be exercised in using the results.

Most statistical program packages and many reliability packages include programs for data analysis based on the proportional hazards model and other models used in accelerated tests.

Example 14.5 (Case 2.1)

We consider a simple example based on a regression analysis of the data given in Table 2.1b of Case 2.1. Four temperature–humidity combinations were used in the study, with temperatures ranging from 27 °C to 32 °C and relative humidity ranging from 50 to 100%. The output of a multiple regression analysis of the data (without transformation) is given in Example 10.12. Suppose that nominal conditions in a particular application are 21 °C, 30% RH. If the regression model is used to estimate to average bond strength at these conditions, the result is $\hat{y} = 358$ pounds with a 95% confidence interval for the true mean strength of (288, 428). ∎

Additional discussion of analysis of accelerated test data, including graphical analyses and analyses based on specific models in many types of applications, may be found in the references cited.

14.6 RELIABILITY ASSESSMENT

As noted, models and methods of reliability prediction and assessment have been many and varied. The evolution of reliability evaluation as advanced theory and methods have been developed is depicted in Figure 14.3. Reliability analysis began with a "seat-of-the-pants" approach, amounting basically to overdesign, and evolved to the current state of the art through a number of steps. Initially, designs that could withstand far more than any anticipated stress were the norm. These were based primarily on engineering experience and judgment. Prior to the development of mathematical models and DOE, a relatively crude ("build 'em and bust 'em") approach to testing was used. With the development of experi-

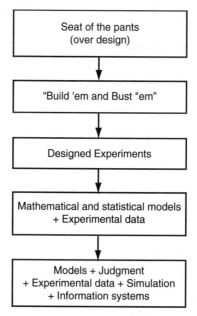

Figure 14.3. Evolution of reliability assessment.

mental design methodology, carefully designed experiments were conducted and the resulting data were analyzed in the context of newly developed mathematical and statistical models. The current state of the art employs, in addition, judgmental and other information,
computer simulation, and information systems. In the remainder of this chapter, we briefly
address a number of these areas.

By reliability assessment is meant determining the reliability of an actual physical system,
based on test data. There are a number important reasons for careful reliability assessment.
First and foremost is to verify that the predicted reliability is attained or is attainable. An important problem nearly always faced is the fact that actual reliability is found to be less than
that predicted by engineering design and analysis and prototype construction and testing.
There are many factors that contribute to this, including unforeseen difficulties and uncertainties in production, manufacturing errors, defects in components or parts received from
suppliers, improper specifications, design changes, inadequate inspection, and so forth. Other reasons for assessment include verification of updated predictions, monitoring of product
quality, determination of reliability growth, and so forth. Testing is required in all of these
endeavors. Testing may be required at more than one level, and it may require the cooperation of suppliers.

The evolution of reliability assessment has been described above. The more important
evolution in practice is that from prediction to assessment. In this evolution, a key principle
is TAAF—test, analyze, and fix. This may involve refinements in design or redesign of product and/or process, alternative materials selection, and so forth, followed by further testing
and analysis, leading to reliability improvements. The process of redesign, modeling, and
testing to improve reliability is often called "reliability growth." This will be discussed in
more detail in Chapter 15. Reliability prediction of designs and testing and assessment of actual components or items are integral parts of the process.

Assessment, then, deals with determining the reliability of an actual physical system. It requires

1. test data at one or more levels from carefully designed experiments (see Chapter 10)
2. statistical estimation and hypothesis testing (see Chapters 5 and 8)

and may require

3. model selection and validation (see Chapter 11)
4. evaluation of the impact of software reliability, if software is a part of the system (see Chapter 9)

Both classical and Bayesian statistical analysis have been used in reliability assessment. If prior information is available, the Bayesian approach is preferable because it is a methodology for incorporating this information into the assessment process and provides a well-developed means for updating as more information, primarily from tests, becomes available. Problems are that modeling of information from many sources may be difficult, subjective information may not be reliable or may not be agreed upon, and analytical difficulties are encountered. The classical method would lead to assessments based primarily on the likelihood function, and hence would be "objective," although many decisions that are at least partially subjective, e.g., choice of the life distribution of a part or item, are involved here as well. It is important, whichever approach is used, to provide interval estimates of system reliability (confidence intervals or Bayesian intervals) at each stage.

Both reliability predictions and assessment at any given point in time may significantly affect management decisions regarding continued product development, reliability improvement, production, marketing, and other activities requiring capital expenditures. Thus it is important to track reliability and to assess product performance as precisely as possible. (Note, however, that reliability considerations are only one of many inputs to the management decision process, and that this process, too, is at least partially subjective.) The following example looks at reliability prediction and assessment in a simplified scenario. More complete analyses of cases will be discussed in Chapter 19.

Example 14.6

In Example 14.1, the reliability of a liquid rocket engine for use in a specific mission was predicted, at the design stage, to be 0.998983. In this context, reliability is defined to be the probability that the engine will achieve at least 1.05 times nominal thrust for a specified duration in a specified environment. The prediction was based on materials properties, prior experience, handbook information, and similarities to the proposed design. After further analysis, component testing, design refinements, and limited testing of the entire system, an updated reliability prediction of $R = 0.998543$ was obtained. Suppose that a propulsion system consists of three such engines, and that the standard deviation of the prediction, based on prior experience and engineering judgment, is determined to be 0.0015. This information may be used in a Bayesian reliability analysis (see Chapter 8).

We take the prior distribution to be a beta distribution with mean $\mu = 0.998543$ and standard deviation $\sigma = 0.0015$. The relationship between the mean and variance and the parameters α and β of the beta distribution is given in Equation (8.47). From (8.50), we obtain

$$\alpha = \frac{1}{2-\mu}\left(\frac{(1-\mu)\mu}{\mu} - 1\right) = 644.7 \qquad \beta = \alpha(1-\mu) = 0.9393.$$

From (8.60), with $n = x = 0$, a 95% Bayesian lower prediction interval is found to be

$$\frac{\alpha}{\alpha + \beta F_{0.95;2\beta,2\alpha}} = \frac{644.7}{644.7 + 0.9393 F_{0.95;2,1289}} = 0.995364$$

where $F_{p;a,b}$ is the p-fractile of the F-distribution with a, b df, obtained in this case from Minitab. Thus the true reliability is predicted, with 95% certainly, to be at least 0.995364.

Suppose that a propulsion system consists of three engines (plus solid rocket motors, guidance system, etc.) and that we are interested in assessing the reliability of the liquid engines after five missions with no engine failures. Here $n = x = 15$. The assessed reliability is

$$\hat{p}_b = \frac{644.7 + 15}{644.7 + 0.9393 + 15} = 0.998578$$

The posterior distribution of R is a beta distribution with $\alpha = 644.7 + 15 = 659.7$, and $\beta = 0.9393$. The lower one-sided 95% Bayesian limit is $959.7/(659.7 + 3.0025) = 0.995469$. The limit has increased only slightly from the predicted value. ∎

NOTES

1. For additional information and other approaches to prediction, see Cruse et al. (1994), Basu (1995), and Singh (1995). Prediction based on incomplete data is discussed by Escobar and Meeker (1999).

2. There is a large literature on testing, analysis of test data, and data-based reliability evaluation. For some additional discussion concerning the role of testing, see Blumenthal et al. (1984), Stochholm (1986), and Raheja (1994).

3. The basic principles of design of accelerated tests, details of traditional test plans, and optimal test procedures are discussed by Nelson (1990) and many additional references are provided. For additional discussion of test plans, selection of test conditions, and the use of environmental factors in specific applications, see Meeker and Escobar (1998) and Condra (1993).

Modeling and analysis of the inverse power law for multiple tests run at two or more stress levels are discussed by Mann et al. (1974).

For more on proportional hazards models and additional multivariate models, see Kalbfleisch and Prentice (1980), Cox and Oakes (1984), and Nelson (1990).

EXERCISES

1. Select two products with which you are familiar, one relatively simple and one complex, and discuss reliability prediction for each. (Think of this from the point of view of the manufacturer and in the conceptual design stage.)

2. Prepare a FMEA for the simple product selected in Exercise 1. Determine the criticality for each component. Give suggestions as to what a company should do to avoid critical failures.

3. Perform a FTA for the simple product selected in Exercise 1. Assign reasonable probabilities to each failure and calculate the product reliability based on these.

4. A series of products on increasing complexity is given in Section 14.2.2. Give another such series, using items ranging from, e.g., an electric iron to an automobile.

5. Merge the lists of the previous exercise and Section 14.2.2, arranging items in your perceived notion of increasing complexity.

6. Use a relatively simple measure such as parts count to evaluate the reliability of the items in the list of the previous exercise. Is the ordinal ranking of reliabilities in perfect correspondence with your ordering based on complexity? Why might it not be?

7. A schematic of a stereo system is given in Figure 7.2. Draw a new block diagram that includes an antenna and six connectors (antenna to tuner; tuner, cassette player and CD to amplifier; and amplifier to both speakers) and 12 connections (one on each end for each connector). Assume that the system is considered to be operational only if all sound sources and both speakers are working.

8. Reanalyze the stereo system of the previous exercise assuming that coherent sound is produced (i.e., if at least one source and one speaker are working).

9. Predict the reliability of the stereo system described in Exercise 4 assuming that components operate independently and lifetimes of each component are exponentially distributed. Reliability is to be calculated as the probability that the system lasts at least one year. The mean lifetimes of the components are

μ(Amplifier) = 14 years

μ(Tuner) = 12 years

μ(CD player) = 15 years

μ(Cassette player) = 8 years

μ(Speaker) = 30 years

The antenna is assumed to have reliability 0.9998 and the connectors and connections a joint reliability of 0.9976.

10. Predict the reliability of the system of the previous exercise if the lifetimes of the components are assumed to follow Weibull distributions with the same means and shape parameters (for amplifier, tuner, etc.) $\alpha = 2, 3, 2, 1$, and 4, and the same reliabilities on connectors and connections.

11. Suppose that the distributions in the previous exercise are gamma rather than Weibull, but have the same mean and variance. Calculate the corresponding values for the gamma parameters, calculate the resulting reliability of the system, and compare the results with those of the previous two exercises.

12. Regression analysis was used to analyze the data of Example 14.5. Test the assumptions required for validity of this analysis, including normality and homogeneity of variance. Do the results of your analysis change your conclusions with regard to the strength of the bond at 21 °C, 30% RH?

13. Reanalyze the data of Example 14.5 using the Cox proportional hazards model (i.e., analyze on the log scale). State your conclusions.

14. Reanalyze the data of Example 14.5 as in the previous exercise, including an interaction term in the model. Does there appear to be an interaction? Do the results change your conclusion?

15. Refer to Exercise 14.6. Suppose that five additional missions have taken place, with no engine failures. Calculate a Bayesian point estimate for the reliability of the liquid rocket engine and a 95% lower confidence limit for its reliability. (Recall that there are three engines on the vehicle.)

16. (Continuation) Suppose that in the next five missions one engine had failed to achieve full thrust. Assess the reliability of the liquid rocket engine.

CHAPTER 15

Reliability Improvement

15.1 INTRODUCTION

As indicated in Chapter 7, the reliability of a system can be expressed in terms of the reliability of its lower-level components, down to the part level. When the reliability of the system during the design phase of the product life cycle is below the target value, it is unacceptable and must be improved. There are two basic approaches to improving system reliability, which are

1. *Use of Redundancy:* This involves the use of replicates rather than a single unit. The replication can be carried out at any level ranging from the system level to the part level.
2. *Reliability Growth:* Here the reliability of a unit (at the assembly or subassembly level) is improved through a development process that involves test–fix cycles.

In this chapter we discuss both of these approaches. The outline of the chapter is as follows. Section 15.2 deals with the use of redundancy for improving system reliability and introduces the three different types of redundancies that can be used. These are hot, cold, and warm redundancies. The next three sections, Sections 15.3–15.5, deal with each of these in more detail. Section 15.6 deals with reliability growth and the classification of the models used for assessing and predicting reliability improvement. These models can be broadly divided into two categories—discrete and continuous models. Section 15.7 discusses some of the discrete models and Section 15.8 deals with continuous models. Some comments on these and related issues are given in Section 15.8.

15.2 REDUNDANCY

Redundancy is a technique whereby one or more of the components of a system are replicated in order to improve the reliability of the system. Redundancy can only be used when the functional design of the system allows for the incorporation of replicated components. It is used extensively in electronic systems to achieve high reliability when individual components have unacceptably low reliability.[1]

Building in redundancy corresponds to using a module consisting of M replications of a component. The manner in which these replicates are put to use depends on the type of re-

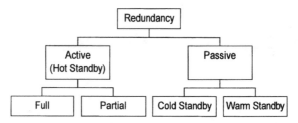

Figure 15.1. Classification of redundancy.

dundancy. A module failure occurs only when some or all of the replicates fail. The decision regarding the use of redundancy always involves a trade-off. Improved reliability is achieved at the cost of designing and manufacturing the system with multiple components, plus the additional operational expense involved. In space systems, for example, this expense can be considerable because of the launch costs due to the extra weight resulting from redundancy, but the added reliability may be well worth the added cost.

Many different types of redundancies are used in practice. The redundancy classification in Figure 15.1 shows the different types. The two main types are (1) active (hot standby) and (2) passive. In active redundancy, all M components of the module are in their operational state, or "fully energized," when put into use. In contrast, in passive redundancy, only one component is in its fully energized state and the remaining are either partially energized (in the case of warm standby) or kept in reserve and energized when put into use (in the case of cold standby). When the fully energized component fails, it is replaced by one of the partially energized components in the case of warm standby, or, in the case of cold standby, by a component from the reserve using a switching mechanism, provided that not all of the components in the module have failed. If all components in the module have failed, then the module has failed.

Redundancy can be either at the system level or at a lower level (e.g., assembly, subassembly or component level). For component-level redundancy, in general, the replicates are statistically similar, but in certain situations they can be different, as well. In the following two sections, we discuss the two types of redundancy and their implications with regard to system reliability in more detail.

15.3   ACTIVE (HOT STANDBY) REDUNDANCY

In the case of active redundancy, all M replicas are in an operational state from the time the module is put in use and the system requires a single operable replica in order to be operable. In other words, the M replicas are connected in parallel. As such, this type of redundancy is also called "parallel redundancy." Polovko (1968) uses the term "constantly connected standby" and Gnedenko et al (1969) use the term "loaded standby" for this type of redundancy.

The active redundancy can be divided into the following two categories: (1) full active and (2) partial active.

15.3.1   Full Active Redundancy

In this case the module fails only when all the M components of the module fail. Let X_i, denote the failure time for component i, $i = 1, 2, \ldots, M$, and Y the time to failure for the module. We assume that the M components are statistically similar and that the component failures are independent. Let $F(x)$ denote the failure distribution for the components, and $G_M(x)$ the failure distribution for the module. Note that

$$Y = \max\{X_1, X_2, \ldots, X_M\} \tag{15.1}$$

If the M components are statistically similar, we have

$$G_M(x) = [F(x)]^M \tag{15.2}$$

and the system reliability $R(x)$ given by

$$R(x) = 1 - G_M(x) = 1 - [F(x)]^M \tag{15.3}$$

It is easily seen that for any fixed x, $R(x)$ increases as M increases.

When the components are not identical, let $F_i(x)$ denote the failure distribution of component i, $i = 1, 2, \ldots, M$. In this case, (15.2) becomes

$$G_M(x) = \prod_{i=1}^{M} F_i(x) \tag{15.4}$$

Example 15.1

Let X be exponentially distributed with parameter λ, so that the mean time to failure is $1/\lambda$. From (15.2) we have

$$G_M(x) = [1 - e^{-(\lambda x)}]^M$$

Suppose that $\lambda = 0.2$ per year, giving a MTTF of 5 years. Table 15.1 gives $R(x)$, obtained from (15.3), for $x = 1, \ldots, 5$ years and $M = 1, \ldots, 5$ components. Note that as x increases, the increase in reliability as M increases becomes progressively more pronounced. Even so, a 5 year MTTF does not provide high reliability over a 5 year period.■

Example 15.2

Let X be Weibull-distributed with shape parameter α and scale parameter β. From (15.2) we have

$$G_M(x) = [1 - e^{-(x/\beta)^\alpha}]^M$$

Table 15.1. $R(x)$ for $\lambda = 0.2$ per year and redundancy M

		M			
x (years)	1	2	3	4	5
1	0.8187	0.9671	0.9940	0.9989	0.9998
2	0.6703	0.8913	0.9642	0.9882	0.9961
3	0.5488	0.7964	0.9082	0.9586	0.9813
4	0.4493	0.6968	0.8330	0.9080	0.9494
5	0.3679	0.6004	0.7474	0.8403	0.8991

Table 15.2. $R(x)$ for the Weibull distribution with $\alpha = 2$, $\mu = 5$ years, and redundancy M

	M				
x (years)	1	2	3	4	5
1	0.9690	0.9990	1.0000	1.0000	1.0000
2	0.8808	0.9858	0.9983	0.9998	1.0000
3	0.7516	0.9383	0.9847	0.9962	0.9991
4	0.6049	0.8439	0.9383	0.9756	0.9904
5	0.4559	0.7040	0.8390	0.9124	0.9523

Suppose that $\alpha = 2.0$ and $\beta = 5.6419$ years. These values also correspond to a MTTF of 5 years, and with $\alpha = 2$, we have an increasing failure rate Table 15.2 gives $R(x)$, obtained from (15.3), for the same values of x and M used in Example 15.1.

Note that the reliabilities here are consistently higher than those for the exponential distribution with the same MTTF. Also note that a considerable amount of redundancy is required to obtain high reliability over periods of time approaching the expected lifetime of the item. The situation is considerably worse if the item has a decreasing failure rate. ∎

15.3.2 Partial Active Redundancy

In the case of full active redundancy, a module failure occurs only when all the components in the module fail. In contrast, in the case of partial active redundancy, the number of components allowed to fail before module failure occurs is less than the full complement of components. Suppose that the module is deemed to be in a failed state when r or more of the M components fail. Then the failure distribution for the module, $G_M(x)$, is given by

$$G_M(x) = \sum_{i=r}^{M} \binom{M}{r} [F(x)]^i [1 - F(x)]^{M-i} \tag{15.5}$$

Note that $r = M$ corresponds to full active redundancy.

15.3.3 Component versus System-Level Redundancy

As mentioned previously, redundancy can be used either at the system, part, or any intermediate level. With system-level redundancy, the system is replicated M times. In contrast, with component level redundancy, the components are replicated M times. Here we contrast the effects of the different types of redundancies.

Let R_s denote the system reliability with no redundancy, and let $R_{sc}(x)$ and $R_{ss}(x)$ denote the system reliability with component and system level redundancy. In general, for any system higher reliability is achieved by using redundancy at the part level as opposed to the system level.

Consider a system with n identical components connected in a series structure. We first consider component-level redundancy where each component is replaced by modules consisting of M replicas, as shown in Figure 15.2. In this case, the reliability of each module is given by (15.3) so that the overall system reliability is

$$R_{sc}(x) = [1 - F(x)^M]^n \tag{15.6}$$

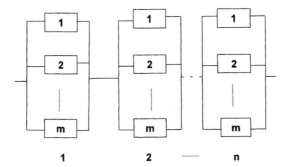

Figure 15.2. Redundancy at the component level.

When redundancy is used at the system level, we have M replicas of the system connected in parallel, as shown in Figure 15.3. In this case, the system reliability is

$$R_{ss}(x) = 1 - \left[1 - \left(1 - F(x)\right)^n\right]^M \qquad (15.7)$$

Example 15.3
Suppose that $F(x)$ is an exponential distribution with failure rate $\lambda = 0.2$ per year, as in Example 15.1, and that $n = 3$. $R_{sc}(x)$ and $R_{ss}(x)$ for the values of x and M used in the previous examples are given in Table 15.3. Note that for $M = 1$, the results for redundancy at the system and component levels are, of course, the same, and that redundancy at the part level provides substantially higher reliability than does redundancy at the system level. ∎

15.3.4   Limitations of Active Redundancy

From an engineering standpoint, some of the limitations of active redundancy are the following:

1. Designing full active standby for mechanical devices is usually a difficult job.
2. The assumption that the failure of one component has no affect on the remaining components is not always true. In reality, when one component fails, the load is distributed to the remaining components. This can lower the reliability of the system because the

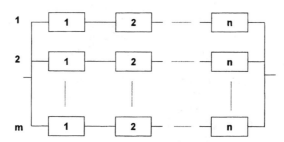

Figure 15.3. Redundancy at the system level.

Table 15.3. $R(x)$ for exponentially distributed items with redundancy at the component and system levels, MTTF = 5, three components

	(a) Redundancy at the component level				
	M				
x (years)	1	2	3	4	5
1	0.5488	0.9046	0.9822	0.9968	0.9994
2	0.3012	0.7081	0.8963	0.9650	0.9884
3	0.1653	0.5052	0.7490	0.8808	0.9449
4	0.0907	0.3383	0.5780	0.7487	0.8557
5	0.0498	0.2165	0.4175	0.5934	0.7268

	(b) Redundancy at the system level				
	M				
x (years)	1	2	3	4	5
1	0.5488	0.7964	0.9082	0.9586	0.9813
2	0.3012	0.5117	0.7615	0.7615	0.8334
3	0.1653	0.3033	0.4148	0.5146	0.5948
4	0.0907	0.1732	0.2482	0.3164	0.3784
5	0.0498	0.0971	0.1420	0.1848	0.2254

load on each component is increased and the multiplicity of redundancy (the number of surviving components in the system) is decreased.

3. For a given component reliability, the incremental gain in reliability with the addition of each additional replicate component decreases, and beyond a certain point any further addition results in a negligible gain in the reliability of the system.

However, examples of this type of redundancy can be found in spacecraft systems (such as dual thrust motors for orbital station keeping), power generating systems, railway signals, bearing assemblies and so forth. In these cases, the benefits of this type of redundancy have been found to or have been projected to more than offset the costs.

15.4 COLD STANDBY REDUNDANCY

In cold standby redundancy, one component is put into use (or energized) and the remaining $(M - 1)$ components of the module are kept in reserve. When the component in use fails, one of the components from the reserve is put into use through a switching mechanism, as shown in Figure 15.4. When this fails, the next in the reserve is put into use and the process continues. The module fails when the last component in the reserve fails after being put into use.

In cold standby, an assumption that is often made is that the components in reserve do not deteriorate with time. Thus, whenever a standby component is called up, it is considered to be a brand new component, irrespective of length of time it has spent waiting in the reserve. This might not be true for certain applications. In this section we assume that there is no deterioration (or if there is deterioration, it is negligible and can be ignored).

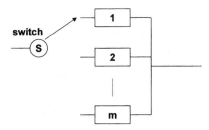

Figure 15.4. Cold standby redundancy involving a switch.

The switch used for disconnecting a failed component and connecting a working one from the reserve can be either perfect or imperfect. In the former case, the switch functions properly whenever it is called on to operate, i.e., the switch is 100% reliable. In the latter case, the switch is unreliable, so that it functions properly on some demand occasions and not on others. We first consider the case where the switch is perfect and then the case where it is not.

15.4.1 Perfect Switch

In this case, the module fails only when all M components of the module fail. As before, let X_i denote the failure time for component i and Y the time to failure for the module. Assume that the M components are statistically identical with failure distribution $F(x)$, and let $G_M(x)$ denote the failure distribution for the module. Y is given by

$$Y = \{X_1, X_2, + \ldots + X_M\} \tag{15.8}$$

This is a sum of M independent and identically distributed random variables. as a result, from Appendix A we have

$$G_M(x) = F^{[M]}(x) = F(x)*F(x)* \cdots *F(x) \tag{15.9}$$

where $F^{[j]}(x)$ is the j-fold convolution of $F(x)$ with itself.

When the components are not identical, let $F_i(x)$ denote the failure distribution of the ith component. In this case,

$$G_M(x) = F_1(x)*F_2(x)**F_M(x) \tag{15.10}$$

Example 15.4
Let $F(x)$ be an exponential distribution with mean μ. In this case, $G_M(x)$ is an Erlangian distribution and the reliability $R_M(x)$ is given by

$$R_M(x) = 1 - G_M(x) = e^{-(x/\mu)}\left[1 + \frac{x}{\mu} + \frac{x^2}{2!\mu^2} + \cdots + \frac{x^{(M-1)}}{(M-1)!\mu^{(M-1)}}\right]$$

Table 15.4 gives $R_M(x)$ for $x/\mu = 1, \ldots, 5$ and $M = 1, \ldots, 5$. ∎

Table 15.4. $R(x)$ for M components with perfect switching

	M				
x/μ	1	2	3	4	5
1	0.3679	0.7358	0.9197	0.9810	0.9963
2	0.1353	0.4060	0.6767	0.8571	0.9473
3	0.0498	0.1991	0.4232	0.6472	0.8153
4	0.0183	0.0916	0.2381	0.4335	0.6288
5	0.0067	0.0404	0.1247	0.2650	0.4405

15.4.2   Imperfect Switch

We model the imperfect switch as follows. When the module is first put into use, the switch is in its working state. Once it is put into use (on failure of a component), it can change to a failed state with probability $(1 - p)$ or continue to be in its working state with probability p. Once the switch fails, it can no longer carry out its function and the module fails. The probability that the switch fails at the kth switching, $k = 1, 2, \ldots, M - 1$, is given by

$$p_k = p^{(k-1)}(1 - p) \tag{15.11}$$

for $k < (M - 1)$ and

$$p_{M-1} = p^{(M-1)} \tag{15.12}$$

Conditional on the switch failing at the kth switching ($k \le M$), Y is a sum of k independent random variables. In other words,

$$G_M(x|\text{switch fails at the } k\text{th switching}) = F^{[k]}(x) \tag{15.13}$$

On removing the conditioning, we have

$$G_M(x) = \sum_{k=1}^{M-1} p^{(k-1)}(1 - p)F^{[k]}(x) + p^{(M-1)}F^{[M]}(x) \tag{15.14}$$

Note that when $p = 1$, (15.14) reduces to (15.9) as expected since this corresponds to a perfect switch.

Example 15.5
Let $M = 2$ and $F(x)$ be an exponential distribution with mean μ. In this case, using (15.14), the density function $g_2(x) (= dG_M(x)/dx)$ is given by

$$g_2(x) + \frac{1}{\mu}\left[1 - p + \frac{px}{\mu}\right]e^{-(x/\mu)}$$

Let $\mu = 1$ year. Plots of the density function $g_2(x)$ for values of p ranging from 0.7 to 1.0 in steps of 0.10 are given in Figure 15.5. As can be seen, as p decreases, the curves move to the left, implying reduced mean time to failure for the module. ∎

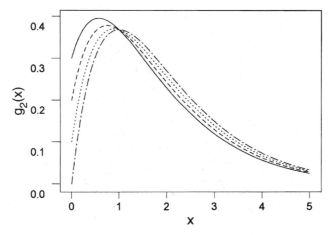

Figure 15.5. Plots of $g_2(x)$, example 15.5, for $p = 0.7$ (—), $p = 0.8$ (– –), $p = 0.9$ (--), and $p = 1.0$ (– -)

15.5 WARM STANDBY REDUNDANCY

With warm standby, the replicas in reserve tend to deteriorate with time. This is equivalent to the situation where those in reserve can be viewed as being in a partially energized state, as opposed to being fully energized in hot standby and not energized in cold standby. The resulting deterioration can lead to the failure of a component in the reserve before it is put into use. In general, the mean time to failure in the partially energized condition is much larger than that when fully energized. This is due to the fact that deterioration under partial energizedzation occurs at a slower rate than when the unit is fully energized. As a result, this is a better representation of the real-world situation than the cold standby model. This type of redundancy has also been called tepid standby (Polovko, 1968). As with cold standby, only one component of the module is fully energized at any given time. When the fully energized component fails, it is replaced by a nonfailed component from the reserve, if one is available.

As in cold standby redundancy, a switching mechanism is necessary to utilize the reserve items. We again consider both a perfect switch and an imperfect switch.

15.5.1 Perfect Switch

We first deal with the case $M = 2$, so that Component 1 is fully energized and Component 2 is partially energized when the module is first put into use. Let X_1 denote the time to failure for Component 1. Since the component is in its fully energized state, the distribution function for X_1 is given by $F(x)$. Let X_{2p} and X_2 denote the time to failure for Component 2 when it is partially energized and fully energized, respectively. Note that we expect that $X_{2p} > X_2$, since the rate of deterioration is less when the unit is partially energizedthan when fully energized. The distribution of X_2 is given by $F(x)$. Let $H(x)$ denote the distribution of X_{2p}, with $F(x) > H(x)$ for all $x > 0$.

If $X_{2p} < X_1$, then Component 2 fails while partially energized before Component 1 fails. In this case, the failure time for the module, Y, is given by $Y = X_1$. On the other hand, if $X_{2p} > X_1$, then Component 2 is switched from partially energized to fully energized at time $x = X_1$. The remaining life of Component 2, at the instant of switching, is less than X_2, since it has been in use for a period X_1 in a partially energized mode. Gnedenko et al. (1969) showed that the remaining life of Component 2 is given by $X_2 - F^{-1}[H(X_1)]$. The reasoning for this is as follows. Component 2 surviving in its partially energized mode for a period X_1 is equivalent to it surviving for a period Z when fully energized, with the two being related by

$$F(Z) = H(X_1) \tag{15.15}$$

This yields an effective age of $Z = F^{-1}[H(X_1)]$ when Component 2 is switched on after being in its working state for a period X_1 in its partially energized mode.

As a result, the time to module failure Y is given by

$$Y = X_1 + \max\{0, X_2 - F^{-1}[H(X_1)]\} \tag{15.16}$$

From this, using standard probabilistic arguments, the distribution function for Y, $G_2(x)$, is found to be

$$G_2(x) = \int_0^x F\{x - x_1 + \overline{F}^{-1}[\overline{H}(x_1)]\}dF(x_1) \tag{15.17}$$

which can be rewritten as

$$G_2(x) = F(x) - \int_0^x \overline{F}\{x - x_1 + \overline{F}^{-1}[\overline{H}(x_1)]\}dF(x_1) \tag{15.18}$$

For $M > 2$, $G_M(x)$ can be obtained using the recursive relationship

$$G_j(x) = \int_0^x \{1 - \overline{F}(x - y + \overline{F}^{-1}[\overline{H}(y)])\}dG_{j-1}(y) \tag{15.19}$$

for $j > 1$ and $G_1(x) = F(x)$.

Example 15.6
Let $M = 2$ and $F(x)$ and $H(x)$ be exponential distributions with means μ and μ_p, respectively. From (15.18), the density function $g_M(x)$ $(= dG_M(x)/dx)$ is found to be

$$g_2(x) = \left[\frac{1}{\mu} + \frac{\mu_p}{\mu^2}\right]\{1 - e^{-(x/\mu_p)}\}e^{-(x/\mu)}$$

Let $\mu = 1$ year. Plots of the density function $g_2(x)$ for values of μ_p ranging from 2 to 5 years in steps of 1.0 are given in Figure 15.6. As can be seen, as μ_p decreases, the curves move to the left, implying a reduction in mean time to failure for the module. ∎

15.5.2 Imperfect Switch

We confine our discussion to the case $M = 2$ and model the operation of the imperfect switch as in Section 15.4.2. Following an approach similar to that used in the case of a perfect switch, we find the module failure time Y to be

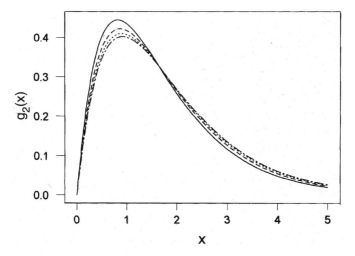

Figure 15.6. Plots of $g_2(x)$, Example 15.6, $\mu_p = 2$ (——), 3 (– –), 4 (- -), and 5 (– -).

$$Y = \begin{cases} X_1 & \text{with probability } (1-p) \\ X_1 + \max\{0, X_2 - F^{-1}[H(X_1)]\} & \text{with probability } p \end{cases} \quad (15.20)$$

Using standard probabilistic arguments, the distribution function for Y, $G_2(x)$, is found to be

$$G_2(x) = (1-p)F(x) + p\int_0^x F\{x - x_1 + \overline{F}^{-1}[\overline{H}(x_1)]\}dF(x_1) \quad (15.21)$$

The extension to arbitrary M is more involved. For further details, see Hussain (1997).

Example 15.7
Let $M = 2$ and $F(x)$ be the exponential distribution with mean μ. In this case, using (15.21), the density function $g_M(x)$ $(= dG_M(x)/dx)$ is found to be

$$g_2(x) = \left[\frac{1}{\mu} + \frac{p\mu_p}{\mu^2} - \frac{p}{\mu^2}\{\mu + \mu_p\}e^{-(x/\mu_p)}\right]e^{-(x/\mu)}$$

Let $\mu = 1$ year and $\mu_p = 10$ years. Plots of the density function $g_2(x)$ for values of p ranging from 0.7 to 1.0 in steps of 0.10 are given in Figure 15.7. As can be seen, as p decreases, the curves move to the left, implying a reduced mean time to failure for the module. ∎

15.6   RELIABILITY GROWTH

In reliability growth, the improvement in reliability is achieved through a test–analyse–and-fix (TAAF) program, discussed briefly in Chapter 14. A TAAF program is carried out in an iterative manner, where during each iteration the various stages are executed sequentially, as shown in Figure 15.8.

The process begins with the testing of prototypes of a product or component, usually un-

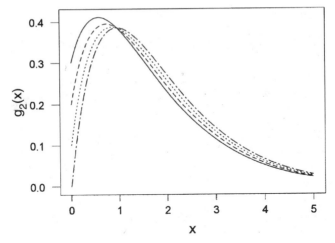

Figure 15.7. Plots of $g_2(x)$, Example 15.7, $p = 0.7$ (—), $p = 0.8$ (— —), p = 0.9 (--), and $p = 1.0$ (— -).

der increasing levels of stress. Should failures occur, the failure data, including modes of failure, TTF, and any other relevant information, are collected and analyzed by engineers to discover the causes of failure. Corrective actions are then taken to reduce the frequency of future failures. This process is repeated until the test results are satisfactory. It is very important that, in such programs, all failures are analyzed fully, and action is taken in design or production to ensure that they do not recur. Tests are usually done at a relatively high level, e.g., the subsystem level, concentrating on those with relatively low predicted reliability, since improvements at this level can be expected to have the maximum effect on system reliability. Subsystems not meeting reliability requirements are subject to redesign.

In TAAF, no failure should be dismissed as being "random" or "nonrelevant," unless it can be demonstrated conclusively that such a failure cannot occur during the normal use of the system. Corrective actions must be taken as soon as possible on all units in the development program. This can cause program delays. However, if faults are not corrected, reliability growth will be delayed, potential failure modes at the "next weakest link" may not be highlighted, and the effectiveness of the corrective action will not be adequately tested.

For further discussion of TAAF, TAAF test design principles, and relationship of TAAF to other testing programs, see Priest (1988, Chapter 9).

Since reliability growth assessment begins fairly early in the development cycle, there is often little prior information available, particularly for new products and components. As a result, modeling of reliability growth is, as in many other situations, a combination of art and science. Good mathematical analyses and sound judgment are both required.

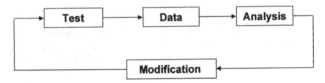

Figure 15.8. Iterative TAAF process.

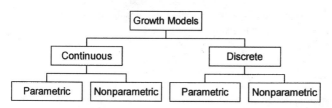

Figure 15.9. Classification of growth models.

A number of reliability growth models have been developed to monitor the progress of the development program and the improvements in reliability. They are broadly categorized into two types of models, continuous and discrete. Each of these can be further subdivided as shown in Figure 15.9. Parametric models are those based on a specified distribution of time to failure, e.g., the exponential or Weibull distribution. Nonparametric models involve specification of a functional form for the reliability improvement relationship apart from the failure distribution. Data analysis for parametric models includes estimation of the parameters of the assumed distribution. In the nonparametric case, curve-fitting techniques, such as regression analysis, are often used.

In general, continuous models are used in the context of continuous (variables) data and attempt to describe the improvement in the failure rate (or mean time between failures) as a function of the total test time and discrete models involve discrete (attribute) data and are concerned with incremental improvements in reliability as a result of design changes. These improvements are expressed as functions of the probability of success in test trials A trial is defined by a period of operation terminated upon successful completion of the test or the occurrence of a failure.[2]

In the following two sections, we look at several models in each of these categories. Estimation procedures will be included for selected models. For further discussion and many additional references, see Lloyd and Lipow (1962), Amstadter (1971), and Dhillon (1983).

15.7 DISCRETE RELIABILITY GROWTH MODELS

The earliest discrete models were the Weiss and Lloyd–Lipow models. Many of the models developed subsequently are special cases of these. The models look at reliability growth in incremental stages, with R_j being the reliability of the item at stage j. In terms of the reliability function, the general form of these models is

$$R_j = R_\infty - \theta g(j) \tag{15.22}$$

where R_∞ is the maximum attainable reliability as $j \to \infty$, θ is a parameter expressing the rate of growth, and $g(\ )$, the growth function, is nonnegative and decreasing.

15.7.1 Weiss Reliability Growth Model

Weiss (1956) proposed a model that basically follows the TAAF philosophy in that reliability growth takes place as a result of corrective actions on failure of an item. The model expresses reliability growth in terms of a decrease in failure rate as corrective actions are undertaken. It is assumed that the failure rate λ_j is constant during the interval between the $(j-1)$st and jth actions, and is given by

$$\lambda_j = \lambda + \frac{\theta}{j} \qquad (15.23)$$

Here λ is the ultimate lowest possible failure rate, and θ/j is maximum remaining improvement (in terms of reduced failure rate) at the jth step.

The model was used under the assumption of exponential time to failure. See Weiss (1956) for estimation and further discussion.

15.7.2 Lloyd–Lipow Reliability Growth Model

Lloyd and Lipow (1962, Section 11.2), formulated a model in which the ultimate reliability attainable is 1, and growth is in exponential increments. The result is

$$R_j = 1 - \alpha e^{-\gamma(j-1)} \qquad (15.24)$$

where α and γ are parameters of the growth function to be estimated from the data. The model is derived under the assumption that there is a single failure mode and that the engineer has a fixed probability of fixing the item after any failure. For derivation and further discussion, see the reference cited.[3]

15.7.3 Hyperbolic Model

Lloyd and Lipow (1962) also consider a model in which reliability improvement takes place in a fixed number M of stages. The model expresses reliability growth as in (15.22), with $g(j) = 1/j$. The result is the hyperbolic function

$$R_j = R_\infty - \frac{\theta}{j} \qquad (15.25)$$

$j = 1, \ldots, M$. At the jth stage, n_j items are tested and the number of successes X_j is observed. It is assumed that at each stage, design changes and other fixes are made.

Parameter Estimation
Lloyd and Lipow (1962) give both least squares and maximum likelihood estimators for the parameters R_∞ and θ. The least squares estimators may be obtained by regressing $r_j = X_j/n_j$ (which is an estimator of R_j) on $(1/j)$. The intercept in the resulting regression equation estimates R_∞ and the negative of the slope estimates θ. Estimates may be obtained by use of a standard regression program or from the equations

$$\hat{R}_\infty = \frac{c_2 \sum_{j=1}^{M} r_j - c_1 \sum_{j=1}^{M} \frac{r_j}{j}}{Mc_2 - c_1^2} \qquad (15.26)$$

and

$$\hat{\theta} = \frac{c_1 \sum_{j=1}^{M} r_j - M \sum_{j=1}^{M} \frac{r_j}{j}}{Mc_2 - c_1^2} \qquad (15.27)$$

where

$$c_1 = \sum_{j=1}^{M} \frac{1}{j} \quad \text{and} \quad c_2 = \sum_{j=1}^{M} \frac{1}{j^2}$$

Estimated standard deviations of the estimates can be obtained from the regression output.

Example 15.8

Suppose that a new product is subjected to $M = 12$ stages in a TAAF process, with a sample of $n_i = 20$ items tested at each stage, and that the resulting data on number of successes (x_j) are 14, 16, 15, 17, 16, 18, 17, 18, 19, 19, 20, and 19. Regressing x_j on $1/j$ provides the following output:

The regression equation is $rsubj = 0.939 - 0.279\ 1/j$

Predictor	Coef	SD	T	P
Constant	0.93870	0.02358	39.81	0.000
$1/j$	−0.27855	0.06529	−4.27	0.002

$S = 0.05701$ $R\text{-}Sq = 64.5\%$ $R\text{-}Sq(\text{adj}) = 61.0\%$

Here the estimated change rate is 0.279 and the maximum reliability is estimated to be 0.939, suggesting that design improvements are necessary. Note that the estimated standard deviation of $\hat{R}_\infty$ is 0.02358. A 95% confidence interval for R_∞ is $0.939 \pm 2.20(0.02358) =$ (0.887, 0.991), where 2.20 is the 0.975 fractile of the Student-t distribution with 11 df. ∎

MLEs, R_∞^* and θ^*, are obtained as the solution of

$$\sum_{j=1}^{M} \left\{ \frac{r_j - R_\infty^* \dfrac{\theta^*}{j}}{\dfrac{1}{n_j}\left(R_\infty^* - \dfrac{\theta^*}{j}\right)\left(1 - R_\infty^* + \dfrac{\theta^*}{j}\right)} \right\} = 0 \tag{15.28}$$

and

$$\sum_{j=1}^{M} \left\{ \frac{\dfrac{1}{j}\left(r_j - R_\infty^* + \dfrac{\theta^*}{j}\right)}{\dfrac{1}{n_j}\left(R_\infty^* - \dfrac{\theta^*}{j}\right)\left(1 - R_\infty^* + \dfrac{\theta^*}{j}\right)} \right\} = 0 \tag{15.29}$$

These equations must be solved numerically. It is suggested that the LS estimates be used as starting values in an iterative procedure. See Lloyd and Lipow (1962) for details and a worked example. The asymptotic covariance matrix of the estimators is also given.

15.7.4   Exponential Model

An alternative form of the exponential model given in (15.24), used by Bonis (1977) in analysis of one-shot devices, is

$$R_j = R_\infty - (1 - R_0)\theta^{j-1} \tag{15.30}$$

where R_0 is the initial reliability of the item and θ is the growth rate. For additional discussion and examples, see Dhillon (1983).

15.7.5 Generalized Hyperbolic Model

A generalization of the hyperbolic model of (15.25) is

$$R_j = R_\infty - \frac{\theta}{j^k} \tag{15.31}$$

This model is discussed by Amstadter (1971) and Dhillon (1983), in both cases with $R_\infty = 1$. As k increases, the model allows for much more rapid convergence to the asymptotic achievable reliability.

Apparently, estimation problems for this model have not been addressed. If k is known, least-squares estimation is straightforward (simply replace j by j^k in the solution for the hyperbolic model). The likelihood equations are also easily modified accordingly. If k is to be estimated, the problem is much more difficult.

15.8 CONTINUOUS RELIABILITY IMPROVEMENT MODELS

The classic and most widely used reliability-growth model is the Duane model. We look at this in some detail and discuss another classic model, the Gompertz reliability growth curve, and a few continuous alternatives.[4]

15.8.1 The Duane Reliability-Growth Model

The model proposed by Duane (1964), also called the "Duane learning curve," has been used extensively to model reliability growth for software (see Chapter 9) as well as all types of hardware. It was first determined empirically in the context of hardware reliability, where it was noted that the cumulative failure rate typically plotted as a linear function of cumulative time on test when plotted on the log–log scale.

The model is particularly appropriate in the TAAF context, with TTF as the variable rather than number of failures as in the discrete models. A typical TAAF cycle in this context would involve testing until failure (under accelerated conditions, if necessary), analysis and fixing, testing again until the next failure, and so forth.

The Duane model is based on total time on test T, and may be written either in terms of cumulative MTBF μ_c, or cumulative failure rate λ_c. In the former case, the resulting model is

$$\mu_c = \alpha T^\beta \tag{15.32}$$

where α is a function of the initial MTBF at the start of testing, and β is the rate of growth. The model in terms of cumulative failure is the reciprocal of (15.32). The instantaneous MTBF μ_i is calculated as

$$\mu_i = \frac{\mu_c}{1 - \beta} \tag{15.33}$$

The Duane model has been shown by Crow (1974) to be equivalent to the assumption that failures times follow a nonhomogeneous Poisson process. The structure and implications of this model and the contribution of Crow and many subsequent results are discussed in detail by Sen (1998).

The model may be used in two ways:

- To estimate the rate of reliability growth
- To estimate the additional time required to attain a specified MTTF μ_i at a given growth rate

We look at the first problem, which is solved by estimating the parameters of the model. The second requires a slight reformulation of the model to include the initial MTBF explicitly. Total required test time is estimated as a function of an estimate of this as well as of the parameters of the model. See Ireson and Coombs (1988, Chapter 10), for details.

Parameter Estimation

A conceptually simple approach to data analysis using the Duane model is to linearize the model, i.e., analyze the data on the log–log scale. The model becomes

$$\log(\mu_c) = \log(\alpha) + \beta \log(T) \tag{15.34}$$

which is in the form of a simple linear regression model. In applications, the values of the predictor variable are logarithms of the observed cumulative times to failure and the values of the response variable are the cumulative observed MTBF at the time of failure.

A problem with the simple least squares regression approach is that the values of the predictor and response variables are highly correlated within each set and with each other, since all are functions of the time to failure of the item currently on test as well as of all earlier failure times. Dhillon (1983) suggests that a weighted regression be used instead to attenuate the effect of this correlation. The weights to be used are the observation number i. The net effect of this is to weight most recent data most heavily. (Weights are easily introduced into the regression analysis in most computer packages.)

Example 15.9.

In his original paper, Duane gave plots of times to failure of an aircraft generator. The data included 14 failures, with the last occurring at 4,596 hours of operating time. The data were tabulated by Black and Rigdon (1996) and analyzed further by Sen (1998). Table 15.5 shows the data from Black and Rigdon and the estimated cumulative MTBFs, $\hat{\mu}_c(i)$, (calculated as t_i/i, where t_i is cumulative operating time at the time of the ith failure), for each data point.

In a simple regression analysis of the data, we regress $\log(\hat{\mu}_c(i))$ on $\log(t_1)$, obtaining the following results:

The regression equation is
C3 = 0.944 + 0.572 C4

Predictor	Coef	SD	T	P
Constant	0.9442	0.1048	9.01	0.000
C4	0.57175	0.01592	35.92	0.000

$S = 0.1010$ $R\text{-Sq} = 99.1\%$ $R\text{-Sq(adj)} = 99.0\%$

Table 15.5. Cumulative time and estimated MTBF, aircraft generator data

i	t_i	$\hat{\mu}_c(i)$
1	10	10.00
2	55	27.50
3	166	55.33
4	205	51.25
5	341	68.20
6	488	81.33
7	567	81.00
8	731	101.50
9	1308	145.33
10	2050	205.00
11	2453	223.00
12	3115	259.58
13	4017	309.00
14	4596	383.00

Analysis of Variance

Source	DF	SS	MS	F	P
Regression	1	13.154	13.154	1290.39	0.000
Error	12	0.122	0.010		
Total	13	13.276			

The estimated regression equation, $y = 0.9442 + 0.57175$ translates to $\hat{\mu}_c = 2.5708t^{0.57175}$ on the original scale.

Note that the results indicate high statistical significance (p-values of zero). This must be interpreted with caution because of the high autocorrelation of the data, as indicated above. For the same reason, the estimated standard errors of the coefficients (SD in the output) are likely to be underestimates of the true values and should not be used in calculating confidence intervals. ∎

15.8.2 Gompertz Curve

The Gompertz reliability growth curve relates reliability to time on test by means of an exponential function of the form

$$R = R_\infty \alpha^{\beta^T} \tag{15.35}$$

where R_∞ is the maximum reliability as $t \to \infty$, as before, and α and β are parameters related to the rate of change in reliability through time, with $0 < \alpha < 1$ and $0 < \beta < 1$. This model is useful in situations in which the rate of change in reliability is nonlinear. The basic form of the curve is due to Gompertz (1825), who postulated the model as the basis of the earliest known probabilistic mortality tables.

Some alternate representations and further discussion of the Gompertz curve are given in

Amstadter (1971) and Dhillon (1983). The model may also be used as a discrete model by re-placing T by the number of tests n in (15.35).

Parameter Estimation

The Gompertz model in (15.35) cannot be linearized; taking logs results in an exponential term in T. Nonlinear regression methods or other numerical techniques are required to obtain the least-squares estimates of the three parameters of the model. Dhillon (1983) suggests, in-stead, discretizing the problem by dividing the total time on test into three equal intervals, counting the number of failures in each, and equating these to their expectations. The result-ing three equations can be solved explicitly for the estimates. The efficiency of this proce-dure is not known. See Dhillon (1983) for computational details.

15.8.3 Other Continuous Reliability-Growth Models

A number of additional continuous reliability-growth models have been suggested. Most are alternate versions of either the Duane or Gompertz curves, involving changes of scale, alter-nate parameterizations, or changes of structure (e.g., from MTBF to failure rate). Several of these, for example, the "exponential model" and the "Weibull model," are discussed by Am-stadter (1971) and Dhillon (1983).

NOTES

1. The literature on redundancy is extensive. For detailed analysis of the three types of redundancy discussed in this chapter, see Gnedenko et al. (1969). The review papers by Osa-ki and Nakagawa (1976), Kumar and Agarwal (1980) and Yearout et al. (1986) provide ref-erences to the vast literature on the subject.

Redundancy in the context of product warranty can be found in Blischke and Murthy (1994) and in Hussain (1997).

2. For further discussion and additional references regarding classification of models by type of data—attribute versus variables data—see Erkanli et al. (1998).

Dhillon (1983) categorizes reliability-growth models in terms of use rather than structure, the categories being assessment, generic models representing historical data on product groups, target models, and projection (prediction) models. A similar categorization is given by Meth (1992).

A general model for reliability growth based on learning curves is given by Jewell (1984). The model includes many of the previous models as special cases. Maximum likelihood and Bayesian estimation are discussed.

3. Barlow and Scheuer (1966) consider the Lloyd–Lipow model in situations where not all failure causes can be detected and corrected. They consider three outcomes—inherent failures (endemic to the system), assignable cause failures (which can be fixed), and success-es. This leads to a trinomial distribution and MLEs are derived for the corresponding para-meters.

Forecasting reliability growth using attribute data in the context of very high reliability requirements is considered by Lloyd (1986). The method is based on accumulating informa-tion in a TAAF program, but reducing failure counts by discounting failures that had previ-ously been fixed. This approach has been used extensively in the aerospace industry.

A survey and extensive discussion of discrete models is given by Fries and Sen (1996).

4. Much has been done in the area of Bayesian estimation of reliability growth models.

See Calabria et al. (1992) and Erkanli et al. (1998) for application to the Duane and many other models.

MLEs for the Duane model are given by McGlone (1984) and in MIL-STD-189. The method is based on discretizing the data by grouping the failure times into k intervals and using counts of cumulative times falling into each. Least squares and maximum likelihood estimators of the parameters of the model (in a reparameterized form) are given by Sen (1998).

McGlone (1984) discusses a number of additional models, including time series models that account for autocorrelation. The models are compared in terms of fit and of long-term predictions. Military and warranty applications are the primary focus.

EXERCISES

1. Redo Example 15.2 with the failure distribution given by a gamma distribution with the mean and variance being the same as in the example.

2. Redo Example 15.2 with the failure distribution given by a lognormal distribution with the mean and variance being the same as in the example.

3. Compare the results of Exercises 15.1 and 15.2 with that of Example 15.2.

4. Redo Example 15.3 with the failure distribution given by a Weibull distribution with scale and shape parameter given by α and β, respectively. Consider the following values for α—0.8, 0.9, 1.1, and 1.2. For each α, the scale parameter is to be selected so that the mean time to failure is 1 year.

5. Discuss the effect of the scale parameter on the results obtained in Example 15.4.

6. Redo Example 15.4 with the failure distribution given by a Weibull distribution. Consider the following values for α—0.8, 0.9, 1.1, and 1.2. For each α, the scale parameter is to be selected so that the mean time to failure is 1 year.

7. Redo Example 15.5 with the failure distribution given by a Weibull distribution. Consider the following values for α—0.8, 0.9, 1.1, and 1.2. For each α, the scale parameter is to be selected so that the mean time to failure is 1 year.

8. Redo Example 15.6 with the failure distribution given by a Weibull distribution. Consider the following values for α—0.8, 0.9, 1.1, and 1.2. For each α, the scale parameter is to be selected so that the mean time to failure is 1 year when fully energized and 10 years when partially energized.

9. Redo Example 15.7 with the failure distribution given by a Weibull distribution. Consider the following values for α—0.8, 0.9, 1.1, and 1.2. For each α, the scale parameter is to be selected so that the mean time to failure is 1 year when fully energized and 10 years when partially energized.

10. The current design of car wheels involves four bolts. The bolt failures can be modelled by a Weibull distribution with shape parameter = 2.0 and scale parameter = 6.0 (years). The wheel is deemed to have failed if three bolts are in failed state. Calculate the probability of a wheel failure in the first 3 years of operation? How does this probability change if the time interval is 5 years?

11. As part of design change, the manufacturer is thinking of increasing the number of bolts from four to five. Redo Exercise 15.10 for this new design configuration.

12. The distribution of time to discharge for a dc battery is exponential with mean = 20 hours. It is used to power an electric bulb. What is the probability that the battery runs out before 20 hours in operation? One way of increasing this probability is to use cold

standby. Assume that the switching is instantaneous and that the battery does not discharge when idle, compute the probability that the system is functional with one, two, and three standbys.

13. How do the results of Exercise 15.12 change if the mission time increases from 20 to 25 hours? How many more standbys would be needed to ensure the same reliability as that with three standbys in Exercise 15.12.

14. The fire alarm system in a building is checked regularly once a year on July 1. The time to failure is exponentially distributed with a mean time to failure = 1 year. Assume that fires occur according to a stationary Poisson process with failure rate 0.5/year. What is the probability that the alarm system fails to detect a fire?

15. In Exercise 15.14, the reliability can be improved by using hot standby. Suppose that the module involves two identical devices connected in parallel. How does the probability of not detecting a fire change?

16. Plot the Lloyd–Lipow reliability growth function of (15.24) for $\alpha = 0.8$, 0.9, and 1.0, and $\gamma = 0.5$, 1.0, and 1.5, and comment on the results.

17. Calculate the MLEs of R_∞ and θ [i.e., solve (15.28) and (15.29)] for the data of Example 15.8 and compare the results with the least squares estimates.

18. Suppose that eight additional stages are run in the TAAF process of Example 15.8, with 20 tests at each stage, resulting in 19, 20, 20, 19, 19, 20, 20, and 20 successes. Restimate the parameters of the model using the data from all 20 stages.

19. A new, ultraefficient model of a refrigerator compressor is being developed. After prototype testing, 50 compressors are built based on the best design then available. These are tested under stress conditions and 8 fail prior to the end of the test period. Failures are analyzed and design changes are made. This process is repeated through 10 cycles, with 50 items tested at each stage. The numbers of successes observed for stages 2 through 10 are 46, 46, 48, 48, 47, 49, 50, 49, and 50, respectively. Use the data to estimate the parameters of the hyperbolic model (15.25). Calculate a 95% confidence interval for R_∞.

20. Use the data of Exercise 15.19 to fit the generalized hyperbolic model of (15.31) with $k = 2$, 3, and 4. Compare the results with those of the previous exercise. Which model gives the best fit? Give a reason for your answer.

21. Use weighted regression, with weights $1/j$ for the tth observation, to analyze the data of Example 15.9 and compare the results with those of the example.

22. Plot the Gompertz curve for $R_\infty = 1$ and all combinations of $\alpha = 0.1$, 0.3, .05, 0.7, 0.9, and $\gamma = 0.1$, 0.3, .05, 0.7, 0.9. Comment on the results.

CHAPTER 16

Maintenance of Unreliable Systems

16.1 INTRODUCTION

As mentioned in Chapter 1, all man-made systems are unreliable in the sense that their performance deteriorates with age and/or usage and they ultimately fail when they are unable to perform their required function under specified operating conditions. Deterioration and failure have a negative consequential effect on the buyer (be it an individual, a business, or a government agency) and are influenced by several factors, some under the control of the manufacturer (e.g., design, manufacture) and others under the control of the buyer (e.g., operational environment, maintenance).

Maintenance can be defined as actions to (1) control the deterioration process leading to failure of a system and (2) restore the system to its operational state through corrective actions after a failure. The former is called "preventive" maintenance and the latter "corrective" maintenance. Carrying out maintenance involves additional costs to the buyer and is worthwhile only if the benefits derived from such actions exceed the costs. From the buyer's viewpoint, this implies that maintenance must be examined in terms of its impact on system performance. Maintenance is of importance to manufacturers as well, since the ease and ability to carry out maintenance actions depends on the inherent characteristics of system design. This notion is defined through the concept of "maintainability."

In this chapter we focus our attention on the maintenance of unreliable systems. Many different approaches to maintenance have evolved over the past few decades. We start with a brief discussion of this in Section 16.2 and consider two approaches (reliability-centered maintenance and total productive maintenance) that highlight the role of reliability in the maintenance context. Section 16.3 deals with a characterization of the different types of maintenance actions at the part and higher (assembly, subsystem, etc.) levels in a system. Evaluation of maintenance actions requires suitable measures to evaluate their impact on system performance. This issue is discussed in Section 16.3. Section 16.4 deals with maintenance policies at the part level. Modeling and analysis of several maintenance policies are discussed. Various cases are considered to illustrate the impact of different information structures on the optimal maintenance actions. The modeling of maintenance at the system level for a multicomponent system is discussed in Section 16.5. Section 16.6 deals with maintainability and logistics. Maintainabiliy links maintenance with design and logistics deals with the support system required to ensure effective maintenance. Section 16.7 deals briefly with maintenance management systems.[1]

16.2   APPROACHES TO MAINTENANCE

The approaches to maintaining unreliable systems have changed significantly in the last 100 years. Up to about 1940, maintenance was viewed as an unavoidable cost and the only maintenance that was usually carried out was corrective maintenance. Whenever a system failure occurred, a specialized maintenance workforce was called upon to make the system operational. Maintenance was neither incorporated into the design of the system, nor was the impact of maintenance on system and business performance duly recognized. The evolution of operations research (OR) from its origin and applications during the Second World War to its subsequent use in industry led to the widespread use of preventive maintenance. Since the 1950's, OR models for maintenance have appeared at an ever-increasing pace. These deal with the effect of different maintenance policies and optimal determination of policy parameters. The models that were developed focus on the operational level by looking at either the cost of maintenance or some operational performance measure of the system. The impact on business, and maintenance viewed in a more strategic sense, is not addressed in these models.

In the 1970's, a more integrated approach to maintenance evolved in both the government and private sectors. New defense acquisitions by the U.S. government required a life cycle costing approach, with maintenance cost being a significant component, and the close linkage between reliability (R) and maintainability (M) was recognized. As a result, the term "R&M" became more widely used in defense-related systems. This concept was also adopted by manufacturers and operators of civilian aircraft through the methodology of reliability-centered maintenance in the United States. Concurrently, the Japanese evolved a concept of total productive maintenance in the context of manufacturing. Both these approaches view maintenance in the broader business context and take into account the strong links between design, operation, and maintenance and the impact of maintenance on the business as a whole. In such a framework, technical and commercial issues are integrated effectively and maintenance is viewed as being an important element of overall business performance. Many other approaches evolved elsewhere—for example, "terotechnology" in the United Kingdom, and the EUT model in the Netherlands.[2]

Figure 16.1 (from Vatn et al., 1996) is an influence diagram showing the effect of maintenance. At Level 1, component quality characterizes the reliability of the component and this is influenced by design and manufacturing decisions. Quality and maintenance affect the number of failures (and failures on demand), as shown in Level 2a. Repair time (at Level 2a) is influenced by the availability of spare parts, which in turn is determined by the maintenance logistics employed. Component performance (Level 2a) has an impact on system performance (Levels 2b and 2c). In a sense, the factors listed in Level 2c show the negative consequences resulting from component failures in a system. Finally, these negative consequences translate into costs or losses, as indicated in Level 3 of the diagram. The aim of maintenance is to reduce these costs and at the same time achieve the business goals and objectives of the company.

In the remainder of the section, we discuss briefly the RCM and TPM approaches to maintenance cost reduction. In the process, we highlight the links between maintenance and other issues such as operations, design, quality, etc.

16.2.1   Reliability-Centered Maintenance

Reliability-centered maintenance (RCM) evolved in the airline industry and was developed by a Maintenance Steering Group (MSG). The first version (called MSG-1) was for the Boe-

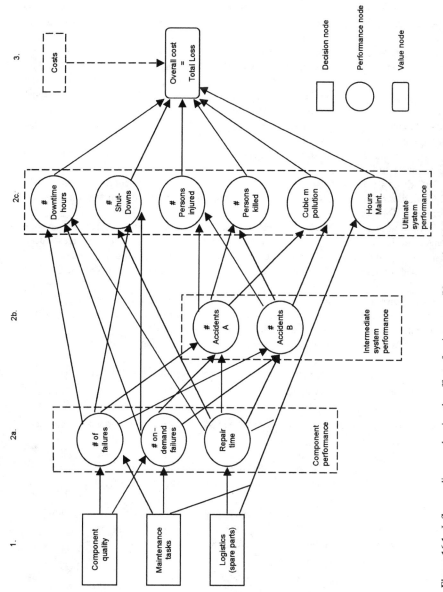

Figure 16.1. Influence diagram showing the effect of maintenance (Vatn et al., 1996). Reprinted from *Reliability Engineering and System Safety*, Vol. 51, p. 243, Copyright 1996, with permission from Elsevier Science.

ing 747. Subsequently, other versions appeared for the maintenance of later aircraft—MSG-2 for the DC-10 and Lockheed L-1011 in 1977; MSG-3 for the Boeing 757 and 767 in 1983; MSG-3-1 (also called EMSG-3) for the Airbus and Concorde. Since then, RCM has been used in other industry sectors, including nuclear energy, utilities, manufacturing, transport, etc.

As mentioned in Chapter 1, every system has an inherent reliability that is determined by design and manufacturing decisions. When the system is put into use, it deteriorates, thus reducing its reliability. System performance is acceptable as long as the actual reliability is at or above the desired level. Once the actual reliability falls below this level, corrective actions are required in order to restore the reliability of the system to at least an acceptable level. Preventive maintenance actions are intended to control the deterioration in reliability and to ensure that the actual reliability is at or above the desired level. Note that if the inherent reliability is below the desired level, then no amount of maintenance effort can increase it to above this level and it is necessary to carry out a major design modification to achieve the needed improvement in reliability.

The following definitions of RCM indicate its goals and objectives. (The first is the original definition of RCM.):

Definition 1 (MSG-1): "RCM is a discipline method logic or methodology used to identify preventive maintenance tasks to realize the inherent reliability of equipment at least expenditures of resources."

Definition 2 (Electric Power Research Institute): "RCM analysis is a systematic evaluation approach for developing and optimizing a maintenance program. RCM utilizes a decision logic tree to identify maintenance requirements of equipment according to the safety and operational consequences of each failure and degradation mechanisms responsible for these failures."

Definition 3 (International Electrotechnical Commission): "RCM is a method of establishing a scheduled preventive maintenance program which will efficiently and effectively achieve the inherent and safety levels of equipment and structures."

The RCM methodology involves consideration of the following:

1. System decomposition in terms of its components to the lowest part level
2. Identification of failure modes for each component
3. Determination of the importance of each component and of its failure
4. Prioritizing preventive maintenance (PM) tasks that effectively reduce failure occurrence

According to Moubray (1991), the RCM process involves asking the following seven questions (either for an existing operational system or a system yet to be built):

Q1: What are the functions and associated performance standards of the system in its present operating context?
Q2: In what way does it fail to fulfill its functions?
Q3: What causes each functional failure?
Q4: What happens when each failure occurs?
Q5: In what way does each failure matter?
Q6: What can be done to prevent each failure?
Q7: What should be done if a suitable preventive task cannot be found?

A methodology to address Q1 and Q2 is discussed in Chapter 1. Concepts and techniques (e.g., FMECA) to address Q3–Q5 are dealt in Chapter 6. Some of the topics covered in the remainder of this chapter are relevant to dealing with Q6 and Q7. All of this clearly shows the role of reliability in the overall maintenance of a system.

The RCM philosophy for PM is "A preventive task is worth doing if it deals successfully with the consequences of failures which it is meant to prevent." This implies that a preventive task must be *technically feasible* (i.e., it can achieve the objective) and is *worth doing*. For further details of the RCM approach and its planning and implementation, see Moubray (1991).

16.2.2 Total Productive Maintenance

The total productive maintenance (TPM) methodology evolved in Japan in the context of manufacturing based on the just-in-time (JIT) approach. The need for avoiding equipment breakdowns during production was critical for the success of the JIT approach since breakdowns resulted in delays, and lowering of the yield and output quality, thus affecting customer satisfaction. These are assessed through the concept of *overall equipment effectiveness* (OEE) and the costs resulting from six important losses.

OEE is given by the simple formula

$$OEE = \text{availability} \times \text{performance rate} \times \text{quality rate} \tag{16.1}$$

where availability, performance rate and quality rate are given by

$$\text{availability} = \{\text{total available time} - \text{actual downtime}\}/\{\text{total available time}\}$$

$$\text{performance rate} = \text{operating speed rate} \times \text{operating rate}$$

with

$$\text{operating rate} = \{\text{actual cycle time} \times \text{output}\}/\{\text{operating time}\}$$

$$\text{operating time} = \text{total available time} - \text{unplanned stoppages}$$

and

$$\text{quality rate} = \{\text{total output} - \text{number of defective items}\}/\{\text{total output}\}$$

Low OEE is due to one or more of the following causes:

1. Availability being affected by (1) breakdowns and (2) setup and adjustments
2. Performance rate being affected by (3) idling and minor stoppages and (4) reduced speeds
3. Quality being affected by (5) defects and rework and (6) startup losses

These result in the following six important losses—(1) breakdown losses, (2) setup and adjustment losses, (3) idling and minor stoppage losses, (4) reduced-speed losses, (5) quality defect and rework losses and, (6) startup losses.

The TPM approach is based on the following five principles:

1. Attack the six losses to improve OEE
2. Set up planned preventive maintenance

3. Establish autonomous maintenance (with operators of equipment maintaining their own equipment and developing the skills for continuous improvement in effective maintenance)

4. Training and education (involving several stages: Innocence → Awareness → Understanding → Competence → Excellence)

5. Equipment improvement and maintenance prevention (requiring design engineers to understand and address maintenance during the design stage)

As can be seen, TPM is based on teamwork and provides a method for achievement of world-class levels of OEE through people, technology, and proper management systems. Maintenance is done at the source and everyone is responsible for maintenance actions and policies. This includes operators, design engineers and managers at all levels. The aim is for continuous improvement, resulting in loss reduction.

TPM uses concepts such as FMEA (discussed in Chapter 6), condition-based assessment, dealing with reliability and maintainability issues during the design stage (discussed in Chapter 13), and formulation of effective maintenance actions.

16.3   MAINTENANCE CLASSIFICATION AND COSTS

The starting point for the study of maintenance is the *maintenance concept.* This consists of statements and illustrations that define the theoretical means of maintaining equipment and relates tasks, tools, techniques, and people in the maintenance process. A proper understanding of the maintenance concept is essential for effective maintenance management. A suitable definition of maintenance is important in this context. One such definition is the following:

Maintenance consists of the different functions (or activities) necessary to keep a system in, or restoring it to, an acceptable state (or operating condition). Maintenance involves one or more of the following actions:

- Servicing
- Testing/Inspection
- Removal/Replacement
- Repair/Overhaul
- Modification

16.3.1   Maintenance Classification

As mentioned earlier, maintenance actions can be divided into the following two broad categories:

1. Corrective maintenance (CM)
2. Preventive maintenance (PM)

As the name implies, corrective maintenance actions are unscheduled actions intended to restore a system from a failed state to a working state. This involves either repair or replacement of failed components. In contrast, preventive maintenance actions are scheduled actions carried out to either reduce the likelihood of a failure or to improve the reliability of the system.

Preventive maintenance (PM) actions are divided into the following categories:

1. **Clock-based maintenance:** Here PM actions are carried out at set times. An example of this is the "block replacement" policy discussed in Section 16.4.
2. **Age-based maintenance:** Here PM actions are based on the age of the component. An example of this is the "age replacement" policy discussed in Section 16.4
3. **Usage-based maintenance:** Here PM actions are based on usage of the product. This is appropriate for items such as tires, components of an aircraft, and so forth.
4. **Condition-based maintenance:** Here PM actions are based on the condition of the component being maintained. This involves monitoring of one or more variables characterizing the wear process (e.g., crack growth in a mechanical component). It is often difficult to measure the variable of interest directly, and in this case, some other variable may be used to obtain estimates of the variable of interest. For example, the wear of bearings can be measured by dismantling the crankcase of an engine. However, measuring the vibration, noise, or temperature of the bearing case provides information about wear since there is a strong correlation between these variables and bearing wear. This is discussed further in Section 16. 6
5. **Opportunity-based maintenance:** This is applicable for multicomponent systems, where maintenance actions (PM or CM) for a component provide an opportunity for carrying out PM actions on one or more of the remaining components of the system.
6. **Design-out maintenance:** This involves carrying out modifications through redesign of the component. As a result, the new component has better reliability characteristics.

16.3.2 Maintenance Costs

The direct costs of maintenance (which are viewed as part of the maintenance budget) are as follows:

- Cost of manpower
- Cost of material and spares
- Cost of tools and equipment needed for carrying out maintenance actions
- Overhead cost

In addition, many other costs are affected either directly or indirectly by maintenance (or, more precisely, by lack of an effective maintenance policy). The costs involved depend on the nature of the business. In the case of a manufacturing operation, some of these costs are as follows:

- Equipment-related
    accelerated wear because of poor maintenance
    excessive spare parts inventory
    unnecessary equipment redundancy
    excessive energy consumption
- Production-related
    rework
    excessive scrap and material losses

idle operators due to breakdowns

delays in fulfilling orders

- Product-related

quality and reliability issues

dissatisfied customers

In the remainder of the chapter, we aggregate the costs and as a result have the following three cost parameters:

1. C_f: cost of a corrective maintenance action involving replacement of a failed component by a new one
2. C_r: cost of a corrective maintenance action through minimal repair
3. C_p: cost of a PM action involving replacing a nonfailed component by a new one

16.3.3 Optimal Maintenance Actions (Policies)

With low level of preventive maintenance (PM) effort, the PM cost is low but the expected corrective maintenance (CM) cost is high. As the PM effort is increased, the CM cost decreases and the PM cost increases as shown in Figure 16.2. Also shown in the figure is the total (PM + CM) cost. This cost decreases initially and then increases with increasing PM effort. This implies that there is an optimum level of PM effort to minimize the total maintenance cost. Minimization of some measure of cost (e.g., total cost, cost per unit time) is one approach to determining the optimal maintenance policy.

Other measures can be used for determining the optimal maintenance actions. These can be either

1. Operational-based (e.g., availability, reliability)
2. A combination involving both operational and cost issues

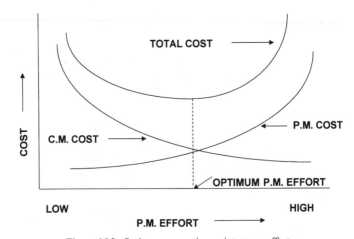

Figure 16.2. Optimum preventive maintenance effort.

In this context, a maintenance policy is a collection of maintenance actions intended to maintain a system. We will study a variety of maintenance policies and their impact on different operational and cost performance measures. A maintenance policy is characterized by one or more parameters. The optimal selection of the parameters of a maintenance policy to optimize a given performance measure is discussed in Chapter 18.

16.4 MAINTENANCE MODELS AT THE PART LEVEL

In this section we discuss a variety of maintenance policies that are appropriate at the part or component level. Failures are modeled using a black-box approach. It is assumed that the component begins in a working state and switches to a failed state after a random length of time. Let $Z(t)$ denote the state of the component at time t. $Z(t) = 1$ if the component is in working state at time t and 0 if it is in a failed state. The failure distribution of component is given by a distribution function $F(x; \theta)$[3] where θ denotes the parameters of the distribution.

Many types of information are relevant for the formulation and/or analysis of maintenance policies. Three of these are:

1. *Information about F(t):* In most cases we assume that this function is known. If this is not the case, the form of the distribution may be deduced by statistical methods (see Chapter 11).
2. *Information about parameters:* If there is an adequate amount of failure data, the parameters of the distribution can be estimated directly by use of any of several methods (see Chapters 5 and 8). An alternate approach, particularly appropriate when there is very little or no failure data, is to model the parameter by means of a prior distribution function and update the function with time as new data becomes available. Depending on the maintenance policy, the data might be either failure data or censored data and a Bayesian framework (see Chapter 8) is used for updating.
3. *Information about component states:* We consider two cases. In the first case, the state, $Z(t)$, is continuously monitored, so that a failure is detected the instant it occurs. In the second case, the state is inspected only at discrete points in time. Inspection points are determined by the policy—they may be periodic or nonperiodic.

Another important factor is the time interval of interest. We denote this by L. We consider finite and infinite time intervals (or horizons). The latter is a valid approximation if the time interval of interest is much greater than the mean time between failures and results in the analysis becoming considerably simpler.

We assume that the times to failure of all items used up to time L (the interval of interest) are identically distributed and that failures are independent. The time to carry out a maintenance (preventive or corrective) action can be either deterministic or random. In the latter case, maintenance time is characterized by a distribution function $G( )$. If the time to repair is much less than the mean time to failure, then it can be approximated by zero, and this simplifies the analysis considerably.

16.4.1 Complete Information

We first consider the case where the failure distribution $F(x; \theta)$ and the parameter θ are known, and the state of the component, $Z(t)$, is continuously monitored, so that a failure is

detected immediately after it occurs. We first consider the case of no preventive maintenance, so that the only maintenance carried out is corrective. Following this, we discuss three
extensively used preventive maintenance policies. We assume that the time to replace a
failed component is negligible, and consider both finite and infinite intervals of interest (L).
We consider the following performance measures:

1. Total expected cost, $J_c(L)$, when L is finite and asymptotic cost per unit time (J_c) when
   L is infinite.
2. Availability, $J_a(L)$, when L is finite and asymptotic availability (J_a) when L is infinite.

No Preventive Maintenance Policy

We first study the case where no preventive maintenance action is used. In this case, the only
maintenance performed is a corrective action, where an item is replaced on failure. We study
several cases in order to compare alternate preventive maintenance actions. This is important
because a preventive maintenance policy is worthwhile only if it is beneficial in the sense of
either lowering the expected cost of maintaining a system or improving its operational performance. Let $N_f(t)$ denote the number of failures over $[0, t)$.

Total Expected Cost, $J_c(L)$. We assume that the time to replace a failed component is negligible, so that it can be approximated as being zero. As a consequence, failures (and replacements) occur according to a renewal process and we have

$$E[N_f(t)] = M(t) \tag{16.2}$$

where $M(t)$ is given by the renewal integral equation associated with the failure distribution
$F(\ )$. From Appendix A, we have

$$M(t) = F(t) + \int_0^t M(t-x)f(x)dx \tag{16.3}$$

As a result, the expected cost of maintaining the component is given by

$$J_c(L) = M(L)C_f \tag{16.4}$$

Note that as L increases, $M(L)$ increases and $M(L)$ approaches infinity as L approaches infinity.

Asymptotic Cost Per Unit Time, J_c. As in Case (1) we assume that the time to replace a failed
item is small and can be ignored. There are two approaches to obtaining the asymptotic cost
per unit of time. The first approach involves taking the limit

$$J_c = \lim_{L \to \infty} \left(\frac{J_c(L)}{L} \right) \tag{16.5}$$

From Appendix B, we have

$$\lim_{t \to \infty} \left\{ \frac{M(t)}{t} \right\} \to \frac{1}{\mu} \tag{16.6}$$

Using (16.4) and (16.6) in (16.5) yields

$$J_c = \frac{C_f}{\mu} \tag{16.7}$$

where μ is the component MTTF.

The second approach is based on the renewal reward theorem (see Appendix D). Note that every failure is a renewal point and the interval between two successive renewal points defines a cycle. As a result,

$$J_c = \frac{ECC}{ECL} \tag{16.8}$$

where ECC is the expected cycle cost and ECL is the expected cycle length. Since the cost involves replacing the failed component by a new one, ECC = C_f and ECL = μ. As a result, we have the same J_c given by (16.7). This approach will be used in many of the succeeding derivations.

An intuitive explanation for (16.7) is as follows. On the average, there is one failure replacement every μ units of time and this costs C_f. The average cost per unit of time is given by the ratio of these two quantities.

Example 16.1

The component under consideration is a valve used in a chemical plant. The failure distribution $F(t)$ is a Weibull distribution [given by Equation (4.44)] with shape parameter α and scale parameter $\beta = 1$ year. We analyze total expected cost under the policy that the component is allowed to operate until it fails. Once a failure occurs, the failed item is replaced by a new item. We assume that the time needed for a replacement is small so that it can be ignored. As a result, failures occur according to a renewal process. The expected number of failures over an interval L is given by $M(L)$. Table 16.1 gives $M(L)$ for values of α varying from 0.5 to 2.0 and L varying from 1 to 5 years. [These were obtained from Blischke and Murthy (1994)]. Note that for $L = 1$, $M(L)$, and hence $J_c(L)$, decreases as α increases and the reverse occurs for $L = 5$. $\alpha = 1$ corresponds to constant failure rate (i.e., the exponential distribution) and in this case the renewal function $M(t)$ is a linear function of t. For $\alpha < 1$ [> 1] $M(t)$ is greater [smaller] than for $\alpha = 1$ for small t and the reverse is true when t becomes large. Total expected cost follows this pattern as well.

In the limit $L \to \infty$, the expected number of failures per unit time (given by 1/MTTF) is

Table 16.1. Expected number of failures, $M(L)$, versus L for the Weibull distribution with $\beta = 1$

$\alpha \to$ $\downarrow L$ (years)	0.5	1.0	1.5	2.0
1	1.3077	1.0000	0.8417	0.7538
2	2.0478	2.0000	1.9456	1.8942
3	2.7018	3.0000	3.0539	3.0129
4	3.3141	4.0000	4.1616	4.1503
5	3.9010	5.0000	5.2693	5.2786

0.5 for $\alpha = 0.5$. This increases to 1.0, 1.1074, and 1.1284 for $\alpha = 1.0$, 1.5, and 2.0, respectively. ■

Availability, $J_a(L)$. In this case, a failure is detected immediately but the time taken to replace a failed component is a random variable with distribution function $G(\ )$. This could occur for a variety of reasons—maintenance personnel not available, replacement parts not available and delivery time random, and so forth. The expected availability is given by

$$J_a(L) = E\left[\int_0^L Z(t)dt\right] \tag{16.9}$$

A conditional approach is used to evaluate this expectation, where the conditioning is done on X_1 (the time to failure for the first item put into use at $t = 0$) and Y_1 (the time to replace the first item on its failure). Conditioning on X_1 and Y_1, we have

$$J_a(L|X_1 = x, Y_1 = y) = \begin{cases} x + J_a(L - x - y) & \text{if } x + y < L \\ x & \text{if } x < L < x + y \\ L & \text{if } x > L \end{cases} \tag{16.10}$$

On removing the conditioning, we have

$$J_a(L) = \int_0^L \left[x + \int_0^{L-x} J_a(L - x - y)g(y)dy\right]f(x)dx + L\bar{F}(L) \tag{16.11}$$

where $f(x)$ and $g(y)$ are the density functions of X_1 and Y_1, respectively.

This is a renewal type integral equation (see Appendix B) which, in general, must be solved by computational methods.

Asymptotic Availability (J_c). The asymptotic availability is given by

$$J_a = \lim_{L\to\infty} \left[\frac{\int_0^L Z(t)dt}{L}\right] \tag{16.12}$$

This can rewritten as

$$J_a = \lim_{n\to\infty} \left[\frac{\sum_{i=1}^n \{X_i\}}{\sum_{i=1}^n \{X_i + Y_i\}}\right] \tag{16.13}$$

From the weak law of large numbers (see Heathcote, 1971), we have

$$\lim_{n\to\infty} \frac{\sum_{i=1}^n X_i}{n} \to E[X_i] \quad \text{and} \quad \lim_{n\to\infty} \frac{\sum_{i=1}^n Y_i}{n} \to E[Y_i] \tag{16.14}$$

and this yields

$$J_a = \frac{E[X_i]}{E[X_i] + E[Y_i]} = \frac{\mu}{\mu + \nu} \tag{16.15}$$

where μ is the mean time to failure and ν is the mean time to replace a failed item.

Note that this also follows from the renewal reward theorem (see Appendix D), where ECC is the expected time in a cycle during which the component is in its operational state (given by μ) and ECL is the expected cycle length (given by $\mu + \nu$).

Age Policy

The age policy is defined as follows:

Replace a component (under preventive maintenance) when it reaches age T (after being put into use) or on failure under corrective maintenance, if the item fails earlier.

This policy is used for components that degrade with age. These include rubber hoses in a pneumatic press, spark plugs and air and oil filters in a car, and so on. When the degradation is due to usage (rather than age) then this policy is used with T representing a measure of usage (e.g., wear of tire tread).

We assume that $T < L$. The time required for a corrective maintenance action (replacing a failed component by a new one) and preventive maintenance action (replacing a nonfailed component of age T by a new one) are sufficiently small so that they can be ignored. The costs of each corrective and preventive maintenance actions are C_f and C_p ($<C_f$), respectively.

Let $\widetilde{X}_i$ denote the age of item i when it is replaced (either under preventive or corrective maintenance). Then it is easily seen that

$$\widetilde{X}_i = \begin{cases} X_i & \text{if } X_i < T \\ T & \text{if } X_i \geq T \end{cases} \tag{16.16}$$

where X_i is the time to failure.

Total Expected Cost, $J_c(L; T)$. Let $J_c(L; T)$ denote the expected cost of maintaining a component over $[0, L)$, starting with a new item at $t = 0$. This cost is given by

$$J_c(L; T) = C_f E[N_f(L)] + C_p E[N_p(L)] \tag{16.17}$$

where $N_f(L)$ and $N_p(L)$ are the number of failure and corrective replacements over the interval L. The probabilistic characterization of these two variables is complex. For solution, we again use a conditional approach, conditioning on the failure time of the component put into use at $t = 0$. Conditional on $X_1 = x$, we have

$$J_c(L; T | X_1 = x) = \begin{cases} C_f + J_c(L - x; T) & \text{if } x < T \\ C_p + J_c(L - T; T) & \text{if } x \geq T \end{cases} \tag{16.18}$$

since a cost of C_f is incurred with a failure replacement and a cost of C_p with a preventive maintenance replacement. On removing the conditioning, we have

$$J_c(L; T) = \int_0^T [C_f + J_c(L - x; T)]f(x)dx + [C_p + J_c(L - T; T)]\overline{F}(T) \qquad (16.19)$$

This is a renewal-type integral equation that must be solved numerically.

Asymptotic Cost Per Unit Time, $J_c(T)$. To obtain $J_c(T)$, the asymptotic cost per unit time, the renewal reward theorem may again be used. Note that any replacement (failure or preventive) is a renewal point. As a result, the expected cost per cycle is given by

$$\text{ECC} = C_f F(T) + C_p \overline{F}(T) \qquad (16.20)$$

and the expected cycle length, ECL, is given by

$$\text{ECL} = E[\tilde{X}_i] = \int_0^T xf(x)dx + T\overline{F}(T) \qquad (16.21)$$

From (16.8), the asymptotic expected maintenance cost per unit time is given by

$$J_c(T) = \frac{C_f F(T) + C_p \overline{F}(T)}{\displaystyle\int_0^T xf(x)dx + T\overline{F}(T)} \qquad (16.22)$$

Comments: (i) $T \to \infty$ implies that no preventive maintenance is carried out. In this case, (16.22) reduces to (16.7) as expected. (ii) An age replacement policy is worthwhile only if $J_c(L; T) > J_c(L; \infty)$ for some $T < \infty$. (iii) The major drawback of the age policy is that it is necessary to keep track of the age of each item once it is put into use. This can be a fairly expensive task if the number of components to be tracked is large (e.g., electric bulbs in hallways of a multistory building; tires on a fleet of trucks or automobiles).

Example 16.2
The component under consideration is a computer chip. We consider the asymptotic cost per unit of time. Let $F(t)$ be an exponential distribution with failure rate λ, so that

$$F(t) = 1 - e^{-\lambda t}$$

We have from (16.22)

$$J_c(T) = \lambda[C_f - C_p] + \frac{\lambda C_p}{F(T)}$$

Since $F(T)$ is increasing in T, we see that $J_c(T)$ decreases with T. As a result, it achieves a minimum values when $T \to \infty$. This implies that no preventive maintenance be carried out and that only corrective maintenance (i.e., replacing an item only on failure) be employed. This is to be expected, since the failure rate λ is a constant and there is no aging effect. ∎

Block Policy

The block replacement policy is defined as follows:

> Replace the component at set times $t = kT$, $k = 1, 2, \ldots$, under preventive maintenance. Any failures in between these times are rectified through corrective maintenance, which involves replacing failed items by new ones.

This policy is used when a large number of identical items are in use. An example of this is electric bulbs in a large building or the street lights in a suburb. In this case, the age policy is not sensible, as it requires tracking the age of each individual item. The block policy does not require tracking the age and hence is more appropriate.

We assume, as with the age policy, that the time to carry out a preventive (or corrective) replacement is negligible and hence approximated as being zero.

Asymptotic Cost Per Unit Time, $J_c(T)$. In this case the time instants $t = kT$, $k = 1, 2, \ldots$, are renewal points and the renewal cycle is the interval between two adjacent renewal points. The cycle length is a deterministic quantity T and the cycle cost is the sum of a preventive replacement (C_p) at the end of the cycle and corrective replacements over the cycle length. These failures occur according to a renewal process so that the expected cycle cost, ECC, is given by

$$\text{ECC} = C_p + C_f M(T) \tag{16.23}$$

where $M(\ )$ is the renewal function associated with $F(\ )$, given by (16.3). Since the expected cycle length is ECC $= T$, we have, from the renewal reward theorem, the asymptotic maintenance cost per unit time, $J_c(T)$, given by

$$J_c(T) = \frac{C_p + C_f M(T)}{T} \tag{16.24}$$

Comments: (i) As $T \to \infty$ (i.e., no preventive maintenance), (6.24) reduces to (6.7) as expected. (ii) For preventive maintenance to be effective, $J_c(T) < J_c(\infty)$ must hold for some $T > 0$. (iii) The main drawback of this policy is that if an item fails very close to kT, $k = 1, 2, \ldots$, it is replaced by a new item, which in turn is replaced under preventive maintenance very soon thereafter. As a result, the replacement item is used for only a very short fraction of its useful life. Several modifications to the block policy have been proposed to overcome this drawback. We describe one such modified policy later in this section.

Example 16.3

The items under consideration are the electric bulbs in a large factory. The failure distribution of the item, $F(t)$, is a Weibull distribution with shape parameter $\alpha = 1.5$ (which corresponds to an increasing failure rate) and scale parameter $\beta = 4.0$ (years). Table 16.2 gives $J_c(T)/C_p$ for a range of T and C_f/C_p. The values for $M(T)$ used in the calculation of the table entries were obtained from Blischke and Murthy (1994). Note that for $C_f/C_p = 1$ or 3, the normalized expected maintenance cost $[J_c(T)/C_p]$ is decreasing with T for the range of values for T considered. In contrast, for $C_f/C_p = 5$ or 10, it is initially decreasing and then increasing. In this case, there is an optimal T (in the interval $1 \leq T \leq 6$) which yields a minimum value for the expected maintenance cost per unit time for infinite operation.

Table 16.2. Normalized expected maintenance cost versus T for block policy

$C_f/C_p \rightarrow$ $\downarrow T$ (years)	1.0	3.0	5.0	10.0
1	1.1219	1.3657	1.6095	2.2190
2	0.6652	0.9955	1.3258	2.1515
3	0.05256	0.9101	1.2947	2.2580
4	0.4604	0.8813	1.3021	2.1063
5	0.4221	0.8689	1.3148	2.4296
6	0.3985	0.8622	1.3258	2.4850

Intuitively, one would expect $J_c(T)$ to be a decreasing function in T for $C_f/C_p = 1$. This follows as there is no penalty associated with failure and hence carrying out any preventive maintenance implies throwing away useful life. When the penalty for failure (C_f/C_p) is high, then preventive maintenance is worthwhile. ■

Modified Block Policy
In the block policy, preventive maintenance actions result in "used items," with ages in the interval $[0, T)$, being scrapped. The modified block policy, defined below, uses both new and used items as part of the preventive maintenance policy, which is characterized by two parameters—T and T_1 ($< T$). The policy is as follows:

Failed items are replaced by new ones (under preventive maintenance actions) at times $t = kT$, $k = 1, 2, \ldots$. Replaced items that have not failed are held in a pool of used items. For failures occurring in the intervals $[kT, kT + T_1)$, failed items are replaced by new ones and for failures occurring in the intervals $[kT + T_1, (k + 1)T)$, failed items are replaced from the pool of used items.

The analysis of this policy is more involved due to the difficulty of obtaining the failure distribution for the items. For details of the analysis, see Murthy and Nguyen (1982).

Periodic Policy
In the age and block replacement policies, failed items are replaced by new ones. This is appropriate for items that are nonrepairable or the cost of repair (relative to the cost of a new item) is large, so that repair is not economical. When it is economical to repair the item, an alternative policy is as follows:

Replace items (under preventive maintenance) at times $t = kT$, $k = 1, 2, \ldots$. Any failure between replacement times is repaired minimally.

As with the earlier age and block policies, we assume that repair and replacement times for each CM and PM are small, so that they can be approximated as zero.

As in the block policy, the time instants at which preventive maintenance occurs are renewal points. Cycle length is a deterministic quantity T and the cycle cost is the sum of a preventive replacement (C_p) at the end of the cycle and corrective repairs over the cycle length. Since failures are rectified through minimal repairs, the failures (and repairs) over a cycle occur according to a nonhomogeneous Poisson process with an intensity function given by the failure rate of the item. As a result, the expected cycle cost (ECC) is given by

$$\text{ECC} = C_p + C_r \int_0^T r(x)dx \tag{16.25}$$

where $r(x)$ is the failure rate associated with the failure distribution $F(x)$. From the renewal reward theorem, we have

$$J_c(T) = \frac{C_p + C_r \int_0^T r(x)dx}{T} \tag{16.26}$$

Example 16.4
The item is a rubber hose of a hydraulic press and a failure occurs whenever there is a leak in the hose. The failure can be fixed by putting on a bandage to stop the leak. This can viewed as minimal repair. As the item ages, the failure rate increases due to the degradation of rubber. The failure distribution, $F(t)$, is a Weibull distribution with shape parameter α and scale parameter β. Then from (16.26) we have

$$J_c(T) = \frac{C_p}{T} + \left(\frac{C_r}{\beta^\alpha}\right)T^{(\alpha-1)}$$

Note that for $\alpha \leq 1$ (decreasing or constant failure rate), $J_c(T)$ is a decreasing function of T, so that it achieves a minimum when $T \to \infty$. This implies that no preventive replacement should be carried out and that the use of only corrective maintenance (repairing the item minimally on failure) is the optimal strategy. When $\alpha > 1$, then $J_c(T)$ is a convex function.

Let $C_p = \$100$ and $C_r = \$10$. The parameters of the failure distribution are $\alpha = 2$, implying an increasing failure rate, and $\beta = 1$ month. $J_c(T)$ for a range of T varying from 1 to 20 months is given in Table 16.3. As can be seen, the expected cost first decreases and then increases with T. ∎

16.4.2   Incomplete Information—I (Component State Unknown)

In Section 16.4.1 it was assumed that the state of the component, $Z(t)$, was known. This implies that the state is continuously monitored. Often this is not possible for economic reasons. In this case, maintenance involves inspection of the component at discrete time instants to determine its state.

Let T_i, $i = 1, 2, 3, \ldots$, denote the time instants, subsequent to an item being put in use, that the item is inspected to detect if it is working [$Z(t) = 1$] or has failed [$Z(t) = 0$]. T_i, $i = 1$, $2. \ldots$, is an increasing sequence in i. Let C_{in} be the cost of each inspection. Thus if an item

Table 16.3. $J_c(T)$ versus T for the periodic policy

T (months)	1	2	3	4	5	10	15	20
$J_c(T)$ ($\$$)	110	70	66.33	65	70	110	156.66	205

failure occurs between $[T_i, T_{i+1})$, then it is detected at time T_{i+1} subsequent to its being put into use. This implies that the component is in an undetected failed state for a fraction of the time interval $[T_i, T_{i+1})$. If X is the time to failure, then the duration for which it in an undetected failed state is given by $(T_{i+1} - X)$. Let C_d denote the penalty per unit of time during which the component is in an undetected failed state.

We look at the impact of inspection on the asymptotic cost for the case of no preventive maintenance and the age policy. We assume that the time needed for a PM or CM action is negligible so that it can be approximated as being zero.

No Preventive Maintenance Policy
In this case, the maintenance action involves inspection and replacement of failed items by new ones.

Asymptotic Cost Per Unit Time (J_c). Note that the inspection point at which an item failure is detected is a renewal point for the process. Using the renewal reward theorem, we have J_c given by (16.8) with ECC, the expected cost per cycle, given by

$$\text{ECC} = C_f + \sum_{i=0}^{\infty} \left[(i+1)C_{\text{in}}\{F(T_{i+1}) - F(T_i)\} + C_d \left\{ \int_{T_i}^{T_{i+1}} (T_{i+1} - x)f(x)dx \right\} \right] \quad (16.27)$$

where $T_0 = 0$, and the expected cycle length, ECL, given by

$$\text{ECL} = \sum_{i=0}^{\infty} [T_{i+1}\{F(T_{i+1}) - F(T_i)\}] \quad (16.28)$$

Age Policy
Here the item is replaced when it reaches an age T, unless it fails earlier. This early failure is detected when the item is inspected. As before, let T_i, $i = 1, 2, \ldots, K$, denote the time instants, subsequent to an item being put into use, that the item is inspected. T_i is an increasing sequence in i. Define $T_0 = 0$ and $T_{K+1} = T$.

Asymptotic Cost Per Unit Time, $J_c(T)$. Since every replacement is a renewal point, the expected cycle cost (ECC) and expected cycle length (ECL) are given by

$$\text{ECC} = C_f + \sum_{i=0}^{K} \left[(i+1)C_{\text{in}}\{F(T_{i+1}) - F(T_i)\} + C_d \left\{ \int_{T_i}^{T_{i+1}} (T_{i+1} - x)f(x)dx \right\} \right] \quad (16.29)$$

and

$$\text{ECL} = \sum_{i=0}^{K} [T_{i+1}\{F(T_{i+1}) - F(T_i)\}] \quad (16.30)$$

respectively. As before, the asymptotic cost per unit of time is given by the ratio of these two quantities.

16.4.3   Incomplete Information—II (Parameters Unknown)

When the true value of θ is not known, one approach is to view θ as a random variable characterized by a prior density function $h_1(u)$. Then, conditional on $\theta = u$, the expected maintenance cost can be evaluated for each maintenance policy. Note that this is a function of u. On removing the conditioning (i.e., carrying out the expectation over θ using $h_1(u)$), we obtain an estimate of the maintenance cost that takes into account the uncertainty in θ.

Each item failure (or replacement under a preventive maintenance policy) provides new information that can be used to update the density function of θ using the Bayesian approach. First consider the case where the information is X, the failure time. Then the density function of X, conditional on $\theta = u$, is given by $f(x|u)$. Let $h_2(u)$ denote the posterior density function of θ given $X = x$. Then from Bayes' theorem (see Chapter 8), we have

$$h_2(u) = \frac{f(x|u)h_1(u)}{\int f(x|u)h_1(u)du} \tag{16.31}$$

This can now be used in the expression for the expectation to obtain an improved estimate of the expected maintenance cost. This process can be repeated with $h_2(u)$ as the new prior density function. By repeated application of the approach, more refined estimates of the maintenance cost are obtained.

When the information is Y, the age of the nonfailed item at preventive replacement, then the analysis must be slightly modified. Note that Y is a censored failure time. Hence, the conditional probability that $Y = y$ given that $\theta = u$ is given by $[1 - F(y|u)]$. This is now used in the Bayesian updating procedure to obtain the posterior distribution.[4]

16.5 MAINTENANCE MODELS AT THE SYSTEM LEVEL

Since a system is comprised of several components, maintenance of a system can be studied by looking at maintenance at the component level and then aggregating over components. This is relatively easy if there is no interaction between components. When there is interaction or dependence between components, this approach cannot be used.

An alternate approach is to model the system as a black box (see Chapter 7). This implies that the degradation in the system and system failures are modeled using suitable models (e.g., Model 7.1, involving a point process formulation) which characterize the changes at the system level. There are many different models that have appeared in the literature. In this section, we consider three simple models that incorporate the impact of the degradation process in distinctly different manners.

16.5.1 Model 16.1 (Discrete Time Formulation)

This model is a discrete time model with the time period being one year. As such, the variables represent the annual aggregates. The variables of the model are:

C: Initial purchase price of the system

R_i: Revenue generated by the system in period i ($i = 1, 2, \ldots$)

O_i: Operating cost (excluding maintenance) in period i ($i = 1, 2, \ldots$)

M_i: Maintenance cost in period i ($i = 1, 2, \ldots$)

S_i: Salvage value of the system at the end of period i ($i = 1, 2, \ldots$)

N: Age (in periods) at which the system is replaced

α: Discount factor

$J(N)$: Discounted profit over the life of the system

Since the system degrades with age and usage, we have the following:

1. R_i is a decreasing sequence in i, implying that as the system ages, its productivity decreases.
2. O_i and M_i are increasing sequences in i, implying that the cost of operating and maintaining the system increases with age.
3. S_i decreases as i increases, implying that the salvage value decreases with age.

The degradation process is uncertain. As a result, the revenue generated and maintenance cost in different time periods and the salvage value are random variables. R_i, O_i and S_i are the expected values of these random variables and hence deterministic. Systems for which this analysis is appropriate include buses and trucks used in commercial operations as well as more complex systems such as power stations, large manufacturing plants and petrochemical plants. The discounted profit over the life of the system is given by

$$J(N) = \left\{ \sum_{i=1}^{N-1} \alpha^i [R_i - M_i - O_i] \right\} + \alpha^N S_N - P \tag{16.32}$$

This allows one to determine the optimal N at which time the system should be replaced.

An estimate of R_i can be obtained by computing the availability of the system in different time periods. This can be done terms of the reliability of the different components and the maintenance actions used. Estimates of O_i and M_i can be obtained similarly.

Example 16.5

Consider a heavy-duty truck. Table 16.4 gives R_i, O_i, and M_i for $i = 1, 2, \ldots$. The salvage value is given by

$$S_N = S_0 e^{-\gamma N}$$

with $S_0 = \$100,000$ and $\gamma = 0.2$.

Table 16.5 shows $J(N)$ versus N for a range of values for N and for three different values of α. As can be seen, the optimal replacement time is at the end of 7 years for all three values of the discount factor.

Table 16.4. Annual revenue, operating, and maintenance costs

i	R_i (\$)	O_i (\$)	M_i (\$)
1	80,000	10,000	1,000
2	75,000	12,000	3,000
3	70,000	14,000	6,000
4	65,000	16,000	9,000
5	60,000	18,000	12,000
6	55,000	20,000	15,000
7	50,000	22,000	20,000
8	50,000	24,000	25,000

Table 16.5. $J(N)$ versus N for three different values of α

N (years)	$\alpha = 1.0$	$\alpha = 0.9$	$\alpha = 0.8$
1	50,873	43,973	37,073
2	96.032	83,032	70,232
3	133.882	115,882	98,082
4	163,933	141,933	120,133
5	185,788	160,788	135,988
6	199,119	172,119	145,319
7	199,660	172,069	144,660
8	191,189	163,989	136,989

16.5.2   Model 16.2 (Continuous Time Formulation)

Maintenance often involves checking the system at fairly short intervals and carrying out minor adjustments; for example, regular checking at short intervals and rectifying any symptoms that can lead to a breakdown. In this case, the effect of maintenance is better characterized through a continuous function $u(t)$ with $0 \le u(t) \le U$. $u(t)$ represents the maintenance effort rate and U is the upper limit on this quantity.

Without any maintenance effort, the system deteriorates with age and usage. The salvage value can be used as a good indicator of the state of the system. Let $S(t)$ denote the salvage value of the system at age t with no maintenance effort. In general, salvage value is a random variable because the degradation process is uncertain. for ease of analysis, we treat $S(t)$ as a deterministic function, representing the expected value of this random process. We model the change in $S(t)$ by a first order differential equation

$$\frac{dS(t)}{dt} = -\theta S(t) \tag{16.33}$$

with the parameter $\theta > 0$ and $S(0) = S_0$. This implies that $S(t)$ decreases exponentially as t increases. The rate of decrease (due to degradation) increases as θ increases.

The productivity of the system is a function of the state of the system. Since the salvage value is a surrogate for the system state, we model productivity by means of a revenue generation rate $R(t)$ which is a linear function of $S(t)$, given by

$$R(t) = aS(t) \tag{16.34}$$

where a is a constant.

The effect of maintenance is to slow down the deterioration. This is modeled as follows:

$$\frac{dS(t)}{dt} = -\theta S(t) + \varphi u(t) \tag{16.35}$$

Even with maximum maintenance [$u(t) = U$], the salvage value decreases over the interval [0, T], where T is the upper limit on the life of the system. This implies that maintenance can never stop the deterioration, but can only slow the process. The maintenance cost rate is given by $bu(t)$ where b is a positive constant.

As a result, the total profit obtained by keeping the system for a period T and using a maintenance effort $u(t)$ over the interval $[0, T)$ is given by

$$J(T; u(t), 0 \le t \le T) = \int_0^T \{aS(t) - bu(t)\}dt + S(T) - P \tag{16.36}$$

where P is the purchase price and $S(t)$ is related to $u(t)$ by (16.35).

For a fixed T, the problem is to determine the optimal maintenance effort that maximizes the total profit. This is a functional optimization problem and will be discussed in Chapter 18. Note that T can also be a decision variable and can also be selected optimally. In this case, T represents the optimal time at which to replace the system.

Consider the following two cases.

Case (i): No maintenance action. In this case $u(t) = 0$. From (16.35) we have

$$S(t) = S_0 e^{-\theta t} \tag{16.37}$$

Using this in (16.36) yields, after simplification,

$$J(T; u(t) = 0) = \left[\frac{a}{\theta}S_0 - P \right] + \left[1 - \frac{a}{\theta} \right] S_0 e^{-\theta T} \tag{16.38}$$

Note that $S_0(a/\theta) > P$ must hold in order for the profit to be positive for all $T > 0$. The system must be kept for at least some length of time to make a positive profit. (*Note:* If the system is replaced after being used for a very short time, the profits can be negative because the revenue generated will be less than $(P - S_0)$, the difference between the purchase price and the salvage value of an almost-new system.) This implies that $(a/\theta) > 1$.

Case (ii): Maximum maintenance used over $[0, T)$. In this case $u(t) = U$. From (16.35) we have

$$S(t) = S_0 e^{-\theta t} + \left[\frac{\varphi U}{\theta} \right] [1 - e^{-\theta t}] \tag{16.39}$$

Note that we need $(\varphi U/\theta) < S_0$ since the system degrades even with a maximum maintenance effort. Using (16.39) in (16.36), we have, after some simplification,

$$J(T; u(t) = U) = \left[\frac{a\varphi}{\theta} - b \right] UT + \left[S_0 - \frac{\varphi U}{\theta} \right] \left[\frac{a}{\theta} \right] - P + \left[S_0 - \frac{\varphi U}{\theta} \right] \left[1 - \frac{a}{\theta} \right] e^{-\theta T} \tag{16.40}$$

When $(a\varphi/\theta) < b$, the plots of $J(\ )$ for the two cases can be as shown in Figure 16.3 for certain parameter values. In this case, there is a T_0 such that, for $T < T_0$, using full maintenance yields a higher value for $J(\ )$ than that for no maintenance. On the other hand, for $T > T_0$, the reverse is true. If $J(T)$ with maximum maintenance is always smaller than $J(T)$ with no maintenance for all T, then $T_0 = 0$.

In the above analysis we are comparing no maintenance with maximum maintenance. The optimal maintenance policy to maximize $J(\ )$ is, in general, more complex.

Comments: (i) A discrete time analog of this continuous time model can be developed. (ii) The variable $S(t)$ can be modeled as a stochastic variable. In this case, the deterministic differential equation in (16.35) is replaced by a stochastic differential equation. The analysis to obtain the total expected profit would be more involved, depending on the nature of the stochastic differential equation.

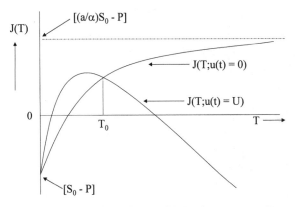

Figure 16.3. $J(T)$ versus T for no and full maintenance over T.

16.5.3 Model 16.3 (Point Process Formulation)

The effect of degradation is that system failures occur in an uncertain manner. When the failures are rectified minimally and the rectification time is negligible, then the failures can be modeled by a point process as indicated in Model 7.2.

Here we consider such a model in which corrective maintenance actions are minimal repairs carried out when the system fails. We also use preventive maintenance in the form of overhauls, as indicated in Model 7.2. This implies that we use both corrective and preventive maintenance actions to control the effect of the degradation processes on the overall profit. The model details are as follows.

The system is subjected to overhauls that rejuvenate the system. Let T_0 denote the age of the system at which the first overhaul is carried out. Failures over the interval $[0, T_0)$ are modeled by an intensity function $\Lambda_0(x)$, where x denotes the time subsequent to the item being put into use, and failures are repaired minimally. It is assumed that the time for each repair is negligibly small so that it can approximated as being zero. As a result, failures occur according to a nonstationary Poisson process with intensity function $\Lambda_0(x)$. After overhaul, failures occur according to the intensity function $\Lambda_1(x)$ (where x denotes the time subsequent to the overhaul) and these are minimally repaired up to time T_1. Subsequent to the first overhaul, the system is subjected to a second overhaul, and the process is repeated until the system is overhauled n times. Failures after the jth overhaul occur according to a Poisson process with intensity function $\Lambda_j(x)$. The time between the jth and the $(j + 1)$st overhauls is given by T_j. After the nth overhaul, the system is kept for a period T_n and is then replaced by a new system.

We make the following additional assumptions:

1. $\Lambda_j(x)$ is an increasing function in x for $j = 0, 1, \ldots, n$. This implies that the system deteriorates with time after each overhaul.
2. $\Lambda_j(x)$ is an increasing sequence in j for a given x, $x \geq 0$. This implies that each overhaul rejuvenates the system but the intensity function for a given x increases with the number of overhauls.
3. The cost of the jth overhaul is $O_j, j = 1, 2, \ldots, n$. O_j is an increasing sequence in j, implying that the cost of overhaul increases with the number of times the system has been overhauled.

4. The cost of each minimal repair is C_r.

5. The acquisition cost is P.

6. The salvage value S is a function of L, the age of the system when it is replaced. L is given by

$$L = \sum_{i=0}^{n} T_i \tag{16.41}$$

Let R denote the revenue generated per unit of time. We assume that the time for each minimal repair is small so that it can be ignored. Similarly, the time for each overhaul is small relative to L, and is also ignored. As a result, the total revenue generated is RL.

Let $\{T\}_0^n$ denote the set $\{T_j: j = 0, 1, \ldots n\}$ and let $J(n, \{T\}_0^n)$ denote the expected profit over the period $[0, L)$. This is calculated as the difference between revenue and the total of purchase and maintenance costs. Maintenance cost is the sum of the costs of minimal repairs and overhauls. The expected cost of minimal repairs between overhauls (j) and $(j + 1)$ $(j = 0$ corresponds to a new item instead of overhaul) is given by

$$C_r \int_0^{T_j} \Lambda_j(x)dx \tag{16.42}$$

This follows since failures occur according to a nonhomogeneous Poisson process. As a result, we have

$$J(n, \{T\}_0^n) = R \sum_{j=0}^{n} T_j - P - \sum_{j=1}^{n} O_j - \sum_{j=0}^{n} C_r \int_0^{T_j} \Lambda_j(x)dx + S(L) \tag{16.43}$$

with L given by (16.41).[5]

Example 16.6

The system under consideration is an electric train used for inner city transport. The various costs are in millions of dollars. The purchase price is $P = 300$ and the revenue generated per year is $R = 20$. The system is designed to last for 45 years so that $L = 45$. We consider the following three cases:

1. No overhaul over L, i.e., $T_0 = 45$ years
2. One overhaul, with $T_0 = 25$ years and $T_1 = 20$ years
3. Two overhauls, with $T_0 = 20$ years, $T_1 = 15$ years, and $T_2 = 10$ years.

The cost of the first overhaul is $O_1 = 30$, and that of the second overhaul is $O_2 = 50$. The average cost of each failure repair is $C_r = 1$. The failure rates per year are given by

$$\Lambda_0(x) = 3 + 0.2x, \Lambda_1(x) = 3 + 0.3x \text{ and } \Lambda_2(x) = 3 + 0.4x$$

Case 1: No Overhaul. In this case, the expected number of failures over L is 337.5. Then from (16.43) we have

$$J(0; T_0 = 45) = 900 - 300 - 337.5 = 262.5$$

Case 2: One Overhaul. In this case, the expected number of failures over L is 257.5. Then from (16.43) we have

$$J(1; T_0 = 25, T_1 = 20) = 900 - 300 - 30 - 257.5 = 312.5$$

Case 3: Two Overhauls. In this case, the expected number of failures over L is 228.75. Then from (16.43) we have

$$J(2; T_0 = 20, T_1 = 15, T_2 = 10) = 900 - 300 - 80 - 228.75 = 291.25$$

Note that Case 2 yields the maximum expected profit. The reason for Case 3 yielding a lower expected profit is that the benefits derived from having an additional overhaul (in terms of the savings in the failure rectification cost) are not worth the cost of the second overhaul. ∎

16.6 MAINTAINABILITY AND LOGISTICS

A goal of any properly designed system (car, stereo, airplane or rocket) is its availability to operate and perform its intended function over its design life. Several notions of availability are discussed in Chapter 13. Availability is affected by many factors. These include (1) reliability, (2) operational requirements, (3) maintainability, and (4) logistics. In this section we deal with the last two factors.

16.6.1 Maintainability

Maintainability for a system or an item has been defined in two ways, namely

1. "A characteristic of design and installation which is expressed as the probability that the item will be retained in or restored to a specified condition within a given time period, when maintenance is performed in accordance with prescribed procedures and resources." (MIL-STD-721)
2. "A system effectiveness concept that measures the ease and rapidity with which a system or equipment is restored to operational state after failing." (Lalli and Packard, 1997)

Maintainability is different from maintenance. Maintainability is the *ability* of a system to be maintained, as opposed to maintenance, which constitutes a series of actions to restore an item to or keep an item in an operational state. As such, maintainability is a design parameter and maintenance is a result of this design.

The reason that maintainability is important is that the usefulness of a system is lost if (1) breakdowns cannot be diagnosed to a level of detail needed to pinpoint the cause in a short time, and (2) repairs require extremely long times for completion. This implies that maintenance of the system must be considered at the design stage.

Effective maintainability requires addressing one or more of the following questions:

1. What parts have high failure rates?
2. How can these failures be diagnosed easily?
3. How quickly can the system be repaired?
4. How much downtime is acceptable?
5. What kind of preventive maintenance needs to be performed?

These questions must be addressed when planning the system and this involves linking reliability, maintenance, and logistics with maintainability. These must be done in the context of the total life cycle of the system, and this has important management implications.[6]

16.6.2   Logistics

Logistics can be defined as the art and science of management, engineering, and technical activities concerned with requirements, design, and planning and maintaining resources to support objectives, plans, and operations. A key issue in logistics is *logistic support.* Some of the key elements of logistic support (Blanchard, 1998) are:

- Maintenance and support planning
- Supply support (spare/repair parts and associated inventories)
- Maintenance and support personnel
- Training and training support
- Test, measurement, handling, and support equipment
- Maintenance facilities
- Technical data, information systems, and databases

We will confine our discussion to spares. This is critical for ensuring availability of the system. The issues that must be addressed include

1. Quantities of spares needed over the life of the system
2. Optimal acquisition policies for spares
3. Location and delivery of spares when needed

The reliability of a system has major implications for the logistic support needed to keep the system operational.

16.7   MAINTENANCE MANAGEMENT INFORMATION SYSTEMS

Maintenance of a system is a complex process and effective maintenance requires proper coordination of many activities. A maintenance management information system (MMIS) is needed to carry out this coordination. The MMIS must be capable of doing the following:

- Tracking components of the system being maintained
- Providing logistic support (e.g., spares inventory)
- Storing maintenance history
- Alerting personnel with regard to predetermined maintenance activities at the correct time
- Performing an analysis of maintenance history
- Determining optimal maintenance policies for items at different levels
- Combining maintenance activities into schedules,
- Updating schedules as and when appropriate (e.g., on occurrence of an unexpected failure or event)
- Planning resources
- Measuring the effectiveness of maintenance activities

An important element of MMIS is the database that contains data regarding component failures. These data provide the information needed, not only for modification of maintenance actions, but also for possible design changes to improve system reliability. Most such databases contain only limited information—for example, time of failure and rectification cost. A more useful database is one that also contains information regarding the cause and mode of failure, the operating environment, and other relevant technical information needed for effective reliability management.

NOTES

1. The literature on maintenance is very large. There are several review papers that have appeared over the last 30 years. These include McCall (1965), Pierskalla and Voelker (1976), Monahan (1982), Jardine and Buzzacot (1985), Sherif and Smith (1986), Thomas (1986), Gits (1986), Valdez-Flores and Feldman (1989), Pintelton and Gelders (1992), and Scarf (1997). Cho and Parlar (1991) and Dekker et al. (1997) deal with the maintenance of mutli-component systems.

There are several books dealing with different aspects of maintenance—see, for example, Gertsbakh (1977), Mann (1983), and Niebel (1985).

2. Many different approaches to maintenance have been proposed. These include the "terotechnology" approach (see Kelly, 1984) and the EUT model (Geraerds, 1992).

For further details of the RCM approach, see Nowlan and Heap (1978), Resnikoff (1978), and Moubray (1991).

For further details of TPM, see Nakijima (1988) for the original Japanese approach and Willmott (1994) for a western version of this approach. The link between reliability and maintenance is recognized at the design stage as stated in the last principle of the TPM approach.

3. Often, we will supress θ for notational ease and write $F(x, \theta)$ as simply $F(x)$.

4. We considered only the basic version of the age and block policies. Many extensions of these have been studied. In many applications, there are discrete time instants (occurring randomly over time) that provide an opportunity to carry out preventive maintenance actions. Dekker and Smeitink (1991) deal with an opportunistic age policy; Dekker and Dijkstra (1992) deal with an opportunistic block policy.

There are several papers dealing with different aspects of maintenance. We give a small illustrative sample of the literature.

Mazzuchi and Soyer (1996) and Percy and Kobbacy (1996) deal with the Bayesian approach to preventive maintenance.

Silver et al. (1992, 1995) deal with preventive maintenance policies based on limited historical failure data.

Berg (1995) deals with the concept of marginal cost analysis.

Ozekici (1995) deals with maintenance policies in random environments.

Christer and Waller (1984) deal with delay time modelling and analysis.

Pham and Wang (1996) deal with imperfect maintenance.

Kobbacy et al. (1997) deal with proportional hazard models and maintenance.

For multicomponent systems, the failure of a component provides an opportunity to carry out preventive maintenance on one or more of the remaining components. For more on this, see the survey paper by Cho and Parlar (1991). Dekker and Roelvink (1995) deal with marginal cost criteria for replacement of a group of components and Christer and Wang (1995) deal with delay time models for multicomponent systems.

5. There are several extensions to Model 16.2. These can be found in the references of Murthy (1984) and Murthy and Hwang (1996). For others, see the review papers mentioned earlier. An extension of Model 16.3 for infinite time operation can be found in Nguyen and Murthy (1981).

6. For more on maintainability, see Blanchard et al. (1995). There are several books on logistics, for example, Green (1991). For more on availability of maintained systems, see Lee et al. (1977).

EXERCISES

1. Compare and contrast RCM and TPM approaches to maintenance.

2. Thirty small electric motors (used in the cooling system for a petrochemical plant) were tested to failure. The sample mean and standard deviation (in years) are 3.2 and 2.8, respectively. Assume that the failures can be modeled by a two-parameter Weibull distribution. Estimate the mean number of failures over a 15-year period assuming that the only action used is corrective maintenance that involves replacing a failed item by a new one. Assume that failure is detected instantaneously.

3. Repeat Exercise 2 assuming that the failure distribution is a lognormal distribution.

4. Repeat Exercise 2 assuming that the failure distribution is a gamma distribution.

5. Compare the results of Exercises 2–4 and comment on the impact of the assumption with regard to the item failure distribution.

6. Suppose that the item in Exercise 2 is preventively maintained using an age policy with parameter T. Calculate the asymptotic expected maintenance cost for $T = 0.5, 1, 1.5$, and 2 years with $C_p = \$200$ and $C_f = \$800$.

7. To compute the correct expected number of replacements in Exercise 6 requires solving the renewal integral equation given by (16.11). Develop a computer program for accomplishing this and carry out the computations for Exercise 6. Compare the exact result with the approximate result.

8. Repeat Exercise 7 assuming a lognormal distribution for item failure times.

9. Repeat Exercise 7 assuming that item failures follow a gamma distribution.

10. Compare the results of Exercises 6–9.

11. Repeat Exercises 6 and 7 with the preventive maintenance involving block replacement rather than age replacement.

12. Repeat Exercise 11 assuming a lognormal distribution.

13. Repeat Exercise 11 assuming a gamma distribution.

14. Compare the results of Exercises 11–13.

15. Repeat Exercises 6 and 7 with the preventive maintenance involving periodic replacement rather than age replacement. The cost of each repair is $50.

16. Repeat Exercise 15 assuming a lognormal distribution.

17. Repeat Exercise 15 assuming a gamma distribution.

18. Compare the results of Exercises 15–17.

19. The failure distribution of an electronic component is exponential with mean time to failure θ unknown. The initial guess (based on the experience of the design engineer) is that it is 3 months. Calculate the expected number of replacements needed per unit time for infinite time operation. The "age-at-failure" for the first three failures are 2.1, 4.8, 0.5 (months), respectively. How would you revise the expected number of replacements per unit time based on these data?

20. Suppose that in Exercise 2 the state of the item (working or failed) is observed only at discrete inspection points equispaced along the time axis. The penalty for the item being in failed state and undetected is $1000/year. Compute the asymptotic cost per unit time with inspection being carried out 4, 8, 12, and 24 times per year.

21. Consider Model 16.2 of Section 16.5.2. Assume the following parameter values: $T = 5$ years, $U = 1$, $S_0 = \$10,000$, $\alpha = 0.3$, $a = 0.5$, $P = \$12,000$, $\beta = 200$, and $b = 200$. Compute $J(T; u(t) = 0)$ and $J(T; u(t) = U)$

22. Consider the following two strategies for the Model in Exercise 21. (1) $u(t) = 0$ for $0 \leq t < U/2$ and $= U$ for $U/2 \leq t \leq U$. (2) $u(t) = U$ for $0 \leq t \leq U/2$ and $= 0$ for $U/2 \leq t \leq U$. Compute $J(T; t(t), 0 \leq t \leq U)$ for the two cases and compare them with the results of exercise 21.

23. Repeat Example 16.6 with $\Lambda_j(x) = a_j + b_j x + c_j x_j^2$, with $a_j = 3$, $b_j = 0.2(1 + j)$ and $c_j = 0.03(1 + j)^2$. Compare the results with those for Example 16.6.

24. Discuss the maintenance requirements for a bicycle. How would this translate into maintainability requirements at the design stage?

25. Suppose that the only maintenance action used is corrective maintenance. The failure distribution is $F(t)$. Failures are detected immediately and a failed item is replaced by a new one. Suppose that n spares are bought at the time of purchase to last for a mission time of L. What is the probability that the spares are not adequate for the mission time? What is the probability that m ($< n$) spares are left unused by the end of the mission time?

26. In Exercise 25, how would you determine the optimal number of spares that should be bought at the time of the purchase. (*Note:* You will need to define a suitable objective function for determining the optimal number.)

CHAPTER 17

Warranties and Service Contracts

17.1 INTRODUCTION

As mentioned in Chapter 1, because of rapid technological changes, new products and systems are appearing at an ever-increasing rate and at the same time they are becoming more and more complex. With the increase in complexity and the use of new materials and design methodologies, the reliability of such products and systems is of concern to buyers. One way for the manufacturer to assure the buyer that the product will perform satisfactorily is to offer a warranty with the sale of the product. In very simple terms, the warranty assures the buyer that the manufacturer will either repair or replace items that do not perform satisfactorily or refund a fraction or the whole of the sale price. For complex systems that are custom built using cutting-edge technologies, buyers often have complex reliability warranty terms included in their contract with the manufacturer.

Offering a warranty results in additional costs to the manufacturer due to the servicing of any problems that arise during the warranty period. We focus our attention on item failures, since these are related to the reliability of the product and are the most common cause of warranty costs. We examine a variety of warranty policies and develop models to obtain the expected cost of servicing the warranty as a function of the reliability of the product. This type of analysis is very important in terms of pricing the product. Since warranty and price play an important role in determining total sales, the implication of reliability on warranty cost is of great importance to manufacturers. Warranty analysis is also relevant for making decisions about reliability targets at the strategic planning level, as discussed in Chapter 12.

One of the ways of reducing warranty cost is through reliability improvement. As indicated in Chapter 15, this can be achieved in two ways. In this chapter, we consider one of these, namely reliability improvement through redundancy and its impact on warranty cost. In Chapter 18 we discuss the impact of reliability growth through development and discuss optimal development in the context of warranty costs.

The outline of the chapter is as follows. We commence with a general discussion on warranties and their role and use in Section 17.2. Following this, in Section 17.3 we discuss some of the many different types of warranty policies that have been studied. Modeling for warranty cost analysis is discussed in Section 17.4, where the results are applied to some of the policies discussed in the previous section. We examine two types of warranty costs: cost per unit and cost over the product life cycle. In Section 17.5, we look at warranty cost to the manufacturer with dynamic sales (sales occurring over time). Section 17.6 deals with improving product reliability through use of redundancy and the impact of this on warranty cost, and Section 17.7 with extended warranties and service contracts. Finally, a brief discussion of some warranty-related management issues is given in Section 17.8.

17.2   WARRANTY—CONCEPT AND ROLE

17.2.1   Warranty Concept

A warranty is a manufacturer's assurance to a buyer that a product or service is or shall be as represented. It may be considered to be a contractual agreement between buyer and manufacturer (or seller) that is entered into upon sale of the product or service. A warranty may be implicit or it may be explicitly stated.

In broad terms, the purpose of a warranty is to establish liability of the manufacturer in the event that an item fails or is unable to perform its intended function when properly used. The contract specifies both the performance that is to be expected and the redress available to the buyer if a failure occurs or the performance is unsatisfactory. The warranty is intended to assure the buyer that the product will perform its intended function under normal conditions of use for a specified period of time.

The terms warranty and guarantee are often used synonymously. The distinction is that a guarantee is defined to be a pledge or assurance of something; a warranty is a particular type of guarantee, namely a guarantee concerning goods or services provided by a seller to a buyer. Another related concept is that of a service contract or "extended warranty." The difference between a warranty and a service contract is that the latter is entered into voluntarily and is purchased separately—the buyer may even have a choice of terms—whereas the basic warranty is a part of the product purchase and is an integral part of the sale.[1]

17.2.2   Role of Warranty

Warranties are an integral part of nearly all consumer and commercial purchases and also of many government transactions that involve product purchases. In such transactions, warranties serve a somewhat different purpose for buyer and manufacturer.

Buyer's Point of View
From the buyer's point of view, the main role of a warranty in these transactions is protectional; it provides a means of redress if the item, when properly used, fails to perform as intended or as specified by the seller. Specifically, the warranty assures the buyer that a faulty item will either be repaired or replaced at no cost or at reduced cost.

A second role is informational. Many buyers infer that a product with a relatively long warranty period is a more reliable and longer-lasting product than one with a shorter warranty period.

Seller's Point of View
One of the main roles of warranty from the seller's point of view is also protectional. Warranty terms may and often do specify the use and conditions of use for which the product is intended and provide for limited coverage or no coverage at all in the event of misuse of the product. The seller may be provided further protection by specification of requirements for care and maintenance of the product.

A second important purpose of warranties for the seller is promotional. Since buyers often infer a more reliable product when a longer warranty is offered, warranty has been used as an effective advertising tool. This is often particularly important when marketing new and innovative products, which may be viewed with a degree of uncertainly by many potential consumers. In addition, warranty has become an instrument, similar to product performance and price, used in competition with other manufacturers in the marketplace.

Warranty In Government Contracting

In simple transactions involving consumer or commercial goods, a government agency may be dealt with in basically the same way as any other customer, providing the standard product warranty for the purchased item. Often, however, the government, as a large entity wielding substantial power as well as a very large consumer, will be dealt with considerably differently, with warranty terms negotiated at the time of purchase rather than specified unilaterally by the seller. The role of warranty in these transactions is usually primarily protectional on the part of both parties.

In some instances, particularly in the procurement of complex military equipment, warranties of a certain type play a very different and important role, that of incentivizing the seller to increase the reliability of the items after they are put into service. This is accomplished by requiring that the contractor service the items in the field and make design changes as failures are observed and analyzed. The incentive is an increased fee paid the contractor if it can be demonstrated that the reliability of the item has, in fact, been increased. Warranties of this type are called reliability improvement warranties (RIW).

17.3   CLASSIFICATION OF WARRANTIES

The first criterion for classification of warranties is whether or not the manufacturer is required to carry out further product development (for example, to improve product reliability) subsequent to the sale of the product as part of the warranty contract. Policies that do not involve such further product development can be further divided into two groups—Group A, consisting of policies applicable for single-item sales, and Group B, policies used in the sale of groups of items (called lot or batch sales). This division and the remainder of a taxonomy of warranties are shown in Figure 17.1.[2]

Policies in Group A can be subdivided into two subgroups, based on whether the policy is renewing or nonrenewing. For renewing policies, the warranty period begins anew with each replacement, whereas for nonrenewing policies, the replacement item assumes the remaining warranty time of the item it replaced. A further subdivision comes about in that warranties may be classified as "simple" or "combination." The free replacement (FRW) and pro-rata (PRW) policies (discussed in the next section) are simple policies. A combination policy is a simple policy combined with some additional features or a policy that combines the terms of two or more simple policies. The resulting four different types of policies under category A are labeled A1–A4 in Figure 17.1. Each of these four groupings can be further subdivided into two subgroups based on whether the policy is one-dimensional or two- (or more) dimensional. A one-dimensional policy is one that is most often based on either time or age of the item, but could instead be based on usage. In contrast, a two-dimensional policy is based on time or age as well as usage. In the case of an automobile, one can either have a one-dimensional policy (e.g., 2 years) or a two-dimensional warranty (e.g., 2 years or 30,000 miles, which ever comes first).

Policies belonging to Group B can also be subdivided into two categories based on whether the policy is "simple" or "combination." These are labeled B1 and B2 in Figure 17.1. As in grouping A, B1 and B2 can be further subdivided based on whether the policy is one-dimensional or two-dimensional.

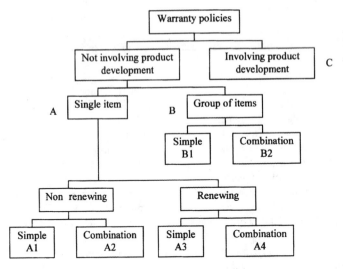

Figure 17.1. Taxonomy of warranty policies.

Finally, policies that involve product development subsequent to the sale are labeled Group C. Warranties of this type are typically part of a service maintenance contract and are used principally in commercial applications and government acquisition of large, complex items—for example, aircraft or military equipment. Nearly all such warranties involve time and/or some function of time as well as a number of characteristics that may not involve time, for example, fuel efficiency.

We describe a few warranty policies from each of the three groups. Some of the policies from Group A (offered with most consumer durables and some industrial and commercial products) will be studied later in the chapter. For a more comprehensive list of policies from Groups A–C, see Blischke and Murthy (1994 and 1996). The following notation is used:

W = length of warranty period

C_b = unit sale price (cost to buyer)

X = time to failure (lifetime) of an item

17.3.1   Group A Policies

We begin with one-dimensional policies. A one-dimensional warranty policy is characterized by an interval, called the warranty period, which is defined in terms of a single variable, e.g., time, age, usage. We define the three most commonly offered one-dimensional warranty policies. Later, we give a two-dimensional warranty policy.

Policy 1: One-Dimensional Nonrenewing Free Replacement Warranty (FRW) Policy
The manufacturer agrees to repair or provide replacements for failed items free of charge up to a time W from the time of the initial purchase. The warranty expires at time W after purchase.

In the case of nonrepairable items, should a failure occur at age X (with $X < W$), under this policy the replaced item has a warranty for a period $(W - X)$, the remaining duration of

the original warranty. Should additional failures occur, this process is repeated until the total service time of the original item and its replacements is at least W. In the case of repairable items, repairs are made free of charge until the total service time of the item is at
least W.

Typical applications of these warranties are consumer products, ranging from inexpensive
items such as photographic film to relatively expensive repairable items such as automobiles,
refrigerators, large-screen color TVs, etc., and expensive nonrepairable items such as mi-
crochips and other electronic components as well.

Policy 2: One-Dimensional Nonrenewing Pro-Rata Rebate Warranty (PRW) Policy

The manufacturer agrees to refund a fraction of the purchase price should the item fail before
time W from the time of the initial purchase. The buyer is not constrained to buy a replacement item.

The refund depends on the age of the item at failure (X) and it can be either a linear or a
nonlinear function of $(W - X)$, the remaining time in the warranty period. Let $q(x)$ denote this
function. This defines a family of pro-rata policies that is characterized by the form of the refund function. Two forms commonly offered are as follows:

1. Linear function: $q(x) = [(W - x)/W]C_b$
2. Proportional function: $q(x) = [\alpha(W - x)/W]C_b$, where $0 < \alpha < 1$

Typically, these policies are offered on relatively inexpensive nonrepairable products
such as batteries, tires, ceramics, etc. The policy is inherently nonrenewing.

Policy 3: One-Dimensional Nonrenewing Combination FRW/PRW

The manufacturer agrees to provide a replacement or repair free of charge up to time W_1
from the time of initial purchase; any failure in the interval W_1 to W (where $W_1 < W$) results in a pro-rated refund. The warranty does not renew. The pro-ration can be either linear or nonlinear.

Again, depending on the form of the pro-ration cost function, we have a family of combined free replacement and pro-rata policies similar to that for the PRW. Warranties of this
type are sometimes used to cover replacement parts or components where the original warranty covers an entire system. They are also widely used in sales of consumer products.

In the case of two-dimensional warranties, a warranty is characterized by a region in a
two-dimensional plane, with one axis representing time or age and the other representing
item usage. As a result, many different types of warranties, based on the shape of the warranty coverage region, may be defined. We describe one such policy.

Policy 4: Two-Dimensional Nonrenewing FRW Policy

The manufacturer agrees to repair or provide a replacement for failed items free of charge up
to a time W or up to a usage U, which ever occurs first, from the time of the initial purchase.
W is called the warranty period and U the usage limit. The warranty region is the rectangle
bounded by the time and usage axes and lines at usage $= U$ and time $= W$.

Note that under this policy, the buyer is provided warranty coverage for a maximum time
period W and a maximum usage U. If usage is heavy, the warranty can expire well before W.
On the other hand, if usage is very light, then the warranty can expire well before the limit U
is reached. Should a failure occur at age X with usage Y, it is covered by warranty only if X is
less than W and Y is less than U. If the item is replaced by a new one, the replacement item is
warrantied for a time period $(W - X)$ and for usage $(U - Y)$. This type of policy is offered by
nearly all auto manufacturers, with usage corresponding to distance driven.

17.3.2 Group B Policies

One-dimensional Group B policies are called cumulative warranties and are applicable only when items are sold as a single lot of n items and the warranty refers to the lot as a whole. The policies are conceptually straightforward extensions of the nonrenewing free replacement and pro-rata warranties discussed previously. Under a cumulative warranty, the lot of n items is warrantied for a total time of nW, with no specific service time guarantee for any individual item. Cumulative warranties would quite clearly be appropriate only for commercial and governmental transactions, since individual consumers rarely purchase items by lot. In fact, warranties of this type have been used by airlines in the purchase of electronic equipment and have been proposed in the United States for use in acquisition of military equipment. The rationale for such a policy is as follows. The advantage to the buyer is that multiple-item purchases can be dealt with as a unit rather than having to deal with each item individually under a separate warranty contract. The advantage to the manufacturer is that fewer warranty claims may be expected because longer-lived items can offset early failures. We describe two such policies from Guin (1984). The following notation will be used:

X_i = service life of item i, $i = 1, 2, \ldots$
$S_n = \sum_{i=1}^{n} X_i$

Policy 5: Cumulative FRW-1
A lot of n items is warrantied for a total (aggregate) period of nW. The n items in the lot are used one at a time. If $S_n < nW$, free replacement items are supplied, also one at a time, until the first instant when the total lifetimes of all failed items plus the service time of the item then in use is at least nW.

Policy 6: Cumulative FRW-2
A lot of n items is warrantied under cumulative warranty for a total period of nW. Of these, k ($< n$) are put into use simultaneously, with the remaining $(n - k)$ items being retained as spares. Spares are used one at a time as failures occur. Upon failure of the nth item, free replacements are supplied as necessary until a total service time of nW is achieved.

These two policies are applicable to components of industrial and commercial equipment bought in lots as spares and used one at a time as items fail. Examples of possible applications are mechanical components such as bearings and drill bits. The policy would also be appropriate for military or commercial airline equipment such as mechanical or electronic modules in airborne units.

17.3.3 Group C Policies

Warranty policies in Group C are also called reliability improvement warranty (RIW) policies. The basic idea is to extend the notion of a basic consumer warranty (usually the FRW) to include guarantees on the reliability of the item and not just on its immediate or short-term performance. This is particularly appropriate in the purchase of complex, repairable equipment that is intended for relatively long use. The intent of reliability improvement warranties is to negotiate warranty terms that will motivate a manufacturer to continue improvements in reliability after a product is delivered.

Under RIW, the contractor's fee is based on his ability to meet the warranty reliability requirements. These often include a guaranteed MTBF as a part of the warranty contract. The following policy from Gandara and Rich (1977) illustrates the concept.

Policy 7: Reliability Improvement Warranty
The manufacturer agrees to repair or provide replacements free of charge for any failed parts or units until time W after purchase. In addition, the manufacturer guarantees the MTBF of the purchased equipment to be at least M. If the computed MTBF is less than M, the manufacturer will provide, at no cost to the buyer, (1) engineering analysis to determine the cause of failure to meet the guaranteed MTBF requirement, (2) engineering change proposals, (3) modification of all existing units in accordance with approved engineering changes, and (4) consignment spares for buyer use until such time as it is shown that the MTBF is at least M.

17.4 WARRANTY COST ANALYSIS

In this section, we first discuss a system characterization that will enable us to build models for carrying out a cost analysis from the perspectives of both the manufacturer and the consumer. Later we discuss the different notions of warranty cost from both of these perspectives.[3]

17.4.1 System Characterization

A system characterization for warranty cost analysis is shown in Figure 17.2. The manufacturer produces products and sells them to consumers with a warranty policy included. Product performance is determined by the interaction between product characteristics (determined primarily by the manufacturer and influenced by design and manufacturing decisions) and product usage (determined by the consumer). When the consumer is not satisfied with product performance, a claim under warranty usually results. The cost of the warranty is the cost incurred by the manufacturer of servicing a claim under warranty. The magnitude of this cost depends on the terms of the warranty policy and product reliability. The characterization of each of the elements in Figure 17.2 can involve many variables, depending on the degree of detail desired. We consider the following simple characterization:

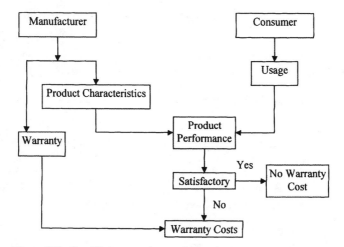

Figure 17.2. Simplified system characterization for warranty cost analysis.

1. All consumers are alike in their usage. One can relax this assumption to divide consumers into two (high- or low-intensity usage) or more groups.

2. All items are statistically similar. One can relax this assumption to include two types of items (conforming and nonconforming) to take into account quality variations in manufacturing.

3. The performance of the product is expressed in terms of a binary characterization, working or failed. The time to first failure is a random variable with failure distribution $F(t)$. The associated density and failure rate functions are $f(t)$ and $r(t)$, respectively. Subsequent failures depend on the type of rectification action (repair or replacement; type of repair, and so on). This is adequate to study the influence of product reliability on the cost of servicing the warranty and to assess the reliability implications at the design stage in order to ensure that warranty costs do not exceed some specified target value.

In order to simplify the building and analysis of the model, we make the following assumptions:

1. Whenever a failure occurs, it results in an immediate claim. Relaxing this assumption involves modeling the delay time between failure and claim.

2. All claims are valid. This can be relaxed by assuming that a fraction of the claims are invalid, either because the item was used in a mode not covered by the warranty or because it was a bogus claim.

3. The time to rectify a failed item (either through repair or replacement) is sufficiently small in relation to the mean time between failures so that it can be approximated as being zero.

4. The manufacturer has the logistic support (spares and facilities) needed to carry out the rectification actions without any delays. Further discussion of logistic related issues can be found in Section 17.8.

17.4.2   Two Notions of Warranty Cost

In the context of warranties, the following costs are of importance to both consumers and manufacturers.

1. Expected warranty cost per unit sale
2. Expected life cycle cost (LCC) of operation over the product life cycle

In this section some issues related to obtaining these cost measures are discussed.

Warranty Cost Per Unit Sale
Whenever an item is returned for rectification action under warranty, the manufacturer incurs various costs (handling, material, labor, facilities, etc). These costs can also be random variables. The total warranty cost (i.e., the cost of servicing all warranty claims for an item over the warranty period) is thus a sum of a random number of such individual costs, since the number of claims over the warranty period is also a random variable.

Expected Life Cycle Cost (LCC)
This cost, also a cost to the buyer, depends on the life cycle of the product, that is, the time interval over which consumers buy the product. After this time period, sales of the product

cease, often because of the introduction of a new and better replacement product. Let L denote the product life cycle. We assume that the consumer continues repeat purchases over this period. The number of repeat purchases is a random variable. The total life cycle cost is the product of this random variable and the warranty cost per item. As a result, the total cost over the product life cycle is also random.

The warranty cost per unit sale is important in the context of pricing the product. The sale price must exceed the manufacturing cost plus the warranty cost or the manufacturer incurs a loss. On the average, warranty cost per item decreases as reliability increases. When a buyer has the option of choosing between different warranty policies, then this cost is of relevance. The life cycle cost of a product is relevant to both buyer and manufacturer, as will be indicated below.

In this section, we confine our attention to obtaining expressions for the expected values for the warranty cost per unit sale for Policies 1, 2, and 4, defined earlier. For Policy 1, we also derive the expected life cycle cost. We make the following additional assumptions:

1. The cost of each rectification under warranty (repair or replacement) is modeled by a single variable that is the aggregate of the different costs (handling, material, labor, facilities, etc.) involved.

2. The product life cycle, L, is modeled as a deterministic variable. One can easily relax this assumption and treat it as a random variable with a specified distribution.

3. All the model parameters (cost and of the various distributions involved) are known.

17.4.3   Policy 1

We first consider the case where the item is nonrepairable, so that any failures during warranty require the replacement of the failed item by a new item. Let C_s denote the manufacturing cost per item (which is also the cost of replacing a failed item).

Expected Cost Per Unit to Manufacturer [Non-repairable Product]
Since failures under warranty are rectified instantaneously through replacement by new items, the number of failures over the period $[0, t)$, $N(t)$, is characterized by a renewal process with time between renewals distributed according to $F(t)$. As a result, the cost $C_m(W)$ to the manufacturer over the warranty period W is a random variable given by

$$C_m(W) = C_s[1 + N(W)] \tag{17.1}$$

where we have included the manufacturing cost C_s of the initial item sold to the buyer.

The expected number of failures during warranty, $E[N(W)]$ is given by

$$E[N(W)] = M(W) \tag{17.2}$$

where $M(.)$ is the renewal function given by

$$M(t) = F(t) + \int_0^t M(t - x)f(x)dx \tag{17.3}$$

As a result, the expected warranty cost per unit to the manufacturer is given by

$$E[C_m(W)] = C_s[1 + M(W)] \tag{17.4}$$

Example 17.1

A color television consists of several components, with the picture tube being the most important. The picture tube typically carries a warranty that is different from that on the remainder of the set. Suppose that the television is sold with an FRW policy with warranty period W for the picture tube. When a picture tube fails, it is replaced by a new tube, since it is not possible to repair one that has failed.

Suppose that the failure time for a picture tube is Weibull distributed, i.e.,

$$F(t) = 1 - e^{-(\lambda t)^\alpha}$$

with $\lambda = 0.5$ (per year) and $\alpha = 1.5$. This implies that the mean time to failure is given by

$$\mu = \frac{1}{\lambda}\Gamma\left(1 + \frac{1}{\alpha}\right) = 2\Gamma(1.666) = 1.805 \text{ years}$$

Since $\alpha = 1.5$, the failure rate is increasing with time. The manufacturer's cost of each item is $C_s = \$80$.

We consider three different values for W. It is not possible to obtain an analytical expression for the renewal function for a Weibull distribution. The values are determined either by the use of tables (Baxter et al., 1981) or by use of a renewal function solver (Blischke and Murthy, 1994). The results are given in Table 17.1(a). If the shape parameter is $\alpha = 1$ (so that failures are exponentially distributed) with MTTF = 2.0 years, the results are shown in Table 17.1(b).

Note that warranty costs are large and become very substantial as W increases. To control these costs either tubes with larger MTTF must be used or a much shorter warranty offered. ■

Expected Life Cycle Cost (Nonrepairable Product)

Here we are interested in the cost over a period L. The first failure after expiration of the warranty results in a new purchase by the buyer and this comes with a new identical warranty. Let $Y_i, i = 1, 2, \ldots$, denote the time interval between successive repeat purchases, with the first purchase occurring at $t = 0$. It is easily seen that the Y's are of the form $Y = W + \gamma(W)$, where $\gamma(W)$ is the remaining life of the item in use at the expiration of the warranty. This is simply the

Table 17.1. Expected warranty costs for Example 17.1

(a) $\alpha = 1.5$		
W (years)	$M(W)$	$E[C_m(W)]$ (\$)
0.5	0.8417	147.33
1.0	1.9456	235.65
2.0	4.1616	412.93
(b) $\alpha = 1.0$		
W (years)	$M(W)$	$E[C_m(W)]$ (\$)
0.5	0.500	120.00
1.0	1.000	160.00
2.0	2.000	240.00

excess age of the renewal process $N(t)$ discussed earlier. Let $F_\gamma(t)$ denote the distribution function for the excess age. This is given by equation (B3.25) of Appendix B as follows:

$$F_\gamma(t) = F(W + t) - \int_0^W [1 - F(W + t - x)]dM(x) \tag{17.5}$$

where $M(t)$ is the renewal function given by (17.3). Since Y is a linear tranaslation of $\gamma(W)$, it is easily shown that the distribution function of Y is given by

$$F_Y(t) = F(t) - \int_0^W [1 - F(t - x)]dM(x) \tag{17.6}$$

Let $C_b(L, W)$ denote the LCC to the buyer. Since the buyer purchases new items according to a renewal process with interval between purchases distributed according to $F_Y(t)$, we have

$$E[C_b(L, W)] = C_b[1 + M_Y(L)] \tag{17.7}$$

where $M_Y(t)$ is the renewal function associated with $F_Y(t)$.

If L is large relative to W and $E[Y]$, then one can use an asymptotic approximation (see Blischke and Murthy, 1996), which gives

$$M_Y(L) \approx \frac{L}{\mu[1 + M(W)]} \tag{17.8}$$

where μ is the MTTF for the product and $M(W)$ is obtained from (17.3).

The LCC, $C_m(L, W)$, to the manufacturer is simply the sum of the unit warranty costs for sales over the life cycle. As a result, we have

$$E[C_m(L, W)] = C_s[1 + M(W)][1 + M_Y(L)] \tag{17.9}$$

Note that the expected profit to the manufacturer over the life cycle, $E[P(L, W)]$, is given by

$$E[P(L, W)] = \{C_b - C_s[1 + M(W)]\}[1 + M_Y(L)] \tag{17.10}$$

Example 17.2

Consider the picture tube of Example 17.1 and let the failure distribution be given by an exponential distribution $F(t) = 1 - e^{-\lambda t}$, with $\lambda = 0.5$ per year, so that the MTTF is 2 years. In this case, $F_\gamma(t)$ is also exponential and, as a result, the distribution of Y is simply the translated exponential:

$$F_Y(t) = 1 - e^{-\lambda(t - W)}$$

The renewal function associated with this can be expressed analytically (see Cinlar, 1975). The result is

$$M_Y(L) = \left[\frac{L}{W}\right] - \sum_{i=1}^{[L/W]} e^{-\lambda(L - iW)} \left\{ \sum_{j=1}^{i-1} \frac{[L - iW]^j}{j!} \right\}$$

where $[x]$ is the largest integer less than x.

Suppose that the life cycle is 7.5 years and $W = 1$. In this case, we have

$$M_Y(7.5) = 7 - \sum_{i=1}^{7} e^{-0.5(7.5-i)} \left\{ \sum_{j=1}^{i-1} \frac{[7.5 - i]^j}{j!} \right\} = 2.222$$

This implies that the expected life cycle cost to the buyer is [from (17.6)] $3.222C_b$.

If $C_b = \$100$, then the expected life cycle cost is \$322.20. Note that if λ is changed to 0.4 (implying an MTTF of 2.5 years), then the expected life cycle cost to the buyer is \$289.90. This is to be expected as the product has become more reliable.

For the manufacturer, the expected life cycle cost (from (17.9) with $\lambda = 0.5$ and $C_s = \$80$) is \$386.64. This changes to \$360.86 when $\lambda = 0.4$. This implies a reduction of \$25.28 over the product life cycle with this improvement in reliability. ■

Buyer's Indifference Price

For unwarranted items, sold at a cost C_u to the buyer, the total expected cost to the buyer over the life cycle L is $C_u[1 + M(W)]$, since the buyer must purchase a replacement item at full price each time an item fails over the life cycle. For warrantied items, the total expected cost to the buyer is $E[C_b(L, W)]$ given by (17.7). The indifference price C_b^* is obtained by equating these two long-term costs and solving for C_b. This is given by

$$C_b^* = \frac{C_u[1 + M(L)]}{M_Y(L)} \tag{17.11}$$

which, for large L, can be approximated as

$$C_b^* = \frac{C_u \mu[1 + M(L)][1 + M(W)]}{\mu[1 + M(L)] + L} \tag{17.12}$$

Example 17.2 (Continued)

With $\lambda = 0.5$, we have from (17.11)

$$C_b^* = \left[\frac{1 + 0.5 \times 7.5}{1 + 2.222} \right] C_u = 1.471 C_u$$

In other words, the buyer should be willing to pay up to 47.1% more for a warrantied item than for an item without warranty. If the manufacturer charges more than this, it is more economical for the buyer to buy the item without warranty. ■

Manufacturer's Indifference Price

If the manufacturer sold the product without warranty, then the expected profit over the life cycle is given by $(C_u - C_s)[1 + M(L)]$. With a sale price of C_b and the items sold with warranty, the expected profit over the life cycle, $E[P(L, W)]$, is given by (17.10). The manufacturer is indifferent between selling with warranty or without warranty when the long run expected profits are the same for both cases. This occurs when the sale price with warranty is given by

$$C_b^{**} = \frac{C_u \mu[1 + M(L)][1 + M(W)]}{\mu[1 + M(L)] + L} \tag{17.13}$$

Note that $C_b^{**} = C_b^*$. However, if one includes an additional administrative cost of C_a for each warranty claim, then we have the indifference price to the seller given by

$$C_b^{**} = \frac{C_u[1 + M(L)] + C_a[M(L) - M_Y(L)]}{1 + M_Y(L)} \qquad (17.14)$$

Comment: The manufacturer prefers to sell an item with warranty at price C_b to selling it without warranty at price C_u if $C_b > C_b^{**}$. The buyer prefers an item with warranty if $C_b < C_b^*$, and *both* prefer an unwarranted item if $C_b^* < C_b < C_b^{**}$.

Expected Cost Per Unit to Manufacturer (Repairable Product)
In this case, the cost depends on the nature of the repair. We first confine our attention to the case where all failures over the warranty period are minimally repaired. We assume that the repair times are small in relation to the MTBF, so that they can be ignored. The average cost of each repair is denoted C_r. As a result, failures over the warranty period occur according to a nonhomogeneous Poisson process with intensity function equal to the failure rate $r(t)$ and the expected warranty cost to the manufacturer is given by

$$E[C_m(W)] = C_s + C_r \int_0^W r(x)dx \qquad (17.15)$$

Example 17.3
The item under consideration is a VCR with a Weibull failure distribution with shape parameter α and scale parameter β. The VCR is sold under FRW with warranty period W and costs \$400 to manufacture. The manufacturer rectifies all failures under warranty using minimal repair at an average cost of $C_r = \$50.00$ per repair. Then, from (17.15), we have

$$E[C_m(W)] = C_s + \frac{C_r W^\alpha}{\beta}$$

Suppose that $\beta = 4.0$ (years). Using this in the above cost expression, we obtain the expected warranty cost for different warranty periods and for $\alpha = 1.5$ and 2.0, as shown in Table 17.2. ∎

17.4.4 Policy 2

We confine our attention to the linear rebate function given by

$$q(x) = \begin{cases} (1 - x/W)C_b & 0 \le x < W \\ 0 & \text{otherwise} \end{cases} \qquad (17.16)$$

Table 17.2. Expected warranty costs for Example 17.3

	(a) $\alpha = 1.5$			
W (years)	1	2	3	4
$E[C_m(W)]$ (\$)	412.50	435.36	469.52	500.00
	(b) $\alpha = 2.0$			
W (years)	1	2	3	4
$E[C_m(W)]$ (\$)	412.50	450.00	512.50	700.00

Expected Cost Per Unit to Manufacturer

The cost per unit to the manufacturer is given by

$$C_m(W) = C_s + q(X) \tag{17.17}$$

where X is the life time of the item supplied. Here X is a random variable with distribution function $F(x)$. As a result, the expected cost per unit to manufacturer is given by

$$E[C_m(W)] = C_s + \int_0^W q(x)f(x)dx \tag{17.18}$$

Using (17.17) in (17.18) and carrying out the integration, we have

$$E[C_m(W)] = C_s + C_b \left[F(W) - \frac{\mu_W}{W} \right] \tag{17.19}$$

where

$$\mu_W = \int_0^W xf(x)dx \tag{17.20}$$

is the partial expectation of X.

The expected profit per unit sale is given by

$$\pi(W) = C_b - E[C_m(W)] = C_b \left[1 - F(W) + \frac{\mu_W}{W} \right] - C_s \tag{17.21}$$

Expected Cost Per Unit to Buyer

The cost per unit to the buyer is given by

$$C_b(W) = C_b - q(X) \tag{17.22}$$

where X is the item failure time and $q(\ )$ is given by (17.16). As a result, the expected per unit cost to the buyer is given by

$$E[C_b(W)] = C_b \left[\frac{\mu_W}{W} + 1 - F(W) \right] \tag{17.23}$$

Example 17.4

Suppose that car battery failures occur according to an exponential distribution with parameter $\lambda = 0.5$ per year. The sale price $C_b = \$50$ and the manufacturing cost is $C_s = \$30$. The partial expectation μ_W is given by

$$\mu_W = \frac{1}{\lambda}(1 + \lambda W)e^{-\lambda W}]$$

Using this in (17.19) yields

$$E[C_m(W)] = C_s + C_b \left[\frac{1 - e^{-\lambda W}}{\lambda W} \right]$$

Suppose that the product is sold with a warranty period of $W = 1$ year. Then the expected cost per unit to the manufacturer is $C_s + 0.2131C_b = \$41.65$. The expected warranty cost to the buyer [using (17.20) in (17.23)] is

$$E[C_b(W)] = C_b - \{E[C_m(W)] - C_s\}$$

As a result the expected per unit cost to the buyer is $\$38.35$. This implies that the manufacturer's profit is $\$8.35$ ($= 38.35 - 30.00$ or $50.00 - 41.65$) per item sold.

If the manufacturer improves the reliability of the product so that $\lambda = 0.4$, then the expected cost per unit to the manufacturer is $C_s + 0.1758C_m = \$38.79$ and average profit per unit increases to $\$11.21$. ∎

Expected Life Cycle Cost

Let $C_b(L, W)$ denote the life cycle cost to the buyer. Then, conditional on X_1, the failure time of the first item, we have (for $L > W$)

$$E[C_b(L, W|X_1 = x)] = \begin{cases} C_b + C_b x/W + E[C_b(L - x, W)] & \text{if } 0 \le x < W \\ 2C_b + E[C_b(L - x, W)] & \text{if } W \le x < L \\ C_b & \text{if } X \ge L \end{cases} \qquad (17.24)$$

On removing the conditioning, this becomes

$$E[C_b(L, W)] = \int_0^L C_b \min\{x/W, 1\} f(x) dx + \int_0^L E[C_b(L - x, W)] f(x) dx \qquad (17.25)$$

This can be rewritten as (for details see Blischke and Murthy, 1994)

$$E[C_b(L, W)] = C_b\{1 + F(L) - [F(W) - \mu_W/W] \times [1 + M(L - W)] \\ + \int_0^{L-W} F(L - x) dM(x) + \int_{L-W}^L \int_0^{L-x} (u/W) dF(u) dM(x)\} \qquad (17.26)$$

The manufacturer's life cycle, $C_m(L, W)$ is the cost of supplying the items over the life cycle. Since failures according to a renewal process, we have

$$E[C_m(L, W)] = C_s[1 + M(L)] \qquad (17.27)$$

The manufacturer's expected profit $E[P(L, W)]$ is the difference between the expected income (which is the expected cost to the buyer) and expected cost. As a result,

$$E[P(L, W)] = E[C_b(L, W)] - C_s[1 + M(L)] \qquad (17.28)$$

Example 17.5

Consider the battery of Example 17.4. Since the failure distribution is exponential, it is possible to obtain an analytical expression for $E[C_b(L, W)]$ given by (17.25). The result is

$$E[C_b(L, W)] = C_b\left\{1 + e^{-\lambda W} + \frac{(\lambda L - 1)(1 - e^{-\lambda W})}{\lambda W}\right\}$$

Table 17.3. Expected warranty costs for Example 17.5

λ (per year)	$E[C_b(L, W)]$	$E[C_m(L, W)]$
0.5	$(0.8195 + 0.3935L)C_b$	$(1 + 0.5L)C_m$
0.4	$(0.8461 + 0.3297L)C_b$	$(1 + 0.4L)C_m$

Let the warranty period be $W = 1$. Then the expected life cycle costs to the buyer and the manufacturer are shown in Table 17.3 for $\lambda = 0.04$ and 0.05. Suppose that $L = 10$ years, $C_b = \$50$, and $C_s = \$30$. Then for $\lambda = 0.5$, the expected life cycle cost to the buyer and manufacturer are \$237.73 and \$180.00, respectively. ∎

One can determine indifference prices for the buyer and manufacturer in a manner similar to that used in analysis of Policy 1. For details, see Blischke and Murthy (1994).

17.4.5 Policy 4

The cost analysis of this policy can be done in two ways. The first is to model failures by a two-dimensional distribution function $F(t, x)$, which jointly characterizes the age and usage at failure. Failures can then be modeled as points in a two-dimensional plane and the analysis carried out in a manner similar to that of Policy 1 using the two-dimensional distribution. An alternate approach is to model usage as a function of age and then model failures as a function of both age and usage. We follow the latter approach.

We assume that the usage by time t, $Y(t)$, is given by a linear function

$$Y(t) = Rt \tag{17.29}$$

with R being a nonnegative random variable with density function $g(r)$. This models the different usage rates across buyers. We assume that the product is repairable and that the manufacturer rectifies all failures under warranty through minimal repair. In this case, the distribution of failures over the warranty period can be characterized by an intensity function $\lambda(\ )$. Conditional on $R = r$, we assume a linear intensity function given by

$$\lambda(t|r) = \theta_0 + \theta_1 t + \theta_2 r + \theta_3 rt \tag{17.30}$$

This implies that the intensity function is a function of the age (t), usage rate (r), and the usage up to t (rt).

Expected Cost Per Unit to Manufacturer
Note that conditional on $R = r$, the warranty expires at time W if the usage rate r is less than γ, where

$$\gamma = U/W \tag{17.31}$$

and at time Z_r given by

$$Z_r = U/r \tag{17.32}$$

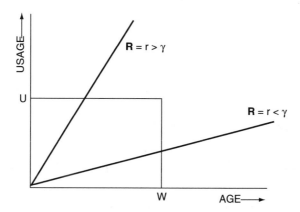

Figure 17.3. Warranty expiration for two usage rates.

if the usage rate $r > \gamma$. (See Figure 17.3.) The expected numbers of failures over the warranty period for these two cases are given by

$$E[N(W, U)|r] = \int_0^W \lambda(t|r)dt \quad \text{and} \quad E[N(W, U)|r] = \int_0^{Z_r} \lambda(t|r)dt \quad (17.33)$$

respectively. On removing the conditioning, we have the expected number of failures during warranty and, using this, we find the expected warranty cost $E[C_m(W, U)]$ to be

$$E[C_m(W, U)] = C_r \left\{ \int_0^\gamma \left[\int_0^W \lambda(t|r)dt \right] g(r)dr + \int_\gamma^\infty \left[\int_0^{Z_r} \lambda(t|r)dt \right] g(r)dr \right\} \quad (17.34)$$

where C_r is the cost of each repair, and $\lambda(t|r)$, γ and Z_r are given by (17.30)–(17.32), respectively.

Example 17.6
We consider an auto warranty for which the unit for usage U is 10^4 miles and for W the unit is years. As a result, $W = 1$ and $U = 2$ corresponds to a time limit of 1 year and a usage limit of 20,000 miles, and the unit for R is 10^4 miles per year. Let $g(r) = 0.2$ for $0 \le r \le 5$ and zero for $r > 5$. This implies that R has a uniform distribution with a mean usage rate of 2.5 (or 25,000 miles per year).

We consider three cases. In the first, the failure intensity is affected by age, usage rate and the total usage. In this case all the four parameters in (17.30) are nonzero. For the second case, we assume that the failure intensity is not affected by the usage rate, so that $\theta_2 = 0$ and only age and usage influence the failure rate. Finally, for the third case, we assume a failure intensity that is not affected by either usage rate or total usage, so that $\theta_2 = \theta_3 = 0$, and only age influences the failure rate. We assume the following parameter values:

Case (i): $\theta_0 = 0.003$, $\theta_1 = 0.007$, $\theta_2 = 0.003$, and $\theta_3 = 0.003$
Case (ii): $\theta_0 = 0.003$, $\theta_1 = 0.007$, $\theta_2 = 0.000$, and $\theta_3 = 0.003$
Case (iii): $\theta_0 = 0.003$, $\theta_1 = 0.007$, $\theta_2 = 0.000$, and $\theta_3 = 0.000$

Table 17.4. Expected warranty costs for Example 17.6

	(a) Case (i)			
↓ U and W →	0.50	1.00	1.50	2.00
1.0	0.0024	0.0034	0.0040	0.0046
1.5	0.0031	0.0048	0.0059	0.0069
2.0	0.0035	0.0061	0.0070	0.0091
	(b) Case (ii)			
↓ U and W →	0.50	1.00	1.50	2.00
1.0	0.0014	0.0021	0.0026	0.0031
1.5	0.0017	0.0027	0.0036	0.0043
2.0	0.0018	0.0033	0.0043	0.0053
	(c) Case (iii)			
↓ U and W →	0.50	1.00	1.50	2.00
1.0	0.0017	0.0022	0.0025	0.0031
1.5	0.0021	0.0029	0.0033	0.0036
2.0	0.0023	0.0034	0.0040	0.0044

Reprinted from Blischke and Murthy (1994), pp. 310–311, by courtesy of Marcel Dekker, Inc.

The ratio of expected warranty cost to repair cost (obtained by solving (17.34) numerically) is shown for each case in Table 17.4 for $U = 1.0$, 1.5, and 2.0 and $W = .50$, 1.00, 1.50, and 2.00.

As can be seen, for all three cases, the expected warranty costs increase with W and/or U increasing. ■

Expected Life Cycle Cost
A life cycle cost analysis can be carried out in a similar manner to that for Policy 1. The results can be found in Blischke and Murthy (1994).

17.5   WARRANTY COST WITH DYNAMIC SALES

For products sold with warranty, the manufacturer is obligated to service all claims made under warranty. The actions taken by the manufacturer to accomplish this depend on the type and the terms of the warranty. Warranty servicing deals with study of such actions and related planning issues.

Under a nonrenewing PRW policy (Policy 2), the manufacturer is required to refund a fraction of the sale price on failure of an item in the warranty period. In order to carry this out, the manufacturer must set aside a fraction of the sale price. This is called warranty reserving. For nonrepairable items sold with an FRW policy (Policy 1), the manufacturer is required to supply a replacement item for failures under warranty. In this case, the number of spares needed is of interest and this is of importance in the context of production and inventory control. For repairable products sold with an FRW policy (Policy 1), planning of repair facilities requires evaluation of the demand for repairs over the warranty period. This depends on the type of repair action and on anticipated sales over the product life cycle.

17.5.1   Product Sales over the Life Cycle

Let L denote the product life cycle and $s(t)$, $0 \leq t \leq L$, denote the sales rate (i.e., sales per unit time) over the life cycle. This includes both first and repeat purchases for the total consuming population. Total sales over the life cycle, S, is given by

$$S = \int_0^L s(t)dt \qquad (17.35)$$

It is assumed that the life cycle L exceeds W, the warranty period, and that items are put into use immediately after they are purchased. Since the manufacturer must provide a refund or replacements for items that fail before reaching age W, and since the last sale occurs at or before time L, the manufacturer has an obligation to service warranty claims over the interval $[0, L + W)$.

17.5.2   Warranty Reserves [PRW Policy]

Suppose that a product is sold with a nonrenewing PRW policy with linear rebate function. The rebate over the interval $[t, t + \delta t)$ is due to failure of items that are sold in the interval $[t - \psi, t)$, where

$$\psi = \max\{0, t - W\}, \qquad (17.36)$$

and fail in the interval $[t, t + \delta t)$. Let $\nu(t)$ denote the expected refund rate (i.e., the amount refunded per unit time) at time t. Then, it is easily shown (for details, see Chapter 9 of Blischke and Murthy, 1994) that

$$\nu(t) = C_b \left[\int_\psi^t s(x) \left(\frac{t-x}{W} \right) f(t-x)dx \right] \qquad (17.37)$$

for $0 \leq t \leq (W + L)$. The expected total reserve needed to service the warranty over the product life cycle, ETR, is given by

$$ETR = \int_0^{W+L} \nu(t)dt \qquad (17.38)$$

Example 17.7
Consider the battery in Example 17.4. Let $L = 7$ years and suppose that the sales rate over the product life cycle is given by

$$s(t) = kte^{-t}$$

$0 \leq t \leq 7$, with $k = S/\{1 - (L + 1)e^{-L}\} = 100,734$. This yields total sales of $S = 100,000$.

As in Example 17.4, the failure distribution $F(t)$ is taken to be exponential with parameter $\lambda = 0.5$/year, so that the mean age of items at failure is two years. Let $W = 1$ year and $C_b = \$50$. The expected refund rate $\nu(t)$, $0 \leq t \leq 5$, obtained from (17.37) is given by

$$\nu(t) = 0.83945 \, t^3 e^{-t} (\times 10^6)$$

for $0 \le t \le 1$, and by

$$v(t) = 0.83945(3t - 2)e^{-t}(\times 10^6)$$

for $1 \le t \le 7$. Plots of $s(t)$, $0 \le t \le 7$, and $v(t)$, $0 \le t \le 8$, are shown in Figure 17.4. Note that $v(t)$ lags $s(t)$. This is because failures occur on the average one year after purchase, since the MTTF is one year. ∎

17.5.3   Demand for Spares (FRW Policy)

When a nonrepairable product is sold with a nonrenewing free replacement warranty, the manufacturer is required to replace all items that fail within warranty period W. The demand for spares in the interval $[t, t + \delta t)$ is due to failure of items sold in the period $[\psi, t)$ where ψ is given by (17.36). It can be shown (details the of derivation can be found in Chapter 9 of Blischke and Murthy, 1994) that the expected demand rate for spares at time t, $\rho(t)$, is given by

$$\rho(t) = \int_{\psi}^{t} s(x)m(t - x)dx \tag{17.39}$$

where $m(t)$ is the renewal density function associated with the failure distribution function $F(t)$, given by

$$m(t) = f(t) + \int_{0}^{t} m(t - x)f(x)dx \tag{17.40}$$

The expected total number of spares required to service the warranty over the product life cycle, ETS, is given by

$$ETS = L + \int_{0}^{L+W} \rho(t)dt \tag{17.41}$$

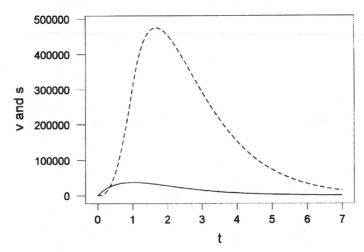

Figure 17.4. Plot of $v(t)$ (– –)and $s(t)$ (——), Example 17.7.

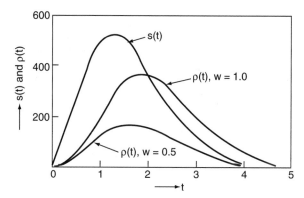

Figure 17.5. Plot of $\rho(t)$ and $s(t)$, Example 17.8.

Example 17.8

The item under consideration is a disk drive used in computers. The technology for disk drives is changing rapidly so that the life cycle L is 4 years. Suppose that the sales rate over the product life cycle is given, for $0 \le t \le 4$, by

$$s(t) = kt^{\beta-1}e^{-\omega t^\beta}$$

Let $k = 722,000$, $\beta = 2.0$, and $\omega = 0.36$. These values were chosen to yield total sales of 100,000.

Suppose that the failure distribution is Erlangian with two stages, that is,

$$F(t) = 1 - (1 + \lambda t)e^{-\lambda t}$$

with parameter λ. The item is nonrepairable and is sold with a FRW policy with warranty period W. Figure 17.5 shows $s(t)$, $0 \le t \le L = 4$ and $\rho(t)$, $0 \le t \le$ L+W, for $W = 0.5$ and 1 year. For both values of W, the peak of $\rho(t)$ lags the peak of $s(t)$, as is to be expected since failures occur after the item has been in use for some time. Table 17.5 gives the expected demand for spares on an annual basis for years 1 through 5 and *ETS*, the total expected demand for spares over the product life cycle, for the two values of W. Both $s(t)$ and $\rho(t)$ initially increase then decrease with time. $\rho(t)$ lags $s(t)$ since the mean time to failure is equal to 1.0 years. The expected number of annual and total replacements increase with the warranty period W, as expected.

Table 17.5. Annual sales and replacements ($\times$ 100) and *ETS* for $W = 0.5$ and 1

	W	0–1	1–2	2–3	3–4	4–5	*ETS*
				Years			
$\int s(t)$		303	463	198	36	0	
$\int \rho(t)$	0.5	46	137	81	19	1	284
$\int \rho(t)$	1.0	59	325	273	87	11	755

Reprinted from Blischke and Murthy (1994), p. 387, by courtesy of Marcel Dekker, Inc.

17.5.4 Demand for Repairs (FRW Policy)

When a repairable product is sold with a FRW policy, the manufacturer has the option of re-pairing items that fail during the warranty period. We consider the case where the manufac-turer always repairs failed items through minimal repair. For each item sold, failures over the warranty period occur according to a nonstationary Poisson process with an intensity func-tion $\lambda(t) = r(t)$, where $r(t)$ is the failure rate associated with the failure distribution function $F(t)$.

Let $\rho_r(t)$ denote the expected repair rate at time t. Then, using an approach similar to that in Section 17.5.3, we have

$$\rho_r(t) = \int_{\psi}^{t} s(x)r(t-x)dx \qquad (17.42)$$

$0 \le t \le L + W$, with ψ given by (17.36). The total expected demand for repair over the war-ranty period, EDR, is given by

$$EDR = \int_{0}^{L+W} \rho_r(t)dt \qquad (17.43)$$

Example 17.9

Consider the disc drive of Example 17.8. The manufacturer decides to minimally repair each failure. The expected repair rate is given by (17.42). Figure 17.6 shows $s(t)$ for $0 \le t \le L = 4$ and $\rho_r(t)$ for $0 \le t \le L + W$, for $W = 0.5$ and 1.0 with the same parameter values as in Exam-ple 17.8. For both values, the peak of $\rho_r(t)$ lags the peak of $s(t)$, which is to be expected since failures occur after the item has been in use for some time. Table 17.6 shows the expected de-mand for spares on an annual basis for years 1 through 5 and the total expected demand for repair (EDR).

On comparing Table 17.6 with Table 17.5, we see that the values (expected annual re-pairs) in Table 17.6 are always greater than the corresponding values (expected number of re-placements) in Table 17.5, as to be expected.

Let C_s, C_r, and C_h denote the unit manufacturing, average cost of each minimal repair, and the handling cost associated with each warranty claim. Then, repair is better than re-

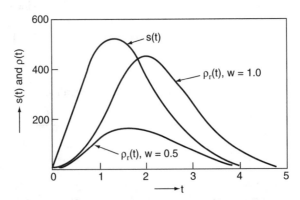

Figure 17.6. Plot of $\rho_r(t)$ and $s(t)$, Example 17.9.

Table 17.6. Annual sales and repairs (× 100) and *EDR* for *W* = 0.5 and 1

	W	0–1	1–2	2–3	3–4	4–5	ETS
				Years			
$\int s(t)dt$		303	463	198	36	0	
$\int \rho_r(t)dt$	0.5	49	148	88	21	1	307
$\int \rho_r(t)dt$	1.0	65	384	331	107	14	901

placement if 307 $(C_r + C_h) < 284 (C_s + C_h)$ for $W = 0.5$. This inequality changes to if 901 $(C_r + C_h) < 755 (C_s + C_h)$ for $W = 1.0$.

17.6 REDUNDANCY AND PRODUCT WARRANTY

17.6.1 Basic Relationship

Most products consist of several components and the reliability of the product is a function of component reliabilities. If a critical component (one whose failure results in product failure) has low reliability, then a large number of failures and claims under warranty can result. This leads to high warranty costs. One way of reducing the warranty cost is to build in redundancy for critical components, thereby improving the overall reliability of the product. Typically, redundancy involves replication of the critical components. This is possible only for certain components where incorporation of such replication is permissible by the functional design of the item. Building in redundancy results in greater manufacturing cost per item and this is justified only if the reduction in warranty costs exceed this increase. In Chapter 15, we discussed different types of redundancy. Two of these are considered in this section.[4]

We confine our attention to the case where the component is replaced by a module consisting of two identical units, with failure distribution $F(x)$ for each. We assume the following:

1. Failures are statistically independent
2. The critical component is nonrepairable so that failed items must be replaced by new ones
3. Time to replace is negligible, so that it can be treated as being zero
4. The product is sold with FRW policy with warranty period W

(*Comment:* The analysis of redundancy for other types of policies, for example PRW or combination, can be carried out in a similar manner.)

We focus our attention on the warranty costs resulting from the failure of the critical component and examine the implications of building in redundancy on the warranty cost and on the total cost.

With no redundancy, the expected number of component failures over the warranty period is given by $M_F(W)$ where $M_F(x)$ is the renewal function associated with the distribution function $F(x)$. If C_s is the component manufacturing cost per unit, then the expected cost to the manufacturer without redundancy is given by

$$E[C_m(W; \text{No Redundancy})] = C_s + (C_s + C_h)M_F(W) \tag{17.44}$$

where C_h is the handling cost associated with each warranty claim.

17.6.2   Hot Standby (Active Redundancy)

From (15.2), the failure distribution for the module, $G(x)$, is given by

$$G(x) = [F(x)]^2 \tag{17.45}$$

As a result, the expected number of warranty claims (due to failures of the critical component) over the warranty period is given by $M_G(W)$, where $M_G(x)$ is the renewal function associated with the distribution function $G(x)$. Each warranty claim costs the manufacturer $2C_s$ as opposed to only C_s in the case of no redundancy. The total expected cost to the manufacturer with redundancy is given by

$$E[C_m(W; \text{Redundancy})] = 2C_s + (2C_s + C_h)M_G(W) \tag{17.46}$$

The use of redundancy is justified only if

$$E[C_m(W; \text{Redundancy})] < E[C_m(W; \text{No Redundancy})] \tag{17.47}$$

where $E[C_m(W; \text{Redundancy})]$ is given by (17.46) and $E[C_m(W; \text{No Redundancy})]$ given by (17.44). Whether the inequality in (17.47) is satisfied or not depends on the failure distribution $F(x)$ and the various cost parameters. The following proposition characterizes whether building in redundancy is worthwhile or not.

Proposition 17.1: Building in redundancy is worthwhile if

$$\frac{C_h}{C_s} > \frac{1 + 2M_G(W) - M_F(W)}{M_F(W) - M_G(W)} \tag{17.48}$$

An implication of this is that redundancy is worthwhile if C_h/C_s is high.

The proof is trivial and follows by using (17.46) and (17.44) in (17.47) and minor rearrangement.

Example 17.10

Consider a complex electronic equipment with a critical component that is relatively cheap. Assume that the failure distribution of the critical component is $F(x)$ is exponential with mean time to failure $1/\lambda = 4$ years. With no redundancy, the expected number of claims over the warranty period is given by

$$M(W) = W\lambda$$

With redundancy, the density function of time between claims is

$$g(x) = 2\lambda(e^{-\lambda x} - e^{-2\lambda x})$$

As a result, the expected number of claims over the warranty period is given by

$$M_G(W) = \frac{2}{3}(W\lambda) - \frac{2}{9}(1 - e^{-3W\lambda})$$

where $G(\ )$ is the distribution function corresponding to g. If the item is relatively cheap to manufacture but the warranty handling (or administrative) cost is high relative to the manufacturing cost, then $C_h/C_s > 1$. Table 17.7 shows $E[C_m(W;\ \text{Redundancy})]/C_s$ for $C_h/C_s = 2$ (which is realistic, since the cost of the critical component is low relative to the handling cost) and a range of warranty periods W (years). For $W = 1$ year, building in redundancy is not economical. The same is true for $W = 2$ years. For $W = 3$ and 4 years, building in redundancy is worthwhile, since it results in a lower overall cost. ∎

17.6.3   Cold Standby (Passive Redundancy)

We consider the case where the switch is imperfect. As a result, from (15.12), the failure density function for the module is given by

$$h(x;\ q) = (1 - q)f(x) + qf(x)*f(x) \tag{17.49}$$

where $*$ is the convolution operator and q is the probability that the switch functions properly when needed ($q = 1$ corresponds to a perfect switch).

The expected number of claims due to critical component failures over the warranty period is given by $M_H(W;\ q)$ where $M_H(x;\ q)$ is the renewal function associated with the distribution function $H(x;\ q)$. The cost associated with each warranty claim is the cost of the module. This consists of two components, each costing C_s, and the switching mechanism, costing δC_s, for a total cost of $(2 + \delta)C_s$. As a result, the manufacturer's expected cost per unit sale is

$$E[C\ (W;\ \text{Redundancy})] = (2 + \delta)C_s + [(2 + \delta)C_s + C_h]M_H(W;\ q) \tag{17.50}$$

As before, building in redundancy is worthwhile if the inequality in (17.47) is satisfied with the left-hand side given by (17.50) and the right-hand side given by (17.44). Note that $M_H(W;\ q)$ decreases as q decreases. The following proposition characterizes whether building in redundancy is worthwhile or not.

Proposition 17.2: Building in redundancy is worthwhile if

$$\frac{C_h}{C_s} > \frac{(2 + \delta)[1 + M_H(W;\ q)] - [1 + M_F(W)]}{M_F(W) - M_H(W;\ q)} \tag{17.51}$$

The proof is trivial and hence omitted. The implication of this is that redundancy is worthwhile if C_h/C_s and q are large and δ is small.

Table 17.7. Expected warranty costs for Example 17.10

Warranty period	Expected warranty cost/C_s	
	No redundancy	Hot standby
1 year	1.750	2.198
2 years	2.500	2.643
3 years	3.250	3.205
4 years	3.822	3.135

Example 17.11

Let $F(x)$ be the exponential distribution with mean time to failure $1/\lambda = 4$ years. With redundancy, $h(x; q)$ is given by

$$h(x; q) = \lambda[(1 - q)(1 - x\lambda) + x\lambda]e^{-\lambda x}$$

Note that if $q = 1$ (perfect switch), then $H(x)$ is a gamma distribution with mean time to failure $= 2/\lambda$ (or 8 years).

The expected number of claims over the warranty period is given by

$$M_H(W; q) = \left(\frac{1}{1+q}\right)(W\lambda) - \left(\frac{q}{q+1}\right)^2(1 - e^{-(1+q)W\lambda})$$

Table 17.8(a) shows $E[C_m(W; \text{Redundancy})]/C_s$, with perfect switching ($q = 1$), for $C_h/C_s = 2$ and a range of warranty periods W (years) and δ. For $W = 1$ year and for all three values of δ, no redundancy is better than building in redundancy. For $W = 2$ and 3 years, we see that building in redundancy is worthwhile for $\delta = 0$ and 0.1 due to the advantage resulting from fewer warranty claims. When $\delta = 0.5$, this advantage is negated by the increase in the module price due to the high cost of the switch. For $W = 4$, the warranty duration is sufficiently large to make building in redundancy worthwhile for all the values of δ considered.

Tables 17.8(b) gives the results for the case $q = 0.8$ (imperfect switch). The results indicate that for $W = 1$ and 2, no redundancy is the better option for the three values of δ considered. For $W = 3$ and 4, building in redundancy is worthwhile for $\delta = 0$ and 0.1 and no redundancy is the better option for $\delta = 0.5$. The reason for this is the same as before, that is, the reductions in the expected number of failures is negated by the high cost of each replacement module.

On comparing the results of Tables 17.8(a) with Table 17.8(b), the interesting thing to note is that the expected warranty costs are larger for the latter case. This is to be expected

Table 17.8. Expected warranty costs for Example 17.11

(a) Perfect switch ($q = 1$)			
		Expected warranty cost/C_s	
Warranty period	δ	0.0	0.1	0.5
1 year		2.107	2.209	2.620
2 years		2.368	2.477	2.939
3 years		2.723	2.841	3.314
4 years		3.135	3.264	3.773

(b) Imperfect switch ($q = 0.8$)			
		Expected warranty cost/C_s	
Warranty period	δ	0.0	0.1	0.5
1 year		2.269	2.376	2.803
2 years		2.642	2.758	3.223
3 years		3.018	3.208	3.717
4 years		3.563	3.702	4.258

for obvious reasons. Also note that for $W = 2$ years and $\delta = 0$ or 0.1, building in redundancy is worthwhile when $q = 1$ (perfect switch) and no redundancy is better when $q = 0.8$ (imperfect switch) In other words, the advantage of redundancy is lost due to uncertainty in the switching operation. ■

17.7   EXTENDED WARRANTIES AND SERVICE CONTRACTS

The warranty that is an integral part of product sale is called the base warranty. It is offered by the manufacturer at no additional cost and is factored into the sale price. Extended warranty provides additional coverage over the base warranty and is obtained by the buyer by paying a premium. Extended warranties are optional warranties that are not tied to the sale process and can be either offered by the manufacturer or a third party (for example, several credit card companies offer extended warranties for products bought using their credit cards and some large merchants offer extended warranties).[5]

The expected cost to the provider of extended warranties can be calculated using models similar to those for the cost analysis of base warranties. The cost of extended warranty is related to product reliability and usage intensity. The reasons for purchase of extended warranties have been analyzed. Padmanabhan (1996) discusses alternate theories and the design of extended warranty policies.

For complex products, the buyer often lacks the expertise for its effective maintenance (both preventive and corrective) after expiration of the warranty. Service contracts allow for the provision of maintenance by an external agent (either the manufacturer or a third party). In a sense, service contracts are similar to warranty contracts, but the scope and coverage can vary and may be negotiated by the buyer and the service provider. The modeling and analysis of service contracts is similar to that given in Chapter 16, with additional features such as buyer options and the pricing of service contracts.

17.8   WARRANTY SERVICING

Offering warranty implies that the manufacturer incurs additional costs in the servicing of warranty. This cost can be minimized through optimal servicing strategies and effective warranty logistic management. In this section, we briefly discuss some of these issues.

Replace versus Repair

When a repairable item is returned to the manufacturer for repair under free replacement warranty, the manufacturer has the option of either repairing it or replacing it with a new one. The optimal strategy is one that minimizes the expected cost of servicing the warranty over the warranty period. In Chapter 18 we study a model that deals with this issue.

Cost Repair Limit Strategy

In general, the cost to repair a failed item is a random variable that can be characterized by a distribution function $H(z)$. Analogous to the notion of a failure rate, one can define a repair cost rate given by $\{h(z)/[1 - H(z)]\}$, where $h(z)$ is the derivative of $H(z)$. Depending on the form of $H(z)$, the repair cost rate can increase, decrease or remain constant with z. A decreasing repair cost rate is usually an appropriate characterization for the repair cost distribution (see, e.g., Mahon and Bailey, 1975). In Chapter 18 we discuss an optimal cost limit strategy to minimize the expected warranty servicing cost.

Inventory of Spares

As seen from Example 17.8, the expected demand for spares varies over the product life cycle. Carrying excess inventory results in higher servicing cost. With a small inventory, servicing under warranty can be delayed and this affects customer satisfaction. Thus the inventory level must be managed in an effective manner. This requires building models and a critical variable in doing so is product reliability. Many models have been developed for inventory management, but very few of these deal with inventory management in the context of product warranty.

Use of Loaners

In certain applications, it is critical that downtime not exceed some specified value. The warranty contract often specifies a penalty should this happen. One way for the manufacturer (or agent) to reduce the probability of this happening is to have a stock of loaners which are issued to the owners of failed items when they are undergoing repair. This implies additional servicing costs and the manufacturer must optimally decide on the number of loaners to be held in stock. Again, because of the complexity of the model needed to study this problem, the optimal number can only be determined by simulation studies. A simple model dealing with loaners can be found in Karmarkar and Kubat (1983).

Product Recall

Thus far the discussion has focused on warranty servicing where a failed item is returned for rectification under warranty. Occasionally, a manufacturer finds it necessary to recall either a fraction or all of the items sold, for some rectification action. The recall of only a fraction of the total production arises when items are produced in batches and some of the batches are defective due to inferior component(s) having been used and this is not detected under quality control. The manufacturer is held responsible for damages resulting from such defective components, either under express or implied warranty. A hypothetical example is the following. Due to poor quality of insulation, certain batches of a domestic appliance (e.g., an electric frying pan) are prone to result in an electrical shock under normal use. In this case, under the terms of implied warranty, the manufacturer can be held liable for damage caused by such defective items and it is more economical for the manufacturer to recall items from such batches than to pay the cost of lawsuits, etc.

A total recall situation usually arises because of poor design specifications that can lead to malfunction and serious damage under certain conditions and is discovered only after the items have been produced and sold. A hypothetical example of this is where the brakes of an automobile malfunction under certain conditions of driving. In such cases, the manufacturer can be held responsible for damages caused under the terms of warranty for fitness. Under the conditions discussed above, the manufacturer has the option of recalling the items, either for replacement of defective components or for replacement of one or more old components by newly designed ones. The optimal decision depends on many factors. Failure to act can result in huge payouts. On the other hand, any recall is not only costly, but can do serious damage to the reputation of the manufacturer and may affect sales for a long time. These all highlight the implications of the lack of adequate reliability in a product.[6]

NOTES

1. The literature on warranty is very large. There are many aspects to warranty. Researchers from different disciplines (engineering, accounting, marketing, statistics, operations research, economics, legal and many others) have examined different issues. Djamaludin et al. (1996) present a survey of this vast literature. Included are chapters written by experts from different disciplines and a bibliography listing 1500 references relating to warranties.

2. For more details of the different warranty policies, see Blischke and Murthy (1994).

3. Blischke and Murthy (1994) deal with the cost analysis of a variety of warranty policies. These include two-dimensional warranties and reliability improvement warranties.

The warranty cost analysis carried out in this chapter is based on expected values. A complete probabilistic characterization of costs is useful for risk analysis. Sahin and Potaloglu (1998) deal with such an analysis for the basic one-dimensional warranty policies.

4. Blischke and Murthy (1996) includes chapters on warranty and design, warranty and manufacturing and warranty servicing; many models that link warranty with product reliability can be found therein. Hussain (1997) deals with reliability improvement (through redundancy and/or reliability growth involving test–fix–test cycles) in the context of product warranty. Optimal improvement strategies are derived to minize the total (development + manufacturing + warranty servicing) costs.

5. Mathematical modeling and analysis of service contracts is a a relatively new area of research. Murthy and Ashgarizadeh (1998 and 1999) deal with a simple game theoretic model that may be used to obtain optimal strategies for the service agent. See also Ashgarizadeh (1997).

6. Blischke and Murthy (1996) include a chapter on warranty servicing along with additional references. Warranty servicing involves many other issues. One of these is the choice between repair and replacement of a failed item or component. A model dealing with this is discussed in Chapter 18.

For more on product recall, see Chandran and Lancioni (1981), Fisk and Chandran (1975), Min (1989), and Dardis and Zent (1982).

The management of reliability for new products requires taking into account the impacts of reliability and warranty costs on each other. The interaction between these implies that the two areas should be closely integrated in the formulation of new product strategy. Murthy and Blischke (1999) deal with this issue and the formulation of an effective warranty strategy that is integrated effectively with other technical and commercial strategies in the context of new product development.

EXERCISES

1. Suppose that item failure is given by a Weibull distribution. The item is covered by an FRW with a warranty period of $W = 1$ year. Compute the expected warranty cost for the following values for the shape parameter: 0.9, 1.0, 1.1, 1.3, 1.5, and 2.0. The scale parameter is to be selected in each case so that the MTTF is 2 years.

2. Repeat Exercise 1 assuming a gamma distribution with the same values for the mean and variance. (You will need to find the values of the shape parameter that will satisfy this condition.)

3. Repeat Exercises 1 and 2 for a linear PRW policy.

4. Summarize the results for the previous three exercises and comment. Would you expect much difference in the results if other distributions (e.g., the lognormal or inverse Gaussian) with the same mean and variance were used?

5. Suppose that a nonrepairable item is sold with the following warranty policy: The buyer receives a total refund should a failure occur in the interval $[0, W_1)$ and a partial refund (given by a linear function) for a failure occurring in the interval $[W_1, W)$. Determine the expected warranty cost.

6. Often, a buyer does not exercise a warranty claim. Suppose that an item is sold with a linear PRW policy with warranty period W. The probability distribution function for failure is given by $F(x)$. Let $q(t)$ denote the probability that a failure occurring at time t is not exercised. This is in general an increasing function of t for $0 \leq t \leq W$. Calculate the expected warranty servicing cost to the manufacturer and compare the results with the case $q(t) = 0$.

7. Formulate a two-dimensional warranty policy that ensures a minimum age and usage (given by W_1 and U_1, respectively) and a maximum age and usage (given by W and U, respectively). Compute the expected warranty cost to the manufacturer.

8. Suppose that the warranty region is a triangle with maximum age W and maximum usage U. Compute the expected warranty cost to the manufacturer.

9. The section on redundancy dealt with the critical component being replaced by a module involving two identical units. Suppose that the module consists of three identical units (in other words, one uses triplication as opposed to duplication). Derive conditions for determine whether or not building in redundancy is worthwhile.

10. In the use of redundancy, we assumed that all units used are identical. Because of manufacturing variability, some of the units may be nonconforming. In this case, the failure distribution of units can be modeled by a mixture of two distributions. Derive the expected warranty cost when the components of the mixture are exponential with an MTTF of 4 years for conforming items and 1 year for nonconforming items.

11. In Exercise 8, one can use burn-in to weed out a fraction of the nonconforming items. Develop a model to determine the effect of burn-in for a period τ and its effect on the expected warranty cost. (*Note:* You need to build a model for the cost of burn-in testing.)

12. Under what conditions is an extended warranty cost effective for the buyer?

13. Discuss the effect of cumulative warranties as opposed to noncumulative on buyer's and seller's costs per unit. How might these costs be modeled?

14. Calculate the buyer's indifference price in Example 17.2 for various values of L (say $L = 5$ to 20) and comment on the results.

15. Class discussion question: Comment on your own experiences with warranties and extended warranties.

CHAPTER 18

Reliability and Optimization

18.1 INTRODUCTION

Product reliability is of great interest to both manufacturers and buyers. From the manufacture's perspective, low product reliability has a detrimental impact on overall business performance. This impact can be minimized by properly managing reliability over the life cycle of the product. In Chapters 12–17, many models that dealt with reliability-related decisions during design, manufacture, and postsale service were developed and their impacts on various goals or objectives (operational and/or economic) were assessed. Lack of reliability in consumer durables affects buyer satisfaction. In the case of industrial and commercial products, it affects business performance of the buyer. Maintenance and service contracts help to reduce these negative impacts.

Decisions by both manufacturers of consumer durables and industrial and commercial products and buyers of industrial and commercial products must be made in a framework that integrates technical and commercial issues with the aim of optimizing a desired goal or objective. Some illustrative examples of such objectives, from the manufacturer's perspective, are as follows:

1. Optimal reliability allocation during the design stage to achieve a prespecified reliability target at minimum cost
2. A quality control program to achieve an optimal trade-off between the cost of testing and the savings in expected warranty cost
3. Optimal servicing strategy under warranty to minimize the expected warranty servicing cost

From the buyer's perspective, some optimization criteria are:

1. Optimum maintenance effort to achieve a trade-off between preventive and failure maintenance cost
2. Optimal spares inventory to ensure a prespecified reliability at minimum cost

The models developed in earlier chapters dealt with decision variables and their impact as defined through different objective functions. In this chapter, we focus our attention on the optimal choice of decision variables to optimize these objective functions. This involves building models for carrying out the optimization effort.[1]

619

The outline of the chapter is as follows. Section 18.2 deals with optimization during product design and development stages. We examine optimal reliability allocation and reliability development. Section 18.3 deals with optimization during the manufacturing stage. We look at optimal testing for effective quality control and optimal lot sizing. Sections 18.4 and 18.5 deal with optimization issues subsequent to the sale of the product. Section 18.4 deals with optimal maintenance at the component and system levels. Section 18.5 deals with optimal warranty decisions. This includes optimal warranty servicing strategies. In Section 18.6, we look at a simple model introduced in Section 12.11 and discuss an integrated approach to optimization at the design, operational and maintenance stages.

18.2   OPTIMIZATION DURING DESIGN AND DEVELOPMENT

One can categorize the optimization problems during design and development into two broad categories.

(a) **Narrow Context:** Here the focus is on technical aspects (reliability, MTTF, etc.) and the resulting economic issues (e.g., unit manufacturing cost). The problem is to optimize some performance measure (technical or economic) subject to any required constraints. Two such problems are discussed in Section 18.2.1.

(b) **Broader Context:** Here the reliability issues dealt with are optimization, taking into account commercial (sales) and postsale service (warranty) implications. It is again important that the optimization be done in a framework that integrates both technical and commercial aspects. We study a variety of such problems in Sections 18.2.2–18.2.5.

18.2.1   Optimal Reliability Allocation[2]

System reliability R is related to component reliabilities R_i, $1 \leq i \leq n$, through a structure function given by

$$R = \phi(R_1, R_2, \ldots, R_n) \tag{18.1}$$

This is the same as Equation (13.5) (with the argument T omitted for notational ease).

Let $C_i(R_i)$ denote the cost of Component i, which has reliability R_i. This is an increasing function of R_i, implying that the cost increases with increasing component reliability. It follows that the total system cost is given by

$$C = \sum_{i=1}^{n} C_i(R_i) \tag{18.2}$$

Define $\{R\}_1^n = \{R_1, R_2, \ldots, R_n\}$.

We discuss two problems. The first is that of determining the optimal reliability allocation so that the system reliability is at least $\overline{R}$ and the total system cost C is minimized. Thus we have the following optimization problem:

$$\min_{\{R\}_1^n} J = \sum_{i=1}^{n} C_i(R_i) \tag{18.3}$$

subject to the constraints

$$\phi(R_1, R_2, \ldots, R_n) \geq \overline{R} \tag{18.4}$$

and

$$0 \le R_i \le 1 \tag{18.5}$$

The second problem is to determine the optimal reliability allocation that maximizes the system reliability R subject to the total cost C not exceeding some pre-specified value $\overline{C}$. This results in the following optimization problem:

$$\max_{\{R\}_1^n} J = \phi(R_1, R_2, \ldots, R_n) \tag{18.6}$$

subject to

$$\sum_{i=1}^{n} C_i(R_i) \le \overline{C} \tag{18.7}$$

and (18.5).

We illustrate the solution to these problems by way of an example

Example 18.1

A four-component series system has the following component reliabilities: $R_1 = 0.90$, $R_2 = 0.95$, $R_3 = 0.93$, and $R_4 = 0.95$. The reliability of the system, gotten as the product of component reliabilities, is 0.7554. The problem is to determine the optimal reliability allocation so as to achieve a system reliability of $\overline{R} = 0.98$ at minimum cost. Suppose that the cost of improving the reliability of component i by an amount x_i is $\alpha_i x_i$, where $\alpha_i = \$10,000$ for each i.

The optimal solution $(x_1^*, x_2^*, x_3^*, x_4^*)$ is obtained by minimizing

$$J = \sum_{i=1}^{4} \alpha_i x_i$$

subject to

$$\prod_{i=1}^{4} (R_i + x_i) = \overline{R}$$

and

$$0 \le x_i \le 1 - R_i$$

for $1 \le i \le 4$.

Using the multiplier approach (see Appendix D), where the equality constraint is adjoined to the objective function J through a multiplier λ, we obtain the optimal solution by minimizing

$$L = \sum_{i=1}^{4} \alpha_i x_i - \lambda \left(\prod_{i=1}^{4} (R_i + x_i) - \overline{R} \right)$$

If the optimal solution is such that $(R_i + x_i) < 1$ (so that it is not on a boundary for the constraints), the optimal solution is obtained by solving

$$\frac{\partial L}{\partial x_i} = \alpha_i - \lambda \left(\prod_{\substack{j=1 \\ j \ne i}}^{4} (R_j + x_j) \right) = 0$$

for $i = 1, 2, 3$ and 4. This yields

$$\lambda = \left\{ \frac{\prod\limits_{i=1}^{4} \alpha_i}{\overline{R}^3} \right\}^{1/4}$$

and

$$x_i^* = \overline{R}^{1/4} \left\{ \prod\limits_{j=1}^{4} \frac{\alpha_j}{\alpha_i} \right\}^{1/4} - R_i$$

If one of the $x_i < 0$ [$> 1 - R_i$] then the value is set to zero [$1 - R_i$] and the optimization is carried out with respect to the remaining decision variables.

Using the numerical values, we have $x_1^* = 0.095$, $x_2^* = 0.045$, $x_3^* = 0.065$, and $x_4^* = 0.045$. Note that the component reliability after optimal allocation is same for all components. This is because the development cost is the same for all components. If the costs are different, then the component reliabilities after optimal allocation need not be the same for all components. ∎

18.2.2   Optimal Reliability Choice

When a product is sold with warranty, product reliability has a significant impact on the total cost. The unit manufacturing cost increases and the expected warranty cost decreases as product reliability increases. Optimal product reliability achieves a balance between these two costs.

Let J denote the number of components in each item. We first consider the single component case ($J = 1$) to illustrate the problem of reliability choice and later consider the multi-component case ($J > 1$).

Optimization for a Single Component
Let $F(x; \theta)$ be the product failure distribution. We first consider the case where the manufacturer has the option of manufacturing the component based on one of K different design choices. For the kth design choice, the parameter θ has a value θ_k, $1 \leq k \leq K$. Let $C_m(\theta_k)$ denote the unit manufacturing cost for the kth design choice. We assume that the K design choices are ordered in terms of decreasing reliability—that is, design choice j implies a less reliable product than design choice i if $j > i$, $1 \leq i, j \leq K$. As a result, $C_m(\theta_k)$ is a decreasing sequence in k, $1 \leq k \leq K$, implying that a less reliable product is also less expensive.

Let the expected warranty cost per unit with design choice k and a warranty period W be given by $\omega(W_k; \theta_k)$. This cost depends on the type of warranty policy and the servicing strategy used by the manufacturer. Expressions for $\omega(W; \theta)$ for different types of one-dimensional warranty policies are given in Chapter 17. The manufacturer's expected total cost per unit is given by

$$TC(\theta_k; W) = \omega(W; \theta_k) + C_m(\theta_k) \tag{18.8}$$

The optimal reliability design choice is given by k^*, the optimal k ($1 \leq k \leq K$), which minimizes $TC(\theta_k; W)$. Obtaining k^* requires an enumerative search, with $TC(\theta_k; W)$ evaluated for all values of k ($1 \leq k \leq K$).

When K is large, one can model the design option as that of choosing θ from a continuum over a specified interval. This allows one to use powerful optimization techniques to determine the optimal design. We propose a model based on the following assumptions:

1. θ is a scalar, with smaller values of θ corresponding to a more reliable product. (The extension to a vector θ is straightforward.) θ can take on any value in the interval $\theta^- \leq \theta \leq \theta^+$ with θ^+ and θ^- representing the limits for allowable and achievable reliability.

2. $C_m(\theta)$, the manufacturing cost per unit, is a continuous function of θ with $dC_m(\theta)/d\theta < 0$, so that θ increasing corresponds to a less reliable and, hence, less expensive item to produce.

As before, $\omega(W; \theta)$, the expected warranty service cost per unit, depends on the design parameter θ, the type of warranty and the warranty period W. The manufacturer's expected total cost per unit sale, $TC(\theta; W)$, is given by

$$TC(\theta; W) = \omega(W; \theta) + C_m(\theta) \tag{18.9}$$

The optimal reliability choice, θ^*, is the value of θ that minimizes $TC(\theta; W)$ subject to the constraint

$$\theta^- \leq \theta \leq \theta^+ \tag{18.10}$$

This is a nonlinear programming problem that can be solved by the usual multiplier method (see Appendix D), where the constraint (18.10) is adjoined to $TC(\theta; W)$ given by (18.9). Depending on the shape of $TC(\theta; W)$, θ^* can be either inside the constraint interval (i.e., $\theta^- \leq \theta \leq \theta^+$) or on the boundary.

Optimization of a Multicomponent System

Let the number of components in the product be J and let $F_j(t, \theta_j)$ denote the failure distribution of the jth component, with θ_j constrained by a relationship of the form given in (18.10). As in the earlier case, we assume that θ_j is a scalar variable with a smaller value implying a more reliable component. Let Θ denote the set $\{\theta_j, 1 \leq j \leq J\}$. The distribution for the product failure time can be obtained, in principle, as a function of Θ. Let $F(t, \Theta)$ denote this distribution. The unit manufacturing cost is a function of Θ. This can be obtained in terms of unit component costs, which, in turn, are functions of component reliabilities. Let $C_{mj}(\theta_j)$ denote the unit manufacturing cost for component j, with reliability parameter θ_j. The production cost per item is given by

$$C_m(\Theta) = \sum_{j=1}^{J} C_{mj}(\theta_j) \tag{18.11}$$

As mentioned earlier, the expected warranty cost depends on whether a failed component can be repaired or not, on the component reliabilities, and on the type and duration of the warranty. As a result, the expected warranty cost per item sold is a function of Θ. Let $\omega(W; \Theta)$ denote this cost. Expressions for $\omega(W; \Theta)$ for different types of one-dimensional warranty policies are given in Chapter 17. The manufacturer's expected total cost per unit sale, $TC(\Theta; W)$, is given by

$$TC(\Theta; W) = \omega(W; \Theta) + C_m(\Theta) \tag{18.12}$$

Θ^*, the optimal Θ, is the value that minimizes $TC(\Theta; W)$] subject to the constraints on the θ_j's.

Components in Series: Exponential Lifetimes

Nguyen and Murthy (1988) discuss the following situation. The product in question consists of J components connected in series. This implies that the item fails whenever one of its components fails. We assume that (1) component failures are statistically independent, (2) the probability of two or more components failing at the same time is zero and, (3) the failure distributions for all components are exponential, with the distribution of time to failure of the jth component given by

$$F_j(x) = 1 - e^{-\theta_j x} \tag{18.13}$$

The mean time to failure for component j is $(1/\theta_j)$. We assume that the θ_j are constrained to satisfy

$$\theta_j^- \le \theta \le \theta_j^+ \tag{18.14}$$

Note that the failure rate of the item is a constant given by

$$r = \sum_{j=1}^{J} \theta_j \tag{18.15}$$

As part of the design specification, we want to ensure that the failure rate r is not smaller than a prespecified value γ. This ensures a minimum reliability for the product.

Suppose the item is sold with a free replacement warranty with warranty period W. We examine two cases.

Case 1: (Repairable Items). Suppose that components are not repairable, but that failed or defective components are replaceable. In this case, the manufacturer can repair failed items by replacing only the failed component(s). If the failure of the item is due to the failure of component j, the failed item is made operational by replacing the failed component. The cost of this is $C_{mj}(\theta_j) + C_h$, where $C_{mj}(\theta_j)$ is the cost of the new component and C_h is the average extra cost incurred in handling the warranty claim. Since components fail independently and component failure distributions are exponential, the expected number of component j failures, $1 \le j \le J$, over the warranty period is given by $W\theta_j$. As a result, the expected cost of servicing the warranty per unit sale is given by

$$\omega(W; \Theta) = \sum_{j=1}^{J} \{W\theta_j[C_{mj}(\theta_j) + C_h]\} \tag{18.16}$$

From (18.12) we have

$$TC(\Theta; W) = \left\{ \sum_{j=1}^{J} W\theta_j[C_{mj}(\theta_j) + C_h] \right\} + C_m(\Theta) \tag{18.17}$$

where $C_m(\Theta)$ is given by 18.11.

Case 2: (Nonrepairable Items). Suppose that failed items cannot be repaired by replacing failed components with new ones, or that such repairs would cost more than a new item. In

Table 18.1. Model parameter values for Example 18.2

j	1	2	3	4
a_j	0.20	0.24	0.30	0.40
d_j	3.00	2.50	2.00	1.50
b_j	8.40	8.64	8.80	8.87
θ_j^-	0.00	0.00	0.00	0.00
θ_j^+	0.50	0.50	0.50	0.50

this case, whenever an item fails, the entire item must be replaced. This is done at a cost $C_m(\Theta)$. The item failures are assumed to occur as in Case 1. As a result, we have

$$TC(\Theta; W) = \left\{ \sum_{j=1}^{J} W\theta_j \right\} [C_m(\Theta) + C_h] + C_m(\Theta) \tag{18.18}$$

where $C_m(\Theta)$ is given by 18.11.

Example 18.2
We consider $J = 4$ and the following form for $C_{mj}(\theta_j)$:

$$C_{mj}(\theta_j) = b_j + a_j(\theta_j)^{-d_j}$$

with $a_j > 0$, $b_j > 0$ and $d_j > 1$. This implies that as θ_j decreases (so that the component becomes more reliable), the unit manufacturing cost increases. The cost parameters and the limits on the θ_j are shown in Table 18.1. The parameters were selected so that (1) the cost per unit with minimum reliability for each component is $50, and, (2) the cost per unit for component j with reliability $(\theta_j^+ - \delta\theta)$, for a given $\delta\theta$, is decreasing in j. The warranty period is $W = 1$ year and C_h is taken to be $50. We consider two values of γ (the prespecified minimum value for the failure rate), $\gamma = 1$ per year and $\gamma = 2$ per year. The former corresponds to a product with greater reliability. Table 18.2 shows the optimal reliability allocation. The optimal allocation of reliability to different components is shown in Figures 18.1 and 18.2 for the repairable and the nonrepairable cases.

Table 18.2. Optimal reliability allocation

			Case 1: Repairable Items					
γ	θ_1^*	θ_2^*	θ_3^*	θ_4^*	$\Sigma\theta_j^*$	$C_m(\Theta^*)$	$\omega(W; \Theta^*)$	$TC(\Theta^*; W)$
2	0.260	0.334	0.398	0.454	1.446	44.9	30.6	75.5
1	0.162	0.224	0.281	0.333	1.000	60.0	22.4	82.4

			Case 2: Nonrepairable Items					
γ	θ_1^*	θ_2^*	θ_3^*	θ_4^*	$\Sigma\theta_j^*$	$C_m(\Theta^*)$	$\omega(W; \Theta^*)$	$TC(\Theta^*; W)$
2	0.219	0.282	0.338	0.387	1.226	49.4	72.9	122.3
1	0.168	0.226	0.279	0.327	1.000	60.0	65.9	125.9

$\Theta = [\theta_1, \theta_2, \theta_3, \theta_4]$.

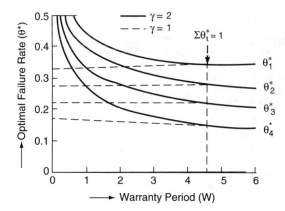

Figure 18.1. Optimal failure rate allocation for repairable products.

The following observations can be made:

1. The optimal cost per unit for $\gamma = 1$ is greater than that for $\gamma = 2$. This is to be expected since the design requirements specify a more reliable product when $\gamma = 1$.

2. For both cases, the constraint is tight when $\gamma = 1$ and W is small. Furthermore, whenever the constraint is not tight, θ_j^* decreases with W. This is to be expected, since W increasing corresponds to a longer warranty period and, to reduce the expected warranty cost, the reliability must improve or equivalently, θ_j^* must decrease.

3. For case (i), individual θ_j^* may increase or decrease with W when the constraint is tight. This is again to be expected since the warranty cost depends on each θ_j. In contrast, for case (ii), the individual θ_j^* do not change with W when the constraint is tight because the warranty cost depends on $\Sigma \theta_j$ as opposed to individual θ_j. ∎

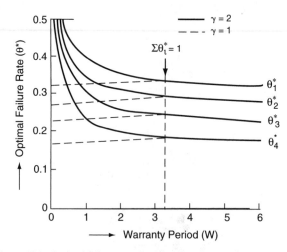

Figure 18.2. Optimal failure rate allocation for nonrepairable products.

18.2.3 Optimal Reliability Development

One way of improving product reliability is through research and development involving a process of test–fix–test–fix iterations. During this process, the product is tested until a failure mode appears. Design and/or engineering modifications are then made as attempts to eliminate the failure mode and the product is tested again. As this continues, the reliability of the product improves and this reduces the expected warranty cost. This development program costs money. Optimal reliability development achieves a trade-off between these two costs.

We consider a model where, during the development period, failures, and subsequent modifications to eliminate them through design improvements, are assumed to occur according to a nonhomogeneous Poisson process with intensity $\nu(t)$. $\nu(t)$ is called the modification rate. $\nu(0)$ is the failure rate before the start of the development program and $\nu(t)$ decreases with increasing development time.

The number of modifications made over $[0, t)$, $N(t)$, is a random variable with

$$E[N(t)] = \Lambda(t) = \int_0^t \nu(x)dx \qquad (18.19)$$

If the development program is stopped at time T_d, then the failure rate for the product is given by

$$\theta = \nu(T_d) \qquad (18.20)$$

Since $\nu(t)$ is a nonincreasing function of t, larger values of T_d imply smaller failure rates. In other words, the longer the duration of product development, the smaller the failure rate for the product at the end of the development program.

Various forms for $\nu(t)$ have been proposed. We consider a model proposed by Crow (1974), namely

$$\nu(t) = \lambda \alpha t^{\alpha-1} \qquad (18.21)$$

where $\lambda > 0$ and $0 < \alpha < 1$. In the literature on reliability growth, this model is called the nonhomogeneous Poisson-process, reliability-growth model. We assume the following costs for development:

1. C_d: The cost per unit time of running the development program
2. C_a: The expected cost of each design modification to fix a failure mode

Let $N(T_d)$ denote the number of design modifications made during the development period. Then the cost of the development program is given by

$$C_d T_d + C_a N(T_d) \qquad (18.22)$$

Note that $N(T_d)$ is a random variable. The failure rate at the end of the development program is given by (18.20).

We assume that: (1) the product is sold with a free replacement (FRW) policy with warranty period W and, (2) the item is repairable and the manufacturer always chooses to minimally repair items which fail under warranty. Let C_r denote the expected cost of each repair.

Table 18.3. Optimal product development—T_d^* versus W and α

α	W (years)	T_d^* (weeks)	$\theta(T_d^*)$ (per year)	$[TC(T_d^*; W)]$ ($\times 100$)
0.4	3	50	0.038	213
	5	71	0.031	281
	7	89	0.027	339
0.5	3	59	0.065	331
	5	87	0.054	448
	7	112	0.047	549

Finally, let Q denote the total number of items sold. Then the expected warranty cost $\omega(W; \theta)$ is given by

$$\omega(W; \theta) = C_r Q \theta W \tag{18.23}$$

Let $TC(T_d; W)$ be the total expected cost to the manufacturer. Then from (18.22) and (18.21) we have

$$TC(T_d; W) = C_d T_d + C_a \Lambda(T_d) + C_r Q W \nu(T_d) \tag{18.24}$$

T_d^*, the optimal T_d, is the value of T_d that minimizes $TC(T_d; W)$.

Example 18.3

Suppose that the product in question is an industrial product sold with a free replacement warranty. Let the modification rate $\nu(t)$ be given by (18.15) with $\lambda = 1$ per week, with time during development measured in weeks. Let $C_a = \$1000$, $C_d = \$100/\text{day}$, $C_r = \$100$ and $Q = 1000$. Table 18.3 gives T_d^*, the optimal development period (in weeks), and the optimal expected cost to the manufacturer, $TC(T_d^*; W)$, for different combinations of W (in years) and α. Note that for a given α, as W increases, the optimal development period also increases. This is to be expected since one needs a more reliable product to reduce the expected warranty cost. For a given W, as α increases, again the development period increases. ∎

18.3 OPTIMIZATION DURING MANUFACTURING

During manufacturing, the reliability of a product generally degrades, in the sense that some of the items do not conform to the design specification. In Section 13.7, we discussed models that characterized the occurrence of nonconforming items for both continuous and batch production. The performance of a nonconforming item is inferior to that for a conforming item. As a result, the warranty and maintenance costs for such an item is higher. In Section 13.8, we discussed different approaches to controlling reliability degradation during manufacturing. In this section, we look at optimal decision making for control of this degradation.[3]

18.3.1 Optimal Testing

In Section 13.10.1, we discussed the use of burn-in to weed out nonconforming items for continuous production manufacturing. We assumed that the process is in steady state. Each item produced is conforming with probability p and nonconforming with probability $(1 - p)$. Let $F(t)$ denote the failure distribution function of a conforming item and $H(t)$ that of a nonconforming item.

The scheme to weed out defective items involves testing items for T units of time. Items that fail are discarded and the rest are released for sale. Let $\omega(W; T)$ denote the expected warranty cost per item when items are released after testing. Note that testing reduces the expected warranty cost but this is done at the expense of increased manufacturing cost per item. The problem is to determine the optimal burn-in time that achieves a balance between these two costs.

Since all items are statistically similar, either one tests all items or none. The choice between the two depends on the savings in expected warranty cost versus the increased manufacturing cost per item due to testing and scrapping or reworking of items which fail during the testing period. Let $J(T)$ denote the asymptotic total (manufacturing + warranty) cost per item when items are tested for a period T.

If no testing is done (i.e., $T = 0$), then the asymptotic total (manufacturing + warranty) cost per item sold is given by

$$J(0) = C_m + \omega(W; 0) \tag{18.25}$$

where C_m is the unit manufacturing cost and $\omega(W; 0)$ is the asymptotic warranty cost per unit. $\omega(W; 0)$ depends on the type of warranty, the warranty duration (W), failure distributions $F(t)$ and $H(t)$ and p, the probability that an item is conforming.

If all items are tested for a period T, then the manufacturing cost per lot of size L is $(C_m + C_i)L$ where $C_i = \phi_0 + \phi_1 T$ is the cost of testing an item for period T. ϕ_0 represents the fixed cost and ϕ_1 the variable cost. The expected number of items per lot which do not fail during testing is given by $[1 - \nu(T)]L$, where

$$\nu(T) = p\{1 - F(T)\} + (1 - p)\{1 - H(T)\} \tag{18.26}$$

is the probability that an item picked randomly will survive the test. Let C_d denote the cost of disposing of each failed item. As a result, the asymptotic total (manufacturing + testing + disposing) cost per item released is given by

$$\gamma(T) = \frac{C_m + C_i + \nu(T)C_d}{1 - \nu(T)} \tag{18.27}$$

The failure distributions $F_1(t)$, for a conforming item, and $H_1(t)$, for a nonconforming item, that survive burn-in for a period T, are given by (13.28) and (13.29) respectively. The asymptotic probability, p_1, that an item released after burn-in is conforming is given by (13.26). Note that $p_1 > p$ and that p_1 increases with T, since $H(T)/F(T)$ decreases as T increases. The asymptotic (manufacturing + testing + disposing + warranty) cost per item is given by

$$J(T) = \gamma(T) + \omega(W; T) \tag{18.28}$$

where $\omega(W; T)$ is the asymptotic warranty cost per item. $\omega(W; T)$ depends on the type of warranty, the warranty duration (W), failure distributions $F_1(t)$ and $H_1(t)$, and p_1, the asymptotic probability that an item released is conforming.

Testing is the optimal strategy if $J(T) > J(0)$ for some $T > 0$ and no testing is the optimal strategy if $J(T) < J(0)$ for all $T > 0$. In the former case, the optimal testing period, T^*, is the value of T that minimizes $J(T)$.

FRW Policy with Minimal Repair

Let C_r denote the cost of each repair. Since failures are repaired minimally and the time to repair is negligible, the expected warranty cost (see Section 17.4.3) is given by

$$\omega(W; 0) = C_r\left[p\int_0^W r_f(t)dt + (1-p)\int_0^W r_h(t)dt\right] \tag{18.29}$$

and

$$\omega(W; T) = C_r\left[p_1\int_0^W r_{1f}(t)dt + (1-p_1)\int_0^W r_{1h}(t)dt\right] \tag{18.30}$$

where $r_f(t)$ and $r_h(t)$ are the failure rates associated with $F(t)$ and $H(t)$, respectively, and, $r_{1f}(t)$ and $r_{1h}(t)$ are the failure rates associated with $F_1(t)$ and $H_1(t)$, respectively.

Consider the special case where $F(t)$ and $H(t)$ are exponential distributions with failure rates λ_1 and λ_2, respectively, with $\lambda_2 > \lambda_1 > 0$. This characterization is often used for modeling failures of electronic systems (with either the whole system or a subsystem being the item under consideration) and here the failure of an item is not dependent on its age. In this case, the choice between testing and no testing is given by the following propositions. We omit the proofs; they can be found in Murthy et al. (1993).

Proposition 18.1

Testing is the optimal strategy if

$$p(1-p)(\lambda_2 - \lambda_1)[H(T) - F(T)] > \frac{[1 - \nu(t)](C_m + C_d) + C_i}{WC_r}$$

for some $T > 0$.
Comments:

1. The inequality is more likely to be satisfied if (1) C_m, C_d, and C_i are small relative to WC_r, (2) $(\lambda_2 - \lambda_1)$ is large and, (3) p is close to 0.5 (i.e., 50% defectives!).
2. If the inequality is satisfied for $p = p_c$, then it is also satisfied for $p_c \le p \le (1 - p_c)$.

Proposition 18.2

No testing is the optimal strategy if

$$C_m + C_d + C_i > WC_r p(1-p)(\lambda_2 - \lambda_1)$$

Comments:

1. The inequality is more likely to be satisfied if (i) C_d, C_m, and C_i are large relative to WC_r, (ii) $(\lambda_2 - \lambda_1)$ is small and, (iii) p is close to 0. This is just the reverse of Proposition 1.
2. The inequality is always satisfied when p is very close to one. In this case, since the majority of the items are defective, there is no advantage in doing 100% testing to weed the defectives out. The more sensible option is to scrap the whole batch.
3. If the inequality is satisfied for $p = p_c$, then it is also satisfied for all $p \le p_c$.

Example 18.4

Suppose that $F(t)$ and $H(t)$ are exponential distributions with parameters $\lambda_1 = 0.2$ and $\lambda_2 = 4.0$ per year, respectively. This implies that the mean time to failure is 5 years for conforming items and 0.25 year for nonconforming items. Let $L = 100$ and the remaining values of the model parameters be as follows: $p = 0.75$, $C_m = \$300$, $\phi_1 = 0.0$, $\phi_2 = \$45/\text{year}$, $C_d = \$5$, and $C_r = \$100$. $p = 0.75$ implies that 25% of the items are defective. This is not an unusual figure for certain electronic systems using integrated chips where the fraction of defectives can be unusually high.

The above situation corresponds to the case where the item is an electronic system composed of three modules, each costing roughly the same. Item failures occur due to one of the modules failing. As a result, a failed item is rectified by replacing the failed module by a new one. Hence, the cost of each replacement is roughly one-third the cost of a new item.

Table 18.4 shows T^* and $J(T^*)$ for four different warranty periods ranging from 1 to 4 years. Note that $T^* = 0$ implies that no testing is the optimal strategy and $T^* > 0$ implies that testing is the optimal strategy. For $W = 1$, no testing is the optimal strategy. For $W \geq 2$, testing is the optimal strategy. Figure 18.3 shows T^* versus W. Note that T^* is zero for $W \leq 1.25$ and increases with W for $W > 1.25$. The reason for this is as follows. Since all items released for sale are rectified through minimal repair, the warranty servicing cost increases rapidly with W for any nonconforming item released. As a result, as the warranty period increases, longer testing is needed to reduce the number of nonconforming items being released. This is seen more clearly in Table 18.5 where, as W increases, T^* increases and the expected number of nonconforming items released decreases.

The optimal testing period varies from 0.2259 year for $W = 2$ years to 0.5343 year for $W = 4$ years. Obviously, it is not possible to test items for such long periods. However, life testing is usually carried out in an accelerated manner, as discussed in Chapter 14. Under accelerated life testing, the item is subjected to a harsher environment that hastens the aging process. As a result, testing for one unit of time in the accelerated mode corresponds to β (> 1) units of testing under normal conditions. This implies that testing for T/β units of time in the accelerated mode corresponds to testing for T units under normal conditions. If the testing in accelerated mode is done with $\beta = 50$, testing for 0.5343 year in the normal mode would require accelerated testing for only approximately 4 days. This practical solution corresponding to the optimal period may be attainable.

Figure 18.4 shows the influence of p on T^*, with the remaining parameters held at their nominal values. Note that $T^* = 0$ for small p (close to 0) and large p (close to 1) as to be expected from Proposition 18.1. For $W = 1$, T^* is zero for all values of p. For $W = 2$, $T^* > 0$ for $0.48 < p < 0.92$. Over this range, T^* first increases and then decreases as shown in the figure. The reason for this is as follows. When $p = 1$, no item is nonconforming and hence there is no need for testing. As p decreases (for $0.92 \leq p \leq 1$), the fraction of nonconforming is sufficiently small so that testing to weed out nonconforming items is not jus-

Table 18.4. T^* and $J(T^*)$ versus W

W	T^*	$J(T^*)$
1	0.0000	415.00
2	0.2259	514.30
3	0.4231	567.13
4	0.5343	606.47

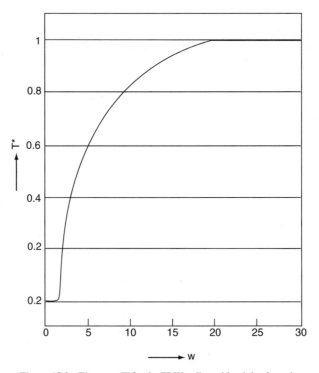

Figure 18.3. T^* versus W for the FRW policy with minimal repair.

tified and, as a result, $T^* = 0$. As p decreases still further (for $0.48 \leq p < 0.92$), the fraction of nonconforming items increases and testing is worthwhile. Note that testing for greater time reduces the warranty cost per item but increases the manufacturing cost per item as a greater fraction of items fail during testing and are discarded. The reason that T^* increases as p decreases in the interval $0.69 \leq p < 0.92$ is that the increase in the manufacturing cost per item is less than the decrease in the warranty cost as p decreases. In the interval $0.48 \leq p < 0.69$, T^* decreases with p decreasing as the increase in the manufacturing cost per item is more than the reduction in the warranty cost per item. Finally, for $0 < p \leq 0.48$, the fraction of nonconforming items is very high. The cost of any testing is not

Table 18.5. Expected number of nondefective and defective items per batch

	Expected number of items per batch of 100 items					
	Weeded out during testing			Released for use		
W	Nondefective	Defective	Total	Nondefective	Defective	Total
1	0.000	0.000	0.000	75.000	25.000	100.000
2	3.317	14.872	18.185	71.687	10.128	81.815
3	6.085	21.397	26.282	68.915	4.603	73.718
4	7.601	22.050	29.651	67.399	2.950	70.349

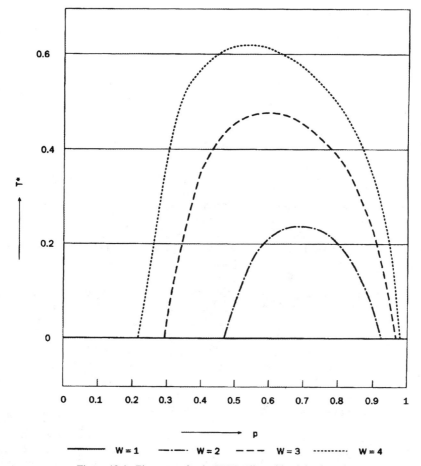

Figure 18.4. T^* versus p for the FRW policy with minimal repair.

worth the reduction in the warranty cost and hence $T^* = 0$. As W increases, the interval over which $T^* > 0$ also increases and T^* has a shape similar to that for $W = 2$ except that the values for any given p are increasing with W. This is to be expected for reasons discussed earlier. ∎

Pro-rata Warranty (Linear Pro-ration)
Under this policy, the manufacturer agrees to refund a fraction of the original sales price S to the consumer if the item sold fails within the warranty period $[0, W]$. From the results from Section 17.4.4, the asymptotic warranty cost per item with linear pro-ration and no testing is

$$\omega(W; 0) = p\left[kS\int_0^W\left(1 - \frac{t}{W}\right)f(t)dt + C_h F(W)\right] + (1 - p)\left[kS\int_0^W\left(1 - \frac{t}{W}\right)h(t)dt + C_h H(W)\right]$$

(18.31)

and with testing, it is given by

$$\omega(W; T) = p_1 \left[kS \int_0^W \left(1 - \frac{t}{W}\right) f_1(t) dt + C_h F_1(W) \right] + (1 - p_1) \left[kS \int_0^W \left(1 - \frac{t}{W}\right) h_1(t) dt + C_h H_1(W) \right]$$

(18.32)

where, $F_1(t)$, $H_1(t)$, and p_1 are given by (13.28), (13.29), and (13.26), respectively.

Example 18.5

As in Example 18.4, let $F(t)$ and $H(t)$ be exponential distributions with failure rates λ_1 and λ_2, respectively. Let $\lambda_1 = 0.2$ per year, $\lambda_2 = 4.0$ per year, $C_h = \$20$, $k = 1.0$ and $S = \$1000$. The remaining parameter values are the same as in Example 18.4.

Table 18.6 shows T^* and $J(T^*)$ for the four different warranty periods. For $W = 1$ to 4, $T^* > 0$, implying that testing is the optimal strategy in all four cases. Figure 18.5 shows T^* as a function of W. For $W < 0.5$, $T^* = 0$. This is to be expected since small W implies smaller warranty service cost, even with all nonconforming items released, and the cost of testing is not justified. For $0.5 < W < 6.0$, $T^* > 0$, in contrast to the FRW policy with minimal repair (Example 18.4), here T^* increases with W for $0.5 < W < 1.7$ and decreases for $1.7 < W < 6$. For $0.5 < W \leq 1.7$, the testing period increases with W because the warranty cost for released nonconforming items increasingly dominates the cost per item released. Thus by reducing the number of released nonconforming items through increased testing, the warranty cost can be decreased by more than the increase in the manufacturing cost per item released. For $1.7 < W < 6.0$, the influence of the warranty cost for released nonconforming items decreases while the influence of the other costs increases, resulting in the testing period decreasing as W increases. This continues until, for $W > 6.0$, testing is again not justified, so that $T^* = 0$. Table 18.7 shows the expected number of nonconforming items released and weeded out in each batch. ∎

18.3.2 Optimal Lot Sizing

Model 2 of Section 13.7.2 dealt with reliability degradation when items are produced in lots. We consider the special case where $p_{in} = 1$ (all items produced are conforming when the process is in control) and $p_{out} = 0$ (all items produced are nonconforming when the process is out of control).

Before a lot production starts, the process is checked to make sure that it is in control. This costs an amount C_f if the process is in-control and $C_f + \eta$ ($\eta \geq 0$) if it is out of control. η is the cost of bringing an out-of-control process to in-control. As a result, the asymptotic manufacturing (fixed + variable) cost per item is given by

$$\gamma(L) = C_m + \frac{C_f + (1 - q^L)\eta}{L}$$

(18.33)

Table 18.6. T^* and $J(T^*)$ versus W

W	T^*	$J(T^*)$
1	0.1539	560.03
2	0.1871	650.45
3	0.1565	720.22
4	0.1090	778.35

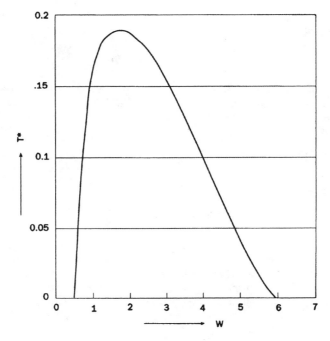

Figure 18.5. T^* versus W for the PRW policy.

where C_m is the variable cost, including material and labor, of producing one unit and q is the probability that the process remains in control.

The expected number of conforming items in a batch of size L is given by (13.23) with $p_{in} = 1$ and $p_{out} = 0$. As a result, the asymptotic probability that an item is conforming is given by

$$\phi = \frac{E[K]}{L} = \frac{q(1 - q^L)}{L(1 - q)} \tag{18.34}$$

The asymptotic warranty cost per item, $\omega(W, L)$, depends on the type of warranty, the warranty duration W, the failure distributions of conforming items $F(t)$ and nonconforming

Table 18.7. Expected number of nondefective and defective items per batch

	Expected number of items per batch of 100 items					
	Weeded out during testing			Released for use		
W	Nondefective	Defective	Total	Nondefective	Defective	Total
1	2.273	11.490	13.763	72.727	13.510	86.237
2	2.755	13.172	15.927	72.245	11.828	84.073
3	2.311	11.633	13.944	72.689	13.367	86.056
4	1.617	8.834	10.451	73.383	16.166	89.549

items $H(t)$, and the probability ϕ that an item is conforming. The asymptotic total (manufacturing + warranty) cost per item is given by

$$J(L) = \gamma(L) + \omega(W, L) \tag{18.35}$$

As L increases, $\gamma(L)$ decreases and ϕ increases. As a consequence, $\omega(W, L)$ increases. L^*, the optimal L, is the value of L that minimizes $J(L)$ and achieves a trade-off between the unit manufacturing cost and the asymptotic warranty cost.

FRW Policy (Minimal Repair)

Suppose that the items are sold with a FRW policy. Whenever an item fails under warranty, it is minimally repaired. In this case, the asymptotic warranty cost per item is given by

$$\omega(W, L) = C_r \left[\phi \int_0^W r_f(x)dx + (1 - \phi) \int_0^W r_h(x)dx \right] \tag{18.36}$$

where ϕ is given by (18.34), C_r is the cost of each repair and $r_f(t)$ and $r_h(t)$ are the failure rates associated with $F(t)$ and $H(t)$ respectively..

L^* is obtained by minimizing $J(L)$ given by (18.35) with $\gamma(L)$ given by (18.33) and $\omega(W, L)$ given by (18.36). In general, a computational scheme must be employed to obtain L^*. However, if q is very close to 1, then q^L can be approximated as

$$q^L \approx 1 + \log(q)L + \{\log(q)L\}^2 \tag{18.37}$$

and in this case, L^{**}, an approximation to L^* obtained by solving $dJ(L)/dL = 0$, is given by

$$L^{**} = q \left(\frac{2C_f}{(1 - q)\left(C_r \left[q \int_0^W r_h(x)dx - \int_0^W r_f(x)dx \right] - \eta \right)} \right)^{1/2} \tag{18.38}$$

For further details, see Djamaludin et al. (1994).

Example 18.6

Suppose that the failure distributions of both conforming and nonconforming items are exponential with failure rates $\lambda_1 = 0.1$ and $\lambda_2 = 1.0$, respectively. This implies that the mean time to failure is 10 years for a conforming item and 1 year for a nonconforming item.

Let L_m denote the upper limit on L. Take $L_m = 100$, and the nominal values for the remaining parameters to be $q = 0.99$, $C_m = \$5.00$, $C_f = \$50.00$, $\eta = \$10.00$, $C_f = \$5.00$. Values for the warranty period W, ranging from 1 to 4 years, are considered. Also included is the case $W = 0$, which corresponds to the product being sold with no warranty.

Table 18.8 shows L^* and $J(L^*)$, obtained by evaluating $J(L)$ for $L = 1, \ldots, L^*$; L^{**}, obtained from (18.38); $J(L^{**})$, and $J(L)$, the cost per unit if the lots are of size $L_m = 100$. The percentage reduction in cost, RC, given by

$$RC = 100 \left(\frac{J(L_m) - J(L^*)}{J(L_m)} \right)$$

is also shown in Table 18.8.

Table 18.8. Exact and approximate optimal lot sizes and related costs (FRW Policy)

W	0	1	2	3	4
L^*	100	58	38	30	26
$J(L^*)$	5.5634	7.8439	9.5526	11.0515	12.4429
L^{**}	100	47	33	27	23
$J(L^{**})$	5.5634	7.8739	9.5758	11.0701	12.4653
$J(L_m)$	5.5634	8.0383	10.5131	12.9879	15.4628
RC (%)	0.00	2.42	9.14	14.91	19.53

For $W = 0$, $L^* = L_m$. As W increases, L^* decreases, since a longer warranty period implies increased warranty costs for nonconforming items released. Hence, smaller lot sizes are required to ensure that the expected fraction of nonconforming items released is smaller. The percentage reduction, RC, obtained by using L^*, increases with W, indicating that lot sizing becomes more critical as the warranty period increases. Note that the error between L^* and L^{**} decreases as W increases and that L^{**} is always less than L^*, indicating that the true optimal lot size is somewhat larger than that given by the approximation. Also note, however, that the increase in cost using the suboptimal result is less than 1%.

The effect of q on L^*, holding the remaining parameters at their nominal values, is shown in Figure 18.6. Note that there is a critical value q_c, dependent on W, such that, for $q < q_c$, $L^* = L_m$ and for $q_c \leq q < 1$, $L^* < L_m$. Firstly, if $q = 1$, then the process is always in control. Re-

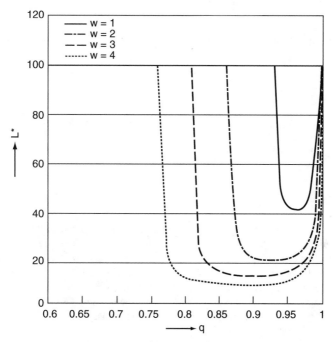

Figure 18.6. L^* versus q for the FRW with minimal repair.

ducing the lot size increases the manufacturing cost per item with no effect on the warranty cost per item and so $L^* = L_m$. Now, as q decreases with L kept at L_m, the expected fraction of nonconforming items in a lot increases, resulting in a higher warranty cost per item. By decreasing L as well, the warranty cost can be reduced but the manufacturing cost per item will then increase. Decreasing L is justified if the reduction in warranty cost compensates for the increase in manufacturing cost. This occurs, so that L^* decreases as q decreases until some value (depending on W) is reached. At this point the increase in manufacturing cost is greater than the decrease in the warranty cost as q decreases, resulting in higher values for L^*. Consequently, L^* increases to the maximum L_m as q decreases from this point and then, for $q <$ q_c, L^* is constrained to be the maximum L_m. Note that q_c decreases as W increases; this is to be expected since the warranty costs increase as W increases. ■

PRW Policy (Linear Rebate)
Here, whenever an item fails under warranty, a fraction of the sale price S is refunded. Under linear rebate, the asymptotic warranty cost per unit is given by

$$\omega(W, L) = \phi \int_0^W \left[S\left(1 - \frac{x}{W}\right) + C_h \right] f(x)dx + (1 - \phi) \int_0^W \left[S\left(1 - \frac{x}{W}\right) + C_h \right] h(x)dx \quad (18.39)$$

where C_h is the handling cost for each warranty claim.

L^* is obtained by minimizing $J(L)$ given by (18.35) with $\gamma(L)$ given by (18.33) and $\omega(W, L)$ given by (18.39). If q is very close to 1, then, using the approximation given in (18.37), L^{**}, an approximation to the optimal lot size, is given by

$$L^{**} = q\left(\frac{2C_f}{(1 - q)[q(A_2 - A_1) - \eta(1 - q)]} \right)^{1/2} \quad (18.40)$$

where

$$A_1 = (S + C_h)F(W) - S\int_0^W \left(\frac{\beta t}{W} \right) f(t)dt \quad (18.41)$$

and

$$A_2 = (S + C_h)H(W) - S\int_0^W \left(\frac{\beta t}{W} \right) h(t)dt \quad (18.42)$$

(For details, see Djamaludin et al., 1994.)

Example 18.7
Suppose the failure distributions of both conforming and nonconforming items are exponential with failure rates λ_1 and λ_2, respectively. Let the parameter values be the same as in Example 18.6 and let $S = \$30$.

Table 18.9 shows L^*, $J(L^*)$ [obtained by evaluating $J(L)$ for $L = 1, \ldots, L_m$], L^{**} [obtained by using (18.40)], $J(L^{**})$, $J(L_m)$, the cost per unit if the lots are of size $L_m = 100$, and RC, the percentage reduction in cost, defined in Example 18.6. The results are similar to those for the FRW policy. Note that for the nominal values used, for $W = 1, 2$, and 3, the optimal lot sizes are smaller than those for the FRW policy and larger for $W = 4$. Also, the percentage reduction, RC, is larger than that for the FRW policy for $W = 1, 2$, and 3 and smaller

Table 18.9. Exact and approximate optimal lot sizes and related costs (PRW Policy)

W	0	1	2	3	4
L^*	100	38	31	28	27
$J(L^*)$	5.5634	9.9418	12.2504	13.9478	15.3331
L^{**}	100	33	27	25	24
$J(L^{**})$	5.5634	9.9658	12.2722	13.9714	15.3620
$J(L_m)$	5.5634	10.8939	14.1262	16.2893	17.8809
RC (%)	0.00	8.74	13.28	14.39	14.25

for $W = 4$. Again, the approximate optimal lot sizes are close to the optimal lot sizes. Finally, the influence of q on L^* is similar to that for FRW policy. ■

18.4   OPTIMAL MAINTENANCE OF UNRELIABLE SYSTEMS

In Chapter 16, we discussed a variety of maintenance models at the component and system levels. In this section we study the optimization of maintenance for some of these models. The goal is to minimize the expected cost of maintaining an unreliable system.[4]

18.4.1   Component Level Models

In Section 16.4.1, three preventive maintenance policies (age, block, and periodic) at the component level were discussed. Each of these policies is characterized by a single parameter T. The asymptotic cost per unit time, $J_c(T)$, for the age, block, and periodic policies are given by (16.22), (16.24), and (16.26), respectively. In this section we focus our attention on the optimal T to minimize $J_c(T)$. A unified approach to all three cases and many others was proposed by Aven and Dekker (1997). Our analysis follows this approach. Let

$$J(T) = \frac{c + \int_0^T K(x)h(x)dx}{d + \int_0^T h(x)dx} \tag{18.43}$$

where c and d are constants. The asymptotic costs for the three policies are special cases of this expression. The results are as follows:

Age Policy. Note that (16.22) can be rewritten as

$$J_c(T) = \frac{C_p + (C_f - C_p)\int_0^T f(x)dx}{\int_0^T \overline{F}(x)dx} \tag{18.44}$$

With $h(x) = 1 - F(x)$, $K(x) = (C_f - C_p)r(x)$, where $r(x)$ is the failure rate associated with $F(x)$, $c = C_p$, and $d = 0$, (18.43) reduces to (18.44).

Block Policy. With $h(x) = 1$, $K(x) = C_f m(x)$, where $m(x)$ is the renewal density associated with $F(x)$, $c = C_p$, and $d = 0$, (18.43) reduces to (16.24).

Periodic Policy. With $h(x) = 1$, $K(x) = C_r r(x)$, $c = C_p$, and $d = 0$, (18.43) reduces to (16.26).

Optimal T

Let T^* denote the optimal T that minimizes $J(T)$ given by (18.43). If T^* exists, it can be obtained from the usual first order condition, i.e.,

$$J'(T) = \frac{dJ(T)}{dT} = 0 \tag{18.45}$$

A characterization of $J(T)$ is useful in determining if T^* exists and is finite. Toward this end, let

$$H(T) = d + \int_0^T h(x)dx \quad \text{and} \quad C(T) = c + \int_0^T K(x)h(x)dx \tag{18.46}$$

so that (18.43) can be written as

$$J(T) = \frac{C(T)}{H(T)} \tag{18.47}$$

Then (18.45) can be written as

$$J'(T) = \frac{\{K(T) - J(T)\}h(T)}{H(T)} = \frac{h(T)}{H^2(T)}[\Psi(T) - c] \tag{18.48}$$

where

$$\Psi(T) = K(T)H(T) - \int_0^T K(x)h(x)dx \tag{18.49}$$

The following proposition (from Aven and Dekker, 1997) gives a characterization of $J(T)$.

Proposition 18.3

Suppose $h(T) > 0$ for all $T > 0$

(i) If $K(T)$ is nonincreasing on $[T_0, T_1]$ and $\Psi(T_0) < c$, then $J(T)$ is decreasing on $[T_0, T_1]$

(ii) If $K(T)$ increases strictly for $T > T_0$, where $\Psi(T_0) < c$, and $\Psi(T_1) > c$ for some $T_1 > T_0$, then $J(T)$ has a minimum, say J^* in T^*, which is unique on $[T_0, \infty)$; moreover

$$K(T) - J(T) \begin{cases} < 0 & \text{for } T_0 < T < T^* \\ > 0 & \text{for } T > T^* \end{cases}$$

and

$$K(T) - J^* \begin{cases} < 0 & \text{for } T_0 < T < T^* \\ > 0 & \text{for } T > T^* \end{cases}$$

If $J(T)$ is differentiable in T then $K(T^*) = J^*$

(iii) If $\psi(T) < c$ for all $T > T_0$, then $J(T)$ is decreasing for $T > T_0$

(iv) If $K(T)$ is increasing for $T > T_0$ and $\psi(T_0) < c$, then $\psi(T_1) > c$ for some $T_1 > T_0$ if one of the following conditions holds

    (a) $\lim_{T \to \infty} K(T) = \infty$
    (b) $\lim_{T \to \infty} K(T) > \lim_{T \to \infty} J(T)$
    (c) $\lim_{T \to \infty} H(T) = \infty$, $\lim_{T \to \infty} K(T) = a$, for some $a > 0$ and $\lim_{T \to \infty} \int_0^T [a - K(x)]h(x)dx$
    $> c - ad$.

The proof is given in the reference cited. The proposition implies that for optimization one only needs to consider those regions where $K(T)$ is increasing.

Age Policy
If $F(x)$ is IFR (see Section 4.7) then we have $K(T) \to \infty$ as $T \to \infty$ so that T^* is finite and unique from (a). The optimal T must be obtained numerically.

Example 18.8
Suppose that component failure times follow a Weibull distribution with shape parameter α and scale parameter β. Define $\rho = C_f/C_p$. Note that $\rho > 1$ since $C_f > C_p$. Let T^* denote the optimal T that yields a minimum value for $J_c(T)$. Tadikamalla (1980) presents tables for T^*/β for a range of values for the shape parameter α and $(1/\rho)$. Huang et al. (1995) present tables for obtaining T^* for a range of values for the scale parameter β and the cost ratio ρ. They also give values for the mean time between failures with no preventive maintenance (i.e., letting $T \to \infty$); mean time between replacements (MTBR) with optimal T^*; cycle reliability $R(T^*)$, which is the probability that a component will not fail before preventive replacement; $J_c(T^*)$, the asymptotic expected cost per unit time with optimal T^*, and η, the percent cost reduction by the use of optimal age replacement. η is given by

$$\eta = [1 - J_c(T^*)/J_c(\infty)] \times 100$$

Numerical results (obtained from Huang et al., 1995) for $\alpha = 2.5$, $\beta = 10$ months, and $C_p = \$100$ are given in Table 18.10.

As can be seen, as the cost ratio ρ increases, T^* decreases so as to reduce the probability of item failure. The benefit of optimal maintenance strategy increases with ρ, as expected. ∎

Table 18.10. Optimal age replacement and other measures for different cost ratios

	$\rho = 2$	$\rho = 4$	$\rho = 8$	$\rho = 10$	$\rho = 20$
T^* (months)	8.8310	5.5494	4.4976	3.5455	2.6254
MTBR	7.3176	5.2087	4.330	3.8208	3.4713
$R(T^*)$	0.4805	0.7950	0.8731	0.9279	0.9653
$J(T^*)$	20.76	31.01	37.74	47.51	68.34
η	0.0788	0.3122	0.4419	0.5785	0.7168

Block Policy

From Appendix B, we have $m(T) \rightarrow 1/\mu$ as $T \rightarrow \infty$, where μ is the mean time to failure. Then from (c) we have a finite and unique T^* if

$$\int_0^\infty \left[\frac{1}{\mu} - m(x) \right] dx > \frac{C_p}{C_f} \qquad (18.50)$$

Example 18.9

Suppose that component failure times follow a Weibull distribution with shape parameter α = 2.5 and scale parameter β = 10 months. Define $\rho = C_f/C_p$. Since there is no analytical expression for the renewal function for a Weibull distribuition, $M(T)$ is obtained from tables (see Blischke and Murthy, 1994). $J_c(T)$ is given by

$$\frac{J_c(T)}{C_p} = \frac{1 + \rho M(T)}{T}$$

Table 18.11 gives $M(T)$ and $J_c(T)/C_r$ for T varying from 0.5 to 10 months in steps of 0.5 months and for four values of ρ. From the table, we see that $T^* > 10$ months for $\rho = 2$. For ρ = 4 we have $T^* = 5.5$ and this decreases with ρ increasing. Also, the expected cost per unit of time with optimal maintenance ($T = T^*$) increases as expected. ■

Table 18.11. $M(T)$ and $J_c(T)/C_r$ versus T for block policy

		$J_c(T)$				
T (months)	$M(T)$	$\rho = 2$	$\rho = 4$	$\rho = 6$	$\rho = 10$	$\rho = 20$
0.5	0.0006	20.024	20.048	20.096	20.120	21.296
1.0	0.0320	10.640	11.280	11.920	13.200	16.4000
1.5	0.0087	6.7827	6.8989	7.0147	7.2467	7.8267
2.0	0.0178	5.1780	5.3560	5.5340	5.8900	6.7800
2.5	0.0309	4.2472	4.4944	4.7416	5.2360	**6.4720**
3.0	0.0483	3.6553	3.9773	4.2993	4.9433	6.5533
3.5	0.0704	3.2594	3.6617	4.0640	**4.8686**	6.8800
4.0	0.0972	2.9860	3.4720	3.9590	4.9300	7.3600
4.5	0.1287	2.7942	3.3662	**3.9382**	5.0822	7.9422
5.0	0.1648	2.6592	3.3184	3.9776	5.2960	8.5920
5.5	0.2053	2.5647	**3.3113**	4.0578	5.5510	9.2836
6.0	0.2500	2.5000	3.3333	4.1667	5.8333	10.0000
6.5	0.2984	2.4566	3.3748	4.2930	6.1292	10.7200
7.0	0.3502	2.4291	3.4297	4.4303	6.4315	11.4343
7.5	0.4048	2.4128	3.4923	4.5717	6.7307	12.1280
8.0	0.4617	2.4043	3.5585	4.7128	7.0213	12.7925
8.5	0.5204	2.4009	3.6254	4.8499	7.2988	13.4212
9.0	0.5805	2.4011	3.6911	4.9811	7.5611	14.0111
9.5	0.6413	2.4027	3.7528	5.1030	7.8032	14.5537
10.0	0.7026	2.4052	3.8104	5.2156	8.0260	15.0520

Periodic Policy
If $F(x)$ is IFR (see Section 4.7), then we have $K(T) \to \infty$ as $T \to \infty$ so that T^* is finite and unique from (a).

Example 18.10
Suppose that the item failure distribution is Weibull with shape parameter $\alpha > 1$ and scale parameter β. From Example 16.4 we have

$$J_c(T) = \frac{C_p}{T} + \left(\frac{C_r}{\beta^\alpha} \right) T^{(\alpha-1)}$$

T^*, the optimal T, is easily obtained from the usual first order necessary condition and is given by

$$T^* = \beta \left(\frac{C_p}{C_r} \right)^{(1/\alpha)}$$

Note that T^*/β is a simple function of the cost ratio (C_p/C_r). It is easily seen that

1. T^* increases as as β increases,
2. T^* decreases as α increases
3. T^* increases as the cost ratio (C_p/C_r) increases

Table 18.12 gives T^*/β for a range of α and (C_p/C_r) values. ∎

18.4.2   System-Level Models

In Section 16.5, three system-level models (Models 16.1–16.3) for maintenance of an unreliable system were discussed. In this section, we consider optimal maintenance actions for Models 16.2 and 16.3.

Model 16.2 (Continuous Preventive Maintenance)
The model is described in Section 16.5.2. The optimization problem is as follows. Maximize total profit given by

$$J(T; u(t), 0 \le t \le T) = \int_0^T \{aS(t) - bu(t)\}dt + S(T) - P \qquad (18.51)$$

Table 18.12. (T^*/β) for periodic policy and Weibull failure distribution

α	C_p/C_r			
	2	5	10	20
1.5	1.5874	2.9242	4.6412	7.3688
2.0	1.4142	2.2361	3.1623	4.4721
2.5	1.3195	1.9037	2.5119	3.3145
3.0	1.2599	1.7099	2.1543	2.7142
5.0	1.1487	1.3979	1.5849	1.8206

by optimally selecting the replacement time T and the maintenance level $u(t)$ over the interval $[0, T)$, subject to the following constraints:

$$S'(T) = \frac{dS(t)}{dt} = -\theta S(t) + \varphi u(t) \tag{18.52}$$

and

$$0 \le u(t) \le U \tag{18.53}$$

This is a standard dynamic optimization problem with one state variable, $S(t)$, one control variable, $u(t)$, free terminal time (as T is a decision variable to be selected optimally), and an inequality constraint on the control variable. By use of the approach outlined in Appendix D, the optimal solution is obtained as follows.

From (D27) we have $H(S, u, \lambda)^5$ given by

$$H(S, \lambda, u) = (aS - bu) + \lambda(-\theta S + \varphi u) \tag{18.54}$$

From (D28) and (D29), we have

$$\lambda'(t) = -\frac{\partial H(S, \lambda, u)}{\partial S} = -a + \theta \lambda \tag{18.55}$$

and

$$\lambda(T) = 1 \tag{18.56}$$

This is a linear first-order differential equation with solution

$$\lambda(t) = e^{-(T-t)} + \left(\frac{a}{\theta}\right)[1 - e^{-\theta(T-t)}] \tag{18.57}$$

Since $(a/\theta) > 1$, we have $\lambda(t) > 1$ for $0 \le t < T$, and it is easily seen that $\lambda(t)$ is a decreasing function of t.

The optimal $u(t)$ is given by (D32). Note that (18.54) can be rewritten as

$$H(S, \lambda, u) = (a - \lambda\theta)S + (-b + \lambda\varphi)u \tag{18.58}$$

Since $0 \le u \le U$, $u^*(t)$, the optimal $u(t)$ is given by

1. $u^*(t) = U$ if $[\lambda(t)\varphi - b] > 0$
2. $u^*(t) = 0$ if $[\lambda(t)\varphi - b] < 0$
3. $u^*(t)$ is undefined if $[\lambda(t)\varphi - b] = 0$

Since $\lambda(t)$ is a decreasing function, (3) can occur at most at one time instant in the interval $[0, T]$ and it can be ignored. As a result, attention can be focused on (1) and (2) to characterize $u^*(t)$.

The optimal maintenance action is as follows:

(A) If $\varphi > b$, then $\lambda(t)\varphi - b > 0$. This implies that $u^*(t) = U$ for $0 \le t \le T$.

(B) If $a\varphi - \theta b < 0$, then $\lambda(t)\varphi - b < 0$. This implies that $u^*(t) = 0$ for $0 \le t \le T$.

(C) If $\varphi < b < \lambda(0)\varphi$, then $\lambda(0)\varphi - b > 0$ and $\lambda(T)\varphi - b < 0$. Since $\lambda(t)$ is a decreasing function of t, this implies that there is a t^* such that

(i) for $0 \le t < t^*$: $\lambda(t)\ \varphi - b > 0$ and $u^*(t) = U$

(ii) for $t^* < t < T$: $\lambda(t)\ \varphi - b < 0$ and $u^*(t) = 0$

t^* is given by $\varphi\lambda(t^*) = b$. Using (18.57) in this relationship, we have, after some simple analysis,

$$t^* = T - \log\left(\frac{(a-\theta)\varphi}{a\varphi - b\theta}\right) \qquad (18.59)$$

As a result, the optimal policy is to use either maximum maintenance or no maintenance at any given time. In (A), maximum maintenance is used over the whole interval; in (B), no maintenance is used over the whole interval; and finally in (C), initially maximum maintenance is used and then no maintenance is used for the remainder of the time interval.

Example 18.11

Consider a system having an exponential failure distribution with parameter θ, and the following parameter values: $\theta = 0.05$ per year, $\varphi = 1000$, $a = 1$, $b = \$4,000$, $P = \$10,000$, and $S(0) = \$8,000$. Let $U = 1$. We consider $T = 1.0, 3.0, 5.0$, and 10.0 years. From, (18.57) we have

$$\lambda(0) = 20 - 19e^{-0.05T}$$

$T = 1.0$: $\lambda(0)\varphi = 1927.2 < b$. As a result, from (A) we have $u^*(t) = 0$ implying that the benefits derived from using maintenance are not worth the additional cost incurred.

$T = 3.0$: $\lambda(0)\varphi = 3647.7 < b$. As a result, from (A) we have $u^*(t) = 0$.

$T = 5.0$: $\lambda(0)\varphi = 5202.8 > b$. As a result, from (C) we have $u^*(t) = U$ for $0 \le t < t^*$ and $= 0$ for $t^* < t < T$. From (18.59), $t^* = 5 - \log(1.1875) = 4.8283$ years.

$T = 10.0$: $\lambda(0)\varphi = 8.4765 > b$. As a result, from (C) we have $u^*(t) = U$ for $0 \le t < t^*$ and $u(t^*) = 0$ for $t^* < t < T$. From (18.59), $t^* = 10 - \log(1.1875) = 9.8283$ years.

Note that when (C) holds, the optimal maintenance strategy is not to use any maintenance for the last 0.1717 years and to use maximum maintenance in the remaining period. ∎

Model 16.3   (Minimal Repairs and Overhauls)

The model is described in Section 16.5.3. We assume that the salvage value is zero, so that it can be ignored. The optimization problem is as follows. Maximize total profit given by

$$J(n, \{T\}_0^n) = R\sum_{j=0}^{n} T_j - P - \sum_{j=1}^{n} O_j - \sum_{j=0}^{n} C_r \int_{0}^{T_j} \Lambda_j(x)dx \qquad (18.60)$$

by optimally selecting n, the number of overhauls, and the set $\{T\}_0^n = \{T_j, j = 0, 1, \ldots, n\}$, which represents the time to first overhaul and time intervals between successive overhauls until the item is replaced. The replacement occurs at age L given by

$$L = \sum_{i=0}^{n} T_i \qquad (18.61)$$

The failure rate $\Lambda_j(x)$ after the jth overhaul (with $j = 0$ corresponding to a new system) and O_j, the cost of the jth overhaul, have the following properties:

1. $\Lambda_j(x)$ is an increasing function in x for $j = 0, 2, \ldots, n$.
2. $\Lambda_j(x)$ is an increasing sequence in j for a given x, $x \geq 0$.
3. O_j is an increasing sequence in j.

This is a mixed nonlinear programming problem as n is an integer and $\{T\}_0^n$ is a set of real variables. The optimal values are obtained as follows:

Step 1: For a fixed n, obtain $T_j^*(n)$, the optimal $T_j(n)$.
Step 2: Obtain n^*, the optimal n, using $T_j^*(n)$. Then T_j^*, the optimal T_j, is given by $T_j^*(n^*)$.

Note that $T_j^*(n)$ can be obtained from the first order condition $\partial J/\partial T_j = 0$.[6] From this, it follows that $T_j^*(n)$ must satisfy

$$\Lambda_j\left(T_j^*(n)\right) = \frac{R}{C_r} \tag{18.62}$$

Note that $T_j^*(n)$ does not depend on n. Hence, $T_j^* = T_j^*(n)$ and from (1) and (2) we have T_j^* to be a decreasing sequence in j as indicated in Figure 18.7.

To determine n^*, we proceed as follows. Increasing the number of overhauls from n to $(n + 1)$ implies that that ΔJ_{n+1}, the change in J, is given by

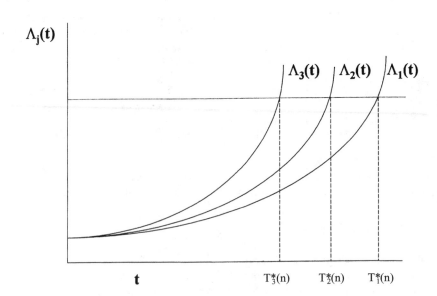

Figure 18.7.

$$\Delta J_{n+1} = RT_{n+1} - O_{n+1} - C_r \int_0^{T_{n+1}} \Lambda_{n+1}(x)dx \qquad (18.63)$$

Since T_j is a decreasing sequence and O_j is an increasing sequence in j, n^* is the smallest n for which $\Delta J_{n+1} < 0$.

Example 18.12
Let $\Lambda_j(x)$ be linear functions of the form

$$\Lambda_j(x) = \zeta(1 + \delta)^j x$$

with $\delta > 0$. This satisfies (1) and (2) above. From (18.62) we have

$$T_j^* = \frac{R}{C_r \zeta(1 + \delta)^j}$$

Let

$$O_{j+1} = \Psi(1 + \gamma)^{(j+1)}$$

with $\alpha > 0$ and $\gamma > 0$. From (18.63), after some simplification, we have

$$\Delta J_{n+1} = \frac{R^2}{2C_r \zeta(1 + \delta)^{(n+1)}} - \Psi(1 + \gamma)^{(n+1)}$$

n^* is the smallest n for which $\Delta J_{n+1} < 0$.

Time is measured in years. Consider the following parameter values: $P = \$100,000$, $R = \$20,000$ per year, $C_r = \$1,000$ per repair, $\zeta = 5$, $\psi = \$7,692$, $\delta = 0.5$, and $\gamma = 0.3$. As a result, the cost of the first overhaul is $\$10,000$, that of the second overhaul is $\$13,000$, and so on. The cost of each repair is $\$1,000$. The optimal time for the first overhaul and the time between successive subsequent overhauls and the corresponding ΔJ_{n+1} are as follows:

j	0	1	2
T_j^* (years)	4.0000	2.6667	1.7778
ΔJ_{n+1} (10^3 \$)	10.63	4.78	-5.05

This suggests that the system should be subjected to two overhauls before it is replaced. The first overhaul occurs after 4 years. The second overhaul occurs when the system is 6.6667 years old (or 2.6667 years after the first overhaul) and finally, the system is replaced at age $4.000 + 2.6667 + 1.7778 = 8.545$ years. ∎

18.5 OPTIMAL WARRANTY DECISIONS

In this section, we discuss a few warranty-related optimization problems. We first consider the joint determination of reliability and warranty duration. Following this, we look at two models for determining the optimal warranty service strategies.[7]

18.5.1 Optimal Reliability and Warranty Duration

As mentioned in Chapter 17, warranty plays an important role in the promotion of the product—better warranty terms conveying the notion of a more reliable product. However, offering a warranty results in additional costs to the manufacturer. Nguyen and Murthy (1988) developed a simple model that links warranty duration and sale price to total sales and hence to the revenue generated. The expected profit is the difference between this and the expected warranty cost, which, in turn, depends on the reliability of the product.

The design decision parameter θ of the failure distribution $F(x, \theta)$, sale price P, and the warranty duration W are selected to maximize the manufacturer's total expected profit over the product life cycle L. The salient features of the model are as follows:

1. The total amount of first purchase sales is given by

$$Q(P, W) = KP^{-\varepsilon}W^{\varphi} \tag{18.64}$$

   where $\varepsilon > 1$ and $0 < \varphi < 1$. This implies that as P increases and/or W decreases, the total sales decrease. ε and φ are the price and warranty period elasticities.

2. The items are nonrepairable and are sold with a linear rebate warranty policy. If the age at failure is X, then the fraction refunded is given by

$$S(X) = (1 - aX/W) \text{ if } X < W \tag{18.65}$$

3. Some of the first purchasers are not happy with the product and cease to buy any replacements. The rest are satisfied customers and continue to buy the product over the life cycle L of the product. Let γ denote the fraction of satisfied first purchasers. As a result, the repeat purchases of satisfied first purchasers occur according to a renewal process associated with the distribution function $F(x, \theta)$.

4. As in earlier models, a smaller value of θ corresponds to a more reliable product. As a result, the manufacturing cost per unit, $C_m(\theta)$, is a decreasing function of θ over a specified interval representing the range of achievable reliability.

Note that over the product life cycle, each satisfied buyer buys items according to a renewal process associated with failure distribution $F(x; \theta)$ and a dissatisfied buyer buys only once. As a result, the expected total purchases by a buyer is $(1 + \gamma M_F(L; \theta))$. From Section (17.4.4), $\omega(W, \theta)$ is given by

$$\omega(W, \theta) = S\frac{(1 - a)WF(W, \theta) + a\int_0^W F(t, \theta)dt}{W} \tag{18.66}$$

Hence the manufacturers total expected profit is given by

$$\Pi(P, W, \theta) = Q(P, W)[P - C_m(\theta) - \omega(W, \theta)][1 + \gamma M_F(L, \theta)] \tag{18.67}$$

The optimal design choice (θ^*) and market choice (P^* and W^*) are given by the values of θ, P, and W, which maximize $\Pi(P, W, \theta)$.

Example 18.13

Let the unit manufacturing cost be given by $C_m(\theta) = \theta^{-0.1}$ with $0 \leq \theta \leq 0.4$ indicating the limits of achievable reliability. Let $K = 1000$, $\varepsilon = 2$, $\varphi = 0.8$, $a = 1$, and $\gamma = 1$. The optimal design choice (θ^*) and market choice (P^* and W^*) are shown in Table 18.13 for $L = 0, 1, \ldots, 5$.

Note that when $L = 0$ (i.e., no repeat purchases), P^* and W^* are high and θ^* low. This is to be expected, since the manufacturer aims to maximize expected profits based solely on first purchase sales. As a result, W^* has to be high to attract more customers. This in turn requires θ^* to be low (implying greater reliability) to reduce warranty costs and forces P^* to be high. This indicates that W^* is an important variable acting as a signal to attract more customers. As L increases, the repeat purchases become more important. Both P^* and W^* decrease with L and θ^* increases. In this case, the joint effect of P^* and W^* is important. As a result, θ^* must increase (implying a less reliable product) to maximize the expected profit. Finally, as γ decreases, for a given L, P^*, and W^* increase, whereas θ^* decreases. ∎

18.5.2   Optimal Repair versus Replace Strategies in Warranty Servicing

When a repairable item is returned to the manufacturer for repair under a free replacement warranty, the manufacturer has the option of either repairing it or replacing it with a new one. The optimal strategy is one that minimizes the expected cost of servicing the warranty over the warranty period. This section examines two simple strategies (involving minimal repair or replacement) that are characterized by a single parameter, which is to be chosen optimally to minimize the expected cost of servicing the warranty.

Strategy 1: A failed item is replaced by a new one if it fails in $(0, W - \tau]$ and subjected to minimal repair if it fails in $(W - \tau, W]$.

Strategy 2: A failed item is subjected to minimal repair if it fails in $(0, \tau]$ and is replaced by a new one if it fails in $(\tau, W]$.

In both cases, the parameter τ $(0 \leq \tau \leq W)$ is to be selected to minimize the expected cost of servicing the warranty.

Comment: Strategy 1 is more appropriate where the initial failure rate is high due to a small fraction of the items being of inferior quality, i.e., not conforming to design specification and having a very high failure rate. As a result, replacing items that fail early by new ones can be viewed as an effective way of weeding out items of inferior quality. Strategy 2 is more appropriate when items have a decreasing failure rate in the early stages of their life and failed items can be repaired relatively cheaply. As the item ages, the failure rate increases and hence it is more sensible to replace failed items by new ones when they are old.

Table 18.13.   Optimal P^*, W, and θ^* for Example 18.13

L	P^*	W^*	θ^*	$\Pi(P^*, W^*, \theta^*)$
0	6.90	7.60	0.125	237
1	6.06	6.38	0.149	269
2	5.23	4.97	0.191	311
3	4.18	2.85	0.333	357
4	3.98	2.38	0.400	422

Let c_m and c_r be the manufacturing cost and the repair cost per unit. From Nguyen and Murthy (1989), the expected cost of servicing the warranty per item sale, $\omega(\tau, W)$, for Strategy 1 is given by

$$\omega(\tau, W) = c_m M(W - \tau) + c_r \log[\overline{F}_\gamma(\tau)] \tag{18.68}$$

where

$$F_\gamma(x) = F(W - \tau + x) - \int_0^{W-\tau} \overline{F}(W - \tau + x - y)dM(y) \tag{18.69}$$

for $x \geq 0$. $M(\ )$ is the renewal function associated with $F(x)$.

Similarly, for Strategy 2, the total expected warranty service cost per item sale, $\omega(\tau, W)$, under this strategy is given by

$$\omega(\tau, W) = c_m M_d(W - \tau) - c_r \log[\overline{F}(\tau)] \tag{18.70}$$

where $M_d(x)$ is the expected value of a modified renewal process with the distribution for the first failure, $F_\gamma(x)$, given by

$$F_\gamma(x) = \frac{F(x + \tau) - F(\tau)}{\overline{F}(\tau)} \tag{18.71}$$

and the distribution for subsequent failures is given by $F(x)$.

τ^*, the optimal τ that minimizes $\omega(\tau, W)$, can be obtained from

$$\frac{d\omega(\tau, W)}{d\tau} = 0 \tag{18.72}$$

provided τ^* exists. From this it follows that for both strategies, τ^* is the solution of the following equation

$$-c_m + \frac{c_r \overline{F}(\tau)}{\overline{F}_\gamma(\tau)} = 0 \tag{18.73}$$

In general, one needs to use a computational scheme to obtain τ^*.

Example 18.14
Let $F(x)$ be an Erlangian distribution (see Section 4.4.2) with two stages and parameter λ, i.e.,

$$F(x) = 1 - (1 + \lambda x)e^{-\lambda x}$$

Let $\lambda = 2.0$ per year. This implies that $F(x)$ is IFR with mean $\mu = 1$ year. Let $c_m = \$100$ and $W = 1$ year.

The existence of τ^* depends on the ratio c_r/c_m, i.e., the ratio of expected repair cost to manufacturing cost. A numerical approach is required to obtain τ^*. Nguyen (1984) uses a search method, in which the left-hand side of (18.73) is evaluated for values of c_r/c_m ranging from 0.00 to 1.00 in steps of 0.01 and for τ ranging from 0.00 to 1.00 in steps of 0.01 and a

Table 18.14. Optimal repair versus replace decisions for strategy 1

c_r	τ_1	$\omega(\tau_1, W)$	τ_2	$\omega(\tau_2, W)$	$\omega(0, W)$	$\omega(W, W)$	τ^*	$\omega(\tau^*, W)$
80	0.39	72.3	0.72	72.7	75.5	72.1	1.00	72.1
85	0.23	73.9	0.84	76.8	75.5	76.6	0.23	73.9
90	0.13	74.9	0.91	81.2	75.5	81.1	0.13	74.9
95	0.06	75.3	0.96	85.6	75.5	85.6	0.06	75.3

finer grid for search when the quantity evaluated is close to zero. The results are as follows:

For Strategy 1, for $c_r/c_m < 0.78$, there is no τ, with $0 \leq \tau \leq 1$, satisfying (18.73); $\omega(\tau, W)$ is minimum for $\tau = 1$. Since repairs are carried out over the interval $(W - \tau, W]$, this implies that the optimal strategy for $c_r/c_m < 0.78$ is that failed items are always repaired. For $0.78 \leq c_r/c_m < 1$, there are two solutions which we denote by τ_1 and τ_2 respectively. τ^* is the one which gives the minimum value for $\omega(\tau, W)$.

Table 18.14 gives results for a range of c_r values. Also included are $\omega(0, W)$ and $\omega(W, W)$. The former corresponds to always replace and the latter to always repair. For $c_r = \$80$, the total expected cost for each of the two solutions is larger than that for $\tau = W$. It follows that in this case $\tau^* = 1$. Note that as c_r increases from \$80 to \$95, τ^*, the optimal period over which minimal repair is to be carried out, decreases.

The corresponding results for Strategy 2 are given in Table 18.15. We see that the values of τ^* are the same as in Table 18.14. However, the resulting optimal expected warranty service cost, $\omega(\tau^*, W)$, is larger.

A comparison between Strategies 1 and 2 for $c_r = \$80$ and \$90 is as follows. For $c_r = \$80$, the optimal decision in both cases is to repair all failures over the warranty period. For $c_r = \$90$, the optimal decision under Strategy 1 is to replace all failures in $(0, 0.87]$ and repair all failures in $(0.87, 1.0]$. In contrast, the optimal decision under Strategy 2 is to repair all failures in $(0, 0.13]$ and replace all failures in $(0.13, 1.0]$. ∎

18.5.3   Cost Repair Limit Strategy

This section deals with the case where the cost to repair a failed unit, C_r, is a random variable characterized by a distribution function $H(z)$. Analogous to the notion of a failure rate, one can define a repair cost rate given by $h(z)/H(z)$, where $h(z)$ is the derivative of $H(z)$. Depending on the form of $H(z)$, the repair cost rate can increase, decrease, or remain constant with z. A decreasing repair cost rate is usually an appropriate characterization for the repair cost distribution (see, e.g., Mahon and Bailey, 1975). Murthy and Nguyen (1988) consider a model where items are sold with a free replacement warranty. Under this model, when an item is returned under warranty, the failed item is inspected and an estimate of the repair cost is deter-

Table 18.15. Optimal repair versus replace decisions for strategy 2

c_r	τ_1	$\omega(\tau_1, W)$	τ_2	$\omega(\tau_2, W)$	$\omega(0, W)$	$\omega(W, W)$	τ^*	$\omega(\tau^*, W)$
80	0.39	74.5	0.72	74.9	75.5	72.1	1.00	72.1
85	0.23	75.1	0.84	78.0	75.5	76.6	0.23	75.1
90	0.13	75.4	0.91	81.7	75.5	81.1	0.13	75.4
95	0.06	75.5	0.96	85.8	75.5	85.6	0.06	75.5

mined. If the estimate is less than a specified limit v, then the failed item is repaired and returned to the owner. If not, the failed item is junked and the customer is supplied with a new
item at no cost. The repair is minimal repair. $\omega(v, W)$, the expected warranty service cost per
unit sale, is given by

$$\omega(v, W) = M_G(W; v)\left[c_m + \left(\frac{\int_0^v z\,dH(z)}{H(v)}\right)\right] \tag{18.74}$$

where $M_G(W; v)$ is the renewal function associated with the distribution function $G(u; v)$, giv-
en by

$$G(u; v) = 1 - [\bar{F}(u)]^{\bar{H}(v)} \tag{18.75}$$

and ζ, the expected cost of each repair carried out, given by

$$\zeta = E[C_r|C_r < v] = \frac{\int_0^v z\,dH(z)}{H(v)} \tag{18.76}$$

(For the details of the derivation, see Murthy and Nguyen (1988)). The optimal v^*, if it exists, is the value of v which minimizes $\omega(v, W)$. This can be obtained from

$$\frac{d\omega(v, W)}{dv} = 0 \tag{18.77}$$

In general, it is difficult to obtain v^* analytically, and a computational scheme to obtain v^*
using (18.77) is required. The following proposition (from Nguyen and Murthy, 1988) gives
bounds for v^*:

Proposition 18. 4

1. If $F(t)$ is IFR, then $0 \leq v^* \leq c_m$.
2. If $F(t)$ is DFR, then $v^* \geq c_m$.

Proof of this result can be found in the reference cited. The proposition states that if the failure rate is decreasing, then it is worth spending more than the replacement cost for repair
since the repaired item has a smaller failure rate and hence is more reliable than a new item.
However, this situation seldom happens in the real world. For increasing failure rate, repaired
items are less reliable than new items and hence the optimal repair limit must be less than the
price of a new unit.

Example 18.15
Let $F(t)$ be a Weibull distribution with parameters λ and α, i.e.,

$$F(t) = 1 - e^{-(\lambda t)^\alpha}$$

with $\alpha = 2$ and $\lambda = 0.886$. This results in $F(t)$ having an increasing failure rate with mean
time to failure of 1.0. The costs are normalized so that $c_m = 1.0$.

Suppose that the repair cost distribution, $H(z)$, is also a Weibull distribution with parameters $\bar{\alpha}$ and $\bar{\lambda}$. Consider the following three sets of parameter values:

(a) $\bar{\alpha} = 0.5$ and $\bar{\lambda} = 2.0$: This implies a decreasing repair cost rate.
(b) $\bar{\alpha} = 1.0$ and $\bar{\lambda} = 1.0$: This implies a constant repair cost rate.
(c) $\bar{\alpha} = 3.0$ and $\bar{\lambda} = 0.883$: This implies an increasing repair cost rate.

The above values of $\bar{\lambda}$ were chosen so that the expected repair cost, c_r, is equal to 1. The corresponding values of ζ will depend on v, but are always less than c_m.

Figure 18.8 shows the optimal $v(v^*)$ as a function of W for these three sets of parameters values. The results were obtained using a computational scheme to solve (18.77) based on an iterative gradient method. Note that v^* is always less than $c_r = 1$, as expected. Also, note that v^* decreases with increasing W. The results imply that the longer the warranty period, the smaller the repair limit, which in turn implies that, as expected, the failed unit is more often replaced by a new unit.

Two limiting cases are as follows:

1. $\zeta = \infty$: This corresponds to always repair and the expected service cost is given by $\omega(\infty; W)$.

2. $\zeta = 0$: This corresponds to always replace and the expected service cost is given by $\omega(0; W)$.

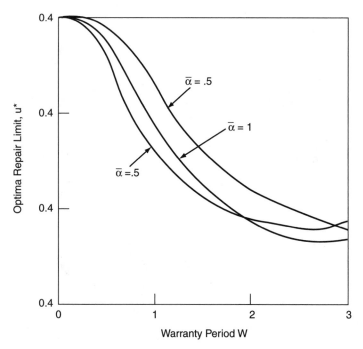

Figure 18.8. Plot of v^* versus W.

If F is IFR, then it is easily seen that $\omega(0; W) < \omega(\infty; W)$ for all W. To compare $\omega(v^*; W)$ with $\omega(0; W)$, define the percentage savings, $\eta(W)$, as

$$\eta(W) = \left(\frac{\omega(0; W) - \omega(v^*; W)}{\omega(0; W)} \right) \times 100$$

Figure 18.9 shows $\eta(W)$ for $\bar{\alpha}$ as a function of W. For a given $\bar{\alpha}$, $\eta(W)$ decreases with W, as is to be expected. For a given W, $\eta(W)$ decreases as $\bar{\alpha}$ increases, as is to be expected for the following reason. When $\bar{\alpha}$ is less than 1, the repair cost rate is decreasing. The smaller the value of $\bar{\alpha}$, the greater is the probability of the repair being carried out at a small cost. When $\bar{\alpha}$ is greater than 1, the repair cost rate is increasing. This implies that the probability of the repair cost being small decreases with increasing $\bar{\alpha}$ and hence the advantage of repair over replacement is reduced. Thus, $\eta(W)$ decreases as $\bar{\alpha}$ increases. ∎

18.6 ANALYSIS OF AN INTEGRATED MODEL

The simple deterministic model given in Section 12.11 integrated three issues—design, operation, and maintenance. The design decision variable is the initial strength (x_0) of the pipe. The strength $x(t)$ decreases with time and rate of change is influenced by the stress $y(t)$, which is related to the pressure at which fluid is pumped through the pipe. Degradation in the strength of the pipe is given by

$$\frac{dx(t)}{dt} = -\theta x(t) - \varphi y(t) \tag{18.78}$$

with the constraint

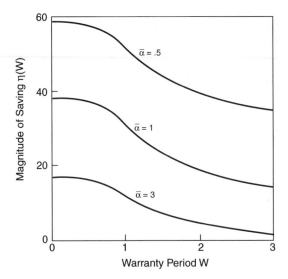

Figure 18.9. Plot of $\eta(W)$ versus W.

$$y_m < y(t) < x(t) - \delta \tag{18.79}$$

In the model, total profit is a function of three decision variables; initial strength x_0; $y(t)$, the stress on the pipe over the life of the pipe; and T, the life of the pipe. The problem is to select these optimally to maximize the total profit. The optimization problem is as follows: Maximize total profit

$$J = \gamma \int_0^T y(t)dt - C(x_0) \tag{18.80}$$

by optimally selecting x_0, the design parameter; $y(t)$, $0 \le t \le T$, the operational variable; and T, the maintenance variable, subject to the constraints given in (18.78) and (18.79).

This is a standard dynamic optimization problem with one state variable $x(t)$, one control variable $y(t)$ with free end point (since T is a decision variable to be selected optimally), and with a control variable inequality. By use of the approach outlined in Appendix D, the optimal solution is obtained as follows.

Define[8]

$$H(x, y, \lambda) = \gamma y + \lambda\{-\theta x - \varphi y\} \tag{18.81}$$

Then from (D28) and (D29) we have

$$\frac{d\lambda}{dt} = \frac{\partial H(x, y, \lambda)}{\partial x} = -\theta\lambda \tag{18.82}$$

with boundary condition

$$\lambda(T) = 0 \tag{18.83}$$

The solution to (18.82) satisfying (18.83) is given by

$$\lambda(t) = 0 \tag{18.84}$$

As a result,

$$H(x, y, \lambda) = \gamma y \tag{18.85}$$

Since $y(t)$ is constrained, we have, from (D30) and (18.85), $y^*(t)$, the optimal $y(t)$, given by

$$y^*(t) = x(t) - \delta \tag{18.86}$$

This implies that the optimal operational variable should be such that the stress on the component is at its limit, which is strength $x(t)$ minus the safety factor δ.

Using (18.86) in (18.78) yields the degradation in the strength over time with optimal operation. The result is

$$x(t) = \left[x_0 - \frac{\varphi\delta}{\theta + \varphi} \right] e^{-(\theta+\varphi)t} + \frac{\varphi\delta}{\theta + \varphi} \tag{18.87}$$

Note that the unit can be kept operational as long as the inequality given by (18.79) is not violated. Let $\overline{T}$ be such that $x(t)$ reaches the limit $y_m + \delta$ when $t = \overline{T}$. Then, from (18.87), after some simplification, we have

$$\overline{T} = \frac{1}{\theta + \varphi} \log\left(\frac{x_0(\theta + \varphi) - \varphi\delta}{y_m(\theta + \varphi) + \theta\delta} \right) \tag{18.88}$$

From (18.80) it can be seen that J increases with T. Hence, T^*, the optimal T, is given by $T^* = \overline{T}$.

Note that both $y^*(t)$ and T^* are functions of x_0. Using $y^*(t)$ and T^* in (18.88), after some simplification, the result can be rewritten as

$$J = Ax_0 - B \log\{(\theta + \varphi)x_0 - \varphi\delta\} - C(x_0) + D \tag{18.89}$$

where

$$A = \frac{\gamma}{\theta + \varphi}, \quad B = \frac{\theta\delta}{(\theta + \varphi)^2} \quad \text{and} \quad D = -\frac{y_m + \delta}{\theta + \varphi} + \frac{\theta\delta}{(\theta + \varphi)^2} \log\{(\theta + \varphi)y_m + \theta\delta\} \tag{18.90}$$

J is a nonlinear function of x_0 and the optimal value of x_0 can be obtained from the usual first order conditions. The expression for it would depend on the form of $C(x_0)$.

NOTES

1. The literature on reliability and optimization is very large. Models dealing with the many different issues are many and varied. The models discussed in this chapter are a very small illustrative sample. In the following notes, we cite some of the more recent references to indicate the issues being addressed and the techniques used in optimization problems.

2. During the years 1970–1980, optimal reliability allocation received considerable attention. See, for example, Tillman et al. (1975, 1980). More recent papers include Coit and Smith (1998), which deals with redundancy allocation to maximize a percentile of the failure distribution; Mi (1998), which deals with maximizing lifetime; and Bland (1998), Painton and Campbell (1995), and Gen and Cheng (1996), which deal with more modern techniques for optimization, such as tabu search and genetic algorithms.

Nguyen and Murthy (1988) deal with optimal reliability allocation, taking into account its impact on warranty costs. Murthy (1990) and Murthy and Hussain (1993) look at optimal reliability choice in new product development. Murthy and Nguyen (1987) deal with optimal development testing policies.

3. Murthy et al. (1993) deal with quality control through burn-in. Optimal burn-in, taking into account the impact of this on warranty, is discussed in Nguyen and Murthy (1982). For a review of optimal burn-in, see Leemis and Benke (1990). Djamaludin (1993) and Djamaludin et al. (1994) deal with optimal lot sizing. Lie and Chun (1987) deal with optimal inspection plans. Mok and Xie (1966) deal with optimal stress screening.

Several papers link quality, design, and product servicing. These include Tapiero and Lee (1989), and Ritchken et al. (1989).

4. There are several survey papers where references to the different maintenance optimization models can be found. These include Pierskalla and Voelker (1976), Sherif and Smith (1976), Thomas (1986), Valdez-Flores and Feldman (1989), Cho and Parlar (1991), Dekker (1996), Vatn (1997), Vatn et al. (1996), Aven and Dekker (1997), and Munoz et al. (1997).

Bergman and Klefsjo (1982) propose a graphical method based on the TTT concept. Berg and Epstein (1978) do a comparison of age, block, and failure replacement policies.

Dohi et al. (1996) and Haneveld and Teunter (1997) are illustrative of papers dealing with the optimization of spares and ordering policies.

5. We supress the argument t for notational ease so that S, λ, u represent $S(t)$, $\lambda(t)$, and $u(t)$, respectively.

6. We supress the arguments for notational ease.

7. Nguyen (1984), Nguyen and Murthy (1986, 1989) and Murthy and Nguyen (1988) deal with optimization in the context of warranty servicing.

Anderson (1977) and Glickman and Berger (1976) deal with optimal selection of price and warranty terms.

8. We supress the argument t. As a result, y represents $y(t)$ and so on.

Acknowledgment: Materials in Examples 18.2, 18.3, 18.4, 18.6, 18.7, 18.13, and 18.14 are based on Blischke and Murthy (1994) and used by permission of Marcel Dekker, Inc.

EXERCISES

1. Consider Example 18.1. Suppose that the cost of development is a nonlinear function, with J given by

$$J = \sum_{i=1}^{4} \alpha_i(x_i)^\gamma$$

   where $\gamma < 1$. Compute the optimal reliability allocation for $\gamma = 0.9$, 0.8, and 0.7 respectively. Compare the results and discuss the effect of γ on the optimal allocations.

2. Redo Example 1 with the following objective—maximize the system reliability subject to the total cost being $\leq \$1500$. How does the result change if the cost limit is increased to $\$2000$?

3. In Section 18.2.3, we assumed that the failure rate [given by equation (18.21)] is deterministic. In real life, the outcome of any development process is uncertain. A stochastic model for $\nu(t)$ is as follows. $\nu(t)$ is modeled as a stochastic process with mean given by the right-hand side of equation (18.21) and standard deviation given by $\sigma(t)$, which is an increasing function of t. How does this affect the optimal development strategy?

4. In Section 18.2.3, the cost of development [given by equation 18.22] is a linear function of the development time and the number of modifications made. How do the results of Example 18.3 change if (18.22) is given by $C_d (T_d)^a + C_a \{k_1 N_a + k_2 (N_a)^2\}$?

5. Prove Proposition 18.1.

6. Prove Proposition 18.2.

7. Suppose that $F(t)$ and $H(t)$ in Example 18.4 are Weibull distributions with shape parameter 1.2 and scale parameters 5.00 and 0.25 years, respectively. Obtain the optimal burn-in time and the expected numbers of conforming and nonconforming items that are weeded out per batch of 100 items.

8. Repeat Exercise 18.7 with $H(t)$ and $F(t)$ being lognormal distributions with the same first and second moments as those in Exercise 18.7.

9. Compare the results of Exercises 18.7 and 18.8 with that of Example 18.4. Discuss the implications of distributional assumptions on the optimal strategies.

10. Suppose that $F(t)$ and $H(t)$ in Example 18.5 are Weibull distributions with shape parameter $= 1.2$ and scale parameters be 5.00 and 0.25 years, respectively. Obtain the optimal burn-in time and the expected numbers of conforming and nonconforming items that are weeded out per batch of 100 items.

11. Repeat Exercise 18.10 with $H(t)$ and $F(t)$ being lognormal distributions with the same first and second moments as those in Exercise 18.10.

12. Compare the results of Exercises 18.10 and 18.11 with that of Example 18.5. Discuss the implications of distributional assumptions on the optimal strategies.

13. Compare the results of Exercises 18.9 and 18.12. Discuss the implication of different warranty policies on the optimal strategies.

14. Suppose that $F(t)$ and $H(t)$ in Example 18.6 are Weibull distributions with shape parameter $= 1.2$ and scale parameters be 5.00 and 0.25 years, respectively. Obtain the optimal lot size.

15. Repeat Exercise 18.14 with $H(t)$ and $F(t)$ being lognormal distributions with the same first and second moments as those in Exercise 18.14.

16. Compare the results of Exercises 18.14 and 18.15 with that of Example 18.6. Discuss the implications of distributional assumptions on the optimal strategies.

17. Suppose that $F(t)$ and $H(t)$ in Example 18.7 are Weibull distributions with shape parameter $= 1.2$ and scale parameters be 5.00 and 0.25 years, respectively. Obtain the optimal lot size.

18. Repeat Exercise 18.17 with $H(t)$ and $F(t)$ being lognormal distributions with the same first and second moments as in Exercise 18.17.

19. Compare the results of Exercises 18.17 and 18.18 with that of Example 18.7. Discuss the implications of distributional assumptions on the optimal strategies.

20. Redo Example 18.8 with $F(t)$ and $H(t)$ being lognormal distributions with the same first and second moments as in Example 18.8. Compare the results of Exercise 18.20 with those of Example 18.8.

21. Redo Example 18.9 with $F(x)$ being lognormal distributions with the same first and second moments as in Example 18.9. Compare the results of Exercise 18.20 with those of Example 18.9.

22. Redo Example 18.10 with $F(x)$ being lognormal distributions with the same first and second moment as those in Example 18.10. Compare the results of Exercise 18.20 with those of Example 18.10.

23. In Model 16.2 (Section 18.4.2), there is no discounting to present value. With discounting, equation (18.51) becomes

$$J(T, u(t), 0 \leq t \leq T) = \int_0^T e^{-\rho t}\{aS(t) - bu(t)\}dt + e^{-\rho T}S(T) - P$$

where ρ is the discounting factor. How does the optimal maintenance strategy change? [Hint: Define new variables $\hat{S}(t) = e^{-\rho t}S(t)$ and $\hat{u}(t) = e^{-\rho t}u(t)$ and reformulate the problem.]

24. Redo Example 18.13 with $F(t; \theta)$ being a Weibull distribution with shape parameter 1.2. The scale parameter is a decision variable (along with the warranty period W and sale price P) to be optimally selected. Determine the optimal solution.

25. Redo Example 18.14 with $F(x)$ being a Weibull distribution with first and second moments the same as in Example 18.14.

PART E

Epilogue

CHAPTER 19

Case Studies

19.1 INTRODUCTION

In our approach to reliability, we have discussed various aspects of the subject from four points of view, attempting to integrate these into a coherent whole that will provide the necessary understanding and tools for analysis and decision making. The four areas are:

- Engineering—product design, manufacturing, maintenance, and optimization
- Mathematical modeling—deterministic and probabilistic models of failure mechanisms and times to failure for components and systems
- Statistical analysis—methods for collection and analysis of reliability data
- Management—strategies and policies for reliability, servicing, warranty, and related areas

In this chapter, we look at two case studies based on data sets presented and briefly discussed in Chapter 2. In the first of these, the problem is well defined and the data are well structured and reliable. In the second, there are many unknowns, the problem is much more nebulous, and many difficulties must be assumed away in order to do even a preliminary analysis. The objective in both cases is to apply the principles of the previous chapters to arrive at reasonable solutions to achieving the management objectives. We consider each of the four areas listed, as appropriate and where feasible, in the context of the available information. A number of the remaining cases from Chapter 2 are assigned as case study exercises, with suggestions for solution provided.[1]

In Section 19.2, the data of Case 2.2—failure times of aircraft windshields—are analyzed and the results applied to the warranty problem posed. The management decision involves the choice of warranty policy (e.g., free replacement or pro rata) and terms (e.g., length of the warranty). The key inputs to the decision making process are predictions of future costs of the various warranties.

In Section 19.3, we look at the electronic components discussed in Case 2.5. Here the data must be interpreted in the light of incomplete information and the data analysis is itself a significant effort. The management decisions involve costs, maintenance, spares provisions, and possible redesign.

19.2   RELIABILITY ANALYSIS OF AIRCRAFT WINDSHIELDS

Data on failures of aircraft windshields were presented and discussed in Case 2.2. The major objective of the study was to obtain information relevant to evaluation of alternative warranties that were being considered. In the following sections, we analyze the data, calculate cost estimates for the various warranty policies, investigate sensitivity of the results to assumptions and parameter estimates, and state conclusions with regard to the management decisions in selection of an optimal strategy.

Note that this case is well structured is the sense that the problems to be addressed are well formulated and data had been routinely collected and are considered to be valid and reliable. This is possible because the number of any given item produced is relatively small—hundreds or thousands, rather than hundreds of thousands or millions or more, as in the case of many consumer products—and because the ultimate customer, the user of the aircraft, provides data on usage and failure times. In fact, complete and accurate data on all major components of importance in operation and safety of both military and commercial aircraft are required for many purposes, and such data are collected extensively by both types of customers.

We begin with some further discussion of the policy issues and the problems to be addressed.

19.2.1   Background and Problem Statement

The manufacturer (hereafter referred to as The Company) of the aircraft windshield in question had been producing this product in this application for a number of years, and was a sole-source supplier to the aircraft manufacturer up to the point of the study. This is a high-technology item and at the time The Company had only one major competitor. Up to this point, the product had been sold without warranty. The customer had now requested that a warranty be provided and that the terms of the warranty be a part of the next contract negotiation. In preparations for this negotiation, The Company compiled information on product performance for several windshields and on warranty policies. In this study, we analyze one windshield.

The product is extremely reliable as far as structural integrity is concerned, with virtually no possibility of a structural failure during normal aircraft operation, even in severe conditions. The windshield consists of several layers of material, including a thin outer layer of extremely strong material with a conductive heating layer just beneath, all laminated under high temperature and pressure. The windshields are subjected to multiple and sometimes intensive stresses in flight, and especially on take-off and landing. Failures consist of damage to the outer layer, failure of the heating mechanism, or product degradation in the form of partial delamination to the point of requiring replacement.

Failed windshields are repairable; they can be rebuilt using much of the original materials for up to three or four failures on each item. Repairs are done by relamination of the component layers and rebuilding of the frame. The result is a good-as-new windshield and the cost is about 60% of the cost of a new item. In the analysis, we assume that a stock of spares is available and that these consist of repaired items. It follows from the previous statement that the life distribution of a repaired item is the same as that of a new item. We thus assume that times to failure are independent and identically distributed. One of the issues to be dealt with is determination of an appropriate failure distribution.

The choice of warranty involves both the policy itself and the length W of warranty to be offered. The warranty policies being considered were the nonrenewing free replacement warranty (FRW) and the nonrenewing pro-rata warranty (PRW). Under the FRW, replacements for failed items are provided free of charge to the buyer until time W from the time of original purchase. The nonrenewing feature means that a replacement item is not warrantied anew for a time period W from the time of replacement; rather, the replacement item is warrantied only for the time remaining in the original warranty period. Under the nonrenewing PRW, a linear pro-rata rebate is given to the buyer for any failure up to time W from the time of original purchase.

The FRW being considered is Policy 1 of Chapter 17. The PRW is Policy 2. Both of these policies and many others are discussed are discussed in detail by Blischke and Murthy (1994) and in the references cited in that source.

Another issue to be addressed is the basis of the warranty. There are many factors that can affect the failure rate and the actual occurrence of a failure. These include number of take-offs and landings, miles flown, flight hours, and age of the windshield. At the time, the air-craft in question was flown, on average, about 3000 to 4000 hours per year, with relatively little variability in the various applications (e.g., different users and national versus interna-tional flights). Warranties of up to two years were being considered. The management deci-sion in this regard was to base the warranty on flight hours alone, and to consider warranties of up to 8,000 flight hours.

19.2.2   Data Structure and Analysis

The basic data on failure and service times are given in Table 2.2. The data are assumed to comprise a random sample from the production of this model windshield in use in normal ap-plications at the time of the study. The sample size is $n = 153$, of which 88 are flight hours (in thousands) for failed windshields and the remaining 65 are service times of windshields that had not failed. Note that a number of windshields had been replaced because of failures due to causes other than failure in normal use. These included accidental failures, for example, damage on installation or during aircraft maintenance, lightning strikes, outer coat shattered by foreign objects (e.g., birds, gravel), and so forth. These are failures that are not considered to be due to windshield faults and hence would not be covered by the proposed warranty. These were therefore listed as service times rather than failure times in the data set.

Information not included in Table 2.2 is the mode of failure for failed windshields. Eight failure modes were listed in the initial data set, two of which are not considered to be failures for our purposes for the reasons just discussed. The remaining six modes are:

1. Upper sill delamination
2. Massive delamination
3. Delamination—other
4. Coating burnout
5. Shattered (other than by foreign objects)
7. Other or unknown

Failure modes for the 88 failed items are given in Table 19.1.

Data Analysis

Previous Results. In a previous look at these data, probability plots were prepared. Figure 11.12(a) is a four-way plot of the entire data set (i.e., ignoring the different failure modes). Prior to the analysis, it had been anticipated that the data would follow a Weibull distribution.

Table 19.1. Times to failure and failure modes for windshield data.

TTF	Mode	TTF	Mode	TTF	Mode	TTF	Mode
0.040	7	1.866	4	2.385	1	3.443	1
0.301	4	1.876	4	2.481	5	3.467	7
0.309	3	1.899	4	2.610	1	3.478	7
0.557	4	1.911	1	2.625	5	3.578	4
0.943	1	1.912	1	2.632	3	3.595	3
1.070	1	1.914	1	2.646	4	3.699	1
1.124	5	1.981	1	2.661	4	3.779	1
1.248	4	2.010	2	2.688	4	3.924	4
1.281	1	2.038	1	2.823	4	4.035	1
1.281	1	2.085	1	2.890	5	4.121	2
1.303	5	2.089	3	2.902	1	4.167	4
1.432	1	2.097	1	2.934	4	4.240	4
1.480	1	2.135	4	2.962	3	4.255	3
1.505	5	2.154	5	2.964	7	4.278	4
1.506	7	2.190	4	3.000	4	4.305	1
1.568	1	2.194	1	3.103	7	4.376	4
1.615	5	2.223	1	3.114	4	4.449	5
1.619	1	2.224	2	3.117	4	4.485	4
1.652	2	2.229	1	3.166	4	4.570	3
1.652	1	2.300	1	3.344	5	4.602	7
1.757	5	2.324	1	3.376	4	4.663	3
1.795	5	2.349	3	3.385	1	4.694	4

It is apparent from the plot that the Weibull fit is not particularly good, especially at the lower end of the distribution, and that the normal distribution seems to provide a better fit. The normal fit shown in Figure 11.12(b) supports this conclusion. The problem, however, may be that the lowest three or four observations are outliers. On the other hand, the Weibull fit shown in Figure 11.12(d), in which the lowest four observations are omitted, is not much better.

We proceed with the analysis of both the normal and, for a priori reasons, the Weibull distributions. The MLEs of the parameters of the two distributions based on the complete data set are obtained from the Minitab output given in Figures 11.12(b) and 11.12(c). The results are $\mu^* = 3.041$ and $\sigma^* = 1.241$ for the normal distribution, and $\alpha^* = 2.443$ and $\beta^* = 3.452$ for the Weibull.

Analysis of Individual Failure Modes. Additional insight into the causes of unreliability may be gained by considering the failure modes separately. A technique for accomplishing this is to prepare a probability plot for each mode, in which failures by that mode alone are considered failures and all other observations are considered service times. The logic of this is that failure by a given mode effectively censors the observation as far as other modes of failure are concerned.

Four-way plots for the six modes are given in Figure 19.1(a)–(f). Tentative conclusions that might be drawn from the graphical results are:

- Mode 1—No fit is really bad, but lognormal and exponential appear to be somewhat better than either the normal or Weibull
- Mode 2—Too few failures to decide

Four-way Probability Plot for Hours
Censoring indicator in Mode 1

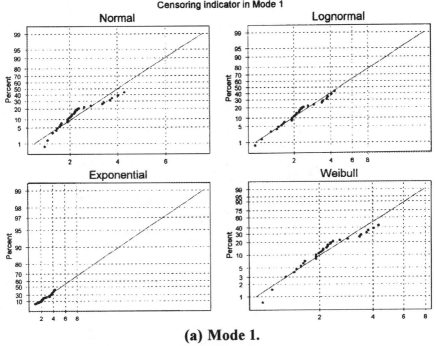

(a) Mode 1.

Four-way Probability Plot for Hours
Censoring indicator in Mode 2

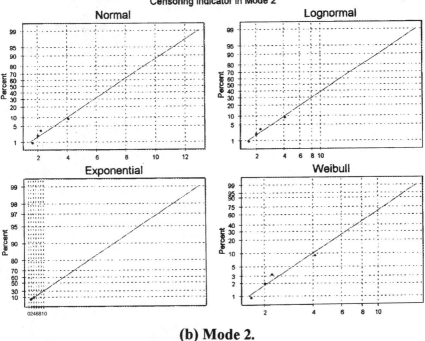

(b) Mode 2.

Figure 19.1. Four-way probability plots for aircraft windshield data, separate failure modes.

Four-way Probability Plot for Hours
Censoring indicator in Mode 3

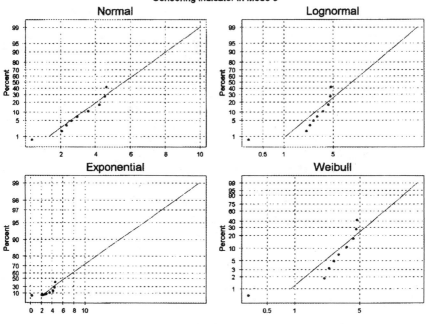

(c) Mode 3.

Four-way Probability Plot for Hours
Censoring indicator in Mode 4

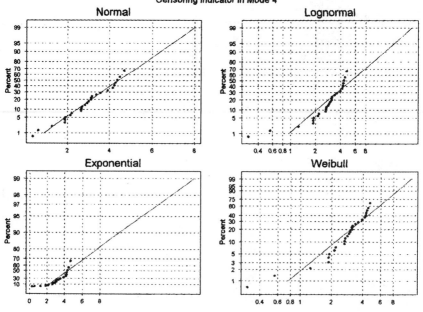

(d) Mode 4.

Figure 19.1. *Continued.*

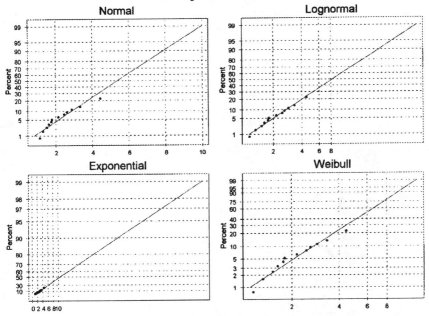

(e) Mode 5.

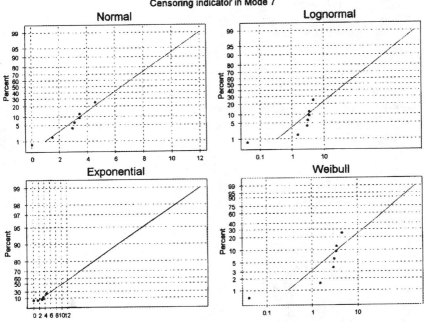

(f) Mode 7.

Figure 19.1. *Continued.*

- Mode 3—The only reasonable choice appears to be the normal
- Mode 4—The normal fits best; the Weibull may be reasonable if a few outliers are removed
- Mode 5—Fits are not bad, but lognormal and exponential appear to be better than either the normal or Weibull.
- Mode 7—Too little data to really tell, but the normal appears to provide the best fit

Overall, the normal distribution appears to provide the best or nearly best fit. The Weibull may also be reasonable, particularly if suspected outliers at the lower end are removed. To compare the modes more precisely, we look at the estimated parameters of the normal and Weibull distributions for each mode. The results are given in Table 19.2. Also included in the table are n_f, the observed number of failures by mode, the estimated coefficient of variation, CV (see Chapter 3) for each mode, and the corresponding statistics for the composite data.

A number of conclusions can be drawn from these results:

- Failure modes 1 and 4—upper sill delamination and burnout—are the most frequently occurring failures (jointly accounting for 63.6% of failures) and, as would therefore be expected, have the lowest estimated MTTF. Any engineering changes in design and/or production that could reduce the failure rates for these modes would have a significant impact on the overall failure rate.
- Note that the estimated MTTF's (μ^*) for individual modes are all higher than that of the composite failure distribution. This is to be expected since the composite is effectively the distribution of the minimum TTF over the different modes. This is also reflected in the Weibull estimates, since for the Weibull distribution $\mu = \beta\Gamma(1 + 1/\alpha)$ (see Chapter 4).
- It is interesting to note that for the Weibull fits, the estimates of α, the shape parameter, are roughly comparable and are approximately equal to that of the composite distribution.
- Since the variability in TTF may be expected to be larger for individual failure modes than for the composite (failures for the individual modes occur less frequently and hence are more spread out), we may expect larger values of σ^* for the individual modes than for the composite, and this is observed.
- By the same reasoning, we may expect the standard deviation to be related to the mean, and, in fact, to increase as the mean increases. This is supported by the sample

Table 19.2. Maximum likelihood estimates of normal and Weibull parameters for different failure modes, windshield data

Mode	μ^*	σ^*	CV (%)	α^*	β^*	n_f
1	4.475	1.799	40.2	2.258	5.578	30
2	7.676	2.442	31.8	2.649	11.075	4
3	5.869	1.791	30.5	2.933	7.507	9
4	4.478	1.487	33.2	2.990	5.189	26
5	6.063	2.247	37.1	2.220	8.504	12
7	6.666	2.185	32.8	1.618	17.163	7
All	3.041	1.241	40.8	2.443	3.452	88

coefficients of variation, defined to be $s/\bar{x}$, given in Table 19.2. These are remarkably consistent, given the small sample sizes, ranging from about 30% to 40%. (See Chapter 5 for a discussion of the CV.)

Goodness-of-Fit. For ease of analysis, it would be desirable to use the composite data set in estimating costs. (The alternative is to employ competing risk models (see Section 4.6), which would lead to significant analytical difficulties in analysis of the warranty cost models to be discussed below.) Based on the discussion in the previous sections, we choose the normal and Weibull models as reasonable alternatives for modeling the composite data.

The next step in the analysis is a test of fit to these distributions. Unfortunately, goodness-of-fit tests, such as the Kolmogorov–Smirnoff and Anderson–Darling tests discussed in Chapter 11, are not applicable to Type I multiply censored data (see Lawless, 1982 for some approaches for censored data). As a compromise, we analyze only the failure data (which, in fact, is all that would be available in many applications). A plot of the failure data on normal probability paper and the value of the Anderson–Darling statistic for the failure data are given in Figure 19.2. The *p*-value of 0.086 indicates that the normal distribution is not rejected at the 5% level.

For testing fit to the Weibull distribution, we use the Anderson–Darling statistic given in (11.12), with the parameters estimated from the data (see Table 19.2). The calculated value is found to be $A^2 = (8251.2/88) - 88 = 5.7636$. The tabulated value for testing at the 1% level is found from Table 11.2 to be 3.857. The Weibull distribution is rejected at the 1% level. Repeating the test after deletion of the lowest three observations, we obtain $A^2 = 0.709$, and the Weibull is not rejected. Possible explanations are that the lowest observations are outliers or that a mixture of Weibulls may be an appropriate model.

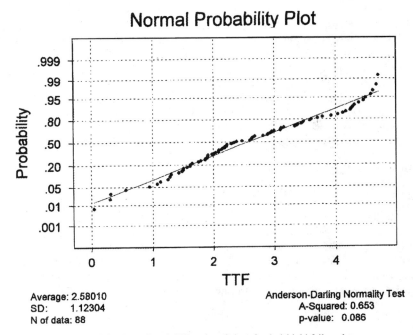

Figure 19.2. Normal probability plot of aircraft windshield failure data.

Since the Weibull distribution was initially assumed and the normal provides a good fit, in our further analyses we will use both the normal and Weibull models, with parameter estimates based on the entire data set for cost estimation. (In studying the sensitivity of the results, we will also consider the Weibull with estimates based on the data omitting the lowest three observations.) Note that neither of these choices is entirely satisfactory. The problems are:

- The Weibull distribution does not provide a good fit to the entire data set. On the other hand, it is difficult to justify deleting the lowest observations just to obtain a better fit, so these are used in estimating the parameters.
- The normal distribution appears to fit quite well. The problem here is that $\mu^* = 2.45\sigma^*$, which gives 0.0071, or nearly 1% probability to the negative region of the time axis. This is not possible for lifetime data. An alternative might be to use a truncated normal distribution. This will be discussed briefly below.

This suggests that further analysis of the data may be desirable. Many other of the distributions discussed in Chapter 4, including mixtures of Weibull distributions or other mixtures, may be applicable.

Some lessons that may be learned from this discussion are:

1. In analyzing data, some compromises between theory and practicalities must sometimes be made. In real problems, a decision must be reached. (On the other hand, it is essential to be aware of such compromises, and their possible consequences.)
2. The possibilities in modeling and analysis are seemingly endless.
3. As noted previously, data analysis often involves art as well as science.

Parameter Estimation and Confidence Intervals. Estimates of the parameters of the normal and Weibull distributions based on the complete data set are given in Table 19.2. Asymptotic normality of the estimators is used to obtain approximate confidence intervals for the parameters of both distributions. We require fractiles of the standard normal distribution, obtained from Appendix C, Table C1, and estimates of the asymptotic standard deviations of the estimators.

For the normal distribution, the estimated standard deviations were obtained in Example 8.7. The results are $\hat{\sigma}_{\mu^*} = 0.0840$ and $\hat{\sigma}_{\sigma^*} = 0.0523$. As a conservative approach, we calculate a lower 95% confidence interval for μ (resulting in the lowest value for the MTTF) and a two-sided 95% confidence interval for σ. The results are

- CI for μ: $3.041 - 1.645(0.0840) = 2.903$.
- CI for σ: $1.241 \pm 1.960(0.0523) = (1.138, 1.344)$

Note that this choice gives us approximately a 90% joint confidence region.

For the Weibull distribution, the maximum likelihood estimates of the parameters are $\alpha^* = 2.443$ and $\beta^* = 3.452$. To determine the asymptotic variances, we transform to the extreme value distribution, analyzing $y_i = \log(x_i)$. (See Chapter 8.) The estimates of the parameters of the extreme value distribution are $\lambda^* = 1/\alpha^* = 0.4093$ and $\mu^* = \log(\beta^*) = 1.2390$. The estimated elements of the information matrix are obtained from (8.18)–(8.20). The results are

$$I_{11} = \frac{r}{\lambda^{*2}} = \frac{88}{0.4093^2} = 525.290$$

$$I_{12} = \frac{1}{0.4093^2} \sum_{i=1}^{153} z_i e^{z_i} = \frac{-6.1781}{0.4093^2} = -36.878$$

and

$$I_{22} = \frac{88}{0.4093^2} + \frac{1}{0.4093^2} \sum_{i=1}^{153} z_i^2 e^{z_i} = 863.020$$

where $z_i = (y_i - \mu^*)/\lambda^*$. The estimated asymptotic covariance matrix of (λ^*, μ^*) is

$$\hat{V}_{\mu^*, \lambda^*} = \begin{pmatrix} 525.290 & -36.878 \\ -36.878 & 863.0202 \end{pmatrix}^{-1} = \begin{pmatrix} 0.001909 & 0.00008159 \\ 0.00008159 & 0.001162 \end{pmatrix}$$

From (8.22) and (8.23), we obtain the approximate variances of the estimators of the parameters of the Weibull distribution as

$$\hat{V}(\alpha^*) = 2.443^4(0.001162) = 0.04139$$

and

$$\hat{V}(\beta^*) = 3.452^2(0.001909) = 0.002275$$

giving estimated standard deviations of $\hat{\sigma}_{\alpha^*} = 0.2034$ and $\hat{\sigma}_{\beta^*} = 0.1508$. As for the normal distribution parameters, we take a conservative approach with regard to the warranty cost estimation to follow and calculate a lower 95% confidence bound for β and a two-sided 95% confidence interval for α. The results are:

- CI for α: $2.443 \pm 1.96(0.2043) = (2.044, 2.842)$
- CI for β: $3.452 - 1.546(0.1508) = 3.204$.

19.2.3 Estimation of Warranty Costs

In estimating warranty costs, we consider the nonrenewing FRW and the nonrenewing PRW for warranty periods of 0.5 to 8.0 thousands of flight hours. This corresponds roughly to periods of 3 to 4 months to a maximum of 2 to 2.67 years. Although warranty lengths at either extreme of this range would probably not be considered, this range provides an overview of the impact of warranty period on estimated costs.

Cost Models

We consider the average per unit cost to the manufacturer, $E[C_m(W)]$, for items sold under Warranty Policies 1 and 2. Cost models for both are given in Chapter 17. For nonrepairable items, the expected cost for the nonrenewing FRW with warranty period W is obtained from (17.4) as

$$E[C_m(W)] = C_s[1 + M(W)] \tag{19.1}$$

where C_s is the seller's (manufacturer's) average per unit cost of providing an item to the buyer (including amortized development costs, manufacturing cost, marketing, distribution, and any other incidental costs), and $M(W)$ is the renewal function associated with the life distribution of the item (see Chapter 17 and Appendix B). If the item is repairable, at an average cost per repair of C_r, (19.1) becomes

$$E[C_m(W)] = C_s + C_r M(W) \tag{19.2}$$

For the rebate form of the PRW, the cost model is obtained from (17.19) as

$$E[C_m(W)] = C_s + C_b\left[F(W) - \frac{\mu_W}{W}\right] \tag{19.3}$$

where C_b is the cost to the buyer (the selling price of the item), and μ_W is the partial expectation defined in (17.20). To evaluate costs for the windshield problem, we require values of the renewal functions and partial expectations for the normal and Weibull distributions, as well as values for the various costs appearing in the models.

For purposes of illustration, we take the manufacturer's cost per unit to be $C_s = \$9,000$ and the selling price to be $C_b = \$17,500$. We will assume that the cost of repair (including trade-in allowance, if any, shipping costs and the cost of rework) is 60% of the cost of building a new item, giving $C_r = \$5,400$. For comparability in the case of the PRW, where a rebate rather than a replacement item is given, we assume that a new windshield can be repaired an average of three times. The average cost C_s' when repaired items are used then becomes $[\$9,000 + 3(5,400)]/4 = \$6,300$. The expected cost of a warrantied windshield is given by (19.3) with C_s' replacing C_s.

Evaluation of Renewal Functions

For the normal distribution, the renewal function is easily evaluated by use of expression (B3.14) in Appendix B. The result is

$$M(W) = \sum_{n=1}^{\infty} F^{(n)}(W) \tag{19.4}$$

where $F^{(n)}(\ )$ is the n-fold convolution of $F(\ )$ with itself, i.e., is the distribution of the sum of n random variables. Since the distribution of the sum of n independent normal random variables is normal with mean $n\mu$ and variance $n\sigma^2$, the sum in (19.4) is easily calculated, usually requiring only a few terms.

For the Weibull distribution, numerical methods or tables are required. The program given in Blischke and Murthy (1994) was used in the calculations below.

Evaluation of Partial Expectations

For the normal distribution, the partial expectation can be determined analytically. The result is

$$\mu_W = \int_{-\infty}^{W} \frac{x}{\sqrt{2\pi\sigma}} e^{-(x-\mu)^2/2\sigma^2} dx = \mu\Phi(z_W) - \frac{\sigma}{\sqrt{2\pi}} e^{-z_W^2/2} \tag{19.5}$$

where $z_W = (W - \mu)/\sigma$ and $\Phi(\ )$ is the CDF of the standard normal distribution.

For the Weibull distribution, the computation involves the incomplete gamma function and is somewhat more tedious. Blischke and Murthy (1994, p. 176), provide an expression for the partial expectation involving an infinite series. The result is

$$\mu_W = \beta\gamma\left(1 + \frac{1}{\alpha}, \left(\frac{W}{\beta}\right)^\alpha\right) \tag{19.6}$$

where $\gamma(a, x)/\Gamma(a)$ is the incomplete gamma function (see Abramowitz and Stegun, 1964.) $\gamma(a, x)$ may be evaluated by summing a sufficient number of terms in the series

$$\gamma(a, x) = a^{-1}x^a e^{-x}\left[1 + \frac{x}{a+1} + \frac{x^2}{(a+1)(a+2)} + \frac{x^3}{(a+1)(a+2)(a+3)} + \cdots\right] \tag{19.7}$$

with $a = 1 + 1/\alpha$ and $x = (W/\beta)^\alpha$.

Estimates of M(W) and μ_W

We consider the Weibull and normal distributions with MLEs of the parameters based on the complete sample of 153 observations, repair versus nonrepair, and the FRW versus the PRW, for a total of eight cases, with warranty coverage of $0.5, 1.0, \ldots, 8.0$ thousand hours.

Estimates of the renewal function are obtained by substitution of the MLEs into the expressions for $M(W)$, as indicated previously. The results are given in Table 19.3. Note that the difference between the normal and Weibull renewal function estimates is not substantial, and appears to decrease as W increases. Note also that long warranty periods are not likely to be considered unless the price of the windshield is raised considerably, since the expected number of replacements becomes quite large as W increases.

For analysis of the PRW, we require estimates of $F(W)$ and μ_W for the normal and Weibull distributions. The estimates are obtained by substitution of the MLEs into (19.5)–(19.7). The results are given in Table 19.4.

In Table 19.4, the results for the normal and Weibull differ somewhat more. For the CDF, the reason is that the normal assigns probability 0.0071 to the negative half of the axis, as not-

Table 19.3. Estimated renewal function for the normal and Weibull distributions

W	$M(W)$—Normal	$M(W)$—Weibull
0.5	0.0211	0.0089
1.0	0.0520	0.0476
1.5	0.1119	0.1240
2.0	0.2113	0.2381
2.5	0.3532	0.3833
3.0	0.5287	0.5491
3.5	0.7196	0.7241
4.0	0.9070	0.8994
4.5	1.0806	1.0700
5.0	1.2412	1.2352
5.5	1.3961	1.3968
6.0	1.5529	1.3968
6.5	1.7148	1.5573
7.0	1.8808	1.8808
7.5	2.0484	2.0443
8.0	2.2152	2.2083

Table 19.4. Estimates of the CDF and partial expectations, normal and Weibull distributions

W	$F(W)$—Normal	$F(W)$—Weibull	μ_W—Normal	μ_W—Weibull
0.5	0.0203	0.0089	0.0009	0.0032
1.0	0.0500	0.0473	0.0241	0.0334
1.5	0.1072	0.1224	0.0969	0.1288
2.0	0.2009	0.2317	0.2623	0.3214
2.5	0.3314	0.3653	0.5577	0.6228
3.0	0.4868	0.5082	0.9856	1.0159
3.5	0.6443	0.6445	1.4968	1.4583
4.0	0.7802	0.7615	2.0052	1.8958
4.5	0.8801	0.8521	2.4284	2.2797
5.0	0.9428	0.9156	2.7246	2.5803
5.5	0.9762	0.9559	2.8992	2.7908
6.0	0.9914	0.9789	2.9861	2.9229
6.5	0.9973	0.9908	3.0227	2.9970
7.0	0.9993	0.9964	3.0358	3.0343
7.5	0.9998	0.9987	3.0397	3.0510
8.0	0.99997	0.9996	3.0407	3.0577

ed, which accounts for the higher cumulative normal probabilities. The partial expectations will converge to the mean μ (or in our case, the estimated mean) as $W \to \infty$. For the normal distribution, the estimated mean is 3.041, and μ_W has very nearly reached this value at $W = 8$. For the Weibull distribution, the mean is $\beta\Gamma(1 + 1/\alpha)$, which is estimated to be 3.452 $\Gamma(1 + 1/2.443) = 3.062$ (where the Γ-function is evaluated using the tables of Abramowitz and Stegun, 1964). Again, μ_W for the Weibull has nearly reached this value at $W = 8$.

Estimates of Warranty Costs

Cost factors C_s, C_r, and C_b were given previously. We compare costs for the FRW and PRW being considered for both the repair and nonrepair options. Results are given for both the normal and Weibull distributions with parameter estimates as before. Given the above computations, the estimated costs per unit to the manufacturer for items under warranty are easily calculated from (19.1)–(19.3).

Estimated costs per unit as functions of warranty period W (in thousands of hours of flight time) for both policies are given in Table 19.5 under the repair option. For the PRW, the costs are calculated using C_s' in place of C_s in (19.3) for reasons indicated previously. Note that the FRW is more costly than the PRW, as expected. The difference between the results for the normal and Weibull distributions is not large, but could be a factor as sales increased. Note that the conservative choice, i.e., the distribution giving the higher cost estimate, depends on W, with the pattern changing alternately as W increases, and that this happens for both the PRW and FRW.

It is apparent from the results that an FRW of longer than 6,000 hours would lead to a loss and that substantially shorter periods (say, 3,000 hours) would be required for acceptable levels of profitability. Under the rebate PRW, longer warranties could be offered. Alternatively, longer warranties could be offered at a higher selling price.

Estimated costs for the option of not repairing failed items (i.e., always replacing failures by new items) are given for the same combination of warranties in Table 19.6. As expected,

Table 19.5. Estimated costs ($) for repairable windshields sold under free replacement and pro-rata warranties, normal and Weibull distributions

W	FRW		PRW	
	Normal	Weibull	Normal	Weibull
0.5	9,114	9,048	6,624	6,345
1.0	9,281	9,257	6,754	6,543
1.5	9,604	9,670	7,045	6,939
2.0	10,141	10,286	7,518	7,542
2.5	10,907	11,070	8,196	8,334
3.0	11,855	11,965	9,070	9,268
3.5	12,885	12,910	10,090	10,288
4.0	13,898	13,857	11,180	11,332
4.5	14,835	14,778	12,258	12,346
5.0	15,703	15,670	13,263	13,292
5.5	16,539	16,543	14,159	14,148
6.0	17,386	17,409	14,941	14,906
6.5	18,260	18,279	15,615	15,571
7.0	19,156	19,156	16,198	16,151
7.5	20,061	20,039	16,704	16,658
8.0	20,962	20,925	17,148	17,104

Table 19.6. Estimated costs for windshields sold under free replacement and pro-rata warranties, nonrepair option, normal and Weibull distributions

W	FRW		PRW	
	Normal	Weibull	Normal	Weibull
0.5	9,190	9,080	9,324	9,045
1.0	9,468	9,428	9,454	9,243
1.5	10,007	10,116	9,745	9,639
2.0	10,902	11,143	10,218	10,242
2.5	12,179	12,450	10,896	11,034
3.0	13,758	13,942	11,770	11,968
3.5	15,476	15,517	12,790	12,988
4.0	17,163	17,095	13,880	14,032
4.5	18,725	18,630	14,958	15,046
5.0	20,171	20,117	15,963	15,992
5.5	21,565	21,571	16,859	16,848
6.0	22,976	23,016	17,641	17,606
6.5	24,433	24,466	18,315	18,271
7.0	25,927	25,927	18,898	18,851
7.5	27,436	27,399	19,404	19,358
8.0	28,937	28,875	19,848	19,804

costs per item for the nonrepair option are substantially higher than under the repair policy. As a result, profitability decreases even more rapidly as the warranty period increases and even shorter warranties must be offered or larger price increases put into effect.

In the next two subsections, we look more carefully at the sensitivity of these results to the distributional and other assumptions and summarize the key management conclusions as a result of the study.

19.2.4 Sensitivity Analysis

We are interested in the sensitivity of the cost analysis results to the selected failure distributions as well to the values of the parameters used in the computations. As noted in the previous section, the results are not overly sensitive to the distribution in the sense that the cost estimates do not vary much under the two assumptions. Here we pursue this further, addressing two key issues:

- Can the normal and Weibull distributions be improved upon in terms of fit, and does this change the results?
- How does uncertainty in the parameter estimates affect the cost estimates?

Modified Normal and Weibull Distributions
Difficulties with the normal and Weibull distributions have been noted previously. The problem with the normal distribution is that μ is not sufficiently large or σ sufficiently small relative to μ so that probability assigned by the distribution to the negative half of the axis is negligible. A possible fix is to use instead the truncated normal distribution, given by

$$f(x) = \frac{1}{a\sigma\sqrt{2\pi}} e^{-(x-\mu)^2/2\sigma^2} \tag{19.8}$$

for $x > 0$, where $a = 1 - \Phi(-\mu/\sigma)$, with $\Phi(\ )$ being the standard normal distribution. For the PRW, the CDF is easily calculated from standard normal tables or by computer, and the partial expectation required in the cost formula can be derived. The renewal function required in the cost equation for the FRW is tabulated by Baxter et al. (1981) and included in the renewal program given in Blischke and Murthy (1994). Thus cost estimates for the two warranties assuming the truncated normal distribution are easily calculated.

As a small indication of sensitivity, we look at the renewal function for the normal and truncated normal, both with mean 3.041 and standard deviation 1.241, for a few values of W. The results are given in Table 19.7.

Note that the normal results are higher in every case than those for the truncated normal. Thus use of the normal distribution would be conservative in the sense that costs would be overestimated. For small values of W, the overestimation is substantial (20% for $W = 1$), but this tails off quickly to 2% for $W = 3$ and about 1% for larger values.

It is easy to see that the probabilities $F(W)$ for the normal and truncated normal will also not differ substantially in this application, nor will the partial expectations. Details of the resulting cost analyses are assigned as an exercise at the end of the chapter.

In the case of the Weibull distribution, the problem is that there are some apparent outliers. In Section 19.2, we considered goodness-of-fit to the Weibull after deleting the lowest three failure times, and found that the fit was acceptable. In the subsequent cost analyses, however, we used parameter estimates based on the complete data set. We now look briefly at the effect of this on the results.

**Table 19.7. Renewal function for normal and truncated
normal distributions**

W	M(W)—Normal	M(W)—Truncated normal
1	0.0520	0.0436
2	0.2113	0.2007
3	0.5287	0.5151
4	0.9070	0.8905
5	1.2412	1.2118
6	1.5529	1.5306
7	1.8808	1.8555
8	2.2152	2.1868

A probability plot of the Weibull CDF with the three observations in question omitted in given in Figure 19.3. The parameter estimates change somewhat but not substantially, α^* changing from 2.443 to 2.827, and β^* from 3.452 to 3.490. The effect of this on the estimated CDF is shown Table 19.8. For purposes of comparison, the normal CDF for the complete data set and MLEs of the parameters is also included. Note that the edited data differ fairly significantly in some respects. The effects of this on the cost estimates is left as an exercise. More significant are the effects of uncertainty in the parameters, which we turn to next.

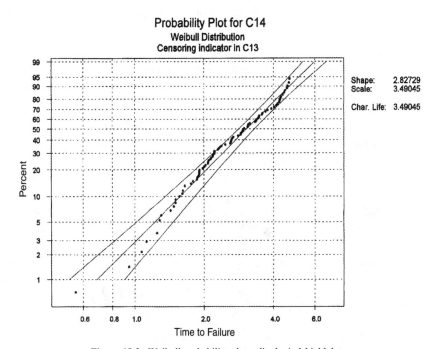

Figure 19.3. Weibull probability plot, edited windshield data.

Table 19.8. Weibull CDF for complete and edited windshield data and normal CDF for complete data

	$F(W)$		
W	Weibull, complete	Weibull, edited	Normal
1	0.0473	0.0407	0.0500
2	0.2317	0.2625	0.2008
3	0.5082	0.6230	0.4868
4	0.7615	0.8923	0.7802
5	0.9156	0.9855	0.9428
6	0.9789	0.9992	0.9914
7	0.9964	1.0000	0.9993
8	0.9996	1.0000	1.0000

Effect of Uncertainty in Parameter Estimates on Estimated Costs

We consider the complete data set. Uncertainty in the estimates of the parameters of the normal and Weibull distributions is expressed in the confidence intervals calculated in the previous section. We wish to determine the cost consequences of changing parameter values. The most important characteristic in this regard is the MTTF. Smaller values of the mean will lead to more replacements under FRW and to a higher rebate under the rebate PRW being considered, and hence to higher costs in either case.

For the normal distribution, we calculate estimated costs using the lower confidence bound for the mean combined with the upper and lower bounds for the standard deviation. The lower bound on the mean will give a conservative cost estimate (i.e., a high estimate of cost and hence a low estimate of profit). The two limits on σ will enable us to assess the effect of uncertainty in that parameter. The results will be compared with those given previously using the point estimates. By the same reasoning, for the Weibull distribution, we use the lower bound on β and the two-sided bounds on α plus the results based on the point estimates. In summary, the estimated parameter values used in the sensitivity study are:

- Normal distribution:
    - Case (i) $(\hat{\mu}, \hat{\sigma}) = (3.041, 1.241)$
    - Case (ii) $(\hat{\mu}, \hat{\sigma}) = (2.903, 1.138)$
    - Case (iii) $(\hat{\mu}, \hat{\sigma}) = (2.903, 1.344)$
- Weibull distribution:
    - Case (iv) $(\hat{\alpha}, \hat{\beta}) = (2.443, 3.452)$
    - Case (v) $(\hat{\alpha}, \hat{\beta}) = (2.044, 3.204)$
    - Case (vi) $(\hat{\alpha}, \hat{\beta}) = (2.842, 3.204)$

We consider the repair option and estimate costs per unit for both the FRW and PRW for the six cases. Warranty periods used are $W = 1, \ldots, 8$ thousand hours. Computations proceed as before and we omit the details.

Results for the FRW are given in Table 19.9. It can be seen that costs do not vary a great

Table 19.9. Estimated per-unit costs ($) for the FRW for the normal and Weibull distributions, Cases (i)–(vi)

W	Case (i)	Case (ii)	Case (iii)	Case (iv)	Case (v)	Case (vi)
1	9,281	9,263	9,457	9,257	9,484	9,194
2	10,141	10,205	10,489	10,286	10,823	10,266
3	11,855	12,113	12,276	11,965	12,644	12,230
4	13,898	14,250	14,335	13,857	14,597	14,332
5	15,703	16,060	16,235	15,670	16,529	16,182
6	17,386	17,838	18,036	17,409	18,430	18,016
7	19,156	19,729	19,884	19,156	20,326	19,930
8	20,962	21,609	21,760	20,925	22,227	21,834

deal across cases, but that for all values of W, the highest estimated costs are associated with Case (v), the Weibull distribution with highest α and lowest β. For this case, the item sold with FRW with $W = 5,000$ hours is still profitable, but with a small profit margin. For higher values of W, a loss is expected.

Results for the PRW for the same six cases are given in Table 19.10. Note that the results are again very comparable across cases. It can be seen that, as expected, all rebate PRW costs to the manufacturer are substantially less than FRW costs for the same warranty length and the same parameter values

For the PRW, there is no consistent worst case, either overall or for the individual distributions. The highest estimated cost depends on W. With a selling price of $17,500, nearly all of the warranties considered would be profitable, although the profit margin would be unacceptably low for the longer warranties.

19.2.5 Summary of Conclusions

In this section, we summarize some of the key results of the previous sections, list some of the engineering and management options suggested by the results, and indicate areas that may require additional analysis.

Table 19.10. Estimated per-unit costs ($) for the PRW for the normal and Weibull distributions, Cases (i)–(vi)

W	Case (i)	Case (ii)	Case (iii)	Case (iv)	Case (v)	Case (vi)
1	6,754	6,690	7,133	6,543	6,818	6,465
2	7,518	7,511	8,062	7,542	8,265	7,409
3	9,070	9,241	9,719	9,268	10,244	9,328
4	11,180	11,543	11,786	11,332	12,297	11,710
5	13,263	13,691	13,760	13,292	14,113	13,853
6	14,941	15,336	15,347	14,906	15,579	15,477
7	16,198	16,543	16,544	16,151	16,715	16,665
8	17,148	17,450	17,450	17,104	17,593	17,557

Statistical Analysis

Some key statistical conclusions and areas needing further investigation are:

- The results obtained appear to be relatively insensitive to the distribution used and the parameter values, changing estimated costs only a few percent except in a few cases. This provides some credibility to the conclusions drawn.
- Many other failure distributions may fit the data as well. This requires further investigation and additional sensitivity studies.
- Tests for outliers should be pursued.
- Models that fit all of the data, including the outliers should be developed.
- Models that take into account other factors for which information may be obtained (e.g., number of landings, flight duration), may be considered. The Cox regression model for the failure intensity function is an example.
- Further analysis of the data, based on competing risk models, would be useful. This may require development of new theory and algorithms for calculation of the associated renewal functions, partial expectations, and other elements of warranty cost models.
- Goodness-of-fit tests for incomplete data should be applied to the data set.

Engineering Analysis

Some key engineering conclusions and areas needing further investigation are:

- The various failure modes require further analysis. As noted, the most frequently occurring and hence most costly are Modes 1 and 4. Any redesign that would reduce the frequency of or eliminate these modes would significantly reduce costs.
- An understanding of the physical processes leading to failure by the various modes would assist is selection of an appropriate failure distribution.
- Warranty costs can be reduced by improving the reliability of the product. This may require redesign or modification of the process. This may include changing raw materials, changing manufacturing conditions such as temperature and pressure during lamination, and so forth.
- A cost–benefit analysis to study the trade-off between the increased cost of improving reliability and the decreased warranty costs should be performed.
- If changes in product or process design are to be considered, properly designed experiments to determine the effect of the various factors and selection of optimal settings for optimization of product reliability should be undertaken.
- The context of the lowest three or four observations (source, conditions, correctness of data, etc.) should be investigated. These appear to be outliers and are listed as having failed by different failure modes. An identified cause of the unusual values would justify their permanent exclusion from the data set.
- Repairability issues that need to be addressed include the number of times that an item may be rebuilt, repairability of the various failure modes, and whether or not repaired items are, indeed, good as new.
- Provision should be made for collection of detailed data on repaired items. (manufacture data, number of times repaired, TTF at each repair, failure mode, etc.)

Management Aspects

Some key management decisions and responsibilities are:

- It is clear that the pro-rata warranty entails significantly less cost to The Company than does the free replacement warranty. Which one is selected is also a function of the actions of competitors, however, and this issue must be addressed.

- The expected per unit cost of items sold with a renewing pro-rata warranty will be even lower. This is because the rebate given under a nonrenewing PRW is a function of the selling price, as is the discount given on a replacement item under a renewing PRW. In the later case, however, the cost to the supplier is simply the cost C_s of supplying a replacement item (see Blischke and Murthy, 1994). An added advantage is that in this case buyers are much more likely to replace a failed item with an item purchased from the same manufacturer rather than from a competitor. This option requires further analysis.

- There are many other warranty policies that might be considered. An alternative often used is a combination FRW/PRW (Policy 3 of Chapter 17 or some form of a renewing version of this policy). This also requires further analysis.

- Whichever warranty policy is selected, warranty costs can be expected to increase significantly as a function of the length of the warranty. Selection of the length of the warranty period must be done in the context of product, product line, and company profit objectives, using the tables of warranty costs provided. This decision also will depend, in part, on the actions of competitors.

- Although the results of the data analysis are relatively insensitive to the assumptions made from a statistical point of view, the cost estimates do change and the worst case should be considered in the decision making process.

- It is clear that repairing items to the extent possible greatly reduces the per-item cost. This does entail additional administrative costs, however, and the true cost of repair/rebuild, including tracking, shipping, etc., needs to be monitored.

- A related issue is whether or not to offer a trade-in on failed windshields returned for rework when failures occur after expiration of the warranty.

- The data of Table 2.2 are early operational data for this particular model windshield. Subsequent engineering changes in the process and improvements in materials have substantially increased the mean time to failure of this product. An ongoing collection of data of this type analysis of the results would involve reliability growth models and periodic updating of the cost estimates. This effort should be supported.

- Another way of reducing warranty costs is improvement of product reliability. This entails an up-front cost to support the required engineering effort. Management involvement and support is essential

19.3   RELIABILITY ANALYSIS OF CASE 2.5

Case 2.5 involves a highly specialized industrial product and user-supplied data on failure times and usage. There are a number of problems with the data, some of which will be discussed below, and many missing data elements that would have been useful in analyzing the data and assessing product reliability. Here we look at the problem of structuring and analyzing the data and using the results for estimating reliability and related quantities.[2]

19.3.1   Statement of the Problem

The product in question is relatively new, with data collection spanning 22 months since the initial installations. All items are tested prior to shipment. The data are operational data, collected under less than rigorous conditions. Although the data were obtained as part of a monitoring effort in connection with an evaluation of warranty costs, the warranty aspects will not be considered in our analysis because of the considerable uncertainty in the basic information. Instead, we focus on the basic aspects of data analysis and reliability assessment.

The key issues to be addressed in the data and reliability analyses are:

1. Interpretation of the structure of the data and completion of an appropriate data analysis
2. Estimation of product reliability
3. Development of a program for obtaining an adequate data base for a proper reliability analysis
4. Formulate recommendations with regard to engineering and management aspects as appropriate

In short, our objective is to perform a preliminary reliability analysis based on the available data and to provide the basis for formulating a plan for future data collection and a more detailed reliability analysis.

19.3.2   Data Analysis

The first issue to be dealt with is structuring of the data. The product is sold in batches and data on product performance are given on the batch as a whole. In some cases, the data are obtained at the time of a warranty claim due to the failure of one or more items in the batch. In other cases, service time of the batch is given, even though no failures have occurred. In order to conduct even a preliminary analysis of the data, assumptions must be made concerning its interpretation. We assume the following:

- All items are put into service immediately on delivery. Thus we assume no storage time and that all data are either actual failure times or service times.
- Reported failures are failures of separate items, except for Batch 6. Thus we assume one failure per item.
- For Batch 6 (two items; three reported failures), we assume that one item failed at 255 days of service, was repaired good as new in a negligibly short time and returned to service, and then failed a second time at 354 days, and that the second item failed at 355 days.

In the above interpretation, all ages of items for which failures have not been reported are considered to be service times. This results in the following data for Batches 1, 2, 5, and 14:

- For Batch 1, we have two items and one failure time. We assume that one failure occurred at 642 days and one item was in service for 679 days.
- Batch 2: two service times of 636 days.

- Batch 5: Failure times of 255 and 350.
- Batch 14: seven failure times (104, 154, etc.) and 93 service times of 469.

Data for the remaining batches are treated similarly.

There are a number of difficulties with this interpretation of the data. One key problem is that if the times reported do include storage time, then the estimate of MTTF will overestimate the average lifetime of the item (unless deterioration of the product takes place at the same rate when in storage as when in use, which is unlikely). A second key problem is the possibility that some items failed more than once. A number of other data issues that must be addressed in dealing with the existing data and providing for improved data in the future will be discussed below. A third problem is that failures may be due to several causes, e.g., damage during shipping, improper installation, errors in manufacturing. We proceed with the data analysis, with the caveat that any conclusions reached must be considered tentative.

Data Analysis

We begin, as usual, with plots of the data. The four-way probability plot is given in Figure 19.4. It is apparent that none of the fits is good and that the problem may be that there may be one or two outliers. Figure 19.5 is a four-way probability plot with the largest two failure times omitted. Here all four distributions appear to give reasonable fits, though it is difficult to tell much about the exponential and the normal plot appears to have roughly a quadratic rather than linear form. We accept the Weibull and lognormal models as tentative

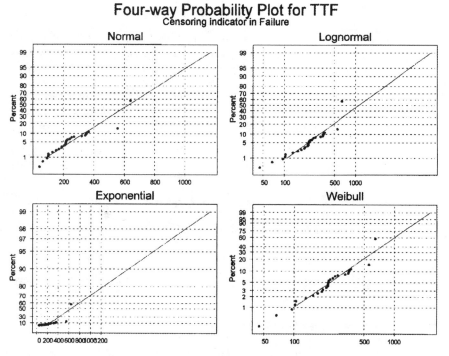

Figure 19.4. Four-way probability plot for Case 2.5 data.

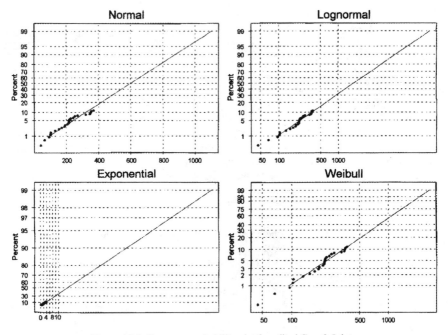

Figure 19.5. Four-way probability plot for edited Case 2.5 data.

failure distributions. A crude test of the lognormal is shown in Figure 19.6, which is a test of normality on the log scale using only failure data (including the possible outliers). Subject to the validity of the test, normality of $y = \log(x)$, i.e., the lognormal distribution, would not be rejected.

MLEs of the parameters of the lognormal and Weibull distributions may be obtained from the Minitab probability plots for these distributions given in Figure 19.7(a) and (b). Here all data are used and the results are

- Lognormal: $\mu^* = 7.6989$; $\sigma^* = 1.3366$
- Weibull: $\alpha^*\ 1.5417$; $\beta^* = 1751.0$

If the highest two observations are deleted, the corresponding estimates are

- Lognormal: $\mu^* = 7.8561$; $\sigma^* = 1.4269$
- Weibull: $\alpha^* = 1.4248$; $\beta^* = 2053.6$

We look next at the sensitivity of reliability estimates and other results to the distributional assumptions.

Estimation of Reliability and Related Quantities
It is instructive to look at plots of the probability distributions for the parameter values given above. The density functions $f(x)$ and CDFs $F(x)$ are given in Figure 19.8(a) and (b), respectively. We see that although the lognormal and Weibull distributions agree quite closely for

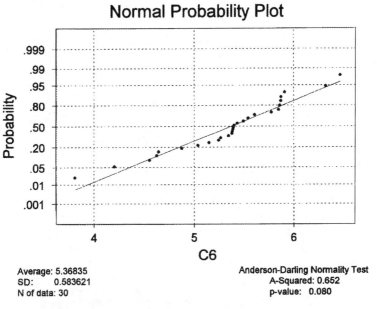

Figure 19.6. Normal probability plot of Case 2.5 failure data.

small values of x, they diverge very significantly as x increases. As a result, the estimated distributions would provide estimates of reliabilities of interest, e.g., 0.95 or 0.99, that are in good agreement, since these correspond to failure probabilities of 0.05 and 0.01, respectively. The disagreement of the distributions as a whole appears to reflect the effect of the relatively large number of censored observations and the associated uncertainty in determining the upper part of the distribution.

This difference is also reflected in the MTTFs of the distributions. The estimates of these, based on the parameter estimates given above, are:

- Lognormal all data: 2.206
- Lognormal, omit two: 2.581
- Weibull, all data: 1.576
- Weibull, omit two: 1.887

Two other quantities that would be of interest in assessing product reliability are estimated fractiles of the failure distribution and tolerance intervals. Point estimates and approximate 95% confidence intervals for fractiles are obtained from the Minitab output. The results for 95% confidence intervals for the lognormal and Weibull distribution, using all data and omitting two observations in both cases, are given in Table 19.11.

Note that these results do not differ dramatically. The estimated 0.05 fractile, corresponding to 0.95 reliability, is about 250, with a 95% confidence interval of roughly 190 to 335. The estimated 0.10 fractile, corresponding to reliability 0.90, is about 410 with a CI of roughly 310 to 525. Based on this preliminary analysis (and the validity of all of the assumptions

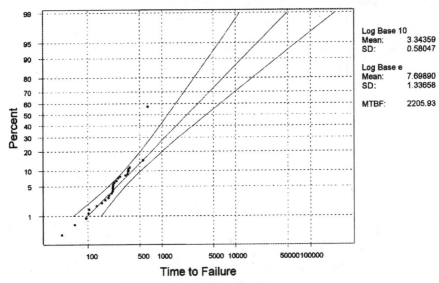

(a) Lognormal Distribution.

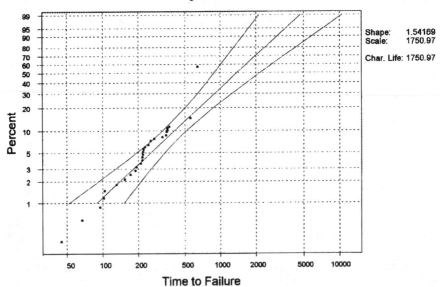

(b) Weibull Distribution.

Figure 19.7. Probability plots and MLEs for Case 2.5 data.

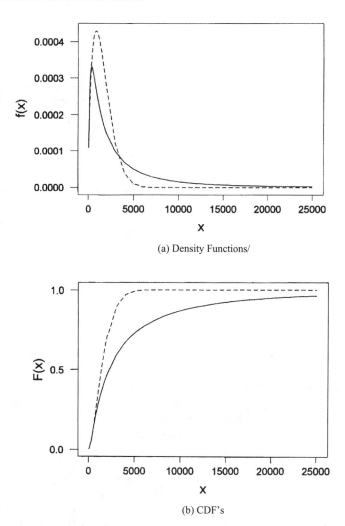

(a) Density Functions/

(b) CDF's

Figure 19.8. Estimated lognormal and Weibull distributions, Case 2.5 (— Lognormal; - -Weibull).

Table 19.11. 95% Confidence intervals for fractiles of estimated lognormal and Weibull distributions

		Reliability			
		0.90		0.95	
		Estimate	95% CI	Estimate	95% CI
All Data	Weibull	407	(323,513)	255	(193,336)
	Lognormal	398	(307,516)	245	(188,319)
Omit 2	Weibull	423	(326,549)	255	(189,345)
	Lognormal	415	(312,552)	247	(186,328)

made regarding the data and failure distribution), we estimate that the reliability of the product over a one-year period to be somewhat above 90%.

19.3.3 Conclusions

It is clear from the discussion of the previous sections that the results are preliminary and that caution must be exercised in interpreting them. This is not uncommon in real applications and it is important to recognize the limitations of both the data and analyses. In the remarks that follow, we indicate some tentative conclusions that may be drawn and areas for future research, including further data collection, engineering and statistical analysis, and some management responsibilities if a serious reliability study is to be undertaken.

Many of the difficulties to be faced in a proper reliability analysis in this case involve engineering (design, maintainability, etc.), statistics (data collection, design of experiments, data analysis), and management (decisions regarding maintenance, etc., and provision of necessary resources for engineering and analysis). In situations such as this, it is essential that an integrated approach, involving all of the relevant disciplines, be taken. With this in mind, we list the issues under the key discipline involved.

Engineering Issues

- Design and production issues have not been addressed in this study. Relevant issues should be analyzed.
- Information on cause of failure is not included in the data. This should be pursued and the resulting information analyzed.
- Information concerning performance of prototypes or any other laboratory data has not been used in the analysis. Any such data that can be obtained should be analyzed.
- Information on similar systems, suppliers, etc., would be useful.
- Block diagrams, fault trees, and other aids to reliability analysis are required.
- Carefully designed experiments to investigate the reliability of the system and its components may be necessary in order to obtain better estimates of system reliability.

Statistical Issues

- There are many alternative interpretations of the data, and the effect of these on the conclusions should be investigated. Two possibilities are:
  1. The item was put into service at time 0, repaired essentially instantaneously upon failure, and then performed its function until the age given in the data. In this case, we have a service time of (age − TTF) in addition to the TTF. How this is to be used in the analysis depends on the assumption made about the distribution of lifetimes of repaired items.
  2. Make some assumptions, albeit arbitrary, about the length of storage of items, particularly when bought in relatively large batches.
- Much more detailed information concerning the data is required. Important details include the following:
  1. Number of installations (and hence the number of items initially stored)
  2. Time at which each item was put into service
  3. Cause of failure
  4. Actual storage time for items stored
  5. Whether failed items were repaired or replaced
  6. Environmental conditions during storage and operation

- The fit of many other distributions to the data as structured should be investigated.
- More complex models, e.g., mixture or competing risk models) that take into account multiple failure modes should be used in analysis of the existing data.
- Models that include age of the item as a covariate and various alternatives concerning operating time would be useful.

Management Issues
- Efforts to obtain more detailed data from buyers should be made.
- If reliability is, indeed, a concern, resources for pursuing the necessary engineering and related analyses must be made available.

NOTES

1. A number of interesting cases are given by Elsayad (1996). Illustrative examples based on real data may be found in Lloyd and Lipow (1962), Kapur and Lamberson (1977), Kalbfleisch and Prentice (1980), Lawless (1982), Nelson (1982), and many of the other sources cited in previous chapters.

2. This case originated in an analysis of warranty issues. The data given were the only data available, and the client had little other hard information. A number of difficulties were mentioned in Sections 19.3. Some other issues not addressed in the analysis were:

- The item is used in regions with very different environments, ranging from hot, dry desert conditions to warm, humid tropics. This may affect the failure rate and that was not taken into consideration in the modeling or analysis.
- Proper installation is very critical for the item to function properly. Installation was often done by personnel who may not have had adequate training or understanding of the product.
- Variable rules were used for reporting failures and for returning items to the manufacturer, e.g., ship when a lot of a certain size is attained or ship when it is convenient to do so.

Another approach to analysis of the data is given in Lyons and Murthy (1996). In their analysis, failures were treated as having occurred according to a nonstationary point process with Weibull intensity function. The results include MLEs of the parameters of the model.

EXERCISES

1. Recalculate the cost per unit of warrantied windshields if the selling price in $18,500 per unit. If the profit goal is $4,000 per unit, what length FRW can be offered with repair? Without repair?
2. Repeat Exercise 1 for the PRW and compare the results.
3. Repeat Exercises 1 and 2 with $C_b = \$20,000$.
4. (Calculus required). Derive an expression for the partial expectation μ_W for the truncated normal distribution with density function given by (19.8).
5. Use the results of the previous exercise and tables or a computer algorithm for the renewal function to calculate cost estimates for items sold under FRW and PRW analogous to those given in Tables 19.5 and 19.6, assuming a truncated normal distribution.

6. The MLEs of the Weibull parameters for the windshield data omitting the lowest three observations are $\alpha^* = 2.827$ and $\beta^* = 3.490$. Recalculate the Weibull results of Tables 19.5 and 19.6 using these values and compare the results with those of the tables and of Exercise 5.

7. Omit the lowest four failure times and the incomplete data and test the remaining for fit to the Weibull distribution, using the Anderson–Darling procedure.

8. Suppose the system discussed in Case 2.5 is intended to last for at least one year. Estimate the reliability of the system using both the lognormal and Weibull models. Use the methods of Chapter 8 to calculate confidence intervals in each case and compare the results.

9. Repeat the previous exercise to determine the two-year reliability of the system.

10. Lawless (1982) suggests the likelihood ratio method for calculating confidence intervals as an alternative to those of the previous two exercises, which are based on asymptotic normality. Look up this source (Lawless, 1982, p. 236) and use the results to determine confidence intervals for reliability. Compare the results with those of the previous two exercises.

CLASS EXERCISES

I. Analyze each of the cases given in Chapter 2 and prepare detailed reports of your analyses, findings, and interpretation. Proceed as follows:

1. Read the case descriptions and formulate objectives (e.g., determine significant factors, estimate reliability, estimate costs, etc.) For some cases, these are stated quite explicitly in Chapter 2. In other cases, formulate what you think would be reasonable engineering or management goals in the context of the case description and the data given.

2. Collect all of the information and results from previous analyses, including exercises and examples in the text. Begin with the tables started in the exercises of Chapter 3 and subsequently expanded. Add any further information obtained from later chapters.

3. Select or construct appropriate reliability models. These may include block diagrams, life distributions, stochastic processes, and so forth. Clearly state the assumptions made in formulating and using the models.

4. Perform any additional analyses as relevant in the context of the experimental objectives and the available data and interpret the results. These may include reliability analyses, statistical procedures such as goodness-of-fit tests, parameter estimates, confidence intervals, and so forth, engineering and cost analyses, etc.

5. Think about additional data that might be useful to have and additional analyses that would be desirable.

6. Prepare a report that includes at least the following:

(i) A statement of experimental objectives.

(ii) A description of the experimental design or the source of the data and method of data collection.

(iii) A summary and description of the results obtained.

(iv) A statistical analysis of the data, including a brief description of the methodology used, the numerical results obtained, the statistical conclusions drawn (e.g., hypotheses rejected, at what level of significance, confidence intervals, etc.).

    (v)   Interpretation of the statistical results in the context of the experimental objectives.

    (vi) A discussion of possible problems or limitations of your conclusions.

    (vii) Suggestions for further research and/or analysis.

II. The data for many of the cases of Chapter 2 have been published previously in articles or reports. For these, sources are indicated. In fact, some of these data sets have been analyzed repeatedly (especially the air conditioning data of Case 2.10 and the ball bearing data of Case 2.14). Look up the sources of the data wherever possible and the various analyses done by finding the references cited and any additional sources that the authors of these publications cite. Summarize the results from the literature and compare them with the results you obtained in Class Exercise I.

CHAPTER 20

Resource Materials

20.1 INTRODUCTION

There are many aspects to reliability. In this book, we have focussed our attention on basic concepts, tools and techniques needed for the study of these different aspects. The literature on reliability is very large and deals many different issues. We have covered many of the important issues and indicated suitable references where further details on the material covered as well as information on related material not covered can be found. Our coverage has necessarily been brief, as it is impossible to provide complete coverage of so broad a topic. Instead, we have attempted to present a solid introduction to the most important aspects of reliability and the linkages between them.

Both the theoretical and application-oriented literature on reliability are growing at an increasing pace and this trend will continue in the future. In addition, the changes in computer and information technology have had a significant impact on the study of reliability. In this chapter, we focus on some of these topics and provide information sources that are useful to both researchers and practitioners.[1]

The outline of the chapter is as follows. In Section 20.2 we give a list of journals of relevance to reliability. Some of these deal exclusively with reliability-related topics and others (from different disciplines) contain papers dealing with issues of relevance to reliability. Reliability standards were discussed in Chapter 12. In Section 20.3 we give a fairly comprehensive list of such standards and illustrate the use of different standards in the different stages of a product life cycle. In Section 20.4 we deal with the topic of reliability databases. This information is very useful for practitioners in the reliability assessment of complex systems during the design stage. Computer and information technologies have made a significant impact on modeling, analysis, and optimization in the context of reliability studies. There are several commercial reliability software packages on the market that provide aids for these activities. Some of these are discussed briefly in Section 20.5. We conclude with some comments on the tools needed for reliability analysis in Section 20.6.

20.2 JOURNALS AND CONFERENCE PROCEEDINGS

Publications in the following lists range from those that publish exclusively articles on reliability to those that print reliability-related articles relatively frequently. The publications are categorized by major subject area.

Journals on Reliability

IEEE Transactions on Reliability
International Journal of Reliability, Quality and Safety Engineering
International Journal of Quality in Maintenance Engineering
International Journal of Quality and Reliability Management
Journal of Applied Reliability
Maintenance Management International
Microelectronics and Reliability
Quality and Reliability Engineering International
Quality and Reliability Management
Reliability Engineering and System Safety
Reliability Review
Total Quality Management

Operations Research Journals

Bulletin of the Australian Society of Operations Research
Asia Pacific Journal of Operations Research
Operations Research (USA)
Engineering Optimization
European Journal of Operational Research
Journal of Operational Research (UK)
Mathematics of Operations Research
OPSEARCH
Naval Research (Logistics) Quarterly
Journal of Operations Research Society of Japan
OR and Computers
Logistics and Transportation
Production and Inventory Management Journal
RAIRO Research Operationalle

Economics Journals

Bell Journal of Economics
RAND Journal of Economics
Industrial Economics
International Journal of Production Economics
Journal of Accounting and Economics
Journal of Cost Management
Journal of Economics and Business
Journal of Economics and Business Administration

Engineering Journals

International Journal of Production Research
Computer and Industrial Engineering
Industrial Engineering
International Journal of Operations and Productions Management
International Journal of Production Research
International Journal of Industrial Engineering—Applications and Practice
Transactions of the Institute of Industrial Engineers
Journal of Erosion and Corrosion

Journal of Engineering for Industry
Journal of Manufacturing Systems

Software Engineering Journals
International Journal of Systems and Software
Computer Science and Informatics

Statistics, Probability, and Stochastic Processes Journals
Advances in Applied Probability
American Statistician
Annals of Probability
Annals of Applied Probability
Annals of Mathematical Statistics
Annals of Statistics
Applied Stochastic Models and Data Analysis
Biometrics
Biometrika
Business Statistics
Communication in Statistics—Series A: Theory and Methods
Communication in Statistics—Series B: Computation and Simulation
Communication in Statistics—Series C: Stochastic Models
Computational Statistics
Computational Statistics and Data Analysis
Electronics Communications in Probability
IMS Bulletin
InterStat
Journal of the American Statistical Association
Journal of Applied Probability
Journal of Applied Statistical Reasoning
Journal of Applied Statistics
Journal of Business and Economics Statistics
Journal of Computational and Graphical Statistics
Journal of the Royal Statistical Society: Series A (Statistics in Society)
Journal of the Royal Statistical Society: Series B (Methodological)
Journal of the Royal Statistical Society: Series C (Applied Statistics)
Journal of the Royal Statistical Society: Series D (The Statistician)
Journal of Statistical Computation and Simulation
Journal of Statistical Planning and Inference
Journal of Statistical Software
Journal of Theoretical Probability
Metrika
Probability and Statistic Letters
Sankhya
Scandinavian Journal of Statistics
Statistical Inference for Stochastic Processes
Statistical Science
Statistica Neerlandica
Statistics—Theoretical and Applied Statistics
Statistics and Computing

Stochastic Processes and their applications
Technometrics

Business and Management Journals
American Journal of Mathematical and Management Sciences
Business Economics
Business Marketing
Decision Sciences
Decision Support Systems
Engineering Management Review
IEEE Transactions on Engineering Management
IMA Journal of Mathematics in Management
Industrial Management
Information Management
Interfaces
Harvard Business Review
Journal of Business
Journal of Business Administration
Journal of Business and Industrial Marketing
Journal of Business Policy
Journal of Business Strategy
Journal of General Management
Journal of Management Case Studies
Management Science
Management and Engineering Manufacture
Management in Engineering
Management Review
Management Sciences
Manufacturing and Management
Omega
Sloan Management Review
Strategic Management Journal

Legal, Safety and Risk Journals
Journal of Product Liability
Journal of Safety Research
Journal of Risk Insurance
Legal Economics
Risk Analysis

Conferences and Symposia
Annual Reliability and Maintainability Symposium
American Society of Quality Control Technical Conference Transactions

Special Issues Dealing with Specific Topics in Reliability
Software Reliability: *IEEE Transactions on Reliability,* Volume 28, Number 3, 1979.
Systems Maintenance: *Quality. and Reliability Engineering International,* Volume 13, Numbers 3-4, 1997.
Maintenance and Reliability: *Reliability Engineering and Systems Safety,* Volume 51, Number 3, 1996.

Failure Analysis: *IEEE Transactions on Reliability,* Volume 29, Number 3, 1980.
Design for Reliability: *IEEE Transactions on Reliability,* Volume 44, Number 2, 1995.
Software Reliability Engineering: *IEEE Transactions on Reliability,* Volume 43, Number 4, 1994.
Maintainability: *IEEE Transactions on Reliability,* Volume 30, Number 3, 1981.
USAF R&M 2000 Initiative: *IEEE Transactions on Reliability,* Volume 36, Number 3, 1987.

20.3   RELIABILITY RELATED STANDARDS

20.3.1   Standards Classified by Application

Reliability standards have been published by many organizations. The following are categorized by type of standard.

Design: MIL-HDBK-251, MIL-HDBK-338, MIL-STD-454, MIL-STD-785, MIL-STD-1556, MIL-STD-1591, MIL-STD-1670, ISO 9000
Environmental Stress Screening: MIL-HDBK-344, MIL-TD-810, MIL-STD-1670, MIL-STD-2164
FMECA: MIL-STD-1629, RADC-TR-83-72
Human Factors: MIL-STD-1472, MIL-H-46855
Logistics: MIL-STD-1388
Maintainability: MIL-STD-470, MIL-STD-472, MIL-STD-721, MIL-STD-741, RADC-TR-83-29
Quality: ISO 9000 series
Reliability (Prediction): MIL-HDBK-217, MIL-STD-756, LC-78-2, NPRD-91, RADC-TR-73-248, RADC-TR-75-22, RADC-TR-83-91
Reliability (General): MIL-HDBK-338, MIL-STD-1543, MIL-STD-2155, DoD-STD-4245.7, RADC-TR-83-29
Reliability Growth: MIL-HDBK-189
Safety: MIL-STD-882, MIL-STD-1574
Sampling: MIL-STD-105, MIL-STD-690
Software: MIL-STD-2167, DoD-STD-2168, MIL-S52779
Testing: MIL-HDBK-781, MIL-STD-781, MIL-STD-785, MIL-STD-790, MIL-STD-810, MIL-STD-883, MIL-STD-2074, MIL-STD-2165

20.3.2   Standards Classified by Origin

In the following lists, standards are categorized by type of organization or agency responsible for issuance and updating.

Commercial
SAE M-110 (1993), *Reliability and Maintainability for Manufacturing Machinery and Equipment, Society of Automotive Engineers, Inc.*

British Standards
BS2011, *Basic Environmental Testing Procedure*
BS4200, *Guide to Reliability of Electronic Equipment and Parts Used Therein*
BS5760, *Reliability of Systems, Equipment and Components*

IEC Standards

IEC 271, *List of Basic Terms, Definitions and Related Mathematics for Reliability*

IEC 300-1, *Dependability Management—Part 1: Dependability Program Management*

IEC 300-2, *Dependability Management—Part 2: Dependability Program Elements and Tasks*

IEC 300-3-1, *Dependability Management Part 3: Application Guide—Section 1: Analysis and Techniques for Dependability*

IEC 300-3-1, *Dependability Management Part 3: Application Guide—Section 2: Collection of Dependability Data from Field*

IEC 605, *Equipment Reliability Testing*

IEC 706, *Guide on Maintainability of Equipment*

IEC 812, *Analysis Techniques for System Reliability—Procedure for Failure Modes and Effects Analysis (FMEA)*

IEC 1014, *Program for Reliability Growth*

IEC 1025, *Fault Tree Analysis*

IEC 1078, *Analysis Techniques for Dependability—Reliability Block Diagram*

IEEE Standards

IEEE Std 352, *Guide for General Principles of Reliability Analysis of Nuclear Power Generating Station Protection Systems*

IEEE Std 500, *IEEE guide to the Collection and Presentation of Electrical, Electronic, Sensing Component, and Mechanical Equipment Reliability Data for Nuclear Power Generating Stations*

ISO Standards

ISO 9004-1, *Quality Management and Quality System Elements—Part 1*

ISO 9004-2, *Quality Management and Quality System Elements—Part 2: Guidelines for Services*

ISO 9004-3, *Quality Management and Quality System Elements—Part 3: Guidelines for Processed Materials*

ISO 9004-4, *Quality Management and Quality System Elements—Part 2: Guidelines for Quality Improvement*

British (Defense)

UK DEF STD 00-41, *MOD R&M Practices and Procedures*

UK DEF STD 00-42, *R&M Design Guides, Part 1—Design and assessment of BIT*

UK DEF STD 00-43, *R&M Assurance Activity, Part 1—In Service Reliability Demonstration*

UK DEF STD 00-44, *Data Collection and Classification, Part 1—Data and Defect Reporting, Part 2—Data Classification.*

UK DEF STD 00-49, *Glossary*

UK DEF STD 07-55, *Environmental Testing*

USA (Defense)

Perry, W. (1994), *Specifications and Standards—A New Way of Doing Business,* Memorandum by Dr. W. Perry, U.S. Secretary of Defense, Dated 29 June 1994.

MIL-STD-105, *Sampling Procedures and Tables for Inspection by Attributes*

MIL-STD-189, *Reliability Growth Management*

MIL-STD-217F, *Reliability Prediction of Electronic Equipment*

MIL-STD-470B, *Maintainability Prediction*

MIL-STD470, *Maintainability Program for Systems and Equipment*

MIL-STD 471, *Maintainability Demonstration*

MIL-STD-690C, *Failure Rate Sampling Plans and Procedures*

MIL-STD-721C, *Definition of Terms for Reliability and Maintainability*

MIL-STD 756B, *Reliability Modeling and Prediction*

MIL-STD 781C, *Reliability Design Qualification and Production Acceptance Tests: Exponential Distribution*

MIL-STD-781D, *Reliability Design Qualification and Production Acceptance Test: Exponential Distribution*

MIL-STD-785B, *Reliability Program for Systems and Equipment, Development and Production*

MIL-STD-790E, *Reliability Assurance Program for Electronic Parts Specification*

MIL-STD-810E, *Environmental Test Methods and Engineering Guidelines*

MIL-STD 839, *Selection and Use of Parts with Established Reliability Levels*

MIL-STD-882C, *System Safety Program Requirements*

MIL-STD-883D, *Test Methods and Procedures for Microelectronics*

MIL-STD-965A, *Parts Control Program*

MIL-STD-1369A, *Integrated Logistics Support Program Requirements*

MIL-STD-1388-1A, *Logistic Support Analysis*

MIL-STD-1390, *Level of Repair*

MIL-STD-1543B, *Reliability Program Requirements for Space Launch and Missile Systems*

MIL-STD-1566B, *Government/Industry Data Exchange Program (GIDEP)*

MIL-STD-1629A, *Procedures for Performing a Failure Mode, Effects, and Criticality Analysis*

MIL-STD-1635, *Reliability Growth Testing*

MIL-STD-2068, *Reliability Development Tests*

MIL-STD-2074, *Failure Classification for Reliability Testing*

MIL-STD-2155, *Failure Reporting, Analysis, and Corrective Action Systems (FRACAS)*

MIL-STD-2164, *Environmental Stress Screening Process for Electronic Equipment*

MIL-STD-2165A, *Testability Program for Systems and Equipment*

MIL-STD-2173AS, *Reliability-Centered Maintenance Requirements for Naval Aircraft, Weapon Systems and Support Equipment,* DoD

NATO

ARMP-1, *NATO Requirements for Reliability and Maintainability*

ARMP-2, *General Application Guidance on the Use of ARMP-1*

ARMP-3, *National R&M Documents*

ARMP-4, *Guidance for Writing R&M Requirements Documents*

ARMP-5, *Guidance on R&M Training*

ARMP-6, *In-service R&M*

ARMP-7, *R&M for Off the Shelf Equipment—Non-developmental Items (NDI)*

2.3.3 Handbooks

Defense

US DOD-HDBK-344, *Environmental Stress Screening (ESS) of Electronic Equipment*

US MIL-HDBK-189, *Reliability Growth Management.*

US MIL-GDBK-189, *Reliability Growth Management*

US MIL-HDBK-217F, *Reliability Prediction for Electronic Equipment.*

US MIL-HDBK-251, *Reliability/Design Thermal Applications*
US MIL-HDBK 259, *Life-Cycle Cost in Navy Acquisitions*
US MIL-HDBK-338-1A, *Electronic Reliability Design Handbook*
US-MIL-HDBK-344, *Environmental Stress screening [ESS] of Electronic Equipment*
US-MIL-HDBK-472, *Maintainability Prediction*
US MIL-HDBK-781, *Reliability Test Methods, Plans and Environments for Engineering Development, Qualifications and Production*
RADC-TR-73-248, *Dormancy and Power On-Off Cycling Effects on Electronic Equipment and Part Reliability*
RADC-TR-75-22, *Non-electronic Reliability Notebook*
RADC-TR-83-29, *Reliability, Maintainability, and Life Cycle Cost Effects of Commercial Off-the-Shelf Equipment*
RADC-TR-83-72, *The Evolution and Practical Applications of Failure Mode and Effects Analysis*
RADC-TR-85-91, *Impact of Non-operating Periods on Equipment Reliability*

Commercial

Ford Motor Company, *Potential Failure Modes and Effects Analysis in Design (Design FMEA) and for Manufacturing and Assembly Processes (Process FMEA) Instruction Manual*
TR-NWT-000332, *Reliability Prediction Procedure for Electronic Equipment,* Bellcore Technical Reference

20.3.4   Sources of Information

For purchase of military specifications, standards and handbooks
　DOSSP-Customer Service
　700 Robbins Ave.
　Building 4D
　Philadelphia, PA 19111-5094
For purchase of DoD Directives and Instructions
　Defense Technical Information Center
　DTIC-FDAC
　8725 John J. Kingman Road, Suite 0944
　Fort Belvoir, VA 22060-3842

20.3.5   Application of Standards

As indicated in Chapter 12, the product life cycle involves several phases. We follow Loll (1998) to illustrate the application of standards to a typical product life cycle using the IEC Standards.

Concept and Definition Phase

IEC 50 (191), International Electrotechnical Vocabulary; Chapter 191: Dependability and Quality of Service
IEC 33-1, Dependability Management
IEC 300-3-3, Life Cycle Costing
IEC 300-3-4, Guide to Specification of Dependability Requirements
IEC 300-3-13, Technical Risk Management

Design and Development Phase
IEC 300-2, Dependability Program Elements and Tasks
IEC 300-3-1, Analysis Techniques for Dependability, Guide on Methodology
IEC 300-3-3, Life Cycle Costing
IEC 300-3-6, Software Aspects of Dependability
IEC 300-3-9, Risk Analysis of Technological Systems
IEC 605-3-2, Design Test Cycles
IEC 600-3-x Series, Preferred Test Conditions
IEC812, Analysis Techniques for System Reliability (FMEA)
IEC 1078, Analysis Techniques for Dependability, Reliability Block Diagram Method
IEC 1014, Program for Reliability Growth
IEC 1025, Fault Tree Analysis
IEC 1160, Formal Design Review
IEC 1165, Application of Markov Techniques
IEC 1709, Electronic Component—Reliability Reference Conditions for Failure Rates and
   Stress Models for Conversion.

Manufacturing and Installation Phase
IEC 300-3-5, Reliability and Test Conditions and Statistical Test Principles
IEC 300-3-7, Reliability Stress Screening of Electronic Hardware
IEC 1163-1, Reliability Stress Screening of Repairable Items Manufactured in Lots
IEC 1163-2, Reliability Stress Screening of Electronic Components
IEC 1164, Reliability Growth—Statistical Tests and Estimation Methods

Operation and Maintenance Phase
IEC 300-3-2, Collection of Dependability Data from the Field
IEC 300-3-11, Reliability Centred Maintenance
IEC 330-3-12, Integrated Logistic Support
IEC 706-5, Diagnostic Testing

Disposal Phase
IEC 300-3-3, Life Cycle Costing

20.4   RELIABILITY DATA BASES[2]

The term reliability data base has two different meanings. The first is an actual computer data base that is set up for storing component data (event, engineering and operating data) for equipment or plants. The second use of the term is in referring to what are also called generic data bases or data handbooks. These usually contain the results of analyses performed on the event data that is found within the computer data bases or, in some cases, based simply on expert opinion. The result that is usually reported in the handbooks is the failure rate for a specific component class (e.g., a pump). Note here that a component can be anything from something as small as a valve to something as large as a pump.[3]

It is interesting to note that component failure rates in the handbooks[4] (and some data bases) are all constant and intended to correspond to the flat region of the bathtub failure-rate curve. However, the data used for estimating this often also include early failures (corresponding to a decreasing failure rate) and late failures (corresponding to an increasing failure rate).

20.4.1   Reliability Data Base Structures

There are three types of reliability computer data bases and the type of data contained in each is quite different. They are as follows:

- Component event data bases (e.g., CREDO, CEDB)
- Abnormal or incident-report data bases (e.g., FACTS, AORS)
- Generic component reliability data bases (e.g., IAEA's Generic Component Reliability Data base)

Component event data bases are compilations of events (e.g., failure, maintenance undertaken, etc.) for use in determining reliability, availability, and maintainability parameters. An incident report data base stores information on incidents or accidents, their probability, causes, and effects. A generic component reliability data base is essentially a compilation of reliability failure rates for various components, based on information gleaned from the literature.

20.4.2   Component Event Data Bases

Component event data bases store three types of data. These are event data, operations data and engineering data. The engineering data can be further subdivided into two categories—basic data that are generic for each component and data particular to a certain type of component.

The components in a modern event data base are generic, as are the modes of failure, as indicated in Table 20.1.

Table 20.2 is a list of the types of generic component data that are usually included in a data base. Each component is given a unique identification (id) which links it with all other data associated with it. Generic data usually do not change unless a component is moved to a different location or undergoes modification.

An example of some of the information that may be gathered for a particular component, in this case a valve, is given in Table 20.3.

Table 20.1. Examples of generic failure modes

Failure mode	Failure mode code
All modes	A
Degraded	B
Fails to change position	C
Fails to remain in position	D
Fails to close	E
. . .	
. . .	
Fails to operate	L
Fails to open	O
Fails to run	R
. . .	

Table 20.2. Generic component data for a component

Unique component identification code
Reporting site and type of plant
Generic component type
Manufacturer, model number, drawing numbers
Date installed and date removed
Components system, subsystem, application designed life
Component's operating and duty factors
Component's scheduled maintenance/inspection data
Components environment (e.g., radiation exposure, humidity, temperature, etc.)
Number and location of all replicate, redundant components

Event information deals with the different possible events for a component. These are as follows:

- An unscheduled maintenance or unscheduled repair is performed on a component (failed or malfunctioning)

- System undergoes an unanticipated change from normal operation (e.g., a spurious operation occurs or a safety system is activated, etc.). The change can be attributed to a component malfunction or failure (e.g., a component malfunction that affects other items)

- Repair or replacement of a component occurs during a scheduled maintenance inspection that was to include a "repair or replacement only when necessary" (e.g., inspection reveals that a component has deteriorated to the point of requiring replacement)

- During a maintenance inspection, a component is found to exhibit abnormal behavior

- Component fails to perform its function for any reason not already covered

The failure of an item depends on its operating mode. The operating modes for equipment can be:

- Continuous operation
- Standby
- Alternating (or intermittent) uasge

Table 20.3. Components for a valve

Type (ball, diaphragm, etc.)
Functional application
Functional characteristics
Connection type to other equipment
Material
Seal type
Operator type
Design characteristics (e.g., temperature, pressure, size, etc.)

Table 20.4. Event or failure data for components

Identification of the reporting site, title and date of the event
Component identification code
A narrative description of the event
Method of detection of the event and what immediate actions where taken
Corrective actions taken
Extent to which human actions were involved in propagating or mitigating the event
Details related to the last maintenance, test, or calibration before the event
Component failure data (eg. system, subsystem, event type, failure mode, failure cause,
    event severity, primary or secondary event)
Downtime, repair time, and cost of repair
Operating mode at the time of an event

The failure modes can vary with the usage mode. Table 20.4 gives a partial list of usage modes. The types of operating data that may be recorded for a component are as shown in Table 20.5.

Because of the ability to sort information in the data bases, the user is able to select a data set according to many different fields, including system type, component type, failure mode, failure cause, operating characteristics, and so on.

Most of the data bases produce reports for the failure rates of different components, typically providing the following information:

- Nnumber of components in the selected population
- Operating hours
- Number of demands
- Number of failures
- Failure rate for each mode (per hour or per demand)
- Combined total failure rate
- Number of failure modes
- Percentage of failures due to different modes and causes
- Mean time between maintenance (or repair rate)
- Confidence limits on one or more of the above

Most data bases and analyses of failure rates treat failures as independent, although this is not necessarily the case. In an attempt to address this issue, a data base DEFEND was developed for the nuclear industry by the National Centre of Systems Reliability (NCSR) to support a wide range of dependent failure models using a cause–effect logic structure.

We briefly describe the salient features of some of the data bases.

Table 20.5. Operating data collected for a component

System or component identification
Operating modes (continuous, hot or cold standby, etc.) and time spent in each mode
Outage times and whether scheduled or unscheduled
Number and times of demands

CREDO

The Centralized Reliability Data Organization (CREDO) is a data bank and data analysis center. The CREDO data base system development was initiated in 1978 at Oak Ridge National Laboratory (ORNL). In 1985, CREDO became a joint venture between the U.S. Department of Energy's (US DOE) Office of Technology Support Programs and Japan's Power Reactor and Nuclear Fuel Development Corporation (PNC).

The main purpose of the reliability data base system was to provide a centralized source of accurate, up-to-date data for use in reliability, availability and maintenance (RAM) analyses required by DOE for advanced reactor safety assessments, design, and licensing. CREDO concentrates on reliability data for liquid metal reactors (LMRs). LMRs use liquid sodium or a mixture of sodium and magnesium in the cooling system.

The CREDO data base management system (DBMS) is a hierarchical structure that allows each data record to have a unique descriptive name. Each name is composed of qualifiers (maximum of 16) separated by periods. The qualifiers correspond to the different levels within the hierarchical framework of the data base. There are three main files in the CREDO data base structure. There are files for the engineering data (ENGR), the operating data (OPER), and the event data (EVENT).

CREDO has concentrated on the mechanical, electrical, and electronic components and devices in LMRs, specifically

- Those whose failure could have an adverse impact upon a unit's operation
- Those that are exposed to or directly associated with a liquid metal environment
- Those components that are liquid-metal-specific
- Those components operating in environments not tracked by other nuclear data bases

OREDA

The Offshore Reliability Data Project (OREDA) was established by eight oil companies to collect data on selected offshore equipment. The project began with a feasibility study in 1980. In 1983 the project began the collection of historical maintenance data. Data collection is now in its fifth phase. The primary objective of the OREDA project was to collect reliability data for improved input to safety and reliability studies. OREDA pioneered the use of diagrams to define component boundaries.

SRDF

Electricité De France (EDF) has developed several data banks for its own and foreign pressurized water reactors (PWRs). They are

- Reliability Data System (SRDF)—a component event data base (published in RPDF handbook)
- Incident data bank (referred to as either INC or FI)—operational information and incidents compiled for foreign PWR units (published in RDE foreign units reports)
- Event data bank (FE)—operational information on French PWR units (published in RDE EDF units reports)

In 1974, EDF developed its own reliability data system, the Système de Recueil de Donnees de Fiabilité (SRDF). The SRDF is a data base containing data collected on mechanical and electro-mechanical devices from nuclear power plants. The data were originally collected from six pressurized water reactors based at Fessenheim and Bugey. In 1983 data collection was extended to all EDF PWR units and later extended to include the fast breeder reactor of Greys Malville (Super Phoenix). Data has been collected on approximately 1100 components per pair of units. Between 1978 and 1992, information on 50,000 failures was collected.

The European Reliability Data System (ERDS) and the Component Event Data Bank (CEBD)

The ERDS project was launched in 1979 as part of the Nuclear Safety program of the JRC (Joint Research Centre) of the Commission of the European Communities at Ispra, Italy. Its aim was to collect and organize information and act as a centralized repository for information related to the operation of nuclear power plants (NPP's), specifically light water reactors (LWR's). The data were gathered for use in probabilistic risk assessment (PRA) and to allow better predictions of component behavior and a better understanding of how faults appear and develop.

The data are gathered from national and international data systems, including those in ECC countries, the U.S.A., Sweden, and Spain. So that the data from the various national data systems could be merged without losing information, a set of Reference Classifications was developed.

ERDS is made up of four data subsystems:

- Component Event Data Bank
- Abnormal Occurrences Reporting System (AORS)
- Operating Unit Status Report
- Reliability Parameter Data Bank

The Component Event Data Bank (CEDB) contains reactor component data from different European countries. The engineering characteristics, operational profiles, and environmental conditions are recorded for each component. The information is coded according to the Common European Reference Classifications.

The Abnormal Occurrences Reporting System (AORS) is a collection of information on safety-related abnormal events from NPP's in Europe and the U.S.A. It is used for safety analysis. The AORS contains information from national data files and is stored according to a common format. It also has a set of satellite data banks in which original data from the different countries are stored unmodified according to their specific classification and format.

The Operating Unit Status Report (OUSR) contains detailed information on outages, and reactor performance information for European NPPs.

The Reliability Parameter Data Bank (RPDB), which is a generic component data base, is a collection of reliability parameters for similar classes of components operating in similar conditions. The data are derived from analysis of CEDB failure information and technical literature.

Shell UK's Equipment Availability Reporting Data Base and Analysis Package

Shell UK Exploration and Production, Northern Operations is responsible for production of oil and gas in the northern North Sea. They have eleven offshore production platforms and two onshore plants. The data base, which began as a pilot study of two production platforms and extended to a full study in 1983, is aimed at collecting operational (availability, utilization, and reliability) data on rotating equipment considered vital to production.

The data base is not a component-event data base but a status-based data base. The downtime data are only collected for outages greater than half an hour and collected on a daily basis and transmitted onshore. The downtime data include

- Date
- Downtime
- Downtime type
- Downtime code

The operating data include

- Running time
- Standby time
- Scheduled downtime
- Unscheduled downtime
- Number of failures

The operating data are compiled into monthly operating reports and then entered into the data base. The only information that can be gleaned from such a data base is the failure rate.

A new package has been developed and will include the Equipment Availability Reporting System as a module. The new Integrated Maintenance Information System (IMIS) will be event-based and extended to include equipment such as pressure vessels, heat exchangers, valves, pumps, and emergency generators.

SRS Data Bank

The Systems Reliability Service (SRS) was incorporated into the National Centre of Systems Reliability (NCSR). The English parliament set up the United Kingdom Atomic Energy Authority (UKAEA) and it was required to minimize harm and danger in the operation of nuclear and chemical plants. This lead to monitoring misfunctions in nuclear reactors and chemical plants. The misfunctions were first logged on cards and then were computerized in the 1960's. About a decade later, the data base was expanded to include reliability in conventional industries.

The data bank consists of several data stores:

- Generic reliability data store—processes reliability data obtained from published information, manufacturers, data collection activities, etc.
- Event data store—raw data from special structured collection schemes set up in association with plant operators
- Accident data store
- Human reliability data store
- Misfunction data store

A large part of the information gathered comes from maintenance job cards, operations log books, permits to work, clearance certificates, stores and requisitions, and stores withdrawals. In 1984, after nearly a quarter of a century, there were 12,000 different records in the data base spread over 400–500 equipment categories. The type of information collected and the sources are

- Electronic—nuclear and chemical industries
- Electrical—electrical generating organizations in the U.K.
- Mechanical data—onshore industries, North Sea operations, and oil production facilities (wide range of operating/environmental conditions)

The SRS data bank collects data on "event" reports. The SRS is not an event data base by definition. The data may be collected on an event basis but are input in a summary form similar to a generic component data base (i.e., constant failure rate information for the various failure modes with confidence intervals and ranges input into the data base). The "event" data are collected and compiled periodically (a year or more apart) from a plant. The SRS data base reports failure rates in terms of calendar time and operating time.

Savannah River Laboratory Data Bank for Nuclear-Fuel Reprocessing Plants (U.S.)
The Savannah River Laboratory maintains a data bank that contains information on operating problems and equipment failures that have occurred in the nuclear fuel reprocessing areas of the Savannah River Plant. The databank has been in operation since 1974 and in 1988 the data bank contained over 175,000 entries on events ranging from minor equipment malfunctions to incidents with potential for injury or contamination of personnel, or for economic loss.

The data are gathered from 66 different sources normally available from the manufacturing plant organizations. The sources include

- Two external to SRP
- 33 internal published reports
- 31 unpublished sources (e.g., shift turnover logbooks)

The information in the data base is stored as entries that are essentially text fields. Data in each entry is specified using key words. Each event or incident can have multiple entries. No failure mode or failure cause is specified, only the fact that a component has failed.

Integrated Information System for Reliability and Maintainability of Transmission Lines (Ontario Hydro, Canada)
This data base is heavily weighted towards maintenance and maintenance logistics. The Ontario hydro system consists of more than 16,000 circuit miles supplying customers in an area of 250,000 square miles. The system is divided into three data bases:

- Technical data base
- Maintenance data base (information on all maintenance activities performed on circuits, including patrols, inspections, tests, preventive maintenance, breakdown maintenance, rehabilitation, and modifications)
- Operating data base

SADE
SADE is the computerized data base of the Centre National d'Études des Telecommincations (CNET) in France. The SADE data bank stores data on electronic components. Failures are recorded at all stages of the lifetime of a component, including factory testing, equipment setup, and operating lifetime. As an example of the amount of data collected, in the period

July, 1987 to June, 1988, 9,290,000 boards (83% switching equipment and 17% transmission equipment) were monitored and resulted in 180,000 failure reports (89% and 11%, respectively). Because climatic conditions affect the reliability of equipment in electronic telephone exchanges, a meteorological data base was set up to record temperature, relative humidity, and atmospheric pressure. The data are gathered from 40 sites distributed in France.

CNET is also collecting data from the French Administrations, public services, industry, and SADE for its Reliability Data Handbook.

ITALTEL

ITALTEL (Società Italianà Telecomunicazioni) is an Italian company that manufactures telecommunication equipment. They have an electronics data base that has been in operation since 1976. It contains information on the reliability of

- Switching systems
- Transmission systems
- Radio mobile networks

ITALGAS—Natural Gas Pressure Regulating Installation (PRI) Reliability Data Base

A reliability data base was created for the natural gas distribution industry. Data on faults and breakdowns was collected from ordinary maintenance operations on pressure regulating installations (PRI). The initial study between 1986 and 1988 involved six sets (or city mains) located in North Italy. A total of 400 installations were surveyed.

Computer Aided Reliability and Risk Assessment (CARARA)

CARARA is a system principally aimed at studying reliability and risk assessment of technical installations that deal with toxic and radioactive substances, but it can also deal with failure data from any source (e.g., the automotive industry). It has been used in the EuReDatA benchmarking exercises. The CARARA system contains several modules, including

- FDB—a component event data base
- FDA—a statistical analysis package for calculating reliability distributions, reliability parameters, failure rates, etc.
- RDB—a data base that stores the failure rates and other items calculated by FDA for components in FDB. A separate data base was created for generic component failure data.
- FTL—a fault tree analysis code. For the top event the annual failure frequency, the unavailability, mean system downtime, and the probability of exceeding tolerable downtime are calculated.
- STAR—module for evaluating risk of release of a toxic or radioactive substance

The FDB and RDB information have been combined in the comparison tables.

Component Event Data Bases

A data base format for collection of reliability data for process plants is presented. A RAM data base, Shipnet, was created as part of the Ship Operations Cooperative Program (SOCP) in the United States. A data base was created for the availability and supportability of new weapons systems.

20.4.3   Incident Data Bases

Incident data bases, as stated earlier, contain information about accident events or abnormal plant operating events. These data bases are concerned with the flow of events rather than failure rates. Typical data required for an abnormal occurrence are shown in Table 20.6

It is important that the data contained in an incident data base conserve the information in the original source reports. The data are stored so as to reflect the chain of occurrences of an incident. Every action in an incident is preceded by a cause. The action and other factors then contribute to causes for the next action in the chain of events.

These data are usually coded through the use of keywords. A data set can then be selected for analysis by searching using any of the keywords. The data in incident data bases are usually used for safety analysis. Analyses can include

- Number of incidents due to different industrial activities
- Number of accidents where different chemicals are involved
- Number of accidents involving a specific piece of equipment
- Number of deaths or injuries
- Number of incidents due to a specific initial cause
- Common factors in accidents

We briefly discuss some of the incident data bases.

FACTS

FACTS (Failure and ACcident Technical information System), created by TNO, the Netherlands Organization for Applied Scientific Research, contains technical information about incidents with hazardous materials that have occurred in the processing industries, during storage and transport, and during use or waste treatment of chemicals.

The aim of FACTS is to enable the cause and course of incidents to be studied so that safety can be improved through the initiation of avoidance and preventative actions. The data collected can be applied to:

Table 20.6.   Abnormal occurrences data bank content

Identification of plant
Identification of the incident:
   date
   status of the plant at the moment of the incident
   initiating event
   time sequence of the events
   systems involved during the events development (whether failed to operate or not)
   failed components
   cause of failure of systems and components
   way of failure(s) discovery
Identification of the consequences:
   on the system/plant operation
   on the people inside and outside the plant
   on the plant itself
   on the environment

- Quantify the risk of particular industrial activities and their justification
- Identify trends, and increase the knowledge on probabilities
- Verify models, scenarios, and probabilities that are used in risk analysis
- Develop event trees
- Investigate causes, circumstances, and consequences of industrial accidents

The selection criteria for an incident to be included in FACTS are

- Danger and/or damage to the nearby population and the environment
- Existence of acute danger of wounding and/or acute danger to property
- Processing, transport, research storage, transshipment, use/application, and waste treatment activities
- Events that can be classified as "near misses"

The data sources for FACTS can be divided into the following groups:

- Organizations with internal reporting systems. Conditions of strict anonymity apply to data retrieved from these sources. They include police, fire brigades, and labor inspectorates
- Companies supplying data under strict conditions of anonymity
- Literature and other publications, annual reports, symposia, etc.
- Magazines and periodicals dealing with industrial safety, risk management, and loss prevention
- Cuttings from newspapers used as signals for obtaining appropriate information
- Information from organizations who represent FACTS in their own country. FACTS agencies exist in the following countries: the U.K., Spain, Italy, the U.S.A., and Norway

The information available on incidents increases with the amount of damage to the environment or equipment or injuries to people so that the data being stored tend to be skewed towards more serious incidents.

The data are encoded as attributes, values and free, text. The attributes and values may be keywords that can be used as a basis of a search. The data in FACTS is stored to reflect the chain of occurrences of an incident. Every action in an incident is preceded by a cause. The action and other factors then contribute to causes for the next action in the chain of events.

The data base used in FACTS is based on a network structure. It contains six different data sets, five master sets and a detailed set that ties all the information together. The different data sets and what is contained in each is shown in Table 20.7

The original documents from which the data are gleaned are stored on microfilm and assigned a unique number, which is stored with the incident information in the data base.

Abnormal Occurrence Reporting System (AORS)

The Abnormal Occurrences Reporting System (AORS) is a collection of information on safety-related abnormal events from nuclear power plants (NPP) in Europe and the U.S.A. It is part of the European Reliability Data System (ERDS) and was developed by the Joint Research Centre of the Commission of the European Communities based at Ispra in Italy as part of their Nuclear Safety program.

Table 20.7. Data sets defined for the FACTS network data base

Data set	Information
1	Accident numbers used
2	Dimensions and conversion coefficients used
3	Attribute numbers used
4	Value numbers used and the relationship between attributes and values
5	Sources used
Detailed	Information for each accident, which consists of a number of lines. Each line consists of
	ACC—accident number
	ATTRIBUTE—attribute number
	VALUE—value number
	SRCE—information source
	TEXT—explanation with regard to items and values
	DIM—dimension of a number or the UN number for a chemical

The AORS contains information from each country's national data files and is stored according to a common format. It also has a set of satellite data banks in which original data from the different countries are stored unmodified according to their specific classification and format.

Information from complementary sources was used to upgrade national event reports. These included

- U.S. Nuclear Regulatory Commission (NRC) "Licensee Event Reports" (LER)
- Nuclear Power Experience (NPE) volumes
- NSAC reports on major accident sequences
- NUREG reports, such as potential precursors studies, etc.
- ENEL event reporting scheme

Data on over 30,000 incidents were collected between 1984 and 1986.

Other Incident Data Bases
Other incident data bases include

- Databank for the Rijnmond Process Industry
- SONATA

20.4.4   Generic Component Reliability Data Bases

An example of this type of data base is IAEA's generic component reliability data base. A generic component reliability data base is essentially a computerized data handbook, containing various component reliability failure rates gleaned from the literature.

Generic reliability data are data that are not specific to the plant (or equipment) that is being analyzed. Generic data are usually used when there is not enough reliability data available to arrive at meaningful reliability statistics, such as failure rates and failure distributions for use in various system reliability studies such as probabilistic safety assessment (PSA) or fault tree analysis. This is can arise when there is limited or no operating experience or a limited number of failures for the plant or equipment being studied.

The generic data in the literature can be divided into three broad categories:

- Data that were used in a previous study, including both plant- (or equipment-) specific data or generic data updated with plant-specific information
- Data compiled from the plant's own operating experience
- Data sources based on expert opinions or a combination of expert opinions and other reliability data experience from sources not necessarily related to the plant (or equipment) of interest

The selection of generic data has several problems associated with it because a large discrepancy exists between failure rate values for 'similar' components in the literature. This affects the use of handbooks and generic data bases alike.

In the compilation of the IAEA's generic reliability data base, the following four problem areas were identified in decreasing order of importance:

- Component boundary definition
- Failure mode definition
- Operating mode definition
- Operating environment definition

The large difference in component failure rates between various generic reliability data bases may be related to component boundaries. There are several ways of defining component (and system) boundaries. They are by

- Indicating component boundaries and points of interface with other systems through the use of a diagram (CREDO, Swedish Reliability Data Book)
- Defining precisely the major interface points (NUREG LER (licensee event report), ERDS)
- Defining the component boundaries by defining the component as off-the-shelf. For some components, this definition is acceptable, but for others, it leads to problems. What may constitute an off-the -shelf component one year may not be an off-the-shelf component the next
- Indirect definition, such as when a fault-tree is constructed as part of a probabilistic safety assessment (PSA)

Mixed-source data bases do not usually provide a detailed component boundary description.

One of the ways in to avoid problems such as those encountered when combining data from different sources is to define "generic" component boundaries as was done for component event data bases. The difficulty, of course, is that other organizations may define different component boundaries and the same problem of combining data then arises.

Failure modes are another source for the difference in failure rates between different data sources. The failure mode description can vary significantly. A way of dealing with this problem is to define generic failure modes.

As stated previously, the three operating modes are continuous, standby, and alternating (or intermittent). There are two different failure-rate definitions. One is time-related (a failure rate per unit time) and the other is demand-related (a failure rate per demand). Some data bases do not distinguish between the two and report both as per-hour rate (NUREG 2815) or combine the two into one time-related failure rate. As also stated previously the failure of a component on standby can be purely demand-related or purely time-related (i.e., related to the time spent in standby) and this also poses problems.

Operating environments that includes such things as temperature, humidity, corrosiveness, stress, and vibration can severely affect the lifetime of a component. The majority of generic data bases do not address this problem at all. The change in environment due to accidents in severe environments can also alter the "normal" lifetimes of components. The WASH 1400 data base gives a separate failure rate for postaccident situations while IEEE Standard 500 and MIL HDBK 217 give multipliers for altering the "normal" operating lifetimes of components for severe environments.

IAEA's Generic Component Reliability Data Base
The Generic Component Reliability Data Base was created by the International Atomic Energy Agency (IAEA) from the perceived need of its member states for generic component reliability data for use in probabilistic safety assessment (PSA).

The data for the Generic Component Reliability Data Base are compiled from the open literature for practically all components relevant to PSAs for NPPs. The data are compiled from 21 different data sources.

The Generic Component Reliability Data base was designed to store failure rates gleaned from these published sources. In an attempt to overcome the issues raised in the preceding section, IAEA defined generic component boundaries, generic component failure modes, and causes.

The IAEA provides a package for browsing and data retrieval and fault/event tree analysis called PASPACK.

HARIS
HARIS is a generic data base system that was developed by RM Consultants in the United Kingdom and contains four basic types of information:

- Incident data—contains reports of major accidents in nuclear, chemical, and offshore plants
- Reliability data—generic reliability data base
- Maintainability—contains information on maintenance and repair processes, including mean repair and waiting times
- Abstracts—contains synopses of relevant reports, articles, standards, codes of practice, etc. Abstracts data sheets provide the link between incident, reliability, and maintainability data sheets and source documents

The failure modes reported are the general terms critical, degraded, incipient, and unknown.

FARADIP.THREE
FARADIP is generic component data base based on over 20 published data sources from offshore and onshore applications including oil, gas, telecom, military, aerospace, and commercial industries.

20.4.5 Data Handbooks

The issues involved in the construction of computerized generic component data bases also affect data handbooks. To reiterate, these are component boundaries, failure modes, failure causes, and the environment.

The information contained in data handbooks has changed over the years as our understanding of reliability has grown. Early handbooks contained reliability data on broad classes of 'components' compiled from many different sources. Because of this, the data held within them must be used with care. Some of the earliest handbooks containing part-failure data were published in the 1950's by Radio Corporation of America (RCA), General Electric, (GE), and Motorola. The data were compiled from life-test and field-failure data.

The early handbooks did not take into account differences in failure rates due to environment and physical differences among subcomponent groups. The exponential distribution with its constant failure rate became the standard for reporting failure rates. The use of K-factors to modify a base failure rate and account for redundancy and operating mode of the component was introduced in the Martin Titan Handbook. These later became known as π-factors in MIL-HNBK-217. The division of component classes by similar environments, thus making the classes more homogeneous, was introduced in the NPRD. FARADA was the first handbook to be supported by a computer data base (i.e., a generic component data base).

Modern computer event data bases were developed to collect reliability information on components in a specific industry such as the nuclear power, offshore oil, and chemical industries. Handbooks were published based on the information compiled in these computer data bases. Since the data in these handbooks are compiled from a particular industry, the problems associated with environmental factors are minimized, and since the components are grouped in smaller subcomponent groups, the physical differences of the components are also minimized. Thus a better understanding of the correspondence between a component's attributes (for example, operating mode, environment, and subtype) and its reliability can be now gained.

A review of handbooks for reliability design analysis can be found in Moss and Strutt (1993).

Martin Titan Handbook

The "Martin Titan Handbook: Procedure and Data for Estimating Reliability and Maintainability" was first published in 1959. It contains generic failure rates for electrical, electronic, electromechanical, and mechanical parts. The Titan handbook was the first source to standardize failure rates in terms of 10^6 hours and it also standardized the use of the exponential distribution with its constant failure rate. It also introduced the use of failure rate adjustment factors or K-factors, which when multiplied by the generic failure rate to adjust for the presence of redundancy and to account for operational mode. It does not provide failure rates for the different failure modes for each component but does provide a list of typical failure modes for each component.

FARADA

FARADA arose out of a program sponsored by the U.S. Army Materiel Command and Air Force Logistics Command. It encouraged the exchange of data on purchased equipment. This program eventually became the Government Industry Data Exchange Program (GIDEP). The data bank that resulted became known as the Reliability and Maintainability Data Bank. The data are from field experience, laboratory accelerated life tests and reliability demonstration tests. The GIDEP data bank includes population information and failure mode information.

MIL-HDBK-217

The requirements for improved reliability in the U.S. armed services lead to the generation of the reliability handbook MIL-HDBK 217. The handbook was first published in 1962 and deals with the reliability prediction of electronic equipment. It is now up to edition F. Contractors to the U.S. Department of Defense are required to use the handbook to estimate the reliability of their products.

Most of the data have come from tests in the laboratory with controlled environmental stresses. The handbook assumes an exponential or constant failure rate model. This base failure rate in 10^6 hours is compiled from a broad range of "similar" components. The data are not related to specific failure modes or causes. No confidence intervals are given nor are the population numbers. The handbook uses π-factors as modifiers to the base failure rate to account for subcategories of the component, quality, environment, temperature, etc.

Several data handbooks that compete with MIL-HDBK-217 are HRD4 (1987), CNET (1983), NTT (1985), and Bellcore (1998).

Reliability Analysis Center (RAC) Handbooks

The U.S. Air Force Rome Air Development Center (RADC) developed a nonelectronic reliability notebook. The data are derived from military field operating experience and test experience. Some of the failure rates are derived from the grouping of failure rates of generic part types that have experienced similar environments. These groups include ground fixed, ship sheltered, airborne, etc.

The failure rates of over 25,000 parts have been collected and cover a wide range of component types. The data base uses factors (K-factors) in the same way that π-factors are used in MIL-HDBK-217. The factors are to account for environment and quality effects. RADC has also developed an electronic reliability notebook. The data are derived from military and commercial experience in the field and are given in the same format as the NRPD. Failure rates are sorted with respect to environment and quality level.

OREDA

The OREDA handbook summarizes the results of the data collection system described earlier. The OREDA handbook was the first data base to define what exactly comprised a component through the use of diagrams. The current handbook available is OREDA 97. This contains data collected up to and including phase III of the OREDA project.

Recueil Periodique de Donnees de Fiabilité (RPDF)

This handbook is published by EDF. It contains the results of analyses on the data contained in their component event data base SRDF that has been described previously. The failures are divided between critical and all (critical, incipient, and degraded) and are further subdivided between the three operating modes (demand, operating, and standby). The RPDF reports the number of failures that have occurred in service, during test, and during maintenance for each of the three operating modes. This handbook shows the percentage of failures due to failure of a component.

IEEE STD 500

The IEEE Guide to the "Collection of and Presentation of Electrical, Electronic, Sensing Components, and Mechanical Equipment Reliability Data for Nuclear Power Generating Stations" contains reliability data applicable to nuclear power stations. The data have been collected from recorded field data, expert judgment, and a technique called the Delphi Technique (described in IEEE Std 500 Appendix B). The IEEE Standard 500 gives multipliers for altering the "normal" operating lifetimes of components to account for severe environments.

WASH-1400

WASH 1400 is a reactor safety study that contains a comprehensive study of U.S. power plants. Appendices 3 and 4 of the report contain failure rates for some electrical and mechanical components in nuclear power plants. The WASH 1400 data base gives separate failure rates for postaccident situations.

20.4.6   Concluding Remarks

The best data base type for collecting reliability data is the component 'event' data base. There are many pitfalls in the search for generic component reliability data particular for a specific application. Care must be taken when selecting the generic component data with regard to physical type, component boundaries, environment, and failure modes and causes.

As knowledge of the factors influencing the reliability of components has improved, so have the reliability data bases. Hopefully, with the knowledge of failure gained from the data bases one will see an even greater increase in the reliability of future equipment and products.

20.5   RELIABILITY SOFTWARE

Reliability software can be grouped into three categories:

- General statistical software that contain packages for reliability analysis
- Software for general reliability analysis
- Software for specific reliability analysis

In this section, we give a brief description of some of the commercially available software for reliability study. For most software, the information given below has been obtained by accessing the producers' websites.

20.5.1   General Statistical Software Packages

Nearly all general statistical packages include programs for descriptive statistics, tables, exploratory data analysis, probability calculations for a wide range of distributions, various plotting routines, regression analysis, ANOVA, ANCOVA, and MANOVA. Here we list additional features of relevance to reliability.

Statistica

- Distribution Fitting: A given data set may be fitted to a wide variety of distributions. These include exponential, gamma, and Weibull, to name a few.
- Survival/Failure Time Analysis: This features a comprehensive implementation of a variety of techniques for analyzing censored data; selection of survival-function fitting procedures (with parameters estimated by several different methods), and implementation of proportional hazard and regression models.

Minitab
Many features of Minitab are mentioned in the text. Reliability-related programs are primarily plotting, including probability, hazard, and survival plots plus distribution identification plots. Goodness-of-fit tests are limited to the normal distribution. Minitab also includes good plotting and analysis routines for quality control.

SAS/STAT Software
SAS/STAT software provides survival analysis methods for both censored and uncensored observations. It compares survival distributions, fits the accelerated failure time model to event-time data, and also performs regression analysis based on the proportional hazards model.

SPSS
In the reliability context, SPSS provides tools for analyzing event history and duration data. These include Kaplan–Meier and the Cox regression fits.

StatView
The Survival Analysis module contains the following:

- Kaplan–Meier and Actuarial estimates, linear rank tests for comparing survival across groups, including logrank (Mantel–Cox or Mantel–Haernszel), Breslow–Gehan–Wilcoxon, Tarone–Ware and others
- Tables: Survival function, quantiles, and summary
- Plots: Cumulative survival, cumulative hazard, hazard (actuarial only), density (actuarial only), and censor/response patterns
- Regression Models: Models include proportional hazards, Cox, Weibull, exponential, lognormal, logistic, stepwise, stratification (proportional hazards only), and quantiles for model distribution (parametric models only)

S-PLUS
S-PLUS is an exploratory data analysis and statistical modeling package. Of relevance to reliability studies are the following: Kaplan–Meier curves, Cox proportional hazards models, parametric survival regression, and expected survival analysis.

20.5.2   General Reliability Software Packages

Relia Comp
This package consists of several programs useful in reliability modeling. It consists of the following modules.

Reliability Block Diagram Editor (RBD): This is an assortment of tools to assist the user in making reliability block diagrams quickly and easily. The skeleton is drawn quickly using various tools. The Shape and Group editors allows the user greater flexibility. Once the diagram is drawn, the defined components can be inserted in their appropriate locations in the diagram.

Shape Editor: This feature enables the user to define the color and size of the various elements, groups, and switches used in RBD.

Group Editor: This allows the user to show or hide the contents of the subcomponents from which it is defined. This allows the user to expand or collapse portions of the RBD.

Switch Editor: This editor is used to study the effect of different types of redundancies with imperfect switches.

Model Development: This includes models from the *Handbook of Reliability Prediction Procedures for Mechanical Equipment.* In the future, it will also contain models for electronic equipment based on MIL-HDBK-217F. In addition, users can define their own models.

Item Software
This system contains the following packages:

1. AvSim: Calculates availability, reliability, and maintainability using Monte Carlo simulation techniques. Uses reliability block diagrams as the logical descriptors of the system.
2. AvSim+: Calculates availability, reliability, and maintainability using Monte Carlo simulation techniques. Reliability block diagrams or fault trees can be used to logically describe the system.
3. BellStress: A reliability prediction program for calculating electronic equipment reliability based on Bellcore standard TR-332.
4. FailMode: Provides a fully interactive graphical environment for performing a FMECA to recognized standards such as MIL-STD-1629A.
5. FaultTree+: Provides an integrated environment for constructing and analyzing fault and event trees.
6. LCCware: Does life cycle costing analysis.
7. MechStress: Calculates failure rates of mechanical equipment using internationally recognized methods for failure-rate evaluations.
8. MilStress: Calculates failure rates of electronic components in accordance with the reliability prediction standard MIL-HDBK-217.
9. MKV: A Markov analysis program that analyzes state transition diagrams, using numerical integration techniques.
10. RCMCost: Uses reliability centred maintenance methods to optimize maintenance strategies with respect to cost, safety, and environmental consequences.
11. RelDraw: Provides facilities to construct and assess large and complex reliability networks. RelDraw uses analytical methods to predict system measures such as unavailability, unreliability, failure frequency, expected downtime, etc.
12. Reliability Workbench: This combines the MilStress, BelStress, MechStress, FailMode, and RelDraw tools in a single integrated program.
13. Weibull-Pro: Analyzes failure and repair data by assigning probability distributions that represent the failure or repair characteristics of a given failure mode.

The Reliability and Maintainability Analyst
This package consists of two modules—(1) life data analysis, and (2) maintenance optimization.

Life Data Analysis Module: This module calculates estimates of the parameters of Weibull, normal, lognormal, and exponential distributions using a variety of techniques (maximum likelihood, probability plotting, hazard plotting, method of moments) for complete, censored, and grouped data. In addition, it carries out Bayesian estimation.

Maintenance Optimization Module: This allows for optimization of preventive maintenance, inspection schedules, and predictive maintenance based on the cost of maintenance and time-to-fail data.

Relia Soft

Weibull++5.0: Designed to perform life data analyses as it applies to reliability engineering. It is designed to work with all types of life data (complete, as well as mutiply censored—right, left, and interval censored) using most applicable distributions, with a special emphasis on the Weibull distribution.

Nonlinear Multi-Parameter Equation Solver: Useful in accelerated testing analyses.

Design of Reliability Tests: Allows one to determine the appropriate sample size, test duration, or any other variable related to reliability demonstration tests. In addition, a Quick Statistical Reference Calculator is included for quick look-ups of values commonly found in tables.

Software for the Weibull Distribution

The Weibull distribution is very widely used and there are several software packages for reliability analysis for the distribution, including:

WinSmith (by Fulton Findings)

BiWeibullSmith (by Fulton Findings). This package contains DOS-based special probability plotting software for analyzing mixed failure modes often encountered with reliability data showing probability plots that are concave upwards.

Weibull Curve Fitter (by Dynacomp, Inc.). Weibull Curve Fitter is an easy to use reliability analysis package that fits the Weibull distribution to the failure data of a sample of test specimens. The user provides the lifetimes at failures, or at removal from test, of each specimen in the test sample. WCF then calculates the mean rank estimates of the cumulative probability of failure. Nonfailed specimens may also be included in the data.

YBath (by BathTub Software Inc.). YBath deals with a variety of models involving mixtures of Weibull distributions. The approach is to separate parts by failure modes as appropriate and analyze them separately. YBath uses a least-squares algorithm to determine the best-fit parameters. Currently, Ybath fits standard Weibull (2-parameter Weibull), minimum life Weibull (3-parameter Weibull), Competing Risk, Competing Risk Mixture (simple mixture), and compound competing risk mixture (producing the bathtub shape).

NCSS Statistical Software (Survival Analysis and Reliability Analysis)

Survival analysis includes several techniques to study data in which the response variable is elapsed time. Procedures include survival distribution analysis (including the Kaplan–Meier survival distribution estimate), log rank tests, and proportional hazards regression.

The reliability procedure estimates the parameters of exponential, extreme value, logistic, log-logistic, lognormal, normal, and Weibull probability distributions by maximum likelihood and least squares. It can fit complete, right censored, left censored, interval censored (readout), and grouped data values. It also computes the nonprameteric Kaplan–Meier and Nelson–Aalen estimates of reliability and associated hazard rates.

Relex Software (Relex, Inc.)

This system consists of three software packages for reliability and maintainability.

Relex Reliability Prediction Software: calculates MTBF (Mean Time between Failures) and the Failure Rate of Electronic and Mechanical Systems.

Relex FMEA/FMECA: Aids in performing failure mode, effects, and criticality analyses.

Relex Maintainability: Useful for performing maintainability predictions. It calculates the mean time to repair and many other related maintenance parameters.

RelCalc (T-Cubed Systems, Inc.)
This automates the reliability prediction procedure of MIL-HDBK-217 or Bellcore.

C-FIT
C-FIT is a tool for fitting probability distributions to statistical data. It provides the best fit to a given data set and produces graphical and textual reports for the results of the fit.

ExpertFit
ExpertFit is a computer program for determining what probability distribution best fits a data set. It is a Window-based successor to UniFit II.

20.5.1 Specific Reliability Software

We list some of the specialist software available for carrying out different reliability-related tasks. For more details, see Web site Http://www.enre.umd.edu/tool.htm.

System Modeling Programs:
Reliability Modeling Programs. ACARA (Availability, Cost and Resource Allocation), ANNE, ANSAR, CARE III (Computer Aided Reliability Estimation), CARMS (Computer-Aided Rate Modeling and Simulation), CARRACE (A Tool for Combined Analysis of Reliability, Redundancy and Cost), CASRE (Computer Aided Software Reliability Estimation), ETARA (Event Time Availability, Reliability Analysis), FASTER (Fault Tolerant Architecture Simulation Tool), HEUR (Redundancy Optimization), IRAT, PC Process Reliability, PC Supportability, PREDICTOR, QARMS (Quality Assurance, R&M, Statistics), &MA2T@ (Reliability, Maintenance, and Availability Analysis, R/M/L CATSOP (Reliability, Maintenance, and Logistic Computer Aided Tailoring, RAM Commanader, RAMP for Windows, RAMP (Reliability, Avaliability, and Maintainability Program), RAMTOOL, RBDA/CS-RAM, REAL (Reliability Effectiveness Analysis Program), RELAV (Reliability/Availability Analysis Program), Relex RBD-Reliability Block Diagram, Reliability Mathematical Modeling, REST (RADC Reliability Simulation Tool), RAILS (Reliability, Availability, and Integrated Logistic Support Simulation System), RKP232: Static and Dynamic Modeling, RPTB (RAM Processor and Product Tree Builder), SRP (System Reliability Program), SYSREL (Reliability Analysis of Series System).

Availability Modeling Programs. ARAM (Automated Reliability/Availability/Maintainability), Avail, AvSIM, ETARA (Event Time Availability, Reliability Analysis), FRANTIC-NRC (Time Dependent System Availability), ICARUS (Redundant System Unavailability Model), PREDICTOR Deployment Analysis, R&MA2T2 (Reliability, Maintainability, and Availability Analysis and Tradeoff Tool), RAMP (Reliability, Availability, and Maintainability Program), RAP (Reliability and Availability Program), RELAV (Reliability/Availability Analysis Program), RAILS (Reliability, Availability, and Integrated Logistic Support Simulation System), TIGER (Reliability, Readiness, and Availability Measures), Markov Modeling Programs.

Reliability Prediction Programs
Detailed Stress Prediction Programs. 217 Predicts, ARM E and F (Advanced Reliability Modeling), CARP (Computer Aided Reliability Program), Milstress for Windows, Milstress, PC MIL-HDBK-217F, PC Stress Analysis, PREVIEW, RAMTOOL, REAL (Reliability Effectiveness Analysis Program), REAPmate (Reliability Effectiveness Analysis Program for

PC), Rel+4, Rel Plus III, RelCalc for Windows—217, RelCalc for Windows—Bellcore, Relex217, Reliability Parts Stress Prediction, Reliability Predictor, RL ORACLE, RDP (Reliability Prediction Program), TECMTBF, TelStress.

Parts Count Prediction Programs. CARP (Computer Aided Reliability Program), Rel+4, Relex Parts Count, Reliability Parts Count Prediction, RPC (Reliability Parts Count Program), TECMTBF2.

Mechanical Prediction Programs. MECHREL, MRP (Mechanical Reliability Prediction Program), PERCEPT, Relex Mechanical.

Maintainability Prediction and Analysis Programs. ARM (Automated Reliability/Availability/Maintainability), CAMP (Computer Aided Maintainability Predictor), ETARA (Event Time Availability, Reliability Analysis), FRANTIC-NRC (Time Dependent System Unavailability), Main Tain, Maintain.

Maintainability Mathematical Modeling
Maintainability Prediction, MEAP (Maintainability Effectiveness Analysis Program), MPP (Maintainability Prediction Program), MTTR Plus, PC Maintainability, PC MIL-HDBK-472, PC Mission Maintainability, PPCM, R&MA2T2 (Reliability, Maintainability, and Availability Analysis and Tradeoff Tool), R/M/L/ CATSOP (Reliability, Maintainability, and Logistics Computer Aided Tailoring), RAM Commander, RAM4, RAMTOOL, RAP (Reliability and Availability Program), RCMCost, Relex Maintainability, RPTB (RAM Processor and Product Tree Builder), TECMTTR, TIME.

Reliability Centered Maintenance Programs
Boeing Computerized Preventive Maintenance Program, PREDICTOR MCR (Maintenance Centered Reliability), Predictor RCM (Reliability Centered Maintenance).

Repair Level Analysis Programs
NRLA (Network Repair Level Analysis Program).

Logistic Programs
CHEAP (Cost Effectiveness Analysis Program), CORIDA, ISLE (Integrated Spares and Logistics Evaluation System), Logistic Support Analysis, R/M/L CATSOP (Reliability, Maintainability, and Logistic Computer Aided Tailoring), SpareCost, STOCKPOT.

Failure Reporting Analysis and Corrective Action Systems (FRACAS) Programs
F.A.R. Manager, FRACAS, FRACAS I.A.I., RT1 (FRACAS).

Reliability Testing/Data Analysis Programs
Espinoza Consulting, Neptune Reliability Analysis, Probability Distribution Plotting (Team Graph Papers), QC-PRO Weibull Reliability Analysis, QT_1 Qualitek 1, RelDraw, Reliability, STAR (Statistical Analysis of Reliability), TP781 (Two Point Sampling Plans for Reliability), Weibull 3.0-Plus, Weibull Curve Fitter, WeibullSMITH, Weibull Trending Toolkit.

Worst-Case Analysis
Software Tool for Worst Case Circuit Analysis Data Base, The Design Master, Worst Case Circuit/Parts Tolerance Analysis, WSI REL.

Root-Cause Analysis Programs
RPOACT Software.

FMECA/FMEA Programs
1629 FMECA, AQUA 2, AQUA-FMEA, Failmode, Failure Modes and Effects (FMEA) software, Failre Modes, Effects, and Criticality Analysis—Maint. Inf., Failmode for Windows, FME (Failure Modes, Effects, and Criticality Analysis Program), FMEA, FMEA/Failsafe, FMEAplus, FMECA (Failure Mode, Effects, and Criticality Analysis), FORMUSER, PC NASA FMECA, PC Process FMECA, PEM (Product/Process Failure Modes and Effects Analysis), PRO-FMECA, RAM (Reliability Analysis Model), RAMTOOL, Relex Complete System, Relex FMECA, RKP648: Failure Mode Effects Criticality Analysis, TECFMEA.

Fault Tree Analysis Programs
RAVO, CAFTA for Windows, ETA-II, Event Tree, FaultTree+for Windows, Fault Tree, FaultrEASE, Formal-FTA, FTRAN, IMPORTANCE (Basic Event Cut Set Ranking), SAPHIRE (formerly known as IRRAS), MFAULT (Fault Tree Analysis Cut Set Production), PREP/KITT (System Fault Tree Evaluation), Result II (Fault Tree Analysis), Risk Spectrum Fault Tree, RK606: Fault Tree Analysis, SAICUT, SAIPLOT, TRACE (Tree Analysis Code), Tree Master, Tree-Master Software Family.

20.6 CONCLUDING COMMENTS

The literature on reliability is growing at an exponential rate. We list a few Web sites that are helpful for keeping in touch with the latest developments.

- The National Information Center for Reliability Engineering: http://www.enre.umd. mainnojs.html
- FAA Center for Aviation System Reliability: http://www.cnde.isstate.edu/faa.html
- Society of Reliability Engineers (SRE) homepage: http://www.sre.org
- Society for Maintenance and Reliability Professionals (SMRP) homepage: http:// www.smrp.org
- World Reliability/Quality Organization (WRO) homepage: http://www.world5000. com/wro

Finally, there are several reliability-oriented professional societies. Some of these are as follows:

- IEE Reliability Society
- ASQC Reliability Division
- SAE Reliability Division
- Society of Reliability Engineers
- Society of Maintenance and Reliability Professionals
- System Safety Society
- Society of Logistic Engineers
- Institute of Environmental Sciences and Technology

- Institute of Industrial Engineers
- World Reliability Organizations

NOTES

1. On the theoretical side, the literature on reliability is vast and continues to grow. On the application side, the literature in the public domain is relatively small. This is mainly for commercial reasons. Any publicity on the lack of adequate reliability has serious commercial implications. However, more papers are appearing on the applications front; these may be found in various reliability-engineering journals and *Proceedings of the Annual Reliability and Maintainability Symposium.*

2. The contribution of Peter Townson to this section is gratefully acknowledged.

3. Data collection is still a weak point. Although there are several large data bases, the type of information being collected is often not appropriate for reliability management (from design to operation). Cannon and Bendell (1991) and Flammm and Luisi (1992) contain several interesting papers on this topic. The Special Issue on the "Design of Reliability Data Bases" (*Reliability Engineering & System Safety,* **51,** 1996) contains more recent papers on the topic.

4. A review of handbooks for reliability design analysis can be found in Moss et al. (1989).

APPENDIX A

Probability

A.1 INTRODUCTION TO PROBABILITY THEORY

In this section we give a brief introduction to elementary probability theory, which is the basis of the mathematical approach to modeling failures. The presentation is nonrigorous. The objective is to develop an intuitive feel for the topic that forms the foundation for most models used in solving reliability-related problems.

1.1 Sample Space and Events

Consider an experiment whose outcome is not known in advance but is such that the set of all outcomes (called the sample space $\Im$) is known. Any subset of the sample space $\Im$ is called an event. The concept is illustrated in the following two examples.

Example 1: A manufactured part may be in working or failed state because of variations in the quality of manufacturing. In this case there are two possible outcomes and the sample space is the set $\Im = \{\text{working, failed}\}$.

Example 2: In the case of an unreliable system, the set E_i may be used to denote the event that the system will fail at age t which lies in the interval $[t_{i-1}, t_i)$ with $t_0 = 0$ and t_i, $i = 1, 2, \ldots$, an increasing sequence. In this case, the sample space is the set of intervals $\Im = \{[t_{i-1}, t_i); i = 1, 2, \ldots\}$.

Events can be viewed as sets in the sample space. Consider two events E_i and E_j. A new event is defined by the union $(E_i \cup E_j)$. This event occurs if either E_i or E_j or both occur. Similarly, the intersection $(E_i \cap E_j)$ defines an event that occurs if both E_i and E_j occur. This event occurs only if there is a common subset between E_i and E_j. If the intersection is the null set (i.e., $E_i \cap E_j = \varnothing$) then the two events are said to be mutually exclusive. In reliability, a component might fail due to more than one cause. Let E_i $[E_j]$ denote the events that the component failure is due to cause i $[j]$. Then $(E_i \cup E_j)$ represents the event that the failure is due either cause i or j and $(E_i \cap E_j)$ represents the event that it is due to both causes occurring simultaneously. Unions and intersections of more than two events are defined as extensions of these notions in an obvious way. In the limiting case we have $\{\cup_{n=1}^{\infty} E_n\}$ and $\{\cap_{n=1}^{\infty} E_n\}$. Finally, for any event E, the complement E^c is characterized by $E \cap E^c = \varnothing$ and $E \cup E^c = \Im$. E^c is the collection of all elements in $\Im$ that are not in E.

1.2   Probabilities Defined on Events

For each event, the probability $P(E)$ satisfies the following conditions (called the Axioms of probability):

1. $0 \leq P(E) \leq 1$
2. $P(\Im) = 1$
3. For any sequence of events E_i, $i = 1, 2, \ldots$, that are mutually exclusive,

$$P(\cup_{n=1}^{\infty} E_n) = \sum_{n=1}^{\infty} P(E_n)$$

From this we have the following:

$$P(\emptyset) = 0$$

$$P(E) + P(E^c) = 1$$

$$P(E_i \cup E_j) = P(E_i) + P(E_j) - P(E_i \cap E_j)$$

and, when E_i and E_j are mutually exclusive,

$$P(E_i \cup E_j) = P(E_i) + P(E_j)$$

This approach to defining probability is the *axiomatic* approach. An alternate approach is the *frequency* or *empirical* approach. In this approach,

$$P(E_i) = \lim_{N \to \infty} \left(\frac{N_i}{N} \right) \tag{A1.1}$$

where N_i is the number of times the event E_i has occurred in N trials of the experiment.

Note that $P(E) = 0$ implies that the event E never occurs and $P(E) = 1$ implies that the event occurs for certain. Since the set of outcomes is given by the sample space, we have $P(\Im) = 1$.

1.3   Conditional Probabilities

$P(E_i|E_j)$ denotes the conditional probability of event E_i given that event E_j has occurred. It is given by

$$P(E_i|E_j) = \frac{P(E_i \cap E_j)}{P(E_j)} \tag{A1.2}$$

provided that $P(E_j) > 0$. Similarly,

$$P(E_j|E_i) = \frac{P(E_i \cap E_j)}{P(E_i)} \tag{A1.3}$$

provided that $P(E_i) > 0$. From the above two equations we have

$$P(E_i \cap E_j) = P(E_j|E_i)P(E_i) = P(E_i|E_j)P(E_j) \tag{A1.4}$$

This is a version of Bayes' theorem.

1.4   Random Variables

A random variable is a function which maps outcomes from the sample space $\mathcal{T}$ to R, the space of real numbers. In other words, for every outcome ω in the sample space $\mathcal{T}$, $X(\omega)$ assigns a real number to ω.

Example 1 [Continued]
In this case, we might have $X(\text{working}) = 1$ and $X(\text{failed}) = 0$. Random variables can be classified into (1) discrete and (2) continuous random variables. A discrete random variable $X(\omega)$ takes on at most a countable number of values (for example, the set of nonnegative integers). In contrast, a continuous random variable can take on values from a set of possible values that is uncountable [for example, values in the interval $(-\infty, \infty)$].

Because the outcomes are uncertain, the value assumed by $X(\omega)$ is uncertain before the event occurs. Once the event occurs, $X(\omega)$ assumes a certain value. The standard convention used is as follows: $X(\omega)$ [often the argument ω is dropped for notational convenience and $X(\omega)$ is simply written as X (upper case)] represents the random variable before the event and the value it assumes after the event is represented by x (lower case)

1.5   Distribution and Density Functions

The distribution function $F(x)$ is defined as the probability that $X \leq x$ and is given by

$$F(x) = \{X \leq x\} = \sum_{i \in I} P(E_i) \tag{A1.5}$$

where I is the set $\{i: X(E_i) \leq x\}$. This links the distribution function to the sample space and the probabilities defined on the sample space. The domain of $F(x)$ is $(-\infty, \infty)$ and the range is $[0, 1]$. $F(x)$ has the following properties:

1. $F(x)$ is a nondecreasing function in x
2. $F(-\infty) = 0$ and $F(\infty) = 1$
3. For $x_1 < x_2$, $P\{x_1 < X \leq x_2\} = F(x_2) - F(x_1)$

When $F(x)$ is differentiable, the density function $f(x)$ is given by

$$f(x) = \frac{dF(x)}{dx} \tag{A1.6}$$

$f(x)$ can be interpreted as

$$P\{x < X \leq x + \delta x\} = f(x)\,\delta x + O(\delta x^2). \tag{A1.7}$$

A.2   MOMENT-GENERATING AND CHARACTERISTIC FUNCTIONS

2.1   Moment-Generating Functions

The jth moment of the random variable X, M_j, is given by

$$M_j = E[X^j] = \begin{cases} \displaystyle\int_{-\infty}^{\infty} x^j f(x)\,dx & \text{if } X \text{ is continuous} \\[2mm] \displaystyle\sum_{x} x^j P(X = x), & \text{if } X \text{ is discrete} \end{cases} \tag{A2.1}$$

The moments are easily obtained from the moment generating function, $\psi(t)$, given by

$$\psi(t) = E[e^{tX}] = \begin{cases} \displaystyle\int_{-\infty}^{\infty} e^{tx} f(x) dx & \text{if } X \text{ is continuous} \\ \displaystyle\sum_x e^{tx} P(X = x), & \text{if } X \text{ is discrete} \end{cases} \tag{A2.2}$$

when it exists. From this, the jth moment can be obtained as

$$M_j = E[X^j] = \left. \frac{d^j \psi(t)}{dt^j} \right|_{t=0} \tag{A2.3}$$

The moment generating functions for some basic distributions are as follows (see Mood et al., 1974):

Binomial Distribution: $\psi(t) = (1 - p + pe^t)^n$

Poisson Distribution: $\psi(t) = e^{\lambda(e^t - 1)}$

Exponential Distribution: $\psi(t) = \dfrac{\lambda}{(\lambda - t)}$ for $t < \lambda$

Gamma Distribution: $\psi(t) = \dfrac{1}{(1 - t\beta)^\alpha}$ for $t < \dfrac{1}{\beta}$

Weibull Distribution: $\psi(t) = \dfrac{\Gamma\left(1 + \dfrac{t}{\alpha}\right)}{\beta^{t/\alpha}}$

2.2 Characteristic Functions

The characteristic function, $\phi(t)$, of a random variable X is defined by

$$\phi(t) = E[e^{itX}] \tag{A2.4}$$

for all real values of t and $i = \sqrt{-1}$. If $\phi(t)$ can be expanded in a power series, i.e.,

$$\phi(t) = \sum_{j=0}^{\infty} \frac{B_j(it)^j}{j!} \tag{A2.5}$$

then $M_j = B_j$ for $j \geq 1$. M_j can be obtained from the following relationship

$$M_j = \left. \frac{1}{(i)^j} \frac{d^j \phi(t)}{dt^j} \right|_{t=0} \tag{A2.6}$$

$\phi(t)$ always exists, whereas the moment generating function $\psi(t)$ may not.

A.3 TWO OR MORE RANDOM VARIABLES

We shall confine our discussion to two continuous random variables, denoted X and Y. The extension to more than two is relatively straightforward.

3.1   Joint, Marginal, and Conditional Distribution and Density Functions

The *joint distribution function F(x, y)* is given by

$$F(x, y) = P\{X \leq x, Y \leq y\} \tag{A3.1}$$

The random variables are said to be jointly continuous if there exists a function $f(x, y)$, called the *joint probability density function*, such that

$$F(x, y) = \int_{-\infty}^{x} \left[\int_{-\infty}^{y} f(u, v)dv \right] du \tag{A3.2}$$

The *marginal distribution functions $F_X(x)$ and $F_Y(y)$* are given by

$$F_X(x) = F(x, \infty) \qquad \text{and} \qquad F_Y(y) = F(\infty, y) \tag{A3.3}$$

The two *marginal density functions* are given by

$$f_X(x) = \frac{dF_X(x)}{dx} \qquad \text{and} \qquad f_Y(y) = \frac{dF_Y(y)}{dy} \tag{A3.4}$$

The *conditional distribution* of X given $Y = y$ is denoted $F(x|y)$ and given by

$$F(x|y) = P\{X \leq x|Y = y\} \tag{A3.5}$$

The conditional distribution of Y given $X = x$, $F(y|x)$, is defined similarly.

For jointly continuous random variables with a joint density function $f(x, y)$, the *conditional probability density* function of X, given $Y = y$, is given by

$$f(x|y) = \frac{f(x, y)}{f_Y(y)} \tag{A3.6}$$

Similarly,

$$f(y|x) = \frac{f(x, y)}{f_X(x)} \tag{A3.7}$$

The random variables X and Y are said to be *independent* (or *statistically independent*) if and only if

$$F(x, y) = F_X(x)F_Y(y) \tag{A3.8}$$

for all x and y.

The results are similar for discrete random variables, with summation replacing integration.

3.2   Moments of Two Random Variables

The covariance of X and Y is defined by

$$\text{Cov}(X, Y) = E[\{X - E[X]\}\{Y - E[Y]\}] = E[XY] - E[X]E[Y] \tag{A3.9}$$

The correlation ρ_{XY} is defined as

$$\rho_{XY} = \frac{\text{Cov}(X, Y)}{\sigma_X \sigma_Y} \qquad \text{(A3.10)}$$

where σ_X and σ_Y are the standard deviations of X and Y, respectively. The random variables X and Y are said to be *uncorrelated* if $\rho_{XY} = 0$. Note that independent random variables are uncorrelated but that the converse is not necessarily true.

3.3 Conditional Expectation

$E[X|Y = y]$ is called the conditional expectation of X given that $Y = y$. The unconditional expectation of X, given by

$$E[X] = \int_{-\infty}^{\infty} x f_X(x) dx \qquad \text{(A3.11)}$$

is related to the conditional expectation by the relation

$$E[X] = \int_{-\infty}^{\infty} E[X|Y = y] f_Y(y) dy \qquad \text{(A3.12)}$$

which is written symbolically as

$$E[X] = E[E[X|Y]] \qquad \text{(A3.13)}$$

Similar results exist for discrete random variables.

3.4 Sums of Independent Random Variables

Let X and Y be two independent random variables with density functions $f_X(x)$ and $f_Y(y)$, respectively, and let $Z = X + Y$. Then the density function of Z, $f_Z(z)$, is given by

$$f_Z(z) = \int_{-\infty}^{\infty} f_Y(t) f_X(z - t) dt \qquad \text{(A3.14)}$$

or

$$f_Z(z) = \int_{-\infty}^{\infty} f_X(t) f_Y(z - t) dt \qquad \text{(A3.15)}$$

This operation is called the convolution operation, indicated by the symbol "*." Thus

$$f_Z(z) = f_X(z)*f_Y(z) = f_Y(z)*f_X(z) \qquad \text{(A3.16)}$$

When Z is the sum of n independent variables, X_i ($i = 1, 2, \ldots, n$), with respective density functions $f_i(x)$, then

$$f_Z(z) = f_1(z)*f_2(z)* \ldots * f_n(z) \qquad \text{(A3.17)}$$

3.5   Central Limit Theorem

If X_i, $1 \le i \le n$, are a sequence of independent and identically distributed random variables with mean μ and finite variance σ^2, then for large n

$$Z = \frac{\sum_{i=1}^{n} X_i}{n} \tag{A3.18}$$

is approximately distributed according to a normal distribution function with mean μ and variance σ^2/n.

A.4   LAPLACE TRANSFORMS

The Laplace transform of a function $f(x)$, provided it exists, is given by

$$\hat{f}(s) = \sum_{-\infty}^{\infty} e^{-sx} f(x) dx \tag{A4.1}$$

$f(x)$ can be obtained from $\hat{f}(s)$ by the inverse Laplace transform given by

$$f(x) = \frac{1}{2\pi} \oint \hat{f}(s) e^{sx} \, ds \tag{A4.2}$$

The Laplace transform and its inverse are very useful in certain analysis. Of particular interest are the following results:

(i) Let $f(x) = [dF(x)/dx]$ and $\hat{F}(s)$ and $\hat{f}(s)$ denote the Laplace transforms of $F(x)$ and $f(x)$, respectively. Then

$$\hat{f}(s) = s\hat{F}(s) - f(0) \tag{A4.3}$$

(ii) Let

$$f_Z(z) = \int_{-\infty}^{\infty} f_X(t) f_Y(z - t) dt \tag{A4.4}$$

Then

$$\hat{f}_Z(s) = \hat{f}_X(s) \hat{f}_Y(s) \tag{A4.5}$$

where $\hat{f}_Z(s)$, $\hat{f}_X(s)$ and $\hat{f}_Y(s)$ are the Laplace transforms of $f_Z(x)$, $f_X(x)$, and $f_Y(x)$, respectively. (iii) If Z is the sum of n independent and identically distributed random variables with density function $f_X(x)$, then $f_Z(x)$ is given by the n-fold convolution

$$f_Z(z) = f_X(x) * f_X(x) * \ldots * f_X(x) \tag{A4.6}$$

In this case, $f_Z(x)$ is also denoted $f^{(n)}(x)$. On taking the Laplace transform, we have

$$\hat{f}_Z(s) = [\hat{f}_X(s)]^n \tag{A4.7}$$

A5   FUNCTIONS OF RANDOM VARIABLES

In reliability analysis, we often deal with estimation problems involving functions of random variables. Here we give some results (exact and asymptotic) that are often useful in these analyses.

5.1   Asymptotic Mean and Variance of a Function of Random Variables

Theorem 5.1. If Y is a random variable with mean μ and finite variance $V(Y)$, and $\tau(Y)$ is a twice-differentiable function of Y, then

$$E[\tau(Y)] \approx \tau(\mu) + \frac{1}{2}\tau''(\mu)V(Y) \tag{A5.1}$$

and

$$V(\tau(Y)\} \approx [\tau'(\mu)]^2 V(Y) \tag{A5.2}$$

In applications, Y is usually an estimator of a parameter, say $\hat{\theta}$ or θ^*, or a sample moment.

These results extend in a natural way to functions of more than one variable (e.g., functions of estimators of several parameters or functions of several moments). The general result is:

Theorem 5.2. Suppose that $Y_1, Y_2, \ldots, Y_k$ are k random variables with means μ_i, finite variances σ_i^2 and covariances σ_{ij}, where

$$\sigma_{ij} = \text{Cov}(Y_i, Y_j) = E[(Y_i - \mu_i)(Y_j - \mu_j)] \tag{A5.3}$$

Let $\tau(Y_1, Y_2, \ldots, Y_k)$ be such that all second order partial derivatives exist. Then

$$E[\tau(Y_1, \ldots, Y_k)] \approx \tau(\mu_1, \ldots, \mu_k) + \sum_{i=1}^{k}\sigma_i^2\left(\frac{\partial^2\tau}{\partial Y_i^2}\right)\bigg|_{\mu_1,\ldots,\mu_k} + 2\sum_{i<j}\sigma_{ij}\left(\frac{\partial^2\tau}{\partial Y_i\partial Y_j}\right)\bigg|_{\mu_1,\ldots,\mu_k} \tag{A5.4}$$

and

$$V(\tau(Y_1, \ldots, Y_k)) \approx \sum_{i=1}^{k}\sigma_i^2\left(\frac{\partial\tau}{\partial Y_i}\right)^2\bigg|_{\mu_1,\ldots,\mu_k} + 2\sum_{i<j}\sigma_{ij}\left(\frac{\partial\tau}{\partial Y_i}\right)\left(\frac{\partial\tau}{\partial Y_j}\right)\bigg|_{\mu_1,\ldots,\mu_k} \tag{A5.5}$$

The expressions in Equations (A5.1)–(A5.5) are estimated by substituting estimates for the parameters in each formula. Note that if the random variables involved are independent, the final terms in (A5.4) and (A5.5) drop out because in that case the covariances are zero.

Comments

1. Theorem 5.1 is easily proven by use of a Taylor's series expansion. This yields

$$\tau(Y) = \tau(\mu) + \tau'(\mu)(Y - \mu) + \tau''(\mu)(Y - \mu)^2/2 + \text{higher-order terms}$$

(A5.1) follows from this by ignoring the higher-order terms and taking the expectation. (A5.2) is obtained by using only the first two terms.

2. In practice, we often use only the first-order approximation for $E[\tau(Y)]$ as well, in which case (A5.1) reduces to $E[\tau(Y)] \cong \tau(\mu)$.

5.2 Sums of Random Variables

The basic result concerning the moments of a sum of random variables is the following:

Theorem 5.3. Suppose that $X_1, X_2, \ldots, X_n$ is a sequence of n random variables with means μ_i and finite variances σ_i^2 and covariances σ_{ij}, and that $c_0, c_1, \ldots, c_n$ is a sequence of constants. Let $Y = c_0 + \sum_{i=1}^n c_i Y_i$. Then Y has mean

$$\mu_Y = c_0 + \sum_{i=1}^n c_i \mu_i \tag{A5.6}$$

and variance

$$\sigma_Y^2 = \sum_{i=1}^n c_i^2 \sigma_i^2 + 2 \sum_{i<j} c_i c_j \sigma_{ij} \tag{A5.7}$$

Proof of Theorem 5.3 requires only straightforward algebra.

Comments

1. If the X_i are independent, the covariance terms in (A5.7) are zero.
2. If the X_i are independent and $c_0 = 0$ and $c_i = 1$, we obtain the result that the mean is the sum of the means and the variance of a sum is the sum of the variances.
3. The mean and variance of a difference are obtained by putting $n = 2$, $c_0 = 0$, $c_1 = 1$, and $c_2 = -1$.
4. If $c_0 = 0$ and $c_i = 1/n$ for $i = 1, \ldots, n$, then $Y = \overline{X}$. If the X_i are independent and identically distributed, we see that $E(\overline{X}) = \mu$ and $V(\overline{X}) = \sigma^2/n$.
5. If the X_i are normally distributed, so is Y (not deduced from the theorem).
6. If the X_i have an exponential distribution, then Y has a gamma distribution (also not deduced from the theorem).

5.3 Distribution of a Product

Products are particularly important in many reliability calculations. If $X_1, X_2, \ldots, X_n$ are independent and $Y = X_1, X_2, \ldots, X_n$, then $\mu_Y = \mu_1 \cdot \mu_2 \cdots \mu_n$. The variance is a bit messier. For any two variables, we have

$$V(X_i X_j) = \mu_i^2 \sigma_j^2 + \mu_j^2 \sigma_i^2 + \sigma_i^2 \sigma_j^2 \tag{A5.8}$$

For three variables, X_i, X_j, X_h, the variance is

$$V(X_i X_j X_h) = \mu_i^2 \mu_j^2 \sigma_h^2 + \mu_i^2 \mu_h^2 \sigma_j^2 + \mu_j^2 \mu_h^2 \sigma_i^2 + \mu_i^2 \sigma_j^2 \sigma_h^2 + \mu_j^2 \sigma_i^2 \sigma_h^2 + \mu_h^2 \sigma_i^2 \sigma_j^2 + \sigma_i^2 \sigma_j^2 \sigma_h^2 \tag{A5.9}$$

The result extends in an obvious manner.

5.4   Ratio of Two Random Variables

The approximate the mean and variance of a ratio of two random variables, say X_i/X_j, is

$$E\left(\frac{X_i}{X_j}\right) \approx \frac{\mu_i}{\mu_j} - \frac{\sigma_{ij}}{\mu_j^2} + \frac{\mu_j}{\mu_j^3}\sigma_j^2 \tag{A5.10}$$

$$V\left(\frac{X_i}{X_j}\right) \approx \left(\frac{\mu_i}{\mu_j}\right)^2 \left\{\frac{\sigma_i^2}{\mu_j^2} + \frac{\sigma_j^2}{\mu_i^2} - \frac{2\sigma_{ij}}{\mu_i\mu_j}\right\} \tag{A5.11}$$

APPENDIX B

Introduction to Stochastic Processes

B.1 BASIC CONCEPTS

In this Appendix, we give a brief introduction to stochastic processes and discuss some of the processes that are used in the book. Our presentation will be intuitive and nonrigorous and will highlight the important concepts. Readers interested in a deeper understanding of the underlying theory should consult the references given at the end of the book.

1.1 Stochastic Processes

In Appendix A.1 we defined a random variable, $X(\omega)$, as function that maps outcomes from the sample space to real numbers. A stochastic process $X(t, \omega)$, $t \in T$, where T is a set of nonnegative numbers, can be viewed as an extension of $X(\omega)$ in the following sense: t represents a time instant in the set T, which may be either finite or infinite. For a fixed $t \in T$, $X(t, \omega)$ is a random variable in the usual sense. For a fixed ω (outcome), $X(t, \omega)$ can be viewed as a function of t. $X(t, \omega)$ denotes the state of the process at time t. If T is countable, then $X(t, \omega)$ is called a discrete time stochastic process and if T is a continuum, then it is called a continuous time stochastic process. Henceforth, we omit ω and represent $X(t, \omega)$ as simply $X(t)$.

Let $t_i(i = 1, 2, \ldots, n)$ denote n different time instants. The probabilistic characterization of the process $X(t)$ at these n points can be done through the joint probability distribution

$$F(t_1, x_1; t_2, x_2; \ldots ; t_n, x_n) = P\{X(t_1) \le x_1; X(t_2) \le x_2; \ldots ; X(t_n) \le x_n\} \quad \text{(B1.1)}$$

As n increases, this becomes cumbersome and is of limited use in modeling real-world problems.

1.2 Markov Property

A stochastic process $X(t)$ is said to have the Markov property if

$$P\{X(t + \tau) \le x | X(u) = x(u), -\infty < u \le t\} = P\{X(t + \tau) \le x | X(t) = x(t)\} \quad \text{(B1.2)}$$

In other words, the probabilistic characterization of $X(t + \tau)$ (a future event) given $\{X(u) = x(u), -\infty < u \le t\}$ (past history and present value of the process) depends only on $x(t)$ [the present value of $X(t)$] and not its past values. This simplifies the mathematical characterization of the process considerably. Using conditional probability, we have, for an increasing sequence in t_i,

$$P\{X(t_1) \leq x_1; X(t_2) \leq x_2; \ldots; X(t_n) \leq x_n\} = P\{X(t_n) \leq x_n | X(t_{n-1}) \leq x_{n-1}\} \ldots$$

$$P\{X(t_2) \leq x_2 | X(t_1) \leq x_1\} P\{X(t_1) \leq x_1\} \tag{B1.3}$$

Thus the joint probability distribution for $X(t)$ at n different points along the time axis can be obtained in terms of the conditional distribution of $X(t)$ involving two different values of t. In other words, the probabilistic characterization of the process can be done as a function of sets of four variables $F(t_i, x_i; t_j, x_j)$ with

$$F(t_i, x_i; t_j, x_j) = P\{X(t_i) \leq x_i \ \text{and} \ X(t_j) \leq x_j\} \tag{B1.4}$$

for all t_i and t_j over the interval T and all x_i and x_j over the real line.

Stochastic processes which have the Markov property are called *Markov processes*. Processes which do not have this property are called non-Markov processes. Often, a *non-Markov process* can be transformed into a Markov process by enlarging the state space.

Most, but not all, of the stochastic processes that are used in this book have the Markov property.

1.3 Classification of Stochastic Processes

Stochastic processes can be divided into four categories depending on:

1. Whether the values assumed by the process $X(t)$ are discrete or continuous
2. Whether the values assumed by the time variable t are discrete or continuous

We briefly discuss each of these four categories.

(i) Discrete State/Discrete Time Process
Here both $X(t)$ and t assume only discrete values. Let the values assumed by $X(t)$ be denoted by $s_i, i = 1, 2, \ldots, s_r$. r may be either finite or infinite. The values assumed by $t_i, i = 1, 2, \ldots$, form an increasing sequence. If the process is Markovian and r is finite, then it is called a Markov chain. If $r \to \infty$, then it is called a Markov chain with infinite states.

(ii) Discrete State/Continuous Time Process
Here $X(t)$ assumes only discrete values as in (i) with r either finite or infinite and t assumes a continuous range of values in the interval $(-\infty, \infty)$. If the process is Markovian, it is called a continuous chain Markov formulation. A special type of this is a point process which is discussed in Appendix B.2.

(iii) Continuous State/Discrete Process
In this case, $X(t)$ assumes a continuous range of values and t assumes discrete values. If the process is Markovian, it is called a discrete time Markov process.

(iv) Continuous State/Continuous Time Process
In this process, both $X(t)$ and t assume continuous ranges of values. If the process is Markovian, it is called a continuous time Markov process (or simply a Markov process).

A further subclassification is *stationary* and *nonstationary* stochastic processes. A stochastic process is said to be stationary if the joint distribution function is invariant under a shift in t, i.e., if

$$F(t_1', x_1; t_2', x_2; \ldots, t_n', x_n) = F(t_1, x_1; t_2, x_2; \ldots; t_n, x_n) \tag{B1.5}$$

with $t_i' = t_i + \tau$ $(1 \leq i \leq n)$ for all τ and n.

B.2   MARKOV CHAINS

2.1   Discrete Time Markov Chains

Let r denote the number of states [i.e., the different values assumed by $X(t)$], with r either finite or infinite. The discrete states are s_i, $i = 1, 2, \ldots, r$. The discrete values of t_i $(i \geq 0)$ correspond to the set of integers $i = 0, 1, 2, \ldots$. We shall use X_i to denote $X(t_i)$, $i \geq 0$.

A *nonstationary* r-state Markov chain is characterized by an $r \times r$ one-step transition matrix $P(t)$, $t = 0, 1, 2, \ldots$. The entries of $P(t)$ are $P_{ij}(t)$, $1 \leq i, j \leq r$, the one step transition probabilities (for $i \neq j$) or no transition probabilities (for $i = j$) at time t, i.e.,

$$P_{ij}(t) = P\{X_{t+1} = s_j | X_t = s_i\} \tag{B2.1}$$

Note that $0 \leq P_{ij}(t) \leq 1$ and

$$\sum_{j=1}^{r} P_{ij}(t) = 1 \tag{B2.2}$$

for $1 \leq i \leq r$.

For a *stationary* Markov chain, $P(t) = P$, implying that the elements of the matrix do not change with t.

2.2   Continuous Time Markov Chains

As above, r denotes the number of states and may be either finite or infinite. The discrete states are s_i, $i = 1, 2, \ldots, r$. The values assumed by t are continuous. Once $X(t)$ enters state s_i at time t, it moves to state s_j with probability P_{ij}, $j \neq i$, after spending a random amount of time in state s_i.

Note that $P_{ii} = 0$ for $1 \leq i \leq r$. Hence the transition between states is given by a transition matrix P with diagonal elements zero. The elements of P are constrained to satisfy (B2.2). The duration for which $X(t)$ stays in state s_i before moving out is a random variable τ_i which is exponentially distributed with mean time $= 1/\lambda_i$ $(1 \leq i \leq r)$.

The continuous time Markov chain can be characterized in an alternate way involving transition rates λ_{ij} $(1 \leq i, j \leq r)$. $\lambda_{ij}\, \delta t$ is the probability that the state will move from i to j in a small interval $[t, t + \delta t)$ given that $X(t)$ was in state i at time t. For more on the connection between the two, see Ross (1992).

B.3   POINT PROCESSES

A *point process* is a continuous time stochastic process characterized by events that occur randomly along the time continuum. Examples of such events, in the context of reliability, are an item being put into operation or an item failing. The theory of point processes is very rich, as a variety of such processes have been formulated and studied. Of particular interest to reliability modeling is the counting process.

3.1 Counting Processes

A point process $\{N(t), t \geq 0\}$ is a counting process if it represents the number of events that have occurred until time t. It must satisfy:

1. $N(t) \geq 0$
2. $N(t)$ is integer valued
3. If $s < t$, then $N(s) \leq N(t)$
4. For $s < t$, $\{N(t) - N(s)\}$ is the number of events in the interval $(s, t]$

We shall confine ourselves to $t \geq 0$. The behavior of $N(t)$, for $t \geq 0$, depends on whether of not $t = 0$ corresponds to the occurrence of an event. The analysis of the case with $t = 0$ corresponding to the occurrence of an event is simpler than the alternate case. Also, we assume that $N(0) = 0$.

A counting process $\{N(t), t \geq 0\}$ is said to have *independent increments* if, for all choices $0 \leq t_1 < t_2 < t_3 \ldots < t_n$, the $(n-1)$ random variables

$$\{N(t_2) - N(t_1)\}, \{N(t_3) - N(t_2)\}, \ldots, \{N(t_n) - N(t_{n-1})\}$$

are independent. A counting process $\{N(t), t \geq 0\}$ is said to have *stationary independent increments* if, for each $s > 0$, $\{N(t_1 + s) - N(t_1)\}$ and $\{N(t_2 + s) - N(t_2)\}$ have the same distribution function, i.e., if the distribution function of $\{N(t + s) - N(t)\}$ does not depend on t.

Two special counting processes of particular importance to reliability modeling are (1) Poisson process, and (2) renewal process.

3.2 Poisson Processes

We first consider the stationary Poisson process and later discuss some extensions.

3.2.1 Stationary Poisson Process

Definition 1: A counting process $N(t)$, $t \geq 0$, is a stationary Poisson process if:

1. $N(0) = 0$
2. The process has independent increments
3. The number of events in any interval of length t is distributed according to Poisson distribution with parameter λt, i.e.,

$$P\{N(t + s) - N(s)\} = \frac{e^{-\lambda t}(\lambda t)^n}{n!} \tag{B3.1}$$

$n = 0, 1, 2, \ldots$, and for all s and $t \geq 0$.

It can be shown through simple analysis (see, e.g., Ross, 1970) that for a stationary Poisson process, the times between events (also called interevent times) are independent and identically distributed exponential random variables with mean $(1/\lambda)$. This is the basis of a second definition for a stationary Poisson process.

Definition 2: Consider a counting process. Let X_1 denote the time instant of the first event occurrence, and for $j \geq 2$, let X_j denote the time interval between the $(j-1)$st and jth events. The counting process is a stationary Poisson process with parameter λ if the sequence X_j, $j \geq 1$, are independent and identically distributed exponential random variables with mean $(1/\lambda)$.

For a stationary Poisson process it can be shown (see Ross, 1970) that

1. The probability of an event occurring in $[t, t + \delta t)$ is $\lambda \delta t + o(\delta t)$
2. The probability of two or more events occurring in $[t, t + \delta t)$ is $o(\delta t)$
3. The occurrence of an event in $[t, t + \delta t)$ is independent of the number of events in $[0, t)$

As a result, we have a third definition for a stationary Poisson process.
Definition 3: A counting process $\{N(t), t \geq 0\}$ is a stationary Poisson process if

1. The probability of an event occurring in $[t, t + \delta t)$ is $\lambda \delta t + o(\delta t)$
2. The probability of two or more events occurring in $[t, t + \delta t)$ is $o(\delta t)$
3. The occurrence of an event in $[t, t + \delta t)$ is independent of the number of events in $[0, t)$

λ is called the intensity of the process.

Comment: The above discussion illustrates the point that there is more than one way of characterizing a counting process. In the context of reliability modeling, a particular characterization may be more appropriate than alternate, equivalent characterizations. For example, in the case of nonrepairable items, Definition 2 is more appropriate; in the case of repairable items with the item being subjected to minimal repair after each failure, Definition 3 is more appropriate.

Let $M(t)$ denote the expected number of events in $[0, t)$. Since $N(t)$ is distributed according to Poisson distribution with parameter λt, we have

$$M(t) = E[N(t)] = \lambda t \tag{B3.2}$$

3.2.2   *Nonstationary Poisson Process*

In a stationary Poisson process, the probability of an event occurring in $[t, t + \delta t)$ is $\lambda \delta t + o(\delta t)$, with λ a constant. A nonstationary Poisson process is a natural extension in which λ changes with time.

A counting process $\{N(t), t \geq 0\}$ is a nonstationary Poisson process if

1. $N(0) = 0$
2. $\{N(t), t \geq 0\}$ has independent increments
3. $P\{N(t + \delta t) - N(t) = 1\} = \lambda(t)\, \delta t + o(\delta t)$
4. $P\{N(t + \delta t) - N(t) \geq 2\} = o(\delta t)$

$\lambda(t)$ is called the intensity function. Let

$$\Lambda(t) = \int_0^t \lambda(x)dx \tag{B3.3}$$

Then it can be shown (see Ross, 1970) that

$$P\{N(t + s) - N(t) = j\} = \frac{e^{-\{\Lambda(t+s)-\Lambda(t)\}}\{\Lambda(t + s) - \Lambda(t)\}^j}{j!} \tag{B3.4}$$

for $j \geq 0$.

This result may be used to define a nonstationary Poisson process in a manner similar to Definition 1 for a stationary Poisson process.

Expected Number of Events in [0, t)
Since the probability of j events ($j \geq 0$) in $[0, t)$ is given by

$$P\{N(t) = j\} = \frac{e^{-\Lambda(t)}\{\Lambda(t)\}^j}{j!} \tag{B3.5}$$

the expected number of events in $[0, t)$, $M(t)$, is given by

$$M(t) = E[N(t)] = \Lambda(t) \tag{B3.6}$$

3.2.3 Conditional Poisson Process
For a stationary Poisson process, the intensity λ is a deterministic quantity. For a conditional Poisson process the intensity is a random variable Λ with a distribution $G(\ )$.

Definition: A counting process $\{N(t), t \geq 0\}$ is a conditional Poisson process if, conditional on the event $\Lambda = \lambda$, $\{N(t), t \geq 0\}$ is a stationary Poisson process with intensity λ.

In this case, the probability of the event $\{N(t + s) - N(t) = n\}$ is given by

$$P\{N(t + s) - N(t) = n\} = \int_0^\infty \frac{e^{-\lambda s}\{\lambda s\}^n}{n!} dG(\lambda) \tag{B3.7}$$

Note that $\{N(t), t > 0\}$ is *not* a Poisson process except for trivial measures dG.

3.2.4 Doubly Stochastic Poisson Process
This is an extension of the conditional Poisson process. Here the intensity function is itself a stochastic process. Conditional on the intensity function, the process is a nonstationary Poisson process. The analysis of such processes is done by first conditioning on the intensity function and then removing the conditioning. In general, the analysis is involved, and often intractable, even for simple stochastic characterizations of the intensity function.

3.3 Renewal Processes

We first consider the ordinary renewal process and then discuss some extensions.

3.3.1 Ordinary Renewal Processes
As indicated earlier, a counting process characterized in terms of interevent times is a stationary Poisson process if these times are independent and identically distributed exponential random variables. A natural generalization is one where the interevent times are independent and identically distributed with an *arbitrary* distribution.

A counting process $\{N(t), t \geq 0\}$ is an ordinary renewal process if

1. $N(0) = 0$
2. X_1, the time to occurrence of the first event (from $t = 0$) and $X_j, j \geq 2$, the time between the $(j - 1)$st and jth events, are a sequence of independent and identically distributed random variables with distribution function $F(x)$
3. $N(t) = \text{Sup} \{n: s_n \leq t\}$, where

$$S_0 = 0, \qquad S_n = \sum_{i=1}^{n} X_i, n \geq 1 \tag{B3.8}$$

[*Note:* The stationary Poisson process is a special case of the ordinary renewal process with $F(x)$ an exponential distribution function.]

Analysis of Ordinary Renewal Processes

Note that S_n is the time instant for the nth renewal (or event) and is the sum of n independent and identically distributed random variables. Since the X_i's are distributed with distribution function $F(x)$, from the result of Appendix A.3, the distribution of S_n is given by the n-fold convolution of F with itself, i.e.,

$$P(S_n \leq x\} = F^{(n)}(x) = F(x)*F(x)*, \ldots, *F(x) \tag{B3.9}$$

Distribution of N(t)

It is easily seen that $N(t) \geq n$ if and only if $S_n \leq t$. As a result,

$$P\{N(t) = n\} = P\{N(t) \geq n\} - P\{N(t) \geq (n + 1)\} = P\{S_n \leq t\} - P(S_{n+1} \leq t\} \tag{B3.10}$$

for $n = 0, 1, \ldots$, where $S_0 \equiv 0$. Since

$$P\{S_n \leq t\} = F^{(n)}(t) \tag{B3.11}$$

where $F^{(0)} \equiv 1$, we have

$$P\{N(t) = n\} = F^{(n)}(t) - F^{(n+1)}(t) \tag{B3.12}$$

From this, expressions for the moments of $N(t)$ can be obtained. Of particular interest in warranty analysis is the first moment, the expected number of renewals in $[0, t)$.

Expected Number of Renewals in [0, t)

The expected number of renewals in $[0, t)$, $M(t)$, is given by

$$M(t) = E[N(t)] = \sum_{n=0}^{\infty} nP\{N(t) = n\} \tag{B3.13}$$

Using (B3.12), this can be written as

$$M(t) = \sum_{n=0}^{\infty} n\{F^{(n)}(t) - F^{(n+1)}(t)\} = \sum_{n=1}^{\infty} F^{(n)}(t) \tag{B3.14}$$

Let $\hat{M}(s)$ denote the Laplace transform (see Appendix A.4) of $M(t)$. Then we have,

$$\hat{M}(s) = \frac{1}{s}\sum_{n=0}^{\infty}\{\hat{f}(s)\}^n \tag{B3.15}$$

This follows from the result of Appendix A.4 for the Laplace transform of $F^{(n)}(t)$. As a result, $\hat{M}(s)$ can be written as

$$\hat{M}(s) = \frac{\hat{f}(s)}{s[1-\hat{f}(s)]}$$

or, after rearranging, as

$$\hat{M}(s)[1-\hat{f}(s)] = \hat{F}(s) \tag{B3.16}$$

as $\hat{F}(s) = \hat{f}(s)/s$. On taking the inverse Laplace Transform, we have

$$M(t) - M(t)*f(t) = F(t) \tag{B3.17}$$

or

$$M(t) = F(t) + \int_0^t M(t-x)f(x)dx \tag{B3.18}$$

An alternate derivation for $M(t)$, based on conditional expectation, is as follows. Conditioned on X_1, the time to first failure, $M(t)$ can be written as

$$M(t) = \int_0^{\infty} E[N(t)|X_1 = x]dF(x) \tag{B3.19}$$

But,

$$E[N(t)|X_1 = x] = \begin{cases} 0 & \text{if } x > t \\ 1 + M(t-x) & \text{if } x \le t \end{cases} \tag{B3.20}$$

Note that we have used the renewal property in deriving the above expression. If the first failure occurs at $x \le t$, then the renewals over $(t-x)$ occur according to an identical renewal process and hence the expected number of renewals over this period is $M(t-x)$. Using (B3.20) in (B3.19), we have (B3.18).

Equation (B3.18) is called the *renewal integral equation* and $M(t)$ is called the *renewal function* associated with the distribution function $F(t)$. $M(t)$ plays an important role in reliability analysis. In general, it is difficult to obtain $M(t)$ analytically.

The renewal density function, $m(t)$, is given by

$$m(t) = \frac{dM(t)}{dt} \tag{B3.21}$$

and satisfies the equation

$$m(t) = f(t) + \int_0^t m(t-x)f(x)dx \tag{B3.22}$$

where $f(t)$ is the density function associated with $F(t)$.

Variance of the Number of Renewals in [0, t)
From (B3.12)) it can be shown that the variance of $N(t)$ is given by

$$\text{Var}[N(t)] = \sum_{n=1}^{\infty}(2n-1)F^{(n)}(t) - [M(t)]^2 \qquad \text{(B3.23)}$$

Excess (or Residual) Life at t
Let $B(t)$ denote the time from t until the next renewal, that is,

$$B(t) = S_{N(t)+1} - t \qquad \text{(B3.24)}$$

$B(t)$ is called the excess or residual life at t. It represents the remaining life of the item in use at time t. The distribution function for $B(t)$ is given by

$$P\{B(t) \le x\} = F(t+x) - \int_0^t [1 - F(t+x-y)]dM(y), \qquad x \ge 0 \qquad \text{(B3.25)}$$

where $M(t)$ is the renewal function associated with the distribution $F(x)$ and is given by (B3.18). In general, it is difficult to solve (B3.25) analytically. In the limit as $t \to \infty$, this becomes

$$\lim_{t \to \infty} P\{B(t) \le x\} = \frac{\int_0^x [1 - F(y)]dy}{\mu} \qquad \text{(B.26)}$$

where μ is $E[X]$.
    The details of the derivations of (B3.25) and (B3.26) can be found in Ross (1970).

Age at Time t
Let $A(t)$ be the time from t since the last renewal, that is,

$$A(t) = t - S_{N(t)} \qquad \text{(B3.27)}$$

$A(t)$ is the age of the item in use at time t and hence is called the age at t. The distribution function for $A(t)$ is given by

$$P\{A(t) \le x\} = \begin{cases} F(t) - \int_0^{t-x} [1 - F(t-y)]dM(y), & 0 \le x \le t \\ 1, & x > t \end{cases} \qquad \text{(B3.28)}$$

where $M(t)$ is the renewal function associated with $F(t)$. The derivation of (B3.28) can be found in Ross (1970).

3.3.2   Delayed Renewal Process
A counting process $\{N(t), t \ge 0\}$ is a delayed renewal process if

1. $N(0) = 0$
2. X_1, the time to the first event, is a nonnegative random variable with distribution function $F(x)$

3. $X_j, j \geq 2$, the time intervals between the jth and $(j-1)$st events, are independent and identically distributed random variables with a distribution function $G(x)$ different from $F(x)$

4. $N(t) = \text{Sup } \{n: S_n \leq t\}$ where S_n is given by (B3.8)

Note that when $G(x)$ equals $F(x)$, then the delayed renewal process reduces to an ordinary renewal process.

Expected Number of Renewals in [0, t)

Let $M_d(t)$ denote the expected number of renewals over $[0, t)$ for the delayed renewal process, that is,

$$M_d(t) = E[N(t)] = \int_0^\infty E[N(t)|X_1 = x] dF(x) \tag{B3.29}$$

An expression for this can easily be obtained using the conditional expectation approach used for obtaining $M(t)$ for the ordinary renewal process. Conditioning on X_1, the time to the first renewal, we have

$$E[N(t)|X_1 = x] = \begin{cases} 0 & \text{if } x > t \\ 1 + M_g(t-x) & x \leq t \end{cases} \tag{B3.30}$$

where $M_g(t)$ is the renewal function associated with the distribution function $G(t)$. This follows from the fact that, if the first event occurs at $x \leq t$, then over the interval (x, t), the events occur according to a renewal process with distribution G. As a result, $M_d(t)$ is given by

$$M_d(t) = F(t) + \int_0^t M_g(t-x) f(x) dx \tag{B3.31}$$

Excess (or Residual) Life at Time t

Let $B_d(t)$ denote the time from t until the next renewal. The distribution function for $B_d(t)$ is given by

$$P\{B_d(t) \leq x\} = F(t+x) - \int_0^t [1 - F(t+x-y)] dM_d(y); \qquad x \leq 0 \tag{B3.32}$$

For a proof of this, see Cinlar (1975).

3.3.3 Alternating Renewal Process

In an ordinary renewal process, the interevent times are independent and identically distributed. In an alternating renewal process, the interevent times are all independent but not identically distributed. More specifically, the odd-numbered interevent times $X_1, X_3, X_5, \ldots$ have a common distribution function $F(x)$ and the even-numbered ones $X_2, X_4, X_6, \ldots$ have a common distribution function $G(x)$.

State of an Item at a Given Time

At any given time the item can be either in its working state or in a failed state and undergoing repair. A variable of interest is the probability $P(t)$ that the item is in its working state at time t. $P(t)$ is given by

$$P(t) = 1 - F(t) + \int_0^t P(t-x)dH(x) \tag{B3.33}$$

where $H(x)$ is the convolution of $F(x)$ and $G(x)$, i.e.,

$$H(x) = F(x)*G(x) \tag{B3.34}$$

3.4   Additional Topics from Renewal Theory

3.4.1   *Renewal Type Equation*
A renewal type equation is an equation of the form

$$g(t) = h(t) + \int_0^t g(t-x)dF(x) \tag{B3.35}$$

where $h(\ )$ and $F(\ )$ are known functions and $g(\ )$ is the unknown function to be obtained as a solution to the integral equation. Then $g(t)$ given by

$$g(t) = h(t) + \int_0^t h(t-x)dM(x) \tag{B3.36}$$

where $M(x)$, the renewal function associated with $F(x)$, is a solution of (B3.35).

3.4.2   *Wald's Equation*
This result is useful in evaluating sums of random numbers of random variables. We need the concept of "Stopping Time." An integer-valued positive random variable N is said to be a stopping time for the sequence $X_1, X_2, \ldots$ if the event $\{N = n\}$ is independent of $X_{n+1}, X_{n+2}, \ldots$ for $n = 1, 2, \ldots$. If $X_1, X_2, \ldots$ are independent and identically distributed random variables with finite mean and N is a stopping time for the sequence such that $E[N] < \infty$, then

$$E\sum_{i=1}^N X_i = E[N]E[X] \tag{B3.37}$$

For a proof, see Ross (1970).

3.4.3   *Renewal Reward Theorem*
Consider an ordinary renewal process with interarrival times $X_1, X_2, \ldots$. Suppose that a reward of Z_i is earned at the time of the ith renewal. Then the total reward earned by time t is given by

$$Z(t) = E\sum_{i=1}^N Z_i \tag{B3.38}$$

where $N(t)$ is the number of renewals in $[0, t)$. $Z(t)$ is a cumulative process (see Section B2.4.2) with $N(t)$ given by a renewal process. If $E[|Y_i|]$ and $E[X_i]$ are finite, then

1. With probability 1, $\lim\limits_{t\to\infty} \dfrac{Z(t)}{t} \to \dfrac{E[Z_i]}{E[X_i]}$

2. $t\to\infty \; E\left[\dfrac{Z(t)}{t}\right] \to \dfrac{E[Z_i]}{E[X_i]}$

For a proof, see Ross (1970).

3.5 Marked Point Process

A marked point process is a point process with an auxiliary variable, called a mark, associated with each event. Let Y_i, $i \geq 1$, denote the mark attached to the ith event. For example, in the case of a mulicomponent item, failure of a component can cause induced failures of one or more of the remaining components. If the number of components that must be replaced at the ith failure of the item is a random variable, then it can be viewed as a mark attached to an underlying point process characterizing item failures.

3.5.1 A Simple Marked Point Process

A *simple marked* point process is characterized by

1. $\{N(t), t \geq 0\}$, a stationary Poisson process with intensity λ
2. A sequence of independent and identically distributed random variables $\{Y_i\}$, called marks, which are independent of the Poisson process

This point process is also called a *compound Poisson process*. Various extensions (e.g., a nonstationary point process, and marks constituting a dependent sequence, to name two) yield more complex marked point processes.

B.4 MARKOV PROCESSES

4.1 Weiner Process

The Weiner process, $W(t)$, is a continuous-time Markov process with the following properties:

1. $W(t) = 0$
2. $W(t)$ has zero mean, i.e., $E[W(t)] = 0$
3. $W(t)$ has independent increments, i.e., for $t_1 < t_2 < t_3 < t_4$, $E[\{W(t_2) - W(t_1)\}\{W(t_4) - W(t_3)\}] = 0$
4. $E[W(t)\,W(t')] = \min\{t, t'\}$
5. $W(t)$ has a Gaussian distribution
6. $W(t)$ is continuous in t but is nondifferentiable

4.2 Cumulative Process

A cumulative process, $X(t)$, is given by

$$X(t) = \sum_{i=1}^{N(t)} X_i \qquad\qquad (B4.1)$$

with $N(t)$ a marked point process and X_i representing the mark attached to event i, as defined in Section 3.5 of Appendix B.3. The cumulative process is sometimes also called a *mark accumulator process* or *compound Poisson process*.

Note that here $X(t)$ has random jumps at discrete points in time, corresponding to the events for the marked point process, and hence is discontinuous at these points.

4.3   General Markov Process

A general Markov process, $X(t)$, is the solution of a stochastic differential equation of the form

$$dX(t) = f_1(X(t))dt + f_2(X(t))dW(t) + f_3(X(t))dN(t) \qquad (B4.2)$$

where $W(t)$ is a Weiner process and $N(t)$ is a Poisson process. The functions $f_1(\ )$, $f_2(\ )$, and $f_3(\ )$ are suitably constrained to be ensure the existence of the solution (in a stochastic sense). Note that we have $X(t)$ changing continuously as a result of the first two terms on the right-hand side of the equation, and undergoing jumps as a result of the last term.

APPENDIX C

Statistical Tables

Table C1. Fractiles z_p of the standard normal distribution
$(P(Z \leq z_p) = p)$

p	z_p	p	z_p
0.0005	−3.291	0.8000	0.842
0.0010	−3.091	0.8500	1.036
0.0025	−2.807	0.9000	1.282
0.0050	−2.576	0.9500	1.645
0.0100	−2.326	0.9750	1.960
0.0200	−2.054	0.9800	2.054
0.0250	−1.960	0.9900	2.326
0.0500	−1.645	0.9950	2.576
0.1000	−1.282	0.9975	2.807
0.1500	−1.036	0.9990	3.091
0.2000	−0.842	0.9995	3.291

Table C2. Fractiles of the Student-*t* distribution

			p		
df	0.900	0.950	0.975	0.990	0.995
1	3.078	6.314	12.706	31.821	63.657
2	1.886	2.920	4.303	6.965	9.925
3	1.638	2.353	3.182	4.541	5.841
4	1.533	2.132	2.776	3.747	4.604
5	1.476	2.015	2.571	3.365	4.032
6	1.440	1.943	2.447	3.143	3.707
7	1.415	1.895	1.365	2.998	3.499
8	1.397	1.860	2.306	2.896	2.355
9	1.383	1.833	2.262	2.821	3.250
10	1.372	1.812	2.228	2.764	3.169
11	1.363	1.796	2.201	2.718	3.106
12	1.356	1.782	2.179	2.681	3.055
13	1.350	1.771	2.160	2.650	3.012
14	1.345	1.761	2.145	2.624	2.977
15	1.341	1.753	2.131	2.602	2.947
16	1.337	1.746	2.120	2.583	2.921
17	1.333	1.740	2.110	2.567	2.898
18	1.330	1.734	2.101	2.552	2.878
19	1.328	1.729	2.093	2.539	2.861
20	1.325	1.725	2.086	2.528	2.845
21	1.323	1.721	2.080	2.518	2.831
22	1.321	1.717	2.074	2.508	2.819
23	1.319	1.714	2.069	2.500	2.807
24	1.318	1.711	2.064	2.492	2.797
25	1.316	1.708	2.060	2.485	1.787
26	1.315	1.706	2.056	2.479	2.779
27	1.314	1.703	2.052	2.473	2.771
28	1.313	1.701	2.048	2.467	2.763
29	1.311	1.699	2.045	2.462	2.756
30	1.310	1.697	2.042	2.457	2.750
35	1.306	1.690	2.030	2.438	2.715
40	1.303	1.684	2.021	2.423	2.704
45	1.301	1.679	2.014	2.412	2.690
50	1.299	1.676	2.009	2.403	2.678
55	1.297	1.673	2.004	2.396	2.668
60	1.296	1.671	2.000	2.390	2.660
65	1.295	2.669	1.997	2.385	2.654
70	1.294	1.667	1.994	2.381	2.648
75	1.293	2.665	1.992	2.377	2.643
80	1.292	1.664	1.990	2.374	2.639
85	1.292	1.663	1.988	2.371	2.635
90	1.291	1.662	1.987	2.369	2.632
95	1.291	1.661	1.985	2.366	2.629
100	1.290	1.660	1.984	2.364	2.626
200	1.286	1.653	1.972	2.345	2.601
500	1.283	1.648	1.965	2.334	2.586
∞	1.282	1.645	1.960	2.326	2.576

Table C3. Fractiles of the chi-square distribution

					p					
df	0.005	0.010	0.025	0.050	0.010	.900	.950	.975	.990	.995
1	0.0^4393	0.0^3157	0.0^3982	0.0^2393	0.0158	2.71	3.84	5.02	6.63	7.88
2	0.0100	0.0201	0.0506	0.103	0.211	4.61	5.99	7.38	9.21	10.60
3	0.072	0.115	0.216	0.352	0.584	6.25	7.81	9.35	11.34	12.84
4	0.207	0.297	0.484	0.711	1.064	7.78	9.49	11.14	13.28	14.86
5	0.412	0.554	0.831	1.145	1.61	9.24	11.07	12.83	15.09	16.75
6	0.676	0.872	1.24	1.64	2.20	10.64	12.59	14.45	16.81	18.55
7	0.989	1.24	1.69	2.17	2.83	12.02	14.07	16.01	18.48	20.28
8	1.34	1.65	2.18	2.73	3.49	13.36	15.51	17.53	20.09	21.96
9	1.73	2.09	2.70	3.33	4.17	14.68	16.92	19.02	21.87	23.59
10	2.16	2.56	3.25	3.94	4.87	15.99	18.31	20.48	23.21	25.19
11	2.60	3.05	3.82	4.57	5.58	17.28	19.68	21.92	24.72	26.76
12	3.07	3.57	4.40	5.23	6.30	18.55	21.03	23.34	26.22	28.30
13	3.57	4.11	5.01	5.89	7.04	19.81	22.36	24.74	27.69	29.82
14	4.07	4.66	5.63	6.57	7.79	21.06	23.68	26.12	29.14	31.32
15	4.60	5.23	6.26	7.26	8.55	22.31	25.06	27.49	30.58	32.80
16	5.14	5.81	6.91	7.96	9.31	23.54	26.36	28.85	32.00	34.27
17	5.70	6.41	7.56	8.67	10.09	24.77	27.59	30.19	33.41	35.72
18	6.26	7.01	8.23	9.39	10.86	25.99	28.87	31.53	34.81	37.16
19	6.84	7.63	8.91	10.12	11.65	27.20	30.14	32.85	36.19	38.58
20	7.43	8.26	9.59	10.85	12.44	28.41	31.41	34.17	37.57	40.00
21	8.03	8.90	10.28	11.59	13.24	29.62	32.67	35.48	38.93	41.40
22	8.64	9.54	10.98	12.34	14.04	30.81	33.92	36.78	40.29	42.80
23	9.26	10.20	11.69	13.09	14.85	32.01	35.17	38.08	41.64	44.18
24	9.89	10.86	12.40	13.85	15.66	33.20	36.42	39.36	42.98	45.56
25	10.52	11.52	13.12	14.61	16.47	34.38	37.65	40.65	44.31	46.93
26	11.16	12.20	13.84	15.38	17.29	35.56	38.89	41.92	45.64	48.29
27	11.81	12.88	14.57	16.15	18.11	36.74	40.11	43.19	46.96	49.64
28	12.46	13.56	15.31	16.93	18.94	37.92	41.34	44.46	48.28	50.99
29	13.12	14.26	16.05	17.71	19.77	39.09	42.56	45.72	49.59	52.34
30	13.79	14.95	16.79	18.49	20.60	40.26	43.77	46.98	50.89	53.67
40	20.71	22.16	24.43	26.51	29.05	51.80	55.76	59.34	63.69	66.77
50	27.99	29.71	32.36	34.76	37.69	63.17	67.50	71.42	76.15	79.49
60	35.53	37.48	40.48	43.19	46.46	74.40	79.08	83.30	88.38	91.95
70	43.28	45.44	48.76	51.74	55.33	85.53	90.53	95.02	100.4	104.22
80	51.17	53.54	57.15	60.39	64.28	96.58	101.9	106.6	112.3	116.32
90	59.20	61.75	65.65	69.13	73.29	107.6	113.1	118.1	124.1	128.3
100	67.33	70.06	74.22	77.93	82.36	118.5	124.3	129.6	135.8	140.2

Table C4. Factors for two-sided tolerance intervals, normal distribution (confidence γ, coverage p)

s \ p	γ = 0.90				γ = 0.95				γ = 0.99			
	0.75	0.90	0.95	0.99	0.75	0.90	0.95	0.99	0.75	0.90	0.95	0.99
2	11.407	15.978	18.800	24.167	22.858	32.019	37.674	48.430	114.363	160.193	188.491	242.300
3	4.132	5.847	6.919	8.974	5.922	8.380	9.916	12.861	13.378	18.930	22.401	29.055
4	2.932	4.166	4.943	6.440	3.779	5.369	6.370	8.299	6.614	9.398	11.150	14.527
5	2.454	3.494	4.152	5.423	3.002	4.275	5.079	6.634	4.643	6.612	7.855	10.260
6	2.196	3.131	3.723	4.870	2.604	3.712	4.414	5.775	3.743	5.337	6.345	8.301
7	2.034	2.902	3.452	4.521	2.361	3.369	4.007	5.248	3.233	4.613	5.488	7.187
8	1.921	2.743	3.264	4.278	2.197	3.136	3.732	4.891	2.905	4.147	4.936	6.468
9	1.839	2.626	3.125	4.098	2.078	2.967	3.532	4.631	2.677	3.822	4.550	5.966
10	1.775	2.535	3.018	3.959	1.987	2.839	3.379	4.433	2.508	3.582	4.265	5.594
11	1.724	2.463	2.933	3.849	1.916	2.737	3.259	4.277	2.378	3.397	4.045	5.308
12	1.683	2.404	2.863	3.758	1.858	2.655	3.162	4.150	2.274	3.250	3.870	5.079
13	1.648	2.355	2.805	3.682	1.810	2.587	3.081	4.044	2.190	3.130	3.727	4.893
14	1.619	2.314	2.756	3.618	1.770	2.529	3.012	3.955	2.120	3.029	3.608	4.737
15	1.594	2.278	2.713	3.562	1.735	2.480	2.954	3.878	2.060	2.945	3.507	4.605
16	1.572	2.246	2.676	3.514	1.705	2.437	2.903	3.812	2.009	2.872	3.421	4.492
17	1.552	2.219	2.643	3.471	1.679	2.400	2.858	3.754	1.965	2.808	3.345	4.393
18	1.535	2.194	2.614	3.433	1.655	2.366	2.819	3.702	1.926	2.753	3.279	4.307
19	1.520	2.172	2.588	3.399	1.635	2.337	2.784	3.656	1.891	2.703	3.221	4.230
20	1.506	2.152	2.564	3.368	1.616	2.310	2.752	3.615	1.860	2.659	3.168	4.161
21	1.493	2.135	2.543	3.340	1.599	2.286	2.723	3.577	1.833	2.620	3.121	4.100
22	1.482	2.118	2.524	3.315	1.584	2.264	2.697	3.543	1.808	2.584	3.078	4.044
23	1.471	2.103	2.506	3.292	1.570	2.244	2.673	3.512	1.785	2.551	3.040	3.993
24	1.462	2.089	2.489	3.270	1.557	2.225	2.651	3.483	1.764	2.522	3.004	3.947
25	1.453	2.077	2.474	3.251	1.545	2.208	2.631	3.457	1.745	2.494	2.972	3.904
26	1.444	2.065	2.460	3.232	1.534	2.193	2.612	3.432	1.727	2.469	2.941	3.865
27	1.437	2.054	2.447	3.215	1.523	2.178	2.595	3.409	1.711	2.446	2.914	3.828
30	1.417	2.025	2.413	3.170	1.497	2.140	2.549	3.350	1.668	2.385	2.841	3.733
35	1.390	1.988	2.368	3.112	1.462	2.090	2.490	3.272	1.613	2.306	2.748	3.611
40	1.370	1.959	2.334	3.066	1.435	2.052	2.445	3.213	1.571	2.247	2.677	3.518
45	1.354	1.935	2.306	3.030	1.414	2.021	2.408	3.165	1.539	2.200	2.621	3.444
50	1.340	1.916	2.284	3.001	1.396	1.996	2.379	3.126	1.512	2.162	2.576	3.385
55	1.329	1.901	2.265	2.976	1.382	1.976	2.354	3.094	1.490	2.130	2.538	3.335
60	1.320	1.887	2.248	2.955	1.369	1.958	2.333	3.066	1.471	2.103	2.506	3.293
65	1.312	1.875	2.235	2.937	1.359	1.943	2.315	3.042	1.455	2.080	2.478	3.257
70	1.304	1.865	2.222	2.920	1.349	1.929	2.299	3.021	1.440	2.060	2.454	3.225
75	1.298	1.856	2.211	2.906	1.341	1.917	2.285	3.002	1.428	2.042	2.433	3.197
80	1.292	1.848	2.202	2.894	1.334	1.907	2.272	2.986	1.417	2.026	2.414	3.173
85	1.287	1.841	2.193	2.882	1.327	1.897	2.261	2.971	1.407	2.012	2.397	3.150
90	1.283	1.834	2.185	2.872	1.321	1.889	2.251	2.958	1.398	1.999	2.382	3.130
95	1.278	1.828	2.178	2.863	1.315	1.881	2.241	2.945	1.390	1.987	2.368	3.112
100	1.275	1.822	2.172	2.854	1.311	1.874	2.233	2.934	1.383	1.977	2.355	3.096
110	1.268	1.813	2.160	2.839	1.302	1.861	2.218	2.915	1.369	1.958	2.333	3.066
120	1.262	1.804	2.150	2.826	1.294	1.850	2.205	2.898	1.358	1.942	2.314	3.041
130	1.257	1.797	2.141	2.814	1.288	1.841	2.194	2.883	1.349	1.928	2.298	3.019
140	1.252	1.791	2.134	2.804	1.282	1.833	2.184	2.870	1.340	1.916	2.283	3.000
150	1.248	1.785	2.127	2.795	1.277	1.825	2.175	2.859	1.332	1.905	2.270	2.983
160	1.245	1.780	2.121	2.787	1.272	1.819	2.167	2.848	1.326	1.896	2.259	2.968
170	1.242	1.775	2.116	2.780	1.268	1.813	2.160	2.839	1.320	1.887	2.248	2.955
180	1.239	1.771	2.111	2.774	1.264	1.808	2.154	2.831	1.314	1.879	2.239	2.942
190	1.236	1.767	2.106	2.768	1.261	1.803	2.148	2.823	1.309	1.872	2.230	2.931
200	1.234	1.764	2.102	2.762	1.258	1.798	2.143	2.816	1.304	1.865	2.222	2.921
250	1.224	1.750	2.085	2.740	1.245	1.780	2.121	2.788	1.286	1.839	2.191	2.880
300	1.217	1.740	2.073	2.725	1.236	1.767	2.106	2.767	1.273	1.820	2.169	2.850
400	1.207	1.726	2.057	2.703	1.223	1.749	2.084	2.739	1.255	1.794	2.138	2.809

Table C4. *Continued*

$s \backslash p$	γ = 0.90				γ = 0.95				γ = 0.99			
	0.75	0.90	0.95	0.99	0.75	0.90	0.95	0.99	0.75	0.90	0.95	0.99
500	1.201	1.717	2.046	2.689	1.215	1.737	2.070	2.721	1.243	1.777	2.117	2.783
600	1.196	1.710	2.038	2.678	1.209	1.729	2.060	2.707	1.234	1.764	2.102	2.763
700	1.192	1.705	2.032	2.670	1.204	1.722	2.052	2.697	1.227	1.755	2.091	2.748
800	1.189	1.701	2.027	2.663	1.201	1.717	2.046	2.688	1.222	1.747	2.082	2.736
900	1.187	1.697	2.023	2.658	1.198	1.712	2.040	2.682	1.218	1.741	2.075	2.726
1000	1.185	1.695	2.019	2.654	1.195	1.709	2.036	2.676	1.214	1.736	2.068	2.718
∞	1.150	1.645	1.960	2.576	1.150	1.645	1.960	2.576	1.150	1.645	1.960	2.576

Table C5. **Factors for one-sided tolerance intervals, normal distribution (confidence γ, coverage p)**

$n \backslash p$	γ = 0.90					n	γ = 0.95				
	0.900	0.950	0.975	0.990	0.999		0.900	0.950	0.975	0.990	0.999
2	10.253	13.090	15.586	18.500	24.582	2	20.581	26.260	31.257	37.094	49.276
3	4.258	5.311	6.244	7.340	9.651	3	6.155	7.656	8.986	10.553	13.857
4	3.188	3.957	4.637	5.438	7.129	4	4.162	5.144	6.015	7.042	9.214
5	2.744	3.401	3.983	4.668	6.113	5	3.413	4.210	4.916	5.749	7.509
6	2.494	3.093	3.621	4.243	5.556	6	3.008	3.711	4.332	5.065	6.614
7	2.333	2.893	3.389	3.972	5.201	7	2.756	3.401	3.971	4.643	6.064
8	2.219	2.754	3.227	3.783	4.955	8	2.582	3.188	3.724	4.355	5.689
9	2.133	2.650	3.106	3.641	4.771	9	2.454	3.032	3.543	4.144	5.414
10	2.066	2.568	3.011	3.532	4.628	10	2.355	2.911	3.403	3.981	5.204
11	2.012	2.503	2.936	3.444	4.515	11	2.275	2.815	3.291	3.852	5.036
12	1.966	2.448	2.872	3.371	4.420	12	2.210	2.736	3.201	3.747	4.900
13	1.928	2.403	2.820	3.310	4.341	13	2.155	2.670	3.125	3.659	4.787
14	1.895	2.363	2.774	3.257	4.274	14	2.108	2.614	3.060	3.585	4.690
15	1.866	2.329	2.735	3.212	4.215	15	2.068	2.566	3.005	3.520	4.607
16	1.842	2.299	2.700	3.172	4.164	16	2.032	2.523	2.956	3.463	4.534
17	1.819	2.272	2.670	3.137	4.118	17	2.002	2.486	2.913	3.414	4.471
18	1.800	2.249	2.643	3.106	4.078	18	1.974	2.453	2.875	3.370	4.415
19	1.781	2.228	2.618	3.078	4.041	19	1.949	2.423	2.840	3.331	4.364
20	1.765	2.208	2.597	3.052	4.009	20	1.926	2.396	2.809	3.295	4.319
21	1.750	2.190	2.575	3.028	3.979	21	1.905	2.371	2.781	3.262	4.276
22	1.736	2.174	2.557	3.007	3.952	22	1.887	2.350	2.756	3.233	4.238
23	1.724	2.159	2.540	2.987	3.927	23	1.869	2.329	2.732	3.206	4.204
24	1.712	2.145	2.525	2.969	3.904	24	1.853	2.309	2.711	3.181	4.171
25	1.702	2.132	2.510	2.952	3.882	25	1.838	2.292	2.691	3.158	4.143
30	1.657	2.080	2.450	2.884	3.794	30	1.778	2.220	2.608	3.064	4.022
35	1.623	2.041	2.406	2.833	3.730	35	1.732	2.166	2.548	2.994	3.934
40	1.598	2.010	2.371	2.793	3.679	40	1.697	2.126	2.501	2.941	3.866
45	1.577	1.986	2.344	2.762	3.638	45	1.669	2.092	2.463	2.897	3.811
50	1.560	1.965	2.320	2.735	3.604	50	1.646	2.065	2.432	2.863	3.766
60	1.532	1.933	2.284	2.694	3.552	60	1.609	2.022	2.384	2.807	3.695
70	1.511	1.909	2.257	2.663	3.513	70	1.581	1.990	2.348	2.766	3.643
80	1.495	1.890	2.235	2.638	3.482	80	1.560	1.965	2.319	2.733	3.601
90	1.481	1.874	2.217	2.618	3.456	90	1.542	1.944	2.295	2.706	3.567
100	1.470	1.861	2.203	2.601	3.435	100	1.527	1.927	2.276	2.684	3.539
120	1.452	1.841	2.179	2.574	3.402	120	1.503	1.899	2.245	2.649	3.495
145	1.436	1.821	2.158	2.550	3.371	145	1.481	1.874	2.217	2.617	3.455
300	1.386	1.765	2.094	2.477	3.280	300	1.417	1.800	2.133	2.522	3.335
500	1.362	1.736	2.062	2.442	3.235	500	1.385	1.763	2.092	2.475	3.277
∞	1.282	1.645	1.960	2.326	3.090	∞	1.282	1.645	1.960	2.326	3.090

From Owen (1962).

Table C6. Two-sided nonparametric tolerance intervals*

n \ P	$\gamma = 0.75$				$\gamma = 0.90$			
	0.75	0.90	0.95	0.99	0.75	0.90	0.95	0.99
50	5,5	2,1	—	—	5,4	1,1	—	—
55	6,6	2,2	1,1	—	5,5	2,1	—	—
60	7,6	2,2	1,1	—	6,5	2,1	—	—
65	7,7	3,2	1,1	—	6,6	2,2	—	—
70	8,7	3,2	1,1	—	7,6	2,2	—	—
75	8,8	3,3	1,1	—	7,7	2,2	—	—
80	9,8	3,3	2,1	—	8,7	3,2	1,1	—
85	10,9	4,3	2,1	—	8,8	3,2	1,1	—
90	10,10	4,3	2,1	—	9,8	3,2	1,1	—
95	11,10	4,3	2,1	—	9,9	3,3	1,1	—
100	11,11	4,4	2,1	—	10,10	3,3	1,1	—
110	12,12	5,4	2,2	—	11,11	4,3	2,1	—
120	14,13	5,5	2,2	—	12,12	4,4	2,1	—
130	15,14	6,5	3,2	—	13,13	5,4	2,1	—
140	16,15	6,6	3,2	—	14,14	5.5	2,2	—
150	17,17	6,6	3,3	—	16,15	5,5	2,2	—
170	20,19	7,7	4,3	—	18,17	6,6	3,2	—
200	23,23	9,8	4,4	—	21,21	8,7	3,3	—
300	35,35	13,13	6,6	1,1	33,32	12,11	5,5	—
400	47,47	18,18	9,8	2,1	45,44	16,16	8,7	1,1
500	59,59	23,22	11,11	2,1	57,56	23,20	10,9	1,1
600	72,71	28,27	13,13	2,2	68,68	26,25	12,11	2,1
700	84,83	33,32	16,15	3,2	80,80	30,30	14,14	2,2
800	96,96	37,27	18,18	3,3	92,92	35,34	16,16	3,2
900	108,108	42,42	21,20	4,3	104,104	40,39	19,18	3,2
1000	121,120	47,47	23,22	4,4	117,116	44,44	21,20	3,3

n \ P	$\gamma = 0.95$				$\gamma = 0.99$			
	0.75	0.90	0.95	0.99	0.75	0.90	0.95	0.99
50	4,4	1,1	—	—	3,3	—	—	—
55	5,4	1,1	—	—	4,3	—	—	—
60	5,5	1,1	—	—	4,4	—	—	—
65	6,5	2,1	—	—	5,4	1,1	—	—
70	6,6	2,1	—	—	5,5	1,1	—	—
75	7,6	2,1	—	—	5,5	1,1	—	—
80	7,7	2,2	—	—	6,5	1,1	—	—
85	8,7	2,2	—	—	6,6	2,1	—	—
90	8,8	3,2	—	—	7,6	2,1	—	—
95	9,8	3,2	1,1	—	7,7	2,1	—	—
100	9,9	3,2	1,1	—	8,7	2,2	—	—
110	10,10	3,3	1,1	—	9,8	2,2	—	—
120	11,11	4,3	1,1	—	10,9	3,2	—	—
130	13,12	4,4	2,1	—	11,10	3,3	1,1	—
140	14,13	4,4	2,1	—	12,11	3,3	1,1	—
150	15,14	5,4	2,1	—	13,13	4,3	1,1	—
170	17,16	6,5	2,2	—	15,15	5,4	2,1	—
200	20,20	7,6	3,2	—	18,18	6,5	2,2	—
300	32,31	11,11	5,4	—	29,29	10,9	4,3	—
400	43,43	15,15	7,6	—	40,40	14,13	6,5	—
500	55,54	20,19	9,8	1,1	52,51	18,17	7,7	—
600	67,66	24,24	11,10	1 1	63,63	22,22	9,9	—

754

Table C6. *Continued*

	$\gamma = 0.95$				$\gamma = 0.99$			
n \ P	0.75	0.90	0.95	0.99	0.75	0.90	0.95	0.99
700	78,78	29,28	13,13	2,1	75,74	26,26	11,11	1,1
800	90,90	33,33	15,15	2,2	86,86	31,30	13,13	1,1
900	102,102	38,37	18,17	2,2	98,97	35,35	15,15	2,1
1000	114,114	43,42	20,19	3,2	110,109	40,39	18,17	2,1

*Values (r, s) such that we may assert with confidence at least γ that $100P$ percent of a population lies between the rth smallest and the sth largest of a random sample of n from that population (no assumption of normality required).

When the values of r and s given in the table are not equal, they are interchangeable; i.e., for $n = 120$ with confidence at least 0.75 we may assert that 75% of the population lies between the 14th smallest and the 13th largest values, or between the 13th smallest and the 14th largest values.

From Somerville, Paul N. (1958). "Tables for obtaining non-parametric tolerance limits," *Annals of Mathematical Statistics* **29**:599–601.

Table C7. One-sided nonparametric tolerance intervals*

	$\gamma = 0.75$				$\gamma = 0.90$				$\gamma = 0.95$				$\gamma = 0.99$			
s \ P	0.75	0.90	0.95	0.99	0.75	0.90	0.95	0.99	0.75	0.90	0.95	0.99	0.75	0.90	0.95	0.99
50	10	3	1	—	9	2	1	—	8	2	—	—	6	1	—	—
55	12	4	2	—	10	3	1	—	9	2	—	—	7	1	—	—
60	13	4	2	—	11	3	1	—	10	2	1	—	8	1	—	—
65	14	5	2	—	12	4	1	—	11	3	1	—	9	2	—	—
70	15	5	2	—	13	4	1	—	12	3	1	—	10	2	—	—
75	16	6	2	—	14	4	1	—	13	3	1	—	10	2	—	—
80	17	6	3	—	15	5	2	—	14	4	1	—	11	2	—	—
85	19	7	3	—	16	5	2	—	15	4	1	—	12	3	—	—
90	20	7	3	—	17	5	2	—	16	5	1	—	13	3	1	—
95	21	7	3	—	18	6	2	—	17	5	2	—	14	3	1	—
100	22	8	3	—	20	6	2	—	18	5	2	—	15	4	1	—
110	24	9	4	—	22	7	3	—	20	6	2	—	17	4	1	—
120	27	10	4	—	24	8	3	—	22	7	2	—	19	5	1	—
130	29	11	5	—	26	9	3	—	25	8	3	—	21	6	2	—
140	31	12	5	1	28	10	4	—	27	8	3	—	23	6	2	—
150	34	12	6	1	31	10	4	—	29	9	3	—	26	7	2	—
170	39	14	7	1	35	12	5	—	33	11	4	—	30	9	3	—
200	46	17	8	1	42	15	6	—	40	13	5	—	36	11	4	—
300	70	26	12	2	65	23	10	1	63	22	9	1	58	19	7	—
400	94	36	17	3	89	32	15	2	86	30	13	1	80	27	11	—
500	118	45	22	3	113	41	19	2	109	39	17	2	103	35	14	1
600	143	55	26	4	136	51	23	3	133	48	21	2	126	44	18	1
700	167	65	31	5	160	60	28	4	156	57	26	3	149	52	22	2
800	192	74	36	6	184	69	32	5	180	66	30	4	172	61	26	2
900	216	84	41	7	208	79	37	5	204	75	35	4	195	70	30	3
1000	241	94	45	8	233	88	41	6	228	85	39	5	219	79	35	3

*Largest values of m such that we may assert with confidence at least γ that $100P$ percent of a population lies below the mth largest (or above the mth smallest) of a random sample of n from that population (no assumption of normality required).

From Somerville, Paul N. (1958). "Tables for obtaining non-parametric tolerance limits," *Annals of Mathematical Statistics* **29**:599–601.

Table C8. Fractiles F_{1-p} of the F Distribution (p = upper-tail probability)

					Degrees of freedom in the numerator				
p	1	2	3	4	5	6	7	8	9
0.100	39.86	49.50	53.59	55.83	57.24	58.20	58.91	59.44	59.86
0.050	161.45	199.50	215.71	224.58	230.16	233.99	236.77	238.88	240.54
1 0.025	647.79	799.50	864.16	899.58	921.85	937.11	948.22	956.66	963.28
0.010	4052.2	4999.5	5403.4	5624.6	5763.6	5859.0	5928.4	5981.1	6022.5
0.001	405284	500000	540379	562500	576405	585937	592873	598144	602284
0.100	8.53	9.00	9.16	9.24	9.29	9.33	9.35	9.37	9.38
0.050	18.51	19.00	19.16	19.25	19.30	19.33	19.35	19.37	19.38
2 0.025	38.51	39.00	39.17	39.25	39.30	39.33	39.36	39.37	39.39
0.010	98.50	99.00	99.17	99.25	99.30	99.33	99.36	99.37	99.39
0.001	998.50	999.00	999.17	999.25	999.30	999.33	999.36	999.37	999.39
0.100	5.54	5.46	5.39	5.34	5.31	5.28	5.27	5.25	5.24
0.050	10.13	9.55	9.28	9.12	9.01	8.94	8.89	8.85	8.81
3 0.025	17.44	16.04	15.44	15.10	14.88	14.73	14.62	14.54	14.47
0.010	34.12	30.82	29.46	28.71	28.24	27.91	27.67	27.49	27.35
0.001	167.03	148.50	141.11	137.10	134.58	132.85	131.58	130.62	129.86
0.100	4.54	4.32	4.19	4.11	4.05	4.01	3.98	3.95	3.94
0.050	7.71	6.94	6.59	6.39	6.26	6.16	6.09	6.04	6.00
4 0.025	12.22	10.65	9.98	9.60	9.36	9.20	9.07	8.98	8.90
0.010	21.20	18.00	16.69	15.98	15.52	15.21	14.98	14.80	14.66
0.001	74.14	61.25	56.18	53.44	51.71	50.53	49.66	49.00	48.47
0.100	4.06	3.78	3.62	3.52	3.45	3.40	3.37	3.34	3.32
0.050	6.61	5.79	5.41	5.19	5.05	4.95	4.88	4.82	4.77
5 0.025	10.01	8.43	7.76	7.39	7.15	6.98	6.85	6.76	6.68
0.010	16.26	13.27	12.06	11.39	10.97	10.67	10.46	10.29	10.16
0.001	47.18	37.12	33.20	31.09	29.75	28.83	28.16	27.65	27.24
0.100	3.78	3.46	3.29	3.18	3.11	3.05	3.01	2.98	2.96
0.050	5.99	5.14	4.76	4.53	4.39	4.28	4.21	4.15	4.10
6 0.025	8.81	7.26	6.60	6.23	5.99	5.82	5.70	5.60	5.52
0.010	13.75	10.92	9.78	9.15	8.75	8.47	8.26	8.10	7.98
0.001	35.51	27.00	23.70	21.92	20.80	20.03	19.46	19.03	18.69
0.100	3.59	3.26	3.07	2.96	2.88	2.83	2.78	2.75	2.72
0.050	5.59	4.74	4.35	4.12	3.97	3.87	3.79	3.73	3.68
7 0.025	8.07	6.54	5.89	5.52	5.29	5.12	4.99	4.90	4.82
0.010	12.25	9.55	8.45	7.85	7.46	7.19	6.99	6.84	6.72
0.001	29.25	21.69	28.77	17.20	16.21	15.52	15.02	14.63	14.33
0.100	3.46	3.11	2.92	2.81	2.73	2.67	2.62	2.59	2.56
0.050	5.32	4.46	4.07	3.84	3.69	3.58	3.50	3.44	3.39
8 0.025	7.57	6.06	5.42	5.05	4.82	4.65	4.53	4.43	4.36
0.010	11.26	8.65	7.59	7.01	6.63	6.37	6.18	6.03	5.91
0.001	25.41	18.49	15.83	14.39	13.48	12.86	12.40	12.05	11.77
0.100	3.36	3.01	2.81	2.69	2.61	2.55	2.51	2.47	2.44
0.050	5.12	4.26	3.86	3.63	3.48	3.37	3.29	3.23	3.18
9 0.025	7.21	5.71	5.08	4.72	4.48	4.32	4.20	4.10	4.03
0.010	10.56	8.02	6.99	6.42	6.06	5.80	5.61	5.47	5.35
0.001	22.86	16.39	13.90	12.56	11.71	11.13	10.70	10.37	10.11
0.100	3.29	2.92	2.73	2.61	2.52	2.46	2.41	2.38	2.35
0.050	4.96	4.10	3.71	3.48	3.33	3.22	3.14	3.07	3.02
10 0.025	6.94	5.46	4.83	4.47	4.24	4.07	3.95	3.85	3.78
0.010	10.04	7.56	6.55	5.99	5.64	5.39	5.20	5.06	4.94
0.001	21.04	14.91	12.55	11.28	10.48	9.93	9.52	9.20	8.96
0.100	3.23	2.86	2.66	2.54	2.45	2.39	2.34	2.30	2.27
0.050	4.84	3.98	3.59	3.36	3.20	3.09	3.01	2.95	2.90
11 0.025	6.72	5.26	4.63	4.28	4.04	3.88	3.76	3.66	3.59
0.010	9.65	7.21	6.22	5.67	5.32	5.07	4.89	4.74	4.63
0.001	19.69	13.81	11.56	10.35	9.58	9.05	8.66	8.35	8.12
0.100	3.18	2.81	2.61	2.48	2.39	2.33	2.28	2.24	2.21
0.050	4.75	3.89	3.49	3.26	3.11	3.00	2.91	2.85	2.80
12 0.025	6.55	5.10	4.47	4.12	3.89	3.73	3.61	3.51	3.44
0.010	9.33	6.93	5.95	5.41	5.06	4.82	4.64	4.50	4.39
0.001	18.64	12.97	10.80	9.63	8.89	8.38	8.00	7.71	7.48

Degrees of freedom in the denominator

Table C8. *Continued*

			Degrees of freedom in the numerator							
10	12	15	20	25	30	40	50	60	120	1000
60.19	60.71	61.22	61.74	62.05	62.26	62.53	62.69	62.79	63.06	63.30
241.88	243.91	245.95	248.01	249.26	250.10	251.14	251.77	252.20	253.25	254.19
968.63	976.71	984.87	993.10	998.08	1001.4	1005.6	1008.1	1009.8	1014.0	1017.7
6055.8	6106.3	6157.3	6208.7	6239.8	6260.6	6286.8	6302.5	6313.0	6339.4	6362.7
605621	610668	615764	620908	624017	626099	628712	630285	631337	633972	636301
9.39	9.41	9.42	9.44	9.45	9.46	9.47	9.47	9.47	9.48	9.49
19.40	19.41	19.43	19.45	19.46	19.46	19.47	19.48	19.48	19.49	19.49
39.40	39.41	39.43	39.45	39.46	39.46	39.47	39.48	39.48	39.49	39.50
99.40	99.42	99.43	99.45	99.46	99.47	99.47	99.48	99.48	99.49	99.50
999.40	999.42	999.43	999.45	999.46	999.47	999.47	999.48	999.48	999.49	999.50
5.23	5.22	5.20	5.18	5.17	5.17	5.16	5.15	5.15	5.14	5.13
8.79	8.74	8.70	8.66	8.63	8.62	8.59	8.58	8.57	8.55	8.53
14.42	14.34	14.25	14.17	14.12	14.08	14.04	14.01	13.99	13.95	13.91
27.23	27.05	26.87	26.69	2~.58	26.50	26.41	26.35	26.32	26.22	26.14
129.25	128.32	127.37	126.42	125.84	125.45	124.96	124.66	124.47	123.97	123.53
3.92	3.90	3.87	3.84	3.83	3.82	3.80	3.80	3.79	3.78	3.76
5.96	5.91	5.86	5.80	5.77	5.75	5.72	5.70	5.69	5.66	5.63
8.84	8.75	8.66	8.56	8.50	8.46	8.41	8.38	8.36	8.31	8.26
14.55	14.37	14.20	14.02	13.91	13.84	13.75	13.69	13.65	13.56	13.47
48.05	47.41	46.76	46.10	45.70	45.43	45.09	44.88	44.75	44.40	44.09
3.30	3.27	3.24	3.21	3.19	3.17	3.16	3.15	3.14	3.12	3.11
4.74	4.68	4.62	4.56	4.52	4.50	4.46	4.44	4.43	4.40	4.37
6.62	6.52	6.43	6.33	6.27	6.23	6.18	6.14	6.12	6.07	6.02
10.05	9.89	9.72	9.55	9.45	9.38	9.29	9.24	9.20	9.11	9.03
26.92	26.42	25.91	25.39	25.08	24.87	24.60	24.44	24.33	24.06	23.82
2.94	2.90	2.87	2.84	2.81	2.80	2.78	2.77	2.76	2.74	2.72
4.06	4.00	3.94	3.87	3.83	3.81	3.77	3.75	3.74	3.70	3.67
5.46	5.37	5.27	5.17	5.11	5.07	5.01	4.98	4.96	4.90	4.86
7.87	7.72	7.56	7.40	7.30	7.23	7.14	7.09	7.06	6.97	6.89
18.41	17.99	17.56	17.12	16.85	16.67	16.44	16.31	16.21	15.98	15.77
2.70	2.67	2.63	2.59	2.57	2.56	2.54	2.52	2.51	2.49	2.47
3.64	3.57	3.51	3.44	3.40	3.38	3.34	3.32	3.30	3.27	3.23
4.76	4.67	4.57	4.47	4.40	4.36	4.31	4.28	4.25	4.20	4.15
6.62	6.47	6.31	6.16	6.06	5.99	5.91	5.86	5.82	5.74	5.66
14.08	13.71	13.32	12.93	12.69	12.53	12.33	12.20	12.12	11.91	11.72
2.54	2.50	2.46	2.42	2.40	2.38	2.36	2.35	2.34	2.32	2.30
3.35	3.28	3.22	3.15	3.11	3.08	3.04	3.02	3.01	2.97	2.93
4.30	4.20	4.10	4.00	3.94	3.89	3.84	3.81	3.78	3.73	3.68
5.81	5.67	5.52	5.36	5.26	5.20	5.12	5.07	5.03	4.95	4.87
11.54	11.19	10.84	10.48	10.26	10.11	9.92	9.80	9.73	9.53	9.36
2.42	2.38	2.34	2.30	2.27	2.25	2.23	2.22	2.21	2.18	2.16
3.14	3.07	3.01	2.94	2.89	2.86	2.83	2.80	2.79	2.75	2.71
3.96	3.87	3.77	3.67	3.60	3.56	3.51	3.47	3.45	3.39	3.34
5.26	5.11	4.96	4.81	4.71	4.65	4.57	4.52	4.48	4.40	4.32
9.89	9.57	9.24	8.90	8.69	8.55	8.37	8.26	8.19	8.00	7.84
2.32	2.28	2.24	2.20	2.17	2.16	2.13	2.12	2.11	2.08	2.06
2.98	2.91	2.85	2.77	2.73	2.70	2.66	2.64	2.62	2.58	2.54
3.72	3.62	3.52	3.42	3.35	3.31	3.26	3.22	3.20	3.14	3.09
4.85	4.71	4.56	4.41	4.31	4.25	4.17	4.12	4.08	4.00	3.92
8.75	8.45	8.13	7.80	7.60	7.47	7.30	7.19	7.12	6.94	6.78
2.25	2.21	2.17	2.12	2.10	2.08	2.05	2.04	2.03	2.00	1.98
2.85	2.79	2.72	2.65	2.60	2.57	2.53	2.51	2.49	2.45	2.41
3.53	3.43	3.33	3.23	3.16	3.12	3.06	3.03	3.00	2.94	2.89
4.54	4.40	4.25	4.10	4.01	3.94	3.86	3.81	3.78	3.69	3.61
7.92	7.63	7.32	7.01	6.81	6.68	6.52	6.42	6.35	6.18	6.02
2.19	2.15	2.10	2.06	2.03	2.01	1.99	1.97	1.96	1.93	1.91
2.75	2.69	2.62	2.54	2.50	2.47	2.43	2.40	2.38	2.34	2.30
3.37	3.28	3.18	3.07	3.01	2.96	2.91	2.87	2.85	2.79	2.73
4.30	4.16	4.01	3.86	3.76	3.70	3.62	3.57	3.54	3.45	3.37
7.29	7.00	6.71	6.40	6.22	6.09	5.93	5.83	5.76	5.59	5.44

(continued)

Table C8. Fractiles F_{1-p} of the F Distribution (p = upper-tail probability)

			Degrees of freedom in the numerator							
	p	1	2	3	4	5	6	7	8	9
	0.100	3.14	2.76	2.56	2.43	2.35	2.28	2.23	2.20	2.16
	0.050	4.67	3.81	3.41	3.18	3.03	2.92	2.83	2.77	2.71
13	0.025	6.41	4.97	4.35	4.00	3.77	3.60	3.48	3.39	3.31
	0.010	9.07	6.70	5.74	5.21	4.86	4.62	4.44	4.30	4.19
	0.001	17.82	12.31	10.21	9.07	8.35	7.86	7.49	7.21	6.98
	0.100	3.10	2.73	2.52	2.39	2.31	2.24	2.19	2.15	2.12
	0.050	4.60	3.74	3.34	3.11	2.96	2.85	2.76	2.70	2.65
14	0.025	6.30	4.86	4.24	3.89	3.66	3.50	3.38	3.29	3.21
	0.010	8.86	6.51	5.56	5.04	4.69	4.46	4.28	4.14	4.03
	0.001	17.14	11.78	9.73	8.62	7.92	7.44	7.08	6.80	6.58
	0.100	3.07	2.70	2.49	2.36	2.27	2.21	2.16	2.12	2.09
	0.050	4.54	3.68	3.29	3.06	2.90	2.79	2.71	2.64	2.59
15	0.025	6.20	4.77	4.15	3.80	3.58	3.41	3.29	3.20	3.12
	0.010	8.68	6.36	5.42	4.89	4.56	4.32	4.14	4.00	3.89
	0.001	16.59	11.34	9.34	8.25	7.57	7.09	6.74	6.47	6.26
	0.100	3.05	2.67	2.46	2.33	2.24	2.18	2.13	2.09	2.06
	0.050	4.49	3.63	3.24	3.01	2.85	2.74	2.66	2.59	2.54
16	0.025	6.12	4.69	4.08	3.73	3.50	3.34	3.22	3.12	3.05
	0.010	8.53	6.23	5.29	4.77	4.44	4.20	4.03	3.89	3.78
	0.001	16.12	10.97	9.01	7.94	7.27	6.80	6.46	6.19	5.98
	0.100	3.03	2.64	2.44	2.31	2.22	2.15	2.10	2.06	2.03
	0.050	4.45	3.59	3.20	2.96	2.81	2.70	2.61	2.55	2.49
17	0.025	6.04	4.62	4.01	3.66	3.44	3.28	3.16	3.06	2.98
	0.010	8.40	6.11	5.19	4.67	4.34	4.10	3.93	3.79	3.68
	0.001	15.72	10.66	8.73	7.68	7.02	6.56	6.22	5.96	5.75
	0.100	3.01	2.62	2.42	2.29	2.20	2.13	2.08	2.04	2.00
	0.050	4.41	3.55	3.16	2.93	2.77	2.66	2.58	2.51	2.46
18	0.025	5.98	4.56	3.95	3.61	3.38	3.22	3.10	3.01	2.93
	0.010	8.29	6.01	5.09	4.58	4.25	4.01	3.84	3.71	3.60
	0.001	15.38	10.39	8.49	7.46	6.81	6.35	6.02	5.76	5.56
	0.100	2.99	2.61	2.40	2.27	2.18	2.11	2.06	2.02	1.98
	0.050	4.38	3.52	3.13	2.90	2.74	2.63	2.54	2.48	2.42
19	0.025	5.92	4.51	3.90	3.56	3.33	3.17	3.05	2.96	2.88
	0.010	8.18	5.93	5.01	4.50	4.17	3.94	3.77	3.63	3.52
	0.001	15.08	10.16	8.28	7.27	6.62	6.18	5.85	5.59	5.39
	0.100	2.97	2.59	2.38	2.25	2.16	2.09	2.04	2.00	1.96
	0.050	4.35	3.49	3.10	2.87	2.71	2.60	2.51	2.45	2.39
20	0.025	5.87	4.46	3.86	3.51	3.29	3.13	3.01	2.91	2.84
	0.010	8.10	5.85	4.94	4.43	4.10	3.87	3.70	3.56	3.46
	0.001	14.82	9.95	8.10	7.10	6.46	6.02	5.69	5.44	5.24
	0.100	2.96	2.57	2.36	2.23	2.14	2.08	2.02	1.98	1.95
	0.050	4.32	3.47	3.07	2.84	2.68	2.57	2.49	2.42	2.37
21	0.025	5.83	4.42	3.82	3.48	3.25	3.09	2.97	2.87	2.80
	0.010	8.02	5.78	4.87	4.37	4.04	3.81	3.64	3.51	3.40
	0.001	14.59	9.77	7.94	6.95	6.32	5.88	5.56	5.31	5.11
	0.100	2.95	2.56	2.35	2.22	2.13	2.06	2.01	1.97	1.93
	0.050	4.30	3.44	3.05	2.82	2.66	2.55	2.46	2.40	2.34
22	0.025	5.79	4.38	3.78	3.44	3.22	3.05	2.93	2.84	2.76
	0.010	7.95	5.72	4.82	4.31	3.99	3.76	3.59	3.45	3.35
	0.001	14.38	9.61	7.80	6.81	6.19	5.76	5.44	5.19	4.99
	0.100	2.94	2.55	2.34	2.21	2.11	2.05	1.99	1.95	1.92
	0.050	4.28	3.42	3.03	2.80	2.64	2.53	2.44	2.37	2.32
23	0.025	5.75	4.35	3.75	3.41	3.18	3.02	2.90	2.81	2.73
	0.010	7.88	5.66	4.76	4.26	3.94	3.71	3.54	3.41	3.30
	0.001	14.20	9.47	7.67	6.70	6.08	5.65	5.33	5.09	4.89
	0.100	2.93	2.54	2.33	2.19	2.10	2.04	1.98	1.94	1.91
	0.050	4.26	3.40	3.01	2.78	2.62	2.51	2.42	2.36	2.30
24	0.025	5.72	4.32	3.72	3.38	3.15	2.99	2.87	2.78	2.70
	0.010	7.82	5.61	4.72	4.22	3.90	3.67	3.50	3.36	3.26
	0.001	14.03	9.34	7.55	6.59	5.98	5.55	5.23	4.99	4.80

Degrees of freedom in the denominator

Table C8. *Continued*

10	12	15	20	25	30	40	50	60	120	1000
2.14	2.10	2.05	2.01	1.98	1.96	1.93	1.92	1.90	1.88	1.85
2.67	2.60	2.53	2.46	2.41	2.38	2.34	2.31	2.30	2.25	2.21
3.25	3.15	3.05	2.95	2.88	2.84	2.78	2.74	2.72	2.66	2.60
4.10	3.96	3.82	3.66	3.57	3.51	3.43	3.38	3.34	3.25	3.18
6.80	6.52	6.23	5.93	5.75	5.63	5.47	5.37	5.30	5.14	4.99
2.10	2.05	2.01	1.96	1.93	1.91	1.89	1.87	1.86	1.83	1.80
2.60	2.53	2.46	2.39	2.34	2.31	2.27	2.24	2.22	2.18	2.14
3.15	3.05	2.95	2.84	2.78	2.73	2.67	2.64	2.61	2.55	2.50
3.94	3.80	3.66	3.51	3.41	3.35	3.27	3.22	3.18	3.09	3.02
6.40	6.13	5.85	5.56	5.38	5.25	5.10	5.00	4.94	4.77	4.62
2.06	2.02	1.97	1.92	1.89	1.87	1.85	1.83	1.82	1.79	1.76
2.54	2.48	2.40	2.33	2.28	2.25	2.20	2.18	2.16	2.11	2.07
3.06	2.96	2.86	2.76	2.69	2.64	2.59	2.55	2.52	2.46	2.40
3.80	3.67	3.52	3.37	3.28	3.21	3.13	3.08	3.05	2.96	2.88
6.08	5.81	5.54	5.25	5.07	4.95	4.80	4.70	4.64	4.47	4.33
2.03	1.99	1.94	1.89	1.86	1.84	1.81	1.79	1.78	1.75	1.72
2.49	2.42	7.35	2.28	2.23	2.19	2.15	2.12	2.11	2.06	2.02
2.99	2.89	2.79	2.68	2.61	2.57	2.51	2.47	2.45	2.38	2.32
3.69	3.55	3.41	3.26	3.16	3.10	3.02	2.97	2.93	2.84	2.76
5.81	5.55	5.27	4.99	4.82	4.70	4.54	4.45	4.39	4.23	4.08
2.00	1.96	1.91	1.86	1.83	1.81	1.78	1.76	1.75	1.72	1.69
2.45	2.38	2.31	2.23	2.18	2.15	2.10	2.08	2.06	2.01	1.97
2.92	2.82	2.72	2.62	2.55	2.50	2.44	2.41	2.38	2.32	2.26
3.59	3.46	3.31	3.16	3.07	3.00	2.92	2.87	2.83	2.75	2.66
5.58	5.32	5.05	4.78	4.60	4.48	4.33	4.24	4.18	4.02	3.87
1.98	1.93	1.89	1.84	1.80	1.78	1.75	1.74	1.72	1.69	1.66
2.41	2.34	2.27	2.19	2.14	2.11	2.06	2.04	2.02	1.97	1.92
2.87	2.77	2.67	2.56	2.49	2.44	2.38	2.35	2.32	2.26	2.20
3.51	3.37	3.23	3.08	2.98	2.92	2.84	2.78	2.75	2.66	2.58
5.39	5.13	4.87	4.59	4.42	4.30	4.15	4.06	4.00	3.84	3.69
1.96	1.91	1.86	1.81	1.78	1.76	1.73	1.71	1.70	1.67	1.64
2.38	2.31	2.23	2.16	2.11	2.07	2.03	2.00	1.98	1.93	1.88
2.82	2.72	2.62	2.51	2.44	2.39	2.33	2.30	2.27	2.20	2.14
3.43	3.30	3.15	3.00	2.91	2.84	2.76	2.71	2.67	2.58	2.50
5.22	4.97	4.70	4.43	4.26	4.14	3.99	3.90	3.84	3.68	3.53
1.94	1.89	1.84	1.79	1.76	1.74	1.71	1.69	1.68	1.64	1.61
2.35	2.28	2.20	2.12	2.07	2.04	1.99	1.97	1.95	1.90	1.85
2.77	2.68	2.57	2.46	2.40	2.35	2.29	2.25	2.22	2.16	2.09
3.37	3.23	3.09	2.94	2.84	2.78	2.69	2.64	2.61	2.52	2.43
5.08	4.82	4.56	4.29	4.12	4.00	3.86	3.77	3.70	3.54	3.40
1.92	1.87	1.83	1.78	1.74	1.72	1.69	1.67	1.66	1.62	1.59
2.32	2.25	2.18	2.10	2.05	2.01	1.96	1.94	1.92	1.87	1.82
2.73	2.64	2.53	2.42	2.36	2.31	2.25	2.21	2.18	2.11	2.05
3.31	3.17	3.03	2.88	2.79	2.72	2.64	2.58	2.55	2.46	2.37
4.95	4.70	4.44	4.17	4.00	3.88	3.74	3.64	3.58	3.42	3.28
1.90	1.86	1.81	1.76	1.73	1.70	1.67	1.65	1.64	1.60	1.57
2.30	2.23	2.15	2.07	2.02	1.98	1.94	1.91	1.89	1.84	1.79
2.70	2.60	2.50	2.39	2.32	2.27	2.21	2.17	2.14	2.08	2.01
3.26	3.12	2.98	2.83	2.73	2.67	2.58	2.53	2.50	2.40	2.32
4.83	4.58	4.33	4.06	3.89	3.78	3.63	3.54	3.48	3.32	3.17
1.89	1.84	1.80	1.74	1.71	1.69	1.66	1.64	1.62	1.59	1.55
2.27	2.20	2.13	2.05	2.00	1.96	1.91	1.88	1.86	1.81	1.76
2.67	2.57	2.47	2.36	2.29	2.24	2.18	2.14	2.11	2.04	1.98
3.21	3.07	2.93	2.78	2.69	2.62	2.54	2.48	2.45	2.35	2.27
4.73	4.48	4.23	3.96	3.79	3.68	3.53	3.44	3.38	3.22	3.08
1.88	1.83	1.78	1.73	1.70	1.67	1.64	1.62	1.61	1.57	1.54
2.25	2.18	2.11	2.03	1.97	1.94	1.89	1.86	1.84	1.79	1.74
2.64	2.54	2.44	2.33	2.26	2.21	2.15	2.11	2.08	2.01	1.94
3.17	3.03	2.89	2.74	2.64	2.58	2.49	2.44	2.40	2.31	2.22
4.64	4.39	4.14	3.87	3.71	3.59	3.45	3.36	3.29	3.14	2.99

(*continued*)

Table C8. Fractiles F_{1-p} of the F Distribution (p = upper-tail probability)

					Degrees of freedom in the numerator					
	p	1	2	3	4	5	6	7	8	9
	0.100	2.92	2.53	2.32	2.18	2.09	2.02	1.97	1.93	1.89
	0.050	4.24	3.39	2.99	2.76	2.60	2.49	2.40	2.34	2.28
25	0.025	5.69	4.29	3.69	3.35	3.13	2.97	2.85	2.75	2.68
	0.010	7.77	5.57	4.68	4.18	3.85	3.63	3.46	3.32	3.22
	0.001	13.88	9.22	7.45	6.49	5.89	5.46	5.15	4.91	4.71
	0.100	2.91	2.52	2.31	2.17	2.08	2.01	1.96	1.92	1.88
	0.050	4.23	3.37	2.98	2.74	2.59	2.47	2.39	2.32	2.27
26	0.025	5.66	4.27	3.67	3.33	3.10	2.94	2.82	2.73	2.65
	0.010	7.72	5.53	4.64	4.14	3.82	3.59	3.42	3.29	3.18
	0.001	13.74	9.12	7.36	6.41	5.80	5.38	5.07	4.83	4.64
	0.100	2.90	2.51	2.30	2.17	2.07	2.00	1.95	1.91	1.87
	0.050	4.21	3.35	2.96	2.73	2.57	2.46	2.37	2.31	2.25
27	0.025	5.63	4.24	3.65	3.31	3.08	2.92	2.80	2.71	2.63
	0.010	7.68	5.49	4.60	4.11	3.78	3.56	3.39	3.26	3.15
	0.001	13.61	9.02	7.27	6.33	5.73	5.31	5.00	4.76	4.57
	0.100	2.89	2.50	2.29	2.16	2.06	2.00	1.94	1.90	1.87
	0.050	4.20	3.34	2.95	2.71	2.56	2.45	2.36	2.29	2.24
28	0.025	5.61	4.22	3.63	3.29	3.06	2.90	2.78	2.69	2.61
	0.010	7.64	5.45	4.57	4.07	3.75	3.53	3.36	3.23	3.12
	0.001	13.50	8.93	7.19	6.25	5.66	5.24	4.93	4.69	4.50
	0.100	2.89	2.50	2.28	2.15	2.06	1.99	1.93	1.89	1.86
	0.050	4.18	3.33	2.93	2.70	2.55	2.43	2.35	2.28	2.22
29	0.025	5.59	4.20	3.61	3.27	3.04	2.88	2.76	2.67	2.59
	0.010	7.60	5.42	4.54	4.04	3.73	3.50	3.33	3.20	3.09
	0.001	13.39	8.85	7.12	6.19	5.59	5.18	4.87	4.64	4.45
	0.100	2.88	2.49	2.28	2.14	2.05	1.98	1.93	1.88	1.85
	0.050	4.17	3.32	2.92	2.69	2.53	2.42	2.33	2.27	2.21
30	0.025	5.57	4.18	3.59	3.25	3.03	2.87	2.75	2.65	2.57
	0.010	7.56	5.39	4.51	4.02	3.70	3.47	3.30	3.17	3.07
	0.001	13.29	8.77	7.05	6.12	5.53	5.12	4.82	4.58	4.39
	0.100	2.84	2.44	2.23	2.09	2.00	1.93	1.87	1.83	1.79
	0.050	4.08	3.23	2.84	2.61	2.45	2.34	2.25	2.18	2.12
40	0.025	5.42	4.05	3.46	3.13	2.90	2.74	2.62	2.53	2.45
	0.010	7.31	5.18	4.31	3.83	3.51	3.29	3.12	2.99	2.89
	0.001	12.61	8.25	6.59	5.70	5.13	4.73	4.44	4.21	4.02
	0.100	2.81	2.41	2.20	2.06	1.97	1.90	1.84	1.80	1.76
	0.050	4.03	3.18	2.79	2.56	2.40	2.29	2.20	2.13	2.07
50	0.025	5.34	3.97	3.39	3.05	2.83	2.67	2.55	2.46	2.38
	0.010	7.17	5.06	4.20	3.72	3.41	3.19	3.02	2.89	2.78
	0.001	12.22	7.96	6.34	5.46	4.90	4.51	4.22	4.00	3.82
	0.100	2.79	2.39	2.18	2.04	1.95	1.87	1.82	1.77	1.74
	0.050	4.00	3.15	2.76	2.53	2.37	2.25	2.17	2.10	2.04
60	0.025	5.29	3.93	3.34	3.01	2.79	2.63	2.51	2.41	2.33
	0.010	7.08	4.98	4.13	3.65	3.34	3.12	2.95	2.82	2.72
	0.001	11.97	7.77	6.17	5.31	4.76	4.37	4.09	3.86	3.69
	0.100	2.76	2.36	2.14	2.00	1.91	1.83	1.78	1.73	1.69
	0.050	3.94	3.09	2.70	2.46	2.31	2.19	2.10	2.03	1.97
100	0.025	5.18	3.83	3.25	2.92	2.70	2.54	2.42	2.32	2.24
	0.010	6.90	4.82	3.98	3.51	3.21	2.99	2.82	2.69	2.59
	0.001	11.50	7.41	5.86	5.02	4.48	4.11	3.83	3.61	3.44
	0.100	2.73	2.33	2.11	1.97	1.88	1.80	1.75	1.70	1.66
	0.050	3.89	3.04	2.65	2.42	2.26	2.14	2.06	1.98	1.93
200	0.025	5.10	3.76	3.18	2.85	2.63	2.47	2.35	2.26	2.18
	0.010	6.76	4.71	3.88	3.41	3.11	2.89	2.73	2.60	2.50
	0.001	11.15	7.15	5.63	4.81	4.29	3.92	3.65	3.43	3.26
	0.100	2.71	2.31	2.09	1.95	1.85	1.78	1.72	1.68	1.64
	0.050	3.85	3.00	2.61	2.38	2.22	2.11	2.02	1.95	1.89
1000	0.025	5.04	3.70	3.13	2.80	2.58	2.42	2.30	2.20	2.13
	0.010	6.66	4.63	3.80	3.34	3.04	2.82	2.66	2.53	2.43
	0.001	10.89	6.96	5.46	4.65	4.14	3.78	3.51	3.30	3.13

Degrees of freedom in the denominator

Table C8. *Continued*

				Degrees of freedom in the numerator						
10	12	15	20	25	30	40	50	60	120	1000
1.87	1.82	1.77	1.72	1.68	1.66	1.63	1.61	1.59	1.56	1.52
2.24	2.16	2.09	2.01	1.96	1.92	1.87	1.84	1.82	1.77	1.72
2.61	2.51	2.41	2.30	2.23	2.18	2.12	2.08	2.05	1.98	1.91
3.13	2.99	2.85	2.70	2.60	2.54	2.45	2.40	2.36	2.27	2.18
4.56	4.31	4.06	3.79	3.63	3.52	3.37	3.28	3.22	3.06	2.91
1.86	1.81	1.76	1.71	1.67	1.65	1.61	1.59	1.58	1.54	1.51
2.22	2.15	2.07	1.99	1.94	1.90	1.85	1.82	1.80	1.75	1.70
2.59	2.49	2.39	2.28	2.21	2.16	2.09	2.05	2.03	1.95	1.89
3.09	2.96	2.81	2.66	2.57	2.50	2.42	2.36	2.33	2.23	2.14
4.48	4.24	3.99	3.72	3.56	3.44	3.30	3.21	3.15	2.99	2.84
1.85	1.80	1.75	1.70	1.66	1.64	1.60	1.58	1.57	1.53	1.50
2.20	2.13	2.06	1.97	1.92	1.88	1.84	1.81	1.79	1.73	1.68
2.57	2.47	2.36	2.25	2.18	2.13	2.07	2.03	2.00	1.93	1.86
3.06	2.93	2.78	2.63	2.54	2.47	2.38	2.33	2.29	2.20	2.11
4.41	4.17	3.92	3.66	3.49	3.38	3.23	3.14	3.08	2.92	2.78
1.84	1.79	1.74	1.69	1.65	1.63	1.59	1.57	1.56	1.52	1.48
2.19	2.12	2.04	1.96	1.91	1.87	1.82	1.79	1.77	1.71	1.66
2.55	2.45	2.34	2.23	2.16	2.11	2.05	2.01	1.98	1.91	1.84
3.03	2.90	2.75	2.60	2.51	2.44	2.35	2.30	2.26	2.17	2.08
4.35	4.11	3.86	3.60	3.43	3.32	3.18	3.09	3.02	2.86	2.72
1.83	1.78	1.73	1.68	1.64	1.62	1.58	1.56	1.55	1.51	1.47
2.18	2.10	2.03	1.94	1.89	1.85	1.81	1.77	1.75	1.70	1.65
2.53	2.43	2.32	2.21	2.14	2.09	2.03	1.99	1.96	1.89	1.82
3.00	2.87	2.73	2.57	2.48	2.41	2.33	2.27	2.23	2.14	2.05
4.29	4.05	3.80	3.54	3.38	3.27	3.12	3.03	2.97	2.81	2.66
1.82	1.77	1.72	1.67	1.63	1.61	1.57	1.55	1.54	1.50	1.46
2.16	2.09	2.01	1.93	1.88	1.84	1.79	1.76	1.74	1.68	1.63
2.51	2.41	2.31	2.20	2.12	2.07	2.01	1.97	1.94	1.87	1.80
2.98	2.84	2.70	2.55	2.45	2.39	2.30	2.25	2.21	2.11	2.02
4.24	4.00	3.75	3.49	3.33	3.22	3.07	2.98	2.92	2.76	2.61
1.76	1.71	1.66	1.61	1.57	1.54	1.51	1.48	1.47	1.42	1.38
2.08	2.00	1.92	1.84	1.78	1.74	1.69	1.66	1.64	1.58	1.52
2.39	2.29	2.18	2.07	1.99	1.94	1.88	1.83	1.80	1.72	1.65
2.80	2.66	2.52	2.37	2.27	2.20	2.11	2.06	2.02	1.92	1.82
3.87	3.64	3.40	3.14	2.98	2.87	2.73	2.64	2.57	2.41	2.25
1.73	1.68	1.63	1.57	1.53	1.50	1.46	1.44	1.42	1.38	1.33
2.03	1.95	1.87	1.78	1.73	1.69	1.63	1.60	1.58	1.51	1.45
2.32	2.22	2.11	1.99	1.92	1.87	1.80	1.75	1.72	1.64	1.56
2.70	2.56	2.42	2.27	2.17	2.10	2.01	1.95	1.91	1.80	1.70
3.67	3.44	3.20	2.95	2.79	2.68	2.53	2.44	2.38	2.21	2.05
1.71	1.66	1.60	1.54	1.50	1.48	1.44	1.41	1.40	1.35	1.30
1.99	1.92	1.84	1.75	1.69	1.65	1.59	1.56	1.53	1.47	1.40
2.27	2.17	2.06	1.94	1.87	1.82	1.74	1.70	1.67	1.58	1.49
2.63	2.50	2.35	2.20	2.10	2.03	1.94	1.88	1.84	1.73	1.62
3.54	3.32	3.08	2.83	2.67	2.55	2.41	2.32	2.25	2.08	1.92
1.66	1.61	1.56	1.49	1.45	1.42	1.38	1.35	1.34	1.28	1.22
1.93	1.85	1.77	1.68	1.62	1.57	1.52	1.48	1.45	1.38	1.30
2.18	2.08	1.97	1.85	1.77	1.71	1.64	1.59	1.56	1.46	1.36
2.50	2.37	2.22	2.07	1.97	1.89	1.80	1.74	1.69	1.57	1.45
3.30	3.07	2.84	2.59	2.43	2.32	2.17	2.08	2.01	1.83	1.64
1.63	1.58	1.52	1.46	1.41	1.38	1.34	1.31	1.29	1.23	1.16
1.88	1.80	1.72	1.62	1.56	1.52	1.46	1.41	1.39	1.30	1.21
2.11	2.01	1.90	1.78	1.70	1.64	1.56	1.51	1.47	1.37	1.25
2.41	2.27	2.13	1.97	1.87	1.79	1.69	1.63	1.58	1.45	1.30
3.12	2.90	2.67	2.42	2.26	2.15	2.00	1.90	1.83	1.64	1.43
1.61	1.55	1.49	1.43	1.38	1.35	1.30	1.27	1.25	1.18	1.08
1.84	1.76	1.68	1.58	1.52	1.47	1.41	1.36	1.33	1.24	1.11
2.06	1.96	1.85	1.72	1.64	1.58	1.50	1.45	1.41	1.29	1.13
2.34	2.20	2.06	1.90	1.79	1.72	1.61	1.54	1.50	1.35	1.16
2.99	2.77	2.54	2.30	2.14	2.02	1.87	1.77	1.69	1.49	1.22

Table C9. Upper percentage points of the Studentized range

$\alpha = 0.05$

df\\j	2	3	4	5	6	7	8	9	10	11	12	13	14	15	16	17	18	19	20
5	3.64	4.60	5.22	5.67	6.03	6.33	6.58	6.80	6.99	7.17	7.32	7.47	7.60	7.72	7.83	7.93	8.03	8.12	8.12
6	3.46	4.34	4.90	5.31	5.63	5.89	6.12	6.32	6.49	6.65	6.79	6.92	7.03	7.14	7.24	7.34	7.43	7.51	7.59
7	3.34	4.16	4.68	5.06	5.36	5.61	5.82	6.00	6.16	6.30	6.43	6.55	6.66	6.76	6.85	6.94	7.02	7.09	7.17
8	3.26	4.04	4.53	4.89	5.17	5.40	5.60	5.77	5.92	6.05	6.18	6.29	6.39	6.48	6.57	6.65	6.73	6.80	6.87
9	3.20	3.95	4.42	4.76	5.02	5.24	5.43	5.60	5.74	5.87	5.98	6.09	6.19	6.28	6.36	6.44	6.51	6.58	6.64
10	3.15	3.88	4.33	4.65	4.91	5.12	5.30	5.46	5.60	5.72	5.83	5.93	6.03	6.11	6.20	6.27	6.34	6.40	6.47
11	3.11	3.82	4.26	4.57	4.82	5.03	5.20	5.35	5.49	5.61	5.71	5.81	5.90	5.99	6.06	6.14	6.20	6.26	6.33
12	3.08	3.77	4.20	4.51	4.75	4.95	5.12	5.27	5.40	5.51	5.62	5.71	5.80	5.88	5.95	6.03	6.09	6.15	6.21
13	3.06	3.73	4.15	4.45	4.69	4.88	5.05	5.19	5.32	5.43	5.53	5.63	5.71	5.79	5.86	5.93	6.00	6.05	6.11
14	3.03	3.70	4.11	4.41	4.64	4.83	4.99	5.13	5.25	5.36	5.46	5.55	5.64	5.72	5.79	5.85	5.92	5.98	6.03
15	3.01	3.67	4.08	4.37	4.60	4.78	4.94	5.08	5.20	5.31	5.40	5.49	5.58	5.65	5.72	5.79	5.85	5.90	5.96
16	3.00	3.65	4.05	4.33	4.56	4.74	4.90	5.03	5.15	5.26	5.35	5.44	5.52	5.59	5.66	5.72	5.79	5.84	5.90
17	2.98	3.63	4.02	4.30	4.52	4.71	4.86	4.99	5.11	5.21	5.31	5.39	5.47	5.55	5.61	5.68	5.74	5.79	5.84
18	2.97	3.61	4.00	4.28	4.49	4.67	4.82	4.96	5.07	5.17	5.27	5.35	5.43	5.50	5.57	5.63	5.69	5.74	5.79
19	2.96	3.59	3.98	4.25	4.47	4.65	4.79	4.92	5.01	5.14	5.23	5.32	5.39	5.46	5.53	5.59	5.65	5.70	5.75
20	2.95	3.58	3.96	4.23	4.45	4.62	4.77	4.90	5.01	5.11	5.20	5.28	5.36	5.43	5.49	5.55	5.61	5.66	5.71
24	2.92	3.53	3.90	4.17	4.37	4.54	4.68	4.81	4.92	5.01	5.10	5.18	5.25	5.32	5.38	5.44	5.50	5.54	5.59
30	2.89	3.49	3.84	4.10	4.30	4.46	4.60	4.72	4.83	4.92	5.00	5.08	5.15	5.21	5.27	5.33	5.38	5.43	5.48
40	2.86	3.44	3.79	4.04	4.23	4.39	4.52	4.63	4.74	4.82	4.91	4.98	5.05	5.11	5.16	5.22	5.27	5.31	5.36
60	2.83	3.40	3.74	3.98	4.16	4.31	4.44	4.55	4.65	4.73	4.81	4.88	4.94	5.00	5.06	5.11	5.16	5.20	5.24
120	2.80	3.36	3.69	3.92	4.10	4.24	4.36	4.48	4.56	4.64	4.72	4.78	4.84	4.90	4.95	5.00	5.05	5.09	5.13
∞	2.77	3.31	3.63	3.86	4.03	4.17	4.29	4.39	4.47	4.55	4.62	4.68	4.74	4.80	4.85	4.89	4.93	4.97	5.01

$\alpha = 0.01$

df\\j	2	3	4	5	6	7	8	9	10	11	12	13	14	15	16	17	18	19	20
1	90.0	135	164	186	202	216	227	237	246	253	260	266	272	227	282	286	290	294	298
2	14.0	19.0	22.3	24.7	26.6	28.2	29.5	30.7	31.7	32.6	33.4	34.1	34.8	35.4	36.0	36.5	37.0	37.5	37.9
3	8.26	10.6	12.2	13.3	14.2	15.0	15.6	16.2	16.7	17.1	17.5	17.9	18.2	18.5	18.8	19.1	19.3	19.5	19.8
4	6.51	8.12	9.17	9.96	10.6	11.1	11.5	11.9	12.3	12.6	12.8	13.1	13.3	13.5	13.7	13.9	14.1	14.2	14.4
5	5.70	6.97	7.80	8.42	8.91	9.32	9.67	9.97	10.24	10.48	10.70	10.89	11.08	11.24	11.40	11.55	11.68	11.81	11.93
6	5.24	6.33	7.03	7.56	7.97	8.32	8.61	8.87	9.10	9.30	9.49	9.65	9.81	9.95	10.08	10.21	10.32	10.43	10.54
7	4.95	5.92	6.54	7.01	7.37	7.68	7.94	8.17	8.37	8.55	8.71	8.86	9.00	9.12	9.24	9.35	9.46	9.55	9.65
8	4.74	5.63	6.20	6.63	6.96	7.24	7.47	7.68	7.87	8.03	8.18	8.31	8.44	8.55	8.66	8.76	8.85	8.94	9.03
9	4.60	5.43	5.96	6.35	6.66	6.91	7.13	7.32	7.49	7.65	7.78	7.91	8.03	8.13	8.23	9.32	8.41	8.49	8.57
10	4.48	5.27	5.77	6.14	6.43	6.67	6.87	7.05	7.21	7.36	7.48	7.60	7.71	7.81	7.91	7.99	8.07	8.15	8.22
11	4.39	5.14	5.62	5.97	6.25	6.48	6.67	6.84	6.99	7.13	7.25	7.36	7.46	7.56	7.65	7.73	7.81	7.88	7.95
12	4.32	5.04	5.50	5.84	6.10	6.32	6.51	6.67	6.81	6.94	7.06	7.17	7.26	7.36	7.44	7.52	7.59	7.66	7.73
13	4.26	4.96	5.40	5.73	5.98	6.19	6.37	6.53	6.67	6.79	6.90	7.01	7.10	7.19	7.27	7.34	7.42	7.48	7.55
14	4.21	4.89	5.32	5.63	5.88	6.08	6.26	6.41	6.54	6.66	6.77	6.87	6.96	7.05	7.12	7.20	7.27	7.33	7.39
15	4.17	4.83	5.25	5.56	5.80	5.99	6.16	6.31	6.44	6.55	6.66	6.76	6.84	6.93	7.00	7.07	7.14	7.20	7.26
16	4.13	4.78	5.19	5.49	5.72	5.92	6.08	6.22	6.35	6.46	6.56	6.66	6.74	6.82	6.90	6.97	7.03	7.09	7.15
17	4.10	4.74	5.14	5.43	5.66	5.85	6.01	6.15	6.27	6.38	6.48	6.57	6.66	6.73	6.80	6.87	6.94	7.00	7.05
18	4.07	4.70	5.09	5.38	5.60	5.79	5.94	6.08	6.20	6.31	6.41	6.50	6.58	6.65	6.72	6.79	6.85	6.91	6.96
19	4.05	4.67	5.05	5.33	5.55	5.73	5.89	6.02	6.14	6.25	6.34	6.43	6.51	6.58	6.65	6.72	6.78	6.84	6.89
20	4.02	4.64	5.02	5.29	5.51	5.69	5.84	5.97	6.09	6.19	6.29	6.37	6.45	6.52	6.59	6.65	6.71	6.76	6.82
24	3.96	4.54	4.91	5.17	5.37	5.54	5.69	5.81	5.92	6.02	6.11	6.19	6.26	6.33	6.39	6.45	6.51	6.56	6.61
30	3.89	4.45	4.80	5.05	5.24	5.40	5.54	5.65	5.76	5.85	5.93	6.01	6.08	6.14	6.20	6.26	6.31	6.36	6.41
40	3.82	4.37	4.70	4.93	5.11	5.27	5.39	5.50	5.60	5.69	5.77	5.84	5.90	5.96	6.02	6.07	6.12	6.17	6.21
60	3.76	4.28	4.60	4.82	4.99	5.13	5.25	5.36	5.45	5.53	5.60	5.67	5.73	5.79	5.84	5.89	5.93	5.98	6.02
120	3.70	4.20	4.50	4.71	4.87	5.01	5.12	5.21	5.30	5.38	5.44	5.51	5.56	5.61	5.66	5.71	5.75	5.79	5.83
∞	3.64	4.12	4.40	4.60	4.76	4.88	4.99	5.08	5.16	5.23	5.29	5.35	5.40	5.45	5.49	5.54	5.57	5.61	5.65

APPENDIX D

Basic Results in Optimization

In this appendix, some basic results for deterministic optimization problems are presented. We first consider static optimization and present results for both the unconstrained and constrained cases. Following this, we look at dynamic optimization, where both multistage and continuous time formulations are considered.

D.1   UNCONSTRAINED STATIC OPTIMIZATION

The simplest case is that in which x is a scalar decision variable to be selected optimally to maximize a scalar function $L(x)$[1]. Let x^* denote the local optimal x. x^* may or may not exist and when it does, there can one or several local or global maxima.

Scalar x

If x is a real variable, then to be a local maximum, x^* must satisfy the first order necessary condition given by

$$L'(x) = \frac{dL(x)}{dx} = 0 \tag{D1}$$

as well as the condition

$$L''(x) = \frac{d^2L(x)}{dx^2} \leq 0 \tag{D2}$$

Any solution to D1 yields a stationary point. A sufficient condition for x^* to yield a local maximum is given by (D1) and

$$L''(x) = \frac{d^2L(x)}{dx^2} < 0 \tag{D3}$$

In general, it is necessary to solve (D1) computationally to obtain the stationary points and then check (D2) to determine which yield a local maximum. Many different techniques

763

for accomplishing this have been developed. One of the simplest is the first order gradient method (used extensively in the Chapter 18 examples) which computes $x*$ by an iterative process as indicated below.

Select an arbitrary value for x_0. Compute x_i, $i = 1, 2, \ldots$ as follows:

$$x_i = x_{i-1} + \varepsilon \frac{dL(x)}{dx}\bigg|_{x=x_{i-1}} \tag{D4}$$

where ε is a small positive number. The iteration is stopped when the first derivative becomes sufficiently small.

If $L(x)$ is a concave function (or a convex function in the case of minimization), then $x*$ exists and is unique. However, the difficulty then is proving the concavity (or convexity) of the function.

If x takes on only discrete values s_i, $i = 1, 2, \ldots$ (e.g., the set of positive integers), then $x* = s_m$ is a local maximum if

$$L(s_m) > L(s_{m \pm 1}) \tag{D5}$$

Vector x

The results easily generalize to the case where $L(x)$ is a scalar function and x is an m-dimensional vector given by

$$x^T = [x_1, x_2, \ldots, x_m] \tag{D6}$$

In this case, the above results hold with the first derivative being an m-dimension vector given by

$$L'(x) = \left[\frac{\partial L(x)}{\partial x_1}, \frac{\partial L(x)}{\partial x_2}, \ldots, \frac{\partial L(x)}{\partial x_m} \right] \tag{D7}$$

and the second derivative is a $m \times m$ matrix with the element corresponding to row i and column j ($1 \leq i, j \leq m$) given by

$$[L''(x)]_{ij} = \frac{\partial^2 L(x)}{\partial x_i \partial x_j} \tag{D8}$$

In this case, (D1) is a set of n coupled equations and (D3) implies that the matrix is negative definite, i.e., that all of the eigenvalues of the second derivative matrix are negative.

One can use first-order gradient methods to obtain the stationary point. However, there are many other methods that are more efficient. These can be found in most textbooks on optimization, for example, Fletcher (1980) and Rao (1996).

D.2   CONSTRAINED STATIC OPTIMIZATION

There are two different types of constraints—equality and inequality constraints.

Equality Constraint

Let x be an $(n + m)$-dimensional vector. Let the first n components be represented by an n-dimensional vector z (called the *state vector*), with $n \geq 1$, and the last m components by an m-dimensional vector $u(m \geq 1)$.[2] The problem is to maximize $L(z, u)$ subject to a n-dimensional constraint of the form

$$f(z, u) = 0 \tag{D9}$$

where $f(z, u)$ is an n-dimensional vector. Note that u is the decision variable to be optimally selected and z is constrained to ensure that the equality given by (D9) is satisfied.

One way of solving the problem is to adjoin the vector constraint through an n-dimensional set of "undetermined multipliers" and convert the problem into an unconstrained problem with $(2n + m)$ variables and then to carry out the optimization as in the previous section. Let λ be the n-dimensional multiplier and define

$$H(z, u, \lambda) = L(z, u) + \lambda^T f(z, u) \tag{D10}$$

Then the necessary conditions for a stationary point for $L(z, u)$ are given by (D9) and the following equations:

$$\frac{\partial H(z, u, \lambda)}{\partial z} = 0 \tag{D11}$$

and

$$\frac{\partial H(z, u, \lambda)}{\partial u} = 0 \tag{D12}$$

Note that (D11) and (D12) are vector equations of dimension n and m, respectively. As a result, we have $(2n + m)$ equations and solving these yields the stationary values z and u and the associated multiplier λ. Note that this requires that the $(n \times n)$ matrix $L_z = \partial L(z, u)/\partial z$ be nonsingular at the stationary point. This is equivalent to all of the n constraints being independent at the stationary point.

The sufficiency condition involves second-order derivatives. For sufficiency, (D9), (D11), and (D12) must be satisfied and it must be that the $(m \times m)$ matrix

$$H_{uu} - H_{uz} f_z^{-1} f_u - f_u^T (f_z^T)^{-1} H_{zu} + f_u^T (f_z^T)^{-1} H_{zz} f_z^{-1} f_u < 0 \tag{D13}$$

The last condition is equivalent to the matrix being negative definite with all m eigenvalues negative. For further details, see Bryson and Ho (1969).

Inequality Constraint

The problem is to maximize $L(x)$ subject to an inequality constraint of the form

$$f(x) \leq 0 \tag{D14}$$

where $f(\;)$ and x are vectors of different dimensions. Let n and m represent the dimensions of x and $f(\;)$, respectively. Note that the *admissible* x must not violate the constraint given by (D14). This type of problem is also called "nonlinear programming."

The constraint can be adjoined using m multipliers as in the previous case. However, when a constraint is not binding (i.e., for a stationary point with $f_j(x) < 0$), then the associated component of the multiplier λ_j is zero. If a constraint is binding (i.e., for a stationary point with $f_j(x) = 0$) then the associated component of the multiplier λ_j is negative. This implies that the necessary condition is given by (D14) being satisfied and

$$\frac{\partial H(x, \lambda)}{\partial x} = 0 \tag{D15}$$

with

$$H(x, \lambda) = L(x) + \lambda^T f(x) \tag{D16}$$

and

$$\lambda = \begin{cases} < 0, & f(x) = 0 \\ = 0, & f(x) < 0 \end{cases} \tag{D17}$$

(D16) and (D17) are the essence of the "Kuhn–Tucker" condition in nonlinear programming (see Hadley, 1964).

When both $L(x)$ and $f(x)$ are linear in x, the problem is called a "linear programming" problem. In this case, the optimal solution can be obtained using the "simplex" method proposed by Dantzig (1963).

Many algorithmic approaches for optimization have been developed. These can be found in most standard texts, such as Hadley (1964) and Rao (1996). There are several specialized books dealing with this topic. For example, Gen and Cheng (1996) deals with genetic algorithms.

D.3   MULTISTAGE DYNAMIC OPTIMIZATION

Here the optimization is done over K stages, with decisions to be made at each stage. Let u_k (an m-dimensional vector) denote the decision variable at stage k, $k = 0, 1, \ldots, (K - 1)$. These decisions affect the state variables that characterize the linkage between different stages. Let z_k (an n-dimensional vector) denote the state variable corresponding to stage k. The linkage between stages is given by deterministic difference equations of the form

$$z_{k+1} = h_k(z_k, u_k) \tag{D18}$$

for $k = 0, 1, \ldots, (K - 1)$.

The performance index to be maximized is given by

$$J = \phi(z_k) + \sum_{k=0}^{K-1} L_k(z_k, u_k) \tag{D19}$$

The optimization problem is to find the sequence u_k, $k = 0, 1, \ldots, (K - 1)$, to maximize J.

Note that (D18) can be viewed as a set of equality constraints. By adjoining this through a set of n-dimensional multipliers λ_k, $k = 1, \ldots, K$, we effectively convert the problem to an unconstrained optimization problem. This yields[3]

$$J = \phi(z_K) - \lambda_K^T z_K + \sum_{k=0}^{K-1} [H_k - \lambda_k^T z_k] + H_0 \tag{D20}$$

where

$$H_k = L_k(z_k, u_k) + \lambda_{k+1}^T f_k(z_k, u_k) \qquad \text{(D21)}$$

for $k = 0, \ldots, (K-1)$. We give the necessary condition for optimality without the derivation. (Interested readers can find it in Bryson and Ho (1969).) Define

$$\lambda_k = \left[\frac{\partial f_k}{\partial z_k} \right]^T \lambda_{k+1} + \left[\frac{\partial L_k}{\partial z_k} \right]^T \qquad \text{(D22)}$$

for $k = 1, \ldots, (K-1)$;

$$\lambda_K = \frac{\partial \phi(z_K)}{\partial z_K} \qquad \text{(D23)}$$

and

$$\frac{\partial H_k}{\partial u_k} = \frac{\partial L_k}{\partial u_k} + \lambda_{k+1} \frac{\partial f_k}{\partial u_k} = 0 \qquad \text{(D24)}$$

for $k = 0, \ldots, (K-1)$.

The necessary conditions are that

1. (D18) be satisfied for $k = 1, \ldots, (K-1)$, with z_0 given
2. (D22) be satisfied for $k = 1, \ldots, (K-1)$, with λ_K given by (D23)
3. (D24) be satisfied for $k = 0, \ldots, (K-1)$

Note that this is a two-point boundary value problem and, in general, is difficult to solve, requiring an iterative approach. An alternate approach is the "Dynamic Programming" approach of Bellman (1957).

D.4 CONTINUOUS TIME DYNAMIC OPTIMIZATION

A continuous time dynamic optimization problem is also called a *functional optimization* (as opposed to function optimization) problem or a problem in the *calculus of variations*. We first consider fixed terminal time and no constraints on the decision variables. Later, we look at the case where the terminal time is also a decision variable to be selected optimally and the decision variables are constrained.

Fixed Terminal Time

Here the decision variable is an m-dimensional vector $u(t)$, $0 \leq t \leq T$, where T is a specified terminal time. The objective is to select $u(t)$ to maximize a functional J given by

$$J = \Phi(T, x(T)) + \int_0^T L(x(t), u(t), t)dt \qquad \text{(D25)}$$

subject to a dynamic constraint given by an n-dimensional ordinary differential equation

$$\dot{x}(t) = \frac{dx(t)}{dt} = f(x(t), u(t), t) \tag{D26}$$

with initial condition $x(o) = x_0$ specified. $x(t)$ is called the state vector and is n-dimensional.

The approach is to adjoin the dynamic constraint through an n-dimensional multiplier $\lambda(t)$, $0 \le t \le T$, since the constraint is binding for all time instants in the interval of interest. We present the necessary conditions without giving the details of the derivation. These can be found, along with the results for the sufficiency condition, in Bryson and Ho (1969) or Kamien and Schwartz (1991). Define

$$H[x(t), u(t), \lambda(t), t] = L(x(t), u(t), t) + \lambda^T f(x(t), u(t), t) \tag{D27}$$

The necessary conditions are:

1. The multiplier $\lambda(t)$ satisfies

$$\frac{d\lambda^T(t)}{dt} = -\frac{\partial H}{\partial x} = -\frac{\partial L}{\partial x} - \lambda^T \frac{\partial f}{\partial x} \tag{D28}$$

   with boundary condition

$$\lambda^T(T) = \frac{\partial \phi(x(T))}{\partial x(T)} \tag{D29}$$

2. The system equation given by (D26) is satisfied with initial condition given by x_0, and

3. The optimality condition for obtaining the optimal decision variable $u^*(t)$ is given by

$$H_u(T) = \frac{\partial H}{\partial u} = 0 \tag{D30}$$

Again, this requires solving a two-point boundary value problem.

Free Terminal Time

In this case T is also a decision variable to be selected optimally to maximize J. In this case, the necessary conditions are given by (1)–(3) for the fixed terminal case and the additional condition:

4. The optimal T^* is obtained from

$$\frac{\partial \phi(x(t), T)}{\partial T} + H|_{t=T} = 0 \tag{D31}$$

For details of the derivation, see Bryson and Ho (1969).

Inequality Constraint on the Decision Variable

Suppose that $u(t)$ is constrained such that $u(t) \in \Omega$, for example, Ω being the inteval $[0, U]$. In this case condition (3) is replaced by the following:

3a. $H(x(t), u^*(t), \lambda(t), t) \leq H(x(t), u(t), \lambda(t), t)$ for all $u(t) \in \Omega$

[*Note:* If $u(t)$ is unconstrained, then we obtain (3) from (3a).] This is the essence of the "Maximum Principle" of Pontryagin (1962).

D.5   OTHER TOPICS

We have confined our discussion to optimization of a scalar function. For optimization of vector functions, see Steuer (1986).

Stochastic optimization is a topic that has received a great deal of attention. See, for example, Davis (1993).

NOTES

1. If the problem is to find a local minimum for the function $L(x)$, then it is equivalent to finding the local maximum for the function $G(x)$ where $G(x) = -L(x)$. Hence, without loss of generality, we will confine our attention to finding the local maximum for a function.

2. The reason for this will become clearer later in the Appendix.

3. We omit the arguments for notational ease. As a result, H_k is used instead of $H_k(z_k, u_k, \lambda_k)$ and so on.

REFERENCES

Abdushukurov, A. A. (1998). "Nonparametric estimation of the survival function from censored data based on relative risk function," *Communications in Statistics, Theory and Methods* **28:** 1991–2012.

Abernethy, R. B., Breneman, J. E., Medlin, C. H., and Reinman, G. L. (1983). *Weibull Analysis Handbook,* Report No. AFWAL-TR-83–2079, Aero Propulsion Lab., USAF, Wright-Patterson AFB, Ohio.

Abramowitz, M. and Stegun, I. A. (1964). *Handbook of Mathematical Functions,* Applied Mathematics Series No. 55, National Bureau of Standards, Washington, DC.

Aitchison, J. and Brown, J. A. C. (1957). *The Lognormal Distribution,* Cambridge University Press, Cambridge.

Al-Sheikly, M. and Christou, A. (1994). "How radiation affects polymeric materials," *IEEE Transactions on Reliability* **43:** 551–556.

AMPC-706–196 (1976). *Engineering Design Handbook,* Department of Defense, U.S. Army Material Command.

Amstadter, B. L. (1971). *Reliability Mathematics,* McGraw Hill, New York.

Anderson, E. E. (1977). "Product price and warranty terms: An optimization model," *Operations Research Quarterly* **28:** 739–741.

Anderson, R. T. and Neri, L. (1990). *Reliability Centered Maintenance,* Elsevier, London.

Anderson, V. L. (1970). "Restriction errors for linear models (An aid to develop models for designed experiments)," *Biometrics* **26:** 255–268.

Anderson, V. L., and McLean, R. A. (1974). "Restriction errors: another dimension in teaching experimental statistics," *The American Statistician* **28:** 145–152.

Ang, A. H. (1984), *Probability Concepts in Engineering Planning and Design,* Wiley, New York.

Appel, F. C. (1970). "The 747 ushers in a new era," *The American Way,* March 25–29.

Asgharizadeh, E. (1997), *Modelling and Analysis of Maintenance Service Contracts,* Unpublished thesis, The University of Queensland, Australia.

Atwood, C. L. (1986). "The binomial failure rate common cause model," *Technometrics* **28:** 139–148.

Aven, Terje and Dekker, Rommert (1997) "A useful framework for optimal replacement models," *Reliability Engineering and System Safety* **58:** 61–67.

Bain, L. J. and Englehardt, M. (1991). *Statistical Analysis of Reliability and Life-Testing Models,* Dekker, New York.

Balakrishnan, N., Gupta, S. S., and Panchapakesan, S. (1995). "Estimation of the location and scale parameters of the extreme value distribution based on multiply Type-II censored samples," *Communications in Statistics, Theory and Methods* **24:** 2105–2125.

Banarjee, A. K. and Bhattacharyya, G. K. (1979). "Bayesian results for the inverse Gaussian distribution with an application, *Technometrics* **21:** 247–251.

Barlow, R. E. (1984). "Theory of reliability: A historical perspective," *IEEE Transactions on Reliability* **33:** 16–20.

Barlow, R. E. and Campo, R. (1975). "Total time on test processes and applications in failure data analysis," in R. E. Barlow, J. B. Fussell, and N. Singpurwalla (Eds.), *Reliability and Fault Tree Analysis,* SIAM Publications, Philadelphia.

Barlow, R. E. and Hunter, L. (1961). "Optimum preventive maintenance policies," *Operations Research* **8:** 90–100.

Barlow, R. E. and Lambert H. E. (1975). "Introduction to Fault Tree Analysis in Reliability" in *Fault Tree Analysis,* Barlow, R. E., Fussell, J. B. and Singpurwalla, N. D. (Eds.), SIAM, Philadelphia.

Barlow, R. E. and Proschan, F. (1965). *Mathematical Theory of Reliability,* Wiley, New York.

Barlow, R. E. and Proschan, F. (1975). *Statistical Theory of Reliability and Life Testing Probability Models,* Holt, Rinehart and Winston, New York.

Barlow, R. E. and Scheuer, E. M. (1966). "Reliability growth during a development testing program," *Technometrics* **8:** 53–60.

Barndorff-Nielsen, O., and Halgren, C. (1997), "Infinite divisibility of the hyperbolic and generalised inverse Gaussaian distribution," *Z. Wahrsheinlichkeitstheorie verw.* Gebeite **38,** 309–312.

Bâsu, M. (1995). A combined fuzzy-logic & physics-of-failure approach to reliability prediction," *IEEE Transactions on Reliability* **44:** 237–242.

Bates, D. M. and Watts, D. G. (1988).*Nonlinear Regression Analysis and Its Applications,* Wiley, New York.

Baxter, L. A. (1993). "Towards a theory of confidence intervals for system reliability," *Statistics and Probability Letters* **16:** 29–38.

Baxter, L. A., Scheuer, E. M., Blischke, W. R., and McConalogue, D. J. (1981). *Renewal Tables of Functions Arising in Renewal Theory,* Tech. Rep., Decision Systems Dept., University of Southern California, Los Angeles, CA.

Bayes, T. (1763). "An essay towards solving a problem in the doctrine of chances," *Philosophical Transactions Royal Society of London* **53:** 370–418. (Reprinted in *Biometrika* **45**(1958): 293–315 and in Press, J. (1989), Appendix 4.)

Bayley, K. M. and Tabbagh, P. P. (1995). "The assurance of R&M in acquisition programs of the Royal Australian Air Force," *Proceedings of the Annual Reliability and Maintainability Symposium,* 118–124.

Bayraktar, B. A. (1990). "On the concepts of technology and management of technology," *Management of Technology II,* Khalil, T. M. and Bayraktar, B.A. (Eds.), Industrial Engineering & Management Press, 1161–1175.

Beizer, Boris (1990). *Software Testing Techniques,* 2nd Ed., Van Nostrand Reinhold, New York.

Bellman, R. E. (1957). *Dynamic Programming,* Princeton University Press, Princeton, NJ.

Bendell, A. and Humble, S. (1985). "A reliability model with states of partial operation," *Naval Research Logistics Quarterly* **32:** 509–535.

Berg, M. (1995). "The marginal cost analysis and its application to repair and replacement policies," *European Journal of Operational Research* **82:** 214–224.

Berg, M. and Epstein, B. (1978). "Comparison of age, block and failure replacement policies," *IEEE Transactions on Reliability* **27:** 25–29.

Bergman, B. and Klefsjo, B. (1982). "A graphical method applicable to age-replacement problems," *IEEE Transactions on Reliability* **31**: 479–481.

Betz, F. (1993). *Strategic Technology Management,* McGraw-Hill, New York.

Bhat, N. U. (1972). *Elements of Applied Stochastic Processes,* Wiley, New York.

Bhattacharyya, G. K. and Fries, A. (1982). "Fatigue failure models—Birbaum Saunders vs Inverse Gaussian," *IEEE Transactions on Reliability* **31**: 439–440.

Bilikam, J. E. (1985). "Some stochastic stress-strength models," *IEEE Transactions on Reliability* **34**: 269–274.

Birnbaum, , Z. W. (1969). "On the importance of different components in a multi-component system," in *Multivariate Analysis,* R.P. Krishniah (ed.), Academic Press, New York, pp. 581–592.

Birnbaum, Z. W. and Saunders, S. C. (1969). "A new family of life distributions," *Journal of Applied Probability* **6**: 319–327.

Birolini, A. (1997), *Quality and Reliability in Technical Systems (2nd. Ed.),* Kluwer Academic Press, Dordrecht, Holland.

Blache, K. M. and Shrivastava, A. B. (1994). "Defining failure of manufacturing machinery and equipment," *Proceedings Annual Reliability and Maintainability Symposium* 69–75.

Black, S. E. and Rigdon, S. E. (1996). "Statistical inference for a modulated power law process," *Journal of Quality Technology* **28**: 81–90.

Blanchard, B. S. (1981). *Logistic Engineering and Management,* Prentice Hall, Englewood Cliffs, NJ.

Blanchard, B. S. (1998). *Logistics Engineering and Management, 5th Edition,* Prentice-Hall International, Upper Saddle River, NJ.

Blanchard, B. S., Verma, D., and Peterson, E. L. (1995). *Maintainability,* Wiley, New York.

Bland, J A. (1998). "Structural design optimization with reliability constraints using tabu search," *Engineering Optimization* **30**: 55–74.

Blanks, H. S. (1992). *Reliability in Procurement and Use: From Specification to Replacement,* Wiley, New York.

Blischke, W. R. (1994). "Bayesian formulation of the best of liquid and solid reliability methodology," *Journal of Spacecraft and Rockets* **31**: 297–303.

Blischke, W. R. and Murthy, D. N. P. (1994). *Warranty Cost Analysis,* Dekker, New York.

Blischke, W. R., and Murthy, D. N. P. (1996). *Product Warranty Handbook,* Dekker, New York.

Blischke, W. R. and Scheuer, E. M. (1986). "Tabular aids for fitting Weibull moment estimates," *Naval Research Logistics Quarterly* **33**: 145–153.

Blumenthal, S., Greenwood, J. A., and Herbach, L. H. (1984). "Series systems and reliability demonstration tests," *Operations Research* **34**: 641–648.

Blumenthal, S. and Marcus, R. (1975). "Estimating population size with exponential failure," *Journal of the American Statistical Association* **70**: 913–922.

Born, F. and Criscimagna, N. H. (1995). "Translating user diagnostics, reliability, and maintainability needs into specifications," *Proceedings Annual Reliability and Maintainability Symposium,* 106–111.

Bonis, A. J. (1977). "Reliability growth curves for one shot devices," *Annual Reliability and Maintainability Symposium,* 181–185.

Bowles, J. B. (1992), "A survey of reliability-prediction procedures for microelectronic devices," *IEEE Transactions on Reliability* **41**: 2–12.

Box, G. E. P., Hunter, W. G., and Hunter, J. S. (1978). *Statistics for Experimenters,* Wiley, New York.

Box, G. E. P. and Tiao, G. C. (1973). *Bayesian Inference in Statistical Analysis,* Addison-Wesley, Reading, MA.

Brall, A. (1994). "A model for success in implementing an R&M program by a supplier of manufacturing machinery," *Proceedings of the Annual Reliability and Maintainability Symposium,* 59–64.

Brombacher, A. C. (1992). *Reliability by Design, CAE Techniques for Electronic Components and Systems,* Wiley, New York.

Brown, K. A. P. (1985). "In pursuit of reliability," *Reliability Engineering* **10:** 141–150.

Brunelle, R. D. and Kapur, K. C. (1997). "Customer-centered reliability methodology," *Proceedings of the Annual Reliability and Maintainability Symposium,* 268–292.

Bryson, A. E. and Ho, Y. C. (1969). *Applied Optimal Control Theory,* Ginn Waltham, MA.

Calabria, R., Guida, M. M, and Pulcini, G. (1992). "A Bayes procidure for estimation of current system reliability," *IEEE Transactions on Reliability* **41:** 616–620.

Cannon, A. G. and Bendell, A. (Eds.) (1991). *Reliability Data Banks,* Elsevier Applied Science, London.

Carter, A. D. S. (1986). *Mechanical Reliability* (2nd Ed.), Wiley, New York.

Carter, A. D. S. (1997). *Mechanical Reliability and Design,* Macmillan Press, London.

Case, K. E. and Jones, L. L. (1978). *Profit Through Quality. Quality Assurance Programs for Manufacturers,* Institute of Industrial Engineers, New York.

Chandra, M., Singpurwalla, N. D., and Stephens, M. A. (1981). "Kolmogorov statistics for tests of fit for the extreme-value and Weibull distributions," *Journal of the American Statistical Association* **76:** 729–731.

Chandran, R. and Lancioni, R. A. (1981). "Product recall: A challenge for the 1980's," *International Journal of Physical Distribution and Materials Management* **11:** 46–55.

Chao, A. and Huwang, L.-C. (1987). "A modified Monte Carlo technique for confidence limits of system reliability using pass-fail data," *IEEE Transactions on Reliability* **R-36:** 109–112.

Cheng, S. S. (1977). "Optimal replacement rate of devices with lognormal failure distributions," *IEEE Transactions on Reliability* **26:** 174–178.

Chhikara, R. S. and Folks, J. L. (1989). *The Inverse Gaussian Distribution,* Dekker, New York.

Chien, W. K. and Kuo, W. (1995). "Modeling and maximizing burn-in effectiveness," *IEEE Transactions on Reliability* **44:** 19–25.

Cho, D. and Parlar, M. (1991). "A survey of maintenance models for multi-unit systems," *European Journal of Operational Research,* **51:** 1–23.

Christer, A. H. and Waller, W. M. (1994). "Delay time models of industrial inspection maintenance problems," *Journal of Operations Research Society* **35:** 401–406.

Christer, A. H. and Wang, W. (1995). "A delay time based model of a multicomponent system," *IMA Journal of Mathematics Applications In Business and Industry* **6:** 205–222.

Christou, A. (1994), *Integrating Reliability into Microelectronics Manufacturing,* Wiley, New York.

Church, J. D. and Harris, B. (1970). "The estimation of reliability from stress-strength relationships," *Technometrics* **12:** 49–54.

Cinlar, E. (1975). *Introduction to Stochastic Processes,* Prentice Hall, Englewood Cliffs, NJ.

Cohen, H. (1984). "Space reliability technology: A historical perspective," *IEEE Transactions on Reliability* **33:** 36–40.

Coit, D. W. and Smith, A. E. (1998). "Redundancy allocation to maximize a lower percentile of the system time-to-failure distribution," *IEEE Transactions on Reliability* **47:** 79–87.

Condra, Lloyd W. (1993). *Reliability Improvement with Design of Experiments,* Dekker, New York.

Coppola, A. (1984). "Reliability engineering of electronic equipment: A historical perspective," *IEEE Transactions on Reliability* **33:** 29–35.

Court, E. T. (1981). "To specify reliability requirement does not ensure its achievement," *Reliability Engineering* **2**: 243–258.

Coutinho, J. (1973). "Software reliability growth," *IEEE Symposium on Computer Software Reliability.*

Cox, D. R. (1960). *Renewal Theory,* Methuen, London.

Cox, D. R. (1972). "Regression models and life tables (With discussion)," *Journal of the Royal Statistical Society* **34B**: 187–220.

Cox, D. R. and Isham, V. (1980). *Point Processes,* Chapman-Hall, New York.

Cox, D. R. and Miller, H. D. (1965). *The Theory of Stochastic Processes,* Wiley, New York.

Cox, D. R. and Oakes, D. (1984). *Analysis of Survival Data,* Metheun, New York.

Crow, E. L. and Shimizu, K. (1988). *Lognormal Distributions,* Dekker, New York.

Crow, L. H. (1974). "Reliability analysis for complex repairable systems," in *Reliability and Biometry,* F. Proschan and R. J. Serfling (Eds.), pp. 379–410, Society of Industrial and Applied Mathematics, Philadelphia.

Crowder, M. J., Kimber, A. C., Smith, R. L., and Sweeting, R. J. (1991). *Statistical Analysis of Reliability Data,* Chapman and Hall, London.

Cruse, T. A. (Ed.) (1997). *Reliability-Based Mechanical Design,* Dekker, New York.

Cruse, T. A., Mahadevan, S., Huang, Q., and Mehta, S. (1994). "Mechanical system reliability and risk assessment," *AIAA Journal* **32**: 2249–2259.

D'Agostino, R. B. and Stephens, M. A. (1986). *Goodness-of-Fit Techniques,* Dekker, New York.

Dantzig, G. (1963), *Linear Programming and Extensions,* Princeton University Press, Princeton, NJ.

Dardis, R. and Zent, C. (1982). "The economics of the Pinto recall," *Journal of Consumer Affairs* **16**: 261–277.

Dasgupta, A. (1993). "Failure mechanism models for cyclic fatigue," *IEEE Transactions on Reliability* **42**: 548–555.

Dasgupta, A. and Haslach, H. W. (1993). "Failure mechanism models for buckling," *IEEE Transactions on Reliability* **42**: 9–16.

Dasgupta, A. and Hu, J. M. (1992a). "Failure mechanism models for excessive elastic deformation," *IEEE Transactions on Reliability* **41**: 149– 154.

Dasgupta, A. and Hu, J. M. (1992b). "Failure mechanism models for brittle plastic deformation," *IEEE Transactions on Reliability* **41**: 168– 174.

Dasgupta, A. and Hu, J. M. (1992c). "Failure mechanism models for brittle fracture," *IEEE Transactions on Reliability* **41**: pp 328– 337.

Dasgupta, A. and Hu, J. M. (1992d). "Failure models for ductile fracture," *IEEE Transactions on Reliability* **41**: pp 489–495.

Dasgupta, A. and Pecht, M. (1991). "Material failure mechanisms and damage models," *IEEE Transactions on Reliability* **40**: 531–536.

Datsko, J. (1966). *Material Properties and Manufacturing Processes,* Wiley, New York.

David, H. A. and Moeschberger, M. L. (1978). *The Theory of Competing Risks,* Griffin, London.

Davis, D. J. (1952). "An analysis of some failure data," *Journal of the American Statistical Assoc.* **47**: 113–150.

Davis, M. H. A. (1993). *Markov Models and Optimization,* Chapman Hall, London.

De Groot, M. H. (1970). *Optimal Statistical Decisions,* McGraw-Hill, New York.

Dekker, R. (1996). "Applications of maintenance and optimization models," *Reliability Engineering and System Safety* **51**: 229–240.

Dekker, R. and Dijkstra, M. C. (1992), "Opportunity based age replacements: exponentially distributed times between opportunities," *Naval Research Logistics* **39**: 175–190.

Dekker, R. and Roelvink, I. F. K. (1995). "Marginal cost criteria for preventive replacement of a group of components," *European. Journal of Operational Research* **84**: 467–480.

Dekker, R. and Smeitink, E. (1991). "Opportunity based block replacement: the single component case," *European Journal of Operational Research* **53**: 46–63.

Dekker, R., Wildeman, R. E., and van der Duyn Schouten, F. A. (1997), "Review of multi-component models with economic dependence," *Zor/Mathematical Methods of Operations Research* **45**: 411–435.

Dey, A. (1985). *Orthogonal Fractional Factorial Designs,* Wiley, New York.

Dey, L. K. and Lee, T.-M. (1992). "Bayes computation for life testing and reliability estimation," *IEEE Transaction on Reliability* **41**: 621–626.

Dhillon, B. S. (1983). *Reliability Engineering in Systems Design and Operation,* Van Nostrand Reinhold, New York.

Dieter, G. E. (1991). *Engineering Design,* McGraw Hill, New York.

Dixon, W. J., and Massey, F. J., Jr. (1969). *Introduction to Statistical Analysis,* 3rd Ed., McGraw-Hill, New York.

Djamaludin, I. (1993). *Warranty and Quality Control,* Ph.D. thesis, The University of Queensland, Brisbane, Australia.

Djamaludin, I., Murthy, D. N. P. and Blischke, W. R. (1996). "Bibliography on Warranties," in *Product Warranty Handbook,* W. R. Blischke and D. N. P. Murthy (Eds.), Dekker, New York.

Djamaludin, I, Murthy, D. N. P., and Wilson, R. J. (1994). "Quality control through lot sizing for items sold with warranty," *International Journal of Production Economics* **33**: 97–107.

Dohi, T., Kaio, N., and Osaki, S. (1996). "Optimal ordering policies with time-dependent delay structure," *Journal of Quality in Maintenance Engineering* **2**: 50–62.

Dowling, N. E. (1999). *Mechanical Behavior of Materials,* 2nd Ed., Prentice Hall, Upper Saddle River, N.J.

Draper, N. R. and Guttman, I. (1978). "Bayesian analysis of reliability in multicomponent stress-strength models,: " *Communications in Statistics, Theory and Methods* **7A**: 441–451.

Draper, N. R. and Smith, H. (1998). *Applied Regression Aanlysis,* Wiley, New York.

Duane, J. T. (1964). "Learning curve approach to reliability modeling," *IEEE Transactions on Aerospace* **2**: 563–566.

Durham, S. D. and Padgett, W. J. (1997). "Cumulative damage models for system failure with application to carbon fibers and composites," *Technometrics* **39**: 34–44.

Easterling, R. G. (1972). "Approximate confidence limits for system reliability," *Journal of the American Statistical Association* **67**: 220–222.

Edgeman, R. L. (1990). "Assessing the inverse Gaussian distribution assumption," *IEEE Transactions on Reliability* **39**: 352–355.

Edgeman, R. L., Scott, R. C., and Pavur, R. J. (1988). "A modified Kolmogorov-Smirnov test for the inverse Gaussian density with unknown parameters," *Communications in Statistics, Simulation and Computation* **17**: 1203–1212.

Efron, B. (1977). "Efficiency of Cox's likelihood function for censored data," *Journal of the American Statistical Association* **72**: 557–565.

Efron, B. and Tibshirani, R. J. (1993). *An Introduction to the Bootstrap,* Chapman and Hall, New York.

Elandt-Johnson, R. C. and Johnson, N. L. (1980). *Survival Models and Data Analysis,* Wiley, New York.

Elsayed, E. A. (1996). *Reliability Engineering,* Addison-Wesley Longman, Reading, MS.

Engel, P. (1993). "Mechanical failure mechanism models for mechanical wear,"*IEEE Transactions on Reliability* **42:** 262–267.

Engelhardt, M. (1975). "On simple estimation of the parameters of the Weibull or extreme value distribution," *Technometrics* **17:** 369–374.

Engelhardt, M. and Bain, L. J. (1974). "Some results on point estimation for the two-parameter Weibull or extreme value distribution," *Technometrics* **16:** 49–56.

Engelhardt, M. and Bain, L. J. (1977). "Simplified statistical procedures for the Weibull or extreme value distribution," *Technometrics* **19:** 323–331.

English, J. R., Sargent, T., and Landers, T. L. (1996). "A discretizing approach for stress/strength analysis," *IEEE Transactions on Reliability* **45:** 84–89.

Erkanli, A., Mazzuchi, T. A., and Soyer, R. (1998). "Bayesian computations for a class of reliability growth models," *Technometrics* **40:** 14–23.

Esary, J. D., Marshall, A. W. and Proschan, F. (1973). "Shock models and wear processes," *The Annals of Probability* **1:** 627–649.

Escobar, L. A. and Meeker, W. G. (1999). "Statistical prediction based on censored data," *Technometrics* **41:** 113–124.

Evans, J. R. and Lindsay, W. M. (1996), *The Management and Control of Quality,* West Publishing.Company, St. Paul, Minnesota.

Evans, M. W. and Marciniak, J. J. (1987). *Software Quality Assurance and Management,* Wiley, New York.

Evans, R. E. (1997). "Practical reliability engineering and management," *Tutorial Notes, 1997 Annual Reliability and Maintainability Symposium.*

Farewell, V. T. and Prentice, R. L. (1977). "A study of distributional shape in life testing," *Technometrics* **9:** 69–75.

Farr, W. (1995). "Software reliability modeling survey," pp. 71–117 in Lyu, M. R. (Ed.), *Handbook of Software Reliability Engineering,* McGraw-Hill, New York.

Fei, H., Kong, F. and Tang, Y. (1995). "Estimation for two-parameter Weibull distribution and extreme-value distribution under multiply Type-II censoring," *Communications in Statistics, Theory and Methods* **24:** 2087–2104.

Fisher, R. A. (1921). "On the mathematical foundations of theoretical statistics," *Philosophical Transactions Royal Society of London* **222:** 309–368.

Fisher, R. A. and Yates, F. (1957). *Statistical Tables for Biological, Agricultural and Medical Research, Fifth Edition,* Hafner Publishing, New York.

Fisk, G. and Chandran, R. (1975). "How to trace and recall products," *Harvard Business Review* **53:** Number 6, 90–96.

Fitzgerald, R.W. (1982). *Mechanics of Materials,* 2nd Ed., Addison-Wesley, Reading, MA.

Flamm, J. and Luisi, T. (Eds.) (1992). *Reliability Data Collection and Analysis,* Kluwer Academic Publishers, London.

Fletcher, R. (1980). *Practical Methods of Optimization,* Wiley, Chichester, London.

Folks, J. L. and Chhikara, R. S. (1978). "The inverse Gaussian distribution and its statistical applications—A review," *Journal of Royal Statistical Society* **B40:** 263– 289.

Fox, E. P. and Walls, D. P. (1997). "A probabilistic assessment of fiber failure during creep in cotinuous fiber reinforced composites," presented at the 1997 Annual Meeting of the American Statistical Association.

Frechet, M. (1951). "Sur la loi de probabilite de leecart dont les marges sont donnes," *Annals of University of Lyon, Sect. A,* **14:** 53–77.

Freund, J. E. (1961). "A bivariate extension of the exponential distribution," *Journal of the American Statistical Association* **56:** 971–977.

Friedman, A. (1975). *Stochastic Differential Equations and Applications,* Volume 1, Academic Press, New York.

Friedman, A. (1976). *Stochastic Differential Equations and Applications,* Volume 2, Academic Press, New York.

Friedman, M. A. and Voas, J. M. (1995). *Software Assessment: Reliability, Safety, Testability,* Wiley, New York.

Fries, A. and Sen, A. (1996). "A survey of discrete reliability-growth models," *IEEE Transactions on Reliability,* **45**: 582–604.

Fussel, J. B. (1976). "Fault tree analysis—Concepts and techniques" in *Generic Techniques in Reliability Assessment,* Henley, E. J. and J. Lynn (Eds.), Nordoff, Leyden, Holland.

Fussel, J. B. (1994). "Nuclear power system reliability: A historical perspective," *IEEE Transactions on Reliability* **33**: 41–47.

Gan, F. F. and Koehler, K. J. (1990). "Goodness-of-fit tests based on P-P probability plots," *Technometrics* **32**: 289–303.

Gan, F. F., Koehler, K. J., and Thompson, J. C. (1991). "Probability plots and distribution curves for assessing the fit of probability models," *The American Statistician* **45**: 14–21.

Gandara, A. and Rich, M. D. (1977). *Reliability Improvement Warranties for Military Procurement,* Report No. R-2264–AF, RAND Corp., Santa Monica CA.

Garvin, D. A. (1988). *Managing Quality,* The Free Press, New York.

Gen, M. and Cheng, R. (1996a). "Optimal design of system reliability using interval programming and genetic algorithms," *Computers and Industrial Engineering* **31**: 237–240.

Gen, M. and Cheng, R. (1996b). *Genetic Algorithms and Engineering Design,* Wiley, New York.

Genadis, T. C. (1988), "Designing a software reliability programme," *Quality and Reliability Engineering International* **4**: 311–316.

Geraerds, W. M. J. (1992). "The EUT maintenance model," *International Journal Production Economics* **24**: 209–216.

Gertsbakh, I. B. (1977). *Models of Preventive Maintenance,* North Holland, Amsterdam..

Giglmayr, J. (1987). "An age-wear dependent model of failure," *IEEE Transactions on Reliability* **36**: 581–585.

Gits, C. W. (1986). "On the maintenance concept for a technical system: II. Literature review," *Maintenance Management International* **6**: 181–196.

Glaser, R. E. (1980). "Bathtub and related failure rate characterizations," *Journal the American Statistical Association* **75**: 667–672.

Glickman, J. S. and Berger, P. D. (1976). "Optimal price and protection period for a product under warranty," *Management Science* **22**: 1381–1396.

Gnedenko, B. V., Belayaev, Y. K., and Solovyeb, A. D. (1969). *Mathematical Methods of Reliability Theory,* Translation edited by R. E. Barlow, Academic Press, New York.

Goel, A. L. and Okumoto, K. (1979). "Time-dependent error-detection rate. Model for software reliability and other performance measures," *IEEE Transactions on Reliability* **28**: 206–211.

Gompertz, B. (1825). "On the science connected with human mortality," *Transactions of the Royal Society,* June, 1825.

Gooden, R. (1996). "Leadership for product liability prevention," *Proceedings ASQC 50th Annual Quality Congress,* pp. 215–221

Govindarajulu, Z. (1964). "A supplement to Mendenhall's bibliography on life testing and related topics," *Journal of the American Statistical Association* **59**: 1231–1291.

Green, A. E. (1983). *Safety Systems Reliability,* Wiley, New York.

Green, L. L. (1991). *Logistic Engineering,* Wiley, New York.

Green, P. J. (1984). "Iteratively reweighted least squares for maximum likelihood estimation, and some robust and resistant alternatives," *Journal of Royal Statistical Society B* **46:** 149–192.

Gross, A. J. and Rust, P. F. (1987). "Item reliability based on a paucity of item failures," *Communications in Statistics, Theory and Methods* **16:** 2981–2990.

Guilbaud, O. (1988). "Exact Kolmogorov-type tests for left-truncated and/or right-censored data," *Journal of the American Statistical Association* **83:** 213–221.

Guin, L. (1984). *Cumulative Warranties: Conceptualization and Analysis,* Doctoral Dissertation, University of Southern California, Los Angeles, CA.

Gupta, R. C. and Subramanian, S. (1998). "Estimation of reliability in a bivariate normal distribution with equal coefficients of variation," *Communications in Statistics, Simulation and Computation* **27:** 675–698.

Hadley, G. (1964). *Nonlinear and Dynamic Programming,* Addison-Wesley, Reading, MA.

Halstead, M. H. (1997). *Elements of Software Science,* Elsevier-North Holland, New York.

Hamada, M. (1995). "Using statistically designed experiments to improve reliability and to achieve robust reliability," *IEEE Transactions on Reliability* **44:** 206–215.

Hameed, M. S. and Proschan, F. (1973). "Non-Stationary shock models," *Stochastic Processes and Their Applications* **1:** 383–404.

Hanagal, D. D. (1996). "Estimation of system reliability from stress-strength relationship," *Communications in Statistics, Theory and Methods* **25:** 1783–1797.

Haneveld, W. K. K. and Teunter, R. H. (1997). "Optimal provisioning strategies for slow moving spare parts with small lead times," *Journal of the Operational Research Society* **48:** 184–194.

Harr, M. E. (1987), *Reliability-Based Design in Civil Engineering,* McGraw Hill, New York.

Harter, H. L. and Moore, A. H. (1976). "An evaluation of exponential and Weibull test plans," *IEEE Transactions on Reliability* **25:** 100–104.

Haugen, E. B. (1980). *Probabilistic Mechanical Design,* Wiley, New York.

Haviland, R. P. (1964). *Engineering Reliability and Long Life Design,* Van Nostrand, Princeton, NJ.

Healy, J. D., Jain, A. K., and Bennett, J. M. (1997). "Reliability prediction," Tutorial Notes, *1997 Annual Reliability and Maintainability Symposium.*

Heathcote, C. R. (1971). *Probability: Elements and Mathematical Theory,* George Allen and Unwin, London.

Henley, E. J. and Kumamoto, H. (1981). *Reliability Engineering and Risk Assessment,* Prentice Hall, Upper Saddle River, NJ.

Henley, E. J. and Kumamoto, H. (1981). *Probabilistic Risk Assessment,* IEEE Press, New York.

Hicks, C. R. (1982). *Fundamental Concepts in the Design of Experiments, Third Edition,* Holt, Rinehart and Winston, New York.

Hirose, H. and Lai, T. L. (1997). "Inference from grouped data in three-parameter Weibull models with applications to breakdown-voltage experiments," *Technometrics* **39:** 199–210.

Hockley, C. J. and Comer, G. E. (1993). "NATO R&M standards; Blueprint for success," *Proceedings of the Annual Reliability and Maintainability Symposium,* 5–9.

Hollander, N. and Peña, E. A. (1992). "A Chi-Squared goodness-of-fit test for randomly censored data," *Journal of the American Statistical Association* **87:** 458–463.

Hossain, S. A. and Dahiya, R. C. (1993). "Estimating the parameters of a non-homogeneous Poisson process model for software reliability," *IEEE Transactions Reliability* **42:** 604–612.

Hough, J. A. (1997). "Reliability program planning in a commercial environment," *Proceedings of the Annual Reliability and Maintainability Symposium,*Tutorial Notes (23 pages).

Hoyland, A. and Rausand, M. (1994). *System Reliability Theory,* Wiley, New York.

Hu, X. J., Lawless., J. F., and Suzuki, K. (1998). "Nonparametric estimation of a lifetime distribution when censoring times are missing," *Technometrics* **40:** 3–13.

Huang, J., Miller, C. R., and Okogbaa, O. G. (1995). "Optimal preventive replacement intervals for the Weibull life distribution: Solution and applications," *Proceedings Annual Reliability and Maintainability Symposium,* 370–377.

Huet, S. and Kaddour, A. (1994). "Maximum likelihood estimation in survival analysis with grouped data on censored individuals and continuous data on failures," *Applied Statistics* **43:** 325–333.

Hulting, F. L. and Robinson, J. A. (1994). "The reliability of a series system of repairable subsystems: A Bayesian approach," *Naval Research Logistics* **41:** 483–506.

Hussain, A. Z. M. O. (1997). *Warranty and Product Reliability,* Unpublished doctoral thesis, The University of Queensland, Brisbane, Australia.

Hutchinson, T. P. and Lai, C. D. (1990). *Continuous Bivariate Distributions, Emphasizing Applications,* Rumsby Scientific Publishing, Adelaide.

Hwang, C. L., Tillman, F. A., Wei, W. K., and Lie, C. H. (1979). "Optimal scheduled-maintenance policy based on multiple-criteria decision-making," *IEEE Transactions on Reliability* **28:** 395–399.

IEC 50 (191): *International Electrotechnical Vocabulary* (IEV), Chapter 191—Dependability and Quality of Service, International Electrotechnical Commission, Geneva,1990.

IEC Standards, International Electrotechnical Commission, Geneva.

IEEE Std. 352 (1982). *IEEE Guide for General Principles of Reliability Analysis of Nuclear Power Generating Station Protection Systems,* IEEE Inc., New York.

IEEE Standards, The Institute of Electrical and Electronics Engineers, New York.

Ireson, W. G. and Coombs, C. F., Jr. (1988). *Handbook of Reliability Engineering and Management,* McGraw Hill, New York.

Ireson, W. G, Coombs, C. F., and Moss R. Y. (Eds.) (1996). *Handbook of Reliability Engineering and Management,* McGraw-Hill, New York.

ISO 9000 International Standard, *Quality Management and Quality Assurance Standards,* International Standards Organization.

Iyer, R. K. and Lee, I. (1996). "Measurement-Based analysis of software reliability," pp. 303–358 in Lyu, M. R. (Ed.), *Handbook of Software Reliability Engineering,* McGraw-Hill, New York .

Jardine, A. K. S. and Buzacott, J .A. (1985). "Equipment reliability and maintenance," *European Journal of Operational Research* **19:** 285–296.

Jaynes, E. T. (1968). "Prior probabilities," *IEEE Transactions on Systems Science and Cybernetics* **SSC-4:** 227–241.

Jelinski, Z. and Moranda, P. B. (1972). "Software reliability research," *Proceedings of the Statistical Methods for the Evaluation of Computer System Performance,* Academic Press, New York .

Jensen, F. and Peterson, N. E. (1982). *Burn-In,* Wiley, New York.

Jewell, W. S. (1984). "A general framework for learning curve reliability growth models," *Operations Research* **32:** 547–558.

Jiang, R. (1996). *Failure Models Involving Two Weibull Distributions,* Ph.D. Thesis, The University of Queensland, Brisbane, Australia.

Jiang, R. and Murthy, D. N. P. (1995a). "Graphical representation of two mixed Weibull distributions," *IEEE Transactions Reliability* **44:** 477–488.

Jiang, R. and Murthy, D. N. P. (1995b). "Reliability modeling involving two Weibull distributions," *Reliability Engineering & System Safety* **47:** 187–198.

Jiang, R. and Murthy, D. N. P. (1997a). "Two sectional models involving three Weibull distributions," *Quality and Reliability Engineering* **13:** 83–96.

Jiang, R. and Murthy, D. N. P. (1997b). "Mixture of Weibull distributions—Parametric characterization of failure rate function;" *Applied Stochastic Models and Data Analysis,* **14:** 47–65.

Jiang, R. and Murthy, D. N. P. (1997c). "Parametric study of Competing Risk model involving two Weibull distributions," *International Journal of Reliability, Quality and Safety Engineering* **4:** 17–34.

Jiang, R. and Murthy, D. N. P. (1997d). "Parametric study of multiplicative model involving two Weibull distributions," *Reliability Engineering and Systems Safety,* **55:** 217–226.

Jiang, R. and Murthy, D. N. P. (1997e). "The exponentiated Weibull family: A graphical approach," *IEEE Transactions on Reliability,* **48:** 68–72.

Joe, H. and Reid, N. (1985). "Estimating the number of faults in a system," *Journal of the American Statistical Association* **80:** 222–226.

Johns, M. V., Jr. and Lieberman, G. J. (1966). "An exact asymptotically efficient confidence bound for reliability in the case of the Weibull distribution," *Technometrics* **8:** 135–175.

Johnson, N. L. and Kotz, S. (1969). Discrete Distributions, Houghton Mifflin, Boston.

Johnson, N. L. and Kotz, S. (1969). *Distributions in Statistics: Discrete Distributions,* Wiley, New York.

Johnson, N. L. and Kotz, S. (1970a). *Distributions in Statistics: Continuous Univariate Distributions—I,* Wiley, New York.

Johnson, N. L. and Kotz, S. (1970b). *Distributions in Statistics: Continuous Univariate Distributions—II,* Wiley, New York.

Johnson, N. L. and Kotz, S. (1972). *Distributions in Statistics: Continuous Multivairate Distributions,* Wiley, New York.

Jones, W. D. and Vouk, M. A. (1996). "Field data analysis," in Lyu, Michael R. (Ed.), *Handbook of Software Reliability Engineering,* pp. 439–489, McGraw-Hill, New York .

Kalbfleisch, J. D. and Lawless, J. F. (1988). "Estimation of reliability in field-performance studies" (with discussion), *Technometrics* **30:** 365–388.

Kalbfleisch, J. D. and Prentice, R. L. (1980). *The Statistical Analysis of Failure Time Data,* Wiley, New York.

Kamien, M. I. and Schwartz, N. L. (1991). *Dynamic Optimization: The Calculus of Variations and Optimal Control in Economics and Management,* North Holland, Amsterdam.

Kane, V. E. (1989). *Defect Prevention* Dekker, New York.

Kaner, C. (1993). *Testing Computer Software,* Van Nostrand Reinold, New York .

Kapur, K. C. and Lamberson, L. R. (1977). *Reliability in Engineering Design,* Wiley, New York.

Karmarkar, U. S. and Kubat, P. (1983). "Value of loaners in product support," *IIE Transactions* **15:** 5–11.

Kececioglu, D. (1991). *Reliability Engineering Handbook—Vol. 1,* Prentice Hall, Englewoods Ciffs, NJ.

Kececioglu, D. (1994). *Reliability Engineering Handbook,* Vol. 2, Prentice-Hall, Englewood Cliffs, NJ.

Keeney, R. L. and Raiffa, H. (1993). *Decisions with Multiple Objectives: Preference and Value Trade-Offs,* Cambridge University Press, Cambridge.

Kehoe, R. and Jarvis, A. (1996). *ISO 9000-3: A Tool for Software Product and Process Improvement,* Springer-Verlag, New York .

Kelly, A. (1984). *Maintenance Planning and Control,* Butterworths, London.

Kelly, S. (1998). "Loudspeakers," Chapter 15 in *Audio and Hi-Fi Handbook,* Ian R. S. (Editor), Newnes, Oxford.

Kindree, J. D., Melton, R. R., Raiffa, H. E., and Reynolds, R. J. (1994). "An overview of an R&M guideline for manufacturing machinery and equipment," *Proceedings of the Annual Reliability and Maintainability Symposium,* 65–68.

King, James R. (1971). *Probability Charts for Decision Making,* Industrial Press, New York.

Klefsjö, B. and Kumar, U. (1992). "Goodness-of-fit tests for the power-law process based on the TTT-plot," *IEEE Transactions on Reliability* **41:** 593–597.

Kline, M. B. (1984). "Suitability of the lognormal distribution for corrective maintenance repair times," *Reliability Engineering* **9:** 65–80.

Knight, C. R. (1991). "Four decades of reliability progress," *Proceedings Annual Reliability and Maintainability Symposium,* 156–159.

Knowles, I., Malhotra, A., Stadterman, T. J., and Munmarty, R. (1995). "Framework for dual-use reliability program standard," *Proceedings of the Annual Reliability and Maintainability Symposium* 102–105.

Kobbacy, K. A. H., Fawzi, B. B., Percy, D. F. and Ascher, H. E. (1997). "Full history proportional hazards model for preventive maintenance scheduling," *Quality and Reliability Engineering International* **3:** 187–198.

Kohoutek, H. J. (1982). "Establishing reliability goals for new technology products," *Proceedings Annual Reliability and Maintainability Symposium,* 460–465.

Kohoutek, H. J. (1996). "Reliability Specification and Goal Setting," in *Handbook of Reliability Engineering and Management,* Ireson, W. G, Coombs, C. F. and Moss R. Y. (Eds.), McGraw-Hill, New York.

Kordonsky, Kh. B. and Gertsbakh, I. B. (1995). "System state monitoring and lifetime scales" *Reliability Engineering and System Safety* **47:** 1–14.

Kotler, P. and Armstrong, G. (1966). *Principles of Marketing, Prentice Hall,* Englewood Cliffs, NJ.

Kumar, A. and Agarwal, M. (1980), "A review of standby redundant systems," *IEEE Transactions on Reliability,* **29:** 290–294.

Kumar, D. and Klefsjö, B. (1994). "Proportional hazards model: A review," *Reliability Engineering and System Safety* **44:** 177–188.

Kumar, U.and Klefsjö, B. (1992). "Reliability analysis of hydraulic systems of LHD machines using the power law process model," *Reliability Engineering and System Safety* **35:** 217–224.

Kunchur, S. H. and Munoli, S. B. (1993). "Estimation of reliability for a multicomponent survival stress-strength model based on exponential distributions," *Communications in Statistics, Theory and Methods* **22:** 769–779.

Kuo, W. (1985). "Bayesian availability using gamma distributed priors," *IEE Transactions* **17:** 132–140.

Lakey, M. J. (1991). "Statistical analysis of field data for aircraft warranties," *Proceedings Annual Reliability and Maintainability Symposium,* 340–344.

Lall, P. (1996). "Temperature as an input to micro-electronics-reliability models," *IEEE Transactions on Reliability* **45:** 3–9.

Lalli, V. R. and Packard, M. (1997), System Maintainability, *Tutorial Notes, 1997 Annual Reliability and Maintainability Symposium,* Philadelphia.

Lalli, V. R., Packard, M., and Ziemianski, T. (1998). "Software design improvements," *Tutorial Notes, 1998 Annual Reliability and Maintainability Symposium,* Anaheim, CA.

Lambert, H. E. (1975). "Measures of importance of events and cut sets in fault trees," in *Re-*

liability and Fault Tree Analysis (Eds. R. E. Barlow, J. B. Fussel, and N. D. Singpurwalla), SIAM, Philadelphia.

Lawless, J. F. (1982). *Statistical Models and Methods for Lifetime Data,* Wiley, New York.

Lee, C. H., Hwang, C. L., and Tillman, F. A. (1977). "Availability of maintained systems: A state-of-the-art survey,"*AIIE Transactions* **9:** 247–259.

Lee, L. (1980). "Testing adequacy of the Weibull and log linear rate models for a Poisson process," *Technometrics* **22:** 195–199.

Leemis, L. M. and Benke, M. (1990). "Burn-in models and methods: A review," *IIE Transactions,* **22:** 172–180.

Lemoine, A. J. and Wencour, M. L. (1985). "On failure modeling," *Naval Research Logistic Quarterly* **22:** 497–508.

Leonard, C. T. and Pecht, M. (1990), "How failure prediction methodology affects electronic equipment design," *Quality and Reliability Engineering International* **6:** 243–249.

Leveson, N. G. (1995). *Safeware: System Safety and Computers,* Addison-Wesley, New York .

Li, M. and Dasgupta, A. (1993). "Failure models for creep," *IEEE Transactions on Reliability* **42:** 339–353.

Li, J. and Dasgupta, A. (1994). "Failure mechanism models for material aging due to interdiffusion," *IEEE Transactions on Reliability* **43:** 2–10.

Lie, C. H. and Chun, Y. H. (1987). "Optimum single-sample inspection plans for products sold under free and rebate warranty," *IEEE Transactions on Reliability* **R-36:** 634–637.

Lieblein, J. and Zelen, M. (1956). "Statistical investigation of the fatigue life of deep-groove ball bearings," *Journal of Research National Bureau of Standards* **57:** 273–316.

Lilliefors, H. W. (1967). "On the Kolmogorov-Smirnov test for normality with mean and variance unknown," *Journal of the American Statistical Association* **62:** 399–402.

Lilliefors, H. W. (1969). Correction to Lilliefors (1967), *Journal of the American Statistical Association* **64:** 1702.

Lin, H.-H. and Chen, H.-H. (1993). "Nonhomogeneous Poisson process software-debugging models with linear dependence," *IEEE Transactions on Reliability* **42:** 613–617.

Lindley, D. V. (1950). "Grouping corrections and maximum likelihood," *Proceedings Cambridge Philosophical Society* **46:** 106–110.

Lindley, D. V. and Singpurwalla, N. D. (1986). "Multivariate distributions for the lifelengths of components of a system sharing a common environment," *Journal of Applied Probability* **23:** 418–431.

Littlewood, B. and Verrall, J. L. (1973). "A Bayesian reliability growth model for computer software," *Applied Statistics* **22:** 322–346.

Littlewood, B. (1980). "A Bayesian differential debugging model for software reliability," *Proceedings IEEE COMPSAC.*

Lloyd, D. K. (1986). "Forecasting reliability growth," *Quality and Reliability Engineering International* **2:** 19–23.

Lloyd, D. K. and Lipow, M. (1962), *Reliability: Management, Methods and Mathematics,* Prentice Hall, Englewood Cliffs, NJ.

Lockhart, R. A. and Stephens, M. A. (1994). "Estimation and tests of fit for the three-parameter Weibull distribution," *Journal of Royal Statistical Society B* **56:** 491–500.

Loll, V. (1998). "Dependability standards: An international cooperation," *Proceedings Annual Reliability and Maintainability Symposium,* 27–28.

Lorenzen, T. J. and Anderson, V. L. (1993). *Design of Experiments, A No-Name Approach,* Dekker, New York.

Lu, J. C. (1989). "Weibull extensions of the Freund and Marshall-Olkin bivariate exponential models," *IEEE Transactions on Reliability* **38**: 615–619.

Luxhoj, J. T. and Shyur, H.-J. (1995). "Reliability curve fitting for aging helicopter components," *Reliability Engineering and System Safety* **38**: 229–234.

Lyons, K. F. (1998). *Warranty Management System,* Master of Engineering Thesis, The University of Queensland, Brisbane Australia.

Lyons, K. F. and Murthy, D. N. P. (1996). "Warranty data analysis: A case study," pp. 396–405 in *Stochastic Models in Engineering, Technology and Management,* R. J. Wilson, D. N. P. Murthy, and S. Osaki (Eds.), The University of Queensland Printery, Brisbane, Australia.

Lyu, M. R. (1996). *Handbook of Software Quality Engineering,* McGraw-Hill, New York .

Mahon, B. H. and Bailey, R. J. M. (1975). "A proposed improved replacement policy for army vehicles," *Operational Research Quarterly* **26**: 477–494.

Majeske, K. D. and Herrin, G. D. (1995). "Assessing mixture-model goodness-of-fit with an application to automobile warranty data," *Proceedings Annual Reliability and Maintainability Symposium,* 378–383.

Mann, L. (1983). *Maintenance Management,* Lexington Press, Lexington, MA.

Mann, N. R. and Fertig, K. W. (1975). "Simplified efficient point and interval estimators for Weibull parameters," *Technometrics* **17**: 361–368.

Mann, N. R. and Grubbs, F. E. (1974). "Approximately optimum confidence bounds for system reliability based on component test data," *Technometrics* **16**: 335–347.

Mann, N. R., Schafer, R. E., and Singpurwalla, N. D. (1974). *Methods for Statistical Analysis of Reliability and Life Test Data,* Wiley, New York.

Mann, N. R., Scheuer, E. M., and Fertig, K. W. (1973). "A new goodness-of-fit test for the two-parameter Weibull or extreme-value distribution with unknown parameters," *Communications in Statistics* **2**: 383–400.

Marshall, A. W. and Olkin, I. (1972). "A generalized bivariate exponential model," *Journal of the American Statistical Association* **62**: 30–44.

Martin, P., Strutt, J. E. and Kinkead, N. (1983). "A review of mechanical reliability modelling in relation to failure mechanisms," *Reliability Engineering* **6**: 13–42.

Martz, H. F. and Duran, B. S. (1985). "A comparison of three methods for calculating lower confidence limits on system reliability using binomial component data," *IEEE Transactions on Reliability* **R-34**: 113–120.

Martz, H. F. and Waller, R. A. (1982). *Bayesian Reliability Analysis,* Wiley, New York.

Martz, H. F. and Waller, R. A. (1990). "Bayesian reliability analysis of complex series/parallel systems of binomial subsystems and components," *Technometrics* **32**: 407–416.

Martz, H. F., Waller, R. A., and Fickas, E. T. (1988). "Bayesian reliability analysis of series systems of binomial subsystems and components," *Technometrics* **30**: 143–159.

Mason, A. L. and Bell, C. B. (1986). "New Lilliefors and Srinivasan tables with applications," *Communications in Statistics, Simulation and Computation* **15**: 451–477.

Massey, Frank J. (1951). "The Kolmogorov-Smirnov test for goodness of fit," *Journal of the American Statistical Association* **6**: 68–78.

Mastran, D. D. and Singpurwalla, N. D. (1978). "A Bayesian estimation of the reliability of coherent structures," *Operations Research* **26**: 663–672.

Mazzuchi, T. A. and Soyer, R. (1988). "A Bayes empirical-Bayes model for software reliability," *IEEE Transactions on Reliability* **37**: 248–254.

Mazzuchi, T. A. and Soyer, R. (1996). "A Bayesian perspective on some replacement strategies," *Reliability Engineering and Systems Safety* **51**: 295–303.

McCall, J. J. (1965). "Maintenance policies for stochastically failing equipment: A survey," *Management Science* **11**: 493–524.

McClave, J. T. and Benson, P. G. (1994). *Statistics for Business and Economics,* Macmillan College Publishing, New York.

McConnell, S. (1997). "Gauging software readiness with defect tracking," *IEEE Software.*

McDonald, J. B. and Richards, D. O. (1987a). "Hazard rates and generalized Beta distributions," *IEEE Transactions on Reliability* **36:** 463–466.

McDonald, J. B. and Richards, D. O. (1987b). "Model selection: Some generalized distributions," *Communications in Statistics—Theory and Methods.* **16:** 1049–1074.

McGlone, M. E. (1984). *Reliability—Growth, Assessment, Prediction, and Control for Electronic Engine Control* (GAPCEED), *Rept., No. AFWAL-TR-84-2024,* Aero Propulsion Laboratory, Wright-Patterson AFB, Ohio.

Meeker, W. Q. and Escobar, L. A. (1998). *Statistical Methods for Reliability Data,* Wiley, New York.

Meeker, W. Q. and Hamada, M. (1995). "Statistical tools for the rapid development & evaluation of high-reliability products," *IEEE Transactions on Reliability* **44:** 187–198.

Meth, M. (1992). "Reliability growth myths and methodologies: A critical view," *Proceedings Annual Reliability and Maintainability Symposium,* 337–342.

Mi, J. (1998). "Bolstering components for maximizing system lifetime," *Naval Research Logistics* **45:** 497–509.

MIL-STD 882. *System Safety Program Requirement,* U.S. Department of Defense, Washington, DC, 1984.

Min, H. (1989). "A bicriterion reverse distribution model for product recall," *Omega,* **17:** 483–490.

Misra, K. (1992). *Reliability Analysis and Prediction,* Elsevier, Amsterdam.

MMR (1989). *National Survey of Early New Car Buyers,* Vol. 4, Maritz Marketing Research, OH.

Mok, Y. L. and Xie, M. (1996). "Optimizing environmental stress screening using mathematical programming," *Journal of the Institute of Environmental Sciences* **39:** 37–43.

Möller, K. (1991). "Increasing software quality by objectives and residual fault prognosis," Chapter 10 in Ince, Darrell (Ed.), *Software Quality and Reliability,* Chapman and Hall, London.

Monahan, G. E. (1982). "A survey of partially observable Markov decision processes: Theory, models and algorithms," *Management Science* **28:** 1–16.

Montgomery, D. C. (1985). *Introduction to Statistical Quality Control,* Wiley, New York.

Montgomery, D. C. (1997). *Design and Analysis of Experiments,* Fourth Ed., Wiley, New York.

Mood, A. M., Graybill, F. A., and Boes, D. C. (1974). *Introduction to the Theory of Statistics,* Third Edition, McGraw-Hill, New York.

Moore, A. H., Hobbs, J. R., and Hasaballa, M. S. B. (1985). "A Monte Carlo method for determining confidence bounds on reliability and availability of maintained systems," *IEEE Transactions on Reliability* **R-34:** 497–498.

Moore, N., Ebbeler, D., and Creager, M. (1990). "A methodology for probabilistic prediction of structural failures of launch vehicle propulsion systems," 31st Annual Structures, Structural Dynamics and Materials Conference, American Institute of Aeronautics and Astronautics.

Moranda, P. B. (1975). "Predictions of software reliability during debugging," pp. 327–332 in *Proceedings of the Annual Reliability and Maintainability Symposium.*

Moss, L. L. C., Taylor, M. S., and Tingey, H. B. (1990). "Quantiles of the Anderson-Darling statistic," *Communications in Statistics, Simulation and Computation* **19:** 1007–1014.

Moss, R. Y. (1996). "Design for Reliability," in *Handbook of Reliability Engineering and*

Management, Ireson, W. G, Coombs, C. F. and Moss R. Y. (Eds.), McGraw-Hill, New York.

Moss, T. R., Sandtrov, H., and Pratella, C. (1989). "Phase II data collection experience," *Proceedings of the Sixth EuREData Conference,* 51–67.

Moss, T. R. and Strutt, J. E. (1993). "Data sources for reliability design and analysis," *Journal of Process Mechanical Engineering* **207**: 13–19.

Moubray, J. (1991). *Reliability Centred Maintenance,* Butterworth-Hinemann, Oxford, UK.

Mudholkar, G. S. and Shrivastava, D. K. (1993), "Exponentiated Weibull family for analyzing bathtub failure-rate data;" *IEEE Transactions on Reliability* **42**: 299–302.

Mudholkar, G. S., Srivastava, D. K., and Freimer, M. (1995). "The exponentiated Weibull family: A reanalysis of the bus-motor-failure data," *Technometrics* **37**: 436–445.

Munoz, A., Martorell, S., and Serradell, V. (1997). "Genetic algorithms in optimizing surveillance and maintenance of components," *Reliability Engineering and System Safety* **57**: 107–120.

Munson, J. C. (1996). "Software faults, software failures and software reliability modeling," *Information and Software Technology* **38**: 687–699.

Murthy, D. N. P. (1983). "Analysis and design of unreliable multi-component system with modular structure," *Large Scale Systems* **5**: 245–254.

Murthy, D. N. P. (1984). "Optimal maintenance and sale date of a machine," *International J. Systems Science* **15**: 277–292.

Murthy, D. N. P. (1990). "Optimal reliability choice in product design," *Engineering Optimization* **15**: 281–294.

Murthy, D. N. P. (1991a). "A note on minimal repair," *IEEE Transactions on Reliability* **40**: 245–246.

Murthy, D. N. P. (1991b). "A new warranty costing model," *Mathematical and Computer Modelling* **13**: 56–69.

Murthy, D. N. P. (1992). "A usage dependent model for warranty costing," *European J. Operations Research* **62**: 89–99.

Murthy, D. N. P. and Asgharizadeh, E. (1998). "A stochastic model for service contract," *International Journal of Reliability, Quality and Safety Engineering,* **5**: 29–45.

Murthy, D. N. P. and Asgharizadeh, E. (1999). "Optimal decision making in maintenance service operation," *European Journal of Operational Research,* **116**: 259–273.

Murthy, D. N. P. and Blischke, W. R. (1999). "Strategic warranty management—A life cycle approach," accepted for publication in *IEEE Transactions on Engineering Management.*

Murthy, D. N. P., Djamaludin, I., and Wilson, R. J. (1993). "Product warranty and quality control," *Quality and Reliability Engineering* **9**: 431–443.

Murthy, D. N. P. and Hussain, A. Z. M. O. (1994). "Warranty and optimal redundancy design," *Engineering Optimization* **23**: 301–314.

Murthy, D. N. P. and Hwang, M. C. (1996). "Optimal discrete and continuous maintenance policy for a complex unreliable machine" *International J. Systems Science* **27**: 483–495.

Murthy, D. N. P. and Jiang, R. (1997). "Parametric study of sectional models involving two Weibull distributions," *Reliability Engineering and Systems Safety* **56**: 151–160.

Murthy, D. N. P. and Nguyen, D. G. (1982). "A note on extended block replacement policy with used items," *Journal of. Applied Probability* **19**: 481–485.

Murthy, D. N. P. and Nguyen, D. G. (1984). "Study of multi-component system with failure interaction," *European Journal of Operational Research* **21**: 330–338.

Murthy, D. N. P. and Nguyen, D. G. (1985). "Study of two component system with failure interaction," *Naval Research Logistics Quarterly* **32**: 239–248.

Murthy, D. N. P. and Nguyen, D. G. (1987). "Optimal development testing policies for products sold with warranty," *Reliability Engineering* **19**: 113–123.

Murthy, D. N. P. and Nguyen, D. G. (1988). "An optimal repair cost limit policy for servicing warranty," *Mathematical and Computer Modelling* **11**: 595–599.

Murthy, D. N. P., Page, N. W., and Rodin, Y. (1990). *Mathematical Modelling,* Pergamon Press, Oxford, UK.

Murthy, D. N. P. and Wilson, R. J: (1994). "Parameter estimation in multi-component system with failure interaction;" *Stochastic Models and Data Analysis* **10**: 47–60.

Musa, J. D. and Okumoto, K. (1983). "Software reliability models: Concepts, classification, comparisons, and practice," pp. 395–424 in *Electronic Systems Effectiveness and Life Cycle Costing,* Skwirzynski, J. K. (Ed.), Springer-Verlag, Heidelberg.

Musa, J. D. and Okumoto, K. (1984). "A logarithmic Poisson execution time model for software reliability measurement, " pp. 230–237 in *Proceedings 7th International Conference on Software Engineering,* Orlando FL.

Musa, J. D., Iannino, A., and Okumoto, K. (1990). *Software Reliability,* McGraw-Hill, New York .

Myers, G. J. (1976).*Software Reliability: Principles and Practices,* Wiley, New York .

Myhre, J. (1978). "Determining confidence bounds for highly reliable coherent systems," *Proceedings Annual Reliability and Maintainability Symposium,* pp. 519–524.

Myhre, J. M. and Rennie, M. W. (1986). "Confidence bounds for reliability of coherent systems based on binomially distributed component data," pp. 265–280 in Basu, A. P. (Ed.), *Reliability and Quality Control,*Elsevier Science Publishers B. V. (North Holland), New York.

Myhre, J. M., Rosenfeld, A. M., and Saunders, S. C. (1978). "Determining confidence bounds for highly reliable coherent systems based on a paucity of component failures," *Naval Research Logistics Quarterly* **25**: 213–227.

Nakagawa, T. and Kowada, M. (1983). "Analysis of a system with minimal repair and its application to a replacement policy," *European Journal of Operational Research* **12**: 176–182.

Nakagawa, T. and Osaki, S. (1974). "Some aspects of damage models," *Micro-electronics and Reliability* **13**: 253–257.

Nakajima, S. (1988). *Introduction to TPM: Total Productive Maintenance,* Productivity Press, Cambridge, MA.

Natvig, B. (1979). "A suggestion of a new measure of importance of system components," *Stochastic Processes and Their Applications* **9**: 319–330.

Nayak, T. K. (1986). "Software reliability: Statistical modeling & estimation," *IEEE Transactions on Reliability* **35**: 566–570.

Nelson, W. (1969). "Hazard plotting for incomplete failure data," *Journal of Quality Technology* **1**: 27–52.

Nelson, W. (1982). *Applied Life Data Analysis,* Wiley, New York.

Nelson, W. (1990). *Accelerated Testing,* Wiley, New York.

Neter, J., Kutner, M. H., Nachtsheim, C. J., and Wasserman, W. (1996). *Applied Linear Statistical Models, Fourth Edition,* Richard D. Irwin, Homewood, IL.

Newlin, L., Sutharshana, S., Ebbeler, D., and Moore, N. (1990). "Probabilistic low cycle fatigue failure analysis with application to liquid propellant rocket engines," 31st Annual Structures, Structural Dynamics and Materials Conference, American Institute of Aeronautics and Astronautics.

Nguyen, D. G. (1984). *Studies in Warranty Policies and Product Reliability;* Unpublished Ph.D. Thesis, The University of Queensland, Australia.

Nguyen, D. G. and Murthy, D. N. P. (1981). "Optimal preventive maintenance policies for repairable systems," *Operations Research* **29**: 1181–1194.

Nguyen, D. G. and Murthy, D. N. P. (1982). "Optimal burn-in time to minimize cost for products sold under warranty," *IIE Transactions* **14**: 167–174.

Nguyen, D. G. and Murthy, D. N. P. (1986). "An optimal policy for servicing warranty," *Journal of Operations Research Society* **37**: 1081–1088.

Nguyen, D. G. and Murthy, D. N. P. (1988). "Optimal reliability allocation for products sold under warranty" *Engineering Optimization* **13**: 35–45.

Nguyen, D. G. and Murthy, D. N. P. (1989). "Optimal replace-repair strategy for servicing warranty sold under warranty," *European Journal of Operations Research* **39**: 206–212.

Niebel, B. (1985). *Engineering Maintenance Management,* Dekker, New York.

Nieuwhof, G. W. E. (1984). "The concept of failure in reliability engineering," *Reliability Engineering* **7**: 53–59.

Nikora, A. P and Lyu, M. R. (1995). "Software reliability measurement experience," pp. 255–301 in Lyu, M. R. (Ed.), *Handbook of Software Reliability Engineering,* McGraw-Hill, New York .

Nishida, S. I. (1992). *Failure Analysis in Engineering Applications,* Butterworth-Hinemann, Oxford, UK.

NORSOK Standards, Norwegian Technology Standards Institute, Oslo.

Nowlan, F. S. and Heap, H. (1978),. *Reliability-centered Maintenance,* National Technical Information Service, US Department of Commerce, Springfield, VA.

O'Leary, D. E. (1998). "Knowledge acquisition for multiple experts: An empirical study," *Management Science* **44**: 1049–1058.

O'Leary, D. J. (1996). "International standards: Their new role in the global economy," *Proceedings of the Annual Reliability Symposium.* 17–23.

Oakes, D. (1983). "Survival analysis,"*European Journal of Operational Research* **12**: 3–14.

Ohba, M. (1984). "Software reliability analysis models," *IBM Journal of Research and Development* **21**: 428–443.

Osaki, S. and Nakagawa. T. (1976). "Bibliography for reliability and availability of stochastic systems," *IEEE Transactions on Reliability,* **25**: 284–286.

Oskarsson, Ö. and Glass, R. L. (1996). *ISO 9000 Approach to Building Quality Software,* Prentice-Hall, Englewood Cliffs, NJ.

Owen, D. B. (1962). *Handbook of Statistical Tables,* Addison-Wesley, Reading, MA.

Ozekici, S. (1995), "Optimal maintenance policies in random environments," *European Journal of Operational Research* **82**: 283–294.

Padgett, W. J. and Wei, L. J. (1978). "Bayesian lower bounds on reliability for the lognormal distribution," *IEEE Transactions on Reliability* **R27**: 161–165.

Padmanabhan, V. (1996). "Extended Warranties," in *Product Warranty Handbook,* W. R. Blischke and D. N. P. Murthy (Eds.), Dekker, New York.

Painton, L. and Campbell, J. (1995). "Genetic algorithms in optimization of system reliability," *IEEE Transactions on Reliability* **44**: 172–178.

Papoulis, A. (1984). *Probability, Random Variables and Stochastic Processes,* 2nd Ed., McGraw-Hill, New York.

Patel, J. K., Kapadia, C. H. and Owen, D.B. (1976). *Handbook of Statistical Distributions,* Dekker, New York.

Pavur, R. J., Edgeman, R. L., and Scott, R. C. (1992). "Quadratic statistics for the goodness-of-fit test of the inverse Gaussian distribution," *IEEE Transactions on Reliability* **41**: 118–123.

Pecht, M. G. (1993). "Design for qualification," *Proceedings Annual Reliability and Maintainability Symposium,* 1–4.

Peck, D. S. (1986). "Comprehensive model in humidity testing correlation," *Proceedings of the 24th IEEE International Reliability Physics Symposium,* 44–50.

Percy, D. F. and Kobbacy, K. A. H. (1996). "Preventive maintenance modelling—A Bayesian perspective," *Journal of Quality and Maintenance Engineering* **2:** 15–24.

Pham, H. and Wang, H. (1996). "Imperfect maintenance." *European Journal of Operations Research* **94:** 425–438.

Pham-Gia, T. and Turkkan, N. (1995). "Comment on Weerahandi and Johnson (1992)," *Technometrics* **37:** 130–131.

Phillipson, L. L. (1996). "The failure of Bayes system reliability inference based on data with multi-level applicability," *IEEE Transactions on Reliability* **45:** 66–68.

Pierskalla, W. P. and Voelker, J. A. (1976). "A survey of maintenance models: The control and surveillance of deteriorating systems," *Naval Research Logistics Quarterly* **23:** 353–388.

Pintelton, L. M. and Gelders, L. (1992). "Maintenance management decision making," *European Journal of Operations Research* **58:** 301–317.

Ploe, R. J. and Skewis, W. H. (1990). *Handbook of Reliability Prediction Procedures for Mechanical Equipment,* David Taylor Research Center, Bethesda, MD.

Polovko, A. M. (1968), *Fundamentals of Reliability Theory,* Academic Press, New York.

Pontryagin, L. S., Bollyanski, V. G., Gamkrelidze, R. V., and Mischchenko, E. F. (1962). *The Mathematical Theory of Optimal Processes,* Wiley Interscience, New York.

Porter, J. E. III, Coleman, J. W., and Moore, A. H. (1992). "Modified KS, AD, and C-vM tests for the Pareto distribution with unknown location & scale parameters," *IEEE Transactions on Reliability* **41:** 112–117.

Press, J. S. (1989). *Bayesian Statistics: Principles, Models, and Applications,* Wiley, New York.

Priest, J. W. (1988). *Engineering Design for Producibility and Reliability,* Dekker, New York .

Proctor and Gamble (1992). *Report to the Total Quality Leadership Steering Committee and Working Councils,* Cincinnati, OH.

Proschan, F. (1963). "Theoretical explanation of observed decreasing failure rate," *Technometrics* **5:** 375–383.

Pullum, L. L. and Doyle, S. A. (1998). "Software Testing," *Tutorial Notes, 1998 Annual Reliability and Maintainability Symposium,* Anaheim, CA.

Raghavarao, D. (1971). *Constructions and Combinatorial Problems in Design of Experiments,* Dover Publications, New York.

Raheja, D. G. (1991). *Assurance Technologies: Principles and Practices,* McGraw-Hill, New York .

Raheja, D. G. (1994). "Reliability testing—the way it should be," *ASQC Technical Conference Transactions,* pp. 901–903.

Rahman, M. S. and King, M. L. (1999). "Improved model selection criterion," *Communications in Statistics, Simulation and Computation* **28:** 51–71.

Raiffa, H. and Schlaifer, R. (1961). *Applied Statistical Decision Theory,* Harvard University Press, Boston.

Rajashri, S. and Rajashri, M. B. (1988). "Bathtub distributions: A review," *Communications in Statistics—Theory and Methods* **17:** 2597–2621.

Rao, S. S. (1992), *Reliability-Based Design,* McGraw Hill, New York.

Rao, S. S. (1996). *Engineering Optimization Theory and Practice,* Wiley, New York.

Rausand, M. and Oien, K. (1996). "The basic concept of failure analysis," *Reliability Engineering and System Safety* **53**: 73–83.

RDH376 (1975). *Reliability Design Handbook,* No. RDH376, Reliability Analysis Center, IIT Research Institute, Chicago.

Rekab, K. (1993). "A sampling scheme for estimating the reliability of a series system," *IEEE Transactions on Reliability* **42**: 287–293.

Resnikoff, H. L. (1978). *Mathematical Aspects of Reliability-centered Maintenance,* Dolby Access Press, Los Altos, CA.

Rice, R. E. and Moore, A. H. (1983). "A Monte-Carlo technique for estimating lower confidence limits on system reliability using pass-fail data," *IEEE Transactions on Reliability* **R-32**: 366–369.

Richards, D. O. and McDonald, J. B. (1987). "A general methodology for determining distributional forms with applications in reliability," *Journal of Statistical Planing and Inference* **16**: 365–376.

Rink, D. R. and Swan, J. E. (1979). "Product life cycle research: A literature review," *Journal of Business Research* **78**: 219–242.

Ritchken, P. H., Chandramohan, J., and Tapiero, C. S. (1989). "Servicing, quality design and control," *IIE Transactions* **21**: 213–220.

Rohrbach, K. V. (1996), "Managing Reliability as a Process," in *Handbook of Reliability Engineering and Management,* Ireson, W. G, Coombs, C. F., and Moss R. Y. (Eds.), McGraw-Hill, New York.

Rooney, J. P. (1994). "Customer satisfaction," *Proceedings Annual Reliability and Maintainability Symposium,* 376–381.

Ross, P. J. (1988). *Taguch Techniques for Quality Engineering,* McGraw-Hill, New York.

Ross, S. M. (1970). *Applied Probability Models with Optimization Applications,* Holden-Day, San Francisco.

Ross, S. M. (1980). *Stochastic Processes,* Wiley, New York.

Ross, S. M. (1983). *Introduction to Probability Models,* Academic Press, New York.

Ryan, T. P. (1989). *Statistical Methods for Quality Improvement,* Wiley, New York.

SAE M-110 (1993). *Reliability and Maintainability Guideline for Manufacturing Machinery and Equipment,* Society of Automotive Engineers Inc.

Sagola, G. and Albin, J. L. (1984). "Reliability models and software development: A practical approach," in Girard, E. (Ed.), *Software Engineering: Practice and Experience,* Oxford Pub., Oxford, UK.

Sahin. I. and Potaloglu, H. (1998) *Quality, Warranty and Preventive Maintenance,* Kluwer, Boston.

Sandtorv, H. A., Hokstad, P., and Tompson, D. W. (1996). "Practical experiences with a data collection project," *Reliability Engineering and System Safety* **51**: 159–167.

Scarf, P. S. (1997). "On the application of mathematical models to maintenance," *European Journal of Operations Research* **63**: 493–506.

Schaeffer, R. L, Mendenhall, W. III, and Ott, R. L. (1996). *Elementary Survey Sampling,* 5th Edition, Duxbury Press, New York.

Schick, G. J. and Wolverton, R. W. (1978). "Assessment of software reliability," pp. 395–422 in *Proceedings of Operations Research,* Physica-Verlag, Vienna, Austria.

Schilling, E. G. (1982). *Acceptance Sampling in Quality Control,* Dekker, New York.

Schlager, N. (1994) (Ed.). *When Technology Fails,* Gale Research Inc., Detroit, MI.

Schroeder, G. J. and Johnson, M. M. (1990). "Complex availability: The new availability problem;" *Proceedings Annual Reliability and Maintainability Symposium,* 268–273

Searle, S. R. (1971). *Linear Models,* Wiley, New York.

Sen, A. (1998). "Estimation of current reliability in a Duane-based reliability growth model," *Technometrics* **40:** 334–344.

Sengupta, D. (1994). "Another look at the moment bounds on reliability," *Journal of Applied Probability* **31:** 777–787.

Sengupta, D. (1996). "Graphical tools for censored survival data," pp. 193–217 in Koul, H. L. and Deshpande, J. V. (Eds.), *Analysis of Censored Data,* Institute of Mathematical Statistics, Hayward, CA.

Shaked, M. (1984). Wear and damage processes from shock models in reliability theory, in *The Theory and Applications of Reliability,* Vol. 1, C. P. Tsokos and I. N. Shimi (Eds.), Academic Press, New York.

Shaked, M. and Shantikumar, J. G. (1986). "Multivariate imperfect repair," *Operations Research* **34:** 437–448.

Shaked, M. and Shantikumar, J. G. (1990). "Reliability and Maintainability" in *Handbooks in OR and MS,* D. P. Heyman and M. J. Sobel (Eds.), North Holland, New York.

Sherif, Y. S. and Smith, M. L. (1976). "Optimal maintenance models for systems subject to failure—A review," *Naval Logistics Research Quarterly* **23:** 47–74.

Shewart, W. A. (1931). *Economic Control of Quality of Manufactured Product,* Van Nostrand, Princeton, NJ. (Reprinted by American Society for Quality Control, Milwaukee, WI.)

Shigley, J. E. and Mischke, C. R. (1986), *Standard Handbook of Machine Design,* McGraw-Hill, New York.

Shirahata, S. (1987). "A goodness of fit test based on some graphical representation when parameters are estimated," *Computational Statistics and Data Analysis* **5:** 127–136.

Shooman, M. L. (1984). "Software reliability: A historical perspective," *IEEE Transasctions on Reliability* **33:** 48–55.

Silver, E. A. and Fiechter, C.-N. (1992). "A simple case of preventive maintenance decision making with limited historical data," *International Journal of Production Economics* **27:** 241–250.

Silver, E. A. and Fiechter, C.-N. (1995). "Preventive maintenance with limited historical data," *European Journal of Operations Research* **82:** 125–144.

Sinclair, C. D., Spurr, B. D., and Ahmad, M. I. (1990). "Modified Anderson Darling test," *Communications in Statistics, Theory and Methods* **19:** 3677–3686.

Singh, N. (1995). "Stochastic modeling of aggregates & products of variable failure rates" *IEEE Transactions on Reliability* **44:** 279–284.

Singpurwalla N. D. (1988). "An interactive PC-based procedure for reliability assessment incorporating expert opinion and survival data," *Journal of the American Statistical Association* **83:** 43–51.

Singpurwalla, N. D. (1995a), Survival in dynamic environments, *Statistical Science,* **10:** 86–103.

Singpurwalla, N. D. (1995b). "The failure rate of software: Does it exist?" *IEEE Transactions on Reliability* **44:** 463–469.

Singpurwalla, N. D. and Wilson, S. P. (1992/1993). "Models for assessing reliability of computer software," *Naval Research Reviews* **44/45:** 21–29.

Sinha, M. N. and Willborn, W. O. (1985). *The Management of Quality Assurance,* Wiley, New York.

Sinha, S. K. (1986). *Reliability and Life Testing,* Wiley, New York.

Sinha, S. K. and Guttman, I. (1988). "Bayesian analysis of life-testing problems involving the Weibull distribution," *Communications in Statistics, Theory and Methods* **17:** 343–356.

Smith, R. L. (1991). "Weibull regression models for reliability data," *Reliability Engineering and System Safety* **34**: 55–77.

Sobczyk, K. (1987). "Stochastic models for fatigue damage of material," *Advances in Appied. Probability* **19**: 652–673.

Soms, A. P. (1989). "Some recent results for series system reliability," *Communications in Statistics, Theory and Methods* **18**: 4211–4227.

Spencer, F. W. and Easterling, R. G. (1986). "Lower confidence bounds on system reliability using component data. The Maximus method," pp. 353–367 in *Reliability and Quality Control*, A. P. Basu (Ed.), Elsevier Science Publishers B. V., North Holland.

Springer, M. D. and Thompson, W. E. (1966). "Bayesian confidence limits for the product of N binomial parameters," *Biometrika* **53**: 611–613.

STATOIL Technical Standard, Center for Statoil Teckniske Standarder, Stavenger, Norway.

Steffens, P. (1991). Modelling Product Life Cycles for Consumer Durables, Doctoral Dissertation, The University of Queensland, Brisbane, Australia.

Stephens, M. A. (1974). "EDF statistics for goodness-of-fit and some comparisons," *Journal of the American Statistical Association* **69**: 730–737.

Stephens, M. A. (1977). "Goodness-of-fit for the extreme value distribution," *Biometrika* **64**: 583–588.

Steuer, R. E. (1986). *Multiple Criteria Optimization Theory and Applications,* Wiley, New York.

Stochholm, K. (1986). "Correlation between reliability prediction, reliability testing methods and field data," in *Reliability Methodology—Theory and Applications,* pp. 153–159, J. Møltoft and F. Jensen (Eds.), Elsevier Science Publishers, New York.

Stuart, A. and Ord, J. K. (1991 *Kendall's Advanced Theory of Statistics,* Vol. 2, 5th Edition, Oxford University Press, New York.

Sutharshana, S., Newlin, L., Ebbeler, D., and Moore, N. (1990). "Probabilistic high-cycle fatigue failure analysis with application to liquid propellant rocket engines," 31st Annual Structures, Structural Dynamics and Materials Conference, American Institute of Aeronautics and Astronautics.

Tadikmalla, P. R. (1980). "Age replacement policies for Weibull failure rates," *IEEE Transactions on Reliability* **29**: 88–90.

Taguchi, G. (1981). *On Line Quality Control during Production,* Japanese Standards Association, Tokyo.

Taguchi, G., Elsayed, E. A., and Hsiang, T. (1989). *Quality Engineering in Production Systems,* McGraw-Hill Publishing Co., New York.

Taguchi, G. and Wu, Y. (1979). *Introduction to Off-Line Quality Control,* American Supplies Institute, Romulus, MI.

Tang, J., Tang, K., and Moskowitz, H. (1994). "Bayes confidence interval for reliability of large series systems with highly reliable components," *IEEE Transactions on Reliability* **43**: 132–137.

Tang, J., Tang, K., and Moskowitz, H. (1997). "Bayesian estimation of system reliability from component test data," *Naval Research Logistics* **44**: 127–146.

Tapiero, C. S. and Lee, H. L. (1989). "Quality control and product servicing: A decision framework," *European Journal of Operational Research* **39**: 61–73.

Taylor, H. M. (1987). "A model for the failure process of semicrystalline polymer materials under static fatigue," *Probability in Engineering & Information Sciences* **1**: 133–162.

Thomas, L. C. (1986). "A survey of maintenance and replacement models for maintainability and reliability of multi-item systems," *Reliability Engineering* **16**: 297–309.

Thompson, A. A. and Strickland, A. J. (1990). *Strategic Management: Concepts and Cases,* BPI Irwin, Homewood, IL.

Thompson, S. K. (1992). *Sampling,* Wiley, New York.

Thompson, W. E. and Chang, E. Y. (1975). "Bayes confidence limits for reliability of redundant systems," *Technometrics* **17:** 89–93.

Thompson, W. E. and Haynes, R. D. (1980). "On the reliability, availability and Bayes confidence intervals for multicomponent systems," *Naval Research Logistics Quarterly* **27:** 345–358.

Tillman, F. A., Hwang, C. L., and Kuo, W. (1975). "Optimization techniques for system reliability with redundancy—A review," *IEEE Transactions on Reliability* **24:** 148–155.

Tillman, F. A., Hwang, C. L., and Kuo, W. (Eds.) (1980), *Optimization of System Reliability,* Dekker, New York.

Tillman, F. A., Kuo, W., Hwang, C. L., and Grosh, D. L. (1982). "Bayesian reliability and availability—A review," *IEEE Transactions on Reliability* **R-31:** 362–372.

Titterington, D. M., Smith, A. F. M., and Makov, U. E. (1985). *Statistical Analysis of Finite Mixture Distributions,* Wiley, New York.

Tufte, E. R. (1983). *The Visual Display of Quantitative Information,* Graphics Press, Cheshire, CT.

Tufte, E. R. (1989). *Envisioning Information,* Graphics Press, Cheshire, CT.

Tufte, E. R. (1997). *Visual Explanations,* Graphics Press, Cheshire, CT.

Tummala, V. M. R. and Sathe, P. T. (1978). "Minimum expected loss estimators of reliability and parameters of certain lifetime distributions," *IEEE Transactions on Reliability* **R-27:** 283–285.

Ueno, K. (1995). "From the product-oriented development to technology-oriented development," *IEEE Transactions on Reliability* **44:** 220–224.

Ury, H. K. (1972). "On distribution-free confidence bounds for $\Pr\{Y < X\}$," *Technometrics* **14:** 577–581.

Valdez-Flores, C. and Feldman, R. M. (1989). "A survey of preventive maintenance models for stochastically deteriorating single-unit systems," *Naval Research Logistics Quarterly* **36:** 419–446.

Vatn, J. (1997). "Maintenance optimisation from a decision theoretical point of view," *Reliability Engineering and System Safety* **58:** 119–126.

Vatn, J., Hokstad, P., and Bodsberg, L. (1996). "An overall model for maintenance optimization," *Reliability Engineering and Systems Safety* **51:** 241–257.

Vesely, W. E., Goldberg, F. F., Roberts, N. H., and Haasl, D. F. (1981). *Fault Tree Handbook, U.S. Nuclear Reliability, maintainability and Safety.* Vol 1: *Methods and Techniques.* Vol 2: *Assessment, Hardware, Software and Human Factors,* Wiley, New York.

Vincent, J., Waters, A., and Sinclair, J. (1988) *Software Quality Assurance,* Prentice-Hall, Englewood Cliffs, NJ.

Vinogradov, A. (1991). *Introduction to Mechanical Reliability: A Designer's Approach,* Hemisphere, New York.

Vouk, M. A (1998). "Introduction to software reliability engineering," *Tutorial Notes, 1998 Annual Reliability and Maintainability Symposium,* Anaheim, CA.

Wallmüller, E. (1994). *Software Quality Assurance: A Practical Approach,* Hemel Hempstead Herfordshire, UK.

Walters, G. F. and McCall, J. A. (1979). "Software quality metrics for life-cycle cost-reduction," *IEEE Transactions on Reliability* **R-28:** 212–220.

Wang, C. J. (1990), "Concept of durability index in product assurance planning," *Proceedings Annual Reliability and Maintainability Symposium,* 221–227.

Watson, A. S. and Smith, R. L. (1985). "An examination of statistical theories for fibrous materials in the light of experimental data," *Journal of Material Science* **20**: 3260–3270.

Weerahandi, S. and Johnson, R. A. (1992). "Testing reliability in a stress-strength model when X and Y are normally distributed," *Technometrics* **34**: 83–91.

Weibull, W. (1951). "A statistical distribution of wide applicability," *Journal of Applied Mechanics* **18**: 293–297.

Weiss, H. K. (1956). "Estimation of reliability growth in a complex system with Poisson type failure," *Operations Research* **4**: 532–545.

Winer, B. J. (1971). *Statistical Principles in Experimental Design,* McGraw-Hill, New York.

Wilmot, P. (1994), *Total Productive Maintenance—The Western Way,* Butterworth-Hinemann, Oxford.,UK.

Winkler, R. L. (1967). "The assessment of prior distributions in Bayesian analysis," *Journal of the American Statistical Association* **62**: 776–800.

Winkler, R. L. (1981). "Combining probability distributions from dependent information sources," *Management Science* **27**: 479–488.

Winkler, R. L. (1986). "Expert resolution," *Management Science* **32**: 298–303.

Winterbottom, A. (1974). "Lower confidence limits for series system reliability from binomial subsystem data," *Journal of the American Statistical Association* **69**: 782–788.

Winterbottom, A. (1984). "The interval estimation of system reliability from component test data," *Operations Research* **32**: 628–640.

Witherell, C. E. (1994). *Mechanical Failure Avoidance,* McGraw-Hill, New York.

Wolstenholme, L. C. (1995). "A nonparametric test of the weakest-link principle," *Technometrics* **37**: 169–175.

Wolstenholme, L. C. (1996). "An alternative to the Weibull distribution," *Communications in Statistics, Simulation and Computation* **25**: 119–137.

Wong, E. (1971). *Stochastic Processes in Information and Dynamical Systems,* McGraw Hill, New York.

Woodruff, B. W. and Moore, A. H. (1988). "Application of goodness-of-fit tests in reliability," pp. 113–120 in P. R. Krishnaiah and C. R. Rao (Eds.), *Handbook of Statistics, Volume 7. Quality Control and Reliability,* North-Holland, Amsterdam.

Woodruff, B. W., Moore, A. H., Dunne, E. J., and Cortes, R. (1983). "A modified Kolmogorov-Smirnov test for Weibull distributions with unknown location and scale parameters," *IEEE Transactions on Reliability* **R-32**: 209–213.

Woodruff, B. W., Viviano, P. J., Moore, A. H., and Dunne, E. J. (1984). "Modified goodness-of-fit tests for gamma distributions with unknown location and scale parameters," *IEEE Transactions on Reliability* **R-33**: 241–245.

Woodward, M. R. (1991). "Aspects of path testing and mutation testing," pp. 1–15 in Ince, D. (Ed.), *Software Quality and Reliability: Tools and Methods,* Chapman and Hall, London.

Wong, K. M. (1989). "The roller-coaster curve is in," *Quality and Reliability Engineering International* **5**: 29–36.

Xie, M. (1991). *Software Reliability Modelling,* World Scientific Press, Singapore.

Xie, M. (1993). "Software reliability models—A selected annotated bibliography," *Software Testing, Verification and Reliability* **3**: 3–28.

Xie, M. and Lai, C. D. (1996). "Reliability analysis using an additive Weibull model with bathtub-shaped failure rate function," *Reliability Engineering and System Safety* **52**: 87–93.

Yamada, S., Ohba, M., and Osaki, S. (1984). "s-shaped software reliability growth models and their applications," *IEEE Transactions on Reliability* **33**: 289–292.

Yang, K. and Kapur, K. C. (1997). "Customer driven reliability: Integration of QFD and robust design," *Proceedings of the Annual Reliability and Maintainability Symposium,* 339–345.

Yearout, R. D., Reddy, P. and Grosh, D. L. (1986). "Standby redundancy in reliability—A review," *IEEE Transactions on Reliability* **35:** 285–292

Yoo, J. N. and Smith, G. (1991) "Implementing total quality into reliability and maintainabilty," *Proceedings Annual Reliability and Maintainability Symposium,* 547–549.

Young, D. and Christou, A. (1994). "Failure mechanism models for electromigration," *IEEE Transactions on Reliability* **43:** 186–192.

Zhao, M. and Xie, M. (1992)."Applications of the log-power NHPP software reliability model," *Proceedings of the International Symposium on Software Reliability Engineering.*

Zok, F. W., Chen, X., and Weber, C. H. (1995). "Tensile strength of SiC fibers," *Journal of American Ceramic Soc.* **78:** 1965–1968.

Zonnenshain, A. and Haim, M. (1984). "Assessment of reliability prior distributions," *1984 Proceedings Annual Reliability and Maintainability Symposium,* pp. 44–47.

Author Index

Subject Index

WILEY SERIES IN PROBABILITY AND STATISTICS
ESTABLISHED BY WALTER A. SHEWHART AND SAMUEL S. WILKS

Editors
Vic Barnett, Noel A. C. Cressie, Nicholas I. Fisher,
Iain M. Johnstone, J. B. Kadane, David G. Kendall, David W. Scott,
Bernard W. Silverman, Adrian F. M. Smith, Jozef L. Teugels;
Ralph A. Bradley, Emeritus, J. Stuart Hunter, Emeritus

Probability and Statistics Section

*ANDERSON · The Statistical Analysis of Time Series
ARNOLD, BALAKRISHNAN, and NAGARAJA · A First Course in Order Statistics
ARNOLD, BALAKRISHNAN, and NAGARAJA · Records
BACCELLI, COHEN, OLSDER, and QUADRAT · Synchronization and Linearity:
  An Algebra for Discrete Event Systems
BASILEVSKY · Statistical Factor Analysis and Related Methods: Theory and
  Applications
BERNARDO and SMITH · Bayesian Statistical Concepts and Theory
BILLINGSLEY · Convergence of Probability Measures, *Second Edition*
BOROVKOV · Asymptotic Methods in Queuing Theory
BOROVKOV · Ergodicity and Stability of Stochastic Processes
BRANDT, FRANKEN, and LISEK · Stationary Stochastic Models
CAINES · Linear Stochastic Systems
CAIROLI and DALANG · Sequential Stochastic Optimization
CONSTANTINE · Combinatorial Theory and Statistical Design
COOK · Regression Graphics
COVER and THOMAS · Elements of Information Theory
CSÖRGŐ and HORVÁTH · Weighted Approximations in Probability Statistics
CSÖRGŐ and HORVÁTH · Limit Theorems in Change Point Analysis
*DANIEL · Fitting Equations to Data: Computer Analysis of Multifactor Data,
  Second Edition
DETTE and STUDDEN · The Theory of Canonical Moments with Applications in
  Statistics, Probability, and Analysis
DEY and MUKERJEE · Fractional Factorial Plans
*DOOB · Stochastic Processes
DRYDEN and MARDIA · Statistical Shape Analysis
DUPUIS and ELLIS · A Weak Convergence Approach to the Theory of Large Deviations
ETHIER and KURTZ · Markov Processes: Characterization and Convergence
FELLER · An Introduction to Probability Theory and Its Applications, Volume 1,
  Third Edition, Revised; Volume II, *Second Edition*
FULLER · Introduction to Statistical Time Series, *Second Edition*
FULLER · Measurement Error Models
GHOSH, MUKHOPADHYAY, and SEN · Sequential Estimation
GIFI · Nonlinear Multivariate Analysis
GUTTORP · Statistical Inference for Branching Processes
HALL · Introduction to the Theory of Coverage Processes
HAMPEL · Robust Statistics: The Approach Based on Influence Functions
HANNAN and DEISTLER · The Statistical Theory of Linear Systems
HUBER · Robust Statistics
HUSKOVA, BERAN, and DUPAC · Collected Works of Jaroslav Hajek—
  with Commentary

*Now available in a lower priced paperback edition in the Wiley Classics Library.

Applied Probability and Statistics Section

*Now available in a lower priced paperback edition in the Wiley Classics Library.

*Now available in a lower priced paperback edition in the Wiley Classics Library.

*Now available in a lower priced paperback edition in the Wiley Classics Library.

*Now available in a lower priced paperback edition in the Wiley Classics Library.

Texts and References Section

*Now available in a lower priced paperback edition in the Wiley Classics Library.

Texts and References (Continued)

HOEL · Introduction to Mathematical Statistics, *Fifth Edition*

HOLLANDER and WOLFE · Nonparametric Statistical Methods, *Second Edition*

HOSMER and LEMESHOW · Applied Survival Analysis: Regression Modeling of Time to Event Data

JOHNSON and BALAKRISHNAN · Advances in the Theory and Practice of Statistics: A Volume in Honor of Samuel Kotz

JOHNSON and KOTZ (editors) · Leading Personalities in Statistical Sciences: From the Seventeenth Century to the Present

JUDGE, GRIFFITHS, HILL, LÜTKEPOHL, and LEE · The Theory and Practice of Econometrics, *Second Edition*

KHURI · Advanced Calculus with Applications in Statistics

KOTZ and JOHNSON (editors) · Encyclopedia of Statistical Sciences: Volumes 1 to 9 wtih Index

KOTZ and JOHNSON (editors) · Encyclopedia of Statistical Sciences: Supplement Volume

KOTZ, REED, and BANKS (editors) · Encyclopedia of Statistical Sciences: Update Volume 1

KOTZ, REED, and BANKS (editors) · Encyclopedia of Statistical Sciences: Update Volume 2

LAMPERTI · Probability: A Survey of the Mathematical Theory, *Second Edition*

LARSON · Introduction to Probability Theory and Statistical Inference, *Third Edition*

LE · Applied Categorical Data Analysis

LE · Applied Survival Analysis

MALLOWS · Design, Data, and Analysis by Some Friends of Cuthbert Daniel

MARDIA · The Art of Statistical Science: A Tribute to G. S. Watson

MASON, GUNST, and HESS · Statistical Design and Analysis of Experiments with Applications to Engineering and Science

MURRAY · X-STAT 2.0 Statistical Experimentation, Design Data Analysis, and Nonlinear Optimization

PURI, VILAPLANA, and WERTZ · New Perspectives in Theoretical and Applied Statistics

RENCHER · Linear Models in Statistics

RENCHER · Methods of Multivariate Analysis

RENCHER · Multivariate Statistical Inference with Applications

ROSS · Introduction to Probability and Statistics for Engineers and Scientists

ROHATGI · An Introduction to Probability Theory and Mathematical Statistics

RYAN · Modern Regression Methods

SCHOTT · Matrix Analysis for Statistics

SEARLE · Matrix Algebra Useful for Statistics

STYAN · The Collected Papers of T. W. Anderson: 1943–1985

TIERNEY · LISP-STAT: An Object-Oriented Environment for Statistical Computing and Dynamic Graphics

WONNACOTT and WONNACOTT · Econometrics, *Second Edition*

*Now available in a lower priced paperback edition in the Wiley Classics Library.

WILEY SERIES IN PROBABILITY AND STATISTICS

ESTABLISHED BY WALTER A. SHEWHART AND SAMUEL S. WILKS

Editors
Robert M. Groves, Graham Kalton, J. N. K. Rao, Norbert Schwarz, Christopher Skinner

Survey Methodology Section

*Now available in a lower priced paperback edition in the Wiley Classics Library.